FUZZY SETS
AND APPLICATIONS

FUZZY SETS AND APPLICATIONS:

Selected Papers by L.A. Zadeh

Edited by

R.R. Yager
Iona College

S. Ovchinnikov
San Francisco State University

R.M. Tong
Advanced Decision Systems

H.T. Nguyen
New Mexico State University

A Wiley-Interscience Publication

JOHN WILEY & SONS

New York   Chichester   Brisbane   Toronto   Singapore

Library of Congress Cataloging in Publication Data:

Zadeh, Lotfi Asker.
  Fizzy sets and applications.

  "A Wiley-Interscience publication."
  Bibliography: p.
  1. Fuzzy sets—Collected works. I. Yager, Ronald R.,
1941-   .   II. Title.
QA248.Z32   1987   511.3′2   87-6088
ISBN 0-471-85710-6

Printed in the United States of America

10 9 8 7 6 5 4 3 2 1

PREFACE

At the beginning of 1985 the four of us decided that the 20th anniversary of the publication of the first paper on fuzzy sets would be a good time to undertake the project that resulted in this collection of papers. Our original conception was to be a personal tribute to Professor Zadeh who had in each case been instrumental in encouraging our efforts in the area of fuzzy sets. However, we soon realized that a selection of Professor Zadeh's papers would be of interest to a wide audience, and so by degrees our small project evolved into this volume.

The order of our names in the list of editors was determined in a way that we hope will appeal to all those who have had the fortune to be entertained by Professor Zadeh at any one of his favorite Chinese restaurants in Berkeley. During one of our dinner meetings to discuss the form and content of the book we decided that we should take the messages from those ubiquitous cookies and use them to define a lexical ordering based on the first word of the fortune. So:

"He loves you as much as he can, but he cannot love you very much."

"To open a book brings profit."

"You love sports, horses and gambling but not to excess."

"You will triumph over your enemy."

We leave the reader to decide whether these have any special significance!

All efforts of this kind require contributions from others. Accordingly, we thank Maria Taylor of Wiley for her support in putting together this volume. We also wish to give a special thank you to Margaret Tong, who achieved the nearly impossible by retrieving the paper reprints from the Zadeh archive. Then finally, of course, we must thank Professor Zadeh himself for providing us with the opportunity and the encouragement that he alone could give.

R.R. YAGER
Iona College

R.M. TONG
Advanced Decision Systems

S. OVCHINNIKOV
San Francisco State University

H.T. NGUYEN
New Mexico State University

April 1987

Acknowledgments

Pp. 9–28: "**Coping with the Imprecision of the Real World: An Interview with Lotfi A. Zadeh,**" *Comm. ACM,* 27 (April 1984), pp. 304–311. Reprinted with permission.

Pp. 29–44: L.A. Zadeh, "Fuzzy Sets," *Information and Control,* Vol. 8, New York: Academic Press, 1965, pp. 338–353. Reprinted with permission. Copyright 1965 by Academic Press.

Pp. 45–51: L.A. Zadeh, "Probability Measures of Fuzzy Events," *J. Math. Analysis and Appl.,* 10 (1968), pp. 421–427. Reprinted with permission. Copyright 1968 by Academic Press.

Pp. 53–79: R.E. Bellman and L.A. Zadeh, "Decision-Making in a Fuzzy Environment," *Management Science,* 17:4 (December 1970), pp. 141–164. Reprinted by permission. Copyright 1970 by the Institute of Management Sciences, 290 Westminster Street, Providence, RI 02903.

Pp. 81–104: L.A. Zadeh, "Similarity Relations and Fuzzy Orderings," *Information Sciences,* 3 (1971), pp. 177–200. Reprinted by permission of the publisher. Copyright 1977 by Elsevier Science Publishing Co., Inc.

Pp. 105–146: L.A. Zadeh, "Outline of a New Approach to the Analysis of Complex Systems and Decision Processes," *IEEE Trans. Systems, Man, and Cybernetics,* SMC-3 (1973), pp. 28–44. Reprinted with permission. Copyright 1973 IEEE.

Pp. 147–192: L.A. Zadeh, "A Fuzzy-Algorithmic Approach to the Definition of Complex or Imprecise Concepts" *Int. J. Man-Machine Studies,* 8 (1976), pp. 249–291. Reprinted with permission. Copyright 1976 by Academic Press Inc. (London) Ltd.

Pp. 193–218: L.A. Zadeh, "Fuzzy Sets as a Basis for a Theory of Possibility," *Fuzzy Sets and Systems,* 1 (1978), pp. 3–28. Reprinted with permission.

Pp. 219–327: L.A. Zadeh, "The Concept of a Linguistic Variable and its Application to Approximate Reasoning," Parts 1 and 2, *Information Sciences,* 8 (1975), pp. 199–249, 301–357. Reprinted with permission. Copyright 1975 by Elsevier Science Publishing Co., Inc.

Pp. 329–366: L.A. Zadeh, "The Concept of a Linguistic Variable and its Application to Approximate Reasoning," Part 3, *Information Sciences,* 9 (1976), pp. 43–80. Reprinted by permission. Copyright 1976 by Elsevier Science Publishing Co., Inc.

Pp. 367–412: L.A. Zadeh, "A Theory of Approximate Reasoning," in J. Hayes, D. Michie, and L.I. Mikulich, Eds., *Machine Intelligence,* Vol. 9, New York: Halstead Press, 1979, pp. 149–194. Reprinted with permission.

Pp. 413–441: L.A. Zadeh, "The Role of Fuzzy Logic in the Management of Uncertainty in Expert Systems," *Fuzzy Sets and Systems,* 11 (1983), pp. 199–227. Reprinted with permission.

Pp. 443–466: L.A. Zadeh, "Syllogistic Reasoning in Fuzzy Logic and its Application to Usuality and Reasoning with Dispositions," *IEEE Trans. Systems, Man, and Cybernetics,* SMC-15 (1985), pp. 754–763. Reprinted with permission. Copyright 1985 IEEE.

CONTENTS

FUZZY SETS
AND APPLICATIONS

INTRODUCTION

Selecting from Professor Zadeh's extensive list of publications proved to be a challenging task. In the end we were limited more by the format of the book than by our ability to select important papers. The final choice was guided by our desire both to provide some historical perspective on his work and to assemble a collection that could be used as a reference by those working in the field.

The works are unedited, and furthermore, since we believe each is a classic in its own way, we have refrained from adding any significant, interpretative comments of our own. In particular, we have not tried to perform any scholarly analysis of the impact of these papers on the work of others, nor have we made any attempt to disentangle the many themes that run through them. We leave this to those who in the future may wish to publish "The Collected Works."

We start with a recent interview with Professor Zadeh that appeared in the Communications of the ACM [1]. This illustrates his basic philosophy, and so gives us a perspective from which to interpret this significant body of scientific work. Following that are eighteen papers that are representative of his enormous intellectual output in the years 1965 to 1986. Although we have grouped them into a number of presentational categories, they form a consistent and remarkably comprehensive approach to the problem of reasoning and problem solving under uncertainty.

FORMAL FOUNDATIONS

We have chosen seven papers in this category and present them chronologically. They contain formal developments of issues in fuzzy sets and possibility theory, and provide the framework within which to interpret the more application-oriented papers. As we might expect, they also tend to be the earlier publications.

We start, naturally, with the very first published paper on fuzzy sets [2] in which Professor Zadeh introduces the seminal idea and defines inclusion, union, intersection, complement, relation and convexity. Although all the material is familiar by now, it is especially remarkable for its recognition that a large part of human reasoning is concerned with problems "... in which the source of imprecision is the absence of sharply defined criteria of class membership rather than the

1

presence of random variables." This is the central theme in all of Professor Zadeh's work and we see it in many forms in the papers in this collection.

Our second selection contains the first development of the idea of fuzzy events and their associated probability measures [3]. In this paper Professor Zadeh is again concerned with extensions of conventional mathematical ideas to what he considers to be everyday experience. In this case it is the interpretation of probabilistic statements about events such as "It is a warm day." The paper defines fuzzy event, the probability of a fuzzy event, independence, conditional probability and entropy.

The third selection is the highly influential paper coauthored with Richard Bellman that appeared in 1970 [4]. In it Professors Bellman and Zadeh develop a decision theory based on fuzzy goals and constraints. This paper provides the first explication of goals, constraints and decisions in environments that are fuzzy rather than random. With these concepts defined, the authors show how they can be applied to multistage decision problems in which the system under control is either deterministic or stochastic.

The fourth selection is concerned with the concept of fuzzy binary relations [5]. Here we see Professor Zadeh defining such important concepts as similarity relations and fuzzy orderings and also developing some valuable mathematical techniques for the resolution of fuzzy relations and the description of the transitivity properties of similarity relations.

The next paper is included because in many ways it is a summary of the results and ideas developed in 1973 [6]. This paper from *IEEE Transactions on Systems, Man, and Cybernetics* also has the distinction of being amongst the most often cited of Professor Zadeh's early works. Indeed for many of us it was our first introduction to the concepts in fuzzy sets theory. The paper contains a statement of the "Principle of Incompatibility" and presents an approach to the description of the behavior of complex systems using linguistic variables and fuzzy algorithms.

The sixth paper is also concerned with mechanisms for the description and definition of complex and imprecise concepts [7]. In it Professor Zadeh extends his ideas on fuzzy algorithms, focussing especially on those that can be structured as branching questionnaires.

The final paper in this section is another landmark. In it Professor Zadeh introduces for the first time the concept of a possibility distribution [8]. Published in 1978, it represents his next major step in the formal development of a theory of uncertain reasoning. Here we see clearly stated the premise that "... the imprecision that is intrinsic in natural languages is, in the main, possibilistic rather than probabilistic in nature." The mathematics of possibility distributions is thus intended as a unifying principle around which to address many questions of natural language interpretation. According to Professor Zadeh this is particularly important in the context of trying to develop machines "... which can simulate the remarkable human ability to attain imprecisely defined goals in a fuzzy environment."

APPROXIMATE REASONING

With this next category the focus of the papers changes. The overall vision remains constant, but Professor Zadeh is now primarily concerned with modelling the human ability to reach conclusions when the information available is imprecise, incomplete and not totally reliable. For this Professor Zadeh uses the especially apt phrase "approximate reasoning." The six papers included in this category are still formal—and especially logical—approaches to approximate reasoning, but we begin to see how the ideas may be transformed into computer-based reasoning systems. The work thus parallels, and in many cases anticipates, the efforts of workers in artificial intelligence.

The first three papers are to be read together. In them Professor Zadeh defines the concept of linguistic variables, performs a complete analysis of their properties and shows how they are used in approximate reasoning [9, 10, 11]. Among the many important topics discussed are the extension principle, fuzzy sets of type 2, marginal and conditional restrictions, separability and non-interaction, and the compositional rule of inference.

The next paper in this category is in the nature of a reprise for the work in approximate reasoning, while at the same time setting the scene for new developments [12]. In summarizing what he now describes as a theory of approximate reasoning, Professor Zadeh draws together the results from several earlier papers showing how the ideas of fuzzy logic, linguistic variables and possibility distributions all work together to provide a system for reasoning from imprecise premises to imprecise conclusions.

The fifth paper is an exposition on the role of fuzzy logic in expert systems [13]. Drawing on the attempts by various workers in artificial intelligence to construct expert systems that manipulate imprecise information, Professor Zadeh shows how the rules of inference in fuzzy logic can be used to model many of the problems that arise.

The last paper contains the most recent developments in the use of syllogisms in approximate reasoning systems [14]. In this work Professor Zadeh identifies the notion of a fuzzy syllogism as being a central unifying concept around which to formalize our ideas on commonsense reasoning. In addition, the computational framework which the syllogisms provide can be used to support a more satisfactory scheme for combining evidence than that used in the current generation of expert systems.

MEANING REPRESENTATION

The third and final category contains five papers that address directly the problem of meaning representation in natural language, with the primary effort now being directed towards a mathematically oriented interpretation of commonsense reasoning. In Professor Zadeh's conception of this problem, the key is to provide an interpretation of natural language through the representational mechanisms of fuzzy sets and possibility theory. These papers overlap with those in the preceding categories, but a distinguishing characteristic is their focus on the direct translation of various propositional forms of natural language, rather than on the representational mechanisms themselves.

The first paper is an early attempt to use fuzzy sets to capture the meaning of hedges in natural language [15]. In this paper Professor Zadeh shows how a linguistic hedge such as "very," "more or less," "much," "essentially," or "slightly," may be viewed as an operator which acts on the fuzzy set representing the meaning of its operand. More complex hedges whose effect is strongly context-dependent require, in Professor Zadeh's view, the use of a fuzzy-algorithmic mode of characterization.

The second selection is the landmark paper on PRUF [16] which provides many of the basic elements in Professor Zadeh's approach to meaning representation. PRUF departs from the conventional approaches to meaning representation in several important ways. First, it assumes that imprecision in language is possibilistic rather than probabilistic. Second, the underlying logic is fuzzy rather than two-valued. Finally, it allows quantifiers to be linguistic. The paper contains formal definitions of four translation rules: modification, composition, quantification and qualification.

The third selection contains a discussion of what Professor Zadeh considers to be the crucial role of fuzzy quantifiers in meaning representation [17]. In this paper, fuzzy quantifiers are treated as fuzzy numbers that can be manipulated through the use of fuzzy arithmetic and fuzzy logic. The computational approach can be viewed as a derivative of fuzzy logic and test-score semantics and makes extensive use of the concept of the cardinality of a fuzzy set.

The next selection contains the beginnings of a theory of commonsense knowledge [18] that is based on the idea that this form of knowledge can be viewed as a collection of dispositions, that is, prepositions with implied fuzzy quantifiers. So for example, a typical dispositional statement that we may consider as being part of a commonsense knowledge base is "Snow is white." Professor Zadeh shows that if we assume the intended meaning of this is "Usually snow is white" then it has a much simple interpretation in terms of his test-score semantics than that provided by truth-conditional semantics.

Finally, we include a recent paper that shows how the test-score semantics can provide the basis of a computational approach to the representation of meaning [19]. Here Professor Zadeh illustrates the use of test-score semantics in the representation of the meaning of propositions, predicates, dispositions, and commands. He places much emphasis on the idea of a canonical form which "... may be viewed as a possibilistic analog of an assignment statement" and notes that semantic networks are actually special cases of this concept.

REFERENCES

[1] Coping with the Imprecision of the Real World: An Interview with Lotfi A. Zadeh. *Comm. ACM.* 27 (1984):304–311.

[2] Fuzzy Sets. *Information and Control* 8 (1965):338–353.

[3] Probability Measures of Fuzzy Events. *J. Math. Analysis and Appl.* 23 (1968):421–427.

[4] With R.E. Bellman. Decision-Making in a Fuzzy Environment. *Management Science* 17 (1970):B-141–164.

[5] Similarity Relations and Fuzzy Orderings. *Information Sciences* 3 (1971):177–200.

[6] Outline of a New Approach to the Analysis of Complex Systems and Decision Processes. *IEEE Trans. Systems, Man, and Cybernetics.* SMC-3 (1973):28–44.

[7] A Fuzzy-Algorithmic Approach to the Definition of Complex or Imprecise Concepts. *Int. J. Man-Machine Studies* 8 (1976):249–291.

[8] Fuzzy Sets as a Basis for a Theory of Possibility. *Fuzzy Sets and Systems* 1 (1978):3–28.

[9] The Concept of a Linguistic Variable and its Application to Approximate Reasoning. Part 1. *Information Sciences* 8 (1975):199–249.

[10] The Concept of a Linguistic Variable and its Application to Approximate Reasoning. Part 2. *Information Sciences* 8 (1975):301–357.

[11] The Concept of a Linguistic Variable and its Application to Approximate Reasoning. Part 3. *Information Sciences* 9 (1975):43–80.

[12] A Theory of Approximate Reasoning. In: *Machine Intelligence*, vol. 9, edited by J. Hayes, D. Michie, and L.I. Mikulich, 149–194. New York: Halstead Press, 1979.

[13] The Role of Fuzzy Logic in the Management of Uncertainty in Expert Systems. *Fuzzy Sets and Systems* 11 (1983):199–227.

[14] Syllogistic Reasoning in Fuzzy Logic and its Application to Usuality and Reasoning with Dispositions. *IEEE Trans. Systems, Man, and Cybernetics* SMC-15 (1985):754–763.

[15] A Fuzzy-Set-Theoretic Interpretation of Linguistic Hedges. *Journal of Cybernetics* 2 (1972):4–34.

[16] PRUF—A Meaning Representation Language for Natural Languages. *Int. J. Man-Machine Studies* 10 (1978):395–460.

[17] A Computational Approach to Fuzzy Quantifiers in Natural Language. *Comp. and Maths with Appls* 9 (1983):149–184.

[18] A Theory of Commonsense Knowledge. In: *Aspects of Vagueness*, edited by H.J. Skala, S. Termini, and E. Trillas, 257–296. Dordrecht, Holland: D. Reidel, 1984.

[19] Test-Score Semantics as a Basis for a Computational Approach to the Representation of Meaning. *Literary and Linguistic Computing* 1 (1986):24–35.

The Selected Papers

Coping with the Imprecision of the Real World: an Interview with Lotfi A. Zadeh

The tools researches use to probe certain AI problems, says this Berkeley professor, are sometimes too precise to deal with the "fuzziness" of the real world.

Q. **Professor Zadeh, in this interview today, you agreed to talk mainly about the limits of traditional logic in dealing with many of the problems in the field of artificial intelligence (AI) and your approach toward helping to overcome those difficulties. Before getting into those issues, though, could you first give our readers a brief overview of what you see as the major areas for computer applications in the years ahead?**

ZADEH. In the years ahead, there will be three major areas of computer applications. One, in the traditional vein, is the use of computers for purposes of numerical analysis. Numerical analysis will be very important in a number of fields—particularly in scientific computations and simulation of large-scale systems.

For such purposes, there will be a need for larger and larger computers. This is especially true for applications in meteorology, in nuclear physics, in modeling of large-scale economic systems, in the solution of partial differential equations, and in the simulation of complex phenomena like turbulence, fluid flow, etc.

Area number two will be concerned with masses of data—large databases. This is the sort of thing that is playing and will be playing an important role in banking, insurance, records processing, information retrieval, etc. What will be important in these areas is not so much number-crunching capabilities as the capability to store massive amounts of data and to access whatever data are needed rapidly and at a reasonably low cost.

Furthermore, in these areas, computer networking, of course, will be playing an essential role. For you will

9

Lotfi A. Zadeh, a professor of electrical engineering and computer science, has been a member of the faculty at the University of California at Berkeley since 1959. Born in the Soviet Union of Iranian parents, Zadeh came to the U.S. in 1944 and studied at MIT and Columbia, from which he got his Ph.D. in 1949. After serving on the Columbia faculty during the 1950s, Zadeh went to Berkeley. Dissatisfied with the use of very precise mathematics to describe the sometimes highly imprecise real world, Zadeh, during the 1960s, developed the theory of fuzzy sets. In recent years, his controversial ideas on "fuzzy thinking" have begun to win a following in some quarters of the world.

have to have access not just to a single database but to a collection of interconnected databases. In response to this need, we will see many advances in computer networking during the next several years.

The third major area for computer applications is what has come to be known as *knowledge engineering*. This area has received considerable publicity during the past few years, particularly since the Japanese have highlighted it as an area of prime concern. This is a rapidly growing field in terms of importance and breadth of applications.

Knowledge engineering is one of the major areas of AI. And within knowledge engineering, a field of primary importance is that of expert systems. True, there may be exaggerated expectations of what expert systems can accomplish at this juncture, but as Jules Verne observed at the turn of the century, scientific progress is driven by exaggerated expectations.

So we have these three major areas for computer applications in the years ahead. All will be *growing* in importance. But knowledge engineering, I think, will be growing in importance more rapidly than the other two, because it is the youngest and, in a sense, the most pervasive of the three.

In saying that knowledge engineering is going to become very important, I don't want to imply that the other two will become less important. They will become more important also. But in relative terms, knowledge engineering will certainly be much more important than it is today.

Now, what I'm going to say will relate to this third area, rather than the first two.

Q. **Do you see supercomputers playing an important role only in the area of numerical analysis?**
ZADEH.   Supercomputers pertain to all three areas: *numerical analysis, large databases,* and *knowledge engineering.* But they apply primarily to the first area: numerical analysis. There is at this point some controversy as to how the available research funds should be distributed between the efforts to build supercomputers and to build machines that will be AI oriented.

These are somewhat distinct efforts. The Japanese are pushing both of them. And in the United States, the emphasis on AI-oriented types of computers is just beginning to become strong, largely as a reaction to the Japanese effort. As you may know, Edward Feigenbaum of Stanford University is a leading advocate of the establishment of a U.S. National Center for Computer Technology as a rallying point for the U.S. effort.

Q. **I'm not perfectly clear on the distinction between a supercomputer and an AI-oriented device, which I believe is sometimes referred to as a fifth generation computer. Could you clarify?**
ZADEH.   Fifth generation has become somewhat of a misnomer. Basically, the Fifth Generation Project is perhaps the most publicized of the several ambitious research programs undertaken by the Japanese. Another one of those programs is called the Supercomputer Project.

The Fifth Generation Project is not hardware oriented. Rather it is intended to exploit the advances in hardware that might be achieved under the other projects in the overall program.

For supercomputers, the emphasis is on large-scale computations relating to scientific and technological ap-

plications. For AI-oriented computers, on the other
hand, the emphasis is shifted away from "data" to
"knowledge."

What matters in the case of these AI-oriented computers is their ability *to infer* from the information resident in a large knowledge base—especially when this
information is imprecise, incomplete, or not totally reliable.

We can expect that advances in the design of supercomputers—on both the hardware and the software
levels—will have a significant impact on the architecture of AI-oriented machines.

Of course, the supercomputer is not a unique concept. It can take a variety of forms. Parallel processing
may play a major role in AI-oriented applications—
especially in pattern recognition, natural language processing, and inference from large knowledge bases.
Whether this will actually happen within the next decade is a matter of conjecture.

During the past 20 years, parallel processing has been
a highly promising area for research. Yet what could
actually be accomplished with parallel processing has
always lagged behind expectations. At this juncture,
however, it's possible that important breakthroughs
may be around the corner.

But there are many problems in AI that will not be
helped to an appreciable extent by the availability of
supercomputers, whether they will be von Neumann-
type supercomputers or some other kind. The reason
why this is so is because the limitation is not so much
computing power, but our lack of understanding of
some of the processes required to perform even simple
cognitive tasks.

Q. **You say that there are many problems that won't
be solved by the availability of supercomputers. Could
you give an example of such a problem?**
ZADEH.   Let me start with a problem basic to most
other problems: the problem of summarization. Now,
when I talk about summarization, I am not talking
about summarizing a short stereotypical story. That capability we have, thanks to Roger Schank, his associates
at Yale University, and others within the AI community.

But what we have no understanding of, whatsoever,
is how to summarize a nonstereotypical story that is
not a short story.

Q. **Could you give an example of a stereotypical story?**
ZADEH. An example would be accounts of automobile accidents. They tend to be stereotypical. In other words, in the story there is an indication of what kind of accident it was, when it occurred, where it occurred, whether there were injuries, etc.

Q. **I see. It has a predetermined structure?**
ZADEH. Yes. When I say stereotypical, I mean it has a predetermined structure. So if you have a predetermined structure, then you can understand the story and you can summarize it.

Now, the reason why summarization is so difficult— it is far more difficult than machine translation from one language to another language—is because summarization requires understanding. And the ability of a computer to summarize a short, stereotypical story would be a little bit like the ability of a person who sees a story of that kind in the newspaper and doesn't understand completely the language in which the story is written. Nonetheless, he can discern a few words here and there and, on that basis, summarize the story. Many people could do that if they have some minimal competence in the language in which the story is written. But that minimal competence is completely inadequate when it comes to summarizing something that is not stereotypical and not short.

Q. **You said that it is very difficult and often impossible at present to write a program to summarize a story. So why is that an important point?**
ZADEH. The ability to summarize is an acid test of the ability to understand, which in turn is a test of intelligence and competence. Suppose I asked a person not familiar with mathematics to summarize a paper in a mathematical journal. It would be impossible for him to summarize it, because he doesn't understand what that paper is about, what the results are, what the significance is, and so forth.

So in a situation like that, it wouldn't help us to have a supercomputer. It wouldn't help us to have all the supercomputers in the world put together. That's not where the problem lies.

Q. **Professor Zadeh, could you give me a few other examples of problems that will still defy solution even with major advances in supercomputer design?**

ZADEH.   Take the problem of identification of ethnic origin. Humans can do that. You can look at a person and say, "Gee, he looks Irish," or whatever. Now, it would be impossible at this point to write a program that would look at somebody's picture and identify the ethnic origin of that person. I don't want to say that it's impossible period. I'm merely saying that at this moment it's impossible.

Another problem is estimation of age. Assume that you look at somebody, and you say, "Well, this person must be around 35." Again, we cannot write a program that would enable a computer to do it at this point. And we can't put our finger on subtle differences, like the difference between a person who is 20 years old and somebody who is 10.

Q.   **Why is that? Why is it impossible to write a program to estimate a person's age from an analysis of physical features?**
ZADEH.   Because we don't understand too well how we arrive at assessments of that kind. In other words, in order to write a program, we have to have an understanding of how we do it. The limitation in problems of this kind is that we cannot articulate the rules that we employ subconsciously to make that kind of an assessment.

Of course, you know, there are certain things that might not be so difficult. It's not so difficult to differentiate between a person who is 70 years old and a person who is, for instance, 5 years old. But I'm talking about kinds of problems in the estimation of age that are not as trivial as that. And we know that you cannot base it entirely on wrinkles or color of hair. It's the totality of these things put together that enable us to make an assessment.

Still another problem of that same type is the problem of the identification of a musical tune. People can identify a tune if they hear just a few bars. They generally can guess who the composer is, even though they may never have heard this piece before. In other words, there's something about the way the music composed by a particular composer sounds that makes it possible for us to say, "Well, this is Mozart," even though we may never have heard that piece before.

Again, if somebody asked how you guessed that it was Mozart, you would not be able to put down on paper the criteria that you have employed.

Q. **You're saying that this process of recognizing that a short burst of music is from Mozart, Beethoven, or someone else is something going on unconsciously or intuitively?**

ZADEH. It's something that we can do without being able to articulate the rules. In other words, the decisional algorithms that we employ for this purpose are opaque rather than transparent.

The problems that I mentioned—ethnic origin identification, age estimation, composer identification, or tune recognition—all of these are problems in pattern recognition. Many of these problems are far from solution at this point. And they present right now a stumbling block to such applications as speech recognition for connected speech.

Q. **What is the point of all these examples? What is the lesson here?**

ZADEH. What I have said so far is intended merely to give the reasons for my feeling that the availability of supercomputers will not help us much in solving problems of that kind. This is the issue that I was really addressing myself to.

But at this point, all I'm trying to say is that the supercomputer effort and the AI-oriented type of computer effort to develop machines that can perform nontrivial cognitive tasks are not quite the same. There is some interaction between them, but they are qualitatively different.

Let us return now to area number three, *knowledge engineering*. What I have to say here will be at considerable variance with the widely held positions within the AI community.

AI, as we know it today, is based on two-valued logic—that is, the classical Aristotelian logic. And it is generally assumed that all you need as a foundation of AI is first-order logic.

Q. **You say that AI today is based on two-valued logic. Is that a yes-or-no kind of logic?**

ZADEH. Yes, that's right. It's a yes-or-no kind of logic. Actually, two-valued logic encompasses a variety of logical systems, all of which share the basic assumption that truth is two-valued. One of these logical systems is what is called first-order logic. And so the assumption that many people make is that first-order logic, perhaps with some modifications, is sufficient.

Q. **Does a simple example come to mind of first-order logic?**

ZADEH. Well, suppose you say, "All men are mortal. Socrates is a man. Therefore Socrates is mortal." This would be a very simple example of reasoning in first-order logic.

Briefly, within the AI community at this point, there are two camps. One camp, the conservative camp, takes the position that AI, and more generally knowledge engineering, should be based on logic, and in particular on first-order logic. One of the main proponents of this view is John McCarthy of Stanford University. Other prominent proponents include Nils Nilsson of SRI, Wolfgang Bibel of the University of Munich, Robert Kowalski of London, and Alain Colmerauer of Marseilles, France.

Now the other camp takes the position that logic is of limited or no relevance to AI. They believe that first-order logic is too limited to be able to deal effectively with the complexity of human cognitive processes. Instead of systematic, logical methods, this second camp relies on the use of ad hoc techniques and heuristic procedures. The prime exponents of this position are Roger Schank of Yale University and, more recently and less emphatically, Marvin Minsky of MIT.

Q. **How does your position, Professor Zadeh, differ from those in the two AI camps you mentioned—the conservatives who believe in first-order logic and the other camp that believes logic has only limited relevance?**

ZADEH. The position that I take—and this is really what differentiates me from most of the people in AI— is that we need logic in AI. But the kind of logic we need is not *first-order logic*, but *fuzzy logic*—that is, the logic that underlies inexact or approximate reasoning.

I feel this way because most of human reasoning— almost all of human reasoning—is imprecise. Much of it is what might be called common sense reasoning. And first-order logic is much too precise and much too confining to serve as a good model for common sense reasoning.

The reason why humans can do many things that present-day computers cannot do well or perhaps even at all is because existing computers employ two-valued logic.

To put it another way, the inability of today's computers to solve some of those problems I mentioned earlier is not that we don't have enough computing capacity. Rather, the computers we have today—in terms of both hardware and software—are not oriented toward the processing of fuzzy knowledge and common sense reasoning. This is where the problem lies in my view.

Q. Doesn't that view—the idea that computers don't mimic human thought processes very well—imply that there is something sacred about the way human beings think? Isn't it possible that the way humans think about things is not very good and that it might be possible to conceive an artificial way of thinking that is superior?

ZADEH. Of course, one could take the position, as some workers in AI do, that it is not essential to mimic the human mind in the design of AI systems. One argument is that when we design an aircraft, so goes one of the arguments, we don't design it like a bird. But somehow this plane-design analogy doesn't seem to pertain to the design of AI systems. For when we actually attempt to build AI systems that can perform humanlike cognitive tasks we invariably seem to come back to the human model. The human model is a pretty good one—better than many people thought.

Q. Could you elaborate a bit more on just what fuzzy logic is?

ZADEH. Let me do that. I will explain the main differences between fuzzy logic and classical two-valued logic. In classical two-valued systems, all classes are assumed to have sharply defined boundaries. So either an object is a member of a class or it is not a member of a class.

Now, this is okay if you are talking about something like mortal or not mortal, dead or alive, male or female, and so forth. These are examples of classes that have sharp boundaries.

But most classes in the real world do not have sharp boundaries. For example, if you consider characteristics or properties like tall, intelligent, tired, sick, and so forth, all of these characteristics lack sharp boundaries. Classical two-valued logic is not designed to deal with properties that are a matter of degree. This is the first point.

Now, there is, of course, a generalization of two-valued logic. And these generalized logical systems are called multivalued logics. So in multivalued logical systems, a property can be possessed to a degree.

Q. I'm not perfectly clear here. Consider the word "tall." Are you saying "tall" can take on multiple values?

ZADEH. Yes, tallness becomes a matter of degree, as does intelligence, tiredness, and so forth. Usually you have degrees between zero and one. So you can say, for example, that a person is tall to the degree 0.9. These degrees are grades of membership that may be interpreted as truth values.

In classical logic, there are just two truth values: true/false (or one and zero). In multivalued logical systems, there are more than two truth values. There may be a finite or even an infinite number of truth values, that is, an infinite number of degrees to which a property may be possessed.

In a three-valued system, for instance, something can be true, false, or on the boundary. Or you can have systems in which one has a continuum of truth values from zero to one.

Q. Who first developed multivalued logic?

ZADEH. The person best known in that connection is a Polish mathematician by the name of J. Lukasiewiecz. He first developed the concept of multivalued logic during the 1920s.

Q. Could you give an example or two of a situation that requires multivalued logic?

ZADEH. Well, you would need a multivalued system to be able to say something like "John is tall." For tall is a property that requires an infinity of truth values to describe it. So something as simple as "John is tall" would require multivalued logic—unless you arbitrarily establish a threshold by saying "somebody over 6 feet tall is tall, and those who are less than 6 feet tall are not tall." In other words, unless you artificially introduce some sort of a threshold like that, you will need multivalued logic.

But even though these multivalued logical systems have been available for some time, they have not been used to any significant extent in linguistics, in psychology, and in other fields where human cognition plays an important role.

Q. Why hasn't multivalued logic been used?

ZADEH. The reason such systems haven't been used is that multivalued logic doesn't go far enough. And this is where *fuzzy logic* enters the picture.

What differentiates fuzzy logic from multivalued logic is that in fuzzy logic you can deal with fuzzy quantifiers, like "most," "few," "many," and "several."

Fuzzy quantifiers have something to do with enumeration, that is, with counting. But they are fuzzy because they don't give you the count exactly, but fuzzily. For instance, you say "many" or "most."

In multi-valued logic you have only two quantifiers, "all" and "some," whereas in fuzzy logic you have all the fuzzy quantifiers. This is one of the important differences.

Q. **Now are those the only fuzzy quantifiers?**
ZADEH.   Well, the ones that I mentioned are merely examples. In reality, there is an infinite number of fuzzy quantifiers. For example, you can say "not very many," "quite a few," "many more than 10," "a large number," "many," "few," or "very many." There is an infinite number of ways in which you can describe in an approximate fashion a count of objects.

Q. **Besides the fuzzy quantifiers, what else distinguishes fuzzy logic from multivalued logic?**
ZADEH.   Another key difference is that in fuzzy logic *truth* itself is allowed to be fuzzy. So it is okay to say that something is "quite true." You can say "it's more or less true." You can also use fuzzy probability like "not very likely," "almost impossible," or "rarely." In this way, fuzzy logic provides a system that is sufficiently flexible and expressive to serve as a natural framework for the semantics of natural languages.

Furthermore, it can serve as a basis for reasoning with common sense knowledge, for pattern recognition, decision analysis, and other application areas in which the underlying information is imprecise. Within the restricted framework of two-valued and even multivalued systems, these problem areas have proved to be difficult to deal with systematically.

The crux of the problem, really, is the excessively wide gap between the precision of classical logic and the imprecision of the real world.

Q. **Does fuzzy logic, then, provide a good match with the imprecise real world?**
ZADEH.   I don't wish to imply that fuzzy logic is in any sense an ultimate system. I do believe, however, that it is far better suited for dealing with real-world problems than the traditional logical systems.

Q. **At this juncture, Professor Zadeh, could you encapsulate what you feel is your most important point so far?**

ZADEH. Yes. It is that the limitation in knowledge engineering is not the unavailability of supercomputers. But it is the fact that computers—both their hardware and software—are based on a kind of logic that is not a good model for human reasoning.

Ultimately, the problem lies at the hardware level. For computers are basically digital devices: they deal with discrete bits of information. Fuzzy information, on the other hand, is not discrete. With fuzzy information, one thing merges into another.

Now an important point is this: even though present-day computers are based on two-valued logic, they can be programmed to process fuzzy information using fuzzy logic. But doing this does not represent an efficient use of the computational capabilities of present-day computers.

Q. **You're saying there are problems with both the hardware and the software of existing computer systems, but that you can overcome them so that traditional computers can still handle fuzzy information?**

ZADEH. Yes, you can overcome the hardware limitations of current computers with software. But doing that involves an inefficient use of computers.

The ability of the human mind to reason in fuzzy terms is actually a great advantage. Even though a tremendous amount of information is presented to the human senses in a given situation—an amount that would choke a typical computer—somehow the human mind has the ability to discard most of this information and to concentrate only on the information that is task relevant. This ability of the human mind to deal only with the information that is task relevant is connected with its ability to process fuzzy information. By concentrating only on the task-relevant information, the amount of information the brain has to deal with is reduced to a manageable level.

Q. **So you can work with traditional computers, but they are inefficient because they are fundamentally incompatible with fuzzy information. What kind of computer could you use then?**

ZADEH. At some point, we may be able to conceive computers that are radically different from existing computers in that the operations they perform are rooted in fuzzy rather than two-valued logic. In other words, they may need a different kind of hardware.

There has been talk about "chemical," "biological," or "molecular" computers.

Some imaginative thinkers are talking about computers of that kind, but we don't have them yet.

*Q. **What is a molecular computer?***
ZADEH. It's difficult to say. But if you compare the way the human brain works with the way a modern computer works, I think that you will find there are some fundamental differences. The human brain, in a way that we don't understand too well at present, uses fuzzy logic.

So the hardware—if you may call it that—of the human brain is the kind of hardware that is effective for manipulating imprecise information. When one uses the term molecular computer, or biological computer or chemical computer, what one has in mind is something that approximates the way in which the human brain processes information. And therein lies a fundamental challenge: how to develop a better understanding of how the human brain processes this fuzzy information so effectively.

*Q. **How much is presently understood about how the brain processes information?***
ZADEH. Scientists know a lot about the functioning of the brain at the neuron level. But how does the activity at that level aggregate into thinking processes? Trying to understand the brain at the neuron level is like trying to understand the functioning of a telephone system in a large city by examining the wiring of a telephone set. We can understand something on the microlevel but are unable to integrate that into an understanding of functioning on higher levels.

*Q. **I don't completely understand your reasons for writing off traditional computers. Maybe people haven't tried hard enough or long enough, as yet, to make them work for certain AI applications?***
ZADEH. In fact, I'm not writing off traditional computers in regard to their ability to process fuzzy information. Rather, my position is that we do not have a good understanding at this point as to how to use them efficiently for handling fuzzy information.

In fact, I believe that there will be a growing number of applications of fuzzy logic in a wide variety of fields

using present-day computers. But, ultimately, to
achieve a higher level of efficiency, it may be necessary
to employ computers that are specially designed for
dealing with fuzzy information.

To give one example, speech recognition is a problem
far from a satisfactory solution. We do have speech
recognition systems with limited capability to understand speech. But all of these systems do not scale up.
That is, they can not be merely modified and improved
in an evolutionary way until they come close to the
human ability to understand speech. So it's obvious
that what is needed is an altogether different approach.

Q. **I'm not perfectly clear when you say these systems won't scale up. Could you expand on that point?**
ZADEH. Yes. In some situations we have a system
that has a limited capability. But we see clearly how,
by improving that system, we can raise its level of performance to a point where it can compete with humans
in terms of certain abilities.

Now, within AI, this is generally not the case. That
is, many AI systems do not scale up: They reach very
quickly the limit of their ability. In other words, you
cannot push them beyond that point.

Q. **Could you give an example of a system that does
not scale up?**
ZADEH. A good example, to go back to what I said
previously, are the programs that can summarize.
These programs reach the limit of their ability very
quickly—in terms of the length of the story they can
summarize or in terms of the degree to which the story
is nonstereotypical. And you cannot go beyond those
limits without radically altering the approaches used.

Another example. Before we had integrated circuits,
we depended on vacuum tubes in the design of our
computers, and you could push the capabilities of those
early computers only up to a point. We had to come up
with something new, the concept of an integrated cir-
cuit, and eventually very large-scale integration (VLSI).
Those breakthroughs greatly increased our capability to
compute, to store, and more generally, to process information.

That was a situation that called for something radically different. And it wasn't a matter of evolution—but
of revolution.

And so I think that we are faced with a somewhat
similar situation in the case of computers that can per-

form high-level cognitive tasks. That is, we cannot hope to be able to solve the problems of the kind I mentioned earlier, in particular the problem of summarization, through evolutionary improvements in present computer hardware or software.

Q. In other words, some of the more difficult problems in AI won't be solved by innovations in super-computer architecture—innovations like parallel processing?
ZADEH.   That's right, such innovations aren't going to help much.

But there is more to this fuzzy logic than simply the enhancement of the ability of computers to solve various problems—to perform nontrivial cognitive tasks. Accepting fuzzy logic will also call for a certain fundamental shift in attitudes, particularly in theoretical computer science. At this point theoretical computer science is mathematical in spirit, in the sense that it is oriented toward the discovery and proof of results that can be stated as theorems.

Unfortunately, there is an incompatibility between precision and complexity. As the complexity of a system increases, our ability to make precise and yet non-trivial assertions about its behavior diminishes. For example, it is very difficult to prove a theorem about the behavior of an economic system that is of relevance to real-world economics.

What I anticipate in the future is a growing recognition of the necessity to find an accommodation with the pervasive imprecision of the real world. This change is needed to be able to make assertions that are not just nontrivial theorems, but something of relevance to practice. In computer science today, people use two-valued logic to establish certain results. But such results are often limited in their relevance to the real world—because they are excessively precise. In other words, we have to accord acceptance to assertions that do not adhere to high standards of precision.

This accommodation with imprecision will require the use of fuzzy logic. Gradually and perhaps rather slowly, there will be a growing acceptance of fuzzy logic as a conceptual framework for computer science.

Now, it is a little bit more difficult to articulate this particular position than some of the earlier things that I said. For this gets into issues that relate not just to computer science but, more generally, to science itself. Science at this point is based on two-valued logic. So what I'm talking about is a significant shift in attitude, not just in computer science, but more generally in scientific thinking.

At this point there is a long-standing and deep-seated tradition of according respectability to what is mathematical and precise. We may have to retreat from this tradition in order to be able to say something useful about complex systems and in particular about systems in which human reasoning plays an important role.

Q. Okay, then, there have been instances in the past where scientists have been too preoccupied with mathematics and precision and, as a result, have failed to come up with useful results. Does an example come to mind?

ZADEH. Yes. Take economics. Time and again, it has been demonstrated that what actually happens in the realm of economics is very different from what the experts predicted. These experts might be using large-scale econometric models, sophisticated mathematics, large-scale computers, and the like. Despite all that, the forecasts turn out to be wrong—very wrong.

Why? Two reasons. One is that economic systems are very complex. Second, and more important, human psychology plays an essential role in the behavior of such systems. And this complexity, together with human reasoning, makes the classical mathematical approaches, based on two-valued logic, ineffective.

So, again, to approximate the way humans can sort through large masses of data and arrive at some sort of a qualitative conclusion, it might be necessary to use fuzzy logic.

Q. Has fuzzy logic been able to solve some of the difficult problems in AI you mentioned earlier? Or is it still just a promise?

ZADEH. These problems are intrinsically complex, and fuzzy logic by itself does not provide a solution to them. Rather, it merely enhances our ability to do so without guaranteeing success. It's a little like finding a cure for cancer. You may develop a technique that may help in finding a cure but it doesn't guarantee a cure will be found.

Fuzzy logic, then, is a necessary but not sufficient condition to finding solutions to these problems. It is a tool that enhances our ability to deal with problems that are too complex and too ill-defined to be susceptible to solution by conventional means. It will be an ingredient of the tools that will eventually be used to solve these problems.

Q. **Have you made any headway in persuading people that they needn't always be superprecise, that in fact such an approach may be an inappropriate approach for attacking certain types of problems?**

ZADEH. It will be a slow process. It's not very easy to change some of the basic attitudes people have been educated with, like the attitude that we must be very precise and that we have to try to come up with results that can be stated as theorems. It's difficult to change these attitudes.

Let me draw an analogy with the way people dress. Classical logic is like a person who comes to a party dressed in a black suit, a white, starched shirt, a black tie, shiny shoes, and so forth. And fuzzy logic is a little bit like a person dressed informally, in jeans, tee shirt, and sneakers. In the past, this informal dress wouldn't have been acceptable. Today, it's the other way around. Somebody who comes dressed to a party in the way I described earlier would be considered funny.

Changes in attitude may take place not only in dress but also in science, music, art, and many other fields. And, in science, there may be an increasing willingness to realize that excessively high degree of formalism, rigor, and precision is counterproductive.

Freedom of expression in science could exhibit itself as a movement away from two-valued logic and toward fuzzy logic. Fuzzy logic is much more general and it gives you much more flexibility.

Q. **How long will it take for traditional scientific attitudes about precision to change and fuzzy logic to take hold?**

ZADEH. Well, I think it will take something on the order of perhaps a couple of decades. Fuzzy logic is making inroads, but it is not something that has coalesced into a broad movement. In other words, there are pockets. These pockets exist in various fields, and of course, there are some people who view these pockets with suspicion and hostility—just as some people who are conservative look with suspicion on those who dress informally.

The difficulty of persuading people has to do also with the question of where does respectability lie. Traditionally, respectability went along with being more mathematical, more precise and more quantitative. And these attitudes go back to Lord Kelvin who said that it's not really a science if it's not quantitative.

But fuzzy logic now challenges that. There are many things that cannot be expressed in numbers, for example, probabilities that have to be expressed as "very likely," or "unlikely." Such linguistic probabilities may be viewed as fuzzy characterizations of conventional numeric probabilities.

And so in that sense fuzzy logic represents a retreat. It represents a retreat from standards of precision that are unrealistic.

There are many parallels to that sort of thing in the history of human thought, where people didn't realize that the objectives they set were unrealizable.

Q.   Does an example or two come to mind of a situation where scientists had to retreat from standards of precision that were not attainable?
ZADEH.   Well, a good example of that sort of thing is statistical mechanics. People in the beginning of the nineteenth century were firm believers in the possibility of using the mechanics that were developed at that time by people like Lagrange and applying those mechanics to the solution of all sorts of problems involving the motion of bodies. But then they encountered the "two-body," "three-body," and "n-body" problems, and it became clear that they could not push this too far. That's where the groundwork was laid for statistical mechanics.

So statistical mechanics represented a retreat, a retreat in the sense that you say, "Well, I cannot say something precisely, but I'll say it statistically."

Now, the same thing happened in the case of the solution of differential equations. Today we freely accept numerical solutions. It is hard to realize that the idea of a numerical solution was not acceptable even as recently as perhaps 30–40 years ago.

Q.   The rise of numerical analysis, then, constituted a retreat. Was it more of a brute force approach rather than an elegant, logical approach to the solution of differential equations?
ZADEH.   Effectively, yes. People were simply not willing to say that, if you use the computer to come up with a numerical solution, you have really done something worthwhile. Somehow we tend to forget that things that are acceptable today were not acceptable 20 to 30 years ago.

Q.   I can remember reading books on science of a few decades ago that always spelled science with a capital S.

ZADEH.   Yes. It's that kind of veneration or worship I'm talking about. I sometimes use a word that offends people who take the more traditional view, and that word is *fetishism*—fetishism of precision and rigor in the context of classical logic.

There is also what might be referred to as "the curse of respectability in science." In trying to be respectable, scientists deny themselves the use of more flexible logical systems in which truth is a matter of degree.

Q. Is there anything that could be done to get certain people to stop worshiping precision?
ZADEH.   I think it has to be a natural process. But because of the current emphasis on AI, and in particular on expert systems, there is a rapidly growing interest in inexact reasoning and processing of knowledge that is imprecise, incomplete, or not totally reliable. And it is in this connection that it will become more and more widely recognized that classical logical systems are inadequate for dealing with uncertainty and that something like fuzzy logic is needed for that purpose.

Q. Since you first developed the concept of fuzzy logic in the 1960s, Professor Zadeh, has there been much of a growth in interest? Have others picked up the banner?
ZADEH.   Between then and now, somewhere between 3,000 and 4,000 papers have been written worldwide on fuzzy sets and their applications. And there are two regular journals: *Fuzzy Sets and Systems*, in English, and *Fuzzy Mathematics*, in Chinese. In addition, a quarterly entitled *Bulletin on Fuzzy Sets and their Applications* is published in France. The countries where most activity is taking place at this point are the Soviet Union, China, Japan, France, Great Britain, West Germany, East Germany, Poland, Italy, Spain and India. There has been less activity in the United States.

There is growing acceptance, but there is also considerable skepticism and in some instances hostility. At this point the largest number of researchers working on fuzzy sets is in China.

There appears to be more sympathy for other than two-valued systems in oriental countries, perhaps because their logic is not like Western, Cartesian logic. There is a greater acceptance of truth that is neither perfect truth nor perfect falsehood. This is particularly characteristic of Hindu, Chinese, and Japanese cultures.

Q. **Professor Zadeh, that's the end of our questions. We on the editorial staff of *Communications* thank you warmly for giving our readers some of your views.**
ZADEH. It was my pleasure.

RECOMMENDED READING
- Dubois and Prade. *Fuzzy Sets and Systems: Theory and Applications.* Academic Press, New York, 1980. This is a good, broad introduction to the subject.
- *Fuzzy Information and Decision Processes.* Elsevier North-Holland, New York, 1982. This is a recent volume of edited papers dealing with both theory and applications.
- *Fuzzy Sets and Possibility Theory.* Pergamon Press, New York, 1982. This is also a recent volume of edited papers dealing with both theory and applications.
- *Fuzzy Reasoning and Its Applications.* Academic Press, New York, 1981. A volume of edited papers.
- *Fuzzy Sets.* Plenum Press, New York, 1980.
- There are three periodicals devoted to fuzzy sets: *Fuzzy Sets and Systems,* Elsevier North-Holland, New York; *Bulletin on Fuzzy Sets and Their Application,* published in Toulouse, France; and *Fuzzy Mathematics* (in Chinese), published in Wuhan, China.

Fuzzy Sets*

L.A. ZADEH

*Department of Electrical Engineering and Electronics Research
Laboratory, University of California, Berkeley, California*

A fuzzy set is a class of objects with a continuum of grades of membership. Such a set is characterized by a membership (characteristic) function which assigns to each object a grade of membership ranging between zero and one. The notions of inclusion, union, intersection, complement, relation, convexity, etc., are extended to such sets, and various properties of these notions in the context of fuzzy sets are established. In particular, a separation theorem for convex fuzzy sets is proved without requiring that the fuzzy sets be disjoint.

I. INTRODUCTION

More often than not, the classes of objects encountered in the real physical world do not have precisely defined criteria of membership. For example, the class of animals clearly includes dogs, horses, birds, etc. as its members, and clearly excludes such objects as rocks, fluids, plants, etc. However, such objects as starfish, bacteria, etc. have an ambiguous status with respect to the class of animals. The same kind of ambiguity arises in the case of a number such as 10 in relation to the "class" of all real numbers which are much greater than 1.

Clearly, the "class of all real numbers which are much greater than 1," or "the class of beautiful women," or "the class of tall men," do not constitute classes or sets in the usual mathematical sense of these terms. Yet, the fact remains that such imprecisely defined "classes" play an important role in human thinking, particularly in the domains of pattern recognition, communication of information, and abstraction.

The purpose of this note is to explore in a preliminary way some of the basic properties and implications of a concept which may be of use in

* This work was supported in part by the Joint Services Electronics Program (U.S. Army, U.S. Navy and U.S. Air Force) under Grant No. AF-AFOSR-139-64 and by the National Science Foundation under Grant GP-2413.

dealing with "classes" of the type cited above. The concept in question is that of a *fuzzy set*,[1] that is, a "class" with a continuum of grades of membership. As will be seen in the sequel, the notion of a fuzzy set provides a convenient point of departure for the construction of a conceptual framework which parallels in many respects the framework used in the case of ordinary sets, but is more general than the latter and, potentially, may prove to have a much wider scope of applicability, particularly in the fields of pattern classification and information processing. Essentially, such a framework provides a natural way of dealing with problems in which the source of imprecision is the absence of sharply defined criteria of class membership rather than the presence of random variables.

We begin the discussion of fuzzy sets with several basic definitions.

II. DEFINITIONS

Let X be a space of points (objects), with a generic element of X denoted by x. Thus, $X = \{x\}$.

A *fuzzy set* (*class*) A in X is characterized by a *membership* (*characteristic*) *function* $f_A(x)$ which associates with each point[2] in X a real number in the interval [0, 1],[3] with the value of $f_A(x)$ at x representing the "grade of membership" of x in A. Thus, the nearer the value of $f_A(x)$ to unity, the higher the grade of membership of x in A. When A is a set in the ordinary sense of the term, its membership function can take on only two values 0 and 1, with $f_A(x) = 1$ or 0 according as x does or does not belong to A. Thus, in this case $f_A(x)$ reduces to the familiar characteristic function of a set A. (When there is a need to differentiate between such sets and fuzzy sets, the sets with two-valued characteristic functions will be referred to as *ordinary sets* or simply *sets*.)

Example. Let X be the real line R^1 and let A be a fuzzy set of numbers

---

[1] An application of this concept to the formulation of a class of problems in pattern classification is described in RAND Memorandum RM-4307-PR, "Abstraction and Pattern Classification," by R. Bellman, R. Kalaba and L. A. Zadeh, October, 1964.

[2] More generally, the domain of definition of $f_A(x)$ may be restricted to a subset of X.

[3] In a more general setting, the range of the membership function can be taken to be a suitable partially ordered set P. For our purposes, it is convenient and sufficient to restrict the range of f to the unit interval. If the values of $f_A(x)$ are interpreted as truth values, the latter case corresponds to a multivalued logic with a continuum of truth values in the interval [0, 1].

which are much greater than 1. Then, one can give a precise, albeit subjective, characterization of A by specifying $f_A(x)$ as a function on R^1. Representative values of such a function might be: $f_A(0) = 0; f_A(1) = 0;$ $f_A(5) = 0.01; f_A(10) = 0.2; f_A(100) = 0.95; f_A(500) = 1.$

It should be noted that, although the membership function of a fuzzy set has some resemblance to a probability function when X is a countable set (or a probability density function when X is a continuum), there are essential differences between these concepts which will become clearer in the sequel once the rules of combination of membership functions and their basic properties have been established. In fact, the notion of a fuzzy set is completely nonstatistical in nature.

We begin with several definitions involving fuzzy sets which are obvious extensions of the corresponding definitions for ordinary sets.

A fuzzy set is *empty* if and only if its membership function is identically zero on X.

Two fuzzy sets A and B are *equal*, written as $A = B$, if and only if $f_A(x) = f_B(x)$ for all x in X. (In the sequel, instead of writing $f_A(x) = f_B(x)$ for all x in X, we shall write more simply $f_A = f_B$.)

The *complement* of a fuzzy set A is denoted by A' and is defined by

$$f_{A'} = 1 - f_A . \tag{1}$$

As in the case of ordinary sets, the notion of containment plays a central role in the case of fuzzy sets. This notion and the related notions of union and intersection are defined as follows.

Containment. A is *contained in* B (or, equivalently, A is a *subset of B*, or A is *smaller than or equal to B*) if and only if $f_A \leqq f_B$. In symbols

$$A \subset B \Leftrightarrow f_A \leqq f_B . \tag{2}$$

Union. The *union* of two fuzzy sets A and B with respective membership functions $f_A(x)$ and $f_B(x)$ is a fuzzy set C, written as $C = A \cup B$, whose membership function is related to those of A and B by

$$f_C(x) = \text{Max} [f_A(x), f_B(x)], \qquad x \in X \tag{3}$$

or, in abbreviated form

$$f_C = f_A \vee f_B . \tag{4}$$

Note that $\cup$ has the associative property, that is, $A \cup (B \cup C) = (A \cup B) \cup C$.

Comment. A more intuitively appealing way of defining the union is the following: The union of A and B is the smallest fuzzy set containing both A and B. More precisely, if D is any fuzzy set which contains both A and B, then it also contains the union of A and B.

To show that this definition is equivalent to (3), we note, first, that C as defined by (3) contains both A and B, since

$$\text{Max} \, [f_A \, , f_B] \geqq f_A$$

and

$$\text{Max} \, [f_A \, , f_B] \geqq f_B \, .$$

Furthermore, if D is any fuzzy set containing both A and B, then

$$f_D \geqq f_A$$

$$f_D \geqq f_B$$

and hence

$$f_D \geqq \text{Max} \, [f_A \, , f_B] = f_C$$

which implies that $C \subset D$. Q.E.D.

The notion of an intersection of fuzzy sets can be defined in an analogous manner. Specifically:

Intersection. The *intersection* of two fuzzy sets A and B with respective membership functions $f_A(x)$ and $f_B(x)$ is a fuzzy set C, written as $C = A \cap B$, whose membership function is related to those of A and B by

$$f_C(x) \, = \, \text{Min} \, [f_A(x), f_B(x)], \qquad x \in X, \qquad (5)$$

or, in abbreviated form

$$f_C = f_A \wedge f_B \, . \qquad (6)$$

As in the case of the union, it is easy to show that the intersection of A and B is the *largest* fuzzy set which is contained in both A and B. As in the case of ordinary sets, A and B are *disjoint* if $A \cap B$ is empty. Note that $\cap$, like $\cup$, has the associative property.

The intersection and union of two fuzzy sets in R^1 are illustrated in Fig. 1. The membership function of the union is comprised of curve segments 1 and 2; that of the intersection is comprised of segments 3 and 4 (heavy lines).

Comment. Note that the notion of "belonging," which plays a fundamental role in the case of ordinary sets, does not have the same role in

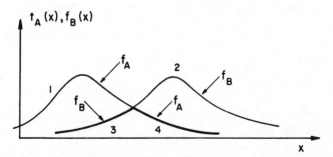

$t_A(x), f_B(x)$

FIG. 1. Illustration of the union and intersection of fuzzy sets in R^1

the case of fuzzy sets. Thus, it is not meaningful to speak of a point x "belonging" to a fuzzy set A except in the trivial sense of $f_A(x)$ being positive. Less trivially, one can introduce two levels α and β $(0 < \alpha < 1,$ $0 < \beta < 1,$ $\alpha > \beta)$ and agree to say that (1) "x belongs to A" if $f_A(x) \geqq \alpha$; (2) "x does not belong to A" if $f_A(x) \leqq \beta$; and (3) "x has an indeterminate status relative to A" if $\beta < f_A(x) < \alpha$. This leads to a three-valued logic (Kleene, 1952) with three truth values: T $(f_A(x) \geqq \alpha)$, F $(f_A(x) \leqq \beta)$, and U $(\beta < f_A(x) < \alpha)$.

III. SOME PROPERTIES OF ∪, ∩, AND COMPLEMENTATION

With the operations of union, intersection, and complementation defined as in (3), (5), and (1), it is easy to extend many of the basic identities which hold for ordinary sets to fuzzy sets. As examples, we have

$$(A \cup B)' = A' \cap B' \atop (A \cap B)' = A' \cup B' \Big\} \text{De Morgan's laws} \qquad (7) \atop (8)$$

$$C \cap (A \cup B) = (C \cap A) \cup (C \cap B) \qquad \text{Distributive laws.} \quad (9)$$

$$C \cup (A \cap B) = (C \cup A) \cap (C \cup B) \qquad (10)$$

These and similar equalities can readily be established by showing that the corresponding relations for the membership functions of A, B, and C are identities. For example, in the case of (7), we have

$$1 - \text{Max}\,[f_A, f_B] = \text{Min}\,[1 - f_A, 1 - f_B] \qquad (11)$$

which can be easily verified to be an identity by testing it for the two possible cases: $f_A(x) > f_B(x)$ and $f_A(x) < f_B(x)$.

Similarly, in the case of (10), the corresponding relation in terms of f_A, f_B, and f_C is:

$$\text{Max}\,[f_C\,,\,\text{Min}\,[f_A\,,f_B] = \text{Min}\,[\text{Max}\,[f_C\,,f_A],\,\text{Max}\,[f_C\,,f_B]] \quad (12)$$

which can be verified to be an identity by considering the six cases:

$$f_A(x) > f_B(x) > f_C(x), f_A(x) > f_C(x) > f_B(x), f_B(x) > f_A(x) > f_C(x),$$

$$f_B(x) > f_C(x) > f_A(x), f_C(x) > f_A(x) > f_B(x), f_C(x) > f_B(x) > f_A(x).$$

Essentially, fuzzy sets in X constitute a distributive lattice with a 0 and 1 (Birkhoff, 1948).

An Interpretation for Unions and Intersections

In the case of ordinary sets, a set C which is expressed in terms of a family of sets $A_1\,,\,\cdots\,,\,A_i\,,\,\cdots\,,\,A_n$ through the connectives $\cup$ and $\cap$, can be represented as a network of switches $\alpha_1\,,\,\cdots\,,\,\alpha_n$, with $A_i \cap A_j$ and $A_i \cup A_j$ corresponding, respectively, to series and parallel combinations of α_i and α_j. In the case of fuzzy sets, one can give an analogous interpretation in terms of sieves. Specifically, let $f_i(x)$, $i = 1,\,\cdots\,,\,n$, denote the value of the membership function of A_i at x. Associate with $f_i(x)$ a sieve $S_i(x)$ whose meshes are of size $f_i(x)$. Then, $f_i(x) \vee f_j(x)$ and $f_i(x) \wedge f_j(x)$ correspond, respectively, to parallel and series combinations of $S_i(x)$ and $S_j(x)$, as shown in Fig. 2.

More generally, a well-formed expression involving $A_1\,,\,\cdots\,,\,A_n\,,\,\cup$, and $\cap$ corresponds to a network of sieves $S_1(x)\,,\,\cdots\,,\,S_n(x)$ which can be found by the conventional synthesis techniques for switching circuits. As a very simple example,

$$C = [(A_1 \cup A_2) \cap A_3] \cup A_4 \quad (13)$$

corresponds to the network shown in Fig. 3.
Note that the mesh sizes of the sieves in the network depend on x and that the network as a whole is equivalent to a single sieve whose meshes are of size $f_C(x)$.

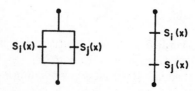

Fig. 2. Parallel and series connection of sieves simultating $\cup$ and $\cap$

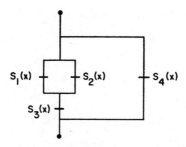

F$_{\mathrm{IG}}$. 3. A network of sieves simultating $\{[f_1(x) \ \vee \ f_2(x)] \ \wedge \ f_3(x)\} \ \vee \ f_4(x)$

IV. ALGEBRAIC OPERATIONS ON FUZZY SETS

In addition to the operations of union and intersection, one can define a number of other ways of forming combinations of fuzzy sets and relating them to one another. Among the more important of these are the following.

Algebraic product. The *algebraic product* of A and B is denoted by AB and is defined in terms of the membership functions of A and B by the relation

$$f_{AB} = f_A f_B \, . \tag{14}$$

Clearly,

$$AB \subset A \cap B. \tag{15}$$

Algebraic sum.[4] The *algebraic sum* of A and B is denoted by $A + B$ and is defined by

$$f_{A+B} = f_A + f_B \tag{16}$$

provided the sum $f_A + f_B$ is less than or equal to unity. Thus, unlike the algebraic product, the algebraic sum is meaningful only when the condition $f_A(x) + f_B(x) \leqq 1$ is satisfied for all x.

Absolute difference. The *absolute difference* of A and B is denoted by $|A - B|$ and is defined by

$$f_{|A-B|} = |f_A - f_B| \, .$$

Note that in the case of ordinary sets $|A - B|$ reduces to the relative complement of $A \cap B$ in $A \cup B$.

[4] The dual of the algebraic product is the *sum* $A \oplus B = (A'B')' = A + B - AB$. (This was pointed out by T. Cover.) Note that for ordinary sets $\cap$ and the algebraic product are equivalent operations, as are $\cup$ and $\oplus$.

Convex combination. By a convex combination of two vectors f and g is usually meant a linear combination of f and g of the form $\lambda f + (1 - \lambda)g$, in which $0 \leq \lambda \leq 1$. This mode of combining f and g can be generalized to fuzzy sets in the following manner.

Let A, B, and Λ be arbitrary fuzzy sets. The *convex combination of A, B, and Λ* is denoted by $(A, B; \Lambda)$ and is defined by the relation

$$(A, B; \Lambda) = \Lambda A + \Lambda' B \tag{17}$$

where Λ' is the complement of Λ. Written out in terms of membership functions, (17) reads

$$f_{(A,B;\Lambda)}(x) = f_\Lambda(x)f_A(x) + [1 - f_\Lambda(x)]f_B(x), \qquad x \in X. \tag{18}$$

A basic property of the convex combination of A, B, and Λ is expressed by

$$A \cap B \subset (A, B; \Lambda) \subset A \cup B \qquad \text{for all } \Lambda. \tag{19}$$

This property is an immediate consequence of the inequalities

$$\text{Min } [f_A(x), f_B(x)] \leq \lambda f_A(x) + (1 - \lambda)f_B(x)$$

$$\leq \text{Max } [f_A(x), f_B(x)], \qquad x \in X \tag{20}$$

which hold for all λ in [0, 1]. It is of interest to observe that, given any fuzzy set C satisfying $A \cap B \subset C \subset A \cup B$, one can always find a fuzzy set Λ such that $C = (A, B; \Lambda)$. The membership function of this set is given by

$$f_\Lambda(x) = \frac{f_C(x) - f_B(x)}{f_A(x) - f_B(x)}, \qquad x \in X. \tag{21}$$

Fuzzy relation. The concept of a *relation* (which is a generalization of that of a *function*) has a natural extension to fuzzy sets and plays an important role in the theory of such sets and their applications—just as it does in the case of ordinary sets. In the sequel, we shall merely define the notion of a fuzzy relation and touch upon a few related concepts.

Ordinarily, a relation is defined as a set of ordered pairs (Halmos, 1960); e.g., the set of all ordered pairs of real numbers x and y such that $x \geq y$. In the context of fuzzy sets, a *fuzzy relation in X* is a fuzzy set in the product space $X \times X$. For example, the relation denoted by $x \gg y$, $x, y \in R^1$, may be regarded as a fuzzy set A in R^2, with the membership function of $A, f_A(x, y)$, having the following (subjective) representative values: $f_A(10, 5) = 0; f_A(100, 10) = 0.7; f_A(100, 1) = 1$; etc.

More generally, one can define an *n-ary fuzzy relation* in X as a fuzzy set A in the product space $X \times X \times \cdots \times X$. For such relations, the membership function is of the form $f_A(x_1, \cdots, x_n)$, where $x_i \in X$, $i = 1, \cdots, n$.

In the case of binary fuzzy relations, the *composition* of two fuzzy relations A and B is denoted by $B \circ A$ and is defined as a fuzzy relation in X whose membership function is related to those of A and B by

$$f_{B \circ A}(x, y) = \text{Sup}_v \text{ Min } [f_A(x, v), f_B(v, y)].$$

Note that the operation of composition has the associative property

$$A \circ (B \circ C) = (A \circ B) \circ C.$$

Fuzzy sets induced by mappings. Let T be a mapping from X to a space Y. Let B be a fuzzy set in Y with membership function $f_B(y)$. The inverse mapping T^{-1} induces a fuzzy set A in X whose membership function is defined by

$$f_A(x) = f_B(y), \qquad y \in Y \tag{22}$$

for all x in X which are mapped by T into y.

Consider now a converse problem in which A is a given fuzzy set in X, and T, as before, is a mapping from X to Y. The question is: What is the membership function for the fuzzy set B in Y which is induced by this mapping?

If T is not one-one, then an ambiguity arises when two or more distinct points in X, say x_1 and x_2, with different grades of membership in A, are mapped into the same point y in Y. In this case, the question is: What grade of membership in B should be assigned to y?

To resolve this ambiguity, we agree to assign the larger of the two grades of membership to y. More generally, the membership function for B will be defined by

$$f_B(y) = \text{Max}_{x \in T^{-1}(y)} f_A(x), \qquad y \in Y \tag{23}$$

where $T^{-1}(y)$ is the set of points in X which are mapped into y by T.

V. CONVEXITY

As will be seen in the sequel, the notion of convexity can readily be extended to fuzzy sets in such a way as to preserve many of the properties which it has in the context of ordinary sets. This notion appears to be particularly useful in applications involving pattern classification, optimization and related problems.

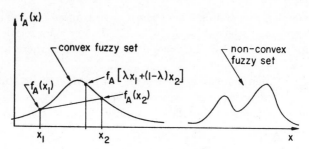

FIG. 4. Convex and nonconvex fuzzy sets in E^1

In what follows, we assume for concreteness that X is a real Euclidean space E^n.

DEFINITIONS

Convexity. A fuzzy set A is *convex* if and only if the sets Γ_α defined by

$$\Gamma_\alpha = \{x \mid f_A(x) \geq \alpha\} \tag{24}$$

are convex for all α in the interval $(0, 1]$.

An alternative and more direct definition of convexity is the following[5]: A is *convex* if and only if

$$f_A[\lambda x_1 + (1 - \lambda)x_2] \geq \mathrm{Min}\,[f_A(x_1), f_A(x_2)] \tag{25}$$

for all x_1 and x_2 in X and all λ in $[0, 1]$. Note that this definition does not imply that $f_A(x)$ must be a convex function of x. This is illustrated in Fig. 4 for $n = 1$.

To show the equivalence between the above definitions note that if A is convex in the sense of the first definition and $\alpha = f_A(x_1) \leq f_A(x_2)$, then $x_2 \in \Gamma_\alpha$ and $\lambda x_1 + (1 - \lambda)x_2 \in \Gamma_\alpha$ by the convexity of Γ_α. Hence

$$f_A[\lambda x_1 + (1 - \lambda)x_2] \geq \alpha = f_A(x_1) = \mathrm{Min}\,[f_A(x_1), f_A(x_2)].$$

Conversely, if A is convex in the sense of the second definition and $\alpha = f_A(x_1)$, then Γ_α may be regarded as the set of all points x_2 for which $f_A(x_2) \geq f_A(x_1)$. In virtue of (25), every point of the form $\lambda x_1 + (1 - \lambda)x_2$, $0 \leq \lambda \leq 1$, is also in Γ_α and hence Γ_α is a convex set. Q.E.D.

A basic property of convex fuzzy sets is expressed by the
THEOREM. *If A and B are convex, so is their intersection.*

[5] This way of expressing convexity was suggested to the writer by his colleague, E. Berlekamp.

Proof: Let $C = A \cap B$. Then

$$f_C[\lambda x_1 + (1 - \lambda)x_2]$$
$$= \text{Min } [f_A[\lambda x_1 + (1 - \lambda)x_2], f_B[\lambda x_1 + (1 - \lambda)x_2]]. \quad (26)$$

Now, since A and B are convex

$$f_A[\lambda x_1 + (1 - \lambda)x_2] \geqq \text{Min } [f_A(x_1), f_A(x_2)]$$
$$f_B[\lambda x_1 + (1 - \lambda)x_2] \geqq \text{Min } [f_B(x_1), f_B(x_2)] \quad (27)$$

and hence

$$f_C[\lambda x_1 + (1 - \lambda)x_2]$$
$$\geqq \text{Min } [\text{Min } [f_A(x_1), f_A(x_2)], \text{Min } [f_B(x_1), f_B(x_2)]] \quad (28)$$

or equivalently

$$f_C[\lambda x_1 + (1 - \lambda)x_2]$$
$$\geqq \text{Min } [\text{Min } [f_A(x_1), f_B(x_1)], \text{Min } [f_A(x_2), f_B(x_2)]] \quad (29)$$

and thus

$$f_C[\lambda x_1 + (1 - \lambda)x_2] \geqq \text{Min } [f_C(x_1), f_C(x_2)]. \quad \text{Q. E. D.} \quad (30)$$

Boundedness. A fuzzy set A is *bounded* if and only if the sets $\Gamma_\alpha = \{x \mid f_A(x) \geqq \alpha\}$ are bounded for all $\alpha > 0$; that is, for every $\alpha > 0$ there exists a finite $R(\alpha)$ such that $\| x \| \leqq R(\alpha)$ for all x in Γ_α.

If A is a bounded set, then for each $\epsilon > 0$ then exists a hyperplane H such that $f_A(x) \leqq \epsilon$ for all x on the side of H which does not contain the origin. For, consider the set $\Gamma_\epsilon = \{x \mid f_A(x) \geqq \epsilon\}$. By hypothesis, this set is contained in a sphere S of radius $R(\epsilon)$. Let H be any hyperplane supporting S. Then, all points on the side of H which does not contain the origin lie outside or on S, and hence for all such points $f_A(x) \leqq \epsilon$.

LEMMA. *Let A be a bounded fuzzy set and let $M = \text{Sup}_x f_A(x)$. (M will be referred to as the* maximal grade *in A.) Then there is at least one point x_0 at which M is essentially attained in the sense that, for each $\epsilon > 0$, every spherical neighborhood of x_0 contains points in the set $Q(\epsilon) = \{x \mid f_A(x) \geqq M - \epsilon\}$.*

Proof.[6] Consider a nested sequence of bounded sets $\Gamma_1, \Gamma_2, \cdots$, where $\Gamma_n = \{x \mid f_A(x) \geqq M - M/(n + 1)\}, n = 1, 2, \cdots$. Note that

[6] This proof was suggested by A. J. Thomasian.

Γ_n is nonempty for all finite n as a consequence of the definition of M as $M = \text{Sup}_x f_A(x)$. (We assume that $M > 0$.)

Let x_n be an arbitrarily chosen point in Γ_n, $n = 1, 2, \cdots$. Then, $x_1, x_2, \cdots$, is a sequence of points in a closed bounded set Γ_1. By the Bolzano-Weierstrass theorem, this sequence must have at least one limit point, say x_0, in Γ_1. Consequently, every spherical neighborhood of x_0 will contain infinitely many points from the sequence $x_1, x_2, \cdots$, and, more particularly, from the subsequence $x_{N+1}, x_{N+2}, \cdots$, where $N \geq M/\epsilon$. Since the points of this subsequence fall within the set $Q(\epsilon) = \{x \mid f_A(x) \geq M - \epsilon\}$, the lemma is proved.

Strict and strong convexity. A fuzzy set A is *strictly convex* if the sets Γ_α, $0 < \alpha \leq 1$ are strictly convex (that is, if the midpoint of any two distinct points in Γ_α lies in the interior of Γ_α). Note that this definition reduces to that of strict convexity for ordinary sets when A is such a set.

A fuzzy set A is *strongly convex* if, for any two distinct points x_1 and x_2, and any λ in the open interval $(0, 1)$

$$f_A[\lambda x_1 + (1 - \lambda)x_2] > \text{Min}\,[f_A(x_1), f_A(x_2)].$$

Note that strong convexity does not imply strict convexity or vice-versa. Note also that if A and B are bounded, so is their union and intersection. Similarly, if A and B are strictly (strongly) convex, their intersection is strictly (strongly) convex.

Let A be a convex fuzzy set and let $M = \text{Sup}_x f_A(x)$. If A is bounded, then, as shown above, either M is attained for some x, say x_0, or there is at least one point x_0 at which M is essentially attained in the sense that, for each $\epsilon > 0$, every spherical neighborhood of x_0 contains points in the set $Q(\epsilon) = \{x \mid M - f_A(x) \leq \epsilon\}$. In particular, if A is strongly convex and x_0 is attained, then x_0 is unique. For, if $M = f_A(x_0)$ and $M = f_A(x_1)$, with $x_1 \neq x_0$, then $f_A(x) > M$ for $x = 0.5x_0 + 0.5x_1$, which contradicts $M = \text{Max}_x f_A(x)$.

More generally, let $C(A)$ be the set of all points in X at which M is essentially attained. This set will be referred to as the *core* of A. In the case of convex fuzzy sets, we can assert the following property of $C(A)$.

THEOREM. *If A is a convex fuzzy set, then its core is a convex set.*

Proof: It will suffice to show that if M is essentially attained at x_0 and x_1, $x_1 \neq x_0$, then it is also essentially attained at all x of the form $x = \lambda x_0 + (1 - \lambda)x_1$, $0 \leq \lambda \leq 1$.

To the end, let P be a cylinder of radius ϵ with the line passing through x_0 and x_1 as its axis. Let x_0' be a point in a sphere of radius ϵ centering on x_0' and x_1' be a point in a sphere of radius ϵ centering on x_1 such that $f_A(x_0') \geqq M - \epsilon$ and $f_A(x_1') \geqq M - \epsilon$. Then, by the convexity of A, for any point u on the segment $x_0'x_1'$, we have $f_A(u) \geqq M - \epsilon$. Furthermore, by the convexity of P, all points on $x_0'x_1'$ will lie in P.

Now let x be any point in the segment $x_0 x_1$. The distance of this point from the segment $x_0'x_1'$ must be less than or equal to ϵ, since $x_0'x_1'$ lies in P. Consequently, a sphere of radius ϵ centering on x will contain at least one point of the segment $x_0'x_1'$ and hence will contain at least one point, say w, at which $f_A(w) \geqq M - \epsilon$. This establishes that M is essentially attained at x and thus proves the theorem.

COROLLARY. *If $X = E^1$ and A is strongly convex, then the point at which M is essentially attained is unique.*

Shadow of a fuzzy set. Let A be a fuzzy set in E^n with membership function $f_A(x) = f_A(x_1, \cdots, x_n)$. For notational simplicity, the notion of the *shadow* (projection) of A on a hyperplane H will be defined below for the special case where H is a coordinate hyperplane, e.g., $H = \{x \mid x_1 = 0\}$.

Specifically, the *shadow* of A on $H = \{x \mid x_1 = 0\}$ is defined to be a fuzzy set $S_H(A)$ in E^{n-1} with $f_{S_H(A)}(x)$ given by

$$f_{S_H(A)}(x) = f_{S_H(A)}(x_2, \cdots, x_n) = \operatorname{Sup}_{x_1} f_A(x_1, \cdots, x_n).$$

Note that this definition is consistent with (23).

When A is a convex fuzzy set, the following property of $S_H(A)$ is an immediate consequence of the above definition: If A is a convex fuzzy set, then its shadow on any hyperplane is also a convex fuzzy set.

An interesting property of the shadows of two convex fuzzy sets is expressed by the following implication

$$S_H(A) = S_H(B) \text{ for all } H \Rightarrow A = B.$$

To prove this assertion,[7] it is sufficient to show that if there exists a point, say x_0, such that $f_A(x_0) \neq f_B(x_0)$, then their exists a hyperplane H such that $f_{S_H(A)}(x_0^*) \neq f_{S_H(B)}(x_0^*)$, where x_0^* is the projection of x_0 on H.

Suppose that $f_A(x_0) = \alpha > f_B(x_0) = \beta$. Since B is a convex fuzzy set, the set $\Gamma_\beta = \{x \mid f_B(x) > \beta\}$ is convex, and hence there exists a hyperplane F supporting Γ_β and passing through x_0. Let H be a hyperplane orthogonal to F, and let x_0^* be the projection of x_0 on H. Then, since

---

[7] This proof is based on an idea suggested by G. Dantzig for the case where A and B are ordinary convex sets.

$f_B(x) \leqq \beta$ for all x on F, we have $f_{S_H(B)}(x_0^*) \leqq \beta$. On the other hand, $f_{S_H(A)}(x_0^*) \geqq \alpha$. Consequently, $f_{S_H(B)}(x_0^*) \neq f_{S_H(A)}(x_0^*)$, and similarly for the case where $\alpha < \beta$.

A somewhat more general form of the above assertion is the following: Let A, but not necessarily B, be a convex fuzzy set, and let $S_H(A) = S_H(B)$ for all H. Then $A = \text{conv } B$, where conv B is the convex hull of B, that is, the smallest convex set containing B. More generally, $S_H(A) = S_H(B)$ for all H implies conv $A = \text{conv } B$.

Separation of convex fuzzy sets. The classical separation theorem for ordinary convex sets states, in essence, that if A and B are disjoint convex sets, then there exists a separating hyperplane H such that A is on one side of H and B is on the other side.

It is natural to inquire if this theorem can be extended to convex fuzzy sets, without requiring that A and B be disjoint, since the condition of disjointness is much too restrictive in the case of fuzzy sets. It turns out, as will be seen in the sequel, that the answer to this question is in the affirmative.

As a preliminary, we shall have to make a few definitions. Specifically, let A and B be two bounded fuzzy sets and let H be a hypersurface in E^n defined by an equation $h(x) = 0$, with all points for which $h(x) \geqq 0$ being on one side of H and all points for which $h(x) \leqq 0$ being on the other side.[8] Let K_H be a number dependent on H such that $f_A(x) \leqq K_H$ on one side of H and $f_B(x) \leqq K_H$ on the other side. Let M_H be Inf K_H. The number $D_H = 1 - M_H$ will be called the *degree of separation of A and B by H.*

In general, one is concerned not with a given hypersurface H, but with a family of hypersurfaces $\{H_\lambda\}$, with λ ranging over, say, E^m. The problem, then, is to find a member of this family which realizes the highest possible degree of separation.

A special case of this problem is one where the H_λ are hyperplanes in E^n, with λ ranging over E^n. In this case, we define the *degree of separability* of A and B by the relation

$$D = 1 - \bar{M} \tag{31}$$

where

$$\bar{M} = \text{Inf}_H M_H \tag{32}$$

with the subscript λ omitted for simplicity.

[8] Note that the sets in question have H in common.

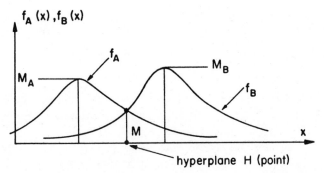

FIG. 5. Illustration of the separation theorem for fuzzy sets in E^1

Among the various assertions that can be made concerning D, the following statement[9] is, in effect, an extension of the separation theorem to convex fuzzy sets.

THEOREM. *Let A and B be bounded convex fuzzy sets in E^n, with maximal grades M_A and M_B, respectively $[M_A = \text{Sup}_x f_A(x), M_B = \text{Sup}_x f_B(x)]$. Let M be the maximal grade for the intersection $A \cap B$ ($M = \text{Sup}_x \text{Min} \cdot [f_A(x), f_B(x)]$). Then $D = 1 - M$.*

Comment. In plain words, the theorem states that the highest degree of separation of two convex fuzzy sets A and B that can be achieved with a hyperplane in E^n is one minus the maximal grade in the intersection $A \cap B$. This is illustrated in Fig. 5 for $n = 1$.

Proof: It is convenient to consider separately the following two cases: (1) $M = \text{Min} (M_A, M_B)$ and (2) $M < \text{Min} (M_A, M_B)$. Note that the latter case rules out $A \subset B$ or $B \subset A$.

Case 1. For concreteness, assume that $M_A < M_B$, so that $M = M_A$. Then, by the property of bounded sets already stated there exists a hyperplane H such that $f_B(x) \leqq M$ for all x on one side of H. On the other side of H, $f_A(x) \leqq M$ because $f_A(x) \leqq M_A = M$ for all x.

It remains to be shown that there do not exist an $M' < M$ and a hyperplane H' such that $f_A(x) \leqq M'$ on one side of H' and $f_B(x) \leqq M'$ on the other side.

This follows at once from the following observation. Suppose that such H' and M' exist, and assume for concreteness that the core of A (that is, the set of points at which $M_A = M$ is essentially attained) is on the plus side of H'. This rules out the possibility that $f_A(x) \leqq M'$

---

[9] This statement is based on a suggestion of E. Berlekamp.

for all x on the plus side of H', and hence necessitates that $f_A(x) \leq M'$ for all x on the minus side of H', and $f_B(x) \leq M'$ for all x on the plus side of H'. Consequently, over all x on the plus side of H'

$$\text{Sup}_x \text{ Min } [f_A(x), f_B(x)] \leq M'$$

and likewise for all x on the minus side of H'. This implies that, over all x in X, $\text{Sup}_x \text{ Min } [f_A(x), f_B(x)] \leq M'$, which contradicts the assumption that $\text{Sup}_x \text{ Min } [f_A(x), f_B(x)] = M > M'$.

Case 2. Consider the convex sets $\Gamma_A = \{x \mid f_A(x) > M\}$ and $\Gamma_B = \{x \mid f_B(x) > M\}$. These sets are nonempty and disjoint, for if they were not there would be a point, say u, such that $f_A(u) > M$ and $f_B(u) > M$, and hence $f_{A \cap B}(u) > M$, which contradicts the assumption that $M = \text{Sup}_x f_{A \cap B}(x)$.

Since Γ_A and Γ_B are disjoint, by the separation theorem for ordinary convex sets there exists a hyperplane H such that Γ_A is on one side of H (say, the plus side) and Γ_B is on the other side (the minus side). Furthermore, by the definitions of Γ_A and Γ_B, for all points on the minus side of $H, f_A(x) \leq M$, and for all points on the plus side of $H, f_B(x) \leq M$.

Thus, we have shown that there exists a hyperplane H which realizes $1 - M$ as the degree of separation of A and B. The conclusion that a higher degree of separation of A and B cannot be realized follows from the argument given in Case 1. This concludes the proof of the theorem.

The separation theorem for convex fuzzy sets appears to be of particular relevance to the problem of pattern discrimination. Its application to this class of problems as well as to problems of optimization will be explored in subsequent notes on fuzzy sets and their properties.

RECEIVED: November 30, 1964

REFERENCES

BIRKHOFF, G. (1948), "Lattice Theory," Am. Math. Soc. Colloq. Publ., Vol. 25, New York.

HALMOS, P. R. (1960), "Naive Set Theory." Van Nostrand, New York.

KLEENE, S. C. (1952), "Introduction to Metamathematics," p. 334. Van Nostrand, New York.

Probability Measures of Fuzzy Events

L.A. ZADEH*

*Department of Electrical Engineering and Computer Sciences,
University of California, Berkeley, California*

I. Introduction

In probability theory [1], an *event*, A, is a member of a σ-field, $\mathcal{Ot}$, of subsets of a sample space Ω. A *probability measure*, P, is a normed measure over a measurable space $(\Omega, \mathcal{Ot})$; that is, P is a real-valued function which assigns to every A in $\mathcal{Ot}$ a *probability*, $P(A)$, such that (a) $P(A) \geqslant 0$ for all $A \in \mathcal{Ot}$; (b) $P(\Omega) = 1$; and (c) P is countably additive, i.e., if $\{A_i\}$ is any collection of disjoint events, then

$$P\left(\bigcup_{i=1}^{\infty} A_i\right) = \sum_{i=1}^{\infty} P(A_i). \tag{1}$$

The notions of an event and its probability constitute the most basic concepts of probability theory. As defined above, an event is a precisely specified collection of points in the sample space. By contrast, in everyday experience one frequently encounters situations in which an "event" is a fuzzy rather than a sharply defined collection of points. For example, the ill-defined events: "It is a *warm* day," "*x* is *approximately* equal to 5;" "In twenty tosses of a coin there are *several* more heads than tails," are fuzzy because of the imprecision of the meaning of the underlined words.

By using the concept of a fuzzy set [2], the notions of an event and its probability can be extended in a natural fashion to fuzzy events of the type exemplified above. It is possible that such an extension may eventually significantly enlarge the domain of applicability of probability theory, especially in those fields in which fuzziness is a pervasive phenomenon.

The present note has the limited objective of showing how the notion of a fuzzy event can be given a precise meaning in the context of fuzzy sets. Thus, it consists mostly of definitions and is largely preliminary in nature. We make

* Currently on leave as Visiting Professor of Electrical Engineering and Guggenheim Fellow with the Department of Electrical Engineering and Project MAC, M.I.T., Cambridge, Massachusetts. This work was supported in part by Project MAC, sponsored by the Advanced Research Projects Agency, Department of Defense, under Office of Naval Research Contract Nonr-4102(01).

no attempt to formulate our definitions in the most general setting, nor do we attempt to explore in detail any of the paths along which classical probability theory may be generalized through the use of concepts derived from the notion of a fuzzy event.

II. Fuzzy Events

We shall assume for simplicity that Ω is an Euclidean n-space R^n. Thus our probability space will be assumed to be a triplet $(R^n, \mathcal{A}, P)$, where $\mathcal{A}$ is the σ-field of Borel sets in R^n and P is a probability measure over R^n. A point in R^n will be denoted by x.

Let $A \in \mathcal{A}$. Then, the probability of A can be expressed as

$$P(A) = \int_A dP \qquad (2)$$

or equivalently

$$P(A) = \int_{R^n} \mu_A(x) \, dP$$

$$= E(\mu_A), \qquad (3)$$

where μ_A denotes the characteristic function of A $(\mu_A(x) = 0$ or $1)$ and $E(\mu_A)$ is the expectation of μ_A.

Equation (3) equates the probability of an event A with the expectation of the characteristic function of A. It is this equation that can readily be generalized to fuzzy events through the use of the concept of a fuzzy set.

Specifically, a fuzzy set A in R^n is defined by a characteristic function $\mu_A : R^n \to [0, 1]$ which associates with each x in R^n its "grade of membership," $\mu_A(x)$, in A. To distinguish between the characteristic function of a nonfuzzy set and the characteristic function of a fuzzy set, the latter will be referred to as a *membership* function. A simple example of a fuzzy set in R^1 is $A = \{x \mid x \gg 0\}$. A membership function for such a set might be subjectively defined by, say,

$$\mu_A(x) = (1 + x^{-2})^{-1}, \qquad x \geqslant 0$$

$$= 0, \qquad\qquad x < 0. \qquad (4)$$

We are now ready to define a fuzzy event in R^n.

DEFINITION. Let $(R^n, \mathcal{A}, P)$ be a probability space in which $\mathcal{A}$ is the σ-field of Borel sets in R^n and P is a probability measure over R^n. Then, a *fuzzy event* in R^n is a fuzzy set A in R^n whose membership function, $\mu_A(\mu_A : R^n \rightarrow [0, 1])$, is Borel measurable.

The *probability* of a fuzzy event A is defined by the Lebesgue-Stieltjes integral:

$$P(A) = \int_{R^n} \mu_A(x) \, dP$$

$$= E(\mu_A). \tag{5}$$

Thus, as in (3), the probability of a fuzzy event is the expectation of its membership function. The existence of the Lebesgue-Stieltjes integral (5) is insured by the assumption that μ_A is Borel measurable.

The above definitions of a fuzzy event and its probability form a basis for generalizing within the framework of the theory of fuzzy sets a number of the concepts and results of probability theory, information theory and related fields. In many cases, the manner in which such generalization can be accomplished is quite obvious. We shall illustrate this in the sequel by a few simple examples.

There are several basic notions relating to fuzzy sets which we shall need in our discussion. These are summarized below. A more detailed discussion of these and other notions may be found in [2].

Containment $A \subset B \Leftrightarrow \mu_A(x) \leqslant \mu_B(x) \; \forall x$ \hfill (6)

Equality $A = B \Leftrightarrow \mu_A(x) = \mu_B(x) \; \forall x$ \hfill (7)

Complement $A' =$ complement of $A \Leftrightarrow \mu_{A'}(x) = 1 - \mu_A(x) \; \forall x$ \hfill (8)

Union $A \cup B =$ union of A and $B \Leftrightarrow \mu_{A \cup B}(x) = \text{Max}[\mu_A(x), \mu_B(x)] \; \forall x$ \hfill (9)

Intersection $A \cap B =$ intersection of A and B

$$\Leftrightarrow \mu_{A \cap B}(x) = \text{Min}[\mu_A(x), \mu_B(x)] \; \forall x \tag{10}$$

Product $AB =$ product of A and $B \Leftrightarrow \mu_{AB}(x) = \mu_A(x) \, \mu_B(x) \; \forall x$ \hfill (11)

Sum $A \oplus B =$ sum of A and B

$$\Leftrightarrow \mu_{A \oplus B}(x) = \mu_A(x) + \mu_B(x) - \mu_A(x) \, \mu_B(x) \; \forall x \tag{12}$$

We are now ready to draw some elementary conclusions from (5)-(12). First, as an immediate consequence of (6), we have

$$A \subset B \Rightarrow P(A) \leqslant P(B) \tag{13}$$

Similarly, as immediate consequences of (9)-(12), we have the identities

$$P(A \cup B) = P(A) + P(B) - P(A \cap B) \tag{14}$$

$$P(A \oplus B) = P(A) + P(B) - P(AB). \tag{15}$$

A level set, $A(\alpha)$, of a fuzzy set A is a non-fuzzy set defined by

$$A(\alpha) = \{x \mid \mu_A(x) \leqslant \alpha\} \tag{16}$$

A fuzzy set A will be said to be a *Borel fuzzy set* if all of its level sets (for $0 \leqslant \alpha \leqslant 1$) are Borel sets. Since the membership function of a fuzzy event is measurable, it follows that all of the level sets associated with a fuzzy event are Borel sets and hence that a fuzzy event is a Borel fuzzy set.

It is well-known that if μ_A and μ_B are Borel measurable, so are Max $[\mu_A, \mu_B]$, Min $[\mu_A, \mu_B]$, $\mu_A + \mu_B$ and $\mu_A \mu_B$ [3]. The same holds, more generally, for any infinite collection of Borel measurable functions. Consequently, we can assert that, like the Borel sets, Borel fuzzy sets form a σ-field with respect to the operations (8), (9) and (10). In this connection, it should be noted that fuzzy sets obey the distributive law.

$$(A \cup B) \cap C = (A \cap C) \cup (B \cap C) \tag{17}$$

but not

$$(A \oplus B)\, C = AC \oplus BC. \tag{18}$$

Employing induction and making use of (14) and (15), we obtain for fuzzy sets the familiar identities for nonfuzzy sets:

$$P\left(\bigcup_{i=1}^{m} A_i\right) = \sum_i P(A_i) - \sum_{i,j}' P(A_i \cap A_j) + \cdots + (-1)^m P\left(\bigcap_i A_i\right) \tag{19}$$

$$P(A_1 \oplus \cdots \oplus A_m) = \sum_i P(A_i) - \sum_{i,j}' P(A_i A_j) + \cdots + (-1)^m P(A_1 \cdots A_m). \tag{20}$$

In a similar fashion, (14) and (15) yield the generalized Boole inequalities for fuzzy sets

$$P\left(\bigcup_{i=1}^{\infty} A_i\right) \leqslant \sum_{i=1}^{\infty} P(A_i) \tag{21}$$

$$P(A_1 \oplus A_2 \oplus \cdots) \leqslant \sum_{i=1}^{\infty} P(A_i). \tag{22}$$

We turn next to the notion of independence of fuzzy events. Specifically, let A and B be two fuzzy events in a probability space $(R^n, \mathcal{O}\!, P)$. Then A and B will be said to be *independent* if

$$P(AB) = P(A)\, P(B). \tag{23}$$

Note that in defining independence we employ the product AB rather than the intersection $A \cap B$.

An immediate consequence of the above definition is the following: Let $\Omega_1 = R^n$, $\Omega_2 = R^m$ and let P be the product measure $P_1 \times P_2$, where P_1 and P_2 are probability measures on Ω_1 and Ω_2, respectively. Let A_1 and A_2 be events in Ω_1 and Ω_2 characterized by the membership functions $\mu_{A_1}(x_1, x_2) = \mu_{A_1}(x_1)$ and $\mu_{A_2}(x_1, x_2) = \mu_{A_2}(x_2)$, respectively. Then A_1 and A_2 are independent events in the sense of (23). Note that this would not be true if independence were defined in terms of $P(A \cap B)$ rather than $P(AB)$.

To be consistent with (23), the *conditional probability* of A given B is defined by

$$P(A \mid B) = \frac{P(AB)}{P(B)}, \tag{24}$$

provided $P(B) > 0$. Note that if A and B are independent, then $P(A \mid B) = P(A)$, as in the case of nonfuzzy independent events.

Many of the basic notions in probability theory, such as those of the mean, variance, entropy, etc., are defined as functionals of probability distributions. The concept of a fuzzy event suggests that it may be of interest to define these notions in a more general way which relates them to both a fuzzy event and a probability measure. For example, the mean of a fuzzy event A relative to a probability measure P may be defined as follows:

$$m_P(A) = \frac{1}{P(A)} \int_{R^u} x \mu_A(x) \, d P \tag{25}$$

where μ_A is the membership function of A and $P(A)$ serves as a normalizing factor. Similarly, the variance of a fuzzy event in R^1 relative to a probability measure P may be defined as

$$G_P^2(A) = \frac{1}{P(A)} \int_{R^1} (x - m_P(A))^2 \, \mu_A(x) \, d P \tag{26}$$

The subscript P in (25) and (26) may be omitted when the dependence on P of the quantities in question is implied by the context.

Turning to the notion of entropy, we note that its usual definition in information theory is as follows: Let x be a random variable which takes the values $x_1, \ldots, x_n$ with respective probabilities $p_1, \ldots, p_n$. Then, the entropy of x—or, more properly, the entropy of the distribution $P = \{p_1, \ldots, p_n\}$—is given by

$$H(x) = - \sum_{i=1}^{n} p_i \log p_i. \tag{27}$$

This definition suggests that the entropy of a fuzzy subset, A, of the finite set $\{x_1, ..., x_n\}$ with respect to a probability distribution $P = \{p_1, ..., p_n\}$ be defined as follows

$$H^P(A) = - \sum_{i=1}^{n} \mu_A(x_i)\, p_i \log p_i , \qquad (28)$$

where μ_A is the membership function of A. Note that whereas (27) expresses the entropy of a distribution P, (28) represents the entropy of a fuzzy event A with respect to the distribution P. Thus, (28) does not reduce to (27) when A is nonfuzzy, unless A is taken to be the whole space $\{x_1, ..., x_n\}$. Intuitively, $H^P(A)$ may be interpreted as the uncertainty associated with a fuzzy event.

Let x and y be independent random variables with probability distributions $P = \{p_1, ..., p_n\}$ and $Q = \{q_1, ..., q_m\}$, respectively. One of the basic properties of the joint entropy of x and y is that when x and y are independent, we can write

$$H(x, y) = H(x) + H(y). \qquad (29)$$

It is easy to verify that for fuzzy events this identity generalizes to

$$H^{PQ}(AB) = P(A)\, H^P(A) + P(B)\, H^Q(B), \qquad (30)$$

where

$$PQ = \{p_i q_j\}, \qquad i = 1,..., n, \qquad j = 1,..., m$$

$$P(A) = \sum_{i=1}^{n} \mu_A(x_i)\, p_i$$

$$P(B) = \sum_{j=1}^{m} \mu_B(y_j)\, q_j$$

$$H^P(A) = - \sum_{i=1}^{n} \mu_A(x_i)\, p_i \log p_i$$

and

$$H^Q(B) = - \sum_{j=1}^{m} \mu_B(y_j)\, q_j \log q_j .$$

Note that (30) reduces to (29) when $A = \{x_1, ..., x_n\}$ and $B = \{y_1, ..., y_m\}$.

The foregoing examples are intended merely to demonstrate possible ways of defining some of the elementary concepts of probability theory in a more general setting in which fuzzy events are allowed. It appears that there are many concepts and results in probability theory, information theory and related fields which admit of such generalization.

REFERENCES

1. H. G. TUCKER. "A Graduate Course in Probability." Academic Press, New York, 1967.
2. L. A. ZADEH. Fuzzy Sets. *Inform. Control* 8 (1965).
3. W. RUDIN. "Principles of Mathematical Analysis." McGraw-Hill, New York, 1967.

Decision-Making in a Fuzzy Environment*†

R.E. BELLMAN‡ and L.A. ZADEH§

By decision-making in a fuzzy environment is meant a decision process in which the goals and/or the constraints, but not necessarily the system under control, are fuzzy in nature. This means that the goals and/or the constraints constitute classes of alternatives whose boundaries are not sharply defined.

An example of a fuzzy constraint is: "The cost of A should not be *substantially* higher than α," where α is a specified constant. Similarly, an example of a fuzzy goal is: "x should be in the *vicinity* of x_0," where x_0 is a constant. The italicized words are the sources of fuzziness in these examples.

Fuzzy goals and fuzzy constraints can be defined precisely as fuzzy sets in the space of alternatives. A fuzzy decision, then, may be viewed as an intersection of the given goals and constraints. A maximizing decision is defined as a point in the space of alternatives at which the membership function of a fuzzy decision attains its maximum value.

The use of these concepts is illustrated by examples involving multistage decision processes in which the system under control is either deterministic or stochastic. By using dynamic programming, the determination of a maximizing decision is reduced to the solution of a system of functional equations. A reverse-flow technique is described for the solution of a functional equation arising in connection with a decision process in which the termination time is defined implicitly by the condition that the process stops when the system under control enters a specified set of states in its state space.

1. Introduction

Much of the decision-making in the real world takes place in an environment in which the goals, the constraints and the consequences of possible actions are not known precisely. To deal quantitatively with imprecision, we usually employ the concepts and techniques of probability theory and, more particularly, the tools provided by decision theory, control theory and information theory. In so doing, we are tacitly accepting the premise that imprecision—whatever its nature—can be equated with randomness. This, in our view, is a questionable assumption.

* Received October 1969; revised February 1970.
† Research sponsored by National Institute of Health Grant 16197 to the University of Southern California, and the National Aeronautics and Space Administration, Grant NSG-354 (Sup. 5) to the University of California, Berkeley.
‡ University of Southern California.
§ University of California, Berkeley.

Specifically, our contention is that there is a need for differentiation between *ran-domness* and *fuzziness*, with the latter being a major source of imprecision in many decision processes. By fuzziness, we mean a type of imprecision which is associated with *fuzzy sets*, [20], [21] that is, classes in which there is no sharp transition from membership to nonmembership. For example, the class of *green objects* is a fuzzy set. So are the classes of objects characterized by such commonly used adjectives as large, small, substantial, significant, important, serious, simple, accurate, approximate, etc. Actually, in sharp contrast to the notion of a class or a set in mathematics, most of the classes in the real world do not have crisp boundaries which separate those objects which belong to a class from those which do not. In this connection, it is important to note that, in the discourse between humans, fuzzy statements such as "John is *several* inches taller than Jim," "*x* is *much larger* than *y*," "Corporation *X* has a *bright future*," "the stock market has suffered a *sharp decline*," convey information despite the imprecision of the meaning of the italicized words. In fact, it may be argued that the main distinction between human intelligence and machine intelligence lies in the ability of humans—an ability which present-day computers do not possess—to manipulate fuzzy concepts and respond to fuzzy instructions.

What is the distinction between randomness and fuzziness? Essentially, randomness has to do with uncertainty concerning membership or nonmembership of an object in a nonfuzzy set. Fuzziness, on the other hand, has to do with classes in which there may be grades of membership intermediate between full membership and nonmembership. To illustrate the point, the fuzzy assertion "Corporation *X* has a modern outlook" is imprecise by virtue of the fuzziness of the terms "modern outlook." On the other hand, the statement "The probability that Corporation *X* is operating at a loss is 0.8," is a measure of the uncertainty concerning the membership of Corporation *X* in the nonfuzzy class of corporations which are operating at a loss. Similarly, "The grade of membership of John in the class of tall men is 0.7," is a nonprobabilistic statement concerning the membership of John in the fuzzy class of tall men, whereas "The probability that John will get married within a year is 0.7," is a probabilistic statement concerning the uncertainty of the occurrence of a nonfuzzy event (marriage).

Reflecting this distinction, the mathematical techniques for dealing with fuzziness are quite different from those of probability theory. They are simpler in many ways because to the notion of probability measure in probability theory corresponds the simpler notion of membership function in the theory of fuzziness. Furthermore, the correspondents of $a + b$ and ab, where a and b are real numbers, are the simpler operations $\text{Max}(a, b)$ and $\text{Min}(a, b)$. For this reason, even in those cases in which fuzziness in a decision process can be simulated by a probabilistic model, it is generally advantageous to deal with it through the techniques provided by the theory of fuzzy sets rather than through the employment of the conceptual framework of probability theory.

Decision processes in which fuzziness enters in one way or another can be studied from many points of view. [22], [9], [14] In the present note, our main concern is with introducing three basic concepts: fuzzy goal, fuzzy constraint and fuzzy decision, ıd exploring the application of these concepts to multistage decision processes in hich the goals or the constraints may be fuzzy, while the system under control may e either deterministic or stochastic—but not fuzzy. This, however, is not an intrinsic striction on the applicability of the concepts and techniques described in the follow-g sections.

Roughly speaking, by a fuzzy goal we mean an objective which can be character-ed as a fuzzy set in an appropriate space. To illustrate, a simple example of a fuzzy al involving a real-valued variable x would be: "x should be *substantially* larger than 0." Similarly, a simple example of a fuzzy constraint would be: "x should be *approxi-ately* in the range 20–25." The sources of fuzziness in these statements are the itali-zed words.

A less trivial example is provided by a deterministic discrete-time system charac-rized by the state equations

$$x_{n+1} = x_n + u_n, \qquad n = 0, 1, 2, \cdots,$$

here x_n and u_n denote, respectively, the state and input at time n and in which for mplicity x_n and u_n are assumed to be real-valued. Here a fuzzy constraint on the ıput may be

$$-1 \underset{\sim}{\leq} u_n \underset{\sim}{\leq} 1$$

here the wavy bar under a symbol plays the role of a *fuzzifier*, that is, a transforma-on which takes a nonfuzzy set into a fuzzy set which is approximately equal to it. ı this instance, $u_n \underset{\sim}{\leq} 1$, would read "$u_n$ should be *approximately* less than or equal to , and the effect of the fuzzifier is to transform the nonfuzzy set $-1 \leq u_n \leq 1$ into fuzzy set $-1 \underset{\sim}{\leq} u_n \underset{\sim}{\leq} 1$. The way in which the latter set can be given a precise ıeaning will be discussed in §2.

Assume that the fuzzy goal is to make x_3 approximately equal to 5, starting with ıe initial state $x_0 = 1$. Then, the problem is to find a sequence of inputs u_0, u_1, u_2 hich will realize the specified goal as nearly as possible, subject to the specified ınstraints on u_0, u_1, u_2.

In what follows, we shall consider in greater detail a few representative problems of ıis type. It should be stressed that our limited objective in the present paper is to raw attention to problems involving multistage decision processes in a fuzzy en-ironment and suggest tentative ways of attacking them, rather than to develop a eneral theory of decision processes in which fuzziness and randomness may enter in a ariety of ways and combinations. In particular, we shall not concern ourselves with ıe application to decision-making of the concept of a fuzzy algorithm [22]—a con-ept which may be of use in problems which are less susceptible to quantitative analysis ıan those considered in the sequel.

For convenience of the reader, a brief summary of the basic properties of fuzzy ets is provided in the following section.

2. A Brief Introduction to Fuzzy Sets

Informally, a fuzzy set is a class of objects in which there is no sharp boundary between those objects that belong to the class and those that do not. A more precise definition may be stated as follows.

Definition. Let $X = \{x\}$ denote a collection of objects (points) denoted generically by x. Than a *fuzzy set A in X* is a set of ordered pairs

$$(1) \qquad A = \{(x, \mu_A(x))\}, \qquad x \in X$$

where $\mu_A(x)$ is termed the *grade of membership of x in A*, and $\mu_A : X \to M$ is a function from X to a space M called the *membership space*. When M contains only two points, 0 and 1, A is nonfuzzy and its membership function becomes identical with the characteristic function of a nonfuzzy set.

In what follows, we shall assume that M is the interval [0, 1], with 0 and 1 representing, respectively, the lowest and highest grades of membership. (More generally, M can be a partially ordered set or, more particularly, a lattice [15], [6].) Thus, our basic assumption is that a fuzzy set A—despite the unsharpness of its boundaries—can be defined *precisely* by associating with each object x a number between 0 and 1 which represents its grade of membership in A.

Example. Let $X = \{0, 1, 2, \cdots\}$ be the collection of nonnegative integers. In this space, the fuzzy set A of "*several* objects" may be defined (subjectively) as the collection of ordered pairs

$$(2) \qquad A = \{(3, 0.6), (4, 0.8), (5, 1.0), (6, 1.0), (7, 0.8), (8, 0.6)\}$$

with the understanding that in (2) we list only those pairs $(x, \mu_A(x))$ in which $\mu_A(x)$ is positive.

Comment. It should be noted that in many practical situations the membership function, μ_A, has to be estimated from partial information about it, such as the values which it takes over a finite set of sample points $x_1, \cdots, x_N$. When A is defined incompletely—and hence only approximately—in this fashion, we shall say that it is partially defined by *exemplification*. The problem of estimating μ_A from the knowledge of the set of pairs $(x_1, \mu_A(x_1)), \cdots, (x_N, \mu_A(x_N))$ is the problem of *abstraction*—a problem that plays a central role in pattern recognition. [4], [18] We shall not concern ourselves with the solution of this problem in the present paper and will assume throughout—except where explicitly stated to the contrary—that $\mu_A(x)$ is given for all x in X.

For notational purposes, it is convenient to have a device for indicating that a fuzzy set A is obtained from a nonfuzzy set $\bar{A}$ by fuzzifying the boundaries of the latter set. For this purpose, we shall employ a wavy bar under a symbol (or symbols) which define $\bar{A}$. For example, if A is the set of real numbers between 2 and 5, i.e., $\bar{A} = \{x \mid 2 \le x \le 5\}$, then $A = \{x \mid 2 \underset{\sim}{\le} x \underset{\sim}{\le} 5\}$ is a fuzzy set of real numbers which are approximately between 2 and 5. Similarly, $A = \{x \mid x \underset{\sim}{=} 5\}$ or simply $\underset{\sim}{5}$ will denote the set of numbers which are approximately equal to 5. The symbol $\sim$ will be referred to as a *fuzzifier*.

We turn next to the definition of several basic concepts which we shall need in later sections.

Normality. A fuzzy set A is *normal* if and only if $\text{Sup}_x\,\mu_A(x) = 1$, that is, the supremum of $\mu_A(x)$ over X is unity. A fuzzy set is *subnormal* if it is not normal. A nonempty subnormal fuzzy set can be normalized by dividing each $\mu_A(x)$ by the factor $\text{Sup}_x\,\mu_A(x)$. (A fuzzy set A is *empty* if and only if $\mu_A(x) \equiv 0$.)

Support. The *support* of a fuzzy set A is a set $S(A)$ such that $x \in S(A) \Leftrightarrow \mu_A(x) > 0$. If $\mu_A(x) = $ constant over $S(A)$, then A is *nonfuzzy*. Note that a nonfuzzy set may be subnormal.

Equality. Two fuzzy sets are *equal*, written as $A = B$, if and only if $\mu_A = \mu_B$, that is, $\mu_A(x) = \mu_B(x)$ for all x in X. (In the sequel, to simplify the notation we shall omit the argument x when an equality or inequality holds for all values of x in X.)

Containment. A fuzzy set A is *contained in* or is a *subset of* a fuzzy set B, written as $A \subset B$, if and only if $\mu_A \leq \mu_B$. In this sense, the fuzzy set of very large numbers is a subset of the fuzzy set of large numbers.

Complementation. A' is said to be the *complement* of A if and only if $\mu_{A'} = 1 - \mu_A$. For example, the fuzzy sets: $A = \{\text{tall men}\}$ and $A' = \{\text{not tall men}\}$ are complements of one another if the negation "not" is interpreted as an operation which replaces $\mu_A(x)$ with $1 - \mu_A(x)$ for each x in X.

Intersection. The *intersection* of A and B is denoted by $A \cap B$ and is defined as the largest fuzzy set contained in both A and B. The membership function of $A \cap B$ is given by

$$(3) \qquad \mu_{A \cap B}(x) = \text{Min}\,(\mu_A(x), \mu_B(x)), \qquad x \in X$$

where $\text{Min}\,(a, b) = a$ if $a \leq b$ and $\text{Min}\,(a, b) = b$ if $a > b$. In infix form, using the conjunction symbol $\wedge$ in place of Min, (3) can be written more simply as

$$(4) \qquad \mu_{A \cap B} = \mu_A \wedge \mu_B.$$

The notion of intersection bears a close relation to the notion of the connective "and." Thus, if A is the class of tall men and B is the class of fat men, then $A \cap B$ is the class of men who are both tall *and* fat.

Comment. It should be noted that our identification of "and" with (4) implies that we are interpreting "and" in a "hard" sense, that is, we do not allow any tradeoff between $\mu_A(x)$ and $\mu_B(x)$ so long as $\mu_A(x) > \mu_B(x)$ or vice-versa. For example, if $\mu_A(x) = 0.8$ and $\mu_B(x) = 0.5$, then $\mu_{A \cap B}(x) = 0.5$ so long as $\mu_A(x) \geq 0.5$. In some cases, a softer interpretation of "and" which corresponds to forming the algebraic product of $\mu_A(x)$ and $\mu_B(x)$—rather than the conjunction $\mu_A(x) \wedge \mu_B(x)$—may be closer to the intended meaning of "and." From the mathematical as well as practical points of view, the identification of "and" with $\wedge$ is preferable to its identification with the product, except where $\wedge$ clearly does not express the sense in which one wants "and" to be interpreted. For this reason, in what follows "and" will be understood to be a *hard* "and" unless explicitly stated that it should be interpreted as a *soft* "and" (in the sense of corresponding to the algebraic product of membership functions).

Union. The notion of the *union* of A of B is dual to the notion of intersection. Thus, the union of A and B, denoted as $A \cup B$, is defined as the smallest fuzzy set containing both A and B. The membership function of $A \cup B$ is given by

(5) $$\mu_{A \cup B}(x) = \text{Max}\,(\mu_A(x), \mu_B(x)), \qquad x \in X$$

where Max $(a, b) = a$ if $a \geq b$ and Max $(a, b) = b$ if $a < b$. In infix form, using the disjunction symbol $\vee$ in place of Max, we can write (5) more simply as

(6) $$\mu_{A \cup B} = \mu_A \vee \mu_B .$$

As in the case of the intersection, the union of A and B bears a close relation to the connective "or." Thus, if $A = \{\text{tall men}\}$ and $B = \{\text{fat men}\}$, then $A \cup B = \{\text{tall } or \text{ fat men}\}$. Also, we can differentiate between a *hard* "or", which corresponds to (6), and a *soft* "or", corresponding to the *algebraic sum* of A and B, which is denoted by $A \oplus B$ and is defined by (9).

It is easy to verify that $\cup$ and $\cap$ are related to one another by the identity

(7) $$A \cup B = (A' \cap B')'.$$

Algebraic product. The *algebraic product* of A and B is denoted by AB and is defined by

(8) $$\mu_{AB}(x) = \mu_A(x)\mu_B(x), \qquad x \in X.$$

Algebraic sum. The *algebraic sum* of A and B is denoted by $A \oplus B$ and is defined by

(9) $$\mu_{A \oplus B}(x) = \mu_A(x) + \mu_B(x) - \mu_A(x)\mu_B(x), \qquad x \in X.$$

It is easy to verify that

(10) $$A \oplus B = (A'B')'.$$

Comment. It should be noted that the operations $\vee$ and $\wedge$ are associative and distributive over one another. On the other hand, $\cdot$ (product) and $\oplus$ (sum) are associative but not distributive. Note also that $\cdot$ (product) distributes over $\vee$ but not vice-versa. This property is possessed, more generally, by any operation $*$ which is monotone nondecreasing in each of its arguments. More specifically, if $b \geq b' \Rightarrow a * b \geq a * b'$ and $a \geq a' \Rightarrow a * b \geq a' * b$, then $a * (b \vee c) = (a * b) \vee (a * c)$. Many of the results described in the following sections remain valid when $\wedge$ is replaced by an operation $*$ which is associative and distributes over $\vee$.

Convexity and concavity. Let A be a fuzzy set in $X = R^n$. Then A is *convex* if and only if for every pair of points x, y in X, the membership function of A satisfies the inequality

(11) $$\mu_A(\lambda x + (1 - \lambda)y) \geq \text{Min}\,(\mu_A(x), \mu_A(y)),$$

for $0 \leq \lambda \leq 1$. Dually, A is *concave* if its complement A' is convex. It is easy to show that if A and B are convex, so is $A \cap B$. Dually, if A and B are concave, so is $A \cup B$.

Relation. A *fuzzy relation*, R, in the product space $X \times Y = \{(x, y)\}, x \in X, y \in Y$, is a fuzzy set in $X \times Y$ characterized by a membership function μ_R which associates with each ordered pair (x, y) a grade of membership $\mu_R(x, y)$ in R. More generally, an n-ary fuzzy relation in a product space $X = X_1 \times X_2 \times \cdots \times X_n$ is a fuzzy set in X characterized by an n-variate membership function $\mu_R(x_1, \cdots, x_n)$, $x_i \in X_i$, $i = 1, \cdots, n$.

Example. Let $X = Y = R'$, where R' is the real line $(-\infty, \infty)$. Then $x \gg y$ is a fuzzy relation in R^2. A subjective expression for μ_R in this case might be: $\mu_R(x, y) = 0$ for $x \leq y$; $\mu_R(x, y) = (1 + (x - y)^{-2})^{-1}$ for $x > y$.

Fuzzy sets induced by mappings. Let $f : X \to Y$ be a mapping from $X = \{x\}$ to $Y = \{y\}$, with the image of x under f denoted by $y = f(x)$. Let A be a fuzzy set in X. Then, the mapping f induces a fuzzy set B in Y whose membership function is given by

$$(12) \qquad \mu_B(y) = \mathrm{Sup}_{x \epsilon f^{-1}(y)} \, \mu_A(x),$$

where the supremum is taken over the set of points $f^{-1}(y)$ in X which are mapped by f into y.

Conditioned fuzzy sets. A fuzzy set $B(x)$ in $Y = \{y\}$ is *conditioned on* x if its membership function depends on x as a parameter. This dependence is expressed by $\mu_B(y \mid x)$.

Suppose that the parameter x ranges over a space X, so that to each x in X corresponds a fuzzy set $B(x)$ in Y. Thus, we have a mapping—characterized by $\mu_B(y \mid x)$—from X to the space of fuzzy sets in Y. Through this mapping, any given fuzzy set A in X induces a fuzzy set B in Y which is defined by

$$(13) \qquad \mu_B(y) = \mathrm{Sup}_x \, \mathrm{Min} \, (\mu_A(x), \mu_B(y \mid x))$$

where μ_A and μ_B denote the membership functions of A and B, respectively. In terms of $\wedge$ and $\vee$ (13) may be written more simply as

$$(14) \qquad \mu_B(y) = \vee_x (\mu_A(x) \wedge \mu_B(y \mid x)).$$

Note that this equation is analogous—but not equivalent—to the expression for the marginal probability distribution of the joint distribution of two random variables, with $\mu_B(y \mid x)$ playing a role analogous to that of a conditional distribution.

Decomposability. Let $X = \{x\}$, $Y = \{y\}$ and let C be a fuzzy set in the product space $Z = X \times Y$ defined by a membership function $\mu_C(x, y)$. Then C is *decomposable along* X and Y if and only if C admits of the representation $C = A \cap B$ or equivalently

$$(15) \qquad \mu_C(x, y) = \mu_A(x) \wedge \mu_B(y)$$

where A and B are fuzzy sets with membership functions of the form $\mu_A(x)$ and $\mu_B(y)$, respectively. (Thus, A and B are cylindrical fuzzy sets in Z.) The same holds for a fuzzy set in the product of any finite number of spaces.

Probability of fuzzy events. Let P be a probability measure on R^n. A *fuzzy event* [23] A in R^n is defined to be a fuzzy subset A of R^n whose membership function, μ_A, is measurable. Then, the *probability of* A is defined by the Lebesgue-Stieltjes integral

$$(16) \qquad P(A) = \int_{R^n} \mu_A(x) \, dP.$$

Equivalently, $P(A) = E\mu_A$ where E denotes the expectation operator. In the case of a normal nonfuzzy set, (16) reduces to the conventional definition of the probability of a nonfuzzy event.

This concludes our brief introduction to some of the basic concepts relating to fuzzy sets. In the following section, we shall use these concepts as a basis for defining the basic notions of goal, constraint and decision in a fuzzy environment.

3. Fuzzy Goals, Constraints and Decisions

In the conventional approach to decision-making, the principal ingredients of a decision process are (a) a set of alternatives; (b) a set of constraints on the choice between different alternatives; and (c) a performance function which associates with each alternative the gain (or loss) resulting from the choice of that alternative.

When we view a decision process from the broader perspective of decision-making in a fuzzy environment, a different and perhaps more natural conceptual framework suggests itself. The most important feature of this framework is its symmetry with respect to goals and constraints—a symmetry which erases the differences between them and makes it possible to relate in a relatively simple way the concept of a decision to those of the goals and constraints of a decision process.

More specifically, let $X = \{x\}$ be a given set of alternatives. Then, a *fuzzy goal* or simply a *goal*, G, in X will be identified with a given fuzzy set G in X. For example, if $X = R^1$ (the real line), then the fuzzy goal expressed in words as "x should be substantially larger than 10" might be represented by a fuzzy set in R^1 whose membership function is (subjectively) given by

$$(17) \qquad \begin{aligned} \mu_G(x) &= 0, \qquad x < 10, \\ &= (1 + (x - 10)^{-2})^{-1}, \qquad x \geq 10. \end{aligned}$$

Similarly, the goal "x should be in the vicinity of 15" might be represented by a fuzzy set whose membership function is of the form

$$(18) \qquad \mu_G(x) = (1 + (x - 15)^4)^{-1}.$$

Note that both of these sets are convex in the sense of (11).

In the conventional approach, the performance function associated with a decision process serves to define a linear ordering on the set of alternatives. Clearly, the membership function, $\mu_G(x)$, of a fuzzy goal serves the same purpose[1] and, in fact, may be derived from a given performance function by a normalization which leaves the linear ordering unaltered. In effect, such normalization provides a common denominator for the various goals and constraints and thereby makes it possible to treat them alike. This, as we shall see, is one of the significant advantages of regarding the concept of a goal—rather than that of a performance function—as one of the principal components of a conceptual framework for decision-making in a fuzzy environment.

In a similar manner, a *fuzzy constraint* or simply a *constraint*, C, in X is defined to be a fuzzy set in X. For example, in R^1, the constraint "x should be approximately between 2 and 10," could be represented by a fuzzy set whose membership function might be of the form

$$\mu_C(x) = (1 + a(x - 6)^m)^{-1}$$

where a is a positive number and m is a positive even integer chosen in such a way

---

[1] Assuming, of course, that μ_G takes values in a linearly ordered set.

as to reflect the sense in which the approximation to the interval [2, 10] is to be understood. For example, if we set $m = 4$ and $a = 5^{-4}$, then at $x = 2$ and $x = 10$ we have
approximately $\mu_C(x) = 0.71$, while at $x = 1$ and $x = 11$, $\mu_C(x) = 0.5$; and at $x = 0$
and $x = 12$, $\mu_C(x)$ is approximately equal to 0.32.

An important aspect of the above definitions of the concepts of goal and constraint
is that both are defined as fuzzy sets in the space of alternatives and thus, as will be
elaborated upon below, can be treated identically in the formulation of a decision.
By contrast, in the conventional approach to decision-making, a constraint set is
taken to be a nonfuzzy set in the space of alternatives X, whereas a performance function is a function from X to some other space. Nevertheless, even in the case of the
conventional approach, the use of Lagrangian multipliers and penalty functions makes
it apparent that there is an intrinsic similarity between performance functions and
constraints [17, Chapter 15]. This similarity—indeed identity—is made explicit in
our formulation.

As an illustration, suppose that we have a fuzzy goal G and a fuzzy constraint C
expressed as follows:

G: x should be substantially larger than 10, with $\mu_G(x)$ given by (17) *and*

C: x should be in the vicinity of 15, with $\mu_C(x)$ expressed by (18).

Note that G and C are connected to one another by the connective *and*. Now, as
was pointed out in §2, *and* corresponds to the intersection of fuzzy sets. This implies
that in the example under consideration the combined effect of the fuzzy goal G and
the fuzzy constraint C on the choice of alternatives may be represented by the intersection $G \cap C$. The membership function of the intersection is given by

$$\mu_{G \cap C}(x) = \mu_G(x) \wedge \mu_C(x)$$

or more explicitly

$$\mu_{G \cap C}(x) = \text{Min } ((1 + (x - 10)^{-2})^{-1}, (1 + (x - 15)^4)^{-1}) \quad \text{for} \quad x \geq 10,$$

$$= 0 \quad \text{for} \quad x < 10.$$

Note that $G \cap C$ is a convex fuzzy set since both G and C are convex fuzzy sets.

Turning to the concept of a *decision*, we observe that, intuitively, a *decision* is basically a choice or a set of choices drawn from the available alternatives. The preceding
example suggests that a *fuzzy decision* or simply a *decision* be defined as the fuzzy set
of alternatives resulting from the intersection of the goals and constraints. We formalize this idea in the following definition.

Definition. Assume that we are given a fuzzy goal G and a fuzzy constraint C in a
space of alternatives X. Then, G and C combine to form a *decision*, D, which is a
fuzzy set resulting from intersection of G and C. In symbols,

(19) $D = G \cap C$

and correspondingly $\mu_D = \mu_G \wedge \mu_C$. The relation between G, C and D is depicted in
Figure 1.

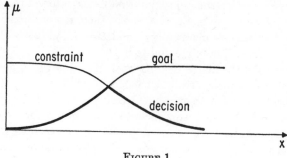

FIGURE 1

More generally, suppose that we have n goals $G_1, \cdots, G_n$ and m constraints $C_1, \cdots, C_m$. Then, the resultant decision is the intersection of the given goals $G_1, \cdots, G_n$ and the given constraints $C_1, \cdots, C_m$. That is,

$$(20) \qquad D = G_1 \cap G_2 \cap \cdots \cap G_n \cap C_1 \cap C_2 \cap \cdots \cap C_m$$

and correspondingly

$$(21) \qquad \mu_D = \mu_{G_1} \wedge \mu_{G_2} \wedge \cdots \wedge \mu_{G_n} \wedge \mu_{C_1} \wedge \mu_{C_2} \wedge \cdots \wedge \mu_{C_m}.$$

Note that in the above definition of a decision, the goals and the constraints enter into the expression for D in exactly the same way. This is the basis for our earlier statement concerning the identity of the roles of goals and constraints in our formulation of decision processes in a fuzzy environment.

Comment. The definition of a decision as the intersection of the goals and constraints reflects our interpretation of "and" in the "hard" sense of (4). If the interpretation of "and" is left open, we shall say that a decision—viewed as a fuzzy set—is a *confluence* of the goals and the constraints. Thus, "confluence" acquires the meaning of "intersection" when "and" is interpreted in the sense of (4); the meaning of "algebraic product" when "and" is interpreted in the sense of (8); and may be assigned other concrete meanings when a need for a special interpretation of "and" arises. (See *Comment* following (10).) In short, a broad definition of the concept of *decision* may be stated as:

Decision = Confluence of Goals and Constraints.

As an illustration of (21), we shall consider a very simple example in which $X = \{1, 2, \cdots, 10\}$ and G_1, G_2, C_1 and C_2 are defined below:

x	1	2	3	4	5	6	7	8	9	10
μ_{G_1}	0	0.1	0.4	0.8	1.0	0.7	0.4	0.2	0	0
μ_{G_2}	0.1	0.6	1.0	0.9	0.8	0.6	0.5	0.3	0	0
μ_{C_1}	0.3	0.6	0.9	1.0	0.8	0.7	0.5	0.3	0.2	0.1
μ_{C_2}	0.2	0.4	0.6	0.7	0.9	1.0	0.8	0.6	0.4	0.2

Forming the conjunction of μ_{G_1}, μ_{G_2}, μ_{C_1} and μ_{C_2}, we obtain the following table of values for $\mu_D(x)$:

x	1	2	3	4	5	6	7	8	9	10
μ_D	0	0.1	0.4	0.7	0.8	0.6	0.4	0.2	0	0

Thus the decision in this case is the fuzzy set

$$D = \{(2, 0.1),\ (3, 0.4),\ (4, 0.7),\ (5, 0.8),\ (6, 0.6),\ (7, 0.4),\ (8, 0.2)\}.$$

Note that no x in X has full (that is, unity grade) membership in D. This reflects, of course, the fact that the specified goals and constraints conflict with one another, ruling out the existence of an alternative which fully satisfies all of them.

The concept of a decision as a fuzzy set in the space of alternatives may appear at first to be somewhat artificial. In fact it is quite natural, since a fuzzy decision may be viewed as an instruction whose fuzziness is a consequence of the imprecision of the given goals and constraints. Thus, in our example, G_1, G_2, C_1 and C_2 may be respectively expressed in words as: "x should be close to 5," "x should be close to 3," "x should be close to 4" and "x should be close to 6". The decision, then, is to choose x to be close to 5. The exact meaning of "close" in each case is given by the values of the corresponding membership function.

How should a fuzzy instruction such as "x should be close to 5" be executed? Although there does not appear to be a universally valid answer to questions of this type,[2] it is reasonable in many instances to choose that x or x's which have maximal grade of membership in D. In the case of our example, this would be $x = 5$.

More generally, let D be a fuzzy decision represented by a membership function μ_D. Let K be the set of points in X on which μ_D attains its maximum, if it exists. Then, the nonfuzzy, but, in general, subnormal, subset D^M of D defined by

$$\mu_{D^m}(x) = \text{Max } \mu_D(x) \quad \text{for} \quad x \in K,$$

$$= 0 \text{ elsewhere}$$

will be said to be the *optimal decision* and any x in the support of D^M will be referred to as a *maximizing decision*. In other words, a maximizing decision is simply any alternative in X which maximizes $\mu_D(x)$, e.g., $x = 5$ in the foregoing example. Note that in R^n a sufficient condition for the uniqueness of a maximizing decision is that D be a strongly convex fuzzy set, i.e., that D be convex and have a unimodal membership function.

In defining a fuzzy decision D as the intersection—or, more generally, as the confluence—of the goals and constraints, we are tacitly assuming that all of the goals and constraints that enter into D are, in a sense, of equal importance. There are some situations, however, in which some of the goals and perhaps some of the constraints are of

[2] The execution of fuzzy instructions is discussed in [22].

greater importance than others. In such cases, D might be expressed as a convex combination of the goals and the constraints, with the weighting coefficients reflecting the relative importance of the constituent terms. More explicitly, we may express $\mu_D(x)$ as

$$(22) \qquad \mu_D(x) = \sum_{i=1}^{n} \alpha_i(x)\mu_{G_i}(x) + \sum_{j=1}^{m} \beta_j(x)\mu_{C_j}(x)$$

where the α_i and β_j are membership functions such that

$$\sum_{i=1}^{n} \alpha_i(x) + \sum_{j=1}^{m} \beta_j(x) \equiv 1.$$

Subject to this constraint, then, the values of $\alpha_i(x)$ and $\beta_j(x)$ can be chosen in such a way as to reflect the relative importance of $G_1, \cdots, G_n$ and $C_1, \cdots, C_m$. In particular, if $m = n = 1$, it is easy to verify that (22) can generate any fuzzy set which is contained in $G \cup C$ and contains $G \cap C$. Note that (22) resembles the familiar artifice of transforming a vector-valued criterion into a scalar-valued criterion by forming a linear combination of the components of the vector-valued objective function.

So far, we have restricted our attention to situations in which the goals and the constraints are fuzzy sets in X, the space of alternatives. A more general case which is of practical interest is one in which the goals and the constraints are fuzzy sets in different spaces. Specifically, let f be a mapping from $X = \{x\}$ to $Y = \{y\}$, with x representing an input (cause) and y, $y = f(x)$, representing the corresponding output (effect).

Suppose that the goals are defined as fuzzy sets $G_1, \cdots, G_n$ in Y while the constraints $C_1, \cdots, C_m$ are defined as fuzzy sets in X. Now, given a fuzzy set G_i in Y, one can readily find a fuzzy set $\bar{G}_i$ in X which induces G_i in Y. Specifically, the membership function of $\bar{G}_i$ is given by the equality

$$(23) \qquad \mu_{\bar{G}_i}(x) = \mu_{G_i}(f(x)), \qquad i = 1, \cdots, n.$$

The decision D, then, can be expressed as the intersection of $\bar{G}_1, \cdots, \bar{G}_n$ and $C_1, \cdots, C_m$. Using (23), we can express $\mu_D(x)$ more explicitly as

$$(24) \qquad \mu_D(x) = \mu_{G_1}(f(x)) \wedge \cdots \wedge \mu_{G_n}(f(x)) \wedge \mu_{C_1}(x) \wedge \cdots \wedge \mu_{C_m}(x),$$

where $f: X \rightarrow Y$. In this way, the case where the goals and the constraints are defined as fuzzy sets in different spaces can be reduced to the case where they are defined in the same space. We shall find (24) of use in the analysis of multistage decision processes in the following section.

4. Multistage Decision Processes

As an application of the concepts introduced in the preceding sections, we shall consider a few basic types of problems involving multistage decision-making in a fuzzy environment. It should be stressed that, in what follows, our main purpose is to illustrate the use of the concepts of fuzzy goal, fuzzy constraint and fuzzy decision, rather than to develop a general theory of multistage decision processes in which fuzziness enters in one way or another.

For simplicity we shall assume that the system under control, A, is a time-invariant finite-state deterministic system in which the state, x_t, at time t, $t = 0, 1, 2, \cdots$, ranges over a finite set $X = \{\sigma_1, \cdots, \sigma_n\}$, and the input, u_t, ranges over a finite set $U = \{\alpha_1, \cdots, \alpha_m\}$. The temporal evolution of A is described by the state equation

$$(25) \qquad x_{t+1} = f(x_t, u_t), \qquad t = 0, 1, 2, \cdots$$

in which f is a given function from $X \times U$ to X. Thus, $f(x_t, u_t)$ represents the *successor state* of x_t for input u_t. Note that if f is a random function, then A is a *stochastic* system whose state at time $t + 1$ is a probability distribution over X, $P(x_{t+1} \mid x_t, u_t)$, which is conditioned on x_t and u_t. Analogously, if f is a fuzzy function, then A is a *fuzzy* system [21] whose state at time $t + 1$ is a fuzzy set conditioned on x_t and u_t, which means that it is characterized by a membership function of the form $\mu(x_{t+1} \mid x_t, u_t)$.[3] Since we will not be concerned with such systems in the sequel, it will be understood that f is nonfuzzy unless explicitly stated to the contrary.

We assume that at each time t the input is subjected to a fuzzy constraint C^t, which is a fuzzy set in U characterized by a membership function $\mu_t(u_t)$. Furthermore, we assume that the goal is a fuzzy set G^N in X, which is characterized by a membership function $\mu_{G^N}(x_N)$, where N is the time of termination of the process. These assumptions are common to most of the problems considered in the sequel.

Problem 1. In this case, the system is assumed to be characterized by (25), with f a given nonrandom function. The termination time N is assumed to be fixed and specified. The initial state, x_0, is assumed to be given. The problem is to find a maximizing decision.

Applying (20), the decision—viewed as a decomposable fuzzy set in $U \times U \times \cdots \times U$, may be expressed at once as

$$(26) \qquad R = C^0 \cap C^1 \cap \cdots \cap C^{N-1} \cap \bar{G}^N$$

where $\bar{G}^N$ is the fuzzy set in $U \times U \times \cdots \times U$ which induces G^N in X. More explicitly, in terms of membership functions, we have

$$(27) \qquad \mu_D(u_0, \cdots, u_{N-1}) = \mu_0(u_0) \wedge \cdots \wedge \mu_{N-1}(u_{N-1}) \wedge \mu_{G^N}(x_N)$$

where x_N is expressible as a function of x_0 and $u_0, \cdots, u_{N-1}$ through the iteration of (25).

Our problem, then, is to find a sequence of inputs $u_0, \cdots, u_{N-1}$ which maximizes μ_D as given by (27). As is usually the case in multistage processes, it is expedient to express the solution in the form

$$u_t = \pi_t(x_t), \qquad t = 0, 1, 2, \cdots, N - 1,$$

where π_t is a policy function. Then, we can employ dynamic programming to give us both the π_t and a maximizing decision $u_0^M, \cdots, u_{N-1}^M$.

---

[3] It should be noted that when we speak of a *fuzzy environment*, we mean that the goals and/or the constraints are fuzzy, but not necessarily the system which is under control.

More specifically, using (26) and (25), we can write

$$(28) \quad \mu_D(u_0^M, \cdots, u_{N-1}^M) = \text{Max}_{u_0,\cdots,u_{N-2}} \text{Max}_{u_{N-1}} (\mu_0(u_0) \wedge \cdots \mu_{N-2}(u_{N-2})$$
$$\wedge \mu_{N-1}(u_{N-1}) \wedge \mu_{G^N}(f(x_{N-1}, u_{N-1})),$$

Now, if γ is a constant and g is any function of u_{N-1}, we have the identity

$$\text{Max}_{u_{N-1}} (\gamma \wedge g(u_{N-1})) = \gamma \wedge \text{Max}_{u_{N-1}} g(u_{N-1}).$$

Consequently, (28) may be rewritten as

$$(29) \quad \mu_D(u_0^M, \cdots, u_{N-1}^N) = \text{Max}_{u_0,\cdots,u_{N-1}} (\mu_0(u_0) \wedge \cdots \wedge \mu_{N-2}(u_{N-2}) \wedge \mu_{G^N-1}(x_{N-1}))$$

where

$$(30) \quad \mu_{G^{N-1}} (x_{N-1}) = \text{Max}_{u_{N-1}} (\mu_{N-1}(u_{N-1}) \wedge \mu_{G^N} (f(x_{N-1}, u_{N-1})))$$

may be regarded as the membership function of a fuzzy goal at time $t = N - 1$ which is induced by the given goal G^N at time $t = N$.

On repeating this backward iteration, which is a simple instance of dynamic programming, we obtain the set of recurrence equations

$$(31) \quad \mu_{G^{N-\nu}} (x_{N-\nu}) = \text{Max}_{u_{N-\nu}} (\mu(u_{N-\nu}) \wedge \mu_{G^{N-\nu+1}} (x_{N-\nu+1}))$$
$$x_{N-\nu+1} = f(x_{N-\nu}, u_{N-\nu}), \quad \nu = 1, \cdots, N,$$

which yield the solution to the problem. Thus, a maximizing decision $u_0^M, \cdots, u_{N-1}^M$ is given by the successive maximizing values of $u_{N-\nu}$ in (31), with $u_{N-\nu}^M$ defined as a function of $x_{N-\nu}$, $\nu = 1, \cdots, N$.

Example. As a simple illustration, consider a system with three states σ_1, σ_2, σ_3 and two inputs α_1 and α_2. Assume $N = 2$ for simplicity. Let the fuzzy goal at time $t = 2$ be defined by a membership function μ_{G^2} whose values are given by

$$\mu_{G^2}(\sigma_1) = 0.3; \quad \mu_{G^2}(\sigma_2) = 1; \quad \mu_{G^2}(\sigma_3) = 0.8.$$

Furthermore, let the fuzzy constraints at $t = 0$ and $t = 1$ be defined respectively by

$$\mu_0(\alpha_1) = 0.7, \mu_0(\alpha_2) = 1; \quad \mu_1(\alpha_1) = 1; \quad \mu_1(\alpha_2) = 0.6.$$

The state transition table which defines the function f in (25) is assumed to be

u_t	x_t		
	σ_1	σ_2	σ_3
α_1	σ_1	σ_3	σ_1
α_2	τ_2	σ_1	σ_3

Using (30), the membership function of the fuzzy goal induced at $t = 1$ is found to be

$$\mu_{G^1}(\sigma_1) = 0.6; \quad \mu_{G^1}(\sigma_2) = 0.8; \quad \mu_{G^1}(\sigma_3) = 0.6$$

and the corresponding maximizing decision is given by

$$\pi_1(\sigma_1) = \alpha_2 ; \quad \pi_1(\sigma_2) = \alpha_1 ; \quad \pi_1(\sigma_3) = \alpha_2.$$

Similarly, for $t = 0$

$$\mu_{G^0}(\sigma_1) = 0.8; \quad \mu_{G^0}(\sigma_2) = 0.6; \quad \mu_{G^0}(\sigma_3) = 0.6$$

and

$$\pi_0(\sigma_1) = \alpha_2 ; \quad \pi_0(\sigma_2) = \alpha_1 \text{ or } \alpha_2 ; \quad \pi_0(\sigma_3) = \alpha_1 \text{ or } \alpha_2 .$$

Thus, if the initial state (at $t = 0$) is σ_1, then the maximizing decision is α_2, α_1 and the corresponding value of μ_{G^2} is 0.8.

Next, we turn to a more general multistage decision process in which the system under control is stochastic, while the goal and the constraints are fuzzy.

5. Stochastic Systems in a Fuzzy Environment

As in the preceding problem, assume that the termination time X is fixed and that an initial state x_0 is specified. The system is assumed to be characterized by a conditional probability function $p(x_{t+1} \mid x_t, u_t)$. The problem is to maximize the probability of attainment of the fuzzy goal at time N, subject to the fuzzy constraints $C^0, \cdots, C^{N-1}$.

If the fuzzy goal G^N is regarded as a fuzzy event [23] in X, then the conditional probability of this event given x_{N-1} and u_{N-1} is expressed by

$$(32) \quad \text{Prob } (G_N \mid x_{N-1}, u_{N-1}) = E\mu_{G^N}(x_N) = \sum_{x_N} p(x_N \mid x_{N-1}, u_{N-1})\mu_{G^N}(x_N)$$

where E denotes the conditional expectation and μ_{G^N} is the membership function of the given fuzzy goal.

We observe that (32) expresses Prob $(G_N \mid x_{N-1}, u_{N-1})$ or, equivalently, $E\mu_{G^N}(x_N)$, as a function of x_{N-1} and u_{N-1}, just as in the preceding problem $\mu_G(x_N)$ was expressed as a function of x_{N-1} and u_{N-1} via (25). This implies that $E\mu_{G^N}(x_N)$ can be treated in the same way as $\mu_{G^N}(x_N)$ was treated in the nonstochastic case, thus making it possible to reduce the solution of the problem under consideration to that of the preceding problem.

More specifically, the recurrence equations (31) are replaced by

$$(33) \quad \begin{aligned} \mu_{G^{N-\nu}}(x_{N-\nu}) &= \text{Max}_{u_{N-\nu}} \; (\mu_{N-\nu}(u_{N-\nu}), E\mu_{G^{N-\nu+1}}(x_{N-\nu+1})) \\ E\mu_{G^{N-\nu+1}}(x_{N-\nu+1}) &= \sum_{x_{N-\nu+1}} p(x_{N-\nu+1} \mid x_{N-\nu}, u_{N-\nu})\mu_{G^{N-\nu+1}}(x_{N-\nu+1}) \end{aligned}$$

where, as before, $\mu_{G^{N-\nu}}(x_{N-\nu})$ denotes the membership function of the fuzzy goal at $t = N - \nu$ induced by the fuzzy goal at $t = N - \nu + 1$, $\nu = 1, \cdots, N$. These equations yield a solution to the problem, as is illustrated by the following example.

Example. As in the preceding example, we assume that the system has three states $\sigma_1, \sigma_2, \sigma_3$ and two inputs α_1, α_2. N is assumed to be equal to 2, and the probability function $p(x_{t+1} \mid x_t, u_t)$ is given by the following two tables, corresponding to $u_t = \alpha_1$, and $u_t = \alpha_2$, respectively.

	I. $u_t = \alpha_1$				II. $u_t = \alpha_2$		
		x_{t+1}				x_{t+1}	
x_t	σ_1	σ_2	σ_3	x_t	σ_1	σ_2	σ_3
σ_1	0.8	0.1	0.1	σ_1	0.1	0.9	0
σ_2	0	0.1	0.9	σ_2	0.8	0.1	0.1
σ_3	0.8	0.1	0.1	σ_3	0.1	0	0.9

The entries in these tables are the values of $p(x_{t+1} \mid x_t, u_t)$. Thus, the entry 0.8 in the position (σ_1, σ_2) in the first table signifies that if the system is in state σ_1 at time t and input α_1 is applied, then with probability 0.8 the state at time $t + 1$ will be σ_2.

The fuzzy goal at $t = 2$ is assumed to be the same as in the preceding example, that is

$$\mu_{G^2}(\sigma_1) = 0.3; \quad \mu_{G^2}(\sigma_2) = 1; \quad \mu_{G^2}(\sigma_3) = 0.8.$$

Likewise, the constraints are assumed to be the same. Thus

$$\mu_0(\alpha_1) = 0.7, \quad \mu_0(\alpha_2) = 1; \quad \mu_1(\alpha_1) = 1, \quad \mu_1(\alpha_2) = 0.6.$$

Using (33), we compute $E\mu_{G^2}(x_2)$ as a function of x_1 and u_1. Tabulating the results, we have

		x_1	
u_1	σ_1	σ_2	σ_3
α_1	0.42	0.82	0.42
α_2	0.93	0.42	0.75

Next, using (33) with $\nu = 1$ and computing $\mu_{G^1}(x_1)$ we obtain

$$\mu_{G^1}(\sigma_1) = 0.6; \quad \mu_{G^1}(\sigma_2) = 0.82; \quad \mu_{G^1}(\sigma_3) = 0.6$$

which correspond to the following values of the **maximal policy function**

(33a) $$\pi_1(\sigma_1) = \alpha_2; \quad \pi_1(\sigma_2) = \alpha_1; \quad \pi_1(\sigma_3) = \alpha_2.$$

The final iteration with $\nu = 2$ yields

		x_0	
μ_0	σ_1	σ_2	σ_3
α_1	0.62	0.62	0.62
α_2	0.8	0.62	0.60

(33b) $\mu_{G^0}(\sigma_1) = 0.8;$ $\mu_{G^0}(\sigma_2) = 0.62;$ $\mu_{G^0}(\sigma_3) = 0.62.$

(33c) $\pi_0(\sigma_1) = \alpha_1 ;$ $\pi_0(\sigma_2) = \alpha_1 \text{ or } \alpha_2 ;$ $\pi_0(\sigma_3) = \alpha_1 .$

The values of μ_{G^0} in (33b) represent the probabilities of attaining the given goal at $t = 2$ starting with σ_1, σ_2 and σ_3, respectively, assuming that the inputs are determined by the maximal policy function π_t, that is, $u_t = \pi_t(x_t)$ $(t = 0, 1, x_t = \sigma_1, \sigma_2, \sigma_3, u_t = \alpha_1, \alpha_2)$ whose values are given in (33a) and (33c).

Comment. It should be noted that when the fuzzy goal at time N is defined in such a way that the probability of attaining it is small for all values of x_{N-1} and u_{N-1}, it may be necessary to normalize the fuzzy goal induced at time $N - 1$ before finding its intersection with C_{N-1}, for otherwise the decision would be uninfluenced by the constraints. To be consistent, such normalization may have to be carried out at each stage of the decision process. Although we shall not dwell further upon this aspect of the problem in the present paper, it should be emphasized that it is by no means a trivial one and requires a more thorough analysis.

6. Systems With Implicitly Defined Termination Time

In the preceding cases, we have assumed that the termination time, N, is fixed a priori. In the more general case which we shall consider in this section, the termination time is assumed to be determined implicitly by a subsidiary condition of the form $x_N \in T$, where T is a specified nonfuzzy subset of X termed the *termination set.*[4] Thus, the process terminates when the state of the system under control enters, for the first time, a specified subset of the state space. In this case, the goal is defined as a fuzzy set G in T, rather than in X.

More concretely, assume that the system under control, A, is a deterministic system characterized by a state equation of the form

(34) $x_{t+1} = f(x_t, u_t), \qquad t = 0, 1, 2, \cdots$

where x_t ranges over $X = \{\sigma_1, \cdots, \sigma_l, \sigma_{l+1}, \cdots, \sigma_n\}$, in which $T = \{\sigma_{l+1}, \cdots, \sigma_n\}$ constitutes the termination set. As before, f is assumed to be a given function from $X \times U$ to X, where $U = \{\alpha_1, \cdots, \alpha_m\}$ is the range of u_t, $t = 0, 1, 2, \cdots$. Note that if σ_i is an *absorbing* state, that is, a state in T, then we can write $f(\sigma_i, \alpha_j) = \sigma_i$ for all α_j in U.

The fuzzy goal is assumed to be a subset of T characterized by a membership function $\mu_G(x_N)$, where N is the time at which $x_t \in T$, with $x_t \notin T$ for $t < N$. As for the constraints on the input, we assume for simplicity that they are independent of time but not necessarily the state. Thus, if A is in state σ_i at time t, then the fuzzy constraint on u_t is assumed to be represented by a fuzzy set $C(\sigma_i)$ (or $C(x_t)$) in U which is conditioned on σ_i. The membership function of this set will be denoted by $\mu_C(u_t \mid x_t)$.

Let x_0 be an initial state in T', where $T' = \{\sigma_1, \cdots, \sigma_l\}$ is the complement of T in X. To each such initial state will correspond a decision, $D(x_0)$, given by

---

[4] In its conventional (nonfuzzy) formulation, this case plays an important role in the theory of optimal control and Markovian decision processes. Some of the more relevant papers on this subject are cited in the list of references.

(35) $$D(x_0) = C(x_0) \cap C(x_1) \cap \cdots \cap C(x_{N-1}) \cap G$$

where the successive states $x_1, \cdots, x_{N-1}, x_N$ can be expressed as iterated functions of x_0 and $u_0, \cdots, u_{N-1}$ through the state equation (34). Thus

(36)
$$
\begin{aligned}
x_1 &= f(x_0, u_0) \\
x_2 &= f(x_1, u_1) \\
    &= f(f(x_0, u_0), u_1) \\
x_3 &= f(f(f(x_0, u_0), u_1), u_2)
\end{aligned}
$$

$$\cdots\cdots\cdots\cdots\cdots\cdots\cdots$$

Note that, as in (26), the C's in (35) should be regarded as fuzzy sets in the product space $U \times U \times \cdots \times U \times T$. Another point that should be noted is that $D(x_0)$ is uniquely determined by (35) for each x_0, with the understanding that $D(x_0)$ is empty if there is no finite sequence of inputs $u_0, \cdots, u_{N-1}$ which takes the initial state x_0 into T. In this event, we shall say that T is *not reachable* from the initial state.

From (35), we can readily derive a simpler implicit equation which is satisfied by $D(x_0)$. Specifically, in virtue of the time-invariance of A and the time-independence of the goal and constraint sets, (35) implies

(37) $$D(x_t) = C(x_t) \cap C(x_{t+1}) \cap \cdots \cap C(x_{t+N-1}) \cap G$$

for $t = 0, 1, 2, \cdots$. In particular,

(38) $$D(x_{t+1}) = C(x_{t+1}) \cap \cdots \cap C(x_{t+N-1}) \cap G$$

and hence (37) can be written as

(39) $$D(x_t) = C(x_t) \cap D(x_{t+1})$$

or, using (34),

(40) $$D(x_t) = C(x_t) \cap D(f(x_t, u_t)), \qquad t = 0, 1\ 2, \cdots$$

which is the desired implicit equation. Expressed in terms of the membership functions of the sets in question, this equation assumes the following form (for $t = 0$)

(41) $$\mu_D(u_0, \cdots, u_{N-1} \mid x_0) = \mu_C(u_0 \mid x_0) \wedge \mu_D(u_1, \cdots, u_{N-1} \mid f(x_0, u_0))$$

where the termination time N is also a function of x_0 and $u_0, u_1, u_2, \cdots$ through the state equation (34) and the termination condition $x_N \in T$, with $x_0 \notin T, \cdots, x_{N-1} \notin T$.

Now suppose that the successive inputs $u_0, u_1, \cdots, u_{N-1}$ are determined by a stationary (time-invariant) policy function π, $\pi: T' \to \mathcal{U}$, which associates with each state x_t in T' an input u_t which should be applied to A when it is in state x_t. Thus,

(42) $$u_t = \pi(x_t), \qquad t = 0, \cdots, N-1, \quad x_t \in T'.$$

Since $u_0, \cdots, u_{N-1}$ are determined by x_0 and π through (42) and the state equation (34), the membership function of $D(x_0)$ can be written as $\mu_D(x_0 \mid \pi)$. Similarly, $\mu_C(u_0 \mid x_0)$ can be written as $\mu_C(\pi(x_0) \mid x_0)$, and $\mu_D(u_1, \cdots, u_{N-1} \mid f(x_0, u_0))$ as $\mu_D(f(x_0, \pi(x_0)) \mid \pi)$. With these substitutions, (41) assumes the more compact form

$$(43) \qquad \mu_D(x_0 \mid \pi) = \mu_C(\pi(x_0) \mid x_0) \wedge \mu_D(f(x_0, \pi(x_0)) \mid \pi), \qquad x_0 \in T',$$

which in effect is a system of l equations (one for each value of x_0) in the μ_D. This system of equations determines μ_D as a function of x_0 for each π, with the understanding that $\mu_D = 0$ if under π the process does not terminate, that is, there does not exist a finite N such that $x_N \in T$. Furthermore, it is understood that $\mu_D = \mu_G$ for states in T.

It is easy to demonstrate that (43) has a unique solution. Specifically, by decomposing the set of states $T' = \{\sigma_1, \cdots, \sigma_l\}$ into disjoint subsets $T'_1, \cdots, T'_\kappa$, where T'_λ, $\lambda = 1, \cdots, \kappa$, represents the set of states from which T is reachable in λ steps, it is readily seen that the equations in (43) corresponding to the x_0 which are in T_1 yield uniquely the respective values of μ_D. In terms of these, the equations in (43) corresponding to the x_0 in T_2 yield uniquely the values of μ_D for x_0 in T_2. Continuing in this manner, all the μ_D's can be determined uniquely by successively solving subsets of the system of equations (43) for the blocks of variables in $T'_1, \cdots, T'_\kappa$.

For our purposes, it will be convenient to represent a policy π as a *policy vector*

$$(44) \qquad \pi = (\pi(\sigma_1), \cdots, \pi(\sigma_l))$$

whose ith component, $i = 1, \cdots, l$, is the input which must be applied when A is in state σ_i. Note that $\pi(\sigma_i)$ ranges over the set $U = \{\alpha_1, \cdots, \alpha_m\}$ and thus that there are m^l distinct policies in the policy space.

With reference to the system of equations (43), let

$$(45) \qquad \mu_D(\pi) = (\mu_D(\sigma_1 \mid \pi), \cdots, \mu_D(\sigma_n \mid \pi))$$

be an n-vector, termed the *goal attainment vector*, whose components are the values of the membership function of D at $\sigma_1, \cdots, \sigma_n$ (corresponding to policy π). It is natural to define a preordering in the policy space by

$$(46) \qquad \pi' \geq \pi'' \Leftrightarrow \mu_D(\pi') \geq \mu_D(\pi'')$$

which means that a policy π' is better than or equal to a policy π'' if and only if $\mu_D(\sigma_i \mid \pi') \geq \mu_D(\sigma_i \mid \pi'')$ for $i = 1, \cdots, n$. Then, a policy π will be said to be *optimal* if and only if π is better than or equal to every policy in the policy space.

Does there exist an optimal policy for the problem under consideration? The answer to this question is in the affirmative. This assertion can be proved rigorously,[5] but it will suffice for our purposes to regard it as a consequence of the alternation principle [13]—a principle of broad validity which in concrete cases can be asserted as a provable theorem.

Specifically, let π' and π'' be two arbitrary policy vectors, with $\mu_D(\pi')$ and $\mu_D(\pi'')$ being the corresponding goal attainment vectors. Using π' and π'', let us construct a policy vector π in accordance with the following rules:

$$(47) \qquad \begin{aligned} \pi_i &= \pi'_i \quad \text{if} \quad \mu_D(\sigma_i \mid \pi') \geq \mu_D(\sigma_i \mid \pi'') \\ &= \pi''_i \quad \text{if} \quad \mu_D(\sigma_i \mid \pi') < \mu_D(\sigma_i \mid \pi'') \end{aligned}$$

---

[5] A proof for the case of a stochastic finite-state system is given in [12].

for each component π_i of π, $i = 1, \cdots, l$. Then, according to the alternation principle, $\pi \geq \pi'$ and $\pi \geq \pi''$, that is, π is better than or equal to both π' and π''. From this and the finiteness of the policy space it follows at once that there exists an optimal policy.

From (43) it is a simple matter to derive a functional equation satisfied by the goal attainment vector corresponding to the optimal policy. Thus, let

$$(48) \qquad \mu_D{}^M = \mathrm{Max}_\pi \, \mu_D(\pi)$$

and let $P(\pi)$ be an $n \times n$ matrix of zeros and ones whose ijth element is one if and only if $\sigma_j = f(\sigma_i, \pi(\sigma_i))$, that is, the state σ_j is the immediate successor of σ_i under policy π.

Furthermore, let $\mu_C(\pi)$ denote a vector whose ith component is $\mu_C(\pi(\sigma_i) \mid \sigma_i)$. Then, on taking the maximum of both sides of (43), we obtain

$$(49) \qquad \mu_D{}^M = \mathrm{Max}_\pi \, (\mu_C(\pi) \wedge P(\pi)\mu_D{}^M)$$

which is the desired functional equation for $\mu_D{}^M$. Although different in detail, equation (49) is of the same general form as the functional equations arising in the theory of Markovian decision processes [17]. Its solution, however, is considerably simpler to obtain because of the distributivity of Max and $\wedge$.

Specifically, let $\pi^1, \cdots, \pi^r$, where $r = m^l$, denote the m^l distinct policy vectors. Then, on using $\vee$ in place of Max, (49) becomes

$$(50) \qquad \mu_D{}^M = (\mu_C(\pi^1) \wedge P(\pi^1)\mu_D{}^M) \vee \cdots \vee (\mu(\pi^r) \wedge P(\pi^r)\mu_D{}^M).$$

Taking advantage of the distributivity of $\vee$ and $\wedge$, and factoring like terms, we can put (50) into a much simpler form which, written as a system of equations in the components of $\mu_D{}^M$, reads

$$(51) \quad \mu_D{}^M(\sigma_i) = \vee_j(\mu_C(\alpha_j \mid \sigma_i) \wedge \mu_D{}^M(f(\sigma_i, \alpha_j))), \quad i = 1, \cdots, n, \quad j = 1, \cdots, m$$

where $\alpha_j = \pi(\sigma_i) = $ input under policy π in state σ_i; $\mu_D{}^M(\sigma_i) = i$th component of the optimal goal attainment vector; $f(\sigma_i, \alpha_j) = $ successor state[6] of σ_i for input α_j, with $f(\sigma_i, \alpha_j) = \sigma_i$ for $i = l + 1, \cdots, n$ (that is, for σ_i in the termination set T); $\mu_C(\alpha_j \mid \sigma_i) = $ value of the membership function of the constraint C in state σ_i for input α_j, with $\mu_C(\alpha_j \mid \sigma_i) = 1$ for $i = l + 1, \cdots, n$; and for $i = l + 1, \cdots, n$, $\mu_D{}^M(\sigma_i) = \mu_G(\sigma_i) = $ value of the membership function of the given goal G at σ_i. Thus, the $\mu_D{}^M(\sigma_i)$, $i = 1, \cdots, l$, are the unknowns in (51), while the $\mu_D{}^M(\sigma_i)$, $i = l + 1 \cdots, n$, and the $\mu_C(\alpha_j \mid \sigma_i)$, $i = 1, \cdots, n, j = 1, \cdots, m$, are given constants.

To make the solution of (51) more transparent, it is helpful to simplify the notation in (51) by letting the unknowns in (51) be denoted by ω_i, that is, $\omega_i = \mu_D{}^M(\sigma_i)$ for $i = 1, \cdots, l$. Furthermore, let the product and plus symbols denote $\wedge$ and $\vee$, respectively. Then, (51) can be written more compactly in matrix form as

$$(52) \qquad \omega = B\omega + \gamma$$

---

[6] Note that the successor states in (49) are defined by $P(\pi)$.

where $\omega = (\omega_1, \cdots, \omega_l)$, $\gamma = (\gamma_1, \cdots, \gamma_l)$, $B = (b_{ik})$; furthermore,

$\qquad b_{ik} = 0$ if σ_k is not an immediate successor of σ_i ;

$\qquad b_{ik} = \bigvee_{\alpha_p} \mu_c(\alpha_p \mid \sigma_i)$, where the α_p are inputs which take σ_i into σ_k ;

and

$$(53) \qquad \gamma_i = \bigvee_j \left(\mu_c(\alpha_j \mid \sigma_i) \wedge \mu_G(f(\sigma_i, \alpha_j)) \right)$$

with the understanding that $\mu_G(\sigma_i) = 0$ for states outside the termination set T.

Having put (51) into the form of a linear equation (52), it is easy to show that (53) and hence (51) can be solved by iteration. Specifically, let $\omega^0 = (0, \cdots, 0)$ and

$$(54) \qquad \omega^{s+1} = B\omega^s + \gamma, \qquad s = 0, 1, 2, \cdots .$$

Then, by induction, the sequence $\omega^0, \omega^1, \omega^2, \cdots$ is monotone nondecreasing. For, assume that $\omega^{k+1} \geq \omega^k$ for some k. Using (54), we have

$$(55) \qquad \omega^{k+2} = B\omega^{k+1} + \gamma \geq B\omega^k + \gamma = \omega^{k+1},$$

and noting that $\omega^1 \geq \omega^0 = 0$, it follows that $\omega^{s+1} \geq \omega^s$ for $s = 0, 1, 2, \cdots$.

Since the sequence $\omega^0, \omega^1, \cdots$ is monotone nondecreasing and bounded from above by $\omega = (1, \cdots, 1)$, it follows that it converges to the solution of (52), that is, to the first l components[7] of the optimal goal attainment vector $\mu_D{}^M$. Actually, a more detailed argument shows that (54) yields the solution of (52) in not more than l iterations. This is an immediate consequence of the following lemma.

LEMMA. *Let $B = [b_{ij}]$ be a matrix of order l with real-valued elements. Let B^s denote the sth power of B with the operations $\vee$ and $\wedge$ replacing the sum and product, respectively. Then, for all integral $s \geq l$,*

$$(56) \qquad B + B^2 + \cdots + B^s = B + B^2 + \cdots + B^l$$

and

$$(57) \quad I + B + B^2 + \cdots + B^s = I + B + B^2 + \cdots + B^{l-1}, \qquad s \geq l - 1$$

where I is the identity matrix.

PROOF. The validity of this lemma becomes rather evident when (56) is interpreted in graph-theoretic terms. Specifically, let $G(B)$ denote a graph with l nodes in which b_{ij}, $i, j = 1, \cdots$, represents the "strength" of the link between node i and node j.

Let $\gamma^s_{i,j,\mu}$ denote a chain of s links in $G(B)$,

$$\gamma^s_{i,j,\mu} = (b_{i\lambda_1}, b_{\lambda_1\lambda_2}, \cdots, b_{\lambda_{s+1},j})$$

starting at node i and ending at node j. The subscript μ serves as a label for the chain in question, with μ ranging from 1 to M, where M is the number of distinct chains of length s linking i to j.

---

[7] The remaining $n - l$ components of $\mu_D{}^M$ are given by the corresponding components of μ_G.

Define the *strength* of $\gamma_{i,j,\mu}^{s}$, $\sigma(\gamma_{i,j,\mu}^{s})$, as the strength of its weakest link, that is,

$$(58) \qquad \sigma(\gamma_{i,j,\mu}^{s}) = b_{i\lambda_1} \wedge b_{\lambda_1\lambda_2} \wedge \cdots \wedge b_{\lambda_{s+1},j}.$$

By the definition of matrix product (with plus and product replaced by $\vee$ and $\wedge$, respectively) it is evident that b_{ij}^{s}, the (i, j) element of B^{s}, $s \geq 1$, may be expressed as

$$(59) \qquad b_{ij}^{s} = \sigma(\gamma_{i,j,1}^{s}) \vee \sigma(\gamma_{i,j,2}^{s}) \vee \cdots \vee \sigma(\gamma_{i,j,M}^{s})$$

or more compactly

$$(60) \qquad b_{ij}^{s} = \vee_{\mu}\, \sigma(\gamma_{i,j,\mu}^{s})$$

where $\vee_{\mu}$ denotes the supremum over all chains of length s linking i to j. Thus, in words,

b_{ij}^{s} = strength of the strongest chain among all chains of length s linking node i to node j.

From this interpretation of the elements of B^{s}, it follows that the (i, j) element of $B + B^{2} + \cdots + B^{s}$ can be expressed as

(i, j) element of $B + \cdots + B^{s}$ = strength of the strongest chain among all chains of length $\leq s$ linking node i to node j.

Thus, in words, the statement of the lemma implies and is implied by: If B is a matrix of order l and $s \geq l$, then:

strength of the strongest chain among all chains of length $\leq s$ linking node i to node j = strength of the strongest chain among all chains of length $\leq l$ linking node i to node j.

Stated in this form, the lemma is very easy to establish. In the first place, it is evident that, for $s \geq l$,

$$(61) \qquad B + \cdots + B^{s} \geq B + \cdots + B^{l}.$$

Thus, it suffices to establish the reverse inequality $B + \cdots + B^{s} \leq B + \cdots + B^{l}$ to complete the proof.

Let $\gamma_{i,j}^{s}$ be a chain from i to j of length $s > l$. Clearly, in any such chain at least one node must appear more than once, implying that every chain of length $s > l$ must have one or more loops. The deletion of these loops results in a chain $\gamma_{i,j}^{r}$ of length $r \leq l$. Now, from the definition of the strength of a chain, (58), it follows that

$$(62) \qquad \sigma(\gamma_{i,j}^{s}) \leq \sigma(\gamma_{i,j}^{r})$$

and hence the supremum of $\sigma(\gamma_{i,j}^{s})$ over chains of length s $(s > l)$ is less than or equal to the supremum of $\sigma(\gamma_{i,j}^{r})$ over chains of length $\leq l$. Thus

$$(63) \qquad B^{s} \leq B + B^{2} + \cdots + B^{l}, \qquad s \geq l$$

and hence

$$(64) \qquad B + B^{2} + \cdots + B^{s} \leq B + B^{2} + \cdots + B^{l}, \qquad s \geq l$$

which, in conjunction with (61), establishes (56).

As for (57), note that if $i \neq j$, then (62) is true for $s \geq l - 1$ and $r \leq l - 1$. Without this restriction $(i \neq j)$, (62) is true with $s \geq l - 1$ and $r \leq l - 1$ if $b_{ii} \geq b_{ij}$

for $i, j = 1, \cdots, l$. The latter condition is satisfied if B is replaced by $I + B$. This implies that the exponent l in (56) may be replaced by $l - 1$ if B is replaced by $I + B$. The result is (57).

Returning to the solution of (52), we note that the expression for sth iterate is given by

$$(65) \qquad W^s = (B^{s-1} + \cdots + B + I)\gamma.$$

Making use of the lemma, we see that

$$(66) \qquad W^s = W^l, \qquad s > l$$

which implies that (54) yields the solution to (52) in not more than l iterations.

To gain an intuitive insight into the solution of (52), it is helpful to interpret the transition from (49) to (51) with the aid of the state diagram of A. Thus, for concreteness assume that A has five states, with transitions corresponding to various inputs shown in Figure 2. In this diagram, the number associated with the branch leading from σ_i to its successor state via input α_j is the value of $\mu_c(\alpha_j \mid \sigma_i)$. States σ_4 and σ_5 are in the termination set and the corresponding values of $\mu_G(\sigma_i)$ are shown alongside. The indicated values of the $\mu_c(\alpha_j \mid \sigma_i)$ correspond to the constraint sets

$$C(\sigma_1) = \{(\alpha_1, 0.6), (\alpha_2, 1)\},$$
$$C(\sigma_2) = \{(\alpha_1, 0.8), (\alpha_2, 1)\},$$
$$C(\sigma_3) = \{(\alpha_1, 1), (\alpha_2, 0.7)\}.$$

For the system in question, the state transition function $f(\sigma_i, \alpha_j)$ is given by the following table:

	σ_i				
α_j	σ_1	σ_2	σ_3	σ_4	σ_5
α_1	σ_4	σ_3	σ_5	σ_4	σ_5
α_2	σ_2	σ_2	σ_1	σ_4	σ_5

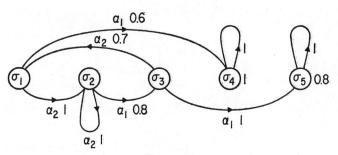

FIGURE 2

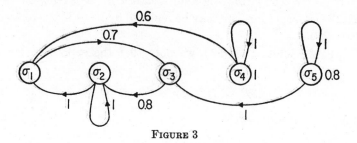

FIGURE 3

From this table, it is easy to construct the matrix $P(\pi)$ for any given policy. For example, for $\pi = (\alpha_2, \alpha_1, \alpha_2)$, we have

$$P(\alpha_2, \alpha_1, \alpha_2) = \begin{bmatrix} 0 & 1 & 0 & 0 & 0 \\ 0 & 0 & 1 & 0 & 0 \\ 1 & 0 & 0 & 0 & 0 \\ 0 & 0 & 0 & 1 & 0 \\ 0 & 0 & 0 & 0 & 1 \end{bmatrix}.$$

The system of equations (51) is obtained by reversing the direction of flow in each branch (see Figure 3) and treating the states in T, that is, σ_4 and σ_5 as sources, with the states in T', that is, σ_1, σ_2 and σ_3, playing the role of receptors (sinks). From the diagram shown in Figure 3, the equations in (51) can be written by inspection. Thus,

$$\mu_D^M(\sigma_1) = (0.6 \wedge \mu_D^M(\sigma_4)) \vee (1 \wedge \mu_D^M(\sigma_2)),$$
$$\mu_D^M(\sigma_2) = (0.8 \wedge \mu_D^M(\sigma_3)) \vee (1 \wedge \mu_D^M(\sigma_2)),$$
(67) $\qquad \mu_D^M(\sigma_3) = (1 \wedge \mu_D^M(\sigma_5)) \vee (0.7 \wedge \mu_D^M(\sigma_1)),$
$$\mu_D^M(\sigma_4) = \mu_G(\sigma_4) = 1,$$
$$\mu_D^M(\sigma_5) = \mu_G(\sigma_5) = 0.8.$$

Employing the simplified notation in which $\wedge$ and $\vee$ are replaced by the product and sum, respectively, and $\omega_i = \mu_D^M(\sigma_i)$, $i = 1, 2, 3$, the system of equations (67) becomes

(68) $\qquad\qquad\qquad\qquad \omega = B\omega + \gamma$

where

$$B = \begin{bmatrix} 0 & 1 & 0 \\ 0 & 1 & 0.8 \\ 0.7 & 0 & 0 \end{bmatrix}, \qquad \gamma = \begin{bmatrix} 0.6 \\ 0 \\ 0.8 \end{bmatrix}.$$

Letting $\omega^0 = (0, 0, 0)$, we obtain on first interation $\omega^1 = (0.6, 0, 0.8)$. Subsequent iterations yield

$$\omega^2 = (0.6, 0.8, 0.8), \qquad \omega^3 = (0.8, 0.8, 0.8), \qquad \omega^4 = (0.8, 0.8, 0.8).$$

Thus, $\omega^3 = (0.8, 0.8, 0.8)$ is the solution of (68).

To visualize the iteration process, imagine that each of the sources in Figure 3 (which are the absorbing states in Figure 2) generates balls of various diameters, with σ_i, $i = l + 1, \cdots, n$, generating balls of diameters ranging from 0 to $\mu_G(\sigma_i)$. Furthermore, imagine that a branch in Figure 2 which leaves state σ_i via input α_j, is a pipe of diameter $\mu_C(\alpha_j \mid \sigma_i)$ which can carry balls of diameter $\leq \mu_C(\alpha_j \mid \sigma_i)$ along the reverse direction, that is, along the direction shown in Figure 3. Thus, the diagram of Figure 3 may be visualized as a network of pipes whose diameters are indicated in the diagram and which can carry balls of lesser or equal diameter in the indicated directions. The states in the termination set (σ_4 and σ_5) play the role of sources of balls of diameters up to $\mu_G(\sigma_4)$ and $\mu_G(\sigma_5)$ respectively, while the remaining states (σ_1, σ_2 and σ_3) act as receptors. Because the absorbing states act as sources, we shall refer to the method of solution described above as a *reverse-flow* technique.

Now assume that it takes one unit of time for the balls to travel from a node of the network of Figure 3 to another node. If we start with no balls at σ_1, σ_2 and σ_3 at time 0, then at time $t = 1$ the maximum diameters of balls at σ_1, σ_2 and σ_3 will be, respectively, ω_1^1, ω_2^1 and ω_3^1, where $\omega^1 = (\omega_1^1, \omega_2^1, \omega_3^1)$ is the first iterate of (68). At time $t = 2$, the maximum diameters of balls will be given by ω^2 and at time $t = 3$ by ω^3. Since it takes no more than three units of time for any ball to travel from its source to any node in the network, there will be no further increase in the size of balls at each source upon further iteration. Thus, ω^3 gives the maximum diameter of balls at each receptor node and hence is the desired solution of (68).

Turning to the illustration of (43) and the alternation principle, consider the policy vector $\pi = (\alpha_1, \alpha_1, \alpha_1)$. For this π, the system of equations (43) becomes

$$\mu_D(\sigma_1 \mid \pi) = 0.6 \wedge \mu_D(\sigma_4 \mid \pi),$$

(69) $$\mu_D(\sigma_2 \mid \pi) = 0.8 \wedge \mu_D(\sigma_3 \mid \pi),$$

$$\mu_D(\sigma_3 \mid \pi) = 1 \wedge \mu_D(\sigma_5 \mid \pi).$$

In this case, σ_1 and σ_3 are in T'_1 and σ_2 is in T'_2. Noting that $\mu_D(\sigma_4 \mid \pi) = \mu_G(\sigma_4) = 1$ and $\mu_D(\sigma_5 \mid \pi) = \mu_G(\sigma_5) = 0.8$, we find at once $\mu_D(\sigma_1 \mid \pi) = 0.6$; $\mu_D(\sigma_2 \mid \pi) = 0.8$; $\mu_D(\sigma_3 \mid \pi) = 0.8$ which is the desired solution.

Carrying out the same computation for other policy vectors, we obtain the results tabulated below

π	σ		
	σ_1	σ_2	σ_3
$(\alpha_1, \alpha_1, \alpha_1)$	0.6	0.8	0.8
$(\alpha_1, \alpha_1, \alpha_2)$	0.6	0.6	0.6
$(\alpha_1, \alpha_2, \alpha_1)$	0.6	0	0.8
$(\alpha_1, \alpha_2, \alpha_2)$	0.6	0	0.6
$(\alpha_2, \alpha_1, \alpha_1)$	0.8	0.8	0.8
$(\alpha_2, \alpha_1, \alpha_2)$	0	0	0
$(\alpha_2, \alpha_2, \alpha_1)$	0	0	0.8
$(\alpha_2, \alpha_2, \alpha_2)$	0	0	0

As a check on the alternation principle, let us take $\pi' = (\alpha_1, \alpha_1, \alpha_2)$ and $\pi'' = (\alpha_1, \alpha_2, \alpha_1)$. Using (47) leads to $\pi = (\alpha_1, \alpha_1, \alpha_1)$. Note that $\pi \geq \pi'$ and $\pi \geq \pi''$. From inspection of the table, the maximal policy is seen to be $(\alpha_2, \alpha_1, \alpha_1)$, which agrees with the result obtained by iteration.

The approach to the solution of problems involving implicitly defined termination time which we have described in this section can be extended to more complex decision processes in a fuzzy environment. In particular, the technique employed for solving the functional equation (49) can readily be extended to fuzzy systems in a fuzzy environment. Furthermore, (43) and (49) can be extended also—as in §4—to stochastic finite-state systems. Because of limitations on space, we shall not consider these cases in the present paper.

7. Concluding Remarks

The task of developing a general theory of decision-making in a fuzzy environment is one of very considerable magnitude and complexity. Thus, the results presented in this paper should be viewed as merely a first attempt at constructing a conceptual framework for such a theory.

There are many facets of the theory of decision-making in a fuzzy environment which require more thorough investigation. Among these are the question of execution of fuzzy decisions; the way in which the goals and the constraints must be combined when they are of unequal importance or are interdependent; the control of fuzzy systems and the implementation of fuzzy algorithms; the notion of fuzzy feedback and its effect on decision-making; control of systems in which the fuzzy environment is partially defined by exemplification; and decision-making in mixed environments, that is, in environments in which the imprecision stems from both randomness and fuzziness.

References

1. ATHANS, M. AND FALB, P., *Optimal Control*, McGraw-Hill Book Co., Inc., New York, N.Y., 1966.
2. BELLMAN, R. E., *Dynamic Programming*, Princeton Univ. Press, Princeton, N.J., 1957.
3. ——, "A Markoffian Decision Process," *Jour. of Math. and Mechanics*, vol. 6, pp. 679–684, (1957).
4. ——, KALABA, R. AND ZADEH, L. A., "Abstraction and Pattern Classification," *Jour. Math. Anal. and Appl.*, Vol. 13 (January 1966), pp. 1–7.
5. BOLTYANSKII, V. G., GAMKRELIDZE, R. V. AND PONTRYAGIN, L. S., "On the Theory of Optimal Processes," *Izv. Akad. Nauk SSSR*, Vol. 24 (1960) pp. 3–42.
6. BROWN, J. G., "Fuzzy Sets on Boolean Lattices," Rep. No. 1957, Ballistic Research Laboratories, Aberdeen, Maryland, January, 1969.
7. BRYSON, A. E., JR., AND HO, Y. C., *Applied Optimal Control*, Blaisdell Co., Waltham, Mass., 1969.
8. CHANG, C. L., "Fuzzy Topological Spaces," *Jour. Math. Analysis and Appl.*, Vol. 24 (1968), pp. 182–190.
9. CHANG, S. S. L., "Fuzzy Dynamic Programming and the Decision Making Process," Proc. 3d Princeton Conference on Information Sciences and Systems, 1969, pp. 200–203.
10. DERMAN, C., "Markoffian Sequential Control Processes,—Denumerable State Space," *Jour. Math. Anal. and Appl.*, Vol. 10 (1965), pp. 295–302.

11. —— AND KLEIN, M., "Some Remarks on Finite Horizon Markoffian Decision Models," *Operations Research*, Vol. 13 (1965), pp. 272–278.
12. EATON, J. H. AND ZADEH, L. A., "Optimal Pursuit Strategies in Discrete-State Probabilistic Systems," *Jour. Basic Engineering (ASME)*, Vol. 84, Series D (March 1962), pp. 23–29.
13. —— AND ——, "An Alternation Principle for Optimal Control," *Automation and Remote Control*, Vol. 24 (March 1963), pp. 328–330.
14. FU, K. S. AND LI, T. J., "On the Behavior of Learning Automata and its Applications," Tech. Rep. TR-EE 68-20, Purdue University, Lafayette, Indiana, Aug. 1968.
15. GOGUEN, J., "L-fuzzy Sets," *Jour. Math. Anal. and Appl.*, Vol. 18 (April 1967), pp. 145–174.
16. HOWARD, R. A., *Dynamic Programming and Markoff Processes*, M.I.T. Press and J. Wiley, Inc., Cambridge, Mass. and New York, N.Y., 1960.
17. WAGNER, H. M., *Principles of Operations Research*, Prentice-Hall, Inc., 1969.
18. WEE, W. G., "On Generalization of Adaptive Algorithms and Application of the Fuzzy Set Concept to Pattern Classification," Technical Report TR-EE-67-7, Purdue University, Lafayette, Indiana, July, 1967.
19. WOLFE, P. AND DANTZIG, G. B., "Linear Programming in a Markoff Chain," *Oper. Res.*, Vol. 10 (1962), pp. 702–710.
20. ZADEH, L. A., "Fuzzy Sets," *Information and Control*, Vol. 8 (June 1965), pp. 338–353.
21. ——, "Toward a Theory of Fuzzy Systems," ERL Report No. 69-2, Electronics Research Laboratories, University of California, Berkeley, June, 1969.
22. ——, "Fuzzy Algorithms," *Information and Control*, Vol. 12 (February 1968), pp. 99–102.
23. ——, "Probability Measures of Fuzzy Events," *Jour. Math. Anal. and Appl.*, vol. 10 (August 1968), pp. 421–427.

Similarity Relations and Fuzzy Orderings†

L.A. ZADEH

*Department of Electrical Engineering and Computer Sciences
and Electronics Research Laboratory, University of California,
Berkeley, California*

ABSTRACT

The notion of "similarity" as defined in this paper is essentially a generalization of the notion of equivalence. In the same vein, a fuzzy ordering is a generalization of the concept of ordering. For example, the relation $x \gg y$ (x is much larger than y) is a fuzzy linear ordering in the set of real numbers.

More concretely, a *similarity relation*, S, is a fuzzy relation which is reflexive, symmetric, and transitive. Thus, let x, y be elements of a set X and $\mu_S(x,y)$ denote the grade of membership of the ordered pair (x,y) in S. Then S is a similarity relation in X if and only if, for all x, y, z in X, $\mu_S(x,x) = 1$ (reflexivity), $\mu_S(x,y) = \mu_S(y,x)$ (symmetry), and $\mu_S(x,z) \geqslant \bigvee_y (\mu_S(x,y) \wedge \mu_S(y,z))$ (transitivity), where $\vee$ and $\wedge$ denote max and min, respectively.

A *fuzzy ordering* is a fuzzy relation which is transitive. In particular, a *fuzzy partial ordering*, P, is a fuzzy ordering which is reflexive and antisymmetric, that is, $(\mu_P(x,y) > 0$ and $x \neq y) \Rightarrow \mu_P(y,x) = 0$. A *fuzzy linear ordering* is a fuzzy partial ordering in which $x \neq y \Rightarrow \mu_S(x,y) > 0$ or $\mu_S(y,x) > 0$. A *fuzzy preordering* is a fuzzy ordering which is reflexive. A *fuzzy weak ordering* is a fuzzy preordering in which $x \neq y \Rightarrow \mu_S(x,y) > 0$ or $\mu_S(y,x) > 0$.

Various properties of similarity relations and fuzzy orderings are investigated and, as an illustration, an extended version of Szpilrajn's theorem is proved.

---

1. INTRODUCTION

The concepts of equivalence, similarity, partial ordering, and linear ordering play basic roles in many fields of pure and applied science. The classical theory of relations has much to say about equivalence relations and various types of orderings [1]. The notion of a distance, $d(x,y)$, between objects x and y has long been used in many contexts as a measure of similarity or dissimilarity between elements of a set. Numerical taxonomy [2], factor

† This work was supported in part by a grant from the National Science Foundation, NSF GK-10656X, to the Electronics Research Laboratory, University of California, Berkeley, California.

81

analysis [3], pattern classification [4–7], and analysis of proximities [8–10] provide a number of concepts and techniques for categorization and clustering. Preference orderings have been the object of extensive study in econometrics and other fields [11, 12]. Thus, in sum, there exists a wide variety of techniques for dealing with problems involving equivalence, similarity, clustering, preference patterns, etc. Furthermore, many of these techniques are quite effective in dealing with the particular classes of problems which motivated their development.

The present paper is not intended to add still another technique to the vast armamentarium which is already available. Rather, its purpose is to introduce a unifying point of view based on the theory of fuzzy sets [13] and, more particularly, fuzzy relations. This is accomplished by extending the notions of equivalence relation and ordering to fuzzy sets, thereby making it possible to adapt the well-developed theory of relations to situations in which the classes involved do not have sharply defined boundaries.[1] Thus, the main contribution of our approach consists in providing a unified conceptual framework for the study of fuzzy equivalence relations and fuzzy orderings, thereby facilitating the derivation of known results in various applied areas and, possibly, stimulating the discovery of new ones.

In what follows, our attention will be focused primarily on defining some of the basic notions within this conceptual framework and exploring some of their elementary implications. Although our approach might be of use in areas such as cluster analysis, pattern recognition, decision processes, taxonomy, artificial intelligence, linguistics, information retrieval, system modeling, and approximation, we shall make no attempt in the present paper to discuss its possible applications in these or related problem areas.

2. NOTATION, TERMINOLOGY, AND PRELIMINARY DEFINITIONS

In [13], a fuzzy (binary) relation R was defined as a fuzzy collection of ordered pairs. Thus, if $X = \{x\}$ and $Y = \{y\}$ are collections of objects denoted generically by x and y, then a *fuzzy relation* from X to Y or, equivalently, a fuzzy relation in $X \cup Y$, is a fuzzy subset of $X \times Y$ characterized by a membership (characteristic) function μ_R which associates with each pair (x, y) its "grade of membership," $\mu_R(x, y)$, in R. We shall assume for simplicity that the range of μ_R is the interval $[0, 1]$ and will refer to the number $\mu_R(x, y)$ as the *strength* of the relation between x and y.

---

[1] In an independent work which came to this writer's attention [14], S. Tamura, S. Higuchi, and K. Tanaka have applied fuzzy relations to pattern classification, obtaining some of the results described in Section 3.

In the following definitions, the symbols $\vee$ and $\wedge$ stand for max and min, respectively.

The *domain* of a fuzzy relation R is denoted by dom R and is a fuzzy set defined by

$$\mu_{\text{dom } R}(x) = \bigvee_y \mu_R(x, y), \qquad x \in X, \tag{1}$$

where the supremum, $\bigvee_y$, is taken over all y in Y. Similarly, the *range* of R is denoted by ran R and is defined by

$$\mu_{\text{ran } R}(y) = \bigvee_x \mu_R(x, y), \qquad x \in X, y \in Y. \tag{2}$$

The *height* of R is denoted by $h(R)$ and is defined by

$$h(R) = \bigvee_x \bigvee_y \mu_R(x, y). \tag{3}$$

A fuzzy relation is *subnormal* if $h(R) < 1$ and *normal* if $h(R) = 1$.

The *support* of R is denoted by $S(R)$ and is defined to be the non-fuzzy subset of $X \times Y$ over which $\mu_R(x, y) > 0$.

The *containment* of a fuzzy relation R in a fuzzy relation Q is denoted by $R \subset Q$ and is defined by $\mu_R \leqslant \mu_Q$, which means, more explicitly, that $\mu_R(x, y) \leqslant \mu_Q(x, y)$ for all (x, y) in $X \times Y$.

The *union* of R and Q is denoted by $R + Q$ (rather than $R \cup Q$) and is defined by $\mu_{R+Q} = \mu_R \vee \mu_Q$, that is

$$\mu_{R+Q}(x, y) = \max(\mu_R(x, y), \mu_Q(x, y)), \qquad x \in X, y \in Y. \tag{4}$$

Consistent with this notation, if $\{R_\alpha\}$ is a family of fuzzy (or non-fuzzy) sets, we shall write $\Sigma_\alpha R_\alpha$ to denote the union $\cup_\alpha R_\alpha$.

The *intersection* of R and Q is denoted by $R \cap Q$ and is defined by $\mu_{R \cap Q} = \mu_R \wedge \mu_Q$.

The *product* of R and Q is denoted by RQ and is defined by $\mu_{RQ} = \mu_R \mu_Q$. Note that, if R, Q, and T are any fuzzy relations from X to Y, then

$$R(Q + T) = RQ + RT.$$

The *complement* of R is denoted by R' and is defined by $\mu_R' = 1 - \mu_R$.

If $R \subset X \times Y$ and $Q \subset Y \times Z$, then the *composition*, or, more specifically, the *max-min composition*, of R and Q is denoted by $R \circ Q$ and is defined by

$$\mu_{R \circ Q}(x, z) = \bigvee_y (\mu_R(x, y) \wedge \mu_Q(y, z)), \qquad x \in X, z \in Z. \tag{5}$$

The *n-fold composition* $R \circ R \ldots \circ R$ is denoted by R^n.

From the above definitions of the composition, union, and containment it follows at once that, for any fuzzy relations $R \subset X \times Y$, Q, $T \subset Y \times Z$, and $S \subset Z \times W$, we have

$$R \circ (Q \circ S) = (R \circ Q) \circ S, \tag{6}$$

$$R \circ (Q + T) = R \circ Q + R \circ T, \tag{7}$$

and

$$Q \subset T \Rightarrow R \circ Q \subset R \circ T. \tag{8}$$

Note. On occasion it may be desirable to employ an operation $*$ other than $\wedge$ in the definition of the composition of fuzzy relations. Then (5) becomes

$$\mu_{R * Q}(x, z) = \bigvee_y (\mu_R(x, y) * \mu_Q(y, z)) \tag{9}$$

with $R * Q$ called the *max-star* composition of R and Q.

In order that (6), (7), and (8) remain valid when $\wedge$ is replaced by $*$, it is sufficient that $*$ be associative and monotone non-decreasing in each of its arguments, which assures the distributivity of $*$ over $+$.[2] A simple example of an operation satisfying these conditions and having the interval $[0,1]$ as its range is the product. In this case, the definition of the composition assumes the form

$$\mu_{R \cdot Q}(x, z) = \bigvee_y (\mu_R(x, y) \cdot \mu_Q(y, z)), \tag{10}$$

where we use the sumbol $\cdot$ in place of $\wedge$ to differentiate between the max-min and max-product compositions. In what follows, in order to avoid a confusing multiplicity of definitions, we shall be using (5) for the most part as our definition of the composition, with the understanding that, in all but a few cases, an assertion which is established with (5) as the definition of the composition holds true also for (10) and, more generally, (9) (provided (6), (7), and (8) are satisfied).

Note also that, when X and Y are finite sets, μ_R may be represented by a relation matrix whose (x,y)th element is $\mu_R(x,y)$. In this case, the defining equation (5) implies that the relation matrix for the composition of R and Q is given by the max-min product[3] of the relation matrices for R and Q.

Level Sets and the Resolution Identity

For α in $[0,1]$, an α-*level-set* of a fuzzy relation R is denoted by R_α and is a non-fuzzy set in $X \times Y$ defined by

$$R_\alpha = \{(x,y) | \mu_R(x, y) \geqslant \alpha\}. \tag{11}$$

Thus, the R_α form a nested sequence of non-fuzzy relations, with

$$\alpha_1 \geqslant \alpha_2 \Rightarrow R_{\alpha_1} \subset R_{\alpha_2}. \tag{12}$$

---

[2] An exhaustive discussion of operations having properties of this type can be found in [15].

[3] In the max-min (or quasi-Boolean) product of matrices with real-valued elements, $\wedge$ and $\vee$ play the roles of product and addition, respectively [16, 17].

An immediate and yet important consequence of the definition of a level set is stated in the following proposition:

PROPOSITION 1. *Any fuzzy relation from X to Y admits of the resolution*

$$R = \sum_{\alpha} \alpha R_{\alpha}, \qquad 0 < \alpha \leqslant 1, \tag{13}$$

where $\sum$ stands for the union (see (4)) and αR_{α} denotes a subnormal non-fuzzy set defined by

$$\mu_{\alpha R_{\alpha}}(x, y) = \alpha \mu_{R_{\alpha}}(x, y), \qquad (x, y) \in X \times Y. \tag{14}$$

or equivalently

$$\mu_{\alpha R_{\alpha}}(x, y) = \alpha, \qquad for\ (x, y) \in R_{\alpha},$$

$$= 0, \qquad elsewhere.$$

Proof. Let $\mu_{R_{\alpha}}(x, y)$ denote the membership function of the non-fuzzy set R_{α} in $X \times Y$ defined by (11). Then (11) implies that

$$\mu_{R_{\alpha}}(x, y) = 1, \qquad for\ \mu_{R}(x, y) \geqslant \alpha, \tag{15}$$

$$= 0, \qquad for\ \mu_{R}(x, y) < \alpha,$$

and consequently the membership function of $\sum_{\alpha} \alpha R_{\alpha}$ may be written as

$$\mu_{\sum_{\alpha} \alpha R_{\alpha}}(x, y) = \bigvee_{\alpha} \alpha \mu_{R_{\alpha}}(x, y)$$

$$= \bigvee_{\alpha \leqslant \mu_{R}(x, y)} \alpha$$

$$= \mu_{R}(x, y),$$

which in turn implies (13).

Note. It is understood that in (13) to each R_{α} corresponds a unique α. If this is not the case, e.g., $\alpha_{1} \neq \alpha_{2}$ and $R_{\alpha_{1}} = R_{\alpha_{2}}$, then the two terms are combined by forming their union, yielding $(\alpha_{1} \vee \alpha_{2}) R_{\alpha_{1}}$. In this way, a summation of the form (13) may be converted into one in which to each R_{α} corresponds a unique α. Furthermore, if X and Y are finite sets and the distinct entries in the relation matrix of R are denoted by α_{k}, $k = 1, 2, \ldots, K$, where K is a finite number, then (13) assumes the form

$$R = \sum_{k} \alpha_{k} R_{\alpha_{k}}, \qquad 1 \leqslant k \leqslant K. \tag{16}$$

As a simple illustration of (13), assume $X = Y = \{x_{1}, x_{2}, x_{3}\}$, with the relation matrix μ_{R} given by

$$\mu_{R} = \begin{bmatrix} 1 & 0.8 & 0 \\ 0.6 & 1 & 0.9 \\ 0.8 & 0 & 1 \end{bmatrix}.$$

In this case, the resolution of R reads

$$R = 0.6\{(x_1, x_1), (x_1, x_2), (x_2, x_1), (x_2, x_2), (x_2, x_3), (x_3, x_1), (x_3, x_3)\}$$
$$+ 0.8\{(x_1, x_1), (x_1, x_2), (x_2, x_2), (x_2, x_3), (x_3, x_1), (x_3, x_3)\}$$
$$+ 0.9\{(x_1, x_1), (x_2, x_2), (x_2, x_3), (x_3, x_3)\}$$
$$+ 1\{(x_1, x_1), (x_2, x_2), (x_3, x_3)\}. \tag{17}$$

In what follows, we assume that $X = Y$. Furthermore, we shall assume for simplicity that X is a finite set, $X = \{x_1, x_2, \ldots, x_n\}$.

3. SIMILARITY RELATIONS

The concept of a *similarity* relation is essentially a generalization of the concept of an equivalence relation. More specifically:

Definition. A *similarity* relation, S, in X is a fuzzy relation in X which is (a) *reflexive*, i.e.,

$$\mu_S(x, x) = 1, \qquad \text{for all } x \text{ in dom } S, \tag{18}$$

(b) *symmetric* i.e.,

$$\mu_S(x, y) = \mu_S(y, x), \qquad \text{for all } x, y \text{ in dom } S, \tag{19}$$

and (c) *transitive*, i.e.,

$$S \supset S \circ S, \tag{20}$$

or, more explicitly,

$$\mu_S(x, z) \geqslant \bigvee_y (\mu_S(x, y) \wedge \mu_S(y, z)).$$

Note. If $*$ is employed in place of $\circ$ in the definition of the composition, the corresponding definition of transitivity becomes

$$S \supset S * S \tag{21}$$

or, more explicitly

$$\mu_S(x, z) \geqslant \bigvee_y (\mu_S(x, y) * \mu_S(y, z)).$$

When there is a need to distinguish between the transitivity defined by (20) and the more general form defined by (21), we shall refer to them as *max-min* and *max-star* transitivity, respectively.

An example of the relation matrix of a similarity relation S is shown in Figure 1. It is readily verified that $S = S \circ S$ and also that $S = S \cdot S$.

$$\mu_S = \begin{bmatrix} 1 & 0.2 & 1 & 0.6 & 0.2 & 0.6 \\ 0.2 & 1 & 0.2 & 0.2 & 0.8 & 0.2 \\ 1 & 0.2 & 1 & 0.6 & 0.2 & 0.6 \\ 0.6 & 0.2 & 0.6 & 1 & 0.2 & 0.8 \\ 0.2 & 0.8 & 0.2 & 0.2 & 1 & 0.2 \\ 0.6 & 0.2 & 0.6 & 0.8 & 0.2 & 1 \end{bmatrix}$$

FIGURE 1. Relation matrix of a similarity relation.

Transitivity

There are several aspects of the transitivity of a similarity relation which are in need of discussion. First, note that, in consequence of (18), we have

$$S \supset S^2 \Rightarrow S \supset S^k, \qquad k = 3, 4, \ldots \qquad (22)$$

and hence

$$S \supset S^2 \Leftrightarrow S = \bar{S}, \qquad (23)$$

where

$$\bar{S} = S + S^2 + S^3 + \ldots \qquad (24)$$

is the *transitive* closure of S. Thus, as in the case of equivalence relations, the condition that S be transitive is equivalent to

$$S = \bar{S} = S + S^2 + S^3 + \ldots. \qquad (25)$$

An immediate consequence of (25) is that the transitive closure of any fuzzy relation is transitive. Note also that for any S

$$S = S^2 \Rightarrow S = \bar{S}$$

and, if S is reflexive, then

$$S = S^2 \Leftrightarrow S = \bar{S}.$$

The significance of (25) is made clearer by the following observation. Let $x_{i_1}, \ldots, x_{i_k}$ be k points in X such that $\mu(x_{i_1}, x_{i_2}), \ldots, \mu(x_{i_{k-1}}, x_{i_k})$ are all >0. Then the sequence $C = (x_{i_1}, \ldots, x_{i_k})$ will be said to be a *chain* from x_{i_1} to x_{i_k}, with the strength of this chain defined as the strength of its weakest link, that is,

$$\text{strength of } (x_{i_1}, \ldots, x_{i_k}) = \mu(x_{i_1}, x_{i_2}) \wedge \ldots \wedge \mu(x_{i_{k-1}}, x_{i_k}). \qquad (26)$$

From the definition of the composition (equation (5)), it follows that the (i,j)th element of S^l, $l = 1, 2, 3, \ldots$, is the strength of the strongest chain of length l from x_i to x_j. Thus, the transitivity condition (25) may be stated in words as: for all x_i, x_j in X,

strength of S between x_i and x_j

$$= \text{strength of the strongest chain from } x_i \text{ to } x_j. \qquad (27)$$

Second, if X has n elements, then any chain C of length $k \geqslant n + 1$ from x_{i_1} to x_{i_k} must necessarily have cycles, that is, one or more elements of X must occur more than once in the chain $C = (x_{i_1}, \ldots, x_{i_k})$. If these cycles are removed, the resulting chain, $\tilde{C}$, of length $\leqslant n$, will have at least the same strength as C, by virtue of (26). Consequently, for any elements x_i, x_j in X we can assert that

strength of the strongest chain from x_i to x_j

$$= \text{strength of the strongest chain of length} \leqslant n \text{ from } x_i \text{ to } x_j. \quad (28)$$

Since the (i,j)th element of $\bar{S}$ is the strength of the strongest chain from x_i to x_j, (28) implies the following proposition [18], which is well known for Boolean matrices [16]:

PROPOSITION 2. *If S is a fuzzy relation characterized by a relation matrix of order n, then*

$$\bar{S} = S + S^2 + S^3 + \ldots = S + S^2 + \ldots + S^n. \quad (29)$$

Note. Observe that (29) remains valid when in the definition of the composition and the strength of a chain $\wedge$ is replaced by the product, i.e., S^k, $k = 2, 3, \ldots$ is replaced by the k-fold composition $S \cdot S \cdot \ldots \cdot S$, with $\cdot$ defined by (10), and (26) is replaced by

$$\text{strength of } (x_{i_1}, \ldots, x_{i_k}) = \mu(x_{i_1}, x_{i_2})\mu(x_{i_2}, x_{i_3}) \ldots \mu(x_{i_{k-1}}, x_{i_k}). \quad (30)$$

Since $ab \leqslant a \wedge b$ for $a, b \in [0,1]$, it follows that

$$S \supset SoS \Rightarrow S \supset S \cdot S, \quad (31)$$

that is, max-min transitivity implies max-product transitivity. This observation is useful in situations in which the strength of a chain is more naturally expressed by (30) than by (26).

A case in point is provided by the criticisms [19–21] leveled at the assumption of transitivity in the case of weak ordering. Thus, suppose that X is a finite interval $[a,b]$ and that we wish to define a non-fuzzy preference ordering on X in terms of two relations $>$ and $\approx$ such that

(a) for every x, y in X, exactly one of $x > y$, $y > x$, or $x \approx y$ is true,
(b) $\approx$ is an equivalence relation,
(c) $>$ is transitive.

In many cases, it would be reasonable to assume that

$$x \approx y \Leftrightarrow |x - y| \leqslant \epsilon > 0,$$

where ϵ is a small number (in relation to $b - a$) representing an "indifference" interval. But then, by transitivity of $\approx$, $x \approx y$ for all x, y in X, which is inconsistent with our intuitive expectation that when the difference between x and y is sufficiently large, either $x > y$ or $y > x$ must hold.

This difficulty is not resolved by making $\approx$ a similarity relation in X so long as we employ the max-min transitivity in the definition of $\approx$. For, if we make the reasonable assumption that $\mu_\approx(x, y)$ is continuous at $x = y$, then (20) implies that $\mu_\approx(x, y) = 1$ for all x, y in X.

The difficulty may be resolved by making $\approx$ a similarity relation and employing the max-product transitivity in its definition. As an illustration, suppose that

$$\mu_\approx(x, y) = e^{-\beta|x-y|}, \qquad x, y \in X, \tag{32}$$

where β is any positive number. In this case, $\approx$ may be interpreted as "is not much different from."

Let $x, y, z \in [a, b]$, with $x < z$. Then, substituting (32) in

$$\mu_{\approx^2}(x, z) = \bigvee_y \mu_\approx(x, y) \mu_\approx(y, z),$$

we have

$$\mu_{\approx^2}(x, z) = \bigvee_y e^{-\beta|y-x|} e^{-\beta|z-y|}$$

$$= \bigvee_{y \in [x, z]} e^{-\beta(y-x)} e^{-\beta(z-y)}$$

$$= e^{-\beta(z-x)}$$

$$= \mu_\approx(x, z), \tag{33}$$

which establishes that $\approx^2 = \approx$ and hence that (32) defines a similarity relation which is continuous at $x = y$ and yet is not constant over X.

Finally, it should be noted that the transitivity condition (20) implies and is implied by the ultrameric inequality [22] for distance functions. Specifically, let the complement of a similarity relation S be a *dissimilarity* relation D, with

$$\mu_D(x, y) = 1 - \mu_S(x, y), \qquad x, y \in X. \tag{34}$$

If $\mu_D(x, y)$ is interpreted as a distance function, $d(x, y)$, then (20) yields

$$1 - d(x, z) \geqslant \bigvee_y ((1 - d(x, y)) \wedge (1 - d(y, z))),$$

and since

$$(1 - d(x, y)) \wedge (1 - d(y, z)) = 1 - (d(x, y) \vee (d(y, z)) \tag{35}$$

we can conclude that, for all x, y, z in X,

$$d(x, z) \leqslant d(x, y) \vee d(y, z), \qquad y \in X, \tag{36}$$

which is the ultrameric inequality satisfied by $d(x,y)$. Clearly, (36) implies the triangle inequality

$$d(x,z) \leqslant d(x,y) + d(y,z). \tag{37}$$

Thus, (20) implies (36) and (37), and is implied by (36).

Returning to our discussion of similarity relations, we note that one of their basic properties is an immediate consequence of the resolution identity (13) for fuzzy relations. Specifically,

PROPOSITION 3. *Let*

$$S = \sum_\alpha \alpha S_\alpha, \qquad 0 < \alpha \leqslant 1, \tag{38}$$

be the resolution of a similarity relation in X. Then each S_α in (38) is an equivalence relation in X. Conversely, if the S_α, $0 < \alpha \leqslant 1$, are a nested sequence of distinct equivalence relations in X, with $\alpha_1 > \alpha_2 \Leftrightarrow S_{\alpha_1} \subset S_{\alpha_2}$, S_1 non-empty and $\operatorname{dom} S_\alpha = \operatorname{dom} S_1$, then, for any choice of α's in $(0,1]$ which includes $\alpha = 1$, S is a similarity relation in X.

Proof. $\Rightarrow$ First, since $\mu_S(x,x) = 1$ for all x in the domain of S, it follows that $(x,x) \in S_\alpha$ for all α in $(0,1]$ and hence that S_α is reflexive for all α in $(0,1]$.

Second, for each α in $(0,1]$, let $(x,y) \in S_\alpha$, which implies that $\mu_S(x,y) \geqslant \alpha$ and hence, by symmetry of S, that $\mu_S(y,x) \geqslant \alpha$. Consequently, $(y,x) \in S_\alpha$ and thus S_α is symmetric for each α in $(0,1]$.

Third, for each α in $(0,1]$, suppose that $(x_1,x_2) \in S_\alpha$ and $(x_2,x_3) \in S_\alpha$. Then $\mu_S(x_1,x_2) \geqslant \alpha$ and $\mu_S(x_2,x_3) \geqslant \alpha$ and hence, by the transitivity of S, $\mu_S(x_1,x_3) \geqslant \alpha$. This implies that $(x_1,x_3) \in S_\alpha$ and hence that S_α is transitive for each α in $(0,1]$.

$\Leftarrow$ First, since S_1 is non-empty, $(x,x) \in S_1$ and hence $\mu_S(x,x) = 1$ for all x in the domain of S_1.

Second, expressed in terms of the membership functions of S and S_α, (38) reads

$$\mu_S(x,y) = \bigvee_\alpha \alpha \mu_{S_\alpha}(x,y), \qquad x,y \in \operatorname{dom} S.$$

It is obvious from this expression for $\mu_S(x,y)$ that the symmetry of S_α for each α in $(0,1]$ implies the symmetry of S.

Third, let x_1, x_2, x_3 be some arbitrarily chosen elements of X. Suppose that

$$\mu_S(x_1,x_2) = \alpha \qquad \text{and} \qquad \mu_S(x_2,x_3) = \beta.$$

Then, $(x_1,x_2) \in S_{\alpha \wedge \beta}$ and $(x_2,x_3) \in S_{\alpha \wedge \beta}$, and consequently $(x_1,x_3) \in S_{\alpha \wedge \beta}$ by the transitivity of $S_{\alpha \wedge \beta}$.

From this it follows that, for all x_1, x_2, x_3 in X, we have

$$\mu_S(x_1,x_3) \geqslant \alpha \wedge \beta$$

and hence

$$\mu_S(x_1, x_3) \geqslant \bigvee_{x_2} (\mu_S(x_1, x_2) \wedge \mu_S(x_2, x_3)),$$

which establishes the transitivity of S.

Partition Tree

Let π_α denote the partition induced on X by S_α, $0 < \alpha \leqslant 1$. Clearly, $\pi_{\alpha'}$ is a refinement of π_α if $\alpha' \geqslant \alpha$. For, by the definition of $\pi_{\alpha'}$, two elements of X, say x and y, are in the same block of $\pi_{\alpha'}$ iff $\mu_S(x, y) \geqslant \alpha'$. This implies that $\mu_S(x, y) \geqslant \alpha$ and hence that x and y are in the same block of π_α.

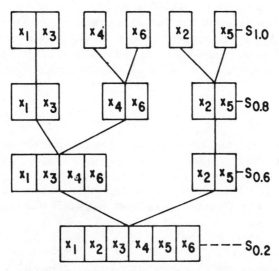

FIGURE 2. Partition tree for the similarity relation defined in Figure 1.

A nested sequence of partitions π_{α_1}, π_{α_2}, ..., π_{α_k} may be represented diagrammatically in the form of a *partition tree*,[4] as shown in Figure 2. It should be noted that the concept of a partition tree plays the same role with respect to a similarity relation as the concept of a quotient does with respect to an equivalence relation.

The partition tree of a similarity relation S is related to the relation matrix of S by the rule: x_i and x_j belong to the same block of π_α iff $\mu_S(x_i, x_j) \geqslant \alpha$.

---

[4] The notion of a partition tree and its properties are closely related to the concept of the hierarchic clustering scheme described in [22].

This rule implies that, given a partition tree of S, one can readily determine $\mu_S(x_i, x_j)$ by observing that

$$\mu_S(x_i, x_j) = \text{largest value of } \alpha \text{ for which } x_i \text{ and } x_j$$

$$\text{are in the same block of } \pi_\alpha. \tag{39}$$

An alternative to the diagrammatic representation of a partition tree is provided by a slightly modified form of the phrase-marker notation which is commonly used in linguistics [23]. Specifically, if we allow recursion and use the notation $\alpha(A, B)$ to represent a partition π_α whose blocks are A and B, then the partition tree shown in Figure 2 may be expressed in the form of a string:

$$0.2(0.6(0.8(1(x_1, x_3)), 0.8(1(x_4), 1(x_6))), 0.6(0.8(1(x_2), 1(x_5)))). \tag{40}$$

This string signifies that the highest partition, π_1, comprises the blocks (x_1, x_3), (x_4), (x_6), (x_2), and (x_5). The next partition, $\pi_{0.8}$, comprises the blocks $((x_1, x_3))$, $((x_4), (x_6))$, and $((x_2), (x_5))$. And so on. Needless to say, the profusion of parentheses in the phrase-marker representation of a partition tree makes it difficult to visualize the structure of a similarity relation from an inspection of (40).

Similarity Classes

Similarity classes play the same role with respect to a similarity relation as do equivalence classes in the case of an equivalence relation. Specifically, let S be a similarity relation in $X = \{x_1, \ldots, x_n\}$ characterized by a membership function $\mu_S(x_i, x_j)$. With each $x_i \in X$, we associate a *similarity class* denoted by $S[x_i]$ or simply $[x_i]$. This class is a fuzzy set in X which is characterized by the membership function

$$\mu_{S[x_i]}(x_j) = \mu_S(x_i, x_j). \tag{41}$$

Thus, $S[x_i]$ is identical with S conditioned on x_i, that is, with x_i held constant in the membership function of S.

To illustrate, the similarity classes associated with x_1 and x_2 in the case of the similarity relation defined in Figure 1 are

$$S[x_1] = \{(x_1, 1), (x_2, 0.2), (x_3, 1), (x_4, 0.6), (x_5, 0.2), (x_6, 0.6)\}$$

$$S[x_2] = \{(x_1, 0.2), (x_2, 1),)x_3, 0.2), (x_4, 0.2), (x_5, 0.8), (x_6, 0.2)\}.$$

By conditioning both sides of the resolution (38) on x_i we obtain at once the following proposition:

PROPOSITION 4. *The similarity class of* x_i, $x_i \in X$, *admits of the resolution*

$$S[x_i] = \sum_\alpha \alpha S_\alpha[x_i], \tag{42}$$

where $S_\alpha[x_i]$ denotes the block of S_α which contains x_i, and $\alpha S_\alpha[x_i]$ is a subnormal non-fuzzy set whose membership function is equal to α on $S_\alpha[x_i]$ and vanishes elsewhere.

For example, in the case of $S[x_1]$, with S defined in Figure 1, we have

$$S[x_1] = 0.2\{x_1, x_2, x_3, x_4, x_5, x_6\} + 0.6\{x_1, x_3, x_4, x_6\} + 1\{x_1, x_3\}$$

and similarly

$$S[x_2] = 0.2\{x_1, x_2, x_3, x_4, x_5, x_6\} + 0.8\{x_2, x_5\} + 1\{x_2\}.$$

The similarity classes of a similarity relation are not, in general, disjoint—as they are in the case of an equivalence relation. Thus, the counterpart of disjointness is a more general property which is asserted in the following proposition:

PROPOSITION 5. *Let $S[x_i]$ and $S[x_j]$ be arbitrary similarity classes of S. Then, the height (see (3)) of the intersection of $S[x_i]$ and $S[x_j]$ is bounded from above by $\mu_S(x_i, x_j)$, that is,*

$$h(S[x_i] \cap S[x_j]) \leqslant \mu_S(x_i, x_j). \tag{43}$$

Proof. By definition of h we have

$$h(S[x_i] \cap S[x_j]) = \bigvee_{x_k} (\mu_S(x_i, x_k) \wedge \mu_S(x_j, x_k)),$$

which in view of the symmetry of S may be rewritten as

$$h(S[x_i] \cap S[x_j]) = \bigvee_{x_k} (\mu_S(x_i, x_k) \wedge \mu_S(x_k, x_j)). \tag{44}$$

Now the right-hand member of (44) is identical with the grade of membership of (x_i, x_j) in the composition of S with S. Thus

$$h(S[x_i] \cap S[x_j]) = \mu_{S \circ S}(x_i, x_j),$$

which, in virtue of the transitivity of S, implies that

$$h(S[x_i] \cap S[x_j]) \leqslant \mu_S(x_i, x_j). \tag{45}$$

Note that, if S is reflexive, then $S^2 = S$ and (45) is satisfied with the equality sign. Thus, for the example of Proposition 4, we have

$$h(S[x_1] \cap S[x_2]) = 0.2 = \mu_S(x_1, x_2),$$

since S is reflexive.

The following corollary follows at once from Proposition 5:

COROLLARY 6. *The height of the intersection of all similarity classes of X is bounded by the infimum of $\mu_S(x_i, x_j)$ over X. Thus*

$$h(S[x_1] \cap \ldots \cap S[x_n]) \leqslant \bigwedge_{x_i} \bigwedge_{x_j} \mu_S(x_i, x_j). \tag{46}$$

We turn next to the consideration of fuzzy ordering relations.

4. FUZZY ORDERINGS

A *fuzzy ordering* is a fuzzy transitive relation.[5] In what follows we shall define several basic types of fuzzy orderings and dwell briefly upon some of their properties.

A fuzzy relation P in X is a *fuzzy partial ordering* iff it is reflexive, transitive, and antisymmetric. By antisymmetry of P is meant that

$$\mu_P(x, y) > 0 \quad \text{and } \mu_P(y, x) > 0 \Rightarrow x = y, \qquad x, y \in X. \tag{47}$$

(On occasion, we may use the notation $x \leqslant y$ to signify that $\mu_P(x, y) > 0$.)

$$\mu_P = \begin{bmatrix} 1 & 0.8 & 0.2 & 0.6 & 0.6 & 0.4 \\ 0 & 1 & 0 & 0 & 0.6 & 0 \\ 0 & 0 & 1 & 0 & 0.5 & 0 \\ 0 & 0 & 0 & 1 & 0.6 & 0.4 \\ 0 & 0 & 0 & 0 & 1 & 0 \\ 0 & 0 & 0 & 0 & 0 & 1 \end{bmatrix}$$

FIGURE 3. Relation matrix for a fuzzy partial ordering.

An example of a relation matrix for a fuzzy partial ordering is shown in Figure 3. The corresponding fuzzy Hasse diagram for this ordering is shown in Figure 4. In this diagram, the number associated with the arc joining x_i to x_j is $\mu_P(x_i, x_j)$, with the understanding that x_j is a *cover* for x_i, that is, there is no x_k

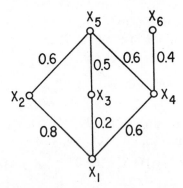

FIGURE 4. Fuzzy Hasse diagram for the fuzzy partial ordering defined in Figure 3.

in X such that $\mu_P(x_i, x_k) > 0$ and $\mu_P(x_k, x_j) > 0$. Note that the numbers associated with the arcs define the relation matrix by virtue of the transitivity identity $P = P^2$.

---

[5] Alternatively, a fuzzy ordering may be viewed as a metrized ordering in which the metric satisfies the ultrametric inequality.

As in the case of a similarity relation, a fuzzy partial ordering may be resolved into non-fuzzy partial orderings. This basic property of fuzzy partial orderings is expressed by

PROPOSITION 7. *Let*

$$P = \sum_{\alpha} \alpha P_{\alpha}, \qquad 0 < \alpha \leqslant 1, \tag{48}$$

be the resolution of a fuzzy partial ordering in X. Then each P_{α} in (48) is a partial ordering in X. Conversely, if the P_{α}, $0 < \alpha \leqslant 1$, are a nested sequence of distinct partial orderings in X, with $\alpha_1 > \alpha_2 \Leftrightarrow P_{\alpha_1} \subset P_{\alpha_2}$, P_1 non-empty, and $\mathrm{dom}\, P_{\alpha} = \mathrm{dom}\, P_1$, then, for any choice of α's in $(0,1]$ which includes $\alpha = 1$, P is a fuzzy partial ordering in X.

Proof. Reflexivity and transitivity are established as in Proposition 3. As for antisymmetry, suppose that $(x,y) \in P_{\alpha}$ and $(y,x) \in P_{\alpha}$. Then $\mu_P(x,y) \geqslant \alpha$, $\mu_P(y,x) \geqslant \alpha$ and hence by antisymmetry of P, $x = y$. Conversely, suppose that $\mu_P(x,y) = \alpha > 0$ and $\mu_P(y,x) = \beta > 0$. Let $\gamma = \alpha \wedge \beta$. Then $(x,y) \in P_{\gamma}$ and $(y,x) \in P_{\gamma}$, and from the antisymmetry of P_{γ} it follows that $x = y$.

In many applications of the concept of a fuzzy partial ordering, the condition of reflexivity is not a natural one to impose. If we allow $\mu_P(x,x)$, $x \in X$, to take any value in $[0,1]$, the ordering will be referred to as irreflexive.

To illustrate the point, assume that X is an interval $[a,b]$, and $\mu_P(x,y) = f(y-x)$, with $f(y-x) = 0$ for $y < x$ and $f(0) = 1$. Then, as was noted in Section 2 ((31) *et seq.*), if $f(x)$ is right-continuous at $x = 0$, the max-min transitivity of μ_P requires that $f(x) = 1$ for $x > 0$. However, if we drop the requirement of reflexivity, then it is sufficient that f be monotone non-decreasing in order to satisfy the condition of transitivity. For, assume that f is monotone non-decreasing and $x \leqslant y \leqslant z$, $x, y, z, \in [a,b]$. Then

$$\begin{aligned} \mu_P(x,z) &= f(z-x) \\ &= f((z-y) + (y-x)), \end{aligned} \tag{49}$$

and, since

$$f((z-y) + (y-x)) \geqslant f(z-y),$$
$$f((z-y) + (y-x)) \geqslant f(y-x),$$

we have

$$f((z-y) + (y-x)) \geqslant f(z-y) \wedge f(y-x),$$

and therefore

$$\mu_P(x,z) \geqslant \bigvee_{y} (\mu_P(x,y) \wedge \mu_P(y,z)),$$

which establishes the transitivity of P.

It should be noted that the condition is not necessary. For example, it is easy to verify that for any

$$\frac{1}{b-a} \leqslant \beta \leqslant \frac{2}{b-a},$$

the function

$$f(x) = \beta x, \qquad 0 \leqslant x \leqslant 1/\beta,$$
$$= 2 - \beta x, \qquad 1/\beta \leqslant x \leqslant b - a,$$

corresponds to a transitive fuzzy partial ordering if $\beta(b-a) \leqslant \frac{4}{3}$.

With each $x_i \in X$, we associate two fuzzy sets: the *dominating* class, denoted by $P_\geqslant[x_i]$ and defined by

$$\mu_{P_\geqslant[x_i]}(x_j) = \mu_P(x_i, x_j), \qquad x_j \in X, \tag{50}$$

and the *dominated* class, denoted by $P_\leqslant[x_i]$ and defined by

$$\mu_{P_\leqslant[x_i]}(x_j) = \mu_P(x_j, x_i), \qquad x_j \in X. \tag{51}$$

In terms of these classes, x_i is *undominated* iff

$$\mu_P(x_i, x_j) = 0, \qquad \text{for all } x_j \neq x_i, \tag{52}$$

and x_i is *undominating* iff

$$\mu_P(x_j, x_i) = 0, \quad \text{for all } x_j \neq x_i. \tag{53}$$

It is evident that, if P is any fuzzy partial ordering in $X = \{x_1, \ldots, x_n\}$, the sets of undominated and undominating elements of X are non-empty.

Another related concept is that of a fuzzy upper-bound for a non-fuzzy subset of X. Specifically, let A be a non-fuzzy subset of X. Then the *upper-bound* for A is a fuzzy set denoted by $U(A)$ and defined by

$$U(A) = \bigcap_{x_i \in A} P_\geqslant[x_i]. \tag{54}$$

For a non-fuzzy partial ordering, this reduces to the conventional definition of an upper-bound. Note that, if the *least* element of $U(A)$ is defined as an x_i (if it exists) such that

$$\mu_{U(A)}(x_i) > 0 \quad \text{and} \quad \mu_P(x_i, x_j) > 0 \quad \text{for all } x_j \text{ in the support of } U(A), \tag{55}$$

then the *least upper-bound* of A is the least element of $U(A)$ and is unique by virtue of the antisymmetry of P.

In a similar vein, one can readily generalize to fuzzy orderings many of the well-known concepts relating to other types of non-fuzzy orderings. Some of these are briefly stated in the sequel.

Preordering

A fuzzy *preordering* R is a fuzzy relation in X which is reflexive and transitive. As in the case of a fuzzy partial ordering, R admits of the resolution

$$R = \sum_{\alpha} \alpha R_{\alpha}, \qquad 0 < \alpha \leqslant 1, \tag{56}$$

where the α-level-sets R_{α} are non-fuzzy preorderings.

For each α, the non-fuzzy preordering R_{α} induces an equivalence relation, E_{α}, in X and a partial ordering, P_{α}, on the quotient X/E_{α}. Specifically,

$$(x_i, x_j) \in E_{\alpha} \Leftrightarrow \mu_{R_{\alpha}}(x_i, x_j) = \mu_{R_{\alpha}}(x_j, x_i) = 1 \tag{57}$$

and

$$([x_i], [x_j]) \in P_{\alpha} \Leftrightarrow \mu_{R\alpha}(x_i, x_j) = 1 \quad \text{and} \quad \mu_{R\alpha}(x_j, x_i) = 0, \tag{58}$$

where $[x_i]$ and $[x_j]$ are the equivalence classes of x_i and x_j, respectively.

As an illustration, consider the fuzzy preordering characterized by the relation matrix shown in Figure 5. The corresponding relation matrices for $R_{0\cdot2}$, $R_{0\cdot6}$, $R_{0\cdot8}$, $R_{0\cdot9}$, and R_1 read as in Figure 6.

$$\mu_R = \begin{bmatrix} 1 & 0.8 & 1 & 0.8 & 0.8 & 0.8 \\ 0.2 & 1 & 0.2 & 0.2 & 0.8 & 0.2 \\ 1 & 0.8 & 1 & 0.8 & 0.8 & 0.8 \\ 0.6 & 0.9 & 0.6 & 1 & 0.9 & 1 \\ 0.2 & 0.8 & 0.2 & 0.2 & 1 & 0.2 \\ 0.6 & 0.9 & 0.6 & 0.9 & 0.9 & 1 \end{bmatrix}$$

FIGURE 5. Relation matrix of a fuzzy preordering.

The preordering in question may be represented in diagrammatic form as shown in Figure 7. In this figure, the broken lines in each level (identified by R_{α}) represent the arcs (edges) of the Hasse diagram of the partial ordering P_{α}, rotated clockwise by 90°. The nodes of this diagram are the equivalence classes of the equivalence relation, E_{α}, induced by R_{α}. Thus, the diagram as a whole is the partition tree of the similarity relation

$$S = 0.2E_{0\cdot2} + 0.6E_{0\cdot6} + 0.8E_{0\cdot8} + 0.9E_{0\cdot9} + 1E_1,$$

with the blocks in each level of the tree forming the elements of a partial ordering P_{α} which is represented by a rotated Hasse diagram.

Linear Ordering

A fuzzy *linear ordering* L is a fuzzy antisymmetric ordering in X in which for every $x \neq y$ in X either $\mu_L(x, y) > 0$ or $\mu_L(y, x) > 0$. A fuzzy linear ordering

$R_{0 \cdot 2}$

$$\begin{bmatrix} 1 & 1 & 1 & 1 & 1 & 1 \\ 1 & 1 & 1 & 1 & 1 & 1 \\ 1 & 1 & 1 & 1 & 1 & 1 \\ 1 & 1 & 1 & 1 & 1 & 1 \\ 1 & 1 & 1 & 1 & 1 & 1 \\ 1 & 1 & 1 & 1 & 1 & 1 \end{bmatrix}$$

$R_{0 \cdot 6}$

$$\begin{bmatrix} 1 & 1 & 1 & 1 & 1 & 1 \\ 0 & 1 & 0 & 0 & 1 & 0 \\ 1 & 1 & 1 & 1 & 1 & 1 \\ 1 & 1 & 1 & 1 & 1 & 1 \\ 0 & 1 & 0 & 0 & 1 & 0 \\ 1 & 1 & 1 & 1 & 1 & 1 \end{bmatrix}$$

$R_{0 \cdot 8}$

$$\begin{bmatrix} 1 & 1 & 1 & 1 & 1 & 1 \\ 0 & 1 & 0 & 0 & 1 & 0 \\ 1 & 1 & 1 & 1 & 1 & 1 \\ 0 & 1 & 0 & 1 & 1 & 1 \\ 0 & 1 & 0 & 0 & 1 & 0 \\ 0 & 1 & 0 & 1 & 1 & 1 \end{bmatrix}$$

$R_{0 \cdot 9}$

$$\begin{bmatrix} 1 & 0 & 1 & 0 & 0 & 0 \\ 0 & 1 & 0 & 0 & 0 & 0 \\ 1 & 0 & 1 & 0 & 0 & 0 \\ 0 & 1 & 0 & 1 & 1 & 1 \\ 0 & 0 & 0 & 0 & 1 & 0 \\ 0 & 1 & 0 & 1 & 1 & 1 \end{bmatrix}$$

R_1

$$\begin{bmatrix} 1 & 0 & 1 & 0 & 0 & 0 \\ 0 & 1 & 0 & 0 & 0 & 0 \\ 1 & 0 & 1 & 0 & 0 & 0 \\ 0 & 0 & 0 & 1 & 0 & 1 \\ 0 & 0 & 0 & 0 & 1 & 0 \\ 0 & 0 & 0 & 0 & 0 & 1 \end{bmatrix}$$

FIGURE 6. Relation matrices for the level sets of the preordering defined in Figure 5.

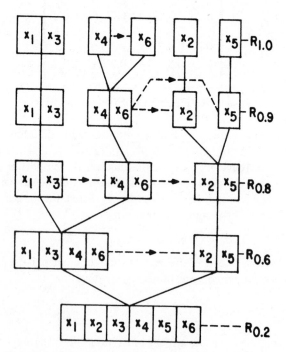

FIGURE 7. Structure of the preordering defined in Figure 5.

admits of the resolution

$$L = \sum_{\alpha} \alpha L_{\alpha}, \qquad 0 < \alpha \leqslant 1, \tag{59}$$

which is a special case of (48) and in which the L_{α} are non-fuzzy linear orderings.

A simple example of an irreflexive fuzzy linear ordering is the relation $y \gg x$ in $X = (-\infty, \infty)$. If we define $\mu_L(x, y)$ by

$$\mu_L(x, y) = (1 + (y - x)^{-2})^{-1}, \qquad \text{for } y - x \geqslant 0,$$

$$= 0, \quad \text{for } y - x < 0,$$

then L is transitive (in virtue of (49)), antisymmetric, and $\mu_L(x, y) > 0$ or $\mu_L(y, x) > 0$ for every $x \neq y$ in $(-\infty, \infty)$. Hence L is a fuzzy linear ordering.

Weak Ordering

If we remove the condition of antisymmetry, then a fuzzy linear ordering becomes a *weak ordering*. Equivalently, a weak ordering, W, may be regarded as a special case of a preordering in which for every $x \neq y$ in X either $\mu_W(x, y) > 0$ or $\mu_W(y, x) > 0$.

Szpilrajn's Theorem

A useful example of a well-known result which can readily be extended to fuzzy orderings is provided by the Szpilrajn theorem [24], which may be stated as follows: Let P be a partial ordering in X. Then, there exists a linear ordering L in a set Y, of the same cardinality as X, and a one-to-one mapping σ from X onto Y (called the Szpilrajn mapping) such that for all x, y in X

$$(x, y) \in P \Rightarrow (\sigma(x), \sigma(y)) \in L.$$

In its extended form, the statement of the theorem becomes:

THEOREM 8. *Let P be a fuzzy partial ordering in X. Then, there exist a fuzzy linear ordering L in a set Y, of the same cardinality as X, and a one-to-one mapping σ from X onto Y such that*

$$\mu_P(x, y) > 0 \Rightarrow \mu_L(\sigma(x), \sigma(y)) = \mu_P(x, y), \qquad x, y \in X. \tag{60}$$

Proof. The theorem can readily be established by the following construction for L and σ: Assume that a fuzzy partial ordering P in $X = \{x_1, \ldots, x_n\}$ is characterized by its relation matrix, which for simplicity will also be referred to as P. In what follows, the relation matrix shown in Figure 3 and the Hasse diagram corresponding to it (Fig. 4) will be used to illustrate the construction for L and σ.

First, we shall show that the antisymmetry and transitivity of P make it possible to relabel the elements of X in such a way that the corresponding relabeled relation matrix P is upper-triangular.

To this end, let C_0 denote the set of undominating elements of X (i.e., $x_i \in C_0 \Leftrightarrow$ column corresponding to x_i contains a single positive element (unity) lying on the main diagonal). The transitivity of P implies that C_0 is non-empty. For the relation matrix of Figure 3, $C_0 = \{x_1\}$.

Referring to the Hasse diagram of P (Fig. 4), it will be convenient to associate with each x_j in X a positive integer $\rho(x_j; C_0)$ representing the *level* of x_j above C_0. By definition

$$\rho(x_j; C_0) = \max_{x_i \in C_0} d(x_i, x_j), \tag{61}$$

where $d(x_i, x_j)$ is the length of the longest upward path between x_i and x_j in the Hasse diagram. For example, in Figure 4, $C_0 = \{x_1\}$ and $d(x_1, x_2) = 1$, $d(x_1, x_3) = 1$, $d(x_1, x_4) = 1$, $d(x_1, x_5) = 2$, $d(x_1, x_6) = 2$.

Now, let C_m, $m = 0, 1, \ldots, M$, denote the subset of X consisting of those elements whose level is m, that is,

$$C_m = \{x_j | \rho(x_j; C_0) = m\}, \tag{62}$$

with the understanding that, if x_j is not reachable (via an upward path) from some element in C_0, then $x_j \in C_0$. For the example of Figure 4, we have $C_0 = \{x_1\}$, $C_1 = \{x_2, x_3, x_4\}$, $C_2 = \{x_5, x_6\}$, $C_3 = \theta$ (empty set). In words,

$x_j \in C_m \Leftrightarrow$ (i) there exists an element of C_0 from which x_j is reachable via a path of length m, and

   (ii) there does not exist an element of C_0 from which x_j is reachable via a path of length $> m$. (63)

From (61) and (62) it follows that $C_0, \ldots, C_M$ have the following properties:

(a)   Every x_j in X belongs to some $C_m, m = 0, \ldots, M$. (64)

Reason. Either x_j is not reachable from any x_i in X, in which case $x_j \in C_0$, or it is reachable from some x_i in X, say x_{i_1} (i.e., $\mu_P(x_{i_1}, x_j) > 0$). Now x_{i_1}, like x_j, either is not reachable from any x_i in X, in which case $x_{i_1} \in C_0$ and hence x_j is reachable from C_0, or is reachable from some x_i in X, say x_{i_2}. Continuing this argument and making use of the antisymmetry of P and the finiteness of X, we arrive at the conclusion that the chain $(x_{i_k}, x_{i_{k-1}}, \ldots, x_i, x_j)$ must eventually originate at some x_{i_k} in C_0. This establishes that every x_j in X which is not in C_0 is reachable from C_0 and hence that $\rho(x_j; C_0) > 0$ and $x_j \in C_{\rho(x_j; c_0)}$.

(b) $C_0, C_1, \ldots, C_M$ are disjoint. Thus, (a) and (b) imply that the collection $\{C_0, \ldots, C_M\}$ is a partition of X.

Reason. Single-valuedness of $\rho(x_j; C_0)$ implies that $x_j \in C_k$ and $x_j \in C_l$ cannot both be true if $k \neq l$. Hence the disjointness of $C_0, \ldots, C_M$.

(c) $$x_i, x_j \in C_m \Rightarrow \mu_P(x_i, x_j) = \mu_P(x_j, x_i) = 0. \tag{65}$$

Reason. Assume $x_i, x_j \in C_m$ and $\mu_P(x_i, x_j) > 0$. Then

$$\rho(x_j; C_0) > \rho(x_i; C_0),$$

which contradicts the assumption that $x_i, x_j \in C_m$. Similarly, $\mu_P(x_j, x_i) > 0$ contradicts $x_i, x_j \in C_m$.

(d) $x_j \in C_l$ and $k < l \Rightarrow x_j$ is reachable from some x_i in C_k.

Reason. If $x_j \in C_l$, then there exists a path T of length l via which x_j is reachable from some x_r in C_0, and there does not exist a longer path via which x_j is reachable from any element of C_0. Now let x_i be the kth node of T (counting in the direction of C_l), with $k < l$. Then $x_i \in C_k$, since there exists a path of length k from x_r to x_i and there does not exist a longer path via which x_i is reachable from any element of C_0. (For, if such a path existed, then x_j would be reachable via a path longer than l from some element of C_0). Thus x_j is reachable from some x_i in C_k.

An immediate consequence of (d) is that the C_m may be defined recursively by

$$C_{m+1} = \{x_j | \rho(x_j; C_m) = 1\}, \qquad m = 0, 1, \ldots, M, \tag{66}$$

with the understanding that $C_M \neq \theta$ and $C_{M+1} = \theta$. More explicitly,

$x_j \in C_{m+1} \Leftrightarrow \mu_P(x_i, x_j) > 0$ for some x_i in C_m, and there does not exist an x_i in C_m and an x_k in X distinct from x_i such that $\mu_P(x_i, x_k) > 0$ and $\mu_P(x_k, x_j) > 0$.

(e) $$x_i \in C_k \quad \text{and} \quad x_j \in C_l \quad \text{and} \quad k < l \Rightarrow \mu_P(x_j, x_i) = 0. \tag{67}$$

Reason. Suppose $\mu_P(x_j, x_i) > 0$. By (d), x_j is reachable from some element of C_k, say x_r. If $x_r = x_i$, then $\mu_P(x_i, x_j) > 0$, which contradicts the antisymmetry of P. If $x_r \neq x_i$, then by transitivity of P, $\mu_P(x_r, x_i) > 0$, which contradicts (c) since $x_r, x_i \in C_k$.

(f) $$x_i \in C_k \quad \text{and} \quad x_j \in C_l \quad \text{and} \quad \mu_P(x_i, x_j) > 0 \Rightarrow l > k. \tag{68}$$

Reason. By negation of (e) and (c).

The partition $\{C_0, \ldots, C_m\}$, which can be constructed from the relation matrix P or by inspection of the Hasse diagram of P, can be put to use in various ways. In particular, it can be employed to obtain the Hasse diagram of P from its relation matrix in cases in which this is difficult to do by inspection. Another application, which motivated our discussion of $\{C_0, \ldots, C_M\}$, relates to the possibility of relabeling the elements of X in such a way as to result in an upper-triangular relation matrix. By employing the properties of $\{C_0,\ldots,C_M\}$ stated above, this can readily be accomplished as follows:

Let n_m denote the number of elements in C_m, $m = 0, \ldots, M$. Let the elements of C_0 be relabeled, in some arbitrary order, as $y_1, \ldots, y_{n_0}$, then the elements of C_1 be relabeled as $y_{n_0+1}, \ldots, y_{n_0+n_1}$, then the elements of C_2 be relabeled as $y_{n_0+n_1+1}, \ldots, y_{n_0+n_1+n_2}$, and so on, until all the elements of X are relabeled in this manner. If the new label for x_i is y_j, we write

$$y_j = \sigma(x_i), \tag{69}$$

where σ is a one-to-one mapping from $X = \{x_1,\ldots,x_n\}$ to $Y = \{y_1,\ldots,y_n\}$. Furthermore, we order the y_j linearly by $y_j > y_i \Leftrightarrow j > i$, $i, j = 1, \ldots, n$.

The above relabeling transforms the relation matrix P into the relation matrix P_r defined by

$$\mu_{P_r}(\sigma(x_i), \sigma(x_j)) = \mu_P(x_i, x_j), \qquad x_i, x_j \in X. \tag{70}$$

To verify that P_r is upper-triangular, it is sufficient to note that, if $\mu_P(x_i, x_j) > 0$, for $x_i \neq x_j$, then by (f) $\sigma(x_j) > \sigma(x_i)$.

It is now a simple matter to construct a linear ordering L in Y which satisfies (60). Specifically, for $x_i \neq x_j$, let

$$
\begin{aligned}
\mu_L(\sigma(x_i), \sigma(x_j)) &= \mu_P(x_i, x_j), \text{ if } \mu_P(x_i, x_j) > 0, \\
&= 0, \text{ if } \mu_P(x_i, x_j) = 0 \text{ and } \mu_P(x_j, x_i) > 0, \\
&= \epsilon, \text{ if } \mu_P(x_i, x_j) = \mu_P(x_j, x_i) = 0 \text{ and } \sigma(x_j) > \sigma(x_i), \\
&= 0, \text{ if } \mu_P(x_i, x_j) = \mu_P(x_j, x_i) = 0 \text{ and } \sigma(x_j) < \sigma(x_i), \quad (71)
\end{aligned}
$$

where ϵ is any positive constant which is smaller than or equal to the smallest positive entry in the relation matrix P.

Note. It is helpful to observe that this construction of L may be visualized as a projection of the Hasse diagram of P on a slightly inclined vertical line Y (Figure 8). The purpose of the inclination is to avoid the possibility that two or more nodes of the Hasse diagram may be taken by the projection into the same point of Y.

All that remains to be demonstrated at this stage is that L, as defined by (71), is transitive. This is insured by our choice of ϵ, for, so long as ϵ is smaller than or equal to the smallest entry in P, the transitivity of P implies the transitivity of L, as is demonstrated by the following lemma:

LEMMA 9. *Let P_r be an upper-triangular matrix such that $P_r = P_r^2$. Let Q denote an upper-triangular matrix all of whose elements are equal to ϵ, where*

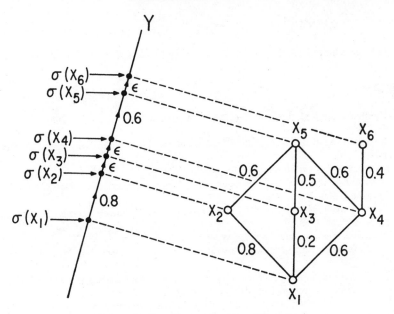

FIGURE 8. Graphic construction of L and σ.

$0 < \epsilon \leqslant$ *smallest positive entry in P_r. Then*

$$P_r \vee Q = (P_r \vee Q)^2. \qquad (72)$$

In other words, if P_r and Q are transitive, so is $P_r \vee Q$.

Proof. We can rewrite (72) as

$$(P_r \vee Q)^2 = P_r{}^2 \vee P_r{\circ}Q \vee Q{\circ}P_r \vee Q^2. \qquad (73)$$

Now $P_r{}^2 = P_r$ and, since Q is upper-triangular, $Q = Q^2$. Furthermore, $P_r{\circ}Q = Q{\circ}P_r = Q$. Hence (72).

To apply this lemma, we note that L, as defined by (71), may be expressed as

$$L = P_r \vee Q, \qquad (74)$$

where P_r and Q satisfy the conditions of the lemma. Consequently, L is transitive and thus is a linear ordering satisfying (60). This completes the proof of our extension of Szpilrajn's theorem.

CONCLUDING REMARK

As the foregoing analysis demonstrates, it is a relatively simple matter to extend some of the well-known results in the theory of relations to fuzzy sets. It appears that such extension may be of use in various applied areas, particularly those in which fuzziness and/or randomness play a significant role in the analysis or control of system behavior.

ACKNOWLEDGMENT

The author is indebted to P. Varaiya, E. T. Lee, and J. Yang for helpful criticisms.

REFERENCES

1 A. Kaufmann and M. Précigout, *Cours de Mathématiques Nouvelles*, Dunod, Paris, 1966.
2 R. R. Sokal and P. H. A. Sneath, *Principles of Numerical Taxonomy*, Freeman, San Francisco, 1963.
3 E. J. Williams, *Regression Analysis*, Wiley, New York, 1959.
4 N. Nilsson, *Learning Machines—Foundations of Trainable Pattern Classification Systems*, McGraw-Hill, New York, 1965.
5 K. S. Fu, *Sequential Methods in Pattern Recognition and Machine Learning*, Academic Press, New York, 1968.
6 S. Watanabe, *Knowing and Guessing*, Wiley, New York, 1969.
7 E. M. Braverman (Ed.), *Automatic Analysis of Complex Patterns*, Mir Press, Moscow, 1969.
8 R. N. Shepard, Analysis of proximities: Multidimensional scaling with an unknown distance function, *Psychometrica* 27 (1962), 125–140, 219–246.
9 J. B. Kruskal, Multidimensional scaling by optimizing goodness of fit to a nonmetric hypothesis, *Psychometrica* 29 (1964), 1–27.
10 J. H. Ward, Jr., Hierarchical grouping to optimize an objective function, *Amer. Statist. Assoc.* 58 (1963), 236–244.
11 G. Debreu, *Theory of Value* (Cowles Commission Monograph 17), Wiley, New York, 1959.
12 R. D. Luce and H. Raiffa, *Games and Decisions*, Wiley, New York, 1958.
13 L. A. Zadeh, Fuzzy sets, *Information and Control* 8 (June, 1965), 338–353.
14 S. Tamura, S. Higuchi, and K. Tanaka, Pattern classification based on fuzzy relations, Osaka University, Osaka, Japan, 1970. To be published in System Science and Cybernetics.
15 J. Aczél, *Lectures on Functional Equations and Their Applications*, Academic Press, New York, 1966.
16. P. L. Hammer (Ivănescu) and S. Rudeanu, *Boolean Methods in Operations Research and Related Areas*, Springer-Verlag, Berlin/New York, 1968.
17 E. G. Santos, Maximin automata, *Information and Control* 13 (Oct. 1968), 363–377.
18 R. E. Bellman and L. A. Zadeh, Decision-making in a fuzzy environment, Electronics Res. Lab. Rept. 69-8, University of California, Berkeley, Nov. 1969 (to appear in *Management Science*).
19 W. E. Armstrong, Uncertainty and the utility function, *Economic J.* 58 (1948), 1–10.
20 K. O. May, Intransitivity, utility and the aggregation of preference patterns, *Econometrica* 22 (Jan. 1954), 1–13.
21 R. D. Luce, Semiorders and a theory of utility discrimination, *Econometrica* 24 (1956), 178 191.
22 S. C. Johnson, Hierarchical clustering schemes, *Psychometrica* 32 (Sept. 1967), 241–254.
23 J. Lyons, *Introduction to Theoretical Linguistics*, Cambridge Univ. Press, Cambridge/New York, 1968.
24 R. M. Baer and O. Osterby, Algorithms over partially ordered sets, *BIT* 9 (1969), 97–118.

Received August 17, 1970

Outline of a New Approach to the Analysis of Complex Systems and Decision Process

LOTFI A. ZADEH§

Abstract—The approach described in this paper represents a substantive departure from the conventional quantitative techniques of system analysis. It has three main distinguishing features: 1) use of so-called "linguistic" variables in place of or in addition to numerical variables; 2) characterization of simple relations between variables by fuzzy conditional statements; and 3) characterization of complex relations by fuzzy algorithms.

A *linguistic variable* is defined as a variable whose values are sentences in a natural or artificial language. Thus, if *tall, not tall, very tall, very very tall*, etc. are values of *height*, then *height* is a linguistic variable. *Fuzzy conditional statements* are expressions of the form IF A THEN B, where A and B have fuzzy meaning, e.g., IF x is *small* THEN y is *large*, where *small* and *large* are viewed as labels of fuzzy sets. A *fuzzy algorithm* is an ordered sequence of instructions which may contain fuzzy assignment and conditional statements, e.g., x = *very small*, IF x is *small* THEN y is *large*. The execution of such instructions is governed by the *compositional rule of inference* and the *rule of the preponderant alternative*.

By relying on the use of linguistic variables and fuzzy algorithms, the approach provides an approximate and yet effective means of describing the behavior of systems which are too complex or too ill-defined to admit of precise mathematical analysis. Its main applications lie in economics, management science, artificial intelligence, psychology, linguistics, information retrieval, medicine, biology, and other fields in which the dominant role is played by the animate rather than inanimate behavior of system constituents.

Manuscript received August 1, 1972; revised August 13, 1972. This work was supported by the Navy Electronic Systems Command under Contract N00039-71-C-0255, the Army Research Office, Durham, N.C., under Grant DA-ARO-D-31-124-71-G174, and NASA under Grant NGL-05-003-016-VP3.

The author is with the Department of Electrical Engineering and Computer Sciences and Electronics Research Laboratory, University of California, Berkeley, Calif. 94720.

I. Introduction

THE ADVENT of the computer age has stimulated a rapid expansion in the use of quantitative techniques for the analysis of economic, urban, social, biological, and other types of systems in which it is the animate rather than inanimate behavior of system constituents that plays a dominant role. At present, most of the techniques employed for the analysis of *humanistic*, i.e., human-centered, systems are adaptations of the methods that have been developed over a long period of time for dealing with *mechanistic* systems, i.e., physical systems governed in the main by the laws of mechanics, electromagnetism, and thermodynamics. The remarkable successes of these methods in unraveling the secrets of nature and enabling us to build better and better machines have inspired a widely held belief that the same or similar techniques can be applied with comparable effectiveness to the analysis of humanistic systems. As a case in point, the successes of modern control theory in the design of highly accurate space navigation systems have stimulated its use in the theoretical analyses of economic and biological systems. Similarly, the effectiveness of computer simulation techniques in the macroscopic analyses of physical systems has brought into vogue the use of computer-based econometric models for purposes of forecasting, economic planning, and management.

Given the deeply entrenched tradition of scientific thinking which equates the understanding of a phenomenon with the ability to analyze it in quantitative terms, one is certain to strike a dissonant note by questioning the growing tendency to analyze the behavior of humanistic systems as if they were mechanistic systems governed by difference, differential, or integral equations. Such a note is struck in the present paper.

Essentially, our contention is that the conventional quantitative techniques of system analysis are intrinsically unsuited for dealing with humanistic systems or, for that matter, any system whose complexity is comparable to that of humanistic systems. The basis for this contention rests on what might be called the *principle of incompatibility*. Stated informally, the essence of this principle is that as the complexity of a system increases, our ability to make precise and yet significant statements about its behavior diminishes until a threshold is reached beyond which precision and significance (or relevance) become almost mutually exclusive characteristics.[1] It is in this sense that precise quantitative analyses of the behavior of humanistic systems are not likely to have much relevance to the real-world societal, political, economic, and other types of problems which involve humans either as individuals or in groups.

An alternative approach outlined in this paper is based on the premise that the key elements in human thinking are not numbers, but labels of fuzzy sets, that is, classes of objects in which the transition from membership to non-membership is gradual rather than abrupt. Indeed, the pervasiveness of fuzziness in human thought processes suggests that much of the logic behind human reasoning is not the traditional two-valued or even multivalued logic, but a logic with fuzzy truths, fuzzy connectives, and fuzzy rules of inference. In our view, it is this fuzzy, and as yet not well-understood, logic that plays a basic role in what may well be one of the most important facets of human thinking, namely, the ability to *summarize* information—to extract from the collections of masses of data impinging upon the human brain those and only those subcollections which are relevant to the performance of the task at hand.

By its nature, a summary is an approximation to what it summarizes. For many purposes, a very approximate characterization of a collection of data is sufficient because most of the basic tasks performed by humans do not require a high degree of precision in their execution. The human brain takes advantage of this tolerance for imprecision by encoding the "task-relevant" (or "decision-relevant") information into labels of fuzzy sets which bear an approximate relation to the primary data. In this way, the stream of information reaching the brain via the visual, auditory, tactile, and other senses is eventually reduced to the trickle that is needed to perform a specified task with a minimal degree of precision. Thus, the ability to manipulate fuzzy sets and the consequent summarizing capability constitute one of the most important assets of the human mind as well as a fundamental characteristic that distinguishes human intelligence from the type of machine intelligence that is embodied in present-day digital computers.

---

[1] A corollary principle may be stated succinctly as, "The closer one looks at a real-world problem, the fuzzier becomes its solution."

Viewed in this perspective, the traditional techniques of system analysis are not well suited for dealing with humanistic systems because they fail to come to grips with the reality of the fuzziness of human thinking and behavior. Thus, to deal with such systems realistically, we need approaches which do not make a fetish of precision, rigor, and mathematical formalism, and which employ instead a methodological framework which is tolerant of imprecision and partial truths. The approach described in the sequel is a step—but not necessarily a definitive step—in this direction.

The approach in question has three main distinguishing features: 1) use of so-called "linguistic" variables in place of or in addition to numerical variables; 2) characterization of simple relations between variables by conditional fuzzy statements; and 3) characterization of complex relations by fuzzy algorithms. Before proceeding to a detailed discussion of our approach, it will be helpful to sketch the principal ideas behind these features. We begin with a brief explanation of the notion of a linguistic variable.

1) *Linguistic and Fuzzy Variables:* As already pointed out, the ability to summarize information plays an essential role in the characterization of complex phenomena. In the case of humans, the ability to summarize information finds its most pronounced manifestation in the use of natural languages. Thus, each word x in a natural language L may be viewed as a summarized description of a fuzzy subset $M(x)$ of a universe of discourse U, with $M(x)$ representing the meaning of x. In this sense, the language as a whole may be regarded as a system for assigning atomic and composite labels (i.e., words, phrases, and sentences) to the fuzzy subsets of U. (This point of view is discussed in greater detail in [4] and [5].) For example, if the meaning of the noun *flower* is a fuzzy subset $M(flower)$, and the meaning of the adjective *red* is a fuzzy subset $M(red)$, then the meaning of the noun phrase *red flower* is given by the intersection of $M(red)$ and $M(flower)$.

If we regard the color of an object as a variable, then its values, *red, blue, yellow, green*, etc., may be interpreted as labels of fuzzy subsets of a universe of objects. In this sense, the attribute *color* is a *fuzzy variable*, that is, a variable whose values are labels of fuzzy sets. It is important to note that the characterization of a value of the variable *color* by a natural label such as *red* is much less precise than the numerical value of the wavelength of a particular color.

In the preceding example, the values of the variable *color* are atomic terms like *red*, *blue*, *yellow*, etc. More generally, the values may be sentences in a specified language, in which case we say that the variable is *linguistic*. To illustrate, the values of the fuzzy variable *height* might be expressible as *tall, not tall, somewhat tall, very tall, not very tall, very very tall, tall but not very tall, quite tall, more or less tall*. Thus, the values in question are sentences formed from the label *tall*, the negation *not*, the connectives *and* and *but*, and the hedges *very*, *somewhat*, *quite*, and *more or less*. In this sense, the variable *height* as defined above is a linguistic variable.

As will be seen in Section III, the main function of linguistic variables is to provide a systematic means for an approximate characterization of complex or ill-defined phenomena. In essence, by moving away from the use of quantified variables and toward the use of the type of linguistic descriptions employed by humans, we acquire a capability to deal with systems which are much too complex to be susceptible to analysis in conventional mathematical terms.

2) *Characterization of Simple Relations Between Fuzzy Variables by Conditional Statements:* In quantitative approaches to system analysis, a dependence between two numerically valued variables x and y is usually characterized by a table which, in words, may be expressed as a set of conditional statements, e.g., IF x is 5 THEN y is 10, IF x is 6 THEN y is 14, etc.

The same technique is employed in our approach, except that x and y are allowed to be fuzzy variables. In particular, if x and y are linguistic variables, the conditional statements describing the dependence of y on x might read (the following italicized words represent the values of fuzzy variables):

IF x is *small*   THEN y is *very large*
IF x is *not very small*   THEN y is *very very large*
IF x is *not small and not large*   THEN y is *not very large*

and so forth.

Fuzzy conditional statements of the form IF A THEN B, where A and B are terms with a fuzzy meaning, e.g., "IF John is *nice* to you THEN you should be *kind* to him," are used routinely in everyday discourse. However, the meaning of such statements when used in communication between humans is poorly defined. As will be shown in Section V, the conditional statement IF A THEN B can be given a precise meaning even when A and B are fuzzy rather than nonfuzzy sets, provided the meanings of A and B are defined precisely as specified subsets of the universe of discourse.

In the preceding example, the relation between two fuzzy variables x and y is *simple* in the sense that it can be characterized as a set of conditional statements of the form IF A THEN B, where A and B are labels of fuzzy sets representing the values of x and y, respectively. In the case of more complex relations, the characterization of the dependence of y on x may require the use of a fuzzy algorithm. As indicated below, and discussed in greater detail in Section VI, the notion of a fuzzy algorithm plays a basic role in providing a means of approximate characterization of fuzzy concepts and their interrelations.

3) *Fuzzy-Algorithmic Characterization of Functions and Relations:* The definition of a fuzzy function through the use of fuzzy conditional statements is analogous to the definition of a nonfuzzy function f by a table of pairs $(x, f(x))$, in which x is a generic value of the argument of f and $f(x)$ is the value of the function. Just as a nonfuzzy function can be defined algorithmically (e.g., by a program) rather than by a table, so a fuzzy function can be defined by a fuzzy algorithm rather than as a collection of fuzzy conditional statements. The same applies to the definition of sets, relations, and other constructs which are fuzzy in nature.

Essentially, a fuzzy algorithm [6] is an ordered sequence of instructions (like a computer program) in which some of the instructions may contain labels of fuzzy sets, e.g.:

Reduce x *slightly* if y is *large*
Increase x *very slightly* if y is *not very large and not very small*
If x is *small* then stop; otherwise increase x by 2.

By allowing an algorithm to contain instructions of this type, it becomes possible to give an approximate fuzzy-algorithmic characterization of a wide variety of complex phenomena. The important feature of such characterizations is that, though imprecise in nature, they may be perfectly adequate for the purposes of a specified task. In this way, fuzzy algorithms can provide an effective means of approximate description of objective functions, constraints, system performance, strategies, etc.

In what follows, we shall elaborate on some of the basic aspects of linguistic variables, fuzzy conditional statements, and fuzzy algorithms. However, we shall not attempt to present a definitive exposition of our approach and its applications. Thus, the present paper should be viewed primarily as an introductory outline of a method which departs from the tradition of precision and rigor in scientific analysis—a method whose approximate nature mirrors the fuzziness of human behavior and thereby offers a promise of providing a more realistic basis for the analysis of humanistic systems.

As will be seen in the following sections, the theoretical foundation of our approach is actually quite precise and rather mathematical in spirit. Thus, the source of imprecision in the approach is not the underlying theory, but the manner in which linguistic variables and fuzzy algorithms are applied to the formulation and solution of real-world problems. In effect, the level of precision in a particular application can be adjusted to fit the needs of the task and the accuracy of the available data. This flexibility constitutes one of the important features of the method that will be described.

II. Fuzzy Sets: A Summary of Relevant Properties

In order to make our exposition self-contained, we shall summarize in this section those properties of fuzzy sets which will be needed in later sections. (More detailed discussions of topics in the theory of fuzzy sets which are relevant to the subject of the present paper may be found in [1]–[17].)

Notation and Terminology

A fuzzy subset A of a universe of discourse U is characterized by a membership function $\mu_A: U \to [0,1]$ which associates with each element y of U a number $\mu_A(y)$ in the interval $[0,1]$ which represents the grade of membership of y in A. The *support* of A is the set of points in U at which $\mu_A(y)$ is positive. A *crossover point* in A is an element of U whose grade of membership in A is 0.5. A *fuzzy singleton* is a fuzzy set whose support is a single point in U. If A is a fuzzy singleton whose support is the point y, we write

$$A = \mu/y \qquad (2.1)$$

where μ is the grade of membership of y in A. To be consistent with this notation, a nonfuzzy singleton will be denoted by $1/y$.

A fuzzy set A may be viewed as the union (see (2.27)) of its constituent singletons. On this basis, A may be represented in the form

$$A = \int_U \mu_A(y)/y \qquad (2.2)$$

where the integral sign stands for the union of the fuzzy singletons $\mu_A(y)/y$. If A has a finite support $\{y_1, y_2, \cdots, y_n\}$, then (2.2) may be replaced by the summation

$$A = \mu_1/y_1 + \cdots + \mu_n/y_n \qquad (2.3)$$

or

$$A = \sum_{i=1}^{n} \mu_i/y_i \qquad (2.4)$$

in which μ_i, $i = 1, \cdots, n$, is the grade of membership of y_i in A. It should be noted that the $+$ sign in (2.3) denotes the union (see (2.27)) rather than the arithmetic sum. In this sense of $+$, a finite universe of discourse $U = \{y_1, y_2, \cdots, y_n\}$ may be represented simply by the summation

$$U = y_1 + y_2 + \cdots + y_n \qquad (2.5)$$

or

$$U = \sum_{i=1}^{n} y_i \qquad (2.6)$$

although, strictly, we should write (2.5) and (2.6) as

$$U = 1/y_1 + 1/y_2 + \cdots + 1/y_n \qquad (2.7)$$

and

$$U = \sum_{i=1}^{n} 1/y_i. \qquad (2.8)$$

As an illustration, suppose that

$$U = 1 + 2 + \cdots + 10. \qquad (2.9)$$

Then a fuzzy subset[2] of U labeled *several* may be expressed as (the symbol $\triangleq$ stands for "equal by definition," or "is defined to be," or "denotes")

several $\triangleq 0.5/3 + 0.8/4 + 1/5 + 1/6 + 0.8/7 + 0.5/8.$

$$(2.10)$$

---

[2] A is a subset of B, written $A \subseteq B$, if and only if $\mu_A(y) \leq \mu_B(y)$, for all y in U. For example, the fuzzy set $A = 0.6/1 + 0.3/2$ is a subset of $B = 0.8/1 + 0.5/2 + 0.6/3$.

Similarly, if U is the interval $[0,100]$, with $y \triangleq age$, then the fuzzy subsets of U labeled *young* and *old* may be represented as (here and elsewhere in this paper we do not differentiate between a fuzzy set and its label)

$$young = \int_0^{25} 1/y + \int_{25}^{100} \left(1 + \left(\frac{y - 25}{5}\right)^2\right)^{-1} /y \quad (2.11)$$

$$old = \int_{50}^{100} \left(1 + \left(\frac{y - 50}{5}\right)^{-2}\right)^{-1} /y. \quad (2.12)$$

(see Fig. 1).

The grade of membership in a fuzzy set may itself be a fuzzy set. For example, if

$$U = \text{TOM} + \text{JIM} + \text{DICK} + \text{BOB} \quad (2.13)$$

and A is the fuzzy subset labeled *agile*, then we may have

$$agile = medium/\text{TOM} + low/\text{JIM}$$

$$+ low/\text{DICK} + high/\text{BOB}. \quad (2.14)$$

In this representation, the fuzzy grades of membership *low*, *medium*, and *high* are fuzzy subsets of the universe V

$$V = 0 + 0.1 + 0.2 + \cdots + 0.9 + 1 \quad (2.15)$$

which are defined by

$$low = 0.5/0.2 + 0.7/0.3 + 1/0.4 + 0.7/0.5 + 0.5/0.6 \quad (2.16)$$

$$medium = 0.5/0.4 + 0.7/0.5 + 1/0.6 + 0.7/0.7 + 0.5/0.8 \quad (2.17)$$

$$high = 0.5/0.7 + 0.7/0.8 + 0.9/0.9 + 1/1. \quad (2.18)$$

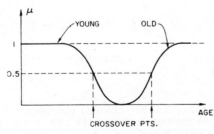

Fig. 1.   Diagrammatic representation of *young* and *old*.

Fuzzy Relations

A *fuzzy relation R* from a set X to a set Y is a fuzzy subset of the Cartesian product $X \times Y$. ($X \times Y$ is the collection of ordered pairs (x,y), $x \in X$, $y \in Y$). R is characterized by a bivariate membership function $\mu_R(x,y)$ and is expressed

$$R \triangleq \int_{X \times Y} \mu_R(x,y)/(x,y). \tag{2.19}$$

More generally, for an *n*ary fuzzy relation R which is a fuzzy subset of $X_1 \times X_2 \times \cdots \times X_n$, we have

$$R \triangleq \int_{X_1 \times \cdots \times X_n} \mu_R(x_1,\cdots,x_n)/(x_1,\cdots,x_n),$$
$$x_i \in X_i, \quad i = 1,\cdots,n. \tag{2.20}$$

As an illustration, if

$$X = \{\text{TOM, DICK}\} \quad \text{and} \quad Y = \{\text{JOHN, JIM}\}$$

then a binary fuzzy relation of *resemblance* between members of X and Y might be expressed as

resemblance = 0.8/(TOM, JOHN) + 0.6/(TOM, JIM)

+ 0.2/(DICK, JOHN) + 0.9/(DICK, JIM).

Alternatively, this relation may be represented as a *relation matrix*

$$\begin{array}{cc} & \text{JOHN} \quad \text{JIM} \\ \begin{array}{c} \text{TOM} \\ \text{DICK} \end{array} & \begin{bmatrix} 0.8 & 0.6 \\ 0.2 & 0.9 \end{bmatrix} \end{array} \tag{2.21}$$

in which the (i,j)th element is the value of $\mu_R(x,y)$ for the ith value of x and the jth value of y.

If R is a relation from X to Y and S is a relation from Y to Z, then the *composition* of R and S is a fuzzy relation denoted by $R \circ S$ and defined by

$$R \circ S \triangleq \int_{X \times Z} \bigvee_y (\mu_R(x,y) \wedge \mu_s(y,z))/(x,z) \tag{2.22}$$

where $\vee$ and $\wedge$ denote, respectively, max and min.[3] Thus, for real a,b,

---

[3] Equation (2.22) defines the max–min composition of R and S. Max–product composition is defined similarly, except that $\wedge$ is replaced by the arithmetic product. A more detailed discussion of these compositions may be found in [2].

$$a \vee b = \max (a,b) \triangleq \begin{cases} a, & \text{if } a \geq b \\ b, & \text{if } a < b \end{cases} \quad (2.23)$$

$$a \wedge b = \min (a,b) \triangleq \begin{cases} a, & \text{if } a \leq b \\ b, & \text{if } a > b \end{cases} \quad (2.24)$$

and $\vee_y$ is the supremum over the domain of y.

If the domains of the variables x, y, and z are finite sets, then the relation matrix for $R \circ S$ is the max–min product[4] of the relation matrices for R and S. For example, the max–min product of the relation matrices on the left-hand side of (2.25) results in the relation matrix $R \circ S$ shown on the right-hand side of

$$\overset{R}{\begin{bmatrix} 0.3 & 0.8 \\ 0.6 & 0.9 \end{bmatrix}} \circ \overset{S}{\begin{bmatrix} 0.5 & 0.9 \\ 0.4 & 1 \end{bmatrix}} = \overset{R \circ S}{\begin{bmatrix} 0.4 & 0.8 \\ 0.5 & 0.9 \end{bmatrix}}. \quad (2.25)$$

Operations on Fuzzy Sets

The negation *not*, the connectives *and* and *or*, the hedges *very*, *highly*, *more or less*, and other terms which enter in the representation of values of linguistic variables may be viewed as labels of various operations defined on the fuzzy subsets of U. The more basic of these operations will be summarized.

The *complement* of A is denoted $\neg A$ and is defined by

$$\neg A \triangleq \int_U (1 - \mu_A(y))/y. \quad (2.26)$$

The operation of complementation corresponds to negation. Thus, if x is a label for a fuzzy set, then *not x* should be interpreted as $\neg x$. (Strictly speaking, $\neg$ operates on fuzzy sets, whereas *not* operates on their labels. With this understanding, we shall use $\neg$ and *not* interchangeably.)

The *union* of fuzzy sets A and B is denoted $A + B$ and is defined by

$$A + B \triangleq \int_U (\mu_A(y) \vee \mu_B(y))/y. \quad (2.27)$$

The union corresponds to the connective *or*. Thus, if u and v are labels of fuzzy sets, then

$$u \text{ or } v \triangleq u + v \quad (2.28)$$

---

[4] In the max–min matrix product, the operations of addition and multiplication are replaced by $\vee$ and $\wedge$, respectively.

The *intersection* of A and B is denoted $A \cap B$ and is defined by

$$A \cap B \triangleq \int_U (\mu_A(y) \wedge \mu_B(y)) / y. \qquad (2.29)$$

The intersection corresponds to the connective *and*; thus

$$u \text{ and } v \triangleq u \cap v. \qquad (2.30)$$

As an illustration, if

$$U = 1 + 2 + \cdots + 10 \qquad (2.31)$$

$$u = 0.8/3 + 1/5 + 0.6/6 \qquad (2.32)$$

$$v = 0.7/3 + 1/4 + 0.5/6 \qquad (2.33)$$

then

$$u \text{ or } v = 0.8/3 + 1/4 + 1/5 + 0.6/6 \qquad (2.34)$$

$$u \text{ and } v = 0.7/3 + 0.5/6. \qquad (2.35)$$

The *product* of A and B is denoted AB and is defined by

$$AB \triangleq \int_U \mu_A(y)\mu_B(y) / y. \qquad (2.36)$$

Thus, if

$$A = 0.8/2 + 0.9/5 \qquad (2.37)$$

$$B = 0.6/2 + 0.8/3 + 0.6/5 \qquad (2.38)$$

then

$$AB = 0.48/2 + 0.54/5. \qquad (2.39)$$

Based on (2.36), A^α, where α is any positive number, is defined by

$$A^\alpha \triangleq \int_U (\mu_A(y))^\alpha / y. \qquad (2.40)$$

Similarly, if α is a nonnegative real number, then

$$\alpha A \triangleq \int_U \alpha\mu_A(y) / y. \qquad (2.41)$$

As an illustration, if A is expressed by (2.37), then

$$A^2 = 0.64/2 + 0.81/5 \qquad (2.42)$$

$$0.5A = 0.4/2 + 0.45/5. \qquad (2.43)$$

In addition to the basic operations just defined, there are other operations that are of use in the representation of linguistic hedges. Some of these will be briefly defined. (A more detailed discussion of these operations may be found in [15].)

The operation of *concentration* is defined by

$$\text{CON } (A) \triangleq A^2. \qquad (2.44)$$

Applying this operation to A results in a fuzzy subset of A such that the reduction in the magnitude of the grade of membership of y in A is relatively small for those y which have a high grade of membership in A and relatively large for the y with low membership.

The operation of *dilation* is defined by

$$\text{DIL } (A) \triangleq A^{0.5}. \qquad (2.45)$$

The effect of this operation is the opposite of that of concentration.

The operation of *contrast intensification* is defined by

$$\text{INT } (A) \triangleq \begin{cases} 2A^2, & \text{for } 0 \le \mu_A(y) \le 0.5 \\ \neg 2(\neg A)^2, & \text{for } 0.5 \le \mu_A(y) \le 1. \end{cases} \qquad (2.46)$$

This operation differs from concentration in that it increases the values of $\mu_A(y)$ which are above 0.5 and diminishes those which are below this point. Thus, contrast intensification has the effect of reducing the fuzziness of A. (An entropy-like measure of fuzziness of a fuzzy set is defined in [16].)

As its name implies, the operation of *fuzzification* (or, more specifically, *support fuzzification*) has the effect of transforming a nonfuzzy set into a fuzzy set or increasing the fuzziness of a fuzzy set. The result of application of a fuzzification to A will be denoted by $F(A)$ or $\tilde{A}$, with the wavy overbar referred to as a *fuzzifier*. Thus $x \approx 3$ means "x is approximately equal to 3," while $x = \tilde{3}$ means "x is a fuzzy set which approximates to 3." A fuzzifier F is characterized by its *kernel* $K(y)$, which is the fuzzy set resulting from the application of F to a singleton $1/y$. Thus

$$K(y) \triangleq \widetilde{1/y}. \qquad (2.47)$$

In terms of K, the result of applying F to a fuzzy set A is given by

$$F(A; K) \triangleq \int_U \mu_A(y)K(y) \qquad (2.48)$$

where $\mu_A(y)K(y)$ represents the product (in the sense of (2.41)) of the scalar $\mu_A(y)$ and the fuzzy set $K(y)$, and $\int_U$ should be interpreted as the union of the family of fuzzy sets $\mu_A(y)K(y)$, $y \in U$. Thus (2.48) is analogous to the integral representation of a linear operator, with $K(y)$ playing the role of impulse response.

As an illustration of (2.48), assume that U, A, and $K(y)$ are defined by

$$U = 1 + 2 + 3 + 4 \qquad (2.49)$$

$$A = 0.8/1 + 0.6/2 \qquad (2.50)$$

$$K(1) = 1/1 + 0.4/2 \qquad (2.51)$$

$$K(2) = 1/2 + 0.4/1 + 0.4/3.$$

Then, the result of applying F to A is given by

$$F(A;K) = 0.8(1/1 + 0.4/2) + 0.6(1/2 + 0.4/1 + 0.4/3)$$

$$= 0.8/1 + 0.32/2 + 0.6/2 + 0.24/1 + 0.24/3$$

$$= 0.8/1 + 0.6/2 + 0.24/3. \qquad (2.52)$$

The operation of fuzzification plays an important role in the definition of linguistic hedges such as *more or less, slightly, much,* etc. Examples of its uses are given in [15].

Language and Meaning

As was indicated in Section I, the values of a linguistic variable are fuzzy sets whose labels are sentences in a natural or artificial language. For our purposes, a language L may be viewed as a correspondence between a set of terms T and a universe of discourse U. (This point of view is described in greater detail in [4] and [5]. For simplicity, we assume that T is a nonfuzzy set.) This correspondence may be assumed to be characterized by a fuzzy *naming relation* N from T to U, which associates with each term x in T and each object y in U the degree $\mu_N(x,y)$ to which x applies to y. For example, if $x = young$ and $y = 23$ years, then $\mu_N(young, 23)$ might be 0.9. A term may be atomic, e.g., $x = tall,$ or composite, in which case it is a concatenation of atomic terms, e.g., $x = very\ tall\ man$.

For a fixed x, the membership function $\mu_N(x,y)$ defines a fuzzy subset $M(x)$ of U whose membership function is given by

$$\mu_{M(x)}(y) \triangleq \mu_N(x,y), \qquad x \in T, \quad y \in U. \qquad (2.53)$$

This fuzzy subset is defined to be the *meaning* of x. Thus, the meaning of a term x is the fuzzy subset $M(x)$ of U for which x serves as a label. Although x and $M(x)$ are different entities (x is an element of T, whereas $M(x)$ is a fuzzy subset of U), we shall write x for $M(x)$, except where there is a need for differentiation between them. To illustrate, suppose that the meaning of the term *young* is defined by

$$\mu_N(young, y) = \begin{cases} 1, & \text{for } y \leq 25 \\ \left(1 + \left(\dfrac{y - 25}{5}\right)^2\right)^{-1}, & \text{for } y > 25. \end{cases}$$

$$(2.54)$$

Then we can represent the fuzzy subset of U labeled *young* as (see (2.11))

$$young = \int_0^{25} 1/y + \int_{25}^{100} \left(1 + \left(\frac{y - 25}{5}\right)^2\right)^{-1}/y \quad (2.55)$$

with the right-hand member of (2.55) representing the meaning of *young*.

Linguistic hedges such as *very, much, more or less*, etc., make it possible to modify the meaning of atomic as well as composite terms and thus serve to increase the range of values of a linguistic variable. The use of linguistic hedges for this purpose is discussed in the following section.

III. LINGUISTIC HEDGES

As stated in Section II, the values of a linguistic variable are labels of fuzzy subsets of U which have the form of phrases or sentences in a natural or artificial language. For example, if U is the collection of integers

$$U = 0 + 1 + 2 + \cdots + 100 \quad (3.1)$$

and *age* is a linguistic variable labeled x, then the values of x might be *young, not young, very young, not very young, old and not old, not very old, not young and not old*, etc.

In general, a value of a linguistic variable is a composite term $x = x_1 x_2 \cdots x_n$, which is a concatenation of atomic terms $x_1, \cdots, x_n$. These atomic terms may be divided into four categories:

1) *primary terms*, which are labels of specified fuzzy subsets of the universe of discourse (e.g., *young* and *old* in the preceding example);
2) the negation *not* and the connectives *and* and *or*;
3) *hedges*, such as *very, much, slightly, more or less*

(although *more or less* is comprised of three words, it is regarded as an atomic term), etc.;

4) *markers*, such as parentheses.

A basic problem P_l which arises in connection with the use of linguistic variables is the following: Given the meaning of each atomic term x_i, $i = 1, \cdots, n$, in a composite term $x = x_1 \cdots x_n$ which represents a value of a linguistic variable, compute the meaning of x in the sense of (2.53). This problem is an instance of a central problem in quantitative fuzzy semantics [4], namely, the computation of the meaning of a composite term. P_l is a special case of the latter problem because the composite terms representing the values of a linguistic variable have a relatively simple grammatical structure which is restricted to the four categories of atomic terms 1)–4).

As a preliminary to describing a general approach to the solution of P_l, it will be helpful to consider a subproblem of P_l which involves the computation of the meaning of a composite term of the form $x = hu$, where h is a hedge and u is a term with a specified meaning; e.g., $x = very\ tall\ man$, where $h = very$ and $u = tall\ man$.

Taking the point of view described in [15], a hedge h may be regarded as an operator which transforms the fuzzy set $M(u)$, representing the meaning of u, into the fuzzy set $M(hu)$. As stated already, the hedges serve the function of generating a larger set of values for a linguistic variable from a small collection of primary terms. For example, by using the hedge *very* in conjunction with *not, and,* and the primary term *tall*, we can generate the fuzzy sets *very tall, very very tall, not very tall, tall and not very tall*, etc. To define a hedge h as an operator, it is convenient to employ some of the basic operations defined in Section II, especially concentration, dilation, and fuzzification. In what follows, we shall indicate the manner in which this can be done for the natural hedge *very* and the artificial hedges *plus* and *minus*. Characterizations of such hedges as *more or less, much, slightly, sort of,* and *essentially* may be found in [15].

Although in its everyday use the hedge *very* does not have a well-defined meaning, in essence it acts as an intensifier, generating a subset of the set on which it operates. A simple operation which has this property is that of concentration (see (2.44)). This suggests that *very x*, where x is a term, be defined as the square of x, that is

$$very\ x \triangleq x^2 \qquad (3.2)$$

or, more explicitly

$$very \ x \ \triangleq \int_U \mu_x^2(y)/y. \qquad (3.3)$$

For example, if (see Fig. 2)

$$x = old \ men \ \triangleq \int_{50}^{100} \left(1 + \left(\frac{y - 50}{5}\right)^{-2}\right)^{-1}/y \quad (3.4)$$

then

$$x^2 = very \ old \ men = \int_{50}^{100} \left(1 + \left(\frac{y - 50}{5}\right)^{-2}\right)^{-2}/y. \quad (3.5)$$

Thus, if the grade of membership of JOHN in the class of *old men* is 0.8, then his grade of membership in the class of *very old men* is 0.64. As another simple example, if

$$U = 1 + 2 + 3 + 4 + 5 \qquad (3.6)$$

and

$$small = 1/1 + 0.8/2 + 0.6/3 + 0.4/4 + 0.2/5 \quad (3.7)$$

then

$$very \ small = 1/1 + 0.64/2 + 0.36/3 + 0.16/4 + 0.04/5.$$
$$(3.8)$$

Viewed as an operator, *very* can be composed with itself. Thus

$$very \ very \ x = (very \ x)^2 = x^4. \qquad (3.9)$$

For example, applying (3.9) to (3.7), we obtain (neglecting small terms)

$$very \ very \ small = 1/1 + 0.4/2 + 0.1/3. \qquad (3.10)$$

In some instances, to identify the operand of *very* we have to use parentheses or replace a composite term by an atomic one. For example, it is not grammatical to write

$$x = very \ not \ exact \qquad (3.11)$$

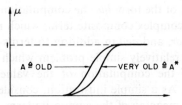

Fig. 2.   Effect of hedge *very*.

but if *not exact* is replaced by the atomic term *inexact*, then

$$x = very\ inexact \qquad (3.12)$$

is grammatically correct and we can write

$$x = (\neg exact)^2. \qquad (3.13)$$

Note that

$$not\ very\ exact = \neg(\text{very exact}) = \neg(exact^2) \quad (3.14)$$

is not the same as (3.13).

The artificial hedges *plus* and *minus* serve the purpose of providing milder degrees of concentration and dilation than those associated with the operations CON and DIL (see (2.44), (2.45)). Thus, as operators acting on a fuzzy set labeled x, *plus* and *minus* are defined by

$$plus\ x \triangleq x^{1.25} \qquad (3.15)$$

$$minus\ x \triangleq x^{0.75}. \qquad (3.16)$$

In consequence of (3.15) and (3.16), we have the approximate identity

$$plus\ plus\ x = minus\ very\ x. \qquad (3.17)$$

As an illustration, if the hedge *highly* is defined as

$$highly = minus\ very\ very \qquad (3.18)$$

then, equivalently,

$$highly = plus\ plus\ very. \qquad (3.19)$$

As was stated earlier, the computation of the meaning of composite terms of the form *hu* is a preliminary to the problem of computing the meaning of values of a linguistic variable. We are now in a position to turn our attention to this problem.

IV. COMPUTATION OF THE MEANING OF VALUES OF A LINGUISTIC VARIABLE

Once we know how to compute the meaning of a composite term of the form *hu*, the computation of the meaning of a more complex composite term, which may involve the terms *not*, *or*, and *and* in addition to terms of the form *hu*, becomes a relatively simple problem which is quite similar to that of the computation of the value of a Boolean expression. As a simple illustration, consider the computation of the meaning of the composite term

$$x = \text{not very small} \qquad (4.1)$$

where the primary term *small* is defined as

$$\text{small} = 1/1 + 0.8/2 + 0.6/3 + 0.4/4 + 0.2/5 \quad (4.2)$$

with the universe of discourse being

$$U = 1 + 2 + 3 + 4 + 5. \qquad (4.3)$$

By (3.8), the operation of *very* on *small* yields

$$\text{very small} = 1/1 + 0.64/2 + 0.36/3 + 0.16/4 + 0.04/5$$
$$(4.4)$$

and, by (2.26),

$$\text{not very small} = \neg(\text{very small})$$
$$= 0.36/2 + 0.64/3 + 0.84/4 + 0.96/5$$
$$\approx 0.4/2 + 0.6/3 + 0.8/4 + 1/5. \quad (4.5)$$

As a slightly more complicated example, consider the composite term

$$x = \text{not very small and not very very large} \qquad (4.6)$$

where *large* is defined by

$$\text{large} = 0.2/1 + 0.4/2 + 0.6/3 + 0.8/4 + 1/5. \quad (4.7)$$

In this case,

$$\text{very large} = \text{large}^2$$
$$= 0.04/1 + 0.16/2 + 0.36/3 + 0.64/4$$
$$+ 1/5 \qquad (4.8)$$
$$\text{very very large} = (\text{large}^2)^2$$
$$\approx 0.1/3 + 0.4/4 + 1/5 \qquad (4.9)$$
$$\text{not very very large} \approx 1/1 + 1/2 + 0.9/3 + 0.6/4 \quad (4.10)$$

and hence

$$\text{not very small and not very very large}$$
$$\approx (0.4/2 + 0.6/3 + 0.8/4 + 1/5)$$
$$\cap (1/1 + 1/2 + 0.9/3 + 0.6/4)$$
$$\approx 0.4/2 + 0.6/3 + 0.6/4. \quad (4.11)$$

An example of a different nature is provided by the values of a linguistic variable labeled *likelihood*. In this case, we assume that the universe of discourse is given by

$$U = 0 + 0.1 + 0.2 + 0.3 + 0.4 + 0.5$$
$$+ 0.6 + 0.7 + 0.8 + 0.9 + 1 \quad (4.12)$$

in which the elements of U represent probabilities. Suppose that we wish to compute the meaning of the value

$$x = highly\ unlikely \quad (4.13)$$

in which *highly* is defined as (see (3.18))

$$highly = minus\ very\ very \quad (4.14)$$

and

$$unlikely = not\ likely \quad (4.15)$$

with the meaning of the primary term *likely* given by

$$likely = 1/1 + 1/0.9 + 1/0.8 + 0.8/0.7$$
$$+ 0.6/0.6 + 0.5/0.5 + 0.3/0.4 + 0.2/0.3. \quad (4.16)$$

Using (4.15), we obtain

$$unlikely = 1/0 + 1/0.1 + 1/0.2 + 0.8/0.3 + 0.7/0.4$$
$$+ 0.5/0.5 + 0.4/0.6 + 0.2/0.7 \quad (4.17)$$

and hence

very very unlikely

$$= (unlikely)^4$$

$$\approx 1/0 + 1/0.1 + 1/0.2 + 0.4/0.3 + 0.2/0.4. \quad (4.18)$$

Finally, by (4.14)

highly unlikely

$$= minus\ very\ very\ unlikely$$

$$\approx (1/0 + 1/0.1 + 1/0.2 + 0.4/0.3 + 0.2/0.4)^{0.75}$$

$$\approx 1/0 + 1/0.1 + 1/0.2 + 0.5/0.3 + 0.3/0.4. \quad (4.19)$$

It should be noted that in computing the meaning of composite terms in the preceding examples we have made implicit use of the usual precedence rules governing the evaluation of Boolean expressions. With the addition of hedges, these precedence rules may be expressed as follows.

Precedence	Operation
First	*h, not*
Second	*and*
Third	*or*

As usual, parentheses may be used to change the precedence order and ambiguities may be resolved by the use of association to the right. Thus *plus very minus very tall* should be interpreted as

$$plus\ (very\ (minus\ (very\ (tall)))).$$

The technique that was employed for the computation of the meaning of a composite term is a special case of a more general approach which is described in [4] and [5]. The approach in question can be applied to the computation of the meaning of values of a linguistic variable provided the composite terms representing these values can be generated by a context-free grammar. As an illustration, consider a linguistic variable x whose values are exemplified by *small, not small, large, not large, very small, not very small, small or not very very large, small and (large or not small), not very very small and not very very large*, etc.

The values in question can be generated by a context-free grammar $G = (V_T, V_N, S, P)$ in which the set of terminals V_T comprises the atomic terms *small, large, not, and, or, very*, etc.; the nonterminals are denoted S, A, B, C, D, and E; and the production system is given by

$S \rightarrow A$	$C \rightarrow D$
$S \rightarrow S\ or\ A$	$C \rightarrow E$
$A \rightarrow B$	$D \rightarrow very\ D$
$A \rightarrow A\ and\ B$	$E \rightarrow very\ E$
$B \rightarrow C$	$D \rightarrow small$
$B \rightarrow not\ C$	$E \rightarrow large$
$C \rightarrow (S).$	(4.20)

Each production in (4.20) gives rise to a relation between the fuzzy sets labeled by the corresponding terminal and nonterminal symbols. In the case of (4.20), these relations are (we omit the productions which have no effect on the associated fuzzy sets)

$$S \rightarrow S\ or\ A \Rightarrow S_L = S_R + A_R$$
$$A \rightarrow A\ and\ B \Rightarrow A_L = A_R \cap B_R$$
$$B \rightarrow not\ C \Rightarrow B_L = \neg C_R$$
$$D \rightarrow very\ D \Rightarrow D_L = D_R^{\ 2}$$

$$E \to very\ E \Rightarrow E_L = E_R{}^2$$

$$D \to small \Rightarrow D_L = small$$

$$E \to large \Rightarrow E_L = large \qquad (4.21)$$

in which the subscripts L and R are used to differentiate between the symbols on the left- and right-hand sides of a production.

To compute the meaning of a composite term x, it is necessary to perform a syntactical analysis of x in terms of the specified grammar G. Then, knowing the syntax tree of x, one can employ the relations given in (4.21) to derive a set of equations (in triangular form) which upon solution yield the meaning of x. For example, in the case of the composite term

$$x = not\ very\ small\ and\ not\ very\ very\ large$$

the solution of these equations yields

$$x = (\neg small^2) \cap (\neg large^4) \qquad (4.22)$$

which agrees with (4.11). Details of this solution may be found in [4] and [5].

The ability to compute the meaning of values of a linguistic variable is a prerequisite to the computation of the meaning of fuzzy conditional statements of the form IF A THEN B, e.g., IF x is *not very small* THEN y is *very very large*. This problem is considered in the following section.

V. Fuzzy Conditional Statements and Compositional Rule of Inference

In classical propositional calculus,[5] the expression IF A THEN B, where A and B are propositional variables, is written as $A \Rightarrow B$, with the implication $\Rightarrow$ regarded as a connective which is defined by the truth table.

A	B	$A \Rightarrow B$
T	T	T
T	F	F
F	T	T
F	F	T

Thus,

$$A \Rightarrow B \equiv \neg A \lor B \qquad (5.1)$$

[5] A detailed discussion of the significance of implication and its role in modal logic may be found in [18].

in the sense that the propositional expressions $A \Rightarrow B$ (*A implies B*) and $\neg A \vee B$ (*not A or B*) have identical truth tables.

A more general concept, which plays an important role in our approach, is a *fuzzy conditional statement*: IF A THEN B or, for short, $A \Rightarrow B$, in which A (the antecedent) and B (the consequent) are fuzzy sets rather than propositional variables. The following are typical examples of such statements:

IF *large* THEN *small*
IF *slippery* THEN *dangerous*

which are abbreviations of the statements

IF x is *large* THEN y is *small*
IF the road is *slippery* THEN driving is *dangerous*.

In essence, statements of this form describe a relation between two fuzzy variables. This suggests that a fuzzy conditional statement be defined as a fuzzy relation in the sense of (2.19) rather than as a connective in the sense of (5.1).

To this end, it is expedient to define first the *Cartesian product* of two fuzzy sets. Specifically, let A be a fuzzy subset of a universe of discourse U, and let B be a fuzzy subset of a possibly different universe of discourse V. Then, the Cartesian product of A and B is denoted by $A \times B$ and is defined by

$$A \times B \triangleq \int_{U \times V} \mu_A(u) \wedge \mu_B(v)/(u,v) \qquad (5.2)$$

where $U \times V$ denotes the Cartesian product of the non-fuzzy sets U and V; that is,

$$U \times V \triangleq \{(u,v) \mid u \in U, v \in V\}.$$

Note that when A and B are nonfuzzy, (5.2) reduces to the conventional definition of the Cartesian product of non-fuzzy sets. In words, (5.2) means that $A \times B$ is a fuzzy set of ordered pairs (u,v), $u \in U$, $v \in V$, with the grade of membership of (u,v) in $A \times B$ given by $\mu_A(u) \wedge \mu_B(v)$. In this sense, $A \times B$ is a fuzzy relation from U to V.

As a very simple example, suppose that

$$U = 1 + 2 \qquad (5.3)$$

$$V = 1 + 2 + 3 \qquad (5.4)$$

$$A = 1/1 + 0.8/2 \qquad (5.5)$$

$$B = 0.6/1 + 0.9/2 + 1/3. \qquad (5.6)$$

Then

$$A \times B = 0.6/(1,1) + 0.9/(1,2) + 1/(1,3)$$
$$+ 0.6/(2,1) + 0.8/(2,2) + 0.8/(2,3). \quad (5.7)$$

The relation defined by (5.7) may be conveniently represented by the relation matrix

$$\begin{array}{ccc} & 1 & 2 & 3 \\ \begin{array}{c} 1 \\ 2 \end{array} & \left[\begin{array}{ccc} 0.6 & 0.9 & 1 \\ 0.6 & 0.8 & 0.8 \end{array} \right]. \end{array} \qquad (5.8)$$

The significance of a fuzzy conditional statement of the form IF A THEN B is made clearer by regarding it as a special case of the conditional expression IF A THEN B ELSE C, where A and (B and C) are fuzzy subsets of possibly different universes U and V, respectively. In terms of the Cartesian product, the latter statement is defined as follows:

$$\text{IF } A \text{ THEN } B \text{ ELSE } C \triangleq A \times B + (\neg A \times C) \quad (5.9)$$

in which $+$ stands for the union of the fuzzy relations $A \times B$ and $(\neg A \times C)$.

More generally, if $A_1, \cdots, A_n$ are fuzzy subsets of U, and $B_1, \cdots, B_n$ are fuzzy subsets of V, then[6]

IF A_1 THEN B_1 ELSE IF A_2 THEN $B_2 \cdots$ ELSE IF A_n THEN B_n

$$\triangleq A_1 \times B_1 + A_2 \times B_2 + \cdots + A_n \times B_n. \quad (5.10)$$

Note that (5.10) reduces to (5.9) if IF A THEN B ELSE C is interpreted as IF A THEN B ELSE IF $\neg A$ THEN C. It should also be noted that by repeated application of (5.9) we obtain

IF A THEN (IF B THEN C ELSE D) ELSE E

$$= A \times B \times C + A \times \neg B \times D + \neg A \times E. \quad (5.11)$$

If we regard IF A THEN B as IF A THEN B ELSE C with unspecified C, then, depending on the assumption made about C, various interpretations of IF A THEN B will result. In particular, if we assume that $C = V$, then IF A THEN B (or $A \Rightarrow B$) becomes[7]

---

[6] It should be noted that, in the sense used in ALGOL, the right-hand side of (5.10) would be expressed as $A_1 \times B_1 + (\neg A_1 \cap A_2) \times B_2 + \cdots + (\neg A_1 \cap \cdots \cap \neg A_{n-1} \cap A_n) \times B_n$ when the A_i and B_i, $i = 1, \cdots, n$, are nonfuzzy sets.

[7] This definition should be viewed as tentative in nature.

$A \Rightarrow B \triangleq$ IF A THEN $B \triangleq A \times B + (\neg A \times V)$.   (5.12)

If, in addition, we set $A = U$ in (5.12), we obtain as an alternative definition

$$A \Rightarrow B \triangleq U \times B + (\neg A \times V).$$ (5.13)

In the sequel, we shall assume that $C = V$, and hence that $A \Rightarrow B$ is defined by (5.12). In effect, the assumption that $C = V$ implies that, in the absence of an indication to the contrary, the consequent of $\neg A \Rightarrow C$ can be any fuzzy subset of the universe of discourse. As a very simple illustration of (5.12), suppose that A and B are defined by (5.5) and (5.6). Then, on substituting (5.8) in (5.12), the relation matrix for $A \Rightarrow B$ is found to be

$$A \Rightarrow B = \begin{bmatrix} 0.6 & 0.9 & 1 \\ 0.6 & 0.8 & 0.8 \end{bmatrix}.$$

It should be observed that when A, B, and C are non-fuzzy sets, we have the identity

IF A THEN B ELSE $C =$ (IF A THEN B) $\cap$ (IF $\neg A$ THEN C)

(5.14)

which holds only approximately for fuzzy A, B, and C. This indicates that, in relation to (5.15), the definitions of IF A THEN B ELSE C and IF A THEN B, as expressed by (5.9) and (5.12), are not exactly consistent for fuzzy A, B, and C. It should also be noted that if 1) $U = V$, 2) $x = y$, and 3) $A \Rightarrow B$ holds for all points in U, then, by (5.12),

$$A \Rightarrow B \quad \text{implies and is implied by} \quad A \subset B \quad (5.15)$$

exactly if A and B are nonfuzzy and approximately otherwise.

As will be seen in Section VI, fuzzy conditional statements play a basic role in fuzzy algorithms. More specifically, a typical problem which is encountered in the course of execution of such algorithms is the following. We have a fuzzy relation, say, R, from U to V which is defined by a fuzzy conditional statement. Then, we are given a fuzzy subset of U, say, x, and have to determine the fuzzy subset of V, say, y, which is induced in V by x. For example, we may have the following two statements.

1) x is *very small*
2) IF x is *small* THEN y is *large* ELSE y is *not very large*

of which the second defines by (5.9) a fuzzy relation R. The question, then, is as follows: What will be the value of y if x is *very small*? The answer to this question is provided by the following rule of inference, which may be regarded as an extension of the familiar rule of *modus ponens*.

Compositional Rule of Inference: If R is a fuzzy relation from U to V, and x is a fuzzy subset of U, then the fuzzy subset y of V which is induced by x is given by the composition (see (2.22)) of R and x; that is,

$$y = x \circ R \qquad (5.16)$$

in which x plays the role of a unary relation.[8]

As a simple illustration of (5.16), suppose that R and x are defined by the relation matrices in (5.17). Then y is given by the max–min product of x and R:

$$\begin{matrix} x & R & y \end{matrix}$$

$$[0.2 \quad 1 \quad 0.3] \circ \begin{bmatrix} 0.8 & 0.9 & 0.2 \\ 0.6 & 1 & 0.4 \\ 0.5 & 0.8 & 1 \end{bmatrix} = [0.6 \quad 1 \quad 0.4]. \quad (5.17)$$

As for the question raised before, suppose that, as in (4.3), we have

$$U = 1 + 2 + 3 + 4 + 5 \qquad (5.18)$$

with *small* and *large* defined by (4.2) and (4.7), respectively. Then, substituting *small* for A, *large* for B and *not very large* for C in (5.9), we obtain the relation matrix R for the fuzzy conditional statement IF *small* THEN *large* ELSE *not very large*. The result of the composition of R with $x = $ *very small* is

$$[1 \quad 0.64 \quad 0.36 \quad 0.16 \quad 0.04] \circ \begin{bmatrix} 0.2 & 0.4 & 0.6 & 0.8 & 1 \\ 0.2 & 0.4 & 0.6 & 0.8 & 0.8 \\ 0.4 & 0.4 & 0.6 & 0.6 & 0.6 \\ 0.6 & 0.6 & 0.6 & 0.4 & 0.4 \\ 0.8 & 0.8 & 0.64 & 0.36 & 0.2 \end{bmatrix}$$

$$= [0.36 \quad 0.4 \quad 0.6 \quad 0.8 \quad 1]. \quad (5.19)$$

[8] If R is visualized as a fuzzy graph, then (5.16) may be viewed as the expression for the fuzzy ordinate y corresponding to a fuzzy abscissa x.

There are several aspects of (5.16) that are in need of comment. First, it should be noted that when $R = A \Rightarrow B$ and $x = A$ we obtain

$$y = A \circ (A \Rightarrow B) = B \qquad (5.20)$$

as an exact identity, when A, B, and C are nonfuzzy, and an approximate one, when A, B, and C are fuzzy. It is in this sense that the compositional inference rule (5.16) may be viewed as an approximate extension of *modus ponens*. (Note that in consequence of the way in which $A \Rightarrow B$ is defined in (5.12), the more different x is from A, the less sharply defined is y.)

Second, (5.16) is analogous to the expression for the marginal probability in terms of the conditional probability function; that is

$$r_j = \sum_i q_i p_{ij} \qquad (5.21)$$

where

$$q_i = \Pr \{X = x_i\}$$
$$r_j = \Pr \{Y = y_j\}$$
$$p_{ij} = \Pr \{Y = y_j \mid X = x_i\}$$

and X and Y are random variables with values $x_1, x_2, \cdots$ and $y_1, y_2, \cdots$, respectively. However, this analogy does not imply that (5.16) is a relation between probabilities.

Third, it should be noted that because of the use of the max–min matrix product in (5.16), the relation between x and y is not continuous. Thus, in general, a small change in x would produce no change in y until a certain threshold is exceeded. This would not be the case if the composition of x with R were defined as max–product composition.

Fourth, in the computation of $x \circ R$ one may take advantage of the distributivity of composition over the union of fuzzy sets. Thus, if

$$x = u \text{ } or \text{ } v \qquad (5.22)$$

where u and v are labels of fuzzy sets, then

$$(u \text{ } or \text{ } v) \circ R = u \circ R \text{ } or \text{ } v \circ R. \qquad (5.23)$$

For example, if x is *small or medium*, and $R = A \Rightarrow B$ reads IF x is *not small and not large* THEN y is *very small*, then we can write

$(small\ or\ medium) \circ (not\ small\ and\ not\ large \Rightarrow very\ small)$
$= small \circ (not\ small\ and\ not\ large \Rightarrow very\ small)\ or\ medium$
$\circ (not\ small\ and\ not\ large \Rightarrow very\ small).$ (5.24)

As a final comment, it is important to realize that in practical applications of fuzzy conditional statements to the description of complex or ill-defined relations, the computations involved in (5.9), (5.10), and (5.16) would, in general, be performed in a highly approximate fashion. Furthermore, an additional source of imprecision would be the result of representing a fuzzy set as a value of a linguistic variable. For example, suppose that a relation between fuzzy variables x and y is described by the fuzzy conditional statement IF *small* THEN *large* ELSE IF *medium* THEN *medium* ELSE IF large THEN *very small*.

Typically, we would assign different linguistic values to x and compute the corresponding values of y by the use of (5.16). Then, on approximating to the computed values of y by linguistic labels, we would arrive at a table having the form shown below:

Given		Inferred	
A	*B*	*x*	*y*
small	large	not small	not very large
medium	medium	very small	very very large
large	very small	very very small	very very large
		not very large	small or medium

Such a table constitutes an approximate linguistic characterization of the relation between x and y which is inferred from the given fuzzy conditional statement. As was stated earlier, fuzzy conditional statements play a basic role in the description and execution of fuzzy algorithms. We turn to this subject in the following section.

VI. FUZZY ALGORITHMS

Roughly speaking, a fuzzy algorithm is an ordered set of fuzzy instructions which upon execution yield an approximate solution to a specified problem. In one form or another, fuzzy algorithms pervade much of what we do. Thus, we employ fuzzy algorithms both consciously and subconsciously when we walk, drive a car, search for an object, tie a knot, park a car, cook a meal, find a number in a telephone directory, etc. Furthermore, there are many instances of uses of what, in effect, are fuzzy algorithms in a wide variety of fields, especially in programming, operations research, psychology, management science, and medical diagnosis.

The notion of a fuzzy set and, in particular, the concept of a fuzzy conditional statement provide a basis for using fuzzy algorithms in a more systematic and hence more effective ways than was possible in the past. Thus, fuzzy algorithms could become an important tool for an approximate analysis of systems and decision processes which are much too complex for the application of conventional mathematical techniques.

A formal characterization of the concept of a fuzzy algorithm can be given in terms of the notion of a fuzzy Turing machine or a fuzzy Markoff algorithm [6]-[8]. In this section, the main aim of our discussion is to relate the concept of a fuzzy algorithm to the notions introduced in the preceding sections and illustrate by simple examples some of the uses of such algorithms.

The instructions in a fuzzy algorithm fall into the following three classes.

1) *Assignment Statements:* e.g.,

$x \approx 5$
$x = small$
x is *large*
x is *not large and not very small.*

2) *Fuzzy Conditional Statements:* e.g.,

IF x is *small* THEN y is *large* ELSE y is *not large*
IF x is positive THEN decrease y *slightly*
IF x is *much greater* than 5 THEN stop
IF x is *very small* THEN go to 7.

Note that in such statements either the antecedent or the consequent or both may be labels of fuzzy sets.

3) *Unconditional Action Statements:* e.g.,

multiply x by x
decrease x *slightly*
delete the first *few* occurrences of 1
go to 7
print x
stop.

Note that some of these instructions are fuzzy and some are not.

The combination of an assignment statement and a fuzzy conditional statement is executed in accordance with the compositional rule (5.16). For example, if at some point in the execution of a fuzzy algorithm we encounter the instructions

1) $x = $ *very small*
2) IF x is *small* THEN y is *large* ELSE y is *not very large*

where *small* and *large* are defined by (4.2) and (4.7), then the result of the execution of 1) and 2) will be the value of y given by (5.19), that is,

$$y = 0.36/1 + 0.4/2 + 0.64/3 + 0.8/4 + 1/5. \quad (6.1)$$

An unconditional but fuzzy action statement is executed similarly. For example, the instruction

$$\text{multiply } x \text{ by itself a } \textit{few} \text{ times} \quad (6.2)$$

with *few* defined as

$$few = 1/1 + 0.8/2 + 0.6/3 + 0.4/4 \quad (6.3)$$

would yield upon execution the fuzzy set

$$y = 1/x^2 + 0.8/x^3 + 0.6/x^4 + 0.4/x^5. \quad (6.4)$$

It is important to observe that, in both (6.1) and (6.4), the result of execution is a fuzzy set rather than a single number. However, when a human subject is presented with a fuzzy instruction such as "take *several* steps," with *several* defined by (see (2.10))

$$several = 0.5/3 + 0.8/4 + 1/5 + 1/6 + 0.8/7 + 0.5/8 \quad (6.5)$$

the result of execution must be a single number between 3 and 8. On what basis will such a number be chosen?

As pointed out in [6], it is reasonable to assume that the result of execution will be that element of the fuzzy set which has the highest grade of membership in it. If such an element is not unique, as is true of (6.5), then a random or arbitrary choice can be made among the elements having the highest grade of membership. Alternatively, an external criterion can be introduced which linearly orders those elements of the fuzzy set which have the highest membership, and thus generates a unique greatest element. For example, in the case of (6.5), if the external criterion is to minimize the number of steps that have to be taken, then the subject will pick 5 from the elements with the highest grade of membership.

An analogous question arises in situations in which a human subject has to give a "yes" or "no" answer to a fuzzy question. For example, suppose that a subject is presented with the instruction

$$\text{IF } x \text{ is } small \text{ THEN stop ELSE go to 7} \qquad (6.6)$$

in which *small* is defined by (4.2). Now assume that $x = 3$, which has the grade of membership of 0.6 in *small*. Should the subject execute "stop" or "go to 7"? We shall assume that in situations of this kind the subject will pick that alternative which is more true than untrue, e.g., "x is *small*" over "x is *not small*," since in our example the degree of truth of the statement "3 is *small*" is 0.6, which is greater than that of the statement "3 is *not small*." If both alternatives have more or less equal truth values, the choice can be made arbitrarily. For convenience, we shall refer to this rule of deciding between two alternatives as the *rule of the preponderant alternative*.

It is very important to understand that the questions just discussed arise only in those situations in which the result of execution of a fuzzy instruction is required to be a single element (e.g., a number) rather than a fuzzy set. Thus, if we allowed the result of execution of (6.6) to be fuzzy, then for $x = 3$ we would obtain the fuzzy set

$$0.6/\text{stop} + 0.4/\text{go to 7}$$

which implies that the execution is carried out in parallel. The assumption of parallelism is implicit in the compositional rule of inference and is basic to the understanding of fuzzy algorithms and their execution by humans and machines.

In what follows, we shall present several examples of fuzzy algorithms in the light of the concepts discussed in the preceding sections. It should be stressed that these examples are intended primarily to illustrate the basic aspects of fuzzy algorithms rather than demonstrate their effectiveness in the solution of practical problems.

It is convenient to classify fuzzy algorithms into several basic categories, each corresponding to a particular type of application: definitional and identificational algorithms; generational algorithms; relational and behavioral algorithms; and decisional algorithms. (It should be noted that an algorithm of a particular type can include algorithms of other types as subalgorithms. For example, a definitional algorithm may contain relational and decisional subalgorithms.) We begin with an example of a definitional algorithm.

Fuzzy Definitional Algorithms

One of the basic areas of application for fuzzy algorithms lies in the definition of complex, ill-defined or fuzzy concepts in terms of simpler or less fuzzy concepts. The following are examples of such fuzzy concepts: sparseness of matrices; handwritten characters; measures of complexity; measures of proximity or resemblance; degrees of clustering; criteria of performance; soft constraints; rules of various kinds, e.g., zoning regulations; legal criteria, e.g., criteria for insanity, obscenity, etc.; and fuzzy diseases such as arthritis, arteriosclerosis, schizophrenia.

Since a fuzzy concept may be viewed as a label for a fuzzy set, a *fuzzy definitional algorithm* is, in effect, a finite set of possibly fuzzy instructions which define a fuzzy set in terms of other fuzzy sets (and possibly itself, i.e., recursively) or constitute a procedure for computing the grade of membership of any element of the universe of discourse in the set under definition. In the latter case, the definational algorithm plays the role of an *identificational algorithm*, that is, an algorithm which identifies whether or not an element belongs to a set or, more generally, determines its grade of membership. An example of such an algorithm is provided by the procedure (see [5]) for computing the grade of membership of a string in a fuzzy language generated by a context-free grammar.

As a very simple example of a fuzzy definitional algorithm, we shall consider the fuzzy concept *oval*. It should be emphasized again that the oversimplified definition that will be given is intended only for illustrative purposes and has no pretense at being an accurate definition of the concept *oval*. The instructions comprising the algorithm OVAL are listed here. The symbol T in these instructions stands for the object under test. The term CALL CONVEX represents a call on a subalgorithm labeled CONVEX, which is a definitional algorithm for testing whether or not T is convex. An instruction of the form IF A THEN B should be interpreted as IF A THEN B ELSE go to next instruction.

Algorithm OVAL:

1) IF T is not closed THEN T is not *oval*; stop.
2) IF T is self-intersecting THEN T is not *oval*; stop.
3) IF T is not CALL CONVEX THEN T is not *oval*; stop.
4) IF T does not have two *more or less* orthogonal axes of symmetry THEN T is not *oval*; stop.

5) IF the major axis of T is not *much* longer than the minor axis THEN T is not *oval*; stop.

6) T is *oval*; stop.

Subalgorithm CONVEX: Basically, this subalgorithm involves a check on whether the curvature of T at each point maintains the same sign as one moves along T in some initially chosen direction.

1) $x = a$ (some initial point on T).

2) Choose a direction of movement along T.

3) $t \approx$ direction of tangent to T at x.

4) $x' \approx x + 1$ (move from x to a neighboring point).

5) $t' \approx$ direction of tangent to T at x'.

6) $\alpha \approx$ angle between t' and t.

7) $x \approx x'$.

8) $t \approx$ direction of tangent to T at x.

9) $x' \approx x + 1$.

10) $t' \approx$ direction of tangent to T at x'.

11) $\beta \approx$ angle between t' and t.

12) IF β does not have the same sign as α THEN T is not convex; return.

13) IF $x' \approx a$ THEN T is convex; return.

14) Go to 7).

Comment: It should be noted that the first three instructions in OVAL are nonfuzzy. As for instructions 4) and 5), they involve definitions of concepts such as "*more or less* orthogonal," and "*much* longer," which, though fuzzy, are less complex and better understood than the concept of *oval*. This exemplifies the main function of a fuzzy definitional algorithm, namely, to reduce a new or complex fuzzy concept to simpler or better understood fuzzy concepts. In a more elaborate version of the algorithm OVAL, the answers to 4) and 5) could be the degrees to which the conditions in these instructions are satisfied. The final result of the algorithm, then, would be the grade of membership of T in the fuzzy set of oval objects.

In this connection, it should be noted that, in virtue of (5.15), the algorithm OVAL as stated is approximately equivalent to the expression

oval = closed $\cap$ non-self-intersecting $\cap$ convex

$\cap$ *more or less* orthogonal axes of symmetry

$\cap$ major axis *much* larger than minor axis   (6.7)

which defines the fuzzy set *oval* as the intersection of the fuzzy and nonfuzzy sets whose labels appear on the right-hand side of (6.7). However, one significant difference is that the algorithm not only defines the right-hand side of (6.7), but also specifies the order in which the computations implicit in (6.7) are to be performed.

Fuzzy Generational Algorithms

As its designation implies, a fuzzy generational algorithm serves to generate rather than define a fuzzy set. Possible applications of generational algorithms include: generation of handwritten characters and patterns of various kinds; cooking recipes; generation of music; generation of sentences in a natural language; generation of speech.

As a simple illustration of the notion of a generational algorithm, we shall consider an algorithm for generating the letter **P**, with the height h and the base b of **P** constituting the parameters of the algorithm. For simplicity, **P** will be generated as a dotted pattern, with eight dots lying on the vertical line.

Algorithm **P**(h,b):

1) $i = 1$.
2) $X(i) = b$ (first dot at base).
3) $X(i + 1) \approx X(i) + h/6$ (put dot *approximately* $h/6$ units of distance above $X(i)$).
4) $i = i + 1$.
5) IF $i = 7$ THEN make right turn and go to 7).
6) Go to 3.
7) Move by $h/6$ units; put a dot.
8) Turn by $45°$; move by $h/6$ units; put a dot.
9) Turn by $45°$; move by $h/6$ units; put a dot.
10) Turn by $45°$; move by $h/6$ units; put a dot.
11) Turn by $45°$; move by $h/6$ units; put a dot; stop.

The algorithm as stated is of open-loop type in the sense that it does not incorporate any feedback. To make the algorithm less sensitive to errors in execution, we could introduce fuzzy feedback by conditioning the termination of the algorithm on an approximate satisfaction of a specified test. For example, if the last point in step 11) does not fall on the vertical part of **P**, we could return to step 8) and either reduce or increase the angle of turn in steps 8)–11) to correct for the terminal error. The flowchart of a cooking recipe for chocolate fudge (Fig. 3), which is

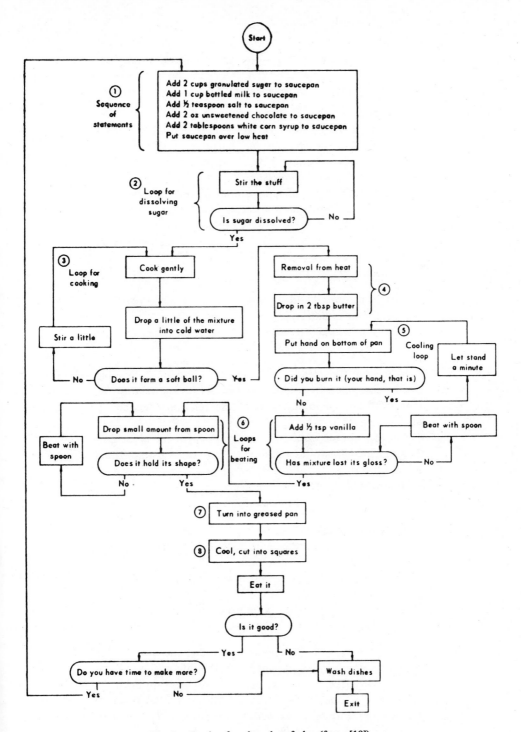

Fig. 3.   Recipe for chocolate fudge (from [19]).

reproduced from [19], is a good example of what, in effect, is a fuzzy generational algorithm with feedback.

Fuzzy Relational and Behavioral Algorithms

A *fuzzy relational algorithm* serves to describe a relation or relations between fuzzy variables. A relational algorithm which is used for the specific purpose of approximate description of the behavior of a system will be referred to as a *fuzzy behavioral algorithm.*

A simple example of a relational algorithm labeled R which involves three parameters x, y, and z is given. This algorithm defines a fuzzy ternary relation R in the universe of discourse $U = 1 + 2 + 3 + 4 + 5$ with *small* and *large* defined by (4.2) and (4.7).

Algorithm $R(x,y,z)$:

1) IF x is *small* and y is *large* THEN z is *very small* ELSE z is *not small.*
2) IF x is *large* THEN (IF y is *small* THEN z is *very large* ELSE z is *small*) ELSE z and y are *very very small.*

If needed, the meaning of these conditional statements can be computed by using (5.9) and (5.11). The relation R, then, will be the intersection of the relations defined by instructions 1) and 2).

Another simple example of a relational fuzzy algorithm $F(x,y)$ which illustrates a different aspect of such algorithms is the following.

Algorithm $F(x,y)$:

1) IF x is *small* and x is increased *slightly* THEN y will increase *slightly.*
2) IF x is *small* and x is increased *substantially* THEN y will increase *substantially.*
3) IF x is *large* and x is increased *slightly* THEN y will increase *moderately.*
4) IF x is *large* and x is increased *substantially* THEN y will increase *very substantially.*

As in the case of the previous example, the meaning of the fuzzy conditional statements in this algorithm can be computed by the use of the methods discussed in Sections IV and V if one is given the definitions of the primary terms *large* and *small* as well as the hedges *slightly, substantially,* and *moderately.*

As a simple example of a behavioral algorithm, suppose

that we have a system S with two nonfuzzy states (see [3]) labeled q_1 and q_2, two fuzzy input values labeled *low* and *high*, and two fuzzy output values labeled *large* and *small*. The universe of discourse for the input and output values is assumed to be the real line. We assume further that the behavior of S can be characterized in an approximate fashion by the algorithm that will be given. However, to represent the relations between the inputs, states, and outputs, we use the conventional state transition tables instead of conditional statements.

Algorithm BEHAVIOR:

u_t \ x_t	x_{t+1}		y_t	
	q_1	q_2	q_1	q_2
low	q_2	q_1	large	small
high	q_1	q_1	small	large

where

u_t input at time t

y_t output at time t

x_t state at time t.

On the surface, this table appears to define a conventional nonfuzzy finite-state system. What is important to recognize, however, is that in the case of the system under consideration the inputs and outputs are fuzzy subsets of the real line. Thus we could pose the question: What would be the output of S if it is in state q_1 and the applied input is *very low*? In the case of S, this question can be answered by an application of the compositional inference rule (5.16). On the other hand, the same question would not be a meaningful one if S is assumed to be a nonfuzzy finite-state system characterized by the preceding table.

Behavioral fuzzy algorithms can also be used to describe the more complex forms of behavior resulting from the presence of random elements in a system. For example, the presence of random elements in S might result in the following fuzzy-probabilistic characterization of its behavior:

u_t \ x_t	x_{t+1}		y_t	
	q_1	q_2	q_1	q_2
low	q_2 *likely*	q_1 *likely*	large likely	small likely²
high	q_1 *likely²*	q_1 *unlikely²*	small likely²	large unlikely²

In this table, the term *likely* and its modifications by *very* and *not* serve to provide an approximate characterization of probabilities. For example, IF the input is *low* and the present state is q_1, THEN the next state is *likely* to be q_2. Similarly, IF the input is *high* and the present state is q_2 THEN the output is *very unlikely* to be *large*. If the meaning of *likely* is defined by (see (4.16))

$$likely = 1/1 + 1/0.9 + 1/0.8 + 0.8/0.7 + 0.6/0.6$$
$$+ 0.5/0.5 + 0.3/0.4 + 0.2/0.3 \quad (6.8)$$

then

$$unlikely = 0.2/0.7 + 0.4/0.6 + 0.5/0.5 + 0.7/0.4$$
$$+ 0.8/0.3 + 1/0.2 + 1/0.1 + 1/0 \quad (6.9)$$
$$very\ likely \approx 1/1 + 1/0.9 + 1/0.8 + 0.6/0.7 + 0.4/0.6$$
$$+ 0.3/0.5 + 0.1/0.4 \quad (6.10)$$
$$very\ unlikely \approx 0.2/0.6 + 0.3/0.5 + 0.5/0.4 + 0.6/0.3$$
$$+ 1/0.2 + 1/0.1 + 1/0. \quad (6.11)$$

Fuzzy Decisional Algorithms

A *fuzzy decisional algorithm* is a fuzzy algorithm which serves to provide an approximate description of a strategy or decision rule. Commonplace examples of such algorithms, which we use for the most part on a subconscious level, are the algorithms for parking a car, crossing an intersection, transferring an object, buying a house, etc.

To illustrate the notion of a fuzzy decisional algorithm, we shall consider two simple examples drawn from our everyday experiences.

Example—Crossing a traffic intersection: It is convenient to break down the algorithm in question into several subalgorithms, each of which applies to a particular type of intersection. For our purposes, it will be sufficient to describe only one of these subalgorithms, namely, the subalgorithm SIGN, which is used when the intersection has a stop sign. As in the case of other examples in this section, we shall make a number of simplifying assumptions in order to shorten the description of the algorithm.

Algorithm INTERSECTION:

1) IF signal lights THEN CALL SIGNAL ELSE IF stop sign THEN CALL SIGN ELSE IF blinking light THEN CALL BLINKING ELSE CALL UNCONTROLLED.

Subalgorithm SIGN:

1) IF no stop sign on your side THEN IF no cars in the intersection THEN cross at *normal* speed ELSE wait for cars to leave the intersection and then cross.
2) IF not *close* to intersection THEN continue approaching at normal speed for a *few* seconds; go to 2).
3) *Slow down.*
4) IF in a *great* hurry and no police cars in sight and no cars in the intersection or its *vicinity* THEN cross the intersection at *slow* speed.
5) IF *very close* to intersection THEN stop; go to 7).
6) Continue *approaching* at *very slow* speed; go to 5).
7) IF no cars *approaching* or in the intersection THEN cross.
8) Wait a *few* seconds; go to 7).

It hardly needs saying that a realistic version of this algorithm would be considerably more complex. The important point of the example is that such an algorithm could be constructed along the same lines as the highly simplified version just described. Furthermore, it shows that a fuzzy algorithm could serve as an effective means of communicating know-how and experience.

As a final example, we consider a decisional algorithm for transferring a blindfolded subject H from an initial position *start* to a final position *goal* under the assumption that there may be an obstacle lying between *start* and *goal* (see Fig. 4). (Highly sophisticated nonfuzzy algorithms of this type for use by robots are incorporated in Shakey, the robot built by the Artificial Intelligence Group at Stanford Research Institute. A description of this robot is given in [20].)

The algorithm, labeled OBSTACLE, is assumed to be used by a human controller C who can observe the way in which H executes his instructions. This fuzzy feedback plays an essential role in making it possible for C to direct H to *goal* in spite of the fuzziness of instructions as well as the errors in their execution by H. The algorithm OBSTACLE consists of three subalgorithms: ALIGN, HUG, and STRAIGHT. The function of STRAIGHT is to transfer H from *start* to an intermediate goal *I-goal$_1$*, and then from *I-goal$_2$* to *goal*. (See Fig. 4.) The function of ALIGN is to orient H in a desired direction; the function of HUG is to guide H along the boundary of the obstacle until the goal is no longer obstructed.

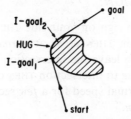

Fig. 4. Problem of transferring blindfolded subject from *start* to *goal*.

Instead of describing these subalgorithms in terms of fuzzy conditional statements as we have done in previous examples, it is instructive to convey the same information by flowcharts, as shown in Figs. 5–7. In the flowchart of ALIGN, ε denotes the error in alignment, and we assume for simplicity that ε has a constant sign. The flowcharts of HUG and STRAIGHT are self-explanatory. Expressed in terms of fuzzy conditional statements, the flowchart of STRAIGHT, for example, translates into the following instructions.

Subalgorithm STRAIGHT:

1) IF not *close* THEN take a step; go to 1).
2) IF not *very close* THEN take a *small* step; go to 2).
3) IF not *very very close* THEN take a *very small* step; go to 3).
4) Stop.

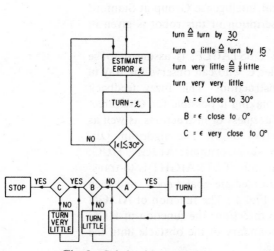

Fig. 5.   Subalgorithm ALIGN.

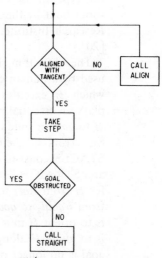

Fig. 6.   Subalgorithm HUG.

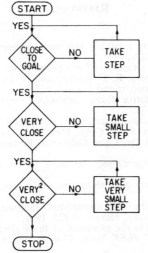

Fig. 7.   Subalgorithm STRAIGHT.

VII. CONCLUDING REMARKS

In this and the preceding sections of this paper, we have attempted to develop a conceptual framework for dealing with systems which are too complex or too ill-defined to admit of precise quantitative analysis. What we have done should be viewed, of course, as merely a first tentative step in this direction. Clearly, there are many basic as well as detailed aspects of our approach which we have treated incompletely, if at all. Among these are questions relating to the role of fuzzy feedback in: the execution of fuzzy algorithms; the execution of fuzzy algorithms by humans; the conjunction of fuzzy instructions; the assessment of the goodness of fuzzy algorithms; the implications of the compositional rule of inference and the rule of the preponderant alternative; and the interplay between fuzziness and probability in the behavior of humanistic systems.

Nevertheless, even at its present stage of development, the method described in this paper can be applied rather effectively to the formulation and approximate solution of a wide variety of practical problems, particularly in such fields as economics, management science, psychology, linguistics, taxonomy, artificial intelligence, information retrieval, medicine, and biology. This is particularly true of those problem areas in these fields in which fuzzy algorithms can be drawn upon to provide a means of description of ill-defined concepts, relations, and decision rules.

REFERENCES

[1] L. A. Zadeh, "Fuzzy sets," *Inform. Contr.*, vol. 8, pp. 338–353, 1965.

[2] ——, "Similarity relations and fuzzy orderings," *Inform. Sci.*, vol. 3, pp. 177–200, 1971.

[3] ——, "Toward a theory of fuzzy systems," in *Aspects of Network and System Theory*, R. E. Kalman and N. DeClaris, Eds. New York: Holt, Rinehart and Winston, 1971.

[4] ——, "Quantitative fuzzy semantics," *Inform. Sci.*, vol. 3, pp. 159–176, 1971.

[5] ——, "Fuzzy languages and their relation to human and machine intelligence," in *Proc. Conf. Man and Computer*, 1970; also Electron. Res. Lab., Univ. California, Berkeley, Memo. M-302, 1971.

[6] ——, "Fuzzy algorithms," *Inform. Contr.*, vol. 12, pp. 94–102, 1968.

[7] E. Santos, "Fuzzy algorithms," *Inform. Contr.*, vol. 17, pp. 326–339, 1970.

[8] L. A. Zadeh, "On fuzzy algorithms," Electron. Res. Lab., Univ. California, Berkeley, Memo. M-325, 1971.

[9] S.-K. Chang, "On the execution of fuzzy programs using finite-state machines," *IEEE Trans. Comput.*, vol. C-21, pp. 241–253, Mar. 1972.

[10] S. S. L. Chang and L. A. Zadeh, "Fuzzy mapping and control," *IEEE Trans. Syst., Man, Cybern.*, vol. SMC-2, pp. 30–34, Jan. 1962.

[11] R. E. Bellman and L. A. Zadeh, "Decision-making in a fuzzy environment," *Management Sci.*, vol. 17, pp. B-141–B-164, 1970.

[12] J. A. Goguen, "The logic of inexact concepts," *Syn.*, vol. 19, pp. 325–373, 1969.

[13] G. Lakoff, "Hedges: a study in meaning criteria and the logic of fuzzy concepts," in *Proc. 8th Reg. Meet. Chicago Linguist. Soc.*, 1972.

[14] L. A. Zadeh, "A system-theoretic view of behavior modification," Electron. Res. Lab., Univ. California, Berkeley, Memo. M-320, 1972.

[15] ——, "A fuzzy-set-theoretic interpretation of hedges," Electron. Res. Lab., Univ. California, Berkeley, Memo. M-335, 1972.

[16] A. De Luca and S. Termini, "A definition of a non-probabilistic entropy in the setting of fuzzy sets theory," *Inform. Contr.*, vol. 20, pp. 301–312, 1972.

[17] R. C. T. Lee, "Fuzzy logic and the resolution principle," *J. Ass. Comput. Mach.*, vol. 19, pp. 109–119, 1972.

[18] G. E. Hughes and M. J. Cresswell, *An Introduction to Modal Logic*. London: Methuen, 1968.

[19] R. S. Ledley, *Fortran IV Programming*. New York: McGraw-Hill, 1966.

[20] B. Raphael, R. Duda, R. E. Fikes, P. E. Hart, N. Nilsson, P. W. Thorndyke, and B. M. Wilbur, "Research and applications—artificial intelligence," Stanford Res. Inst., Menlo Park, Calif., Final Rep., Oct. 1971.

A Fuzzy-Algorithmic Approach to the Definition of Complex or Imprecise Concepts†

L.A. ZADEH‡

Computer Science Division, Department of Electrical Engineering and Computer Sciences and the Electronics Research Laboratory, University of California, Berkeley, California 94720. U.S.A.

(Received 3 January 1975)

It may be argued, rather persuasively, that most of the concepts encountered in various domains of human knowledge are, in reality, much too complex to admit of simple or precise definition. This is true, for example, of the concepts of recession and utility in economics; schizophrenia and arthritis in medicine; stability and adaptivity in system theory; sparseness and stiffness in numerical analysis; grammaticality and meaning in linguistics; performance measurement and correctness in computer science; truth and causality in philosophy; intelligence and creativity in psychology; and obscenity and insanity in law.

The approach described in this paper provides a framework for the definition of such concepts through the use of fuzzy algorithms which have the structure of a branching questionnaire. The starting point is a relational representation of the *definiendum* as a composite question whose constituent questions are either attributional or classificational in nature. The constituent questions as well as the answers to them are allowed to be fuzzy, e.g. the response to: "How large is x?" might be *not very large*, and the response to "Is x large?" might be *quite true*.

By putting the relational representation into an algebraic form, one can derive a fuzzy relation which defines the meaning of the definiendum. This fuzzy relation, then, provides a basis for an interpolation of the relational representation.

To transform a relational representation into an efficient branching questionnaire, the tableau of the relation is subjected to a process of compactification which identifies the conditionally redundant questions. From a maximally compact representation, various efficient realizations which have the structure of a branching questionnaire, with each realization corresponding to a prescribed order of asking the constituent questions, can readily be determined. Then, given the cost of constituent questions as well as the conditional probabilities of answers to them, one can compute the average cost of deducing the answer to the composite question. In this way, a relational representation of a concept leads to an efficient branching questionnaire which may serve as its operational definition.

†To Richard Bellman.

‡This work was supported in part by the Naval Electronics Systems Command under contract N00039-76-C-0022, the Army Research Office, Durham, N.C., under Grant DAHC04-75-G0056 and the National Science Foundation under Grant ENG74-A01. Some of the results were obtained while the author was a visiting member of the International Institute for Applied Systems Analysis in Vienna, Austria.

1. Introduction

The high standards of precision which prevail in mathematics, physics, chemistry, engineering and other "hard" sciences stand in sharp contrast to the imprecision which pervades much of sociology, psychology, political science, history, philosophy, linguistics, anthropology, literature, art and related fields. This marked difference in the standards of precision is due, of course, to the fact that the "hard" sciences are concerned in the main with the relatively simple mechanistic systems whose behavior can be described in quantitative terms, whereas the "soft" sciences deal primarily with the much more complex non-mechanistic systems in which human judgment, perception and emotions play the dominant role.

Although the conventional mathematical techniques have been and will continue to be applied to the analysis of humanistic[†] systems, it is clear that the great complexity of such systems calls for approaches that are significantly different in spirit as well as in substance from the traditional methods—methods which are highly effective when applied to mechanistic systems, but are far too precise in relation to systems in which human behavior plays an important role.

In the *linguistic approach* (Zadeh, 1973, 1975a) which represents one such departure from conventional methods—words or sentences are used in place of numbers to describe phenomena which are too complex or too ill-defined to be susceptible of characterization in quantitative terms. For example, if the probability of an event is not known with precision, then it may be characterized linguistically as, say, *quite likely, not very unlikely, highly unlikely*, etc., where *quite likely, not very unlikely* and *highly unlikely* are interpreted as labels of fuzzy subsets of the unit interval.[‡] Such subsets may be likened to ball-parks without sharply defined boundaries which serve to provide an approximate rather than exact characterization of the value of a variable.

The use of the linguistic approach in the case of humanistic systems is dictated by the fact that as the complexity of a system increases, our ability to make precise and yet significant statements about its behavior diminishes until a threshold is reached beyond which complexity, precision and significance can no longer coexist. The essence of the linguistic approach, then, is that it sacrifices precision to gain significance, thereby making it possible to analyze in an approximate manner those humanistic as well as mechanistic systems which are too complex for the application of classical techniques.

A key feature of the linguistic approach has to do with its use of the notion of a *primary fuzzy set* as a substitute for the basic notion of a *unit of measurement.*[§] More specifically, much of the power of mathematical techniques for dealing with mechanistic systems

---

[†] By a humanistic system we mean a non-mechanistic system in which human behavior plays a major role. Examples of humanistic systems are political systems, economic systems, social systems, religious systems, etc. A single individual and his thought processes may also be viewed as a humanistic system.

[‡] As a fuzzy subset of the unit interval, *quite likely* would be characterized by its compatibility or, equivalently, membership function $\mu_{quite\ likely}: [0,1] \rightarrow [0,1]$. Thus, $\mu_{quite\ likely}(0.8) = 0.9$ means that if the probability of an event is 0.8, then the degree to which 0.8 is compatible with *quite likely* is 0.9. Additional details may be found in the Appendix.

[§] A thorough discussion of the concept of a unit of measurement may be found in Krantz, Luce, Suppes & Twersky (1971).

derives from the existence of a set of units for such basic parameters as length, area, weight, force, current, heat, etc. In general, such units do not exist in the case of humanistic systems, and it is this fact that contributes significantly to the difficulty of analyzing humanistic systems through the use of techniques which depend so essentially on the existence of units of measurement.

In the linguistic approach, a role comparable to that of a unit of measurement is played by one or more primary fuzzy sets from which other sets can be generated through the use of linguistic modifiers such as *very, quite, more or less, extremely, essentially, completely,* etc. To illustrate, consider a property,† say *beautiful,* for which we have neither a unit nor a numerical scale. The meaning of this property may be defined via *exemplification* by associating with each member, *u,* of a subset of objects in a given universe of discourse, U, the grade of membership of *u* in the fuzzy subset labeled *beautiful.* For example, the grade of membership of Fay in the class of beautiful women might be 0·9, that of Jillian 0·85, of Helen 0·8, etc. This set of women, then, would constitute a primary fuzzy set which serves as a reference for defining the meaning of *very beautiful, quite beautiful, more or less beautiful, extremely beautiful,* etc. as fuzzy subsets of U.‡ Thus, in terms of these subsets, an assertion of the form "*Nora is very beautiful*", may be interpreted as the assignment of a linguistic rather than a numerical value to the beauty of Nora. In this way, the linguistic values *beautiful, very beautiful, quite beautiful,* etc. which are generated from the primary fuzzy set *beautiful,* play a role which is roughly similar to that of the multiples of a unit of measurement, when such a unit exists.

Our main purpose in the present paper is to apply the linguistic approach to the definition of concepts which are too complex or too imprecise to be susceptible of exact definition. In general, such concepts are fuzzy in the sense that they correspond to classes of objects or constructs which do not have sharply defined boundaries. For example, the concepts of *oval, in love, young* and *masculine* are fuzzy whereas those of *straight line, married, brother* and *male* are not. Note that *oval* is a more complex concept than *straight line, in love* is more complex than *married, friend* is more complex than *brother,* and *masculine* is more complex than *male.* Indeed, most complex concepts tend to be fuzzy, and it is in this sense that fuzziness may be regarded as a concomitant of complexity.

Note 1.1. In most cases, the question of whether a concept is fuzzy or not may be resolved by examining the applicability of a simple modifier such as *very* to the concept in question. Thus, for example, *very* is applicable to *masculine* but not to *male.* Similarly, *very ill,* where *ill* is a fuzzy concept, is acceptable whereas *very dead* is not. Also, *very much*

---

†At this point we do not differentiate between a property (intension) and the set which it defines (extension). For a discussion of this and other issues relating to concepts, meaning and vagueness see: Carnap (1956), Hempel (1952), Church (1951), Quine (1953), Frege (1952), Martin (1963), Black (1963), Goguen (1969), Fine (1973), van Frassen (1969), Lakoff (1971), Tarski (1956), Scriven (1958), Simon & Siklossy (1972), Hintikka, Moravcsik & Suppes (1973), Minsky (1968), Lukasiewicz (1970), Moisil (1972), and Domotor (1969).

‡The computation of the meaning of a term of the form *mu,* where *m* is a modifier and *u* is a primary term (i.e. a label for a primary fuzzy set) is discussed in Zadeh (1972*a, b*), Lakoff (1972), and, more briefly, in the Appendix.

greater is acceptable (*much greater* is fuzzy), while *very greater* (*greater* is non-fuzzy) is not.

How can a fuzzy concept be defined? The conventional approaches are: (a) giving a dictionary type of definition; (b) writing an essay; and (c) approximating to a fuzzy concept by a non-fuzzy concept and giving a precise definition for the latter. To illustrate, a typical dictionary definition of a fuzzy concept such as *democracy* might read, "A form of government in which the supreme power is vested in the people and exercised by them or by their elected agents under a free electoral system," while a more detailed definition might occupy a chapter in a text on political science. A typical example of (c) is the definition of a *recession* (Silk, 1974; Clark, 1974) as a condition which obtains when the gross national product declines in two successive quarters. In this case, what is in reality a fuzzy concept is defined as one which is both non-fuzzy and simple to understand. The price, of course, is a definition that is oversimplified to a point of uselessness.

An alternative and more systematic approach which is described in the sequel is based on the notion of a fuzzy algorithm (Zadeh, 1968; Santos, 1970; Zadeh, 1971a), that is, an algorithm (or a program or a decision table) in which some of the steps involve the execution of fuzzy instructions, which in turn may require the verification of fuzzy conditions. More specifically, in the *fuzzy-algorithmic* approach the definition of a fuzzy concept F is expressed as a fuzzy recognition algorithm† which acts on a given object u and upon execution yields the degree to which u is compatible with F or, equivalently, the grade of membership of u in the fuzzy set labeled F.

As an illustration, suppose that the concept of an economic recession is defined by a fuzzy algorithm labeled RECESSION. Then, acting on relevant economic data, RECESSION would yield the degree—expressed numerically, e.g., 0·8, or linguistically, e.g. *very true*—to which the data in question are compatible with the concept of *recession* as defined by the algorithm. Similarly, a fuzzy-algorithmic definition of a disease, say *arthritis*, would yield the degree to which a given patient belongs to the class of arthritics. Similarly, a fuzzy-algorithmic definition of the concept of *sparseness* would yield the degree to which a given matrix is sparse. And so on.

As will be seen in the following sections, a fuzzy-algorithmic definition has the form of a branching questionnaire, Q, in which both the questions and the answers are allowed to be fuzzy in nature. For example, to a question such as "Is Valentina *tall*?" (which will be abbreviated as *tall*?) the answer might be "*quite tall*", which may be viewed as being equivalent to the assignment of the linguistic value *quite high* to the grade of membership of Valentina in the class of tall people.

A question, Q_i, in Q may be either *classificational* or *attributional*. In the case of classificational questions, Q_i is concerned with the grade of membership of the subject in a fuzzy set F_i, or, equivalently, with the truth-value of the predicate‡ which corresponds to F_i. For example, Q_i may be "Is Rahim *honest*?" An answer such as *very high*

---

†A recognition algorithm is essentially an algorithmic representation of the membership function of a fuzzy set.

‡The term *predicate* (or, more generally, *fuzzy predicate*) as used here is essentially synonomous with the *membership* (or *compatibility*) *function*. To simplify the notation, the label of a predicate and the label of the set which it defines will be used interchangeably.

would mean that the grade of membership of the subject in the class of *honest* people is *very high*. Equivalently, an answer of the form *very true* would be interpreted as the assignment of the truth-value *very true* to the predicate labeled *honest* evaluated at $x \triangleq$ Rahim. §

In the case of attributional questions, Q_i relates to the value of an attribute of the subject. For example, an instance of Q_i may be "How *old* is Norman?" with the answer being either numerical, e.g. 24 or linguistic, e.g. *quite young*. Thus, in this case the answer may be viewed as the assignment of either a numerical or a linguistic value to an attribute of the subject.

The totality of the questions in Q constitutes a *basis* for Q, or, more specifically, the fuzzy concept defined by Q. If all of the questions in Q are classificational in nature, then the basis for Q defines a collection of fuzzy sets each of which corresponds to a question in Q. In this case, the questionnaire may be viewed as a way of defining the fuzzy set corresponding to Q in terms of the fuzzy sets corresponding to the questions in Q. As a simple illustration, if the predicate *big* is defined as the conjunction of the predicates *long*, *wide*, and *tall*, i.e.,

$$big = long \text{ and } wide \text{ and } tall \tag{1.2}$$

then Q_1, Q_2 and Q_3 may be expressed (in abbreviated form) as

$$Q_1 \triangleq long? \tag{1.3}$$

$$Q_2 \triangleq wide? \tag{1.4}$$

$$Q_3 \triangleq tall? \tag{1.5}$$

and (1.2) is equivalent to

$$big = long \cap wide \cap tall \tag{1.6}$$

where *big*, *long*, *wide* and *tall* are interpreted as the fuzzy sets corresponding to Q, Q_1, Q_2 and Q_3, respectively, and the intersection is defined in the fuzzy-set-theoretic sense. Thus, (1.6) expresses the fuzzy set *big* as a function of the fuzzy sets *long*, *wide* and *tall*, which implies that from the knowledge of the answers to Q_1, Q_2 and Q_3 one can determine the grade of membership of the object under test in the fuzzy set *big*. For example, if the answers to specific instances of Q_1, Q_2 and Q_3 are *true*, *very true* and *very true*, respectively, then from (1.6) it follows that the answer to the question *big?* is *true*. A more detailed discussion of this aspect of fuzzy-algorithmic definitions will be presented in section 3.

By their nature, fuzzy-algorithmic definitions are best suited for the characterization of concepts which are *intrinsically fuzzy*, that is, fuzzy to a degree which makes it unrealistic to approximate to them by non-fuzzy concepts. For example, in law, *insanity* and *obscenity* are intrinsically fuzzy concepts whereas *perjury* is not. Similarly, in system

---

§The symbol $\triangleq$ stands for *denotes* or *is defined to be* or *is equal by definition*.

theory the concepts of *large-scale*, *reliable* and *adaptive* are intrinsically fuzzy, whereas those of *observability* and *controllability* are not. In numerical analysis, the concept of a *sparse* matrix is intrinsically fuzzy while that of a *bounded error* is not. In medicine, most degenerative diseases are intrinsically fuzzy while the infectious diseases, for the most part, are not.

In addition to the intrinsically fuzzy concepts, there are many concepts in various fields which though fuzzy in nature are at present defined in non-fuzzy terms, largely because of a lack of alternative modes of definition. This is true, for example, of the concepts of *recession* and *equilibrium* in economics; *complexity* and *approximation* in mathematics; *structured programming* and *correctness* in computer science; *stability* and *linearity* in system theory; *arthritis* and *hypertension* in medicine, etc. It is very likely that, in time, the use of fuzzy-algorithmic techniques for the characterization of such concepts will become a fairly common practice.

In what follows, our discussion of fuzzy-algorithmic definitions will begin with the notion of an *atomic* question. This notion will serve as a basis for the definition of a *composite* question, which is turn will lead to the concept of a fuzzy-algorithmic *branching questionnaire*. In order to make the discussion self-contained, a brief summary of the relevant aspects of the linguistic approach is presented in the Appendix.

2. Atomic questions

Our focus of attention in this section is the concept of what might be called an *atomic* question, that is, a question which has no constituents other than itself. By contrast, a *composite* question—as its name implies—is composed of a collection of constituent questions. The manner in which the constituent questions are combined to form a composite question as well as other issues relating to the concept of a composite question will be discussed in section 3.

Example 2.1. The question Q $\triangleq$ Is Ruth *tall*? is an atomic question if no other questions have to be asked in order to answer Q.

The question Q $\triangleq$ Is x *big*? where x is some object, is a composite question if *big* is defined as the conjunction of *long*, *wide* and *high* (as in (1.2)), and the answer to Q is deduced from the answers to the constituent questions $Q_1 \triangleq$ Is x *long*?, $Q_2 \triangleq$ Is x *wide*?, and $Q_3 \triangleq$ Is x *high*?

A *questionnaire* is, in effect, a representation of a composite question, and a *branching questionnaire* is a representation in which the order in which the constituent questions are asked is determined by the answers to the previous questions.

In what follows, we shall examine the concept of an atomic question in greater detail with a view to providing a basis for a systematic representation of fuzzy-algorithmic definitions in the form of branching questionnaires.

NOTATION AND TERMINOLOGY

Definition 2.2. An *atomic question*, Q, is characterized by a triple Q $\triangleq$ (X,B,A), where X, the *object-set*, is a set of objects to which Q applies; B, the *body* of Q, is a label of either a class or an attribute; and A, the *answer-set*, is a set of admissible answers to the question. Where necessary, specific instances of Q, X and A will be denoted generically by q, x and a, respectively.† When X and A are implied, Q will be written in an abbreviated form as

$$Q \triangleq B ?$$

and a specific question together with an admissible answer to it will be expressed as

$$Q/A \triangleq B ? a \qquad (2.3)$$

or equivalently

$$q/a \triangleq B ? a.$$

The pair Q/A will be referred to as a *question/answer pair* (or simply *Q/A pair*). Graphically, an atomic question (with implied x) will be represented in the form of a fan as shown in Fig. 1.

Example 2.4. Consider a specific instance of a question Q, e.g. "Is Nancy *well-dressed*?" In this case, with the subject $x \triangleq$ Nancy implied, the specific question may be expressed as

$$q \triangleq well\text{-}dressed? \qquad (2.5)$$

where *well-dressed* is the body of Q. Correspondingly, a specific Q/A pair might be

$$q/a \triangleq well\text{-}dressed? \; true \qquad (2.6)$$

in which *true*, as an admissible answer, is an element of the answer-set A. If the other elements of the answer-set are *false* and *borderline*, then A may be expressed as

$$A = true + borderline + false \qquad (2.7)$$

where $+$ denotes the union rather than the arithmetic sum.

The linguistic truth-values in (2.8) are, in effect, names of fuzzy subsets of the unit interval (Zadeh, 1975b, c). In terms of their respective membership functions, these subsets may be expressed as (see the Appendix)

$$true \quad = \int_0^1 \mu_t(v)/v \qquad (2.8)$$

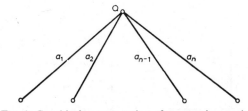

Fig. 1. Graphical representation of an atomic question.

---

†To avoid a proliferation of symbols, Q and q will be used interchangeably when no confusion is likely to arise.

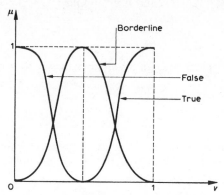

FIG. 2. Membership functions of *true*, *borderline* and *false*.

$$borderline = \int_0^1 \mu_b(v)/v \qquad (2.9)$$

and

$$false = \int_0^1 \mu_f(v)/v \qquad (2.10)$$

where μ_t, μ_b and μ_f are the membership functions of *true*, *borderline* and *false*, respectively, and an expression such as (2.8) means that the fuzzy set labeled *true* is the union of fuzzy singletons $\mu_t(v)/v$ in which the point v in [0,1] has the grade of membership $\mu_t(v)$ in *true*. Typical forms of μ_t, μ_b and μ_f are shown in Fig. 2.

Note 2.11. For the representation of μ_t, μ_b and μ_f it is frequently convenient to employ standardized functions with adjustable parameters, e.g., the S and Π functions which are defined below (see Figs 3(a) and 3(b)).

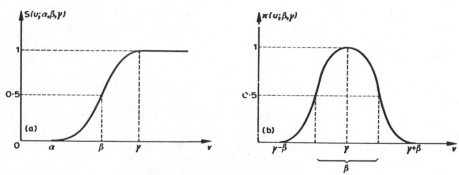

FIG. 3. Plots of S and π functions.

$$S(v;a,\beta,\gamma) = 0 \quad \text{for} \quad v \leqslant a \tag{2.12}$$

$$= 2\left(\frac{v-a}{\gamma-a}\right)^2 \quad \text{for} \quad a \leqslant v \leqslant \beta$$

$$= 1-2\left(\frac{v-\gamma}{\gamma-a}\right)^2 \quad \beta \leqslant v \leqslant \gamma$$

$$= 1 \quad \text{for} \quad v \geqslant \gamma$$

$$\prod(v;\beta,\gamma) = S(v;\gamma-\beta,\gamma-\frac{\beta}{2},\gamma) \quad \text{for} \quad v \leqslant \gamma \tag{2.13}$$

$$= 1-S(v;\gamma,\gamma+\frac{\beta}{2},\gamma+\beta) \quad \text{for} \quad v \geqslant \gamma.$$

In $S(v; a,\beta,\gamma)$, the parameter β, $\beta = (a+\gamma)/2$, is the *crossover point*, that is, the value of v at which S takes the value 0·5. In $\prod(v; \beta, \gamma)$, β is the *bandwidth*, that is, the distance between the crossover points of $\prod$, while γ is the point at which $\prod$ is unity.

In terms of S and $\prod$, μ_t, μ_b and μ_f may be expressed as (suppressing the argument v)

$$\mu_t = S(a, \beta, 1) \tag{2.14}$$
$$\mu_b = \prod(\beta', 0\cdot5) \tag{2.15}$$
$$\mu_f = 1-S(0, \beta, \gamma) \tag{2.16}$$

where the use of the symbol β' in (2.15) signifies that the bandwith of b need not be equal to the value of β in (2.14).

Note 2.17. In cases in which the three linguistic truth-values *true, borderline* and *false* do not offer a sufficiently wide choice, it may be convenient to use, in addition, the truth-values *rather true* and *rather false*, abbreviated as rt and rf, respectively.

As a fuzzy subset of [0,1], *rather true* may be defined approximately as

rather true $\triangleq$ *not very true and not (false or borderline)*

and its membership function may be approximated by a $\prod$ function with γ at, say, the crossover point of *very true*. *Rather false* may be defined similarly in terms of *false* and *borderline*.

CLASSIFICATIONAL AND ATTRIBUTIONAL QUESTIONS

A question, Q, is *classificational* if its body, B, is the label of a fuzzy or non-fuzzy set.

A question, Q, is *attributional* if B is the label of an attribute. In the case of a classificational question, an answer, a, represents the grade of membership of x in the fuzzy set B. The answer might be numerical, e.g., $a \triangleq 0\cdot8$, or linguistic, e.g., $a \triangleq high$. Equivalently, the answer may be expressed as the truth-value of the predicate B(x),† e.g., *true, borderline, false, very true*, etc.

---

†Depending on the circumstances, the arguments of a predicate may be displayed, as in B(x), or suppressed, as in B.

In the case of an attributional question, $Q = B$?, an answer, a, represents the value of the attribute, B, of an object x, e.g. B $\triangleq$ *age* and x $\triangleq$ Haydee. Again, a may be numerical, e.g. a $\triangleq$ 35, or linguistic, e.g. a $\triangleq$ *young*, a $\triangleq$ *very young*, etc.

Comment 2.18. As defined above, a question $Q = (X,B,A)$ may be viewed as a collection of variables $\{B(x)\}$, $x \in X$. From this point of view, answering a classificational question addressed to an x in X corresponds to assigning a value, at x, to the membership function of the fuzzy set B (or, equivalently, assigning a truth-value to the fuzzy predicate $B(x)$). Similarly, answering an attributional question may be interpreted as the assignment of a value to the attribute $B(x)$. In either case, answering a question with body B may be represented as an assignment equation

$$B(x) = a$$

in which a numerical or a linguistic value a is assigned to the variable $B(x)$.

Example 2.19. Suppose that X is the set of objects in a room and $Q = red$? is a fuzzy classificational question. Furthermore, suppose that the set of admissible answers is the interval [0,1], representing the grades of membership of objects in X in the fuzzy subset *red* of X. In this case, an answer such as *true 0·8* to the question "Is the vase *red*?" may be represented as the assignment equation

$$red\ (\text{vase}) = 0·8$$

which implies that the truth-value of the predicate *red* (x) evaluated at x $\triangleq$ vase is 0·8 or, equivalently, that the grade of membership of the object x $\triangleq$ vase in the fuzzy set labeled *red* is 0·8.

Example 2.20. Same as Example 2.19 except that the set of admissible answers, A, is assumed to be expressed by

$$A = low + low^2 + low^{1/2} + medium + medium^2 + medium^{1/2} +$$
$$high + high^2 + high^{1/2} \tag{2.21}$$

where *high* and *medium* and *low* are primary fuzzy subsets of the unit interval which are defined in terms of the S and $\prod$ functions by (2.14), (2.15) and (2.16), and w^2 and $w^{1/2}$ are abbreviations for *very w* and *more or less w*, respectively. Thus, if w is a subset of a universe of discourse U, then

$$w^2 = \int_U (\mu_w(u))^2/u \tag{2.22}$$

and

$$w^{1/2} = \int_U (\mu_w(u))^{1/2}/u, \tag{2.23}$$

which means that the membership functions of w^2 and $w^{1/2}$ are equal, respectively, to the square and square root of the membership function of w.

Example 2.24. Same as Example 2.19, but with the question assumed to be worded as "Is it true that x is *red*?" and the set of admissible answers expressed by

$$A = true + true^2 + true^{1/2} + false + false^2 + false^{1/2} +$$
$$borderline + borderline^2 + borderline^{1/2} \qquad (2.25)$$

where *true, false* and *borderline* are defined in the same way as *high, low* and *medium* and may be used in the same manner. Thus, for example, if the answer to the question "Is it true that the vase is *red*?" is $true^2$ ($\triangleq$ *very true*), then the grade of membership of the vase in the class of red objects is given by the assignment equation.

$$\mu_{red}\ (vase) = true^2 \qquad (2.26)$$

where the right-hand member of (2.26) represents a linguistic truth-value whose meaning is defined by (2.22), and the left-hand member is the membership function of the fuzzy set *red* evaluated at $x \triangleq$ vase.

Example 2.27. As an illustration of an attributional question, suppose that X is the set of employees in a company and Q $\triangleq$ *age*? is an attributional question (e.g. "What is the *age* of Elizabeth?"). If the set of admissible answers is the set of integers

$$A = 20 + 21 + \ldots + 60 \qquad (2.27)$$

then the answer to the question "What is the *age* of Elizabeth?" might be

$$age\ (Elizabeth) = 32$$

On the other hand, if the admissible answers are linguistic in nature, e.g.,

$$A = young + not\ young + very\ young + not\ very\ young +$$
$$old + very\ old + \ldots \qquad (2.28)$$

then an answer might have the form

$$age\ (Elizabeth) = very\ young$$

with the understanding that *very young* is a linguistic value which is assigned to the linguistic variable *age* (Elizabeth). It should be noted that in (2.28) *young* and *old* play the role of primary fuzzy sets which have a specified meaning, e.g.

$$\mu_{young} = 1 - S(20,30,40) \qquad (2.29)$$
$$\mu_{old} = S(50,60,70) \qquad (2.30)$$

where the S and $\prod$ functions are defined by (2.12) and 2.13), and μ_{young} and μ_{old} denote the membership functions of *young* and *old*, respectively. The meaning of the other terms in (2.28) may be computed from the definitions of the modifiers *not* and *very*. Thus,

$$\mu_{not\ young} = 1 - \mu_{young} \qquad (2.31)$$
$$\mu_{very\ young} = (\mu_{young})^2 \qquad (2.32)$$
$$\mu_{not\ very\ young} = 1 - (\mu_{young})^2 \qquad (2.33)$$

and so on. Note that A may be viewed, in effect, as a microlanguage with its own syntax and semantics.

NESTED QUESTIONS

Consider an attributional question of the form "How *old* is Francoise?" to which a linguistic answer might be, "Francoise is *young*", with *young* defined by (2.29).

At this point, one could ask a classificational question concerning the answer "Francoise is *young*", namely, "Is is true that (Francoise is *young*)?" to which a linguistic answer might be *very true*. Continuing this process, one could ask the question "Is it true that ((Francoise is *young*) is *very true*)?" to which a linguistic answer might be *more or less true*. On further repetition, we are led to a *nested question* which, in general terms, may be expressed as

$$\text{Is it true that } (\, . \, . \, (((x \text{ is } w) \text{ is } \tau_1) \text{ is } \tau_2) \ldots \text{ is } \tau_n) \, ? \tag{2.34}$$

in which w is an attribute-value and τ_1, τ_2, ..., τ_n are numerical or linguistic truth-values.

How should the meaning of an answer of the form

$$a \triangleq (\ldots (((x \text{ is } w) \text{ is } \tau_1) \text{ is } \tau_2) \ldots \text{ is } \tau_n) \tag{2.35}$$

be interpreted? A clue is furnished by the following example. Suppose that the answer to the question "Is is true that (Francoise is *young*)?" is a numerical truth-value, say 0·5. As stated earlier, this implies that the grade of membership of Francoise in the class of young women is 0·5, which in turn implies (by (2.29)) that Francoise is 30 years old. Thus, we have

$$(\text{Francoise is } young) \text{ is } 0 \cdot 5 \; true \Rightarrow \text{Francoise is 30 years old.} \tag{2.36}$$

More generally, let u be a base variable for an attribute B and let μ_{young} denote the membership function which defines the answer $a \triangleq young$ as a fuzzy subset of the universe of discourse, U, which is associated with the attribute B (e.g. if B $\triangleq$ *age*, then u is a number in the interval [0,100] and U = [0,100] is the universe of discourse associated with *age*). Now suppose that v is a numerical truth-value of the answer Francoise is *young*. Then, the age of Francoise is given by

$$B(\text{Francoise}) = \mu_B^{-1}(v) \tag{2.37}$$

where μ_B^{-1} is the function inverse to the function μ_B.† Thus, in the particular case where $v = 0 \cdot 5$, (2.29) gives

$$B(\text{Francoise}) = \mu_B^{-1}(0 \cdot 5) \tag{2.38}$$
$$= 30.$$

At this juncture, we can employ the extension principle (see the Appendix) to compute the meaning of the answer $a \triangleq (\text{Francoise is } young)$ is τ, where τ is a linguistic truth-value which is characterized by a membership function μ_τ. (E.g. if τ is *true*, then μ_τ is given by (2.14).) Thus, substituting τ in (2.37), we obtain

$$B(\text{Francoise}) = \mu_B^{-1}(\tau) \tag{2.39}$$
$$= \mu_B^{-1} \circ \tau$$

---

†If the mapping μ_B: U→[0,1] is not 1–1, μ_B^{-1} is the relation (rather than the function) that is inverse to μ_B. In any case, the graph of μ_B^{-1} is the same as that of μ_B, but with the abscissae of μ_B^{-1} being the ordinates of μ_B and vice versa.

which should be interpreted as the composition† of the binary relation μ_B^{-1} and the unary relation τ. In more general terms, this result may be stated as the following proposition.

Proposition 2.40. An answer of the form

$$a \triangleq (x \text{ is } w_1) \text{ is } \tau \tag{2.41}$$

where x is an object in X, w_1 is a fuzzy subset of U, and τ is a truth-value (numerical or linguistic), implies the answer

$$a^\star \triangleq x \text{ is } w_2 \tag{2.42}$$

where w_2 is related to w_1 and τ by

$$w_2 = \mu_{w_1}^{-1} \circ \tau. \tag{2.43}$$

In (2.43), $\mu_{w_1}^{-1}$ is the relation inverse to μ_{w_1}, where μ_{w_1} is the membership function of w_1, and the right-hand member of (2.43) represents the composition of $\mu_{w_1}^{-1}$ with the unary relation (fuzzy set) τ. (See Appendix.)

Repeated application of Proposition 2.40 to an answer of the form (2.25) leads to the general result

$$a \triangleq (\ldots (((x \text{ is } w_1) \text{ is } \tau_1) \text{ is } \tau_2) \ldots \text{ is } \tau_n \Rightarrow a^\star \triangleq x \text{ is } w_{n+1} \tag{2.44}$$

where

$$
\begin{aligned}
w_{n+1} &= \mu_{wn}^{-1} \circ \tau_n \\
w_n &= \mu_{w_{n-1}}^{-1} \circ \tau_{n-1}
\end{aligned} \tag{2.45}
$$

$$-\!-\!-\!-\!-\!-\!-\!-\!-$$

$$w_2 = \mu_{w_1}^{-1} \circ \tau_1$$

and μ_{wi}, $i = 1, \ldots, n$, is the membership function of w_i.

As a simple illustration of (2.43), a graphical representation of the composition $\mu_{w_1}^{-1} \circ \tau_1$, is shown in Fig .4. Here μ_{young} is the membership function of $w_1 = young_1$, with the base variable being the numerical age u. τ_1 is assumed to be *very true*, whose membership function is plotted as shown, with v playing the role of abscissa. The point, a, on $\mu_{very\ true}$ which has the abscissa v has the ordinate $\mu_{very\ true}(v)$, and, correspondingly, the point, β, on $\mu_{young_1}^{-1}$ which has the abscissa v has the ordinate $\mu_{young_1}^{-1}(v)$. Now, from a and β we can construct a point γ on μ_{young_2} with abscissa $\mu_{young_1}^{-1}(v)$ and ordinate $\mu_{very\ true}(v)$. In this way, by varying v from 0 to 1, we can generate the plot of μ_{young_2}, which is the membership function of w_2 as defined by (2.43).

An important conclusion which is implicit in (2.44) is that any nested assertion of the form

$$((x \text{ is } w_1) \text{ is } \tau_1) \ldots \text{ is } \tau_n) \tag{2.46}$$

---

†The composition of a binary relation R in $U_1 \times U_2$ with a unary relation S in U_2 is a unary relation R∘S in U_1 whose membership function is given by $\mu_{R\circ S}(u_1) = V_{u_2} \mu_R(u_1, u_2) \wedge \mu_S(u_2)$, where $V \triangleq$ max and $\wedge \triangleq$ min.

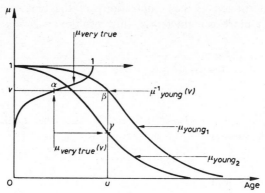

FIG. 4. Composition of $\mu_{young\,1}^{-1}$ with $\tau \triangleq$ *very true*.

may be replaced by an equivalent assertion of the form

$$x \text{ is } w_{n+1} \tag{2.47}$$

which does not contain any truth-values. Thus, the use of truth-values in (2.46) serves indirectly the same function as a linguistic modifier m which transforms w_1 into mw_1.

THE RELATION BETWEEN CLASSIFICATIONAL AND ATTRIBUTIONAL QUESTIONS

In the case of a non-fuzzy classificational question, the answer-set, A, has only two elements which are usually designated as {YES, NO}, {TRUE, FALSE} or {0,1}. By contrast, the answer-set of an attributional question is usually a continuum U or a countable set of linguistic values defined over U. Thus, in general, an answer to an attributional question conveys considerably more information than an answer to a non-fuzzy classificational question.

In the case of fuzzy classificational questions, however, the answer-set may be the unit interval [0,1] or a countable set of linguistic values defined over [0,1]. In such cases, the distinction between classificational and attributional questions is much less pronounced and, in fact, there may be equivalence between them.

To be more specific, let us assume for concreteness that U is the real line and F is a fuzzy subset of U. F will be said to be *amodal* if its membership function μ_F is strictly monotone, which implies that the mapping $\mu_F: U \rightarrow [0,1]$ is one-one. If F is not amodal but is convex† or concave, then F will be said to be *modal*. Typically, the membership function of an amodal fuzzy set has the form shown in Fig. 5, whereas that of a modal set has the appearance of a peak or a valley (Fig. 6).

Let $Q_c \triangleq$ F? be a classificational question which has the same body as an attributional question $Q_a \triangleq$ F? For example, a specific question q_c may be worded as "Is Jeanne *young*?" while the wording of q_a might be "How *young* is Jeanne?" Clearly, if *young* is

---

†A fuzzy set F in U is convex if μ_F satisfies the inequality $\mu_F(\lambda u_1 + (1-\lambda) u_2) \geqslant \min(\mu_F(u_1), \mu_F(u_2))$ for all u_1, u_2 in U and all λ in [0,1]. A fuzzy set F is *concave* if its complement is convex. Additional details may be found in Zadeh (1966).

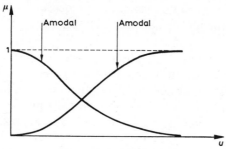

FIG. 5. Amodal fuzzy sets.

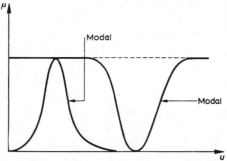

FIG. 6. Compatibility functions of modal fuzzy sets.

an amodal fuzzy set, then from an answer to q_c such as "Jeanne is 0·9 *young*" we can deduce the age of Jeanne and, conversely, from the age of Jeanne, say *age* $\triangleq$ 32, we can deduce her grade of membership in the fuzzy set *young*. Thus, when F is an amodal fuzzy set or, more generally, a fuzzy set whose membership function is a one-one mapping, the answer to a classificational question conveys the same information as the answer to an attributional question.

Now suppose that F is a modal fuzzy set, e.g. F $\triangleq$ *middle-aged*, whose membership function has the form shown in Fig. 7. In this case, from the specification of the grade of membership in *middle-aged*, one cannot deduce the value of the attribute *age* uniquely.

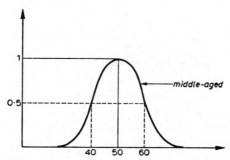

FIG. 7. Representation of *middle-aged* as a modal fuzzy set.

Thus, if F is modal, an answer to the classificational question "Is x F?" e.g., "Is Freda *middle-aged*?" is less informative that an answer to the attributional question "What is the *age* of Freda?"

It should be noted that Comment 2.18 implies that a classificational question $Q \triangleq B$? may always be regarded as an attributional question whose body is the label of the membership function of B. Thus, what the above discussion indicates is that although it is not true in general that an attributional question is equivalent to a classificational question with the same body, this is the case when B is an amodal fuzzy set.

3. Composite questions and their representations

The concept of an atomic question which we discussed in the preceding section provides a basis for the definition of the more general concept of a *composite question*. This concept and its representations will be the focus of our attention in the sequel.

Stated informally, an *n-adic composite question* Q, with body B, is a question composed of n constituent questions $Q_1, \ldots, Q_n$ with bodies $B_1, \ldots, B_n$, respectively, such that the answer to Q is dependent upon the answers to $Q_1, \ldots, Q_n$. Thus, a *monadic* question has a single constituent, a *dyadic* question has two constituents, a *triadic* question has three constituents, etc. A constituent question may be atomic or composite.

An *n-adic composite question* or, simply, an *n-adic question*, Q, is characterized by its *relational representation*, $B(B_1, \ldots, B_n)$ (or simply B, when no confusion with the body, B, of Q can arise), whose tableau has the form shown in Table 1. In this tableau, r_i^j and r_i range over the admissible answers to Q_j and Q, respectively, with A_j and A representing the answer-sets associated with Q_j and Q, and a^j and a denoting their generic elements. Thus, if Q is an *n-adic* question, then B is a non-fuzzy $(n+1)$-ary relation from the cartesian product $A_1 \times \ldots \times A_n$ to A. In particular, if Q is a monadic question, then B is a binary relation, and if Q is atomic then B is a unary relation.

TABLE 1

Relational representation of Q. (*Depending on the circumstances, the columns of* B *may be labeled* $Q_1, \ldots, Q_n$, Q, *or* $B_1, \ldots, B_n$, B.)

Q_1	Q_2	Q_j	Q_n	Q
r_1^1	r_1^2	r_1^j	r_1^n	r_1
r_2^1	r_2^2	r_2^j	r_2^n	r_2
r_i^1	r_i^2	r_i^j	r_i^n	r_i
r_m^1	r_m^2	r_m^j	r_m^n	r_m

Generally, we shall assume that the entries in B are linguistic in nature, i.e. are linguistic attribute-values and/or linguistic truth-values and/or linguistic grades of membership. Thus, if U_j is a universe of discourse associated with A_j, then an answer $a_i^j \in A_j$ will, in general, be a label of a fuzzy subset of U_j. The generic elements of U_j and U will be denoted by u_j and u, respectively, and will be referred to as the *base variables* for A_j and A. When it is necessary to differentiate between attributional and classificational questions, the universes of discourse for the latter will be denoted by V instead of U.

Example 3.1. Consider a composite classificational question Q $\triangleq$ *big*? which is composed of two classificational atomic questions $Q_1 \triangleq$ *wide*? and $Q_2 \triangleq$ *long*?, and one attributional atomic question $Q_3 \triangleq$ *height*? The answer-sets associated with Q_1, Q_2, Q_3 and Q are assumed to be given by (f, b, t, l, m, h are abbreviations for *false, borderline, true, low, medium* and *high*, respectively)

$$A_1 = A_2 = A = f+b+t \tag{3.2}$$

$$A_3 = l+m+h \tag{3.3}$$

where f, b and t are fuzzy subsets of the unit interval defined by (2.8), (2.9) and (2.10), and l, m and h are fuzzy subsets of the real line defined by expressions of the form (2.16), (2.15) and (2.14) with parameters a, β, and γ.

The relational tableau for $B(B_1, \ldots, B_n)$ is assumed to be given by (in partially tabulated form) by Table 2.

TABLE 2

Relational representation of big (wide, long, height)

wide?	long?	height?	big?
t	t	h	t
t	t	m	t
t	t	l	b
t	t	l	f
t	b	h	b
t	f	h	b
t	f	h	f
b	f	l	b
.	.	.	.
f	f	l	f

There are two important observations to be made concerning $B(B_1, \ldots, B_n)$. First in general $B(B_1, \ldots, B_n)$ is a relation rather than a function. In Table 2, this manifests itself by the fact that the entries in the column labeled *big*? are not uniquely determined by the entries in the columns *wide*?, *long*? and *height*?. For example, corresponding to $a^1 = t$, $a^2 = t$ and $a^3 = l$, we have both $a = b$ and $a = f$. This implies that, if the answer to *wide*? is *true*, to *long*? is *true* and to *height*? is *low*, then the answer to *big*? could be either *borderline* or *false*.

Second, the tableau may not be complete, that is, certain combinations of the admissible answers to constituent questions may be missing from the table. For example,

$a^1 = f$, $a^2 = b$ and $a^3 = b$ may not be in the table. This may imply that (a) the particular combination of answers cannot occur, or (b) the answer to Q corresponding to the missing entries is not known—which is equivalent to assuming that the answer is the union of all admissible answers, i.e. is the answer-set A.

Case (a) implies that there is some interdependence between the constituent questions in the sense that the knowledge of answers to some of the constituent questions restricts the possible answers to others. If the Q_i are viewed as variables as in (2.18), then (a) implies that the Q_i are λ-*interactive* in the sense defined in Zadeh (1975a). Unless stated to the contrary, we shall assume that the missing rows imply (a) rather than (b). A more detailed discussion of this issue will be presented in section 4.

ALTERNATIVE REPRESENTATIONS OF B: ALGEBRAIC REPRESENTATION

The relational representation, B, of a composite question Q may in turn be represented in a variety of ways of which the most useful ones are: (a) the *tabular representation*, which we have described already, (b) the *algebraic representation*, which we shall discuss presently, (c) the *analytic representation*, which we shall discuss following (b), and (d) the *branching questionnaire representation*, which will be discussed in section 4.

In the algebraic representation, the ith row, $i = 1, 2, \ldots, m$ of the tableau of B is expressed as a Q/A *sequence* of the form

$$Q_1\, r_i^1\, Q_2\, r_i^2 \ldots Q_i^n\, r_n /\!/ Q\, r_i \tag{3.4}$$

or, more simply as a Q/A *string*

$$r_i^1\, r_i^2 \ldots r_i^n /\!/ r_i \tag{3.5}$$

where it is understood that r_i^j, $j = 1, \ldots, n$ is an admissible answer to the constituent question Q_j, and r_i is an admissible answer to the composite question Q. B as a whole, then, may be expressed algebraically as the summation (i.e. the union) of the Q/A strings corresponding to the rows of the tableau of B. Thus, we may write

$$B = r_1^1 r_1^2 \ldots r_1^n /\!/ r_1 + r_2^1 r_2^2 \ldots r_2^n /\!/ r_2 + \ldots + r_m^1 r_m^2 \ldots r_m^n /\!/ r_m \tag{3.6}$$

or, more compactly,

$$B = \sum_i r_i^1 r_i^2 \ldots r_i^n /\!/ r_i. \tag{3.7}$$

Example 3.8. In the algebraic form, the tableau of the relational representation defined by Table 2 may be expressed as

$$\begin{aligned} B = {}& tth /\!/ t + ttm /\!/ t + ttl /\!/ b \\ & + ttl /\!/ f + tbh /\!/ b + tfh /\!/ b \\ & + tfh /\!/ f + bfl /\!/ b + \ldots + ffl /\!/ f. \end{aligned} \tag{3.9}$$

As in the case of regular expressions, an important advantage of representations of the form (3.9) is that the operations of union ($+$) and string concatenation may be treated in much the same manner as addition and multiplication. Thus, the terms in (3.9) may be combined or expanded in accordance with the replacement rules which are illustrated below by examples.

$$ttf//f \; + \; tff//t \; = \; t(tf//t{+}ff//t) \tag{3.10}$$

$$ttf//t \; + \; ftf//t \; = \; (t+f)tf//t \tag{3.11}$$

$$tfb//t \; + \; ttb//t \; = \; t(f{+}b)b//t \tag{3.12}$$

$$tfb//t \; + \; tfb//b \; = \; tfb//(t{+}b) \tag{3.13}$$

$$(t{+}f)(f{+}b)t//t = tft//t{+}fft//t{+}tbt//t{+}fbt//t. \tag{3.14}$$

For example, using the above identities in (3.9), we can write B in a partially factored form as

$$\begin{aligned} B = \; &tt(h{+}m)//t{+}ttl//(b{+}f){+}t(b{+}f)h//b{+} \\ &{+}(tfh{+}ffl)//f{+} \ldots {+}bfl//b. \end{aligned} \tag{3.15}$$

It should be noted that the replacement of the left-hand member by the right-hand member involves a factorization in (3.10), (3.11), (3.12) and (3.13), and an expansion in (3.14). In general, factorization has the effect of *raising* the *level* of an expression (in the sense of decreasing the number of operations that have to be performed for its evaluation) while an expansion has the opposite effect. For example, the evaluation of the arithmetic expression $xy{+}xz$ requires three operations, while that of the factored form $x(y{+}z)$ requires only two. In this case, the representation of B in the *normal form*† (3.9) has the lowest possible level among all algebraic representations involving the admissible answers to the Q_j and Q.

THE MEANING OF B

The question of what constitutes the meaning of B may be viewed as a special case of the following problem in semantics.‡ Suppose that we are given a string of terms (words) $W_1 W_2 \ldots W_n$ with the meaning of each term defined as a subset of a universe of discourse U. What is the meaning of the composite term $W_1 W_2 \ldots W_n$—that is, what is the subset of U whose label is $W_1 W_2 \ldots W_n$?

As a special instance of this problem consider two finite non-fuzzy sets G and H whose elements are $g_1, \ldots, g_m$ and $h_1, \ldots, h_n$, respectively. When we write

$$G = g_1 + \ldots + g_m \tag{3.16}$$

$$H = h_1 + \ldots + h_n \tag{3.17}$$

the right-hand side of the equation defines the meaning § of the label on the left-hand

---

†This usage of the term *normal form* is consistent with that of Codd (1971) in his work on relational models of data. A related concept is that of *characteristic set* in the Vienna definition language (Lucas, 1968; Wegner, 1972).

‡A more detailed discussion of this problem may be found in Zadeh (1971b, 1972a).

§The term *meaning* is used here in the sense of denotational semantics (Carnap, 1956; Hempel, 1952; Church, 1951; Quine, 1953; Frege, 1952; Martin, 1963).

side. Now, if we write the Cartesian product $G \times H$ as a string GH, then the meaning of $G \times H$ may be obtained very simply by expanding the algebraic product of G and H. Thus,

$$G \times H = GH \qquad (3.18)$$
$$= (g_1 + \ldots + g_m)(h_1 + \ldots + h_n)$$
$$= g_1 h_1 + \ldots + g_m h_n$$

where $g_i h_j$ should be interpreted as the ordered pair (g_i, h_j).

Now suppose that G and H are finite fuzzy sets defined by

$$G = \mu_1/g_1 + \ldots + \mu_m/g_m \qquad (3.19)$$

$$H = v_1/h_1 + \ldots + v_n/h_n \qquad (3.20)$$

where μ_i/g_i means that the grade of membership of g_i in G is μ_i, and likewise for H. Then, for the Cartesian product of G and H we obtain

$$G \times H = (\mu_1/g_1 + \ldots + \mu_m/g_m)(v_1/h_1 + \ldots + v_n/h_n) \qquad (3.21)$$
$$= (\mu_1 \wedge h_1)/g_1 h_1 + \ldots + (\mu_m \wedge v_n)/g_m h_n$$

where

$$\mu_i \wedge v_j \triangleq \min(\mu_i, v_j). \qquad (3.22)$$

More generally, let $G_1, \ldots, G_n$ be fuzzy subsets of $U_1, \ldots, U_n$ defined by

$$G_j = \sum_{i=1}^{m_j} \mu_i^j/u_i^j. \qquad (3.23)$$

Then

$$G_1 \times \ldots \times G_n = G_1 \ldots G_n \qquad (3.24)$$
$$= \sum_i (\mu_{i_1}^1 \wedge \ldots \wedge \mu_{i_n}^n)/u_{i_1}^1 \ldots u_{i_n}^n$$

which implies that the right-hand member of (3.24) constitutes the meaning of the string $G_1 \ldots G_n$ (or, equivalently, $G_1 \times \ldots \times G_n$).

Returning to the question of what constitutes the meaning of B, let us focus our attention on the algebraic representation of B as expressed by (3.6). If the r_i^j and r_i in (3.6) are assumed to be fuzzy subsets of $U_1, \ldots, U_n$, U, then each term in (3.6) is a Cartesian product of fuzzy sets in the sense of (3.24), and B as a whole is the union of such Cartesian products. Thus, upon the expansion of each term in accordance with (3.24) and summing the results, we obtain the expression for a fuzzy $(n+1)$-ary relation from $U_1 \times \ldots \times U_n$ to U which may be viewed as the *denotational meaning* of B.† This fuzzy relation will be denoted by B_β and will be referred to as the *β-representation* of B, with β—standing for *base variable*—serving to signify that B_β is a fuzzy relation from $U_1 \times \ldots \times U_n$ to U whereas B is a non-fuzzy relation from $A_1 \times \ldots \times A_n$ to A.

---

†In performing the expansion and summation of terms in B, we are tacitly assuming that the constituent questions $Q_1, \ldots, Q_n$ are *β-non-interactive* (Zadeh, 1975a) in the sense that the base variables $u_1, \ldots, u_n$ are jointly unconstrained.

In summary, the main points of the foregoing discussion may be stated as follows.

Proposition 3.25. Let B be an $(n+1)$-ary non-fuzzy relation from $A_1 \times \ldots \times A_n$ to A which constitutes a relational representation of a composite question Q. If the answers to Q and the constituent questions in Q are fuzzy subsets of their respective universes of discourse U, $U_1, \ldots, U_n$, then B induces an $(n \times 1)$-ary fuzzy relation B_β which may be derived from B by the process of expansion. The fuzzy relation B_β constitutes the *denotational meaning* of B in the universe of discourse $U_1 \times \ldots \times U_n \times U.\ddagger$

Example 3.26. As a very simple illustration of (3.25), consider a B whose algebraic representation reads

$$B = tt//f^2 + ff//t \tag{3.27}$$

where $t(\triangleq true)$, $f(\triangleq false)$ and $f^2(\triangleq very \ false)$ are fuzzy subsets of the universe of discourse

$$V = 0 + 0\cdot2 + 0\cdot4 + 0\cdot6 + 0\cdot8 + 1 \tag{3.28}$$

and are defined by

$$t = 0\cdot6/0\cdot8 + 1/1 \tag{3.29}$$

$$f = 1/0 + 0\cdot6/0\cdot2 \tag{3.30}$$

and

$$f^2 = 1/0 + 0\cdot36/0\cdot2. \tag{3.31}$$

On substituting (3.29)–(3.31) into $tt//f^2$ and expanding, we have

$$
\begin{aligned}
tt//f^2 = \ & (0\cdot6/0\cdot8 + 1/1)(0\cdot6/0\cdot8 + 1/1)//(1/0 + 0\cdot36/0\cdot2) \tag{3.32}\\
= \ & 0\cdot6/(0\cdot8,0\cdot8,0) + 0\cdot6/0\cdot8,1,0)\\
& + 0\cdot6/(1,0\cdot8,0) + 1/(1,1,0)\\
& + 0\cdot36/(0\cdot8,0\cdot8,0\cdot2) + 0\cdot36/(0\cdot8,1,0\cdot2)\\
& + 0\cdot36/(1,0\cdot8,0\cdot2) + 0\cdot36/(1,1,0\cdot2).
\end{aligned}
$$

Performing the same operation on the other term in (3.27) and summing the results, we obtain the desired expression for B_β

$$
\begin{aligned}
B_\beta = \ & 0\cdot36/((0\cdot8,0\cdot8,0\cdot2) + (0\cdot8,1,0\cdot2) + (1,0\cdot8,0\cdot2) + 1,1,0\cdot2)) \tag{3.33}\\
& + 0\cdot6/((0,0,0\cdot8) + (0,0\cdot2,0\cdot8) + (0\cdot2,0,0\cdot8) + (0\cdot2,0\cdot2,0\cdot8)\\
& + (0,0\cdot2,1) + (0\cdot2,0\cdot2,1)) + 1/((0,0,1) + (1,1,0))
\end{aligned}
$$

as a ternary fuzzy relation in $[0,1] \times [0,1] \times [0,1]$.

INTERPOLATION OF B

Knowledge of B_β is of importance in that it provides a basis for an *interpolation* of B, that is, an approximate way of deducing answers to Q corresponding to entries in B which are not elements of the answer-sets $A_1, \ldots, A_n$.

‡In cases in which the body, B_l, of a classification question, Q_l, is a fuzzy subset of a universe of discourse which does not possess a numerically-valued base variable (e.g. Q_l *beautiful*), it may be necessary to define B_l by exemplification (Bellman, Kalaba & Zadeh, 1966). In general, exemplificational (or ostensive) definitions are human—rather than machine-oriented.

To illustrate, suppose that Q is a dyadic classificational question whose constituent classificational questions Q_1 and Q_2 have the answer-sets

$$A_1 = A_2 = A = t+b+f.$$

Let B be a relational representation of Q and assume that we wish to find the answer to Q when the answers to Q_1 and Q_2 are, respectively,

$$a^1 = not\ very\ true \tag{3.34}$$

and

$$a^2 = rather\ true. \tag{3.35}$$

Since a^1 and a^2 are not among the entries in the Q_1 and Q_2 columns of the tableau of B, we cannot use B to find the corresponding entry in the Q column. On the other hand, if we have B_β as a fuzzy ternary relation in $V_1 \times V_2 \times V$ (which is $[0,1] \times [0,1] \times [0,1]$ in the case under consideration), then by interpolating B we can obtain an approximation to the answer to Q which corresponds to the answers $a^1 = not\ very\ true$ and $a^2 = rather\ true$.

Specifically, the desired approximation is given by the composition of B_β with the fuzzy sets a^1 and a^2, treating a^1 and a^2 as unary fuzzy relations in $[0,1]$. Thus,†

$$\text{Answer to Q} = B_\beta \circ a^1 \circ a^2. \tag{3.36}$$

The significance of (3.36) becomes somewhat clearer if the right-hand member of (3.36) is interpreted as the projection on V of the intersection of B_β with the cylindrical extensions of a^1 and a^2.‡ Thus, if B_β is visualized as a fuzzy surface in $V_1 \times V_2 \times V$, then a^1 and a^2 may be likened to fuzzy points on the coordinate axes V_1 and V_2, and their cylindrical extensions play the role of fuzzy planes passing through these points. The intersection of these planes with the fuzzy surface is a fuzzy point in $V_1 \times V_2 \times V$ which upon projection on V becomes a fuzzy subset of V expressed by the right-hand member of (3.36). A two-dimensional version of this process is shown in Fig. 8.

ANALYTIC REPRESENTATION OF $B(B_1, \ldots, B_n)$

Consider a composite classificational question $Q = B$? whose constituents are classificational questions $Q_1 = B_1$?, $Q_2 = B_2$?, ..., $Q_n = B_n$? in which the body, B_i, of Q_i, $i = 1, \ldots, n$, is a specified fuzzy subset of the universe of discouse V_i. Furthermore, assume that the relation $B(B_1, \ldots, B_n)$ is a function from $A_1 \times \ldots \times A_n$ to A. This implies that an answer to Q—which may be interpreted as a specification of the grade of membership of a given object x in B—is a function of the grades of membership of x in $Q_1, \ldots, Q_n$. In this sense, the B_i form a basis for Q.

When a collection of fuzzy sets $B_1, \ldots, B_n$ forms a basis for Q, it may be convenient to express B, the body of Q, as an explicit function of $B_1, \ldots, B_n$. Such a function may

†It is understood that the right-hand member of (3.36) should be approximated to by an admissible answer to Q.

‡The cylindrical extensions of a^1 and a^2 are, respectively, the ternary fuzzy relations $a^1 \times V \times V$ and $V \times a^2 \times V$. The definition of the projection of a fuzzy relation is given in the Appendix and additional details may be found in Zadeh (1966, 1975a).

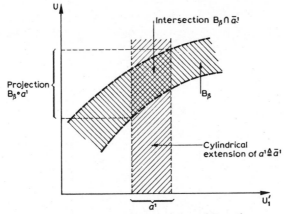

FIG. 8. Graphical interpretation of $B_\beta \bigcirc a^1$.

involve such standard operations as the union, $B_1 + B_2$; intersection, $B_1 \cap B_2$; complement, B_1'; product, $B_1 B_2$; Cartesian product, $B_1 \times B_2$; etc. In addition, it may involve other specified operations—in particular, the interactive versions of $+$ and $\cap$, which will be denoted by $<+>$ and $<\cap>$, respectively.† The expression for B as a function of $B_1, \ldots, B_n$ will be referred to as an *analytic representation* of B.

Example 3.37. Suppose that we wish to define the concept of HIPPIE. To this end, we form the classificational question Q = HIPPIE? and assume that the basis for HIPPIE is the collection of fuzzy sets $B_1 \triangleq$ LONG HAIR, $B_2 \triangleq$ BALD, $B_3 \triangleq$ DRUGS and $B_4 \triangleq$ EMPLOYED, which will be abbreviated as LH, B, D and EMP, respectively.

An analytic representation for B which constitutes the definition of HIPPIE in terms of B_1, B_2, B_3 and B_4 might be:‡

$$\text{HIPPIE} = (\text{LH} + \text{B}) \cap \text{DRUGS} \cap \text{EMP}' \qquad (3.38)$$

or equivalently

$$\text{HIPPIE} = (\text{LH } or \text{ B}) \; and \; \text{DRUGS } and \; not \; \text{EMP} \qquad (3.39)$$

which implies that the grade of membership of a subject x in the fuzzy set HIPPIE is related to the grades of membership of x in the fuzzy set of LONG HAIR subjects, BALD subjects, DRUG TAKING subjects and EMPLOYED subjects by the expression

$$\mu_H(x) = (\mu_{LH}(x) \lor \mu_B(x)) \land \mu_D(x) \land (1 - \mu_{EMP}(x)) \qquad (3.40)$$

where $\lor \triangleq$ max and $\land \triangleq$ min. A representation of (3.39) in the form of a flowchart is shown in Fig. 9, with the understanding that YES and NO are answers of the form

†In general, the angular brackets are used to identify an interactive version of an operation, e.g., $<and>$ is an interactive version of *and*. A brief discussion of interactive operations is given in the Appendix.

‡This definition is used only for illustrative purposes and has no pretense at being a realistic definition of the concept of HIPPIE.

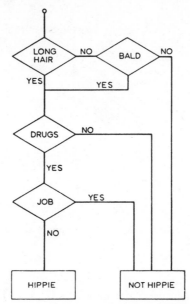

FIG. 9. Flowchart representation of HIPPIE.

YES μ and NO $(1-\mu)$, where μ is the grade of membership of x in the fuzzy set which labels the question.

Note 3.41. If (3.40) does not constitute an acceptable approximation to the expression for $\mu_H(x)$ as a function of $\mu_{LH}(x)$, $\mu_B(x)$, $\mu_D(x)$ and $\mu_{EMP}(x)$, it may be possible to improve on the approximation by employing interactive versions of *and* and/or *or*. For example, we may write

$$\text{HIPPIE} = ((\text{LH } or \text{ B}) <and> \text{DRUGS}) \text{ } and \text{ } not \text{ EMP} \tag{3.41}$$

where $<and>$ is defined by a linguistic relation of the form

u	v	$u <and> v$
t	t	t^2
t	b	b
t	f	f
.	.	.
f	f	f^2

in which t, b, f, t^2 and f^2 are abbreviations for the linguistic truth-values *true, borderline, false, very true,* and *very false.*

Basically, the interactive versions of *and* and *or* serve to extend the usefulness of these connectives by providing a means of taking into account the trade-offs that might exist between their operands. However, it should be noted that, in general, $<and>$ and $<or>$

will not possess such properties as associativity, distributivity, etc., and hence could not be manipulated as conveniently as their non-interactive counterparts.

We turn next to the representation of B by means of branching questionnaires.

4. Branching questionnaires

In one form or another, the concept of a branching questionnaire plays an important role in many fields, especially in taxonomy (Sokal & Sneath, 1963; Cole, 1969; Picard, 1965; Oppenheim, 1966), pattern recognition (Watanabe, 1969; Fu, 1968; Selkow, 1974; Budacker & Saaty, 1965; Fu, 1974; Slagle & Lee, 1971; Hyafil & Rivest, 1973; Meisel & Michalopoulos, 1973), diagnostics and, more particularly, the identification of sequential machines (Gill, 1962; Tal, 1965; Gill, 1969; Kohavi, Rivierre & Kohavi, 1974). In what follows, the term *branching questionnaire* will be used in a more specific sense to refer to a representation of a composite question, $Q \triangleq B$?, in which the constituent questions $Q_1, \ldots, Q_n$ are asked in an order determined by the answers to the previous questions. A branching questionnaire representation of $Q \triangleq B$? will be denoted by $Q^\star$ or, more explicitly, by $B^\star$.

A branching questionnaire, $Q^\star$, may be conveniently represented in the form of a tree as shown in Fig. 10 (or, alternatively, in the form of a block diagram, as in Fig. 11). The root of this tree is labeled with the name of the composite question, Q, or with the name of the body of Q; the leaves are labeled with the admissible answers to Q; and the internal nodes are labeled with the names of the constituent questions or the names of their bodies. Thus, each fan† of the tree represents a constituent question, with each branch of the fan corresponding to an admissible answer to that question. If a branch such as a_1^2 of question Q_2 terminates on Q_1, it means that if the answer to question

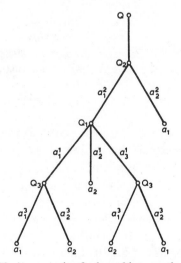

FIG. 10. An example of a branching questionnaire.

†By the *fan* of a tree we mean a node of a tree together with the branches connected to it.

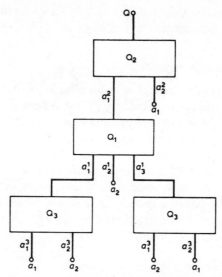

Fɪɢ. 11. Block diagram representation of a branching questionnaire.

Q_2 of a_1^2, then the next question to be asked is Q_1. This implies that if the answer to Q_2 is a_1^2, the answer to Q_1 is a_3^1 and the answer to Q_3 is a_1^3, then the answer to Q is a_2.

Each path from the root of the tree to a leaf represents a particular Q/A sequence, e.g.

$$Q_2\, a_1^2\, Q_1\, a_1^1\, Q_3\, a_1^3//Q\, a_1 \tag{4.1}$$

which may be written more simply as

$$a_1^2\, a_1^1\, a_1^3//a_1 \tag{4.2}$$

if the names of the answers to the constituent questions are labeled in a way that makes it possible to associate each answer in the sequence with a unique constituent question.

It is important to note that the only condition on the structure of a branching questionnaire is that on any path from the root to a leaf each constituent question is encountered at most once. A prescription of the order in which the constituent questions are to be asked (without regard to the answers to Q) is characterized in the manner shown in Fig. 12.

The summation (union) of all Q/A sequences of the form (4.2) constitutes an algebraic representation of $Q^\star$. For example, for the branching questionnaire of Fig. 10, we have the representation (using $Q^\star$ in place of $B^\star$)

$$Q^\star = a_1^2\, a_1^1\, a_1^3//a_1 + a_1^2\, a_1^1\, a_2^3//a_2 + a_1^2\, a_2^1//a_2 \tag{4.3}$$
$$+\ a_1^2\, a_3^1\, a_1^3//a_2 + a_1^2\, a_3^1\, a_2^3//a_1 + a_2^2//a_1.$$

A Q/A sequence which terminates on an internal node of the tree defines an access path to that node and thereby uniquely identifies it. For example, the Q/A sequence a_1^2 identifies the node Q_1 in the tree of Fig. 10. Similarly, the leftmost Q_3 in Fig. 10 is identified by the Q/A sequence $a_1^2\, a_1^1$.‡

‡It should be noted that such Q/A sequences serve a role similar to that of *composite selectors* in the case of a Vienna definition language object (Wegner, 1972).

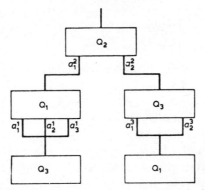

FIG. 12. Specification of the order of questioning.

Each internal node of the tree may be viewed as the root of a subtree which corresponds to a subquestionnaire of $Q^\star$. Thus, on factoring a_1^2 in (4.3), we obtain

$$Q^\star = a_1^2 \, (a_1^1 \, a_1^3//a_1 + a_1^1 \, a_2^3//a_2 + a_2^1//a_2 \qquad (4.4)$$
$$+ \, a_3^1 \, a_1^3//a_2 + a_3^1 \, a_2^3//a_1) + a_2^2//a_1$$

in which the expressions within the parentheses may be regarded as an algebraic representation of a subquestionnaire which has Q_1 and Q_3 as its constituents.

Comment 4.5. By analogy with the concept of a derivative in the case of regular expressions (Tal, 1965; Gill, 1969; Kohavi, Rivierre & Hohavi, 1974; Booth, 1968; Klir & Seidl, 1966; Kohavi, 1970), the coefficients of a_1^2 and a_2^2 in (4.4) may be defined to be the *derivatives* of $Q^\star$ with respect to a_1^2 and a_2^2, respectively. Thus, on denoting these derivatives by $D_{a_1^2} Q^\star$ and $D_{a_1^2} Q^\star$, the expression for $Q^\star$ may be rewritten as

$$Q^\star = a_1^2 \, D_{a1^2} \, Q^\star + a_2^2 \, D_{a2^2} \, Q^\star. \qquad (4.6)$$

More generally, let w denote a Q/A sequence (e.g., $w \triangleq a_1^2 \, a_1^1$), and let S_w denote the subtree of $Q^\star$ which is uniquely determined by w. Then, we may write

$$D_w Q^\star = S_w. \qquad (4.7)$$

Now let $N_1, \ldots, N_r$ be the nodes in a cut† of $Q^\star$ and let $Q/A_1, \ldots, Q/A_r$ denote the Q/A sequences which lead from the root of $Q^\star$ to $N_1, \ldots, N_r$, respectively. Then in consequence of (4.7), we can assert the identity

$$Q^\star = \sum_{i=1}^{r} \, Q/A_i \, D_{Q/A_i} \, Q^\star \qquad (4.8)$$

of which (4.6) may be viewed as a special case.

---

†The *cut* of a tree is a set of nodes with the following properties: (a) no two nodes in the cut are on the same path from the root to a leaf; and (b) no other node of the tree can be added to the cut without violating (a) (Budacker & Saaty, 1965; Aho & Ullman, 1973).

Note 4.9. It should be observed that the constituent questions in Q* may be *λ-interactive* in the sense defined in Zadeh (1973), that is, the answers to, say, $Q_{i_1}, \ldots, Q_{i_k}$, where $(i_1, \ldots, i_k)$ is a subsequence of the index sequence $(1, 2, \ldots, n)$, may restrict the possible answers to $Q_{j_1}, \ldots, Q_{j_l}$, where $(j_1 \ldots, j_l)$ is a subsequence complementary to $(i_1, \ldots, i_k)$ (e.g. for $n = 5$, $(i_1, i_2, i_3) \triangleq (2, 4, 5)$ and $(j_1, j_2) \triangleq (1, 3)$). For example, if the answer to an attributional question $Q_1 \triangleq$ *mother of Julie?* is *Frances*, then the answer to $Q_2 \triangleq$ *sister of Julie?* cannot be *Frances* if there is just one *Frances* in the universes U_1 and U_2. Thus, the answer $a^2 = $ *Frances* is *conditionally impossible given* $a^1 = $ *Frances*.

In the tree representation of a branching questionnaire, the conditional impossibility of an answer to a single question is indicated by associating θ (empty set) with the leaf of the corresponding branch (Fig. 13). Thus, in the example under consideration, a_2 is conditionally impossible given a_1^2. Note that any conditionally impossible answer must of necessity be a leaf of the tree since a Q/A sequence is aborted when a conditionally impossible answer is encountered.

The set of all possible answers to $Q_1, \ldots, Q_n$ constitutes a *restriction* on $Q_1, \ldots, Q_n$. Correspondingly, the conditionally possible answers to $Q_{j_1}, \ldots, Q_{j_l}$ given the answers to $Q_{i_1}, \ldots, Q_{i_k}$ constitute a *conditioned restriction* on $Q_{j_1}, \ldots, Q_{j_l}$ given $Q_{i_1}, \ldots, Q_{i_k}$.† In terms of restrictions, the constituent questions $Q_1, \ldots, Q_n$ are *λ-noninteractive* if the restriction on $Q_1, \ldots, Q_n$ is the Cartesian product of the answer-sets $A_1, \ldots, A_n$. Stated more simply, the non-interaction of $Q_1, \ldots, Q_n$ means that the answers to any subset of constituent questions, say $Q_{i_1}, \ldots, Q_{i_k}$, do not affect the possible answers to the complementary questions $Q_{j_1}, \ldots, Q_{k_l}$. In what follows, we shall assume, unless stated to the contrary, that the constituent questions in Q are *λ-non-interactive*.

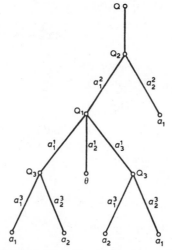

FIG. 13. Illustration of conditional impossibility.

---

† A more detailed discussion of conditioned restrictions may be found in Zadeh (1975a).

CONDITIONAL REDUNDANCE

In comparing the algebraic representations of B and $Q^\star$, (3.6) and (4.3), we observe that every term in B involves the answers to all of the constituent questions in Q, whereas a term in $Q^\star$ involves, in general, a subset of the answers to $Q_1, \ldots, Q_n$.

More specifically, a term such as $a_2^2//a_1$ in $Q^\star$ implies that if the answer to Q_2 is a_2^2, then regardless of the answers to Q_1 and Q_3, the answer to Q is a_1. Thus, in this instance we may say that Q_1 and Q_3 are *conditionally redundant given* a_2^2. Similarly, Q_3 is conditionally redundant given $a_1^2 a_2^1$. By implication, then, a constituent question Q_i, is *unconditionally redundant* if the answers to Q are independent of the answers to Q_i.

A constituent question Q_i will be said to be *conditionally redundant given* $Q_{i_1}, \ldots, Q_{i_k}$ if for every set of possible answers $a_{\lambda_1}^{i1}, \ldots, a_{\lambda_k}^{ik}$, Q_i is conditionally redundant given $a_{\lambda_1}^{i1}, \ldots, a_{\lambda_k}^{ik}$. As we shall see in section 5, the detection of conditional redundancies plays an important role in the construction of efficient branching questionnaires.

Comment 4.10. It should be noted that if the answer to Q_i is uniquely determined by the answers to $Q_{i_1}, \ldots, Q_{i_k}$, then Q_i is conditionally redundant given $Q_{i_1}, \ldots, Q_{i_k}$. However, in general, conditional redundance of Q_i given $Q_{i_1}, \ldots, Q_{i_k}$ is weaker than the dependence of Q_i on $Q_{i_1}, \ldots, Q_{i_k}$.

TABULAR REPRESENTATION OF A BRANCHING QUESTIONNAIRE

As was pointed out already, a term such as $a_2^2//a_1$ in (4.3) signifies that if the answer to Q_2 is a_2^2, then the answer to Q is a_1, no matter what the answers to Q_1 and Q_2 might be. Now, "no matter what" or, equivalently, "don't care" may be interpreted as the answers $a^1 \triangleq a_1^1 + a_2^1 + a_3^1$ to Q_1 and $a^3 \triangleq a_1^3 + a_2^3$ to Q_2. Thus, more generally, a "don't care" answer to Q_i may be interpreted as the answer $a^i \triangleq A_i \triangleq$ answer-set of Q_i.

For simplicity, it is convenient to represent an answer of the form $a^i \triangleq A_i$ by $\star$ or, if necessary, by $\star_i$. With this notation, the tableau of $Q^\star$ (see 4.3) assumes the form shown in Table 3. (The dotted line(s) in this tableau serves to identify the groups of rows which have the same entry in the Q column.)

TABLE 3
Tableau of $Q^\star$

Q_1	Q_2	Q_3	Q
a_1^1	a_1^2	a_1^3	a_1
a_3^1	$\star$	a_2^3	a_1
$\star$	a_2^2	$\star$	a_1
a_3^1	a_1^2	a_1^3	a_2
a_1^1	a_1^2	a_2^3	a_2
a_2^1	a_1^2	$\star$	a_2

A row such as $\star a_2^2 \star$ in this tableau may be represented algebraically as†

$$\begin{aligned}
\star a_2^2 \star &= (a_1^1 + a_2^1 + a_3^1)\, a_2^2\, (a_1^3 + a_2^3) \\
&= a_1^1\, a_2^2\, a_1^3 + a_1^1\, a_2^2\, a_2^3 + a_2^1\, a_2^2\, a_1^3 \\
&\quad + a_2^1\, a_2^2\, a_2^3 + a_3^1\, a_2^2\, a_1^3 + a_3^1\, a_2^2\, a_2^3.
\end{aligned} \tag{4.11}$$

On performing similar expansions for all rows in Table 3 which contain stars, we obtain the complete tableau of $Q\star$, as shown in Table 4.

The preceding discussion indicates that the tableau of Table 3 may be derived from that of Table 4 by a factorization of terms in the algebraic representation of $Q\star$, (4.3), and replacing by $\star$'s those factors which have the form of the sum of all admissible answers to a constituent question. A systematic procedure for carrying out such factorizations will be described in the following section.

TABLE 4
Complete tableau of $Q\star$

Q	Q	Q	Q
a_1^1	a_1^2	a_1^3	a_1
a_3^1	a_1^2	a_2^3	a_1
a_3^1	a_2^2	a_2^3	a_1
a_1^1	a_2^2	a_1^3	a_1
a_1^1	a_2^2	a_2^3	a_1
a_2^1	a_2^2	a_2^3	a_1
a_2^1	a_2^2	a_2^3	a_1
a_3^1	a_2^2	a_1^3	a_1
a_3^1	a_1^2	a_1^3	a_2
a_1^1	a_1^2	a_2^3	a_2
a_2^1	a_1^2	a_1^3	a_2
a_2^1	a_1^2	a_2^3	a_2

Note 4.12. If the constituent questions in Q are λ-interactive, then in a term such as $a_3^1 \star a_2^3$, the star would represent the conditioned restriction on Q_2 given $a_3^1 a_2^3$. More generally, in a term of the form $a_{\lambda_1}^{i_1} \ldots a_{\lambda_k}^{i_k} \star_{j_1} \ldots \star_{j_l}$, the sequence $\star_{j_1} \ldots \star_{j_l}$ would represent the conditioned restriction on $Q_{j_1}, \ldots, Q_{j_l}$ given the Q/A sequence $a_{\lambda_1}^{i_1} \ldots a_{\lambda_k}^{i_k}$.

5. Construction of branching questionnaires

In constructing a fuzzy-algorithmic definition of a concept B, the first step would normally involve a tabulation of the relational representation, $B(Q_1, \ldots, Q_n)$, of a composite question, $Q = B?$, which has B as its body. The second step, then, would

---

†In the terminology of switching theory, the terms on the right-hand side of (4.11) are *covered* by $\star a_2^2 \star$, and $\star a_2^2 \star$ constitutes a *prime implicant* of $Q\star$ (Kohavi, 1970; McCluskey, 1965; Marcovitz & Pugsley, 1971).

involve the construction of a branching questionnaire realization of Q which is efficient in the sense of minimizing a cost function whose components are the costs of answering the constituent questions in Q. In practice, such a cost function would usually be prescribed in a highly approximate fashion.

As an illustration of the first step, suppose that we wish to construct a fuzzy-algorithmic definition of the concept of recession. Using our intuitive knowledge of the factors which enter into this concept and the interrelations between them, we construct in an approximate fashion a linguistic relational representation for RECESSION which might have the form shown in Table 5.† In this table, the observation interval is assumed to be a two-quarter period; GNP ↓ denotes the decline in the gross national product; UNEMP denotes unemployment; BANKR ↑ represents the increase in bankruptcies; and DJ ↓ denotes the decline in the Dow Jones stock average in relation to its maximum value over the observation interval.

TABLE 5
Tableau of relational representation of RECESSION

GNP ↓	UNEMP	BANKR ↑	DJ ↓	RECESSION
small	*low*	*small*	*small*	*false*
moderate	*low*	*small*	*small*	*not true*
high	*low*	*small*	*small*	*borderline*
.	.	.	.	.
high	*moderate*	*moderate*	*large*	*rather true*
high	*high*	*large*	*large*	*very true*

It should be noted that the composite question Q ≙ RECESSION? is treated as a classificational question in Table 5, although all of the constituent questions in RECESSION are attributional in nature. Normally, the meaning of the linguistic values of the attributes would be defined in terms of their compatibility functions, which can be computed from the knowledge of the compatibility functions of the primary fuzzy sets. For example, in the case of unemployment, the compatibility functions of the primary fuzzy sets labeled *low* and *high* might be of the form shown in Fig. 14. From these, one can compute, if needed, the compatibility functions of *very low*, *more or less high*, etc., by the use of (2.22) and (2.23).

There are several basic problems which are ancillary to the transformation of a relational representation of the definiendum (i.e. the concept under definition) into an efficient branching questionnaire. Of these, one is that of determining the conditional redundancies and/or restrictions which may be present in the relational representation. Another is that of determining the order in which the constituent questions must be asked in order to minimize the average cost of finding the answer to Q.

†This representation is used merely for illustrative purposes and should not be taken as a realistic definition of the concept of a recession. A brief but informative discussion of recessions may be found in Silk (1974) and Clark (1974).

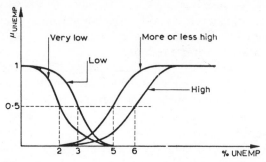

Fig. 14. Compatibility functions of *low, very low, high* and *more or less high* (not to scale).

These and related problems have many features in common with the minimization of switching circuits (McCluskey, 1965; Marcovitz & Pugsley, 1971; Kandel, 1963), optimal encoding (Jelinek, 1968), feature selection in pattern recognition (Chen, 1971; Mucciardi & Gose, 1971; Tou & Heydorn, 1967; Jardine & Sibson, 1971), and the optimization of decision tables (Pollack, Hicks & Harrison, 1971; Montalbano, 1962; Reinwald & Soland, 1967; Bell, 1974). However, the construction of an efficient branching questionnaire for the purpose of defining a concept presents some special problems relating to the fact that the efficiency of a branching questionnaire is influenced not only by the conditional redundancies but also by the cost of the constituent questions as well as by the conditional probabilities of the admissible anwers—probabilities which are conditioned on the answers to the preceding questions in the questionnaire.

In what follows, our discussion of the construction of efficient branching question-naires will be quite restricted in scope. Thus, we shall focus our attention mainly on the determination of the conditional redundancies in a relational representation and an illustration of the computation of the average cost of finding an answer to Q for a given branching questionnaire realization of B.

COMPACTIFICATION OF Q

By the *compactification* of Q (or B) we mean the process of putting the representation of Q (tabular, algebraic or graphical) into a form that places in evidence the conditional redundancies and/or restrictions in the relational representation of Q, and thereby achieves a greater degree of compactness in its mode of representation. Thus, the transition from the tableau of Table 4 to that of Table 3 is an instance of compactification of a tabular representation of a composite question.

If the initial representation has the form of a graph or, more specifically, a tree, then the following rule—which is both general and simple to apply—may be employed to compactify the representation.

Rule 5.1 (merger rule). Let Q^* be a tree representation of a branching questionnaire, and let $S_1, S_2, \ldots, S_l$ be subtrees of Q^* which are identical (i.e. have the same structure as well as branch and node labels)

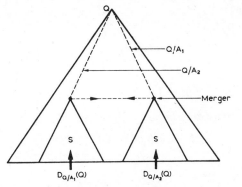

FIG. 15. Illustration of the merger rule.

$$S_1 \equiv S_2 \equiv \ldots \equiv S_l \equiv S. \tag{5.2}$$

Then $S_1, \ldots, S_l$ may be merged into a single subtree S, as shown in Figs 15 and 16. *Comment 5.3.* It should be noted that the structure resulting from a merger is not a tree but an acyclic graph (with the branches oriented downward) which, for convenience, may be referred to as a *semitree*. More generally, then, in the statement of Rule 5.1 the term *tree* should be replaced throughout by *semitree*.

The basis for the merger rule is provided by the following observation. Let $Q/A_1, \ldots,$

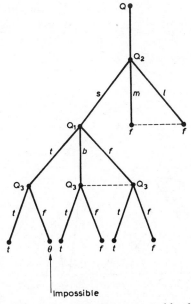

FIG. 16. Illustration of the merger rule—the nodes connected by dashed lines are mergeable.

Q/A_l be the Q/A sequences which terminate on the roots of $S_1, \ldots, S_l$ and let $Q^\star$ be an algebraic representation of Q (see (4.3)). Then by (4.7)

$$S_1 = D_{Q/A_1}Q^\star \qquad (5.4)$$

$$\cdots \cdots \cdots$$

$$S_l = D_{Q/A_l}Q^\star \qquad (5.5)$$

where D_{Q/A_λ} denotes the derivative of $Q^\star$ with respect to Q/A_λ, $\lambda = 1, \ldots, l$.

Now let $N_1, \ldots, N_r$, $r \geq l$, be a set of nodes in $Q^\star$ which form a cut, with the roots of $S_1, \ldots, S_l$ identified with $N_1, \ldots, N_l$, respectively. Then by (4.8), we can express $Q^\star$ as

$$Q^\star = Q/A_1\, D_{Q/A_1}Q^\star + \ldots + Q/A_l\, D_{Q/A_l}Q^\star + \ldots + Q/A_r\, D_{Q/A_r}\, Q^\star. \qquad (5.6)$$

From (5.6) and the assumption that $S_1 \equiv \ldots \equiv S_l \equiv S$, it follows that the common factor $D_{Q/A_1}Q^\star$ may be factored from the first l terms in (5.6), yielding the simpler expression

$$Q^\star = (Q/A_1 + \ldots + Q/A_l)\, D_{Q/A_1}Q^\star + \ldots + Q/A_r\, D_{Q/A_r}Q^\star. \qquad (5.7)$$

The conclusion that follows from (5.7), then, is that the result of application of Rule 5.1 is a semitree whose algebraic representation is expressed by (5.7).

The conditionally redundant questions in $Q^\star$ may readily be deduced by a straightforward application of the merger rule, as illustrated in Fig. 17. Thus, assume that the roots of $S_1, \ldots, S_l$, where $S_1 \equiv \ldots \equiv S_l \equiv S$, are the leaves of a fan which represents a constituent question, say Q_3. Then from (5.7) it follows at once that Q_3 is conditionally redundant given Q/A_3. More generally, if some of the answers to a constituent question are conditionally impossible (e.g. as in Q_2 in Fig. 17), then the condition $S_1 \equiv \ldots \equiv S_l$ need hold only for the conditionally possible answers to Q_2. Thus, in Fig. 17, Q_2 is conditionally redundant given Q/A_2.

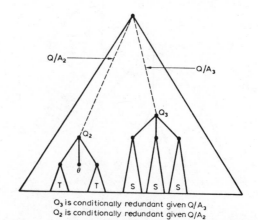

Q₃ is conditionally redundant given Q/A₃
Q₂ is conditionally redundant given Q/A₂

FIG. 17. Application of the merger rule to the identification of conditionally redundant questions.

It is helpful to summarize the foregoing discussion in the form of a proposition.

Proposition 5.8. Let Q_i be a constituent question in $Q^\star$ whose conditionally possible leaves (i.e. the leaves corresponding to conditionally possible answers) are $N_{i_1}, \ldots, N_{i_k}$, and let Q/A_i denote the Q/A sequence terminating on Q_i. Then Q_i is conditionally redundant given Q/A_i if the subtrees (or, more precisely, the semitrees) with roots at $N_{i_1} \ldots, N_{i_k}$ are identical.

COMPACTIFICATION OF A TABULAR REPRESENTATION

Like most graphical procedures, the merger rule discussed above serves to provide a visual and hence more readily comprehensible idea of how it works. For computational purposes, however, it is preferable to employ compactification techniques which operate on tables rather than graphs.

A technique of this type which is described below† is a straightforward adaptation of the well-known Quine–McCluskey algorithm (McCluskey, 1965; Marcovitz & Pugsley, 1971; Kandel, 1973) for the minimization of switching functions. More specifically, suppose that we wish to compactify a given tableau $Q(Q_1, \ldots, Q_n)$, e.g. that of Table 4, in which the rows which have the same entry in the Q column are grouped together as shown. The steps described below, then, would be applied to each such group. (For easier comprehension, the algorithm is expressed in informal terms.)

Algorithm 5.9. The following steps are performed successively for each column in Q, starting with $j = 1$. r_i^j denotes an admissible answer in the ith row of the jth column.

1. Starting with $i = 1$, check if r_1^1 can be replaced by $\star$ (i.e. by the answer-set A_1)·
   (The answer is YES if there are rows in Q which upon addition (treating the rows as strings, as in (3.6), and factoring the common factor $r_1^2 \ldots r_1^n$ yield the term $A_1 r_1^2 \ldots r_1^n$.) If the answer is YES, add the row $\star r_1^2 \ldots r_1^n$ to the tableau, yielding what will be referred to as an *augmented* tableau.

As an illustration, in the tableau of Table 6, the answer is NO for r_1^1 and YES for r_2^1. Consequently, $\star\, a_2^2\, a_1^3$ is added to the tableau as shown in Table 6.

2. Step 1 is applied in succession to all of the entries in column 1 of Q which fall into the group under consideration.

This concludes Pass (1) of the algorithm, yielding an augmented tableau which consists of the original rows together with rows in which the entry in column 1 is a star.

3. Steps 1 and 2 are applied successively to the entries in Columns 2, 3, . . .,n, with the understanding that the initial tableau for Pass $(i+1)$ is the augmented tableau obtained at the conclusion of Pass (i), with $\star$ treated as if it were an element of an answer-set. Furthermore, in applying Step 1 to an entry in column j, all of the rows augmented up to that point must be considered.

4. In the final augmented tableau obtained at the conclusion of Pass (n), each of the

---

†For simplicity, we shall assume that the constituent questions are non-interactive in the sense of Zadeh (1975*a*).

TABLE 6

Intermediate results of Algorithm 5·9 for group 1 of rows of Q

	Q_1	Q_2	Q_3	Q
Group 1 (initial)	a_1^1	a_1^2	a_1^3	a_1
	a_1^1	a_2^2	a_1^3	a_1
	a_2^1	a_2^2	a_1^3	a_1
	a_3^1	a_2^2	a_1^3	a_1
	a_3^1	a_1^2	a_2^3	a_1
	a_1^1	a_2^2	a_2^3	a_1
	a_2^1	a_2^2	a_2^3	a_1
	a_3^1	a_2^2	a_2^3	a_1
Pass (1)	$\star$	a_2^2	a_1^3	a_1
	$\star$	a_2^2	a_2^3	a_1
Pass (2)	a_3^1	$\star$	a_2^3	a_1
Pass (3)	a_1^1	a_2^2	$\star$	a_1
	a_2^1	a_2^2	$\star$	a_1
	a_3^1	a_2^2	$\star$	a_1
	$\star$	a_2^2	$\star$	a_1

rows is checked to see if it is contained as a term in an expansion of a starred term in the final augmented tableau. If the answer is NO, the row in question is transferred to a tableau labeled Q**, with the corresponding answer to Q being the same as for the group under consideration.

As an illustration, in Table 6 $a_1^1 a_2^2 a_1^3$ is contained as a term in the expansion of $\star\ a_2^2 a_1^3$ and hence is not transferred to Q**. The row $a_1^1 a_1^2 a_1^3$ is not contained in the expansion of any starred term and hence is transferred to Q**, with a_1 being the entry in column Q. The row $a_1^1 a_2^2 \star$ is contained in $\star\ a_2^2 \star$ and hence is not transferred to Q**.

5. On applying Steps 1,2,3,4 to each group in the original tableau, we obtain the final form of Q**. The tableau of Q** represents the desired compactified form of Q. It can readily be verified that Q** places in evidence all of the conditionally redundant questions in Q. For this reason, it will be referred to as a *maximally compact* representation of Q.

Note 5.10. The rows in Q** correspond to the prime implicants of a switching function. For our purposes, it is not necessary to compactify Q** still further by deleting the non-essential prime implicants, that is, those terms in Q** which are contained in sums of expansions of some of the starred terms in Q**.

Example 5.11. Intermediate results of the application of Algorithm 5.9 to the tableau of Table 4 are shown in Tables 6 and 7. The final result, Q**, is shown in Table 8.

TABLE 7

Intermediate results of Algorithm 5·9 for group 2 of rows of Q

	Q_1	Q_2	Q_3	Q
Group 2 (initial)	a_2^1	a_1^2	a_1^3	a_2
	a_3^1	a_1^2	a_1^3	a_2
	a_1^1	a_1^2	a_2^3	a_2
	a_2^1	a_1^2	a_2^3	a_2
Pass (3)	a_2^1	a_1^2	$\star$	a_2

COMPUTATION OF THE AVERAGE COST OF FINDING AN ANSWER TO Q

Given a branching questionnaire, $Q^\star$, together with (a) the conditional probabilities of the admissible answers to each constituent question given the answers to the preceding questions; and (b) the cost of each constituent question, it is a simple matter to compute the average cost of finding an answer to Q through the use of $Q^\star$.

Specifically. let $B^\star$ be an algebraic representation for a branching questionnaire $Q^\star$ (see (4.3)), and let $a_{\lambda_1}^{j_1} \ldots a_{\lambda_k}^{j_k} // a_{\lambda_{k+1}}$ be a term in $B^\star$ corresponding to a path from the root to a leaf of $Q^\star$. Let $p_{\lambda_1}^{j_1}, \ldots, p_{\lambda_k}^{j_k}$ be the probabilities associated with the branches $a_{\lambda_1}^{j_1}, \ldots, a_{\lambda_k}^{j_k}$ along this path, and let $C_{J_1}, \ldots, C_{J_k}$ be the costs associated with $Q_{J_1} \ldots, Q_{J_k}$. Then the expected cost of an answer to Q through the use of $Q^\star$ is given by

$$C_{av} = \Sigma p_{\lambda_1}^{j_1} \ldots p_{\lambda_k}^{j_k} (C_{J_1} + \ldots + C_{J_k}) \qquad (5.12)$$

where the summation is taken over all possible paths from the root to the leaves of $Q^\star$.

Example 5.13. Consider the branching questionnaire shown in Fig. 18, in which $C_1 = 2$, $C_2 = 3$, $C_3 = 1$ and the conditional probabilities have the indicated values. (Note that the probabilities associated with the root are not conditional.) Then using (5.12), we have

TABLE 8

Maximally compact representation of Q

	Q_1	Q_2	Q_3	Q
	a_1^1	a_1^2	a_1^3	a_1
	a_3^1	$\star$	a_2^3	a_1
	$\star$	a_2^2	$\star$	a_1
	a_3^1	a_1^2	a_1^3	a_2
	a_1^1	a_1^2	a_2^3	a_2
	a_2^1	a_1^2	$\star$	a_2

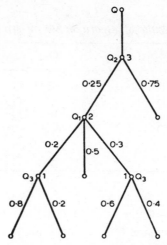

FIG. 18. Conditional probabilities and costs associated with constituent questions.

$$C_{av} = 0{\cdot}04 \times 6 + 0{\cdot}01 \times 6 + 0{\cdot}125 \times 5 + \qquad (5.14)$$
$$0{\cdot}045 \times 6 + 0{\cdot}03 \times 6 + 0{\cdot}75 \times 3$$
$$= 3{\cdot}625$$

Clearly, the determination of a realization of Q** in the form of maximally efficient branching questionnaire—that is, a realization which minimizes C_{av}—is a non-trivial problem. However, since in most situations the conditional probabilities and the costs of constituent questions are likely to be known imprecisely, if at all, highly approximate solutions which yield merely reasonably efficient realizations are likely to be adequate. This may well be the case, for example, in the construction of efficient branching questionnaires for purposes of medical diagnosis, in which both the costs and the probabilities of constituent questions are likely to be both highly variable and poorly defined.

We shall not dwell further upon this problem in the present paper.

6. Concluding remarks

The ideas presented in this paper are merely a first step toward the development of a much more comprehensive theory of fuzzy-algorithmic definitions. We have not considered, for example, fuzzy-algorithmic definitions in which the answers to Q have the form of a probability distribution over an answer-set A. Nor have we considered more complicated types of definitions in which the object, x, is not the same for all constituent questions, or in which the order in which the questions are asked is fuzzy or probabilistic.

Although lacking in complete generality, the relatively simple types of definitions which we have discussed may find useful applications in a variety of fields. Experience with such applications may well suggest many improvements in the approach described in this paper and point to areas requiring further exploration.

The author is indebted to Richard Karp and Jeff Yang for helpful suggestions concerning the optimization of branching questionnaires.

References

AHO, A. V. & ULLMAN, J. D. (1973). *The Theory of Parsing, Translation and Compiling*. Englewood Cliffs, New Jersey: Prentice-Hall.

BELL, D. A. (1974). Decision trees made easy. *Proceedings Second International Joint Conference on Pattern Recognition*, pp. 18–21. Copenhagen, Denmark.

BELLMAN, R. E., KALABA, R. & ZADEH, L. A. (1966). Abstraction and pattern classification. *Journal of Mathematical Analysis and Applications*, **13**, 1.

BLACK, M. (1963). Reasoning with loose concepts. *Dialogue*, **2**, 1.

BOOTH, T. L. (1968). *Sequential Machines and Automata Theory*. New York: John Wiley and Sons.

BUDACKER, R. G. & SAATY, T. L. (1965). *Finite Graphs and Networks*. New York: McGraw-Hill.

CARNAP, R. (1956). *Meaning and Necessity*. Chicago: University of Chicago Press.

CHEN, C. H. (1971). Theoretical comparison of a class of feature selection criteria in pattern recognition. *IEEE Transactions on Computers*, C-20, 1054.

CHURCH, A. (1951). A formulation of the logic of sense and denotation. In *Structure, Method and Meaning*. New York: Liberal Arts Press.

CLARK, L. H., JR. (1974). An economics group is the one that says if there's recession. *The Wall Street Journal*, **1**, 30 August, 1974.

CODD, E. F. (1971). Relational completeness of data base sublanguages. In *Courant Computer Science Symposia*, **6**. Englewood Cliffs, New Jersey, Prentice-Hall.

COLE, E. J., Ed. (1969). *Numerical Taxonomy*. New York: Academic Press.

DOMOTOR, Z. (1969). Probabilistic relational structures and their applications. Technical Report 144, Institute for Mathematical Studies in Social Science, Stanford University.

FINE, K. (1973). Vagueness, truth and logic. Department of Philosophy, University of Edinburgh.

FREGE, G. (1952). In P. GEACH & M. BLACK, Eds. *Philosophical Writings*. Oxford: Blackwell.

FU, K. S. (1968). *Sequential Methods in Pattern Recognition and Machine Learning*. New York: Academic Press.

FU, K. S. (1974). *Syntactic Methods in Pattern Recognition*. New York: Academic Press.

GILL, A. (1962). *Introduction to the Theory of Finite-State Machines*. New York: McGraw-Hill.

GILL, A. (1969). Finite-state systems. In L. A. ZADEH & E. POLAK, Eds, *System Theory*. New York: McGraw-Hill.

GOGUEN, J. A. (1969). The logic of inexact concepts. *Synthese*, **19**, 325.

HEMPEL, C. G. (1952). Fundamentals of concept formation in empirical science. In *International Encyclopedia of Unified Science*, **2**.

HINTIKKA, J., MORAVCSIK, J. & SUPPES, P., Eds (1973). *Approaches to Natural Language*. Dordrecht, The Netherlands.

HYAFIL, L. & RIVEST, R. L. (1973). Graph partitioning and constructing optimal decision trees are polynomial complete problems. IRIA Research Report 33, Le Chesnay, France.

JARDINE, N. & SIBSON, R. (1971). Choice of methods for automatic classification. *The Computer Journal*, **17**, 404.

JELINEK, F. (1968). *Probalistic Information Theory*. New York: McGraw-Hill.

KANDEL, A. (1973). On minimization of fuzzy functions. *IEEE Transactions on Computers*, C-22, 826.

KRANTZ, D. H., LUCE, R. D., SUPPES, P. & TWERSKY, A. (1971). *Foundations of Measurement*. New York: Academic Press.

KAUFMANN, A. (1, 1969; 2, 3, 1975). *An Introduction to the Theory of Fuzzy Sets.* Paris: Masson et Cie. (Also; New York, Academic Press, 1975.)

KLIR, J. & SEIDL, L. (1966). *Synthesis of Switching Circuits.* London: Iliffe Books Ltd.

KOHAVI, Z. (1970). *Swtiching and Finite Automata Theory.* New York: McGraw-Hill.

KOHAVI, A., RIVIERRE, J. A. & KOHAVI, I. (1974). Checking experiments for sequential machines. *Information Sciences*, **7**, 11.

LAKOFF, G. (1971). Linguistic and natural logic. In D. DAVIDSON & G. HARMAN, Eds. *Semantics of Natural Languages*, Dordrecht, The Netherlands: D. Reidel Publishing Company.

LAKOFF, G. (1972). Hedges: A study in meaning criteria and the logic of fuzzy concepts. *Proceedings 8th Regional Meeting of Chicago Linguistic Society.* University of Chicago Linguistics Department.

LUCAS, P. *et al.* (1968). Method and notation for the formal definition of programming languages. Report TR 25.087, IBM Laboratory, Vienna.

LUKASIEWICZ, J. (1970). *Selected Works.* North-Holland Publishing Company.

McCLUSKEY, E. J. (1965). *Introduction to the Theory of Switching Circuits.* New York: McGraw-Hill.

MARCOVITZ, A. B. & PUGSLEY, J. H. (1971). *An Introduction to Switching System Design.* New York: John Wiley and Sons.

MARTIN, R. M. (1963). *Intension and Decision.* Englewood Cliffs, New Jersey: Prentice-Hall Inc.

MEISEL, W. S. & MICHALOPOULOS, D. A. (1973). A partitioning algorithm with application in pattern classification and the optimization of decision trees. *IEEE Transactions on Computers*, **C-22**, 93.

MINSKY, M., Ed. (1968). *Semantic Information Processing.* Cambridge, Mass.: MIT Press.

MOISIL, G. (1972). *Essais sur les Logiques Non-Chrysippiennes.* Bucharest: Collected Works.

MONTALBANO, M. (1962). Tables, flowcharts and program logic. *IBM Systems Journal*, **51.**

MUCCIARDI, A. N. & GOSE, E. E. (1971). A comparison of seven techniques for choosing subsets of pattern recognition properties. *IEEE Transactions on Computers*, **C-20**, 1023.

OPPENHEIM, A. N. (1966). *Questionnaire Design and Attitude Measurement.* New York: Basic Books.

PICARD, C. (1965). *Theorie des Questionnaires.* Paris: Gauthier-Villars.

POLLACK, S. L., HICKS, H. T., JR. & HARRISON, W. J. (1971). *Decision Tables: Theory and Practice.* New York: Wiley-Interscience.

QUINE, W. V. (1953). *From a Logical Point of View.* Cambridge, Mass.: Harvard University Press.

REINWALD, L. T. & SOLAND, R. M. (1967). Conversion of limited-entry decision tables to optimal computer programs. I & II. *Journal of the ACM*, **13**, 339; **14**, 742.

SANTOS, E. (1970). Fuzzy algorithms. *Information and Control*, **17**, 326.

SCRIVEN, M. (1958). Definitions, explanations and theories. In H. FEIGL, M. SCRIVEN & G. MAXWELL, Eds. *Minnesota Studies in the Philosophy of Science*, **2.** Minneapolis.

SELKOW, S. M. (1974). Diagnostic keys as a representation for context in pattern recognition. *IEEE Transactions on Computers*, **C-23**, 970.

SILK, L. (1974). Recession: Some criteria missing, so far. *The New York Times*, **37**, 28 August, 1974.

SIMON, H. A. & SIKLOSSY, L., Eds (1972) *Representation and Meaning Experiments with Information Processing Systems.* Englewood Cliffs, New Jersey, Prentice-Hall.

SLAGLE, J. R. & LEE, R. C. T. (1971). Application of game searching techniques to sequential pattern recognition. *Communications of the ACM*, **14**, 103.

SOKAL, R. & SNEATH, P. (1963). *Principles of Numerical Taxonomy.* San Francisco: W. H. Freeman.

TAL, A. A. (1965). The abstract synthesis of sequential machines from the answers to questions of the first kind in the questionnaire language. *Automation and Remote Control*, **26**, 675.

TARSKI, A. (1956). *Logic, Semantics, Metamathematics.* Oxford: Clarendon Press.

Tou, J. T. & Heydorn, R. P. (1967). Some approaches to optimum feature extraction. J. T. Tou, Ed. *Computer and Information Science*, II. New York: Academic Press.

Van Frassen, B. S. (1969). Presuppositions, supervaluations and free logic. In K. Lambert, Ed. *The Logical Way of Doing Things*. New Haven: Yale University Press.

Watanabe, S., Ed. (1969). *Methodologies of Pattern Recognition*. New York: Academic Press.

Wegner, P. (1972). The Vienna definition language. *ACM Computing Surveys*, **4**, 5.

Zadeh, L. A. (1966). Shadows of fuzzy sets. *Problems in Transmission of Information*, **2**, 37 (in Russian).

Zadeh, L. A. (1968). Fuzzy algorithms. *Information and Control*, **12**, 94.

Zadeh, L. A. (1971a). On fuzzy algorithms. Memorandum No. ERL-M325, Electronics Research Laboratory, University of California, Berkeley.

Zadeh, L. A. (1971b). Quantitative fuzzy semantics. *Information Sciences*, **3**, 159.

Zadeh, L. A. (1972a). Fuzzy languages and their relation to human and machine intelligence. *Proceedings of International Conference on Man and Computer*, 130. Bordeaux, France, S. Karger, Basel.

Zadeh, L. A. (1972b). A fuzzy-set-theoretic interpretation of linguistic hedges. *Journal of Cybernetics*, **2**, 4.

Zadeh, L. A. (1973). Outline of a new approach to the analysis of complex systems and decision processes. *IEEE Transactions on Systems, Man and Cybernetics*, **SMC-3**, 28.

Zadeh, L. A. (1975a). The concept of a linguistic variable and its application to approximate reasoning. *Information Sciences*, **8**, 199 (Part I); **8**, 301 (Part II); **9**, 43 (Part III).

Zadeh, L. A. (1975b). Fuzzy logic and approximate reasoning. *Synthese*, **30**, 407.

Zadeh, L. A. (1975c). Calculus of fuzzy restrictions. In L. A. Zadeh, K. S. Fu, K. Tanaka & M. Shimara, Eds. *Fuzzy Sets and Their Application to Cognitive and Decision Processes*, pp. 1–39. New York: Academic Press.

Appendix

For convenience of the reader, a brief summary of some of the relevant aspects of the theory of fuzzy sets and the linguistic approach is presented in this section. More detailed discussions of the topics touched upon in the sequel may be found in the appended list of references and related publications.

NOTATION AND TERMINOLOGY

The symbol U denotes a universe of discourse, which may be an arbitrary collection of subjects or mathematical constructs.

If A is a finite subset of U whose elements are $u_1, \ldots, u_n$, A is expressed as

$$A = u_1 + \ldots + u_n. \tag{A1}$$

A finite fuzzy subset of U is expressed as

$$F = \mu_1 u_1 + \ldots + \mu_n u_n \tag{A2}$$

or, equivalently, as

$$F = \mu_1/u_1 + \ldots + \mu_n/u_n \tag{A3}$$

where the μ_i, $i = 1, \ldots, n$, represent *grades of membership* of the u_i in F. Unless stated to the contrary, the μ_i are assumed to lie in the interval [0,1], with 0 and 1 denoting *no* membership and *full* membership, respectively.

More generally, a fuzzy subset of U is expressed as

$$F = \int_U \mu_F(u)/u \tag{A4}$$

where $\mu_F \colon U \to [0,1]$ is the *membership* (or *compatibility*) function of F, and $\mu_F(u)/u$ is a *fuzzy singleton*. In effect, (A4) expresses F as the union of its constituent fuzzy singletons.

The points in U at which $\mu_F(u) > 0$ constitute the *support* of F. The points at which $\mu_F(u) = 0.5$ are the *crossover* points of F.

Example A5. Assume

$$U = a+b+c+d. \tag{A6}$$

Then, we may have

$$A = a+b+d \tag{A7}$$

and

$$F = 0.3a+0.9b+d \tag{A8}$$

as non-fuzzy and fuzzy subsets of U, respectively.

If

$$U = 0+0.1+0.2+ \ldots +1 \tag{A9}$$

then a fuzzy subset of U would be expressed as, say,

$$F = 0.3/0.5+0.6/0.7+0.8/0.9+1/1. \tag{A10}$$

If $U = [0,1]$, then F might be expressed as

$$F = \int_0^1 \frac{1}{1+u^2}/u \tag{A11}$$

which means that F is a fuzzy subset of the unit interval [0,1] whose membership function is defined by

$$\mu_F(u) = \frac{1}{1+u^2}. \tag{A12}$$

OPERATION ON FUZZY SETS

If F and G are fuzzy subsects of U, their *union*, F+G, and *intersection*, F∩G, are fuzzy subsets of U defined by

$$F+G \triangleq \int_U \mu_F(u) \vee \mu_G(u)/u \tag{A13}$$

$$F \cap G \triangleq \int_U \mu_F(u) \wedge \mu_G(u)/u \tag{A14}$$

where $\vee$ and $\wedge$ denote max and min, respectively. The *complement* of F is defined by

$$F' = \int_U (1-\mu_F(u))/u. \tag{A15}$$

Example A16. For U defined by (A6) and

$$F \quad = 0{\cdot}4a + 0{\cdot}9b + d \tag{A17}$$

$$G \quad = 0{\cdot}6a + 0{\cdot}5b \tag{A18}$$

we have

$$F + G = 0{\cdot}6 + 0{\cdot}9b + d \tag{A19}$$

$$F \cap G = 0{\cdot}4a + 0{\cdot}5b \tag{A20}$$

$$F' \quad = 0{\cdot}6a + 0{\cdot}1b + c. \tag{A21}$$

The linguistic connectives *and* (conjunction) and *or* (disjunction) are identified with $\cap$ and $+$, respectively. Thus,

$$F \ and \ G \triangleq F \cap G \tag{A22}$$

and

$$F \ or \ G \triangleq F + G. \tag{A23}$$

As defined by (A22) and (A23), *and* and *or* are implied to be *non-interactive* in the sense that there is no "trade-off" between their operands. When this is not the case, *and* and *or* are denoted by $<and>$ and $<or>$, respectively, and are defined in a way that reflects the nature of the trade-off. For example, we may have

$$F <and> G \triangleq \int_U \mu_F(u) \, \mu_G(u)/u \tag{A24}$$

$$F <or> G \triangleq \int_U (\mu_F(u) + \mu_G(u) - \mu_F(u) \, \mu_G(u))/u \tag{A25}$$

whose $+$ denotes the arithmetic sum. In general, the interactive versions of *and* and *or* do not possess the simplifying properties of the connectives defined by (A22) and (A23), e.g. associativity, distributivity, etc.

If a is a real number, then F^α is defined by

$$F^\alpha \triangleq \int_V (\mu_F(u))^\alpha/u. \tag{A26}$$

For example, for the fuzzy set defined by (A17), we have

$$F^2 = 0{\cdot}16a + 0{\cdot}81b + d \tag{A27}$$

and

$$F^{1/2} = 0{\cdot}63a + 0{\cdot}95b + d. \tag{A28}$$

These operations may be used to approximate, very roughly, to the effect of the linguistic modifiers *very* and *more or less*. Thus,

$$very \ F \triangleq F^2 \tag{A29}$$

and

$$more \ or \ less \ F \triangleq F^{1/2}. \tag{A30}$$

If $F_1, \ldots, F_n$ are fuzzy subsets of $U_1, \ldots, U_n$, then the *Cartesian product* of $F_1, \ldots, F_n$

is a fuzzy subset of $U_1 \times \ldots \times U_n$ defined by

$$F_1 \times \ldots \times F_n = \int_{U_1 \times \ldots \times U_n} (\mu_{F_1}(u_1) \wedge \ldots \wedge \mu_{F_n}(u_n))/(u_1, \ldots, u_n). \qquad (A31)$$

As an illustration, for the fuzzy sets defined by (A7) and (A18), we have

$$
\begin{aligned}
F \times G &= (0{\cdot}4a + 0{\cdot}9b + d) \times (0{\cdot}6a + 0{\cdot}5b) \qquad\qquad (A32)\\
&= 0{\cdot}4/(a,a) + 0{\cdot}4/(a,b) + 0{\cdot}6/(b,a)\\
&\quad + 0{\cdot}5/(b,b) + 0{\cdot}6/(d,a) + 0{\cdot}5/(d,b)
\end{aligned}
$$

which is a fuzzy subset of $(a+b+c+d) \times (a+b+c+d)$.

FUZZY RELATIONS

An *n-ary fuzzy relation* R in $U_1 \times \ldots \times U_n$ is a fuzzy subset of $U_1 \times \ldots \times U_n$. The *projection of* R *on* $U_{i_1} \times \ldots \times U_{i_k}$, where $(i_1, \ldots i_k)$ is a subsequence of $(1, \ldots, n)$ is a relation in $U_{i_1} \times \ldots \times U_{i_k}$ defined by

$$\text{Proj } R \text{ on } U_{i_1} \times \ldots \times U_{i_k} \triangleq \int_{U_{i_1} \times \ldots \times U_{i_k}} \vee u_{j_1}, \ldots, u_{j_l} \mu_R(u_1, \ldots, u_n)/(u_1, \ldots, u_n) \qquad (A33)$$

where $(j_1, \ldots, j_l)$ is the sequence complementary to $(i_1, \ldots, i_k)$ (e.g. if $n = 6$ then $(1,3,6)$ is complementary to $(2,4,5)$), and $\vee u_{j_1}, \ldots, u_{j_l}$ denotes the supremum over $U_{j_1} \times \ldots \times U_{j_l}$.

If R is a fuzzy subset of $U_{i_1}, \ldots, U_{i_k}$, then its *cylindrical extension* in $U_1 \times \ldots \times U_n$ is a fuzzy subset of $U_1 \times \ldots \times U_n$ defined by

$$\bar{R} = \int_{U_1 \times \ldots \times U_n} \mu_R(u_{i_1}, \ldots, u_{i_k})/(u_1, \ldots, u_n). \qquad (A34)$$

In terms of their cylindrical extensions, the *composition* of two binary relation R and S (in $U_1 \times U_2$ and $U_2 \times U_3$, respectively) is expressed by

$$R \circ S = \text{Proj } \bar{R} \cap \bar{S} \text{ on } U_1 \times U_3 \qquad (A35)$$

where $\bar{R}$ and $\bar{S}$ are the cylindrical extensions of R and S in $U_1 \times U_2 \times U_3$. Similarly, if R is a binary relation in $U_1 \times U_2$ and S is a unary relation in U_2, their composition is given by

$$R \circ S = \text{Proj } R \cap \bar{S} \text{ on } U_1. \qquad (A36)$$

Example A37. Let R be defined by the right-hand member of (A32) and

$$S = 0{\cdot}4a + b + 0{\cdot}8d. \qquad (A38)$$

Then

$$\text{Proj } R \text{ on } U_1 (\triangleq a+b+c+d) = 0{\cdot}4a + 0{\cdot}6b + 0{\cdot}6d \qquad (A39)$$

and

$$R \circ S = 0 \cdot 4a + 0 \cdot 5b + 0 \cdot 5d. \tag{A40}$$

LINGUISTIC VARIABLES

Informally a linguistic variable, χ, is a variable whose values are words or sentences in a natural or artificial language. For example, if *age* is interpreted as a linguistic variable, then its *term-set*, $T(\chi)$, that is, the set of linguistic values, might be

$$T(age) = young + old + very\ young + not\ young + \tag{A41}$$
$$very\ old + very\ very\ young +$$
$$rather\ young + more\ or\ less\ young + \ldots.$$

where each of the terms in $T(age)$ is a label of a fuzzy subset of a universe of discourse, say $U = [0,100]$.

A linguistic variable is associated with two rules: (a) a *syntactic rule*, which defines the well-formed sentences in $T(\chi)$; and (b) a *semantic rule*, by which the meaning of the terms in $T(\chi)$ may be determined. If X is a term in $T(\chi)$, then its *meaning* (in a denotational sense) is a subset of U. A *primary term* in $T(\chi)$ is a term whose meaning is a *primary fuzzy set*, that is, a term whose meaning must be defined a priori, and which serves as a basis for the computation of the meaning of the non-primary terms in $T(\chi)$. For example, the primary terms in (A41) are *young* and *old*, whose meaning might be defined by their respective compatibility functions μ_{young} and μ_{old}. From these, then, the meaning—or, equivalently, the compatibility functions—of the non-primary terms in (A41) may be computed by the application of a semantic rule. For example, employing (A29) and (A30) we have

$$\mu_{very\ young} = (\mu_{young})^2 \tag{A42}$$

$$\mu_{more\ or\ less\ old} = (\mu_{old})^{1/2} \tag{A43}$$

$$\mu_{not\ very\ young} = 1 - (\mu_{young})^2. \tag{A44}$$

For illustration, plots of the compatibility functions of these terms are shown in Fig. A1.

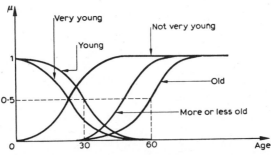

FIG. A1. Compatibility functions of *young, old, very young, more or less old, not very young* (not to scale).

THE EXTENSION PRINCIPLE

Let f be a mapping from U to V. Thus,

$$v = f(u) \tag{A45}$$

where u and v are generic elements of U and V, respectively.

Let F be a fuzzy subset of U expressed as

$$F = \mu_1 u_1 + \ldots + \mu_n u_n. \tag{A46}$$

or more generally

$$F = \int_U \mu_F(u)/u. \tag{A47}$$

By the extension principle (Zadeh, 1975a), the image of F under f is given by

$$f(F) = \mu_1 f(u_1) + \ldots + \mu_n f(u_n) \tag{A48}$$

or, more generally,

$$f(F) = \int_U \mu_F(u)/f(u). \tag{A49}$$

Similarly, if f is a mapping from $U \times V$ to W, and F and G are fuzzy subsets of U and V, respectively, then

$$f(F,G) = \int_W (\mu_F(u) \wedge \mu_G(v))/f(u,v). \tag{A50}$$

Example A51. Assume that f is the operation of squaring. Then, for the set defined by (A10), we have

$$f(0\cdot3/0\cdot5 + 0\cdot6/0\cdot7 + 0\cdot8/0\cdot9 + 1/1) = 0\cdot3/0\cdot25 + 0\cdot6/0\cdot49 + 0\cdot8/0\cdot81 + 1/1. \tag{A51}$$

Similarly, for the binary operation $\vee$ ($\triangleq max$) we have

$$\begin{aligned}
(0\cdot9/0\cdot1 + 0\cdot2/0\cdot5 + 1/1) \vee (0\cdot3/0\cdot2 + 0\cdot8/0\cdot6) = \\
= 0\cdot3/0\cdot2 + 0\cdot2/0\cdot5 + 0\cdot3/1 \\
+ 0\cdot8/0\cdot6 + 0\cdot2/0\cdot6 + 0\cdot8/0\cdot6.
\end{aligned} \tag{A52}$$

Fuzzy Sets as a Basis for a Theory of Possibility*

L.A. ZADEH

Computer Science Division, Department of Electrical Engineering and Computer Sciences and the Electronics Research Laboratory, University of California, Berkeley, CA 94720. U.S.A.

Received February 1977
Revised June 1977

The theory of possibility described in this paper is related to the theory of fuzzy sets by defining the concept of a possibility distribution as a fuzzy restriction which acts as an elastic constraint on the values that may be assigned to a variable. More specifically, if F is a fuzzy subset of a universe of discourse $U = \{u\}$ which is characterized by its membership function μ_F, then a proposition of the form "X is F," where X is a variable taking values in U, induces a possibility distribution Π_X which equates the possibility of X taking the value u to $\mu_F(u)$—the compatibility of u with F. In this way, X becomes a fuzzy variable which is associated with the possibility distribution Π_X in much the same way as a random variable is associated with a probability distribution. In general, a variable may be associated both with a possibility distribution and a probability distribution, with the weak connection between the two expressed as the possibility/probability consistency principle.

A thesis advanced in this paper is that the imprecision that is intrinsic in natural languages is, in the main, possibilistic rather than probabilistic in nature. Thus, by employing the concept of a possibility distribution, a proposition, p, in a natural language may be translated into a procedure which computes the probability distribution of a set of attributes which are implied by p. Several types of conditional translation rules are discussed and, in particular, a translation rule for propositions of the form "X is F is α-possible," where α is a number in the interval $[0, 1]$, is formulated and illustrated by examples.

1. Introduction

. The pioneering work of Wiener and Shannon on the statistical theory of communication has led to a universal acceptance of the belief that information is intrinsically statistical in nature and, as such, must be dealt with by the methods provided by probability theory.

Unquestionably, the statistical point of view has contributed deep insights into the fundamental processes involved in the coding, transmission and reception of data, and played a key role in the development of modern communication, detection and telemetering systems. In recent years, however, a number of other important applications have come to the fore in which the major issues center not on the

To Professor Arnold Kaufmann.
*Research supported by Naval Electronic Systems Command Contract N00039–77–C–0022, U.S. Army Research Office Contract DAHC04–75–G0056 and National Science Foundation Grant MCS76–06693.

transmission of information but on its meaning. In such applications, what matters is the ability to answer questions relating to information that is stored in a database—as in natural language processing, knowledge representation, speech recognition, robotics, medical diagnosis, analysis of rare events, decision-making under uncertainty, picture analysis, information retrieval and related areas.

A thesis advanced in this paper is that when our main concern is with the meaning of information—rather than with its measure—the proper framework for information analysis is possibilistic[1] rather than probabilistic in nature, thus implying that what is needed for such an analysis is not probability theory but an analogous –and yet different—theory which might be called the *theory of possibility*.[2]

As will be seen in the sequel, the mathematical apparatus of the theory of fuzzy sets provides a natural basis for the theory of possibility, playing a role which is similar to that of measure theory in relation to the theory of probability. Viewed in this perspective, a fuzzy restriction may be interpreted as a possibility distribution, with its membership function playing the role of a possibility distribution function, and a fuzzy variable is associated with a possibility distribution in much the same manner as a random variable is associated with a probability distribution. In general, however, a variable may be associated both with a possibility distribution and a probability distribution, with the connection between the two expressible as the *possibility/probability consistency principle*. This principle—which is an expression of a weak connection between possibility and probability—will be described in greater detail at a later point in this paper.

The importance of the theory of possibility stems from the fact that—contrary to what has become a widely accepted assumption—much of the information on which human decisions are based is possibilistic rather than probabilistic in nature. In particular, the intrinsic fuzziness of natural languages—which is a logical consequence of the necessity to express information in a summarized form—is, in the main, possibilistic in origin. Based on this premise, it is possible to construct a universal language[3] in which the translation of a proposition expressed in a natural language takes the form of a procedure for computing the possibility distribution of a set of fuzzy relations in a data base. This procedure, then, may be interpreted as the meaning of the proposition in question, with the computed possibility distribution playing the role of the information which is conveyed by the proposition.

The present paper has the limited objective of exploring some of the elementary properties of the concept of a possibility distribution, motivated principally by the application of this concept to the representation of meaning in natural languages. Since our intuition concerning the properties of possibility distributions is not as yet well developed, some of the definitions which are formulated in the sequel should be viewed as provisional in nature.

---

[1]The term possibilistic—in the sense used here—was coined by Gaines and Kohout in their paper on possible automata [1].

[2]The interpretation of the concept of possibility in the theory of possibility is quite different from that of modal logic [2] in which propositions of the form "It is possible that..." and "It is necessary that..." are considered.

[3]Such a language, called PRUF (Possibilistic Relational Universal Fuzzy), is described in [30].

2. The concept of a possibility distribution

What is a possibility distribution? It is convenient to answer this question in terms of another concept, namely, that of a *fuzzy restriction* [4, 5], to which the concept of a possibility distribution bears a close relation.

Let X be a variable which takes values in a universe of discourse U, with the generic element of U denoted by u and

$$X = u \tag{2.1}$$

signifying that X is assigned the value u, $u \in U$.

Let F be a fuzzy subset of U which is characterized by a membership function μ_F. Then F is a *fuzzy restriction on X* (or *associated with X*) if F acts as an elastic constraint on the values that may be assigned to X—in the sense that the assignment of a value u to X has the form

$$X = u : \mu_F(u) \tag{2.2}$$

where $\mu_F(u)$ is interpreted as the degree to which the constraint represented by F is satisfied when u is assigned to X. Equivalently, (2.2) implies that $1 - \mu_F(u)$ is the degree to which the constraint in question must be stretched in order to allow the assignment of u to X.[4]

Let $R(X)$ denote a fuzzy restriction associated with X. Then, to express that F plays the role of a fuzzy restriction in relation to X, we write

$$R(X) = F. \tag{2.3}$$

An equation of this form is called a *relational assignment equation* because it represents the assignment of a fuzzy set (or a fuzzy relation) to the restriction associated with X.

To illustrate the concept of a fuzzy restriction, consider a proposition of the form $p \triangleq X$ is F,[5] where X is the name of an object, a variable or a proposition, and F is the name of a fuzzy subset of U, as in "Jessie is very intelligent," "X is a small number," "Harriet is blonde is quite true," etc. As shown in [4] and [6], the translation of such a proposition may be expressed as

$$R(A(X)) = F \tag{2.4}$$

where $A(X)$ is an implied attribute of X which takes values in U, and (2.4) signifies that the proposition $p \triangleq X$ is F has the effect of assigning F to the fuzzy restriction on the values of $A(X)$.

As a simple example of (2.4), let p be the proposition "John is young," in which young is a fuzzy subset of $U = [0, 100]$ characterized by the membership function

$$\mu_{young}(u) = 1 - S(u; 20, 30, 40) \tag{2.5}$$

---

[4]A point that must be stressed is that a fuzzy set *per sé* is not a fuzzy restriction. To be a fuzzy restriction, it must be acting as a constraint on the values of a variable.

[5]The symbol $\triangleq$ stands for "denotes" or "is defined to be".

where u is the numerical age and the S-function is defined by [4].

$$S(u;\alpha,\beta,\gamma)=0 \qquad\qquad \text{for } u\leq\alpha \tag{2.6}$$

$$=2\left(\frac{u-\alpha}{\gamma-\alpha}\right)^2 \qquad \text{for } \alpha\leq u\leq\beta$$

$$=1-2\left(\frac{u-\gamma}{\gamma-\alpha}\right)^2 \qquad \text{for } \beta\leq u\leq\gamma$$

$$=1 \qquad\qquad\qquad \text{for } u\geq\gamma,$$

in which the parameter $\beta\overset{\Delta}{=}(\alpha+\gamma)/2$ is the crossover point, that is, $S(\beta;\alpha,\beta,\gamma)=0.5$. In this case, the implied attribute $A(X)$ is Age(John) and the translation of "John is young" assumes the form:

$$\text{John is young}\to R(\text{Age(John)})=\text{young}. \tag{2.7}$$

To relate the concept of a fuzzy restriction to that of a possibility distribution, we interpret the right-hand member of (2.7) in the following manner.

Consider a numerical age, say $u=28$, whose grade of membership in the fuzzy set young (as defined by (2.5)) is approximately 0.7. First, we interpret 0.7 as the degree of *compatibility* of 28 with the concept labeled young. Then, we postulate that the proposition "John is young" converts the meaning of 0.7 from the degree of compatibility of 28 with young to the degree of possibility that John is 28 given the proposition "John is young." In short, the compatibility of a value of u with young becomes converted into the possibility of that value of u given "John is young."

Stated in more general terms, the concept of a possibility distribution may be defined as follows. (For simplicity, we assume that $A(X)=X$.)

Definition 2.1. Let F be a fuzzy subset of a universe of discourse U which is characterized by its membership function μ_F, with the grade of membership, $\mu_F(u)$, interpreted as the compatibility of u with the concept labeled F.

Let X be a variable taking values in U, and let F act as a fuzzy restriction, $R(X)$, associated with X. Then the proposition "X is F," which translates into

$$R(X)=F, \tag{2.8}$$

associates a *possibility distribution*, Π_X, with X which is postulated to be equal to $R(X)$, i.e.,

$$\Pi_X=R(X). \tag{2.9}$$

Correspondingly, the *possibility distribution function associated with X* (or the possibility distribution function of Π_X) is denoted by π_X and is defined to be numerically equal to the membership function of F, i.e.,

$$\pi_X\overset{\Delta}{=}\mu_F. \tag{2.10}$$

Thus, $\pi_X(u)$, the *possibility* that $X=u$, is postulated to be equal to $\mu_F(u)$.

In view of (2.9), the relational assignment equation (2.8) may be expressed equivalently in the form

$$\Pi_X = F, \tag{2.11}$$

placing in evidence that the proposition $p \overset{\Delta}{=} X$ is F has the effect of associating X with a possibility distribution Π_X which, by (2.9), is equal to F. When expressed in the form of (2.11), a relational assignment equation will be referred to as a *possibility assignment equation*, with the understanding that Π_X is *induced by p*.

As a simple illustration, let U be the universe of positive integers and let F be the fuzzy set of small integers defined by ($+ \overset{\Delta}{=}$ union)

$$\text{small integer} = 1/1 + 1/2 + 0.8/3 + 0.6/4 + 0.4/5 + 0.2/6.$$

Then, the proposition "X is a small integer" associates with X the possibility distribution

$$\Pi_X = 1/1 + 1/2 + 0.8/3 + 0.6/4 + 0.4/5 + 0.2/6 \tag{2.12}$$

in which a term such as 0.8/3 signifies that the possibility that X is 3, given that X is a small integer, is 0.8.

There are several important points relating to the above definition which are in need of comment.

First, (2.9) implies that the possibility distribution Π_X may be regarded as an interpretation of the concept of a fuzzy restriction and, consequently, that the mathematical apparatus of the theory of fuzzy sets—and, especially, the calculus of fuzzy restrictions [4]—provides a basis for the manipulation of possibility distributions by the rules of this calculus.

Second, the definition implies the assumption that our intuitive perception of the ways in which possibilities combine is in accord with the rules of combination of fuzzy restrictions. Although the validity of this assumption cannot be proved at this juncture, it appears that there is a fairly close agreement between such basic operations as the union and intersection of fuzzy sets, on the one hand, and the possibility distributions associated with the disjunctions and conjunctions of propositions of the form "X is F." However, since our intuition concerning the behaviour of possibilities is not very reliable, a great deal of empirical work would have to be done to provide us with a better understanding of the ways in which possibility distributions are manipulated by humans. Such an understanding would be enhanced by the development of an axiomatic approach to the definition of subjective possibilities—an approach which might be in the spirit of the axiomatic approaches to the definition of subjective probabilities [7, 8].

Third, the definition of $\pi_X(u)$ implies that the degree of possibility may be any number in the interval [0, 1] rather than just 0 or 1. In this connection, it should be noted that the existence of intermediate degrees of possibility is implicit in such commonly encountered propositions as "There is a slight possibility that Marilyn is very rich," "It is quite possible that Jean-Paul will be promoted," "It is almost impossible to find a needle in a haystack," etc.

It could be argued, of course, that a characterization of an intermediate degree of possibility by a label such as "slight possibility" is commonly meant to be interpreted as "slight probability." Unquestionably, this is frequently the case in everyday discourse. Nevertheless, there is a fundamental difference between probability and possibility which, once better understood, will lead to a more careful differentiation between the characterizations of degrees of possibility vs. degrees of probability—especially in legal discourse, medical diagnosis, synthetic languages and, more generally, those applications in which a high degree of precision of meaning is an important desideratum.

To illustrate the difference between probability and possibility by a simple example, consider the statement "Hans ate X eggs for breakfast," with X taking values in $U = \{1, 2, 3, 4, \ldots\}$. We may associate a possibility distribution with X by interpreting $\pi_X(u)$ as the degree of ease with which Hans can eat u eggs. We may also associate a probability distribution with X by interpreting $P_X(u)$ as the probability of Hans eating u eggs for breakfast. Assuming that we employ some explicit or implicit criterion for assessing the degree of ease with which Hans can eat u eggs for breakfast, the values of $\pi_X(u)$ and $p_X(u)$ might be as shown in Table 1.

Table 1

The possibility and probability distributions associated with X

u	1	2	3	4	5	6	7	8
$\pi_X(u)$	1	1	1	1	0.8	0.6	0.4	0.2
$P_X(u)$	0.1	0.8	0.1	0	0	0	0	0

We observe that, whereas the possibility that Hans may eat 3 eggs for breakfast is 1, the probability that he may do so might be quite small, e.g., 0.1. Thus, a high degree of possibility does not imply a high degree of probability, nor does a low degree of probability imply a low degree of possibility. However, if an event is impossible, it is bound to be improbable. This heuristic connection between possibilities and probabilities may be stated in the form of what might be called the *possibility/probability consistency principle*, namely:

If a variable X can take the values $u_1, \ldots, u_n$ with respective possibilities $\Pi = (\pi_1, \ldots, \pi_n)$ and probabilities $P = (p_1, \ldots, p_n)$, then the degree of consistency of the probability distribution P with the possibility distribution Π is expressed by $(+ \overset{\Delta}{=} \text{arithmetic sum})$

$$\gamma = \pi_1 p_1 + \cdots + \pi_n p_n. \tag{2.13}$$

It should be understood, of course, that the possibility/probability consistency principle is not a precise law or a relationship that is intrinsic in the concepts of possibility and probability. Rather it is an approximate formalization of the heuristic observation that a lessening of the possibility of an event tends to lessen its probability—but not vice-versa. In this sense, the principle is of use in situations in which what is known about a variable X is its possibility—rather than its probability—distribution. In such cases—which occur far more frequently than those in which the reverse is true—the possibility/probability consistency principle provides a basis for the computation of the possibility distribution of the probability distribution of X. Such computations play a particularly important role in decision-making under uncertainty and in the theories of evidence and belief [9–12].

In the example discussed above, the possibility of X assuming a value u is interpreted as the degree of ease with which u may be assigned to X, e.g., the degree of ease with which Hans may eat u eggs for breakfast. It should be understood, however, that this "degree of ease" may or may not have physical reality. Thus, the proposition "John is young" induces a possibility distribution whose possibility distribution function is expressed by (2.5). In this case, the possibility that the variable Age(John) may take the value 28 is 0.7, with 0.7 representing the degree of ease with which 28 may be assigned to Age(John) given the elasticity of the fuzzy restriction labeled young. Thus, in this case "the degree of ease" has a figurative rather than physical significance.

2.1. Possibility measure

Additional insight into the distinction between probability and possibility may be gained by comparing the concept of a *possibility measure* with the familiar concept of a probability measure. More specifically, let A be a nonfuzzy subset of U and let Π_X be a possibility distribution associated with a variable X which takes values in U. Then, the *possibility measure*, $\pi(A)$, of A is defined as a number in $[0, 1]$ given by[6]

$$\pi(A) \triangleq \operatorname{Sup}_{u \in A} \pi_X(u), \tag{2.14}$$

where $\pi_X(u)$ is the possibility distribution function of Π_X. This number, then, may be interpreted as the possibility *that a value of X belongs to A*, that is

$$\operatorname{Poss}\{X \in A\} \triangleq \pi(A) \tag{2.15}$$
$$\triangleq \operatorname{Sup}_{u \in A} \pi_X(u).$$

When A is a fuzzy set, the belonging of a value of X to A is not meaningful. A more general definition of possibility measure which extends (2.15) to fuzzy sets is the following.

Definition 2.2. Let A be a fuzzy subset of U and let Π_X be a possibility distribution associated with a variable X which takes values in U. The *possibility measure*, $\pi(A)$, of A is defined by

$$\operatorname{Poss}\{X \text{ is } A\} \triangleq \pi(A) \tag{2.16}$$
$$\triangleq \operatorname{Sup}_{u \in U} \mu_A(u) \wedge \pi_X(u),$$

---

[6]The measure defined by (2.14) may be viewed as a particular case of the fuzzy measure defined by Sugeno and Terano [20, 21]. Furthermore, $\pi(A)$ as defined by (2.14) provides a possibilistic interpretation for the *scale function*, $\sigma(A)$, which is defined by Nahmias [22] as the supremum of a membership function over a nonfuzzy set A.

where "X is A" replaces "$X \in A$" in (2.15), μ_A is the membership function of A, and $\wedge$ stands, as usual, for min. It should be noted that, in terms of the *height* of a fuzzy set, which is defined as the supremum of its membership function [23], (2.16) may be expressed compactly by the equation

$$\pi(A) \overset{\Delta}{=} \text{Height}(A \cap \Pi_X). \tag{2.17}$$

As a simple illustration, consider the possibility distribution (2.12) which is induced by the proposition "X is a small integer." In this case, let A be the set $\{3, 4, 5\}$. Then

$$\pi(A) = 0.8 \vee 0.6 \vee 0.4 = 0.8,$$

where $\vee$ stands, as usual, for max.

On the other hand, if A is the fuzzy set of integers which are not small, i.e.,

$$A \overset{\Delta}{=} 0.2/3 + 0.4/4 + 0.6/5 + 0.8/6 + 1/7 + \cdots \tag{2.18}$$

then

$$\text{Poss}\{X \text{ is not a small integer}\} = \text{Height}(0.2/3 + 0.4/4 + 0.4/5 + 0.2/6)$$
$$= 0.4. \tag{2.19}$$

It should be noted that (2.19) is an immediate consequence of the assertion

$$X \text{ is } F \Rightarrow \text{Poss}\{X \text{ is } A\} = \text{Height}(F \cap A), \tag{2.20}$$

which is implied by (2.11) and (2.17). In particular, if A is a normal fuzzy set (i.e., $\text{Height}(A) = 1$), then, as should be expected

$$X \text{ is } A \Rightarrow \text{Poss}\{X \text{ is } A\} = 1. \tag{2.21}$$

Let A and B be arbitrary fuzzy subsets of U. Then, from the definition of the possibility measure of a fuzzy set (2.16), it follows that[7]

$$\pi(A \cup B) = \pi(A) \vee \pi(B). \tag{2.22}$$

By comparison, the corresponding relation for probability measures of A, B and $A \cup B$ (if they exist) is

$$P(A \cup B) \leqq P(A) + P(B) \tag{2.23}$$

and, if A and B are disjoint (i.e., $\mu_A(u)\mu_B(u) \equiv 0$),

$$P(A \cup B) = P(A) + P(B), \tag{2.24}$$

---

[7]It is of interest that (2.22) is analogous to the extension principle for fuzzy sets [5], with + (union) in the right-hand side of the statement of the principle replaced by $\vee$.

which expresses the basic additivity property of probability measures. Thus, in contrast to probability measure, possibility measure is not additive. Instead, it has the property expressed by (2.22), which may be viewed as an analog of (2.24) with $+$ replaced by $\vee$.

In a similar fashion, the possibility measure of the intersection of A and B is related to those of A and B by

$$\pi(A \cap B) \leq \pi(A) \wedge \pi(B). \tag{2.25}$$

In particular, if A and B are *noninteractive*,[8] (2.25) holds with the equality sign, i.e.,

$$\pi(A \cap B) = \pi(A) \wedge \pi(B). \tag{2.26}$$

By comparison, in the case of probability measures, we have

$$P(A \cap B) \leq P(A) \wedge P(B) \tag{2.27}$$

and

$$P(A \cap B) = P(A)P(B) \tag{2.28}$$

if A and B are independent and nonfuzzy. As in the case of (2.22), (2.26) is analogous to (2.28), with product corresponding to min.

2.2. Possibility and information

If p is a proposition of the form $p \triangleq X$ is F which translates into the possibility assignment equation

$$\Pi_{A(X)} = F, \tag{2.29}$$

where F is a fuzzy subset of U and $A(X)$ is an implied attribute of X taking values in U, then the *information conveyed by* p, $I(p)$, may be identified with the possibility distribution, $\Pi_{A(X)}$, of the fuzzy variable $A(X)$. Thus, the connection between $I(p)$, $\Pi_{A(X)}$, $R(A(X))$ and F is expressed by

$$I(p) \triangleq \Pi_{A(X)}, \tag{2.30}$$

where

$$\Pi_{A(X)} = R(A(X)) = F. \tag{2.31}$$

For example, if the proposition $p \triangleq$ John is young translates into the possibility assignment equation

$$\Pi_{\text{Age(John)}} = \text{young}, \tag{2.32}$$

---

[8]Noninteraction in the sense defined here is closely related to the concept of noninteraction of fuzzy restrictions [5, 6]. It should also be noted that (2.26) provides a possibilistic interpretation for "unrelatedness" as defined by Nahmias [22].

where young is defined by (2.5), then

$$I(\text{John is young}) = \Pi_{\text{Age(John)}} \tag{2.33}$$

in which the possibility distribution function of Age(John) is given by

$$\pi_{\text{Age(John)}}(u) = 1 - S(u; 20, 30, 40), \qquad u \in [0, 100]. \tag{2.34}$$

From the definition of $I(p)$ it follows that if $p \overset{\Delta}{=} X$ is F and $q \overset{\Delta}{=} X$ is G, then p is at least as informative as q, expressed as $I(p) \geq I(q)$, if $F \subset G$. Thus, we have a partial ordering of the $I(p)$ defined by

$$F \subset G \Rightarrow I(X \text{ is } F) \geq I(X \text{ is } G) \tag{2.35}$$

which implies that the more restrictive a possibility distribution is, the more informative is the proposition with which it is associated. For example, since very tall $\subset$ tall, we have

$$I(\text{Lucy is very tall}) \geq I(\text{Lucy is tall}). \tag{2.36}$$

3. N-ary possibility distributions

In asserting that the translation of a proposition of the form $p \overset{\Delta}{=} X$ is F is expressed by

$$X \text{ is } F \rightarrow R(A(X)) = F \tag{3.1}$$

or, equivalently,

$$X \text{ is } F \rightarrow \Pi_{A(X)} = F, \tag{3.2}$$

we are tacitly assuming that p contains a single implied attribute $A(X)$ whose possibility distribution is given by the right-hand member of (3.2).

More generally, p may contain n implied attributes $A_1(X), \ldots, A_n(X)$, with $A_i(X)$ taking values in U_i, $i = 1, \ldots, n$. In this case, the translation of $p \overset{\Delta}{=} X$ is F, where F is a fuzzy relation in the cartesian product $U = U_1 \times \cdots \times U_n$, assumes the form

$$X \text{ is } F \rightarrow R(A_1(X), \ldots, A_n(X)) = F \tag{3.3}$$

or, equivalently,

$$X \text{ is } F \rightarrow \Pi_{(A_1(X), \ldots, A_n(X))} = F \tag{3.4}$$

where $R(A_1(X), \ldots, A_n(X))$ is an n-ary fuzzy restriction and $\Pi_{(A_1(X), \ldots, A_n(X))}$ is an n-ary possibility distribution which is induced by p. Correspondingly, the n-ary possibility distribution function induced by p is given by

$$\pi_{(A_1(X), \ldots, A_n(X))}(u_1, \ldots, u_n) = \mu_F(u_1, \ldots, u_n), \qquad (u_1, \ldots, u_n) \in U, \tag{3.5}$$

where μ_F is the membership function of F. In particular, if F is a cartesian product of n

unary fuzzy relations $F_1, \ldots, F_n$, then the righthand member of (3.3) decomposes into a system of n unary relational assignment equations, i.e.,

$$X \text{ is } F \rightarrow R(A_1(X)) = F_1 \tag{3.6}$$

$$R(A_2(X)) = F_2$$
$$\vdots \qquad \vdots$$
$$R(A_n(X)) = F_n.$$

Correspondingly,[9]

$$\Pi_{(A_1(X), \ldots, A_n(X))} = \Pi_{A_1(X)} \times \cdots \times \Pi_{A_n(X)} \tag{3.7}$$

and

$$\pi_{(A_1(X), \ldots, A_n(X))}(u_1, \ldots, u_n) = \pi_{A_1(X)}(u_1) \wedge \cdots \wedge \pi_{A_n(X)}(u_n), \tag{3.8}$$

where

$$\pi_{A_i(X)}(u_i) = \mu_{F_i}(u_i), \qquad u_i \in U_i, \qquad i = 1, \ldots, n \tag{3.9}$$

and $\wedge$ denotes min (in infix form).

As a simple illustration, consider the proposition $p \triangleq$ carpet is large, in which large is a fuzzy relation whose tableau is of the form shown in Table 2 (with length and width expressed in metric units).

Table 2

Tableau of large

Large	Width	Length	μ
	250	300	0.6
	250	350	0.7
	.	.	.
	300	400	0.8
	.	.	.
	400	600	1

In this case, the translation (3.3) leads to the possibility assignment equation

$$\Pi_{(\text{width(carpet), length(carpet)})} = \text{large}, \tag{3.10}$$

which implies that if the compatibility of a carpet whose width is, say, 250 cm and length is 350 cm with "large carpet" is 0.7, then the possibility that the width of the carpet is 250 cm and its length is 350 cm—given the proposition $p \triangleq$ carpet is large—is 0.7.

Now, if large is defined as

$$\text{large} = \text{wide} \times \text{long} \tag{3.11}$$

---

[9] If F and G are fuzzy relations in U and V, respectively, then their cartesian product $F \times G$ is a fuzzy relation in $U \times V$ whose membership function is given by $\mu_F(u) \wedge \mu_G(v)$.

where long and wide are unary fuzzy relations, then (3.10) decomposes into the possibility association equations

$$\Pi_{width(carpet)} = wide$$

and

$$\Pi_{length(carpet)} = long$$

where the tableaux of long and wide are of the form shown in Table 3.

Table 3

Tableaux of wide and long

Wide	Width	μ		Long	Length	μ
	250	0.7.			300	0.6
	300	0.8			350	0.7
	350	0.8			400	0.8
	.	.			.	.
	400	1			500	1

3.1. Marginal possibility distributions

The concept of a marginal possibility distribution bears a close relation to the concept of a marginal fuzzy restriction [4], which in turn is analogous to the concept of a marginal probability distribution.

More specifically, let $X = (X_1, \ldots, X_n)$ be an n-ary fuzzy variable taking values in $U = U_1 \times \cdots \times U_n$, and let Π_X be a possibility distribution associated with X, with $\pi_X(u_1, \ldots, u_n)$ denoting the possibility distribution function of Π_X.

Let $q \triangleq (i_1, \ldots, i_k)$ be a subsequence of the index sequence $(1, \ldots, n)$ and let $X_{(q)}$ be the q-ary fuzzy variable $X_{(q)} \triangleq (X_{i_1}, \ldots, X_{i_k})$. The *marginal possibility distribution* $\Pi_{X_{(q)}}$ is a possibility distribution associated with $X_{(q)}$ which is *induced* by Π_X as the projection of Π_X on $U_{(q)} \triangleq U_{i_1} \times \cdots \times U_{i_k}$. Thus, by definition,

$$\Pi_{X_{(q)}} \triangleq \mathrm{Proj}_{U_{(q)}} \Pi_X, \tag{3.12}$$

which implies that the probability distribution function of $X_{(q)}$ is related to that of X by

$$\pi_{X_{(q)}}(u_{(q)}) = \bigvee_{u_{(q')}} \pi_X(u) \tag{3.13}$$

where $u_{(q)} \triangleq (u_{i_1}, \ldots, u_{i_k})$, $q' \triangleq (j_1, \ldots, j_m)$ is a subsequence of $(1, \ldots, n)$ which is complementary to q (e.g., if $n = 5$ and $q \triangleq (i_1 \, i_2) = (2, 4)$, then $q' = (j_1, j_2, j_3) = (1, 3, 5)$, $u_{(q')} \triangleq (u_{j_1}, \ldots, u_{j_m})$ and $\vee_{u_{(q')}}$ denotes the supremum over $(u_{j_1}, \ldots, u_{j_m}) \in U_{j_1} \times \ldots \times U_{j_m}$.

As a simple illustration, assume that $U_1 = U_2 = U_3 = \{a, b\}$ and the tableau of Π_X is given by

Table 4

Tableau of Π_X

Π_X	X_1	X_2	X_3	π
	a	a	a	0.8
	a	a	b	1
	b	a	a	0.6
	b	a	b	0.2
	b	b	b	0.5

Then,

$$\Pi_{(X_1, X_2)} = \text{Proj}_{U_1 \times U_2} \Pi_X = 1/(a, a) + 0.6/(b, a) + 0.5/(b, b) \tag{3.14}$$

which in tabular form reads

Table 5

Tableau of $\Pi_{(X_1, X_2)}$

$\Pi_{(X_1, X_2)}$	X_1	X_2	π
	a	a	1
	b	a	0.6
	b	b	0.5

Then, from Π_X it follows that the possibility that $X_1 = b, X_2 = a$ and $X_3 = b$ is 0.2, while from $\Pi_{(X_1, X_2)}$ it follows that the possibility of $X_1 = b$ and $X_2 = a$ is 0.6.

By analogy with the concept of independence of random variables, the fuzzy variables

$$X_{(q)} \overset{\Delta}{=} (X_{i_1}, \ldots, X_{i_k})$$

and

$$X_{(q')} \overset{\Delta}{=} (X_{j_1}, \ldots, j_m)$$

are *noninteractive* [5] if and only if the possibility distribution associated with $X = (X_1, \ldots, X_n)$ is the cartesian product of the possibility distributions associated with $X_{(q)}$ and $X_{(q')}$, i.e.,

$$\Pi_X = \Pi_{X_{(q)}} \times \Pi_{X_{(q')}} \tag{3.15}$$

or, equivalently,

$$\pi_X(u_1, \ldots, u_n) = \pi_{X_{(q)}}(u_{i_1}, \ldots, u_{i_k}) \wedge \pi_{X_{(q')}}(u_{j_1}, \ldots, u_{j_m}). \tag{3.16}$$

In particular, the variables $X_1, \ldots, X_n$ are noninteractive if and only if

$$\Pi_X = \Pi_{X_1} \times \Pi_{X_2} \times \cdots \times \Pi_{X_n}. \tag{3.17}$$

The intuitive significance of noninteraction may be clarified by a simple example. Suppose that $X \triangleq (X_1, X_2)$, and X_1 and X_2 are noninteractive, i.e.,

$$\pi_X(u_1, u_2) = \pi_{X_1}(u_1) \wedge \pi_{X_2}(u_2).$$

(3.18)

Furthermore, suppose that for some particular values of u_1 and u_2, $\pi_{X_1}(u_1) = \alpha_1$, $\pi_{X_2}(u_2) = \alpha_2 < \alpha_1$ and hence $\pi_X(u_1, u_2) = \alpha_2$. Now, if the value of $\pi_{X_1}(u_1)$ is increased to $\alpha_1 + \delta_1$, $\delta_1 > 0$, it is not possible to decrease the value of $\pi_{X_2}(u_2)$ by a positive amount, say δ_2, such that the value of $\pi_X(u_1, u_2)$ remains unchanged. In this sense, an increase in the possibility of u_1 cannot be compensated by a decrease in the possibility of u_2, and vice-versa. Thus, in essence, noninteraction may be viewed as a form of noncompensation in which a variation in one or more components of a possibility distribution cannot be compensated by variations in the complementary components.

In the manipulation of possibility distributions, it is convenient to employ a type of symbolic representation which is commonly used in the case of fuzzy sets. Specifically, assume, for simplicity, that $U_1, \ldots, U_n$ are finite sets, and let $r^i \triangleq (r_1^i, \ldots, r_n^i)$ denote an n-tuple of values drawn from $U_1, \ldots, U_n$, respectively. Furthermore, let π_i denote the possibility of r^i and let the n-tuple $(r_1^i, \ldots, r_n^i)$ be written as the string $r_1^i \cdots r_n^i$.

Using this notation, a possibility distribution Π_X may be expressed in the symbolic form

$$\Pi_X = \sum_{i=1}^{N} \pi_i r_1^i \ r_2^i \ \cdots \ r_n^i$$

(3.19)

or, in case a separator symbol is needed, as

$$\Pi_X = \sum_{i=1}^{N} \pi_i / r_1^i r_2^i \cdots r_n^i,$$

(3.20)

where N is the number of n-tuples in the tableau of Π_X, and the summation should be interpreted as the union of the fuzzy singletons $\pi_i/(r_1^i, \ldots, r_n^i)$. As an illustration, in the notation of (3.19), the possibility distribution defined in Table 4 reads

$$\Pi_X = 0.8aaa + 1aab + 0.6baa + 0.2bab + 0.5bbb.$$

(3.21)

The advantage of this notation is that it allows the possibility distributions to be manipulated in much the same manner as linear forms in n variables, with the understanding that, if r and s are two tuples and α and β are their respective possibilities, then

$$\alpha r + \beta r = (\alpha \vee \beta) r$$

(3.22)

$$\alpha r \cap \beta r = (\alpha \wedge \beta) r$$

(3.23)

and

$$\alpha r \times \beta s = (\alpha \wedge \beta) r s.$$

(3.24)

where rs denotes the concatenation of r and s. For example, if

$$\Pi_X = 0.8aa + 0.5ab + 1bb \tag{3.25}$$

and

$$\Pi_Y = 0.9ba + 0.6bb \tag{3.26}$$

then

$$\Pi_X + \Pi_Y = 0.8aa + 0.5ab + 0.9ba + 1bb \tag{3.27}$$

$$\Pi_X \cap \Pi_Y = 0.6bb \tag{3.28}$$

and

$$\Pi_X \times \Pi_Y = 0.8aaba + 0.5abba + 0.9bbba + 0.6aabb + 0.5abbb + 0.6bbbb. \tag{3.29}$$

To obtain the projection of a possibility distribution Π_X on $U_{(q)} \triangleq (U_{i_1}, \ldots, U_{i_k})$, it is sufficient to set the values of $X_{j_1}, \ldots, X_{j_m}$ in each tuple in Π_X equal to the null string Λ (i.e., multiplicative identity). As an illustration, the projection of the possibility distribution defined by Table 4 on $U_1 \times U_2$ is given by

$$\mathrm{Proj}_{U_1 \times U_2} \Pi_X = 0.8aa + 1aa + 0.6ba + 0.2ba + 0.5bb \tag{3.30}$$

$$= 1aa + 0.6ba + 0.5bb$$

which agrees with Table 5.

3.2. Conditioned possibility distributions

In the theory of possibilities, the concept of a conditioned possibility distribution plays a role that is analogous—though not completely—to that of a conditional possibility distribution in the theory of probabilities.

More concretely, let a variable $X = (X_1, \ldots, X_n)$ be associated with a possibility distribution Π_X, with Π_X characterized by a possibility distribution function $\pi_X(u_1, \ldots, u_n)$ which assigns to each n-tuple $(u_1, \ldots, u_n)$ in $U_1 \times \cdots \times U_n$ its possibility $\pi_X(u_1, \ldots, u_n)$.

Let $q = (i_1, \ldots, i_k)$ and $s = (j_1, \ldots, j_m)$ be subsequences of the index sequence $(1, \ldots, n)$, and let $(a_{j_1}, \ldots, a_{j_m})$ be an n-tuple of values assigned to $X_{(q')} = (X_{j_1}, \ldots, X_{j_m})$. By definition, the conditioned possibility distribution of

$$X_{(q)} \triangleq (X_{i_1}, \ldots, X_{i_k})$$

given

$$X_{(q')} = (a_{j_1}, \ldots, a_{j_m})$$

is a possibility distribution expressed as

$$\Pi_{X_{(q)}}[X_{j_1} = a_{j_1}; \ldots; X_{j_m} = a_{j_m}]$$

whose possibility distribution function is given by[10]

$$\pi_{X_{(q)}}(u_{i_1},\ldots,u_{i_k}\,|\,X_{j_1}=a_{j_1};\ldots;X_{j_m}=a_{j_m}) \tag{3.31}$$

$$\overset{\Delta}{=}\pi_X(u_1,\ldots,u_n)\Big|_{u_{j_1}=a_{j_1},\ldots,u_{j_m}=a_{j_m}}.$$

As a simple example, in the case of (3.21), we have

$$\Pi_{(X_2,X_3)}[X_1=a]=0.8aa+1ab \tag{3.32}$$

as the expression for the conditioned possibility distribution of (X_2,X_3) given $X_1=a$.

An equivalent expression for the conditioned possibility distribution which makes clearer the connection between

$$\Pi_{X_{(q)}}[X_{j_1}=a_{j_1};\ldots;X_{j_m}=a_{j_m}]$$

and Π_X may be derived as follows.

Let

$$\Pi_X[X_{j_1}=a_{j_1};\ldots;X_{j_m}=a_{j_m}]$$

denote a possibility distribution which consists of those terms in (3.19) in which the j_1th element is a_{j_1}, the j_2th element is $a_{j_2},\ldots$, and the j_mth element is a_{j_m}. For example, in the case of (3.21)

$$\Pi_X(X_1=a]=0.8aaa+1aab. \tag{3.33}$$

Expressed in the above notation, the conditioned possibility distribution of $X_{(q)}=(X_{i_1},\ldots,X_{i_k})$ given $X_{j_1}=a_{j_1},\ldots,X_{j_m}=a_{j_m}$ may be written as

$$\Pi_{X_{(q)}}[X_{j_1}=a_{j_1};\ldots;X_{j_m}=a_{j_m}]$$

$$=\text{Proj}_{U_{(q)}}\Pi_X[X_{j_1}=a_{j_1};\ldots;X_{j_m}=a_{j_m}] \tag{3.34}$$

which places in evidence that $\Pi_{X_{(q)}}$ (conditioned on $X_{(s)}=a_{(s)}$) is a marginal possibility distribution induced by Π_X (conditioned on $X_{(s)}=a_{(s)}$). Thus, by employing (3.33) and (3.34), we obtain

$$\Pi_{(X_2,X_3)}[X_1=a]=0.8aa+1ab \tag{3.35}$$

which agrees with (3.32).

In the foregoing discussion, we have assumed that the possibility distribution of $X=(X_1,\ldots,X_n)$ is conditioned on the values assigned to a specified subset, $X_{(s)}$, of the constituent variables of X. In a more general setting, what might be specified is a

---

[10]In some applications, it may be appropriate to normalize the expression for the conditioned possibility distribution function by dividing the right-hand member of (3.31) by its supremum over $U_{i_1}\times\cdots\times U_{i_k}$.

possibility distribution associated with $X_{(s)}$ rather than the values of $X_{j_1}, \ldots, X_{j_m}$. In such cases, we shall say that Π_X is *particularized*[11] by specifying that $\Pi_{X_{(s)}} = G$, where G is a given m-ary possibility distribution. It should be noted that in the present context $\Pi_{X_{(s)}}$ is a given possibility distribution rather than a marginal distribution that is induced by Π_X.

To analyze this case, it is convenient to assume—in order to simplify the notation—that $X_{j_1} = X_1, X_{j_2} = X_2, \ldots, X_{j_m} = X_m$, $m < n$. Let $\bar{G}$ denote the *cylindrical extension* of G, that is, the possibility distribution defined by

$$\bar{G} \overset{\Delta}{=} G \times U_{m+1} \times \cdots \times U_n \tag{3.36}$$

which implies that

$$\mu_{\bar{G}}(u_1, \ldots, u_n) \overset{\Delta}{=} \mu_G(u_1, \ldots, u_m), \qquad u_j \in U_j, \qquad j = 1, \ldots, n, \tag{3.37}$$

where μ_G is the membership function of the fuzzy relation G.

The assumption that we are given Π_X and G is equivalent to assuming that we are given the intersection $\Pi_X \cap \bar{G}$. From this intersection, then, we can deduce the particularized possibility distribution $\Pi_{X_{(q)}}[\Pi_{X_{(s)}} = G]$ by projection on $U_{(q)}$. Thus

$$\Pi_{X_{(q)}}[\Pi_{X_{(s)}} = G] = \mathrm{Proj}_{U_{(q)}} \Pi_X \cap \bar{G}. \tag{3.38}$$

Equivalently, the left-hand member of (3.38) may be regarded as the composition of Π_X and G [5].

As a simple illustration, consider the possibility distribution defined by (3.21) and assume that

$$G = 0.4aa + 0.8ba + 1bb. \tag{3.39}$$

Then

$$\bar{G} = 0.4aaa + 0.4aab + 0.8baa + 0.8bab + 1bba + 1bbb \tag{3.40}$$

$$\Pi_X \cap \bar{G} = 0.4aaa + 0.4aab + 0.6baa + 0.2bab + 0.5bbb \tag{3.41}$$

and

$$\Pi_{X_3}[\Pi_{(X_1, X_2)} = G] = 0.6a + 0.5b. \tag{3.42}$$

As an elementary application of (3.38), consider the proposition $p \overset{\Delta}{=}$ John is big, where big is a relation whose tableau is of the form shown in Table 6 (with height and weight expressed in metric units).

---

[11]In the case of nonfuzzy relations, particularization is closely related to what is commonly referred to as *restriction*. We are not employing this more conventional term here because of our use of the term "fuzzy restriction" to denote an elastic constraint on the values that may be assigned to a variable.

Table 6

Tableau of big

Big	Height	Weight	μ
	170	70	0.7
	170	80	0.8
	180	80	0.9
	.	.	.
	190	90	1

Now, suppose that in addition to knowing that John is big, we also know that q $\overset{\Delta}{=}$ John is tall, where the tableau of tall is given (in partially tabulated form) by Table 7.

Table 7

Tableau of tall

Tall	Height	μ
	170	0.8
	180	0.9
	190	1

The question is: What is the weight of John? By making use of (3.38), the possibility distribution of the weight of John may be expressed as

$$\Pi_{\text{weight}} = \text{Proj}_{\text{weight}} \, \Pi_{(\text{height, weight})}[\Pi_{\text{height}} = \text{tall}] \tag{3.39}$$

$$= 0.7/70 + 0.9/80 + 1/90.$$

An acceptable linguistic approximation [5], [13] to the right-hand side of (3.39) might be "somewhat heavy," where "somewhat" is a modifier which has a specified effect on the fuzzy set labeled "heavy." Correspondingly, an approximate answer to the question would be "John is somewhat heavy."

4. Possibility distributions of composite and qualified propositions

As was stated in the Introduction, the concept of a possibility distribution provides a natural way for defining the meaning as well as the information content of a proposition in a natural language. Thus, if p is a proposition in a natural language NL and M is its meaning, then M may be viewed as a procedure which acts on a set of relations in a universe of discourse associated with NL and yields the possibility distribution of a set of variables or relations which are explicit or implicit in p.

In constructing the meaning of a given proposition, it is convenient to have a collection of what might be called *conditional translation rules* [30] which relate the meaning of a proposition to the meaning of its modifications or combinations with other propositions. In what follows, we shall discuss briefly some of the basic rules of this type and, in particular, will formulate a rule governing the modification of possibility distributions by the *possibility qualification* of a proposition.

4.1. Rules of type I

Let p be a proposition of the form X is F, and let m be a modifier such as very, quite, rather, etc. The so-called *modifier rule* [6] which defines the modification in the possibility distribution induced by p may be stated as follows.
If

$$X \text{ is } F \rightarrow \Pi_{A(X)} = F \tag{4.1}$$

then

$$X \text{ is } mF \rightarrow \Pi_{A(X)} = F^{+} \tag{4.2}$$

where $A(X)$ is an implied attribute of X and F^{+} is a modification of F defined by m.[12] For example, if $m \overset{\Delta}{=}$ very, then $F^{+} = F^{2}$; if $m \overset{\Delta}{=}$ more or less then $F^{+} = \sqrt{F}$; and if $m \overset{\Delta}{=}$ not then $F^{+} = F' \overset{\Delta}{=}$ complement of F. As an illustration:
If

$$\text{John is young} \rightarrow \Pi_{\text{Age(John)}} = \text{young} \tag{4.3}$$

then

$$\text{John is very young} \rightarrow \Pi_{\text{Age(John)}} = \text{young}^{2}.$$

In particular, if

$$\text{young} = 1 - S(20, 30, 40) \tag{4.4}$$

then

$$\text{young}^{2} = (1 - S(20, 30, 40))^{2},$$

where the S-function (with its argument suppressed) is defined by (2.6).

4.2. Rules of type II

If p and q are propositions, then $r \overset{\Delta}{=} p * q$ denotes a proposition which is a *composition* of p and q. The three most commonly used modes of composition are (i) conjunctive, involving the connective "and"; (ii) disjunctive, involving the connective "or"; and (iii) conditional, involving the connective "if...then." The conditional translation rules relating to these modes of composition are stated below.

---

[12]A more detailed discussion of the effect of modifiers (or hedges) may be found in [15, 16, 17, 8, 6, 13 and 18].

Conjunctive (noninteractive): If

$$X \text{ is } F \rightarrow \Pi_{A(X)} = F \tag{4.5}$$

and

$$Y \text{ is } G \rightarrow \Pi_{B(Y)} = G \tag{4.6}$$

then

$$X \text{ is } F \text{ and } Y \text{ is } G \rightarrow \Pi_{(A(X), B(Y))} = F \times G \tag{4.7}$$

where $A(X)$ and $B(Y)$ are the implied attributes of X and Y, respectively, $\Pi_{(A(X), B(Y))}$ is the possibility distribution of the variables $A(X)$ and $B(Y)$, and $F \times G$ is the cartesian product of F and G. It should be noted that $F \times G$ may be expressed equivalently as

$$F \times G = \bar{F} \cap \bar{G} \tag{4.8}$$

where $\bar{F}$ and $\bar{G}$ are the cylindrical extensions of F and G, respectively.

Disjunctive (noninteractive): If (4.5) and (4.6) hold, then

$$X \text{ is } F \text{ or } Y \text{ is } G \rightarrow \Pi_{(A(X), B(Y))} = \bar{F} + \bar{G} \tag{4.9}$$

where the symbols have the same meaning as in (4.5) and (4.6), and $+$ denotes the union.

Conditional (noninteractive): If (4.5) and (4.6) hold, then

$$\text{If } X \text{ is } F \text{ then } Y \text{ is } G \rightarrow \Pi_{(A(X), B(Y))} = \bar{F}' \oplus \bar{G} \tag{4.10}$$

where F' is the complement of F and $\oplus$ is the bounded sum defined by

$$\mu_{F \oplus G} = 1 \wedge (1 - \mu_F + \mu_G), \tag{4.11}$$

in which $+$ and $-$ denote the arithmetic addition and subtraction, and μ_F and μ_G are the membership functions of F and G, respectively. Illustrations of these rules—expressed in terms of fuzzy restrictions rather than possibility distributions—may be found in [6 and 14].

4.3. Truth qualification, probability qualification and possibility qualification

In natural languages, an important mechanism for the modification of the meaning of a proposition is provided by the adjuction of three types of qualifiers: (i) is τ, where τ is a linguistic truth-value, e.g., true, very true, more or less true, false, etc.; (ii) is λ, where λ is a linguistic probability-value (or likelihood), e.g., likely, very likely, very unlikely, etc.; and (iii) is π, where π is a linguistic possibility-value, e.g., possible, quite possible, slightly possible, impossible, etc. These modes of qualification will be referred to, respectively, as *truth qualification*, *probability qualification* and *possibility qualification*. The rules governing these qualifications may be stated as follows.

Truth qualification: If

$$X \text{ is } F \rightarrow \Pi_{A(X)} = F \tag{4.12}$$

then

$$X \text{ is } F \text{ is } \tau \rightarrow \Pi_{A(X)} = F^{+},$$

where

$$\mu_{F^{+}}(u) = \mu_{\tau}(\mu_{F}(u)), \qquad u \in U; \tag{4.13}$$

μ_{τ} and μ_{F} are the membership functions of τ and F, respectively, and U is the universe of discourse associated with $A(X)$. As an illustration, if young is defined by (4.4); $\tau = $ very true is defined by

$$\text{very true} = S^{2}(0.6, 0.8, 1) \tag{4.14}$$

and

$$\text{John is young} \rightarrow \Pi_{\text{Age(John)}} = \text{young}$$

then

$$\text{John is young is very true} \rightarrow \Pi_{\text{Age(John)}} = \text{young}^{+}$$

where

$$\mu_{\text{young}^{+}}(u) = S^{2}(1 - S(u; 20, 30, 40); 0.6, 0.8, 1), \qquad u \in U.$$

It should be noted that for the *unitary* truth-value, u-true, defined by

$$\mu_{u\text{-true}}(v) = v, \qquad v \in [0, 1] \tag{4.15}$$

(4.13) reduces to

$$\mu_{F^{+}}(u) = \mu_{F}(u), \qquad u \in U$$

and hence

$$X \text{ is } F \text{ is } u\text{-true} \rightarrow \Pi_{X} = F. \tag{4.16}$$

Thus, the possibility distribution induced by any proposition is invariant under unitary truth qualification.

Probability qualification: If

$$X \text{ is } F \rightarrow \Pi_{A(X)} = F$$

then

$$X \text{ is } F \text{ is } \lambda \rightarrow \Pi_{\int_{l} p(u)\mu_{l}(u)du} = \lambda \tag{4.17}$$

where $p(u)du$ is the probability that the value of $A(X)$ falls in the interval $(u, u+du)$; the integral

$$\int_U p(u)\mu_F(u)du$$

is the probability of the fuzzy event F [19]; and λ is a linguistic probability-value. Thus, (4.17) defines a possibility distribution of probability distributions, with the possibility of a probability density $p(\cdot)$ given implicitly by

$$\pi[\int_U p(u)\mu_F(u)du] = \mu_\lambda[\int_U p(u)\mu_F(u)du]. \tag{4.18}$$

As an illustration, consider the proposition $p \stackrel{\Delta}{=} $ John is young is very likely, in which young is defined by (4.4) and

$$\mu_{\text{very likely}} = S^2(0.6, 0.8, 1). \tag{4.19}$$

Then

$$\pi[\int_U p(u)\mu_F(u)du] = S^2[\int_0^1 p(u)(1 - S(u; 20, 30, 40))du; 0.6, 0.8, 1].$$

It should be noted that the probability qualification rule is a consequence of the assumption that the propositions "X is F is λ" and "Prob$\{X$ is $F\} = \lambda$" are *semantically equivalent* (i.e., induce identical possibility distributions), which is expressed in symbols as

$$X \text{ is } F \text{ is } \lambda \leftrightarrow \text{Prob}\{X \text{ is } F\} = \lambda. \tag{4.20}$$

Thus, since the probability of the fuzzy event F is given by

$$\text{Prob}\{X \text{ is } F\} = \int_U p(u)\mu_F(u)du,$$

it follows from (4.20) that we can assert the semantic equivalence

$$X \text{ is } F \text{ is } \lambda \leftrightarrow \int_U p(u)\mu_F(u)du \text{ is } \lambda,$$

which by (2.11) leads to the right-hand member of (4.17).

Possibility qualification: Our concern here is with the following question: Given that "X is F" translates into the possibility assignment equation $\Pi_{A(X)} = F$, what is the translation of "X is F is π," where π is a linguistic possibility-value such as quite possible, very possible, more or less possible, etc.? Since our intuition regarding the behavior of possibility distributions is not well-developed at this juncture, the answer suggested in the following should be viewed as tentative in nature.

For simplicity, we shall interpret the qualifier "possible" as "1-possible," that is, as the assignment of the possibility-value 1 to the proposition which it qualifies. With this understanding, the translation of "X is F is possible" will be assumed to be given by

$$X \text{ is } F \text{ is possible} \rightarrow \Pi_{A(X)} = F^+, \tag{4.21}$$

in which

$$F^+ = F \oplus \Pi \tag{4.22}$$

where Π is a fuzzy set of Type 2[13] defined by

$$\mu_\Pi(u) = [0, 1], \qquad u \in U, \tag{4.23}$$

and $\oplus$ is the bounded sum defined by (4.11). Equivalently,

$$\mu_{F^+}(u) = [\mu_F(u), 1], \qquad u \in U, \tag{4.24}$$

which defines μ_{F^+} as an interval-valued membership function.

In effect, the rule in question signifies that possibility qualification has the effect of weakening the proposition which it qualifies through the addition to F of a possibility distribution Π which represents total indeterminacy[14] in the sense that the degree of possibility which it associates with each point in U may be any number in the interval $[0, 1]$. An illustration of the application of this rule to the proposition $p \overset{\Delta}{=} X$ is small is shown in Fig. 1.

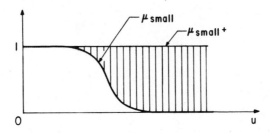

Fig. 1. The possibility distribution of "X is small is possible".

As an extension of the above rule, we have: If

$$X \text{ is } F \to \Pi_{A(X)} = F \tag{4.25}$$

then, for $0 \le \alpha \le 1$,

$$X \text{ is } F \text{ is } \alpha\text{-possible} \to \Pi_{A(X)} = F^+ \tag{4.26}$$

where F^+ is a fuzzy set of Type 2 whose interval-valued membership function is given by

$$\mu_{F^+}(u) = [\alpha \wedge \mu_F(u), \alpha \oplus (1 - \mu_F(u))], \qquad u \in U. \tag{4.27}$$

---

[13]The membership function of a fuzzy set of Type 2 takes values in the set of fuzzy subsets of the unit interval [5, 6].

[14]Π may be interpreted as the possibilistic counterpart of white noise.

As an illustration, the result of the application of this rule to the proposition $p \triangleq X$ is small is shown in Fig. 2. Note that the rule expressed by (4.24) may be regarded as a special case of (4.27) corresponding to $\alpha = 1$.

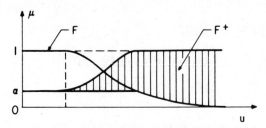

Fig. 2. The possibility distribution of "X is small is α-possible".

A further extension of the rule expressed by (4.25) to linguistic possibility-values may be obtained by an application of the extension principle, leading to the *linguistic possibility qualification rule*:

If

$$X \text{ is } F \rightarrow \Pi_{A(X)} = F$$

then

$$X \text{ is } F \text{ is } \pi \rightarrow \Pi_{A(X)} = F^+ \tag{4.28}$$

where F^+ is a fuzzy set of Type 2 whose membership function is given by

$$\mu_F(u) = \{ \geq \circ (\pi \wedge \mu_F(u)) \cap (\leq \circ (\pi \oplus (1 - \mu_F(u)))) \}, \tag{4.29}$$

where π is the linguistic possibility (e.g., quite possible, almost impossible, etc.) and denotes the composition of fuzzy relations. This rule should be regarded as speculative in nature since the implications of a linguistic possibility qualification are not as yet well understood

An alternative approach to the translation of "X is F is π" is to interpret this proposition as

$$X \text{ is } F \text{ is } \pi \leftrightarrow \text{Poss}\{X \text{ is } F\} = \pi, \tag{4.30}$$

which is in the spirit of (4.20), and then formulate a rule of the form (4.28) in which $\Pi_{A(X)}$ is the largest (i.e., least restrictive) possibility distribution satisfying the constraint $\text{Poss}\{X \text{ is } F\} = \pi$. A complicating factor in this case is that the proposition "X is F is π" may be associated with other implicit propositions such as "X is not F is $[0,1]$-possible," or "X is not F is not impossible," which affect the translation of "X is F is π." In this connection, it would be useful to deduce the translation rules (4.21), (4.26) and (4.29) (or their variants) from a conjunction of "X is F is π" with other implicit propositions involving the negation of "X is F."

An interesting aspect of possibility qualification relates to the invariance of implication under this mode of qualification. Thus, from the definition of implication [6], it follows at once that

$$X \text{ is } F \Rightarrow X \text{ is } G \quad \text{if } F \subset G.$$

Now, it can readily be shown that

$$F \subset G \Rightarrow F^+ \subset G^+ \tag{4.31}$$

where $\subset$ in the right-hand member of (4.31) should be interpreted as the relation of containment for fuzzy sets of Type 2. In consequence of (4.31), then, we can assert that

$$X \text{ is } F \text{ is possible} \Rightarrow X \text{ is } G \text{ is possible} \quad \text{if } F \subset G. \tag{4.32}$$

5. Concluding remarks

The exposition of the theory of possibility in the present paper touches upon only a few of the many facets of this—as yet largely unexplored—theory. Clearly, the intuitive concepts of possibility and probability play a central role in human decision-making and underlie much of the human ability to reason in approximate terms. Consequently, it will be essential to develop a better understanding of the interplay between possibility and probability—especially in relation to the roles which these concepts play in natural languages—in order to enhance our ability to develop machines which can simulate the remarkable human ability to attain imprecisely defined goals in a fuzzy environment.

Acknowledgment

The idea of employing the theory of fuzzy sets as a basis for the theory of possibility was inspired by the paper of B.R. Gaines and L. Kohout on possible automata [1]. In addition, our work was stimulated by discussions with H.J. Zimmermann regarding the interpretation of the operations of conjunction and disjunction; the results of a psychological study conducted by Eleanor Rosch and Louis Gomez; and discussions with Barbara Cerny.

References

[1] B.R. Gaines and L.J. Kohout, Possible automata, Proc. Int. Symp. on Multiple-Valued Logic, University of Indiana, Bloomington, IN (1975) 183–196.
[2] G.E. Hughes and M.J. Cresswell, An Introduction to Modal Logic (Methuen, London, 1968).
[3] A. Kaufmann, Valuation and probabilization, in: A. Kaufmann and E. Sanchez, eds., Theory of Fuzzy Subsets, Vol. 5 (Masson and Co., Paris, 1977).
[4] L.A. Zadeh, Calculus of fuzzy restrictions, in: L.A. Zadeh, K.S. Fu, K. Tanaka and M. Shimura, eds., Fuzzy Sets and Their Applications to Cognitive and Decision Processes (Academic Press, New York, 1975) 1–39.

[5] L.A. Zadeh, The concept of a linguistic variable and its application to approximate reasoning, Part I, Information Sci. 8 (1975) 199–249; Part II, Information Sci. 8 (1975) 301–357; Part III, Information Sci. 9 (1975) 43–80.

[6] R.E. Bellman and L.A. Zadeh, Local and fuzzy logics, ERL Memo. M-584, University of California, Berkeley, CA (1976); Modern Uses of Multiple-Valued Logics, D. Epstein, ed. (D. Reidel, Dordrecht, 1977).

[7] B. DeFinetti, Probability Theory (Wiley, New York, 1974).

[8] T. Fine, Theories of Probability (Academic Press, New York, 1973).

[9] A. Dempster, Upper and lower probabilities induced by multi-valued mapping, Ann. Math. Statist. 38 (1967) 325–339.

[10] G. Shafer, A Mathematical Theory of Evidence (Princeton University Press, Princeton, NJ, 1976).

[11] E.H. Shortliffe, A model of inexact reasoning in medicine, Math. Biosciences 23 (1975) 351–379.

[12] R.O. Duda, P.F. Hart and N.J. Nilsson, Subjective Bayesian methods for rule-based inference systems, Stanford Research Institute Tech. Note 124, Stanford, CA (1976).

[13] F. Wenstop, Deductive verbal models of organization, Int. J. Man-Machine Studies 8 (1976) 293–311.

[14] L.A. Zadeh, Theory of fuzzy sets, Memo. No. UCB/ERL M77/1, University of California, Berkeley, CA (1977).

[15] L.A. Zadeh, A fuzzy-set-theoretic interpretation of linguistic hedges, J. Cybernet. 2 (1972) 4–34.

[16] G. Lakoff, Hedges: a study in meaning criteria and the logic of fuzzy concepts, J. Phil. Logic 2 (1973) 458–508; also paper presented at 8th Regional Meeting of Chicago Linguistic Soc. (1972) and in: D. Hockney, W. Harper and B. Freed, eds., Contemporary Research in Philosophical Logic and Linguistic Semantics (D. Reidel, Dordrecht, 1975) 221–271.

[17] H.M. Hersch and A. Caramazza, A fuzzy set approach to modifiers and vagueness in natural languages, Dept. of Psych., The Johns Hopkins University, Baltimore, MD (1975).

[18] P.J. MacVicar-Whelan, Fuzzy sets, the concept of height and the hedge very, Tech. Memo 1, Physics Dept., Grand Valley State College, Allendale, MI (1974).

[19] L.A. Zadeh, Probability measures of fuzzy events, J. Math. Anal. Appl. 23 (1968) 421–427.

[20] M. Sugeno, Theory of fuzzy integrals and its applications, Ph.D. Thesis, Tokyo Institute of Technology, Tokyo (1974).

[21] T. Terano and M. Sugeno, Conditional fuzzy measures and their applications, in: L.A. Zadeh, K.S. Fu, K. Tanaka and M. Shimura, eds., Fuzzy Sets and Their Applications to Cognitive and Decision Processes (Academic Press, New York, 1975) 151–170.

[22] S. Nahmias, Fuzzy variables, Tech. Rep. 33, Dept. of Ind. Eng., Systems Manag. Eng. and Operations Research, University of Pittsburgh, PA (1976); Presented at the ORSA/TIMS meeting, Miami, FL (November 1976).

[23] L.A. Zadeh, Similarity relations and fuzzy orderings, Information Sci. 3 (1971) 177–200.

[24] W. Rödder, On 'and' and 'or' connectives in fuzzy set theory, Inst. for Oper. Res., Technical University of Aachen, Aachen (1975).

[25] L.A. Zadeh, K.S. Fu, K. Tanaka and M. Shimura, eds., Fuzzy Sets and Their Applications to Cognitive and Decision Processes (Academic Press, New York, 1975).

[26] E. Mamdani, Application of fuzzy logic to approximate reasoning using linguistic synthesis. Proc. 6th Int. Symp. on Multiple-Valued Logic, Utah State University, Logan, UT (1976) 196–202.

[27] V.V. Nalimov, Probabilistic model of language, Moscow State University, Moscow (1974).

[28] J.A. Goguen, Concept representation in natural and artificial languages: axioms, extensions and applications for fuzzy sets, Int. J. Man-Machine Studies 6 (1974) 513–561.

[29] C.V. Negoita and D.A. Ralescu, Applications of Fuzzy Sets to Systems Analysis (Birkhauser, Stuttgart, 1975).

[30] L.A. Zadeh, PRUF–A meaning representation language for natural languages, Memo No. UCB/ERL M77/61, University of California, Berkeley, CA (1977).

The Concept of a Linguistic Variable
and its Application to Approximate
Reasoning-I

L.A. ZADEH‡

Computer Sciences Division, Department of Electrical Engineering and Computer Sciences, and the Electronics Research Laboratory, University of California, Berkeley, California 94720.

ABSTRACT

By a *linguistic variable* we mean a variable whose values are words or sentences in a natural or artificial language. For example, *Age* is a linguistic variable if its values are linguistic rather than numerical, i.e., *young, not young, very young, quite young, old, not very old and not very young,* etc., rather than 20, 21, 22, 23, ... In more specific terms, a linguistic variable is characterized by a quintuple $(\mathcal{X}, T(\mathcal{X}), U, G, M)$ in which $\mathcal{X}$ is the name of the variable; $T(\mathcal{X})$ is the *term-set* of $\mathcal{X}$, that is, the collection of its linguistic values; U is a universe of discourse; G is a *syntactic rule* which generates the terms in $T(\mathcal{X})$; and M is a *semantic rule* which associates with each linguistic value X its *meaning, $M(X)$,* where $M(X)$ denotes a fuzzy subset of U. The meaning of a linguistic value X is characterized by a *compatibility function, $c : U \rightarrow [0,1]$,* which associates with each u in U its compatibility with X. Thus, the compatibility of age 27 with *young* might be 0.7, while that of 35 might be 0.2. The function of the semantic rule is to relate the compatibilities of the so-called *primary* terms in a composite linguistic value—e.g., *young* and *old* in *not very young and not very old*—to the compatibility of the composite value. To this end, the hedges such as *very, quite, extremely,* etc., as well as the connectives *and* and *or* are treated as nonlinear operators which modify the meaning of their operands in a specified fashion. The concept of a linguistic variable provides a means of approximate characterization of phenomena which are too complex or too ill-defined to be amenable to description in conventional quantitative terms. In particular, treating *Truth* as a linguistic variable with values such as *true, very true, completely true, not very true, untrue,* etc., leads to what is called *fuzzy logic.* By providing a basis for *approximate reasoning,* that is, a mode of reasoning which is not exact nor very inexact, such logic may offer a more realistic framework for human reasoning than the traditional two-valued logic. It is shown that probabilities, too, can be treated as linguistic variables with values such as *likely, very likely, unlikely,* etc. Computation with linguistic probabilities requires the solution of

This work was supported in part by the Navy Electronic Systems Command under Contract N00039-71-C-0255, the Army Research Office, Durham, N.C., under Grant DA-ARO-D-31-124-71-G174, and the National Science Foundation under Grant GK-VP-610656X3. The writing of the paper was completed while the author was participating in a Joint Study Program with the Systems Research Department, IBM Research Laboratory, San Jose, California.

nonlinear programs and leads to results which are imprecise to the same degree as the underlying probabilities. The main applications of the linguistic approach lie in the realm of humanistic systems—especially in the fields of artificial intelligence, linguistics, human decision processes, pattern recognition, psychology, law, medical diagnosis, information retrieval, economics and related areas.

---

1. INTRODUCTION

One of the fundamental tenets of modern science is that a phenomenon cannot be claimed to be well understood until it can be characterized in quantitative terms.[1] Viewed in this perspective, much of what constitutes the core of scientific knowledge may be regarded as a reservoir of concepts and techniques which can be drawn upon to construct mathematical models of various types of systems and thereby yield quantitative information concerning their behavior.

Given our veneration for what is precise, rigorous and quantitative, and our disdain for what is fuzzy, unrigorous and qualitative, it is not surprising that the advent of digital computers has resulted in a rapid expansion in the use of quantitative methods throughout most fields of human knowledge. Unquestionably, computers have proved to be highly effective in dealing with *mechanistic* systems, that is, with inanimate systems whose behavior is governed by the laws of mechanics, physics, chemistry and electromagnetism. Unfortunately, the same cannot be said about *humanistic* systems,[2] which—so far at least—have proved to be rather impervious to mathematical analysis and computer simulation. Indeed, it is widely agreed that the use of computers has not shed much light on the basic issues arising in philosophy, psychology, literature, law, politics, sociology and other human-oriented fields. Nor have computers added significantly to our understanding of human thought processes—excepting, perhaps, some examples to the contrary that can be drawn from artificial intelligence and related fields [2, 3, 4, 5, 51].

---

[1] As expressed by Lord Kelvin in 1883 [1], "In physical science a first essential step in the direction of learning any subject is to find principles of numerical reckoning and practicable methods for measuring some quality connected with it. I often say that when you can measure what you are speaking about and express it in numbers, you know something about it; but when you cannot measure it, when you cannot express it in numbers, your knowledge is of a meagre and unsatisfactory kind: it may be the beginning of knowledge but you have scarcely, in your thoughts, advanced to the state of *science,* whatever the matter may be."

[2] By a *humanistic* system we mean a system whose behavior is strongly influenced by human judgement, perception or emotions. Examples of humanistic systems are: economic systems, political systems, legal systems, educational systems, etc. A single individual and his thought processes may also be viewed as a humanistic system.

It may be argued, as we have done in [6] and [7], that the ineffectiveness of computers in dealing with humanistic systems is a manifestation of what might be called the *principle of incompatibility*—a principle which asserts that high precision is incompatible with high complexity.[3] Thus, it may well be the case that the conventional techniques of system analysis and computer simulation—based as they are on precise manipulation of numerical data—are intrinsically incapable of coming to grips with the great complexity of human thought processes and decision-making. The acceptance of this premise suggests that, in order to be able to make significant assertions about the behavior of humanistic systems, it may be necessary to abandon the high standards of rigor and precision that we have become conditioned to expect of our mathematical analyses of well-structured mechanistic systems, and become more tolerant of approaches which are approximate in nature. Indeed, it is entirely possible that only through the use of such approaches could computer simulation become truly effective as a tool for the analysis of systems which are too complex or too ill-defined for the application of conventional quantitative techniques.

In retreating from precision in the face of overpowering complexity, it is natural to explore the use of what might be called *linguistic* variables, that is, variables whose values are not numbers but words or sentences in a natural or artificial language. The motivation for the use of words or sentences rather than numbers is that linguistic characterizations are, in general, less specific than numerical ones. For example, in speaking of age, when we say "John is young," we are less precise than when we say, "John is 25." In this sense, the label *young* may be regarded as a *linguistic value* of the variable *Age*, with the understanding that it plays the same role as the numerical value 25 but is less precise and hence less informative. The same is true of the linguistic values *very young, not young, extremely young, not very young,* etc. as contrasted with the numerical values 20, 21, 22, 23, ...

If the values of a numerical variable are visualized as points in a plane, then the values of a linguistic variable may be likened to ball parks with fuzzy boundaries. In fact, it is by virtue of the employment of ball parks rather than points that linguistic variables acquire the ability to serve as a means of approximate characterization of phenomena which are too complex or too ill-defined to be susceptible of description in precise terms. What is also important, however, is that by the use of a so-called *extension principle,* much of the existing mathematical apparatus of systems analysis can be adapted to the manipulation of linguistic variables. In this way, we may be able to develop an approximate calculus of linguistic variables which could be of use in a wide variety of practical applications.

---

[3] Stated somewhat more concretely, the complexity of a system and the precision with which it can be analyzed bear a roughly inverse relation to one another.

The totality of values of a linguistic variable constitute its *term-set,* which in principle could have an infinite number of elements. For example, the term-set of the linguistic variable *Age* might read

$T(Age) = young + not\ young + very\ young + not\ very\ young + very\ very\ young$
$+ \cdots + old + not\ old + very\ old + not\ very\ old + \cdots + not\ very\ young$
and not very old $+ \cdots +$ *middle-aged* $+$ *not middle-aged* $+ \cdots +$ *not*
old and not middle-aged $+ \cdots +$ *extremely old* $+ \cdots ,$ (1.1)

in which + is used to denote the union rather than the arithmetic sum. Similarly, the term-set of the linguistic variable *Appearance* might be

$T(Appearance) = beautiful + pretty + cute + handsome + attractive + not$
beautiful $+$ *very pretty* $+$ *very very handsome* $+$ *more or less pretty* $+$
quite pretty $+$ *quite handsome* $+$ *fairly handsome* $+$ *not very attractive*
and not very unattractive $+ \cdots$

In the case of the linguistic variable *Age,* the numerical variable *age* whose values are the numbers 0, 1, 2, 3, . . . , 100 constitutes what may be called the *base variable* for *Age.* In terms of this variable, a linguistic value such as *young* may be interpreted as a label for a *fuzzy restriction* on the values of the base variable. This fuzzy restriction is what we take to be the *meaning* of *young.*

A fuzzy restriction on the values of the base variable is characterized by a *compatibility function* which associates with each value of the base variable a number in the interval [0, 1] which represents its *compatibility* with the fuzzy restriction. For example, the compatabilities of the numerical ages 22, 28 and 35 with the fuzzy restriction labeled *young* might be 1, 0.7 and 0.2, respectively. The meaning of *young,* then, would be represented by a graph of the form shown in Fig. 1, which is a plot of the compatibility function of *young* with respect to the base variable *age.*

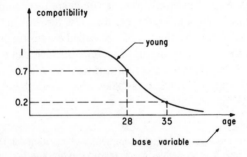

Fig. 1. Compatibility function for *young.*

The conventional interpretation of the statement "John is young," is that John is a member of the class of young men. However, considering that the class of young men is a fuzzy set, that is, there is no sharp transition from being young to not being young, the assertion that John is a member of the class of young men is inconsistent with the precise mathematical definition of "is a member of." The concept of a linguistic variable allows us to get around this difficulty in the following manner.

The name "John" is viewed as a name of a composite linguistic variable whose components are linguistic variables named *Age, Height, Weight, Appearance,* etc. Then the statement "John is young" is interpreted as an *assignment equation* (Fig. 2).

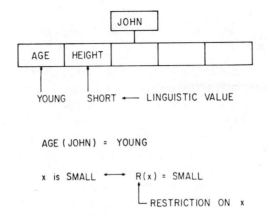

Fig. 2. Assignment of linguistic values to attributes of John and *x*.

$$Age = young$$

which assigns the value *young* to the linguistic variable *Age.* In turn, the value *young* is interpreted as a label for a fuzzy restriction on the base variable *age,* with the meaning of this fuzzy restriction defined by its compatibility function. As an aid in the understanding of the concept of a linguistic variable, Fig. 3 shows the hierarchical structure of the relation between the linguistic variable *Age*, the fuzzy restrictions which represent the meaning of its values, and the values of the base variable *age*.

There are several basic aspects of the concept of a linguistic variable that are in need of elaboration.

First, it is important to understand that the notion of compatibility is distinct from that of probability. Thus, the statement that the compatibility of, say, 28 with *young* is 0.7, has no relation to the probability of the age-value 28. The correct interpretation of the compatibility-value 0.7 is that it

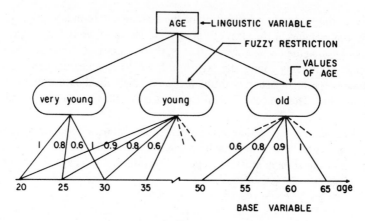

Fig. 3. Hierarchical structure of a linguistic variable.

is merely a subjective indication of the extent to which the age-value 28 fits one's conception of the label *young.* As we shall see in later sections, the rules of manipulation applying to compatibilities are different from those applying to probabilities, although there are certain parallels between the two.

Second, we shall usually assume that a linguistic variable is *structured* in the sense that it is associated with two rules. Rule (i), a *syntactic rule,* specifies the manner in which the linguistic values which are in the term-set of the variable may be generated. In regard to this rule, our usual assumption will be that the terms in the term-set of the variable are generated by a context-free grammar.

The second rule, (ii), is a *semantic rule* which specifies a procedure for computing the meaning of any given linguistic value. In this connection, we observe that a typical value of a linguistic variable, e.g., *not very young and not very old,* involves what might be called the *primary* terms, e.g., *young* and *old,* whose meaning is both subjective and context-dependent. We assume that the meaning of such terms is specified *a priori.*

In addition to the primary terms, a linguistic value may involve connectives such as *and, or, either, neither,* etc.; the negation *not*; and the hedges such as *very, more or less, completely, quite, fairly, extremely, somewhat,* etc. As we shall see in later sections, the connectives, the hedges and the negation may be treated as operators which modify the meaning of their operands in a specified, context-independent fashion. Thus, if the meaning of *young* is defined by the compatibility function whose form is shown in Fig. 1, then the meaning of *very young* could be obtained by squaring the compatibility function of *young,* while that of *not young* would be given by subtracting the compatibility function of *young* from unity (Fig. 4). These two rules are special instances of a more general semantic rule which is described in Part II, Sec. 2.

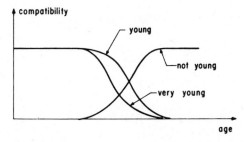

Fig. 4. Compatibilities of *young, not young,* and *very young.*

Third, when we speak of a linguistic variable such as *Age,* the underlying base variable, *age,* is numerical in nature. Thus, in this case we can define the meaning of a linguistic value such as *young* by a compatibility function which associates with each age in the interval [0, 100] a number in the interval [0, 1] which represents the compatibility of that age with the label *young.*

On the other hand, in the case of the linguistic variable *Appearance,* we do not have a well-defined base variable; that is, we do not know how to express the degree of beauty, say, as a function of some physical measurements. We could still associate with each member of a group of ladies, for example, a grade of membership in the class of beautiful women, say 0.9 with Fay, 0.7 with Adele, 0.8 with Kathy and 0.9 with Vera, but these values of the compatibility function would be based on impressions which we may not be able to articulate or formalize in explicit terms. In other words, we are defining the compatibility function not on a set of mathematically well-defined objects, but on a set of labeled impressions. Such definitions are meaningful to a human but not—at least directly—to a computer.[4]

As we shall see in later sections, in the first case, where the base variable is numerical in nature, linguistic variables can be treated in a reasonably precise fashion by the use of the extension principle for fuzzy sets. In the second case, their treatment becomes much more qualitative. In both cases, however, some computation is involved—to a lesser or greater degree. Thus, it should be understood that the linguistic approach is not entirely qualitative in nature. Rather, the computations are performed behind the scene, and, at the end, *linguistic approximation* is employed to convert numbers into words (Fig. 5).

A particularly important area of application for the concept of a linguistic variable is that of *approximate reasoning,* by which we mean a type of reasoning which is neither very precise nor very imprecise. As an illustration, the following inference would be an instance of approximate reasoning:

---

[4]The basic problem which is involved here is that of abstraction from a set of samples of elements of a fuzzy set. A discussion of this problem may be found in [8].

x is small,

x and y are approximately equal;

therefore,

y is more or less small.

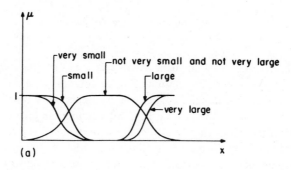

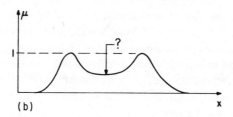

Fig. 5. (a) Compatibilities of *small, very small, large, very large* and *not very small and not very large.* (b) The problem of linguistic approximation is that of finding an approximate linguistic characterization of a given compatibility function.

The concept of a linguistic variable enters into approximate reasoning as a result of treating *Truth* as a linguistic variable whose truth-values form a term-set such as shown below:

$T(Truth) = true + not\ true + very\ true + completely\ true + more\ or\ less\ true$
$\qquad + fairly\ true + essentially\ true + \cdots + false + very\ false + neither\ true$
$\qquad nor\ false + \cdots .$

The corresponding base variable, then, is assumed to be a number in the interval [0, 1], and the meaning of a primary term such as *true* is identified with a fuzzy restriction on the values of the base variable. As usual, such a restriction is characterized by a compatibility function which associates a number in the interval [0, 1] with each numerical truth-value. For example, the compatibility of the numerical truth-value 0.7 with the linguistic truth-value *very true* might be 0.6. Thus, in the case of truth-values, the compatibility function is a mapping from the unit interval to itself. (This will be shown in Part II, Fig. 13.)

Treating truth as a linguistic variable leads to a fuzzy logic which may well be a better approximation to the logic involved in human decision processes than the classical two-valued logic.[5] Thus, in fuzzy logic it is meaningful to assert what would be inadmissibly vague in classical logic, e.g.,

> The truth-value of "Berkeley is close to San Francisco," is *quite true.*
> The truth-value of "Palo Alto is close to San Francisco," is *fairly true.*

Therefore,

the truth-value of "Palo Alto is more or less close to Berkeley," is *more or less true.*

Another important area of application for the concept of a linguistic variable lies in the realm of probability theory. If probability is treated as a linguistic variable, its term-set would typically be:

$T(Probability) = likely + very\ likely + unlikely + extremely\ likely + fairly\ likely$
$+ \cdots + probable + improbable + more\ or\ less\ probable + \cdots.$

By legitimizing the use of linguistic probability-values, we make it possible to respond to a question such as "What is the probability that it will be a warm day a week from today?" with an answer such as *fairly high,* instead of, say, 0.8. The linguistic answer would, in general, be much more realistic, considering, first, that *warm day* is a fuzzy event, and, second, that our understanding of weather dynamics is not sufficient to allow us to make unequivocal assertions about the underlying probabilities.

In the following sections, the concept of a linguistic variable and its applications will be discussed in greater detail. To place the concept of a linguistic variable in a proper perspective, we shall begin our discussion with a formalization of the notion of a conventional (nonfuzzy) variable. For our purposes, it will be helpful to visualize such a variable as a tagged valise with rigid (hard) sides (Fig. 6). Putting an object into the valise corresponds to assigning a value to the variable, and the restriction on what can be put in corresponds to a subset of the universe of discourse which comprises those points which can be assigned as values to the variable. In terms of this analogy, a *fuzzy variable,*

---

[5] Expositions of alternative approaches to vagueness may be found in [9 – 18].

which is defined in Part II, Sec. 1, may be likened to a tagged valise with soft rather than rigid sides (Part II, Fig. 1). In this case, the restriction on what can be put in is fuzzy in nature, and is defined by a compatibility function which associates with each object a number in the interval [0, 1] representing the degree of ease with which that object can be fitted in the valise. For example, given a valise named X, the compatibility of a coat with X would be 1, while that of a record-player might be 0.7.

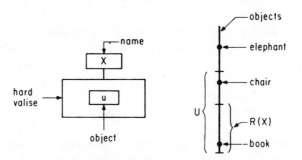

Fig. 6. Illustration of the valise analogy for a unary nonfuzzy variable.

As will be seen in Part II, Sec. 1, an important concept in the case of fuzzy variables is that of *noninteraction,* which is analogous to the concept of independence in the case of random variables. This concept arises when we deal with two or more fuzzy variables, each of which may be likened to a compartment in a soft valise. Such fuzzy variables are *interactive* if the assignment of a value to one affects the fuzzy restrictions placed on the others. This effect may be likened to the interference between objects which are put into different compartments of a soft valise (Part II, Fig. 3).

A linguistic variable is defined in Part II, Sec. 2 as a variable whose values are fuzzy variables. In terms of our valise analogy, a linguistic variable corresponds to a hard valise into which we can put soft valises, with each soft valise carrying a name tag which describes a fuzzy restriction on what can be put into that valise (Part II, Fig. 6).

The application of the concept of a linguistic variable to the notion of *Truth* is discussed in Part II, Sec. 3. Here we describe a technique for computing the conjunction, disjunction and negation for linguistic truth-values and lay the groundwork for fuzzy logic.

In Part III, Sec. 1, the concept of a linguistic variable is applied to probabilities, and it is shown that linguistic probabilities can be used for computational purposes. However, because of the constraint that the numerical probabilities must add up to unity, the computations in question involve the solution of nonlinear programs and hence are not as simple to perform as computations involving numerical probabilities.

The last section is devoted to a discussion of the so-called *compositional rule of inference* and its application to approximate reasoning. This rule of inference is interpreted as the process of solving a simultaneous system of so-called *relational assignment equations* in which linguistic values are assigned to fuzzy restrictions. Thus, if a statement such as "*x* is small" is interpreted as an assignment of the linguistic value *small* to the fuzzy restriction on *x*, and the statement "*x* and *y* are approximately equal" is interpreted as the assignment of a fuzzy relation labeled *approximately equal* to the fuzzy restriction on the ordered pair (x, y), then the conclusion "*y* is more or less small" may be viewed as a linguistic approximation to the solution of the simultaneous equations

$$R(x) = small,$$

$$R(x,y) = approximately\ equal,$$

in which $R(x)$ and $R(x,y)$ denote the restrictions on x and (x, y), respectively (Part III, Fig. 5).

The compositional rule of inference leads to a *generalized modus ponens,* which may be viewed as an extension of the familiar rule of inference: If A is true and A implies B, then B is true. The section closes with an example of a fuzzy theorem in elementary geometry and a brief discussion of the use of fuzzy flowcharts for the representation of definitional fuzzy algorithms.

The material in Secs. 2 and 3 and in Part II, Sec. 1 is intended to provide a mathematical basis for the concept of a linguistic variable, which is introduced in Part II, Sec. 2. For those readers who may not be interested in the mathematical aspects of the theory, it may be expedient to proceed directly to Part II, Sec. 2 and refer where necessary to the definitions and results described in the preceding sections.

2. THE CONCEPT OF A VARIABLE

In the preceding section, our discussion of the concept of a linguistic variable was informal in nature. To set the stage for a more formal definition, we shall focus our attention in this section on the concept of a conventional (nonfuzzy) variable. Then in Sec. 3 we shall extend the concept of a variable to fuzzy variables and subsequently will define a linguistic variable as a variable whose values are fuzzy variables.

Although the concept of a (nonfuzzy) variable is very elementary in nature, it is by no means a trivial one. For our purposes, the following formalization of the concept of a variable provides a convenient basis for later extensions.

DEFINITION 2.1. A variable is characterized by a triple $(X, U, R(X;u))$, in which X is the name of the variable; U is a universe of discourse (finite or infinite set); u is a generic[6] name for the elements of U; and $R(X;u)$ is a subset of U which represents a *restriction*[7] on the values of u imposed by X. For convenience, we shall usually abbreviate $R(X;u)$ to $R(X)$ or $R(u)$ or $R(x)$, where x denotes a generic name for the values of X, and will refer to $R(X)$ simply as the restriction *on u* or the restriction *imposed by X*.

In addition, a variable is associated with an *assignment equation*

$$x = u : R(X) \tag{2.1}$$

or equivalently

$$x = u, \qquad u \in R(X) \tag{2.2}$$

which represents the assignment of a value u to x subject to the restriction $R(X)$. Thus, the assignment equation is *satisfied* iff (if and only if) $u \in R(X)$.

Example 2.1. As a simple illustration consider a variable named *age*. In this case, U might be taken to be the set of integers $0, 1, 2, 3, \ldots$, and $R(X)$ might be the subset $0, 1, 2, \ldots, 100$.

More generally, let $X_1, \ldots, X_n$ be n variables with respective universes of discourse $U_1, \ldots, U_n$. The ordered n-tuple $X = (X_1, \ldots, X_n)$ will be referred to as an *n-ary composite* (or *joint*) *variable*. The universe of discourse for X is the Cartesian product

$$U = U_1 \times U_2 \times \cdots \times U_n, \tag{2.3}$$

and the restriction $R(X_1, \ldots, X_n)$ is an n-ary relation in $U_1 \times \cdots \times U_n$. This relation may be defined by its characteristic (membership) function $\mu_R : U_1 \times \cdots \times U_n \rightarrow \{0, 1\}$, where

$$\mu_R(u_1, \ldots, u_n) = 1 \qquad \text{if } (u_1, \ldots, u_n) \in R(X_1, \ldots, X_n)$$

$$= 0 \qquad \text{otherwise,} \tag{2.4}$$

---

[6] A generic name is a single name for all elements of a set. For simplicity, we shall frequently use the same symbol for both a set and the generic name for its elements, relying on the context for disambiguation.

[7] In conventional terminology, $R(X)$ is the range of X. Our use of the term *restriction* is motivated by the role played by $R(X)$ in the case of fuzzy variables.

and u_i is a generic name for the elements of U_i, $i = 1, \ldots, n$. Correspondingly, the n-ary assignment equation assumes the form

$$(x_1, \ldots x_n) = (u_1, \ldots, u_n) : R(X_1, \ldots, X_n), \tag{2.5}$$

which is understood to mean that

$$x_i = u_i, \qquad i = 1, \ldots, n \tag{2.6}$$

subject to the restriction $(u_1, \ldots, u_n) \in R(X_1, \ldots, X_n)$, with x_i, $i = 1, \ldots, n$, denoting a generic name for values of X_i.

Example 2.2 Suppose that $X_1 \triangleq$ age of father,[8] $X_2 \triangleq$ age of son, and $U_1 \triangleq U_2 = \{1, 2, \ldots, 100\}$. Furthermore, suppose that $x_1 \geqslant x_2 + 20$ (x_1 and x_2 are generic names for values of X_1 and X_2). Then $R(X_1, X_2)$ may be defined by

$$\mu_R(u_1, u_2) = 1 \qquad \text{for } 21 \leqslant u_1 \leqslant 100, \quad u_1 \geqslant u_2 + 20$$

$$= 0 \qquad \text{elsewhere.} \tag{2.7}$$

MARGINAL AND CONDITIONED RESTRICTIONS

As in the case of probability distributions, the restriction $R(X_1, \ldots, X_n)$ imposed by $(X_1, \ldots, X_n)$ induces *marginal* restrictions $R(X_{i_1}, \ldots, X_{i_k})$ imposed by composite variables of the form $(X_{i_1}, \ldots, X_{i_k})$, where the index sequence $q = (i_1, \ldots, i_k)$ is a subsequence of the index sequence $(1, 2, \ldots, n)$.[9] In effect, $R(X_{i_1}, \ldots, X_{i_k})$ is the smallest (i.e., most restrictive) restriction imposed by $(X_{i_1}, \ldots, X_{i_k})$ which satisfies the implication

$$(u_1, \ldots, u_n) \in R(X_1, \ldots, X_n) \Rightarrow (u_{i_1}, \ldots, u_{i_k}) \in R(X_{i_1}, \ldots, X_{i_k}). \tag{2.8}$$

Thus, a given k-tuple $u_{(q)} \triangleq (u_{i_1}, \ldots, u_{i_k})$ is an element of $R(X_{i_1}, \ldots, X_{i_k})$ iff there exists an n-tuple $u \triangleq (u_1, \ldots, u_n) \in R(X_1, \ldots, X_n)$ whose i_1 th, $\ldots$, i_k th components are equal to $u_{i_1}, \ldots, u_{i_k}$, respectively. Expressed in terms of the characteristic functions of $R(X_1, \ldots, X_n)$ and $R(X_{i_1}, \ldots, X_{i_k})$, this statement translates into the equation

---

[8] The symbol $\triangleq$ stands for "denotes" or is "equal by definition."

[9] In the case of a binary relation $R(X_1, X_2)$, $R(X_1)$ and $R(X_2)$ are usually referred to as the *domain* and *range* of $R(X_1, X_2)$.

$$\mu_{R(X_{i_1}, \ldots, X_{i_k})}(u_{i_1}, \ldots, u_{i_k}) = \bigvee_{u_{(q')}} \mu_{R(X_1, \ldots, X_n)}(u_1, \ldots, u_n), \quad (2.9)$$

or more compactly,

$$\mu_{R(X_{(q)})}(u_{(q)}) = \bigvee_{u_{(q')}} \mu_{R(X)}(u), \quad (2.10)$$

where q' is the complement of the index sequence $q = (i_1, \ldots, i_k)$ relative to $(1, \ldots, n)$, $u_{(q')}$ is the complement of the k-tuple $u_{(q)} \stackrel{\triangle}{=} (u_{i_1}, \ldots, u_{i_k})$ relative to the n-tuple $u \stackrel{\triangle}{=} (u_1, \ldots, u_n)$, $X_{(q)} \stackrel{\triangle}{=} (X_{i_1}, \ldots, X_{i_k})$, and $\bigvee_{u_{(q')}}$ denotes the supremum of its operand over the u's which are in $u_{(q')}$. (Throughout this paper, the symbols $\vee$ and $\wedge$ stand for Max and Min, respectively; thus, for any real a, b

$$a \vee b = \text{Max}(a, b) = a \quad \text{if} \quad a \geqslant b$$

$$= b \quad \text{if} \quad a < b \quad (2.11)$$

and

$$a \wedge b = \text{Min}(a, b) = a \quad \text{if} \quad a \leqslant b$$

$$= b \quad \text{if} \quad a > b.$$

Consistent with this notation, the symbol $\bigvee_z$ should be read as "supremum over the values of z.") Since μ_R can take only two values—0 or 1—(2.10) means that $\mu_{R(X_{(q)})}(u_{(q)})$ is 1 iff there exists a $u_{(q')}$ such that $\mu_{R(X)}(u) = 1$.

COMMENT 2.1. There is a simple analogy which is very helpful in clarifying the notion of a variable and related concepts. Specifically, a non-fuzzy variable in the sense formalized in Definition 2.1 may be likened to a tagged valise having rigid (hard) sides, with X representing the name on the tag, U representing a list of objects which could be put in a valise, and $R(X)$ representing a sublist of U which comprises those objects which can be put into valise X. [For example, an object like a boat would not be in U, while an object like a typewriter might be in U but not in $R(X)$, and an object like a cigarette box or a pair of shoes would be in $R(X)$.] In this interpretation, the assignment equation

$$x = u : R(X)$$

signifies that an object u which satisfies the restriction $R(X)$ (i.e., is on the list of objects which can be put into X) is put into X (Fig. 6).

An n-ary composite variable $X \triangleq (X_1, \ldots, X_n)$ corresponds to a valise, carrying the name-tag X, which has n compartments named $X_1, \ldots, X_n$ with adjustable partitions between them. The restrictions $R(X_1, \ldots, X_n)$ corresponds to a list of n-tuples of objects $(u_1, \ldots, u_n)$ such that u_1 can be put in compartment X_1, u_2 in compartment $X_2, \ldots$, and u_n in compartment X_n *simultaneously.* (see Fig. 7.) In this connection, it should be noted that n-tuples on this list could be associated with different arrangements of partitions. If $n = 2$, for example, then for a particular placement of the partition we could put a coat in compartment X_1 and a suit in compartment X_2, while for some other placement we could put the coat in compartment X_2 and a box of shoes in compartment X_1. In this event, both (coat, suit) and (shoes, coat) would be included in the list of pairs of objects which can be put in X simultaneously.

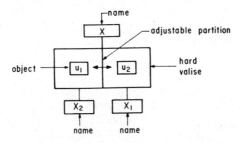

Fig. 7. Valise analogy for a binary nonfuzzy variable.

In terms of the valise analogy, the n-ary assignment equation

$$(x_1, \ldots, x_n) = (u_1, \ldots, u_n) : R(X_1, \ldots, X_n)$$

represents the action of putting u_1 in $X_1, \ldots$, and u_n in X_n simultaneously, under the restriction that the n-tuple of objects $(u_1, \ldots, u_n)$ must be on the $R(X_1, \ldots, X_n)$ list. Furthermore, a marginal restriction such as $R(X_{i_1}, \ldots, X_{i_k})$ may be interpreted as a list of k-tuples of objects which can be put in compartments $X_{i_1}, \ldots, X_{i_k}$ simultaneously, in conjunction with every allowable placement of objects in the remaining compartments.

COMMENT 2.2. It should be noted that (2.9) is analogous to the expression for a marginal distribution of a probability distribution, with $\vee$ corresponding to summation (or integration). However, this analogy should not be construed to imply that $R(X_{i_1}, \ldots, X_{i_k})$ is in fact a marginal probability distribution.

It is convenient to view the right-hand side of (2.9) as the characteristic function of the projection[10] of $R(X_1, \ldots, X_n)$ on $U_{i_1} \times \cdots \times U_{i_k}$. Thus, in symbols,

$$R(X_{i_1}, \ldots, X_{i_k}) = \text{Proj } R(X_1, \ldots, X_n) \text{ on } U_{i_1} \times \cdots \times U_{i_k}, \quad (2.12)$$

or more simply,

$$R(X_{i_1}, \ldots, X_{i_k}) = P_q R(X_1, \ldots, X_n),$$

where P_q denotes the operation of projection on $U_{i_1} \times \cdots \times U_{i_k}$ with $q = (i_1, \ldots, i_k)$.

Example 2.3.   In the case of Example 2.2, we have

$$R(X_1) = P_1 R(X_1, X_2) = \left\{21, \ldots, 100\right\},$$

$$R(X_2) = P_2 R(X_1, X_2) = \left\{1, \ldots, 80\right\}.$$

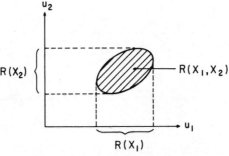

Fig. 8. Marginal restrictions induced by $R(X_1, X_2)$.

Example 2.4.   Fig. 8 shows the restrictions on u_1 and u_2 induced by $R(X_1, X_2)$.

An alternative way of describing projections is the following. Viewing $R(X_1, \ldots, X_n)$ as a relation in $U_1 \times \cdots \times U_n$, let $q' = (j_1, \ldots, j_m)$ denote the index sequence complementary to $q = (i_1, \ldots, i_k)$, and let $R(X_{i_1}, \ldots, X_{i_k} \mid u_{j_1}, \ldots, u_{j_m})$—or, more compactly, $R(X_{(q)} \mid u_{(q')})$—denote a restriction in $U_{i_1} \times \cdots \times U_{i_k}$ which is *conditioned on* $u_{j_1}, \ldots, u_{j_m}$. The characteristic function of this conditioned restriction is defined by

---

[10]The term *projection* as used in the literature is somewhat ambiguous in that it could denote either the operation of projecting or the result of such an operation. To avoid this ambiguity in the case of fuzzy relations, we will occasionally employ the term *shadow* [19] to denote the relation resulting from applying an operation of projection to another relation.

$$\mu_{R(X_{i_1}, \ldots, X_{i_k} | u_{j_1}, \ldots, u_{j_m})} (u_{i_1}, \ldots, u_{i_k}) = \mu_{R(X_1, \ldots, X_n)} (u_1, \ldots, u_n), \tag{2.13}$$

or more simply [see (2.10)],

$$\mu_{R(X_{(q)} | u_{(q')})} (u_{(q)}) = \mu_{R(X)} (u)$$

with the understanding that the arguments $u_{j_1}, \ldots, u_{j_m}$ on the right-hand side of (2.13) are treated as parameters. In consequence of this understanding, although the characteristic function of the conditioned restriction is numerically equal to that of $R(X_1, \ldots, X_n)$, it defines a relation in $U_{i_1} \times \cdots \times U_{i_k}$ rather than in $U_1 \times \cdots \times U_n$.

In view of (2.9), (2.12) and (2.13), the projection of $R(X_1, \ldots, X_n)$ on $U_{i_1} \times \cdots \times U_{i_k}$ may be expressed as

$$P_q R(X_1, \ldots, X_n) = \cup_{u(q')} R(X_{i_1}, \ldots, X_{i_k} | u_{j_1}, \ldots, u_{j_m}), \tag{2.14}$$

where $\cup_{u(q')}$ denotes the union of the family of restrictions $R(X_{i_1}, \ldots, X_{i_k} | u_{j_1}, \ldots, u_{j_m})$ parametrized by $u_{(q')} \triangleq (u_{j_1}, \ldots, u_{j_m})$. Consequently, (2.14) implies that the marginal restriction $R(X_{i_1}, \ldots, X_{i_k})$ in $U_{i_1} \times \cdots \times U_{i_k}$ may be expressed as the union of conditioned restrictions $R(X_{i_1}, \ldots, X_{i_k} | u_{j_1}, \ldots, u_{j_m})$, i.e.,

$$R(X_{i_1}, \ldots, X_{i_k}) = \cup_{u(q')} R(X_{i_1}, \ldots, X_{i_k} | u_{j_1}, \ldots, u_{j_m}), \tag{2.15}$$

or more compactly,

$$R(X_{(q)}) = \cup_{u(q')} R(X_{(q)} | u_{(q')}).$$

Example 2.5. As a simple illustration of (2.15), assume that $U_1 = U_2 \triangleq \{3, 5, 7, 9\}$ and that $R(X_1, X_2)$ is characterized by the following relation matrix. [In this matrix, the (i, j)th entry is 1 iff the ordered pair (ith element of U_1, jth element of U_2) belongs to $R(X_1, X_2)$. In effect, the relation matrix of a relation R constitutes a tabulation of the characteristic function of R.]

R	3	5	7	9
3	0	0	1	0
5	1	0	1	0
7	1	0	1	1
9	1	0	0	1

In this case,

$$R(X_1, X_2 \,|\, u_1 = 3) = \{7\},$$

$$R(X_1, X_2 \,|\, u_1 = 5) = \{3,7\},$$

$$R(X_1, X_2 \,|\, u_1 = 7) = \{3,7,9\},$$

$$R(X_1, X_2 \,|\, u_1 = 9) = \{3,9\},$$

and hence

$$R(X_2) = \{7\} \cup \{3,7\} \cup \{3,7,9\} \cup \{3,9\}$$

$$= \{3,7,9\}.$$

INTERACTION AND NONINTERACTION

A basic concept that we shall need in later sections is that of the *interaction* between two or more variables—a concept which is analogous to the *dependence* of random variables. More specifically, let the variable $X = (X_1, \ldots, X_n)$ be associated with the restriction $R(X_1, \ldots, X_n)$, which induces the restrictions $R(X_1), \ldots, R(X_n)$ on $u_1, \ldots, u_n$, respectively. Then we have

DEFINITION 2.2. $X_1, \ldots, X_n$ are *noninteractive variables under* $R(X_1, \ldots, X_n)$ iff $R(X_1, \ldots, X_n)$ is *separable*, i.e.,

$$R(X_1, \ldots, X_n) = R(X_1) \times \cdots \times R(X_n), \tag{2.16}$$

where, for $i = 1, \ldots, n,$

$$R(X_i) = \text{Proj } R(X_1, \ldots, X_n) \text{ on } U_i$$

$$= \cup_{u_{(q')}} R(X_i \,|\, u_{(q')}), \tag{2.17}$$

with $u_{(q)} \triangleq u_i$ and $u_{(q')} \triangleq$ complement of u_i in $(u_1, \ldots, u_n)$.

Example 2.6. Fig. 9(a) shows two noninteractive variables X_1 and X_2 whose restrictions $R(X_1)$ and $R(X_2)$ are intervals; in this case, $R(X_1, X_2)$ is the Cartesian product of the intervals in question. In Fig. 9(b), $R(X_1, X_2)$ is a proper subset of $R(X_1) \times R(X_2)$, and hence X_1 and X_2 are interactive. Note that in Example 2.3, X_1 and X_2 are interactive.

As will be shown in a more general context in Part II, Sec. 1, if $X_1, \ldots, X_n$ are noninteractive, then an n-ary assignment equation

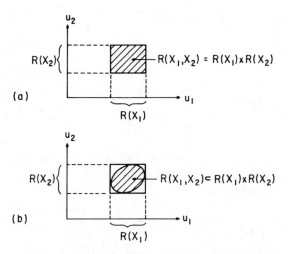

Fig. 9. (a) X_1 and X_2 are noninteractive. (b) X_1 and X_2 are interactive.

$$(x_1, \ldots, x_n) = (u_1, \ldots, u_n) : R(X_1, \ldots, X_n) \qquad (2.18)$$

can be decomposed into a sequence of n unary assignment equations

$$x_1 = u_1 : R(X_1),$$

$$x_2 = u_2 : R(X_2),$$

$$\begin{array}{ccc} \cdot & \cdot & \cdot \\ \cdot & \cdot & \cdot \\ \cdot & \cdot & \cdot \end{array} \qquad (2.19)$$

$$x_n = u_n : R(X_n),$$

where $R(X_i)$, $i = 1, \ldots, n$, is the projection of $R(X_1, \ldots, X_n)$ on U_i, and by Definition 2.2,

$$R(X_1, \ldots, X_n) = R(X_1) \times \cdots \times R(X_n). \qquad (2.20)$$

In the case where $X_1, \ldots, X_n$ are interactive, the sequence of n unary assignment equations assumes the following form [see also Part II, Eq. (1.34)].

$$x_1 = u_1 : R(X_1),$$

$$x_2 = u_2 : R(X_2 | u_1),$$

. . .

. . .   (2.21)

. . .

$$x_n = u_n : R(X_n | u_1, \ldots, u_{n-1}),$$

where $R(X_i | u_1, \ldots, u_{i-1})$ denotes the induced restriction on u_i conditioned on $u_1, \ldots, u_{i-1}$. The characteristic function of this conditioned restriction is expressed by [see (2.13)]

$$\mu_{R(X_i | u_1, \ldots, u_{i-1})}(u_i) = \mu_{R(X_1, \ldots, X_i)}(u_1, \ldots, u_i), \cdot \qquad (2.22)$$

with the understanding that the arguments $u_1, \ldots, u_{i-1}$ on the right-hand side of (2.22) play the role of parameters.

COMMENT 2.3. In words, (2.21) means that, in the case of interactive variables, once we have assigned a value u_1 to x_1, the restriction on u_2 becomes dependent on u_1. Then, the restriction on u_3 becomes dependent on the values assigned to x_1 and x_2, and, finally, the restriction on u_n becomes dependent on $u_1, \ldots, u_{n-1}$. Furthermore, (2.22) implies that the restriction on u_i given $u_1, \ldots, u_{i-1}$ is essentially the same as the marginal restriction on $(u_1, \ldots, u_i)$, with $u_1, \ldots, u_{i-1}$ treated as parameters. This illustrated in Fig. 10.

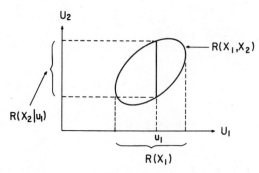

Fig. 10. $R(X_2 | u_1)$ is the restriction on u_2 conditioned on u_1.

In terms of the valise analogy (see Comment 2.1), $X_1, \ldots, X_n$ are noninteractive if the partitions between the compartments named $X_1, \ldots, X_n$ are not adjustable. In this case, what is placed in a compartment X_i has no influence on the objects that can be placed in the other compartments.

In the case where the partitions are adjustable, this is no longer true, and $X_1, \ldots, X_n$ become interactive in the sense that the placement of an object, say u_i, in X_i affects what can be placed in the complementary compartments. From this point of view, the sequence of unary assignment equations (2.21) describes the way in which the restriction on compartment X_i is influenced by the placement of objects $u_1, \ldots, u_{i-1}$ in $X_1, \ldots, X_{i-1}$.

Our main purpose in defining the notions of noninteraction, marginal restriction, conditioned restriction, etc. for nonfuzzy variables is (a) to indicate that concepts analogous to statistical independence, marginal distribution, conditional distribution, etc., apply also to nonrandom, nonfuzzy variables; and (b) to set the stage for similar concepts in the case of fuzzy variables. As a preliminary, we shall turn our attention to some of the relevant properties of fuzzy sets and formulate an extension principle which will play an important role in later sections.

3. FUZZY SETS AND THE EXTENSION PRINCIPLE

As will be seen in Part II, Sec. 1, a fuzzy variable X differs from a nonfuzzy variable in that it is associated with a restriction $R(X)$ which is a fuzzy subset of the universe of discourse.[11] Consequently, as a preliminary to our consideration of the concept of a fuzzy variable, we shall review some of the pertinent properties of fuzzy sets and state an extension principle which allows the domain of a transformation or a relation in U to be extended from points in U to fuzzy subsets of U.

FUZZY SETS–NOTATION AND TERMINOLOGY

A fuzzy subset A of a universe of discourse U is characterized by a *membership function* $\mu_A : U \to [0, 1]$ which associates with each element u of U a number $\mu_A(u)$ in the interval $[0, 1]$, with $\mu_A(u)$ representing the *grade of membership* of u in A.[12] The *support* of A is the set of points in U at which $\mu_A(u)$ is positive. The *height* of A is the supremum of $\mu_A(u)$ over U. A *crossover point* of A is a point in U whose grade of membership in A is 0.5.

Example 3.1. Let the universe of discourse be the interval $[0, 1]$, with u interpreted as *age*. A fuzzy subset of U labeled *old* may be defined by a membership function such as

---

[11] More detailed discussions of fuzzy sets and their properties may be found in the listed references. (A detailed exposition of the fundamentals together with many illustrative examples may be found in the recent text by A. Kaufmann [20]).

[12] More generally, the range of μ_A may be a partially ordered set (see [21], [22]) or a collection of fuzzy sets. The latter case will be discussed in greater detail in Sec. 6.

$$\mu_A(u) = 0 \qquad \text{for } 0 \leqslant u \leqslant 50, \tag{3.1}$$

$$\mu_A(u) = \left[1 + (\tfrac{u-50}{5})^{-2}\right]^{-1} \text{ for } 50 \leqslant u \leqslant 100.$$

In this case, the support of *old* is the interval [50, 100]; the height of *old* is effectively unity; and the crossover point of *old* is 55.

To simplify the representation of fuzzy sets we shall employ the following notation.

A nonfuzzy finite set such as

$$U = \left\{u_1, \dots, u_n\right\} \tag{3.2}$$

will be expressed as

$$U = u_1 + u_2 + \cdots + u_n \tag{3.3}$$

or

$$U = \sum_{i=1}^{n} u_i, \tag{3.4}$$

with the understanding that + denotes the union rather than the arithmetic sum. Thus, (3.3) may be viewed as a representation of U as the union of its constituent singletons.

As an extension of (3.3), a fuzzy subset A of U will be expressed as

$$A = \mu_1 u_1 + \cdots + \mu_n u_n \tag{3.5}$$

or

$$A = \sum_{i=1}^{n} \mu_i u_i, \tag{3.6}$$

where μ_i, $i = 1, \dots, n$, is the grade of membership of u_i in A. In cases where the u_i are numbers, there might be some ambiguity regarding the identity of the μ_i and u_i components of the string $\mu_i u_i$. In such cases, we shall employ a separator symbol such as / for disambiguation, writing

$$A = \mu_1/u_1 + \cdots + \mu_n/u_n \tag{3.7}$$

or

$$A = \sum_{i=1}^{n} \mu_i / u_i . \tag{3.8}$$

Example 3.2. Let $U = \{a, b, c, d\}$ or, equivalently,

$$U = a + b + c + d . \tag{3.9}$$

In this case, a fuzzy subset A of U may be represented unambiguously as

$$A = 0.3a + b + 0.9c + 0.5d. \tag{3.10}$$

On the other hand, if

$$U = 1 + 2 + \cdots + 100, \tag{3.11}$$

then we shall write

$$A = 0.3/25 + 0.9/3 \tag{3.12}$$

in order to avoid ambiguity.

Example 3.3. In the universe of discourse comprising the integers $1, 2, \ldots,$ 10, i.e.,

$$U = 1 + 2 + \cdots + 10, \tag{3.13}$$

the fuzzy subset labeled *several* may be defined as

$$several = 0.5/3 + 0.8/4 + 1/5 + 1/6 + 0.8/7 + 0.5/8. \tag{3.14}$$

Example 3.4. In the case of the countable universe of discourse

$$U = 0 + 1 + 2 + \cdots, \tag{3.15}$$

the fuzzy set labeled *small* may be expressed as

$$small = \sum_{0}^{\infty} \left[1 + \left(\frac{u}{10} \right)^2 \right]^{-1} / u. \tag{3.16}$$

Like (3.3), (3.5) may be interpreted as a representation of a fuzzy set as the union of its constituent fuzzy singletons $\mu_i u_i$ (or μ_i / u_i). From the definition of the union [see (3.34)], it follows that if in the representation of A we have $u_i = u_j$, then we can make the substitution expressed by

$$\mu_i u_i + \mu_j u_i = (\mu_i \vee \mu_j) u_i. \tag{3.17}$$

For example,

$$A = 0.3a + 0.8a + 0.5b \tag{3.18}$$

may be rewritten as

$$A = (0.3 \vee 0.8)a + 0.5b$$
$$= 0.8a + 0.5b. \tag{3.19}$$

When the support of a fuzzy set is a continuum rather than a countable or a finite set, we shall write

$$A = \int_U \mu_A(u)/u, \tag{3.20}$$

with the understanding that $\mu_A(u)$ is the grade of membership of u in A, and the integral denotes the union of the fuzzy singletons $\mu_A(u)/u$, $u \in U$.

Example 3.5. In the universe of discourse consisting of the interval $[0, 100]$, with $u = age$, the fuzzy subset labeled *old* [whose membership function is given by (3.1)] may be expressed as

$$old = \int_{50}^{100} \left[1 + \left(\frac{u-50}{5}\right)^{-2}\right]^{-1} /u. \tag{3.21}$$

Note that the crossover point for this set, that is, the point u at which

$$\mu_{old}(u) = 0.5, \tag{3.22}$$

is $u = 55$.

A fuzzy set A is said to be *normal* if its height is unity, that is, if

$$\mathrm{Sup}_u \, \mu_A(u) = 1. \tag{3.23}$$

Otherwise A is *subnormal.* In this sense, the set *old* defined by (3.21) is *normal,* as is the set *several* defined by (3.14). On the other hand, the subset of $U = 1 + 2 + \cdots + 10$ labeled *not small and not large* and defined by

$$not\ small\ and\ not\ large = 0.2/2 + 0.3/3 + 0.4/4 + 0.5/5$$

$$+ 0.4/6 + 0.3/7 + 0.2/8 \qquad (3.24)$$

is subnormal. It should be noted that a subnormal fuzzy set may be *normalized* by dividing μ_A by $\sup_u \mu_A(u)$.

A fuzzy subset of U may be a subset of another fuzzy or nonfuzzy subset of U. More specifically, A is a *subset of* B or is *contained in* B iff $\mu_A(u) \le \mu_B(u)$ for all u in U. In symbols,

$$A \subset B \Leftrightarrow \mu_A(u) \le \mu_B(u), \qquad u \in U. \qquad (3.25)$$

Example 3.6. If $U = a + b + c + d$ and

$$A = 0.5\,a + 0.8b + 0.3d,$$

$$B = 0.7a + b + 0.3c + d, \qquad (3.26)$$

then $A \subset B$.

LEVEL-SETS OF A FUZZY SET

If A is a fuzzy subset of U, then an α-*level set* of A is a nonfuzzy set denoted by A_α which comprises all elements of U whose grade of membership in A is greater than or equal to α. In symbols,

$$A_\alpha = \left\{ u \mid \mu_A(u) \ge \alpha \right\}. \qquad (3.27)$$

A fuzzy set A may be decomposed into its level-sets through the *resolution identity*[13]

$$A = \int_0^1 \alpha A_\alpha \qquad (3.28)$$

---

[13] The resolution identity and some of its applications are discussed in greater detail in [6] and [24].

or

$$A = \sum_{\alpha} \alpha A_{\alpha}, \qquad (3.29)$$

where αA_{α} is the product of a scalar α with the set A_{α} [in the sense of (3.39)], and $\int_0^1$ (or Σ_{α}) is the union of the A_{α}, with α ranging from 0 to 1.

The resolution identity may be viewed as the result of combining together those terms in (3.5) which fall into the same level-set. More specifically, suppose that A is represented in the form

$$A = 0.1/2 + 0.3/1 + 0.5/7 + 0.9/6 + 1/9. \qquad (3.30)$$

Then by using (3.17), A can be rewritten as

$$A = 0.1/2 + 0.1/1 + 0.1/7 + 0.1/6 + 0.1/9$$
$$+ 0.3/1 + 0.3/7 + 0.3/6 + 0.3/9$$
$$+ 0.5/7 + 0.5/6 + 0.5/9$$
$$+ 0.9/6 + 0.9/9$$
$$+ 1/9$$

or

$$A = 0.1 \, (1/2 + 1/1 + 1/7 + 1/6 + 1/9)$$
$$+ 0.3 \, (1/1 + 1/7 + 1/6 + 1/9)$$
$$+ 0.5 \, (1/7 + 1/6 + 1/9)$$
$$+ 0.9 \, (1/6 + 1/9)$$
$$+ 1(1/9), \qquad (3.31)$$

which is in the form (3.29), with the level-sets given by [see (3.27)]

$$A_{0.1} = 2 + 1 + 7 + 6 + 9,$$

$$A_{0.3} = 1 + 7 + 6 + 9,$$

$$A_{0.5} = 7 + 6 + 9,$$

$$A_{0.9} = 6 + 9,$$

$$A_1 = 9. \tag{3.32}$$

As will be seen in later sections, the resolution identity—in combination with the extension principle—provides a convenient way of generalizing various concepts associated with nonfuzzy sets to fuzzy sets. This, in fact, is the underlying basis for many of the definitions stated in what follows.

OPERATIONS ON FUZZY SETS

Among the basic operations which can be performed on fuzzy sets are the following.

1. The *complement* of A is denoted by $\neg A$ (or sometimes by A') and is defined by

$$\neg A = \int_U [1 - \mu_A(u)]/u. \tag{3.33}$$

The operation of complementation corresponds to negation. Thus, if A is a label for a fuzzy set, then *not A* would be interpreted as $\neg A$. (See Example 3.7 below.)

2. The *union* of fuzzy sets A and B is denoted by $A + B$ (or, more conventionally, by $A \cup B$) and is defined by

$$A + B = \int_U [\mu_A(u) \vee \mu_B(u)]/u. \tag{3.34}$$

The union corresponds to the connective *or*. Thus, if A and B are labels of fuzzy sets, then A *or* B would be interpreted as $A + B$.

3. The *intersection* of A and B is denoted by $A \cap B$ and is defined by

$$A \cap B = \int_U [\mu_A(u) \wedge \mu_B(u)]/u. \tag{3.35}$$

The intersection corresponds to the connective *and*; thus

$$A \text{ and } B = A \cap B. \tag{3.36}$$

COMMENT 3.1. It should be understood that $\vee(\overset{\Delta}{=} \text{Max})$ and $\wedge(\overset{\Delta}{=} \text{Min})$ are not the only operations in terms of which the union and intersection can be defined. (See [25] and [26] for discussions of this point.) In this connection, it is important to note that when *and* is identified with Min, as in (3.36), it represents a "hard" *and* in the sense that it allows no trade-offs between its operands. By contrast, an *and* identified with the arithmetic product, as in (3.37) below, would act as a "soft" *and*. Which of these two and possibly other definitions is more appropriate depends on the context in which *and* is used.

4. The *product* of A and B is denoted by AB and is defined by

$$AB = \int_U \mu_A(u)\mu_B(u)/u. \tag{3.37}$$

Thus, A^α, where α is any positive number, should be interpreted as

$$A^\alpha = \int_U [\mu_A(u)]^\alpha/u. \tag{3.38}$$

Similarly, if α is any nonnegative real number such that $\alpha \operatorname{Sup}_u \mu_A(u) \leqslant 1$, then

$$\alpha A = \int_U \alpha\mu_A(u)/u. \tag{3.39}$$

As a special case of (3.38), the operation of *concentration* is defined as

$$\operatorname{CON}(A) = A^2, \tag{3.40}$$

while that of *dilation* is expressed by

$$\operatorname{DIL}(A) = A^{0.5} \tag{3.41}$$

As will be seen in Part II, Sec. 3, the operations of concentration and dilation are useful in the representation of linguistic hedges.

Example 3.7. If

$$U = 1 + 2 + \cdots + 10,$$

$$A = 0.8/3 + 1/5 + 0.6/6, \tag{3.42}$$

$$B = 0.7/3 + 1/4 + 0.5/6,$$

then

$$\neg A = 1/1 + 1/2 + 0.2/3 + 1/4 + 0.4/6 + 1/7 + 1/8 + 1/9 + 1/10,$$

$$A + B = 0.8/3 + 1/4 + 1/5 + 0.6/6,$$

$$A \cap B = 0.7/3 + 0.5/6,$$

$$AB = 0.56/3 + 0.3/6,$$

$$A^2 = 0.64/3 + 1/5 + 0.36/6,$$

$$0.4A = 0.32/3 + 0.4/5 + 0.24/6,$$

$$\mathrm{CON}(B) = 0.49/3 + 1/4 + 0.25/6,$$

$$\mathrm{DIL}(B) = 0.84/3 + 1/4 + 0.7/6.$$

(3.43)

5. If $A_1, \ldots, A_n$ are fuzzy subsets of U, and $w_1, \ldots, w_n$ are nonnegative weights adding up to unity, then a *convex combination* of $A_1, \ldots, A_n$ is a fuzzy set A whose membership function is expressed by

$$\mu_A = w_1 \mu_{A_1} + \cdots + w_n \mu_{A_n}, \qquad (3.44)$$

where + denotes the arithmetic sum. The concept of a convex combination is useful in the representation of linguistic hedges such as *essentially, typically,* etc., which modify the weights associated with the components of a fuzzy set [27].

6. If $A_1, \ldots, A_n$ are fuzzy subsets of $U_1, \ldots, U_n$, respectively, the *Cartesian product* of $A_1, \ldots, A_n$ is denoted by $A_1 \times \cdots \times A_n$ and is defined as a fuzzy subset of $U_1 \times \cdots \times U_n$ whose membership function is expressed by

$$\mu_{A_1 \times \cdots \times A_n}(u_1, \ldots, u_n) = \mu_{A_1}(u_1) \wedge \cdots \wedge \mu_{A_n}(u_n). \qquad (3.45)$$

Thus, we can write [see (3.52)]

$$A_1 \times \cdots \times A_n = \int_{U_1 \times \cdots \times U_n} [\mu_{A_1}(u_1) \wedge \cdots \wedge \mu_{A_n}(u_n)]/(u_1, \ldots, u_n).$$

(3.46)

Example 3.8. If $U_1 = U_2 = 3 + 5 + 7$, $A_1 = 0.5/3 + 1/5 + 0.6/7$ and $A_2 = 1/3 + 0.6/5$, then

$$A_1 \times A_2 = 0.5/(3,3) + 1/(5,3) + 0.6/(7,3)$$

$$+ 0.5/(3,5) + 0.6/(5,5) + 0.6/(7,5). \qquad (3.47)$$

7. The operation of *fuzzification* has, in general, the effect of transforming a nonfuzzy set into a fuzzy set or increasing the fuzziness of a fuzzy set. Thus, a *fuzzifier F* applied to a fuzzy subset A of U yields a fuzzy subset $F(A;K)$ which is expressed by

$$F(A;K) = \int_U \mu_A(u)K(u), \tag{3.48}$$

where the fuzzy set $K(u)$ is the *kernel* of F, that is, the result of applying F to a singleton $1/u$:

$$K(u) = F(1/u;K); \tag{3.49}$$

$\mu_A(u)K(u)$ represents the product [in the sense of (3.39) of a scalar $\mu_A(u)$ and the fuzzy set $K(u)$; and $\int$ is the union of the family of fuzzy sets $\mu_A(u)K(u)$, $u \in U$. In effect, (3.48) is analogous to the integral representation of a linear operator, with $K(u)$ being the counterpart of the impulse response.

Example 3.9. Assume that U, A and $K(u)$ are defined by

$$U = 1 + 2 + 3 + 4,$$

$$A = 0.8/1 + 0.6/2,$$

$$K(1) = 1/1 + 0.4/2, \tag{3.50}$$

$$K(2) = 1/2 + 0.4/1 + 0.4/3.$$

Then

$$F(A;K) = 0.8(1/1 + 0.4/2) + 0.6(1/2 + 0.4/1 + 0.4/3)$$

$$= 0.8/1 + 0.6/2 + 0.24/3. \tag{3.51}$$

The operation of fuzzification plays an important role in the definition of linguistic hedges such as *more or less, slightly, somewhat, much,* etc. For example, if $A \stackrel{\triangle}{=} positive$ is the label for the nonfuzzy class of positive numbers, then *slightly positive* is a label for a fuzzy subset of the real line whose membership function is of the form shown in Fig. 11. In this case, *slightly* is a fuzzifier which transforms *positive* into *slightly positive.* However, it is not always possible to express the effect of a fuzzifier in the form (3.48), and *slightly* is a case in point. A more detailed discussion of this and related issues may be found in [27].

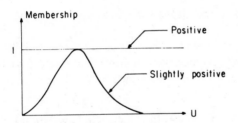

Fig. 11. Membership functions of *positive* and *slightly positive.*

FUZZY RELATIONS

If U is the Cartesian product of n universes of discourse $U_1, \ldots, U_n$, then an n-ary *fuzzy relation*, R, *in* U is a fuzzy subset of U. As in (3.20), R may be expressed as the union of its constituent fuzzy singletons $\mu_R(u_1, \ldots, u_n)$ $/(u_1, \ldots, u_n)$, i.e.,

$$R = \int_{U_1 \times \cdots \times U_n} \mu_R(u_1, \ldots, u_n)/(u_1, \ldots, u_n), \qquad (3.52)$$

where μ_R is the membership function of R.

Common examples of (binary) fuzzy relations are: *much greater than, resembles, is relevant to, is close to,* etc. For example, if $U_1 = U_2 = (-\infty, \infty)$, the relation *is close to* may be defined by

$$\textit{is close to} \triangleq \int_{U_1 \times U_2} e^{-a|u_1 - u_2|} /(u_1, u_2), \qquad (3.53)$$

where a is a scale factor. Similarly, if $U_1 = U_2 = 1 + 2 + 3 + 4$, then the relation *much greater than* may be defined by the relation matrix

R	1	2	3	4
1	0	0.3	0.8	1
2	0	0	0	0.8
3	0	0	0	0.3
4	0	0	0	0

$$(3.54)$$

in which the (i, j)th element is the value of $\mu_R(u_1, u_2)$ for the ith value of u_1 and jth value of u_2.

If R is a relation from U to V (or, equivalently, a relation in $U \times V$) and S is a relation from V to W, then the composition of R and S is a fuzzy relation from U to W denoted by $R \circ S$ and defined by[14]

$$R \circ S = \int_{U \times W} \bigvee_v [\mu_R(u, v) \wedge \mu_S(v,w)] /(u,w). \qquad (3.55)$$

If U, V and W are finite sets, then the relation matrix for $R \circ S$ is the max-min product[15] of the relation matrices for R and S. For example, the max-min product of the relation matrices on the left-hand side of (3.56) is given by the right-hand side of (3.56):

$$
\begin{array}{ccc}
R & S & R \circ S
\end{array}
$$

$$
\begin{bmatrix} 0.3 & 0.8 \\ 0.6 & 0.9 \end{bmatrix} \circ \begin{bmatrix} 0.5 & 0.9 \\ 0.4 & 1 \end{bmatrix} = \begin{bmatrix} 0.4 & 0.8 \\ 0.5 & 0.9 \end{bmatrix} \qquad (3.56)
$$

PROJECTIONS AND CYLINDRICAL FUZZY SETS

If R is an n-ary fuzzy relation in $U_1 \times \cdots \times U_n$, then its *projection (shadow)* on $U_{i_1} \times \cdots \times U_{i_k}$ is a k-ary fuzzy relation R_q in U which is defined by [compare with (2.12)]

$$R_q \triangleq \text{Proj } R \text{ on } U_{i_1} \times \cdots \times U_{i_k}$$

$$\triangleq P_q R$$

$$\triangleq \int_{U_{i_1} \times \cdots \times U_{i_k}} [\bigvee u_{(q')} \mu_R(u_1, \ldots, u_n)] /(u_{i_1}, \ldots, u_{i_k}), \qquad (3.57)$$

---

[14] Equation (3.55) defines the max-min composition of R and S. Max-product composition is defined similarly, except that $\wedge$ is replaced by the arithmetic product. A more detailed discussion of these compositions may be found in [24].

[15] In the max-min matrix product, the operations of addition and multiplication are replaced by $\vee$ and $\wedge$, respectively.

where q is the index sequence $(i_1, \ldots, i_k)$; $u_{(q)} \triangleq (u_{i_1}, \ldots, u_{i_k})$; q' is the complement of q; and $\bigvee_{u_{(q')}}$ is the supremum of $\mu_R(u_1, \ldots, u_n)$ over the u's which are in $u_{(q')}$. It should be noted that when R is a nonfuzzy relation, (3.57) reduces to (2.9).

Example 3.10. For the fuzzy relation defined by the relation matrix (3.54), we have

$$R_1 = 1/1 + 0.8/2 + 0.3/3$$

and

$$R_2 = 0.3/2 + 0.8/3 + 1/4.$$

It is clear that distinct fuzzy relations in $U_1 \times \cdots \times U_n$ can have identical projections on $U_{i_1} \times \cdots \times U_{i_k}$. However, given a fuzzy relation R_q in $U_{i_1} \times \cdots \times U_{i_k}$, there exists a unique *largest*[16] relation $\bar{R}_q$ in $U_1 \times \cdots \times U_n$ whose projection on $U_{i_1} \times \cdots \times U_{i_k}$ is R_q. In consequence of (3.57), the membership function of $\bar{R}_q$ is given by

$$\mu_{\bar{R}_q}(u_1, \ldots, u_n) = \mu_{R_q}(u_{i_1}, \ldots, u_{i_k}), \tag{3.58}$$

with the understanding that (3.58) holds for all $u_1, \ldots, u_n$ such that the $i_1, \ldots, i_k$ arguments in $\mu_{\bar{R}_q}$ are equal, respectively, to the first, second, $\ldots$, kth arguments in μ_{R_q}. This implies that the value of $\mu_{\bar{R}_q}$ at the point $(u_1, \ldots, u_n)$ is the same as that at the point $(u'_1, \ldots, u'_n)$ provided that $u_{i_1} = u'_{i_1}, \ldots, u_{i_k} = u'_{i_k}$. For this reason, $\bar{R}_q$ will be referred to as the *cylindrical extension* of R_q, with R_q constituting the *base* of $\bar{R}_q$. (See Fig. 12.)

---

[16]That is, a relation which contains all other relations whose projection on $U_{i_1} \times \cdots \times U_{i_k}$ is R_q.

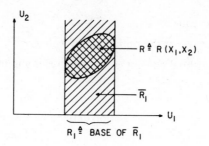

Fig. 12. R_1 is the base of the cylindrical set $\bar{R}_1$.

Suppose that R is an n-ary relation in $U_1 \times \cdots \times U_n$, R_q is its projection on $U_{i_1} \times \cdots \times U_{i_k}$, and $\bar{R}_q$ is the cylindrical extension of R_q. Since $\bar{R}_q$ is the largest relation in $U_1 \times \cdots \times U_n$ whose projection on $U_{i_1} \times \cdots \times U_{i_k}$ is R_q, it follows that R_q satisfies the *containment relation*

$$R \subset \bar{R}_q \tag{3.59}$$

for all q, and hence

$$R \subset \bar{R}_{q_1} \cap \bar{R}_{q_2} \cap \cdots \cap \bar{R}_{q_r} \tag{3.60}$$

for arbitrary $q_1, \ldots, q_r$ [index subsequences of $(1, 2, \ldots, n)$].
  In particular, if we set $q_1 = 1, \ldots, q_r = n$, then (3.60) reduces to

$$R \subset \bar{R}_1 \cap \bar{R}_2 \cap \cdots \cap \bar{R}_n, \tag{3.61}$$

where $R_1, \ldots, R_n$ are the projections of R on $U_1, \ldots, U_n$, respectively, and $\bar{R}_1, \ldots, \bar{R}_n$ are their cylindrical extensions. But, from the definition of the Cartesian product [see (3.45)] it follows that

$$\bar{R}_1 \cap \cdots \cap \bar{R}_n = R_1 \times \cdots \times R_n, \tag{3.62}$$

which leads us to the

PROPOSITION 3.1. *If R is an n-ary fuzzy relation in* $U_1 \times \cdots \times U_n$ *and* $R_1, \ldots, R_n$ *are its projections on* $U_1, \ldots, U_n$, *then (see Fig. 13 for illustration)*

$$R \subset R_1 \times \cdots \times R_n. \tag{3.63}$$

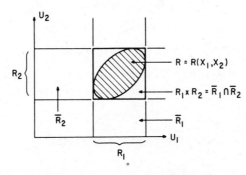

Fig. 13. Relation between the Cartesian product and intersection of cylindrical sets.

The concept of a cylindrical extension can also be used to provide an intuitively appealing interpretation of the composition of fuzzy relations. Thus, suppose that R and S are binary fuzzy relations in $U_1 \times U_2$ and $U_2 \times U_3$, respectively. Let $\bar{R}$ and $\bar{S}$ be the cylindrical extensions of R and S in $U_1 \times U_2 \times U_3$. Then, from the definition of $R \circ S$ [see (3.55)] it follows that

$$R \circ S = \operatorname{Proj} \bar{R} \cap \bar{S} \quad \text{on} \quad U_1 \times U_3. \tag{3.64}$$

If R and S are such that

$$\operatorname{Proj} R \quad \text{on} \quad U_2 = \operatorname{Proj} S \quad \text{on} \quad U_2, \tag{3.65}$$

then $\bar{R} \cap \bar{S}$ becomes the *join*[17] of R and S. A basic property of the join of R and S may be stated as

---

[17]The concept of the join of nonfuzzy relations was introduced by E. F. Codd in [28].

PROPOSITION 3.2. *If R and S are fuzzy relations in $U_1 \times U_2$ and $U_2 \times U_3$, respectively, and $\bar{R} \cap \bar{S}$ is the join of R and S, then*

$$R = \operatorname{Proj} \bar{R} \cap \bar{S} \quad on \quad U_1 \times U_2 \tag{3.66}$$

and

$$S = \operatorname{Proj} \bar{R} \cap \bar{S} \quad on \quad U_2 \times U_3. \tag{3.67}$$

Thus, R and S can be retrieved from the join of R and S.

Proof. Let μ_R and μ_S denote the membership functions of R and S, respectively. Then the right-hand sides of (3.66) and (3.67) translate into

$$\bigvee_{u_3} [\mu_R(u_1,u_2) \wedge \mu_S(u_2,u_3)] \tag{3.68}$$

and

$$\bigvee_{u_1} [\mu_R(u_1,u_2) \wedge \mu_S(u_2, u_3)]. \tag{3.69}$$

In virtue of the distributivity and commutativity of $\vee$ and $\wedge$, (3.68) and (3.69) may be rewritten as

$$\mu_R(u_1,u_2) \wedge [\bigvee_{u_3} \mu_S(u_2,u_3)] \tag{3.70}$$

and

$$\mu_S(u_2,u_3) \wedge [\bigvee_{u_1} \mu_R(u_1,u_2)]. \tag{3.71}$$

Furthermore, the definition of the join implies (3.65) and hence that

$$\bigvee_{u_1} \mu_R(u_1,u_2) = \bigvee_{u_3} \mu_S(u_2,u_3). \tag{3.72}$$

From this equality and the definition of $\vee$ it follows that

$$\mu_R(u_1,u_2) \leqslant \bigvee_{u_1} \mu_R(u_1,u_2) = \bigvee_{u_3} \mu_S(u_2,u_3) \tag{3.73}$$

and

$$\mu_S(u_2, u_3) \leqslant \vee_{u_3} \mu_S(u_2,\ u_3) = \vee_{u_1} \mu_R(u_1, u_2). \qquad (3.74)$$

Consequently

$$\mu_R(u_1, u_2) \wedge [\vee_{u_3} \mu_S(u_2, u_3)] = \mu_R(u_1, u_2) \qquad (3.75)$$

and

$$\mu_S(u_2, u_3) \wedge [\vee_{u_1} \mu_R(u_1, u_3)] = \mu_S(u_2, u_3), \qquad (3.76)$$

which translate into (3.66) and (3.67).   Q.E.D.

A basic property of projections which we shall have an occasion to use in Part II, Sec. 1 is the following.

PROPOSITION 3.3. *If R is a normal relation* [*see (3.23)*]*, then so is every projection of R.*

Proof. Let R be an n-ary relation in $U_1 \times \cdots \times U_n$, and let R_q be its projection (shadow) on $U_{i_1} \times \cdots \times U_{i_k}$, with $q = (i_1, \ldots, i_k)$. Since R is normal, we have by (3.23),

$$\vee_{(u_1, \ldots, u_n)} \mu_R(u_1, \ldots, u_n) = 1, \qquad (3.77)$$

or more compactly

$$\vee_u \mu_R(u) = 1.$$

On the other hand, by the definition of R_q [see (3.57)],

$$\mu_{R_q}(u_{i_1}, \ldots, u_{i_k}) = \vee_{(u_{j_1}, \ldots, u_{j_m})} \mu_R(u_1, \ldots, u_n),$$

or

$$\mu_{R_q}(u_{(q)}) = \vee_{u_{(q')}} \mu_R(u),$$

and hence the height of R_q is given by

$$\vee u_{(q)} \mu_{R_q}(u_{(q)}) = \vee_{u_{(q)}} \vee_{u_{(q')}} \mu_R(u) \qquad (3.78)$$

$$= \vee_u \mu_R(u)$$

$$= 1. \qquad \text{Q.E.D.}$$

THE EXTENSION PRINCIPLE

The *extension principle* for fuzzy sets is in essence a basic identity which allows the domain of the definition of a mapping or a relation to be extended from points in U to fuzzy subsets of U. More specifically, suppose that f is a mapping from U to V, and A is a fuzzy subset of U expressed as

$$A = \mu_1 u_1 + \cdots + \mu_n u_n. \qquad (3.79)$$

Then the extension principle asserts that[18]

$$f(A) = f(\mu_1 u_1 + \cdots + \mu_n u_n) \equiv \mu_1 f(u_1) + \cdots + \mu_n f(u_n). \qquad (3.80)$$

Thus, the image of A under f can be deduced from the knowledge of the images of $u_1, \ldots, u_n$ under f.

Example 3.11. Let

$$U = 1 + 2 + \cdots + 10,$$

and let f be the operation of squaring. Let *small* be a fuzzy subset of U defined by

$$small = 1/1 + 1/2 + 0.8/3 + 0.6/4 + 0.4/5. \qquad (3.81)$$

Then, in consequence of (3.80), we have [19]

$$small^2 = 1/1 + 1/4 + 0.8/9 + 0.6/16 + 0.4/25. \qquad (3.82)$$

If the support of A is a continuum, that is,

---

[18] The extension principle is implicit in a result given in [29]. In probability theory, the extension principle is analogous to the expression for the probability distribution induced by a mapping [30]. In the special case of intervals, the results of applying the extension principle reduced to those of interval analysis [31].

[19] Note that this definition of *small²* differs from that of (3.38).

$$A = \int_U \mu_A(u)/u, \tag{3.83}$$

then the statement of the extension principle assumes the following form:

$$f(A) = f\left(\int_U \mu_A(u)/u\right) \equiv \int_V \mu_A(u)/f(u), \tag{3.84}$$

with the understanding that $f(u)$ is a point in V and $\mu_A(u)$ is its grade of membership in $f(A)$, which is a fuzzy subset of V.

In some applications it is convenient to use a modified form of the extension principle which follows from (3.84) by decomposing A into its constituent level-sets rather than its fuzzy singletons [see the resolution identity (3.28)]. Thus, on writing

$$A = \int_0^1 \alpha A_\alpha, \tag{3.85}$$

where A_α is an α-level set of A, the statement of the extension principle assumes the form

$$f(A) = f\left(\int_0^1 \alpha A_\alpha\right) \equiv \int_0^1 \alpha f(A_\alpha) \tag{3.86}$$

when the support of A is a continuum, and

$$f(A) = f\left(\sum_\alpha \alpha A_\alpha\right) = \sum_\alpha \alpha f(A_\alpha) \tag{3.87}$$

when either the support of A is a countable set or the distinct level-sets of A form a countable collection.

COMMENT 3.2. Written in the form (3.84), the extension principle extends the domain of definition of f from points in U to fuzzy subsets of U. By contrast, (3.86) extends the domain of definition of f from nonfuzzy subsets of U to fuzzy subsets of U. It should be clear, however, that (3.84) and (3.86) are equivalent, since (3.86) results from (3.84) by a regrouping of terms in the representation of A.

COMMENT 3.3. The extension principle is analogous to the superposition principle for linear systems. Under the latter principle, if L is a linear system and $u_1, \ldots, u_n$ are inputs to L, then the response of L to any linear combination

$$u = w_1 u_1 + \cdots + w_n u_n, \tag{3.88}$$

where the w_i are constant coefficients, is given by

$$L(u) = L(w_1 u_1 + \cdots + w_n u_n) = w_1 L(u_1) + \cdots + w_n L(u_n). \tag{3.89}$$

The important point of difference between (3.89) and (3.80) is that in (3.80) + is the union rather than the arithmetic sum, and f is not restricted to linear mappings.

COMMENT 3.4. It should be noted that when $A = u_1 + \cdots + u_n$, the result of applying the extension principle is analogous to that of forming the n-fold Cartesian product of the algebraic system (U, f) with itself. (An extension of the multiplication table is shown in Table 3.1.)

X	I	2	3	4	Iv2	2v4
I	I	2	3	4	Iv2	2v4
2	2	4	6	8	Iv4	4v8
3	3	6	9	12	3v6	6v12
4	I	8	12	16	4v8	8v16
Iv2	Iv2	2v4	3v6	4v8	Iv2v4	2v4v8

$$\begin{array}{c} 3\,\mathrm{v}\,5\,\mathrm{v}\,6 \\ \mathrm{x} \quad\quad\quad \\ 2\,\mathrm{v}\,4\,\mathrm{v}\,6 \\ \hline \end{array}$$

$$6 \vee 10 \vee 12$$
$$12 \vee 20 \vee 24$$
$$18 \vee 30 \vee 36$$

$$\overline{\quad\quad\quad\quad\quad}$$

$$6 \vee 10 \vee 12 \vee 18 \vee 20 \vee 24 \vee 30 \vee 36$$

Table 1. Extension of the multiplication table to subsets of integers. $1\vee2$ means 1 or 2.

In many applications of the extension principle, one encounters the following problem. We have an n-ary function, f, which is a mapping from a Cartesian product $U_1 \times \cdots \times U_n$ to a space V, and a fuzzy set (relation) A in $U_1 \times \cdots \times U_n$ which is characterized by a membership function $\mu_A(u_1, \ldots, u_n)$, with u_i, $i = 1, \ldots, n$, denoting a generic point in U_i. A direct application of the extension principle (3.84) to this case yields

$$f(A) = f \left(\int_{U_1 \times \cdots \times U_n} \mu_A(u_1, \ldots, u_n)/(u_1, \ldots, u_n) \right) \quad (3.90)$$

$$= \int_V \mu_A(u_1, \ldots, u_n)/f(u_1, \ldots, u_n).$$

However, in many instances what we know is not A but its projections $A_1, \ldots,$ A_n on $U_1, \ldots, U_n$, respectively [see (3.57)]. The question that arises, then, is: What expression for μ_A should be used in (3.90)?

In such cases, unless otherwise specified we shall assume that the membership function of A is expressed by

$$\mu_A(u_1, \ldots, u_n) = \mu_{A_1}(u_1) \wedge \mu_{A_2}(u_2) \wedge \cdots \wedge \mu_{A_n}(u_n), \quad (3.91)$$

where μ_{A_i}, $i = 1, \ldots, n$, is the membership function of A_i. In view of (3.45), this is equivalent to assuming that A is the Cartesian product of its projections, i.e.,

$$A = A_1 \times \cdots \times A_n,$$

which in turn implies that A is the largest set whose projections on $U_1, \ldots, U_n$ are $A_1, \ldots, A_n$, respectively. [See (3.63).]

Example 3.12. Suppose that, as in Example 3.11,

$$U_1 = U_2 = 1 + 2 + 3 + \cdots + 10$$

and

$$A_1 = \underset{\sim}{2} \overset{\Delta}{=} approximately\ 2 = 1/2 + 0.6/1 + 0.8/3, \quad (3.92)$$

$$A_2 = \underset{\sim}{6} \overset{\Delta}{=} approximately\ 6 = 1/6 + 0.8/5 + 0.7/7 \quad (3.93)$$

and

$$f(u_1, u_2) = u_1 \times u_2 = arithmetic\ product\ of\ u_1\ and\ u_2.$$

Using (3.91) and applying the extension principle as expressed by (3.90) to this case, we have

$$2 \times 6 = (1/2 + 0.6/1 + 0.8/3) \times (1/6 + 0.8/5 + 0.7/7)$$

$$= 1/12 + 0.8/10 + 0.7/14 + 0.6/6 + 0.6/5 + 0.6/7$$

$$+ 0.8/18 + 0.8/15 + 0.7/21$$

$$= 0.6/5 + 0.6/6 + 0.6/7 + 0.8/10 + 1/12 + 0.7/14$$

$$+ 0.8/15 + 0.8/18 + 0.7/21. \qquad (3.94)$$

Thus, the arithmetic product of the fuzzy numbers *approximately* 2 and *approximately* 6 is a fuzzy number given by (3.94).

More generally, let $*$ be a binary operation defined on $U \times V$ with values in W. Thus, if $u \in U$ and $v \in V$, then

$$w = u * v, \qquad w \in W$$

Now suppose that A and B are fuzzy subsets of U and V, respectively, with

$$A = \mu_1 u_1 + \cdots + \mu_n u_n,$$
$$B = v_1 v_1 + \cdots + v_m v_m. \qquad (3.95)$$

By using the extension principle under the assumption (3.91), the operation $*$ may be extended to fuzzy subsets of U and V by the defining relation

$$A * B = \left(\sum_i \mu_i u_i \right) * \left(\sum_j v_j v_j \right)$$

$$= \sum_{i,j} (\mu_i \wedge v_j)(u_i * v_j). \qquad (3.96)$$

It is easy to verify that for the case where $A = 2$, $B = 6$ and $* = \times$, as in Example 3.12, the application of (3.96) yields the expression for 2×6.

COMMENT 3.5. It is important to note that the validity of (3.96) depends in an essential way on the assumption (3.91), that is,

$$\mu_{(A,B)}(u,v) = \mu_A(u) \wedge \mu_B(v).$$

The implication of this assumption is that u and v are noninteractive in the sense of Definition 2.2. Thus, if there is a constraint on (u,v) which is expressed as a relation R with a membership function μ_R, then the expression for $A * B$ becomes

$$A * B = \left[\left(\sum_i \mu_i u_i \right) * \left(\sum_j v_j v_j \right) \right] \cap R$$

$$= \sum_{i,j} [\mu_i \wedge v_j \wedge \mu_R(u_i, v_j)] (u_i * v_j). \qquad (3.97)$$

Note that if R is a nonfuzzy relation, then the right-hand side of (3.97) will contain only those terms which satisfy the constraint R.

A simple illustration of a situation in which u and v are interactive is provided by the expression

$$w = z \times (x+y), \qquad (3.98)$$

in which $+ \stackrel{\Delta}{=}$ arithmetic sum and $\times \stackrel{\Delta}{=}$ arithmetic product. If x, y and z are noninteractive, then we can apply the extension principle in the form (3.96) to the computation of $A \times (B + C)$, where A, B and C are fuzzy subsets of the real line. On the other hand, if (3.98) is rewritten as

$$w = z \times x + z \times y,$$

then the terms $z \times x$ and $z \times y$ are interactive by virtue of the common factor z, and hence

$$A \times (B + C) \neq A \times B + A \times C. \qquad (3.99)$$

A significant conclusion that can be drawn from this observation is that the product of fuzzy numbers is not distributive if it is computed by the use of (3.96). To obtain equality in (3.99), we may apply the unrestricted form of the extension principle (3.96) to the left-hand side of (3.99), and must apply the restricted form (3.97) to its right-hand side.

REMARK 3.1. The extension principle can be applied not only to functions, but also to relations or, equivalently, to predicates. We shall not discuss this subject here, since the application of the extension principle to relations does not play a significant role in the present paper.

FUZZY SETS WITH FUZZY MEMBERSHIP FUNCTIONS

Our consideration of fuzzy sets with fuzzy membership functions is motivated by the close association which exists between the concept of a linguistic truth with truth-values such as *true, quite true, very true, more or less true,* etc., on the one hand, and fuzzy sets in which the grades of membership are specified in linguistic terms such as *low, medium, high, very low, not low and not high,* etc., on the other.

Thus, suppose that A is a fuzzy subset of a universe of discourse U, and the values of the membership function, μ_A, of A are allowed to be fuzzy subsets of the interval $[0, 1]$. To differentiate such fuzzy sets from those considered previously, we shall refer to them as fuzzy sets of *type* 2, with the fuzzy sets whose membership functions are mappings from U to $[0, 1]$ classified as *type* 1. More generally:

DEFINITION 3.1. A fuzzy set is of type n, $n = 2, 3, \ldots$, if its membership function ranges over fuzzy sets of type n-1. The membership function of a fuzzy set of type 1 ranges over the interval $[0, 1]$.

To define such operations as complementation, union, intersection, etc. for fuzzy sets of type 2, it is natural to make use of the extension principle. It is convenient, however, to accomplish this in two stages: first, by extending the type 1 definitions to fuzzy sets with interval-valued membership functions; and second, generalizing from intervals to fuzzy sets[20] by the use of the level-set form of the extension principle [see (3.86)]. In what follows, we shall illustrate this technique by extending to fuzzy sets of type 2 the concept of intersection—which is defined for fuzzy sets of type 1 by (3.35).

Our point of departure is the expression for the membership function of the intersection of A and B, where A and B are fuzzy subsets of type 1 of U:

$$\mu_{A \cap B}(u) = \mu_A(u) \wedge \mu_B(u), \qquad u \in U.$$

Now if $\mu_A(u)$ and $\mu_B(u)$ are intervals in $[0, 1]$ rather than points in $[0, 1]$ — that is, for a fixed u,

$$\mu_A(u) = [a_1, a_2],$$

$$\mu_B(u) = [b_1, b_2],$$

where a_1, a_2, b_1 and b_2 depend on u—then the application of the extension principle (3.86) to the function $\wedge$(Min) yields

$$[a_1, a_2] \wedge [b_1, b_2] = [a_1 \wedge b_1, a_2 \wedge b_2]. \tag{3.100}$$

Thus, if A and B have interval-valued membership functions as shown in Fig. 14, then their intersection is an interval-valued curve whose value for each u is given by (3.100).

---

[20]We are tacitly assuming that the fuzzy sets in question are convex, that is, have intervals as level-sets (see [29]). Only minor modifications are needed when the sets are not convex.

Next, let us consider the case where, for each u, $\mu_A(u)$ and $\mu_B(u)$ are fuzzy subsets of the interval $[0, 1]$. For simplicity, we shall assume that these subsets are *convex*, that is, have intervals as level-sets. In other words, we shall assume that, for each α in $[0, 1]$, the α-level sets of μ_A and μ_B are interval-valued membership functions. (See Fig. 15).

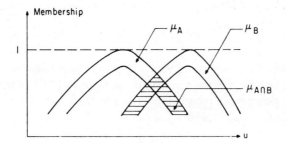

Fig. 14. Intersection of fuzzy sets with interval-valued membership functions.

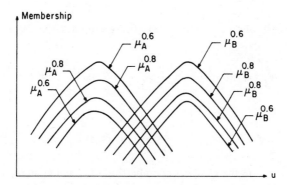

Fig. 15. Level-sets of fuzzy membership functions μ_A and μ_B.

By applying the level-set form of the extension principle (3.86) to the α-level sets of μ_A and μ_B we are led to the following definition of the intersection of fuzzy sets of type 2.

DEFINITION 3.2. Let A and B be fuzzy subsets of type 2 of U such that, for each $u \in U$, $\mu_A(u)$ and $\mu_B(u)$ are convex fuzzy subsets of type 1 of $[0, 1]$, which implies that, for each α in $[0, 1]$, the α-level sets of the fuzzy membership functions μ_A and μ_B are interval-valued membership functions μ_A^α and μ_B^α.

Let the α-level set of the fuzzy membership function of the intersection of A and B be denoted by $\mu_{A \cap B}^\alpha$, with the α-level sets μ_A^α and μ_B^β defined for each u by

$$\mu_A^\alpha \triangleq \{v \mid v_A(v) \geq \alpha\}, \tag{3.101}$$

$$\mu_B^\alpha \triangleq \{v \mid v_B(v) \geq \alpha\}, \tag{3.102}$$

where $v_A(v)$ denotes the grade of membership of a point v, $v \in [0, 1]$, in the fuzzy set $\mu_A(u)$, and likewise for μ_B. Then, for each u,

$$\mu_{A \cap B}^\alpha = \mu_A^\alpha \wedge \mu_B^\alpha \tag{3.103}$$

In other words, the α-level set of the fuzzy membership function of the intersection of A and B is the minimum [in the sense of (3.100)] of the α-level sets of the fuzzy membership functions of A and B. Thus, using the resolution identity (3.28), we can express $\mu_{A \cap B}$ as

$$\mu_{A \cap B} = \int_0^1 \alpha(\mu_A^\alpha \wedge \mu_B^\alpha). \tag{3.104}$$

For the case where μ_A and μ_B have finite supports, that is, μ_A and μ_B are of the form

$$\mu_A = \alpha_1 v_1 + \cdots + \alpha_n v_n, \quad v_i \in [0, 1], \quad i = 1, \ldots, n \tag{3.105}$$

and

$$\mu_B = \beta_1 w_1 + \cdots + \beta_m w_m, \quad w_j \in [0, 1], \quad j = 1, \ldots, m, \tag{3.106}$$

where α_i and β_j are the grades of membership of v_i and w_j in μ_A and μ_B, respectively, the expression for $\mu_{A \cap B}$ can readily be derived by employing the extension principle in the form (3.96). Thus, by applying (3.96) to the operation $\wedge$ ($\triangleq$ Min), we obtain at once

$$\mu_{A \cap B} = \mu_A \wedge \mu_B$$

$$= (\alpha_1 v_1 + \cdots + \alpha_n v_n) \wedge (\beta_1 w_1 + \cdots + \beta_m w_m) \tag{3.107}$$

$$= \sum_{i,j} (\alpha_i \wedge \beta_j)(v_i \wedge w_j)$$

as the desired expression for $\mu_{A \cap B}$.[21]

---

[21] Actually, Definition 3.2 can be deduced from (3.90).

Example 3.13. As a simple illustration of (3.104), suppose that at a point u the grades of membership of u in A and B are labeled as *high* and *medium*, respectively, with *high* and *medium* defined as fuzzy subsets of $V = 0 + 0.1 + 0.2 + \cdots + 1$ by the expressions

$$high = 0.8/0.8 + 0.8/0.9 + 1/1, \qquad (3.108)$$

$$medium = 0.6/0.4 + 1/0.5 + 0.6/0.6. \qquad (3.109)$$

The level sets of *high* and *medium* are expressed by

$$high_{0.6} = 0.8 + 0.9 + 1,$$

$$high_{0.8} = 0.8 + 0.9 + 1,$$

$$high_{1} = 1,$$

$$medium_{0.6} = 0.4 + 0.5 + 0.6,$$

$$medium_{1} = 0.5,$$

and consequently the α-level sets of the intersection are given by

$$\mu_{A \cap B}^{0.6}(u) = high_{0.6} \wedge medium_{0.6}$$

$$= (0.8 + 0.9 + 1) \wedge (0.4 + 0.5 + 0.6)$$

$$= 0.4 + 0.5 + 0.6, \qquad (3.110)$$

$$\mu_{A \cap B}^{0.8}(u) = high_{0.8} \wedge medium_{0.8}$$

$$= (0.8 + 0.9 + 1) \wedge 0.5$$

$$= 0.5 \qquad (3.111)$$

and

$$\mu_{A \cap B}^{1}(u) = high_{1} \wedge medium_{1}$$

$$= 1 \wedge 0.5$$

$$= 0.5 \qquad (3.112)$$

Combining (3.110), (3.111) and (3.112), the fuzzy set representing the grade of membership of u in the intersection of A and B is found to be

$$\mu_{A \cap B}(u) = 0.6/(0.4 + 0.5 + 0.6) + 1/0.5$$

$$= medium, \tag{3.113}$$

which is equivalent to the statement

$$high \wedge medium = medium. \tag{3.114}$$

The same result can be obtained more expeditiously by the use of (3.107). Thus, we have

$$high \wedge medium = (0.8/0.8 + 0.8/0.9 + 1/1) \wedge (0.6/0.4 + 1/0.5 + 0.6/0.6)$$

$$= 0.6/0.4 + 1/0.5 + 0.6/0.6$$

$$= medium. \tag{3.115}$$

In a similar fashion, we can extend to fuzzy sets of type 2 the operations of complementation, union, concentration, etc. This will be done in Part II, Sec. 3, in conjunction with our discussion of a fuzzy logic in which the truth-values are linguistic in nature.

REMARK 3.2. The results derived in Example 3.13 may be viewed as an instance of a general conclusion that can be drawn from (3.100) concerning an extension of the inequality $\leqslant$ from real numbers to fuzzy subsets of the real line. Specifically, in the case of real numbers a, b, we have the equivalence

$$a \leqslant b \quad \leftrightarrow \quad a \wedge b = a. \tag{3.116}$$

Using this as a basis for the extension of $\leqslant$ to intervals, we have in virtue of (3.100),

$$[a_1, a_2] \leqslant [b_1, b_2] \quad \leftrightarrow \quad a_1 \leqslant b_1 \quad \text{and} \quad a_2 \leqslant b_2. \tag{3.117}$$

This, in turn, leads us to the following definition.

DEFINITION 3.3. Let A and B be convex fuzzy subsets of the real line, and let A_α and B_α denote the α-level sets of A and B, respectively. Then an extension of the inequality $\leqslant$ to convex fuzzy subsets of the real line is expressed by[22]

$$A \leqslant B \quad \Leftrightarrow \quad A \wedge B = A \tag{3.118}$$

$$\Leftrightarrow \quad A_\alpha \wedge B_\alpha = A_\alpha \qquad \text{for all } \alpha \text{ in } [0, 1], \tag{3.119}$$

where $A_\alpha \wedge B_\alpha$ is defined by (3.100).

In the case of Example 3.13, it is easy to verify by inspection that

$$medium_\alpha \leqslant high_\alpha \qquad \text{for all } \alpha \tag{3.120}$$

in the sense of (3.119), and hence we can conclude at once that

$$medium \wedge high = medium, \tag{3.121}$$

which is in agreement with (3.114).

---

[22]It can be readily be verified that $\leqslant$ as defined by (3.117) constitutes a partial ordering.

REFERENCES

1. Sir William Thomson, *Popular Lectures and Addresses,* McMillan, London, 1891.
2. E. Feigenbaum, *Computers and Thought,* McGraw-Hill, New York, 1963.
3. M. Minsky and S. Papert, *Perceptrons: An Introduction to Computational Geometry,* M.I.T. Press, Cambridge, Mass., 1969.
4. M. Arbib, *The Metaphorical Brain,* Wiley-Interscience, New York, 1972.
5. A. Newell and H. Simon, *Human Problem Solving,* Prentice-Hall, Englewood Hills, N.J., 1972.
6. L. A. Zadeh, Fuzzy languages and their relation to human and machine intelligence, in *Proc. Int. Conf. on Man and Computer,* Bordeaux, France, S. Karger, Basel, 1972, pp. 130-165.
7. L. A. Zadeh, Outline of a new approach to the analysis of complex systems and decision processes, *IEEE Trans. Syst., Man and Cybern,* SMC-3, 28-44 (January 1973).
8. R. E. Bellman, R. E. Kalaba and L. A. Zadeh, Abstraction and pattern classification, *J. Math. Anal. Appl.* 13, 1-7 (1966).
9. M. Black, Reasoning with loose concepts, *Dialogue* 2, 1-12 (1963).
10. L. Wittgenstein, Logical form, *Proc. Aristotelian Soc.* 9, 162-171 (1929).
11. M. Scriven, The logic of criteria, *J. Philos.* 56, 857-868 (1959).
12. H. Khatchadourian, Vagueness, meaning and absurdity, *Am. Phil. Quart.* 2, 119-129 (1965).
13. R. R. Verma, Vagueness and the principle of excluded middle, *Mind* 79, 67-77 (1970).

14. J. A. Goguen, The Logic of Inexact Concepts, *Synthese* **19**, 325-373 (1969).
15. E. Adams, The logic of "Almost All", *J. Philos. Logic,* to be published.
16. K. Fine, Vagueness, truth and logic, Department of Philosophy, University of Edinburgh, 1973.
17. B.S. van Frassen, Presuppositions, supervaluations and free logic, in *The Logical Way of Doing Things* (K. Lambert, Ed.), Yale U. P., New Haven Conn., 1969.
18. G. Lakoff, Linguistics and natural logic, in *Semantics of Natural Languages,* (D. Davidson and G. Harman, Eds.), D. Reidel, Doudrecht, The Netherlands, 1971.
19. L. A. Zadeh, Shadows of fuzzy sets, *Probl. Transfer Inf.* **2**, 37-44 (1966).
20. A. Kaufmann, *Theory of Fuzzy Sets,* Masson, Paris, 1972.
21. J. Goguen, *L*-fuzzy sets, *J. Math. Anal. Appl.* **18**, 145-174 (1967).
22. J. G. Brown, A note on fuzzy sets, *Inf. Control* **18**, 32-39 (1971).
23. M. Mizumoto, J. Toyoda and K. Tanaka, General formulation of formal grammars, *Inf. Sci.* **4**, 87-100, 1972.
24. L. A. Zadeh, Similarity relations and fuzzy orderings, *Inf. Sci.* **3**, 177-200 (1971).
25. R. E. Bellman and L. A. Zadeh, Decision-making in a fuzzy environment, *Manage. Sci.* **17**, B-141-B-164 (1970).
26. R. E. Bellman and M. Giertz, On the analytic formalism of the theory of fuzzy sets, *Inf. Sci.* **5**, 149-156 (1973).
27. L. A. Zadeh, A fuzzy-set-theoretic intrepretation of linguistic hedges, *J. Cybern.* **2**, 4-34 (1972).
28. E. F. Codd, Relational completeness of data base sublanguages, in *Courant Computer Science Symposia,* Vol. 6, Prentice-Hall, Englewood Hills, N.J., 1971.
29. L. A. Zadeh, Fuzzy sets, *Inf. Control* **8**, 338-353, 1965.
30. A. Thomasian, *The Structure of Probability Theory With Applications,* McGraw-Hill, New York, 1969.
31. R. E. Moore, *Interval Analysis,* Prentice-Hall, Englewood Hills, N.J., 1966.
32. J. A. Brzozowski, Regular expressions for linear sequential circuits, *IEEE Trans. Electron Comput.,* **EC-14**, 148-156 (1965).
33. E. Shamir, Algebraic, Rational and Context-Free Power Series in Noncommuting Variables, in M. Arbib's *Algebraic Theory of Machines, Languages and Semigroups,* Academic, New York, 1968, pp. 329-341.
34. A. Blikle, Equational languages, *Inf. Control* **21**, 134-147 (1972).
35. D. J. Rosenkrantz, Matrix equation and normal forms of context-free grammars, *J. Assoc. Comput. Mach.* **14**, 501-507, 1967.
36. J. E. Hopcroft and J. D. Ullman, *Formal Languages and Their Relations to Automata,* Addison-Wesley, Reading, Mass., 1969.
37. A. V. Aho and J. D. Ullman, *The Theory of Parsing, Translation and Compiling,* Prentice-Hall, Englewood Hills, N.J., 1973.
38. G. Lakoff, Hedges: a study in meaning criteria and the logic of fuzzy concepts, in *Proc. 8th Reg. Meeting of Chic. Linguist. Soc.,* Univ. of Chicago Linguistics Dept., April 1972.
39. L. A. Zadeth, Semantics of context-free languages, *Math. Syst. Theory,* **2**, 127-145 (1968).
39. L. A. Zadeh, Quantitative fuzzy semantics, *Inf. Sci.* **3**, 159-176 (1971).
40. D. Knuth, Semantics of context-free languages, *Math, Syst. Theory,* **2**, 127-145 (1968).
41. P. Lucas et al., Method and notation for the formal definition of programming languages, Rept. TR 25.087, IBM Laboratory, Vienna, 1968.
42. P. Wegner, The Vienna definition language, *ACM Comput. Surv.* **4**, 5-63, (1972).
43. J. A. Lee, *Computer Semantics,* Van Nostrand-Reinhold, New York, 1972.
44. Z. Kohavi, *Switching and Finite Automata Theory,* McGraw-Hill, New York, 1970.

45. G. E. Hughes and M. J. Cresswell, *An Introduction to Modal Logic,* Methuen, London, 1968,

46. N. Rescher, *Many-Valued Logics,* McGraw-Hill, New York, 1969.

47. R. Barkan, A functional calculus of first order based on strict implication, *J. Symbol. Logic* **11**, 1-16 (1946).

48. L. A. Zadeh, Probability measures of fuzzy events, *J. Math. Anal. Appl.* **23**, 421-427 (1968).

49. A. DeLuca and S. Termini, A definition of non-probabilistic entropy in the setting of fuzzy set theory, *Inf. Control* **20**, 201-312 (1972).

50. J. Hintikka and P. Suppes (Eds.), *Aspects of Inductive Logic,* North-Holland, Amsterdam, 1966.

51. T. Winograd, *Understanding Natural Language,* Academic, New York, 1972.

52. A. DeLuca and S. Termini, Algebraic properties of fuzzy sets, *J. Math. Anal. Appl.* **40**, 373-386 (1972).

53. A. Rosenfeld, Fuzzy groups, *J. Math. Anal. Appl.* **35**, 512-517 (1971).

54. L. A. Zadeh, Fuzzy Algorithms, *Inf. Control* **12**, 94-102 (1968).

55. E. Santos, Fuzzy algorithms, *Inf. Control* **17**, 326-339 (1970).

56. S. K. Chang, On the execution of fuzzy programs using finite state machines, *IEEE Trans. Electron. Conput.,* **C-21**, 241-253 (1972).

57. S. S. L. Chang and L. A. Zadeh, Fuzzy mapping and control, *IEEE Trans. Syst., Man and Cybern.,* **SMC-2**, 30-34 (1972).

58. E. T. Lee, Fuzzy languages and their relation to automata, Dissertation, Dept. of Electr. Eng. and Comput. Sci., Univ. of Calif., Berkeley, 1972.

59. R. C. T. Lee, Fuzzy logic and the resolution principle, *J. Assoc. Comput. Mach.* **19**, 109-119 (1972).

60. K. M. Colby, S. Weber and F. D. Hilf, Artificial paranoia, *J. Artif. Intell.* **2**, 1-25 (1971).

Received November, 1973

The Concept of a Linguistic Variable and its Application to Approximate Reasoning—II*

L.A. ZADEH

Computer Sciences Division, Department of Electrical Engineering and Computer Sciences, and the Electronics Research Laboratory, University of California, Berkeley, California 94720

1. THE CONCEPT OF A FUZZY VARIABLE

Proceeding in the development of Part I of this work, we are now in a position to generalize the concepts introduced in Part I, Sec. 2 to what might be called *fuzzy* variables. For our purposes, it will be convenient to formalize the concept of a fuzzy variable in a way that parallels the characterization of a nonfuzzy variable as expressed by Definition 2.1 of Part I. Specifically:

DEFINITION 1.1. A fuzzy variable is characterized by a triple $(X, U, R(X;u))$, in which X is the name of the variable; U is a universe of discourse (finite or infinite set); u is a generic name for the elements of U; and $R(X;u)$ is a fuzzy subset of U which represents a fuzzy *restriction* on the values of u imposed by X. [As in the case of nonfuzzy variables, $R(X;u)$ will usually be abbreviated to $R(X)$ or $R(u)$ or $R(x)$, where x denotes a generic name for the values of X, and $R(X;u)$ will be referred to as the restriction *on u* or the restriction *imposed by X*.] The nonrestricted nonfuzzy variable u constitutes the *base variable* for X.

The *assignment equation* for X has the form

$$x = u : R(X) \tag{1.1}$$

and represents an assignment of a value u to x subject to the restriction $R(X)$.

The degree to which this equation is satisfied will be referred to as the *compatibility of u with R(X)* and will be denoted by $c(u)$. By definition,

---

*This work was supported in part by the Navy Electronic Systems Command under Contract N00039-71-C-0255, the Army Research Office, Durham, N.C., under Grant DA-ARO-D-31-124-71-G174, and the National Science Foundation under Grant GK-10656X3. The writing of the paper was completed while the author was participating in a Joint Study Program with the Systems Research Department, IBM Research Laboratory, San Jose, California.

$$c(u) = \mu_{R(X)}(u), \qquad u \in U \tag{1.2}$$

where $\mu_{R(X)}(u)$ is the grade of membership of u in the restriction $R(X)$.

COMMENT 1.1. It is important to observe that the compatibility of u is not the same as the probability of u. Thus, the compatibility of u with $R(X)$ is merely a measure of the degree to which u satisfies the restriction $R(X)$, and has no relation to how probable or improbable u happens to be.

COMMENT 1.2. In terms of the valise analogy (see Part I, Comment 2.1), a fuzzy variable may be likened to a tagged valise with *soft* sides, with X representing the name on the tag, U corresponding to a list of objects which can be put in a valise, and $R(X)$ representing a sublist of U in which each object u is associated with a number $c(u)$ representing the degree of ease with which u can be fitted in valise X (Fig. 1).

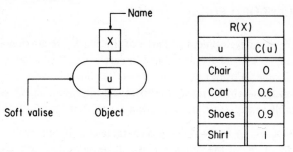

R(X)	
u	C(u)
Chair	0
Coat	0.6
Shoes	0.9
Shirt	1

Fig. 1. Valise analogy for a unary fuzzy variable.

In order to simplify the notation it is convenient to use the same symbol for both X and x, relying on the context for disambiguation. We do this in the following example.

EXAMPLE 1.1. Consider a fuzzy variable named *budget*, with $U = [0, \infty]$ and $R(X)$ defined by (see Fig. 2)

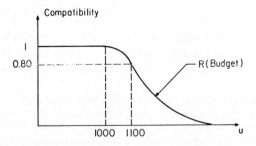

Fig. 2. Compatibility function of *budget*.

$$R(budget) = \int_0^{1000} 1/u + \int_{1000}^{\infty} \left[1 + \left(\frac{u-1000}{200}\right)^2\right]^{-1}/u. \quad (1.3)$$

Then, in the assignment equation

$$budget = 1100 : R(budget), \quad (1.4)$$

the compatibility of 1100 with the restriction imposed by *budget* is

$$c(1100) = \mu_{R\,(budget)}\,(1100)$$

$$= 0.80. \quad (1.5)$$

As in the case of nonfuzzy variables, if $X_1, \ldots, X_n$ are fuzzy variables in $U_1, \ldots, U_n$, respectively, then $X \triangleq (X_1, \ldots, X_n)$ is an *n-ary composite (joint) variable* in $U = U_1 \times \cdots \times U_n$. Correspondingly, in the *n-ary assignment equation*

$$(x_1, \ldots, x_n) = (u_1, \ldots, u_n) : R(X_1, \ldots, X_n), \quad (1.6)$$

x_i, $i = 1, \ldots, n$, is a generic name for the values of X_i; u_i is a generic name for the elements of U_i; and $R(X) \triangleq R(X_1, \ldots, X_n)$ is an *n*-ary fuzzy relation in U which represents the *restriction* imposed by $X \triangleq (X_1, \ldots, X_n)$. The *compatibility of* $(u_1, \ldots, u_n)$ *with* $R(X_1, \ldots, X_n)$ is defined by

$$c(u_1, \ldots, u_n) = \mu_{R\,(X)}\,(u_1, \ldots, u_n), \quad (1.7)$$

where $\mu_{R\,(X)}$ is the membership function of the restriction on $u \triangleq (u_1, \ldots, u_n)$.

EXAMPLE 1.2. Suppose that $U_1 = U_2 = (-\infty, \infty)$, $X_1 \triangleq$ *horizontal proximity;* $X_2 \triangleq$ *vertical proximity;* and the restriction on u is expressed by

$$R(X) = \int_{U_1 \times U_2} (1 + u_1^2 + u_2^2)^{-1}/(u_1, u_2). \quad (1.8)$$

Then the compatibility of the value $u = (2, 1)$ in the assignment equation

$$(x_1, x_2) = (2, 1) : R(X) \quad (1.9)$$

is given by

$$c(2,1) = \mu_{R(X)}(2,1)$$

$$= 0.16. \tag{1.10}$$

COMMENT 1.3. In terms of the valise analogy (see Comment 1.2), an n-ary composite fuzzy variable may be likened to a soft valise named X with n compartments named $X_1, \ldots, X_n$. The compatibility function $c(u_1, \ldots, u_n)$ represents the degree of ease with which objects $u_1, \ldots, u_n$ can be put into respective compartments $X_1, \ldots, X_n$ simultaneously (Fig. 3).

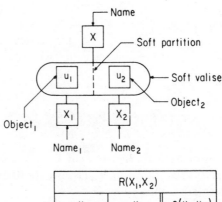

$R(X_1, X_2)$		
u_1	u_2	$c(u_1, u_2)$
Coat	Shoes	0.8
Coat	Shirt	1
Coat	Coat	0.6

Fig. 3. Valise analogy for a binary fuzzy variable.

A basic question that arises in connection with an n-ary assignment equation relates to its decomposition into a sequence of n unary assignment equations, as in Part I, Eq. (2.21). In the case of fuzzy variables, the process of decomposition is somewhat more involved, and we shall take it up after defining marginal and conditioned restrictions.

MARGINAL AND CONDITIONED RESTRICTIONS

In Part I, Sec. 2, the concepts of marginal and conditioned restrictions were intentionally defined in such a way as to make them easy to extend to fuzzy restrictions. Thus, in the more general context of fuzzy variables, these concepts can be formulated in almost exactly the same terms as in Part I, Sec. 2. This is what we shall do in what follows.

NOTE 1.1. As we have seen in our earlier discussion of the notions of marginal and conditioned restrictions in Part I, Sec. 2, it is convenient to simplify the representation of n-tuples by employing the following notation.

Let

$$q \triangleq (i_1, \ldots, i_k) \tag{1.11}$$

be an ordered subsequence of the index sequence $(1, \ldots, n)$. E.g., for $n = 7$, $q = (2, 4, 5)$.

The ordered complement of q is denoted by

$$q' = (j_1, \ldots, j_m). \tag{1.12}$$

E.g., for $q = (2, 4, 5)$, $q' = (1, 3, 6, 7)$.

A k-tuple of variables such as $(v_{i_1}, \ldots, v_{i_k})$ is denoted by $v_{(q)}$. Thus

$$v_{(q)} \triangleq (v_{i_1}, \ldots, v_{i_k}), \tag{1.13}$$

and similarly

$$v_{(q')} \triangleq (v_{j_1}, \ldots, v_{j_m}). \tag{1.14}$$

For example, if

$$v_{(q)} = (v_2, v_4, v_5),$$

then

$$v_{(q')} = (v_1, v_3, v_6, v_7).$$

If $k = n$, we shall write more simply

$$v = (v_1, \ldots, v_n). \tag{1.15}$$

This notation will be used in the following without further explanation.

DEFINITION 1.2. An n-ary restriction $R(X_1, \ldots, X_n)$ in $U_1 \times \cdots \times U_n$ induces a k-ary *marginal restriction* $R(X_{i_1}, \ldots, X_{i_k})$ which is defined as the projection (shadow) of $R(X_1, \ldots, X_n)$ on $U_{i_1} \times \cdots \times U_{i_k}$. Thus, using the definition of projection [see Part I, Eq. (3.57)] and employing the notation of Note 1.1, we can express the membership function of the marginal restriction $R(X_{i_1}, \ldots, X_{i_k})$ as

$$\mu_{R(X_{(q)})}(u_{(q)}) = \bigvee_{u_{(q')}} \mu_{R(X)}(u). \tag{1.16}$$

EXAMPLE 1.3. For the fuzzy binary variable defined in Example 1.2, we have

$$R_1 \overset{\Delta}{=} R(X_1),$$

$$R_2 \overset{\Delta}{=} R(X_2),$$

$$\mu_{R_1}(u_1) = \bigvee_{u_2}(1 + u_1^2 + u_2^2)^{-1}$$

$$= (1 + u_1^2)^{-1},$$

$$\mu_{R_2} = \mu_{R_1}.$$

EXAMPLE 1.4. Assume that

$$U_1 = U_2 = U_3 = 0 + 1 + 2$$

and $R(X_1, X_2, X_3)$ is a ternary fuzzy relation in $U_1 \times U_2 \times U_3$ expressed by

$$R(X_1, X_2, X_3) = 0.8/(0,0,0) + 0.6/(0,0,1) + 0.2/(0,1,0)$$

$$+ 1/(1,0,2) + 0.7/(1,1,0) + 0.4/(0,1,1)$$

$$+ 0.9/(1,2,0) + 0.4/(2,1,1) + 0.8/(1,1,2). \tag{1.17}$$

Applying (1.16) to (1.17), we obtain

$$R(X_1, X_2) = 0.8/(0,0) + 0.4/(0,1) + 1/(1,0)$$

$$+ 0.8/(1,1) + 0.9/(1,2) + 0.4/(2,1) \tag{1.18}$$

and

$$R(X_1) = 0.8/0 + 1/1 + 0.4/2,$$

$$R(X_2) = 1/0 + 0.8/1 + 0.9/2. \tag{1.19}$$

DEFINITION 1.3. Let $R(X_1, \ldots, X_n)$ be a restriction on $(u_1, \ldots, u_n)$, and let $u_{i_1}^0, \ldots, u_{i_k}^0$ be particular values of $u_{i_1}, \ldots, u_{i_k}$, respectively. If in the membership function of $R(X_1, \ldots, X_n)$ the values of $u_{i_1}, \ldots, u_{i_k}$ are set equal to $u_{i_1}^0, \ldots, u_{i_k}^0$, then the resulting function of the arguments $u_{j_1}, \ldots,$ u_{j_m}, where the index sequence $q' = (j_1, \ldots, j_m)$ is complementary to $q = (i_1, \ldots, i_k)$, is defined to be the membership function of a *conditioned restriction* $R(X_{j_1}, \ldots, X_{j_m} | u_{i_1}^0, \ldots, u_{i_k}^0)$ or, more simply, $R(X_{(q')} | u_{(q)}^0)$. Thus

$$\mu_{R(X_{j_1}, \ldots, X_{j_m} | u_{i_1}^0, \ldots, u_{i_k}^0)}(u_{j_1}, \ldots, u_{j_m}) = \mu_{R(X_1, \ldots, X_n)}(u_1, \ldots, u_n | u_{i_1}$$

$$= u_{i_1}^0, \ldots, u_{i_k} = u_{i_k}^0),$$

or more compactly,

$$\mu_{R(X_{(q')} | u_{(q)}^0)}(u_{(q')}) = \mu_{R(X)}(u | u_{(q)} = u_{(q)}^0). \tag{1.20}$$

The simplicity of the relation between conditioned and unconditioned restrictions becomes more transparent if the u_i^0 are written without the superscript. Then, (1.20) becomes

$$\mu_{R(X_{j_1}, \ldots, X_{j_m} | u_{i_1}, \ldots, u_{i_k})}(u_{j_1}, \ldots, u_{j_m}) \overset{\Delta}{=} \mu_{R(X_1, \ldots, X_n)}(u_1, \ldots, u_n),$$

or more compactly,

$$\mu_{R(X_{(q')} | u_{(q)})}(u_{(q')}) \overset{\Delta}{=} \mu_{R(X)}(u). \tag{1.21}$$

NOTE 1.2. In some instances, it is preferable to use an alternative notation for conditioned restrictions. For example, if $n = 4$, $q = (1,3)$ and $q' = (2,4)$, it may be simpler to write $R(u_1^0, X_2, u_3^0, X_4)$ for $R(X_2, X_4 | u_1^0, u_3^0)$. This is particularly true when numerical values are used in place of the subscripted arguments, e.g., 5 and 2 in place of u_1^0 and u_3^0. In such cases, in order to avoid ambiguity we shall write explicitly $R(X_2, X_4 | u_1^0 = 5, u_3^0 = 2)$, or more simply, $R(5, X_2, 2, X_4)$.

EXAMPLE 1.5. In Example 1.4, we have

$$R(X_1, X_2, 0) = 0.8/(0,0) + 0.2/(0,1) + 0.7/(1,1) + 0.9/(1,2),$$

$$R(X_1, X_2, 1) = 0.6/(0,0) + 0.4/(0,1) + 0.4/(2,1), \tag{1.22}$$

$$R(X_1, X_2, 2) = 1/(1,0) + 0.8/(1,1),$$

and, using (1.16),

$$R(X_1,0) = 0.8/0 + 1/1,$$

$$R(X_1,1) = 0.4/0 + 0.8/1 + 0.4/2, \qquad (1.23)$$

$$R(X_1,2) = 0.9/1.$$

It is useful to observe that an immediate consequence of the definitions of marginal and conditioned restrictions is the following

PROPOSITION 4.1. *Let* $R(X_{j_1}, \ldots, X_{j_m})$ *be a marginal restriction induced by* $R(X_1, \ldots, X_n)$, *and let* $R(X_{j_1}, \ldots, X_{j_m} | u_{i_1}, \ldots, u_{i_k})$ *or, more simply,* $R(X_{(q')} | u_{(q)})$ *be a restriction conditioned on* $u_{i_1}, \ldots, u_{i_k}$, *with* $q = (i_1, \ldots, i_k)$ *and* $q' = (j_1, \ldots, j_m)$ *being complementary index sequences. Then, in consequence of* (1.16), (1.21) *and the definition of the union* [*see Part I, Eq.* (3.34)], *we can assert that*

$$R(X_{(q')}) = \sum_{u(q)} R(X_{(q')} | u_{(q)}), \qquad (1.24)$$

where $\sum_{u_{(q)}}$ *stands for the union (rather than the arithmetic sum) over the* $u_{(q)}$.

EXAMPLE 1.6. With reference to Example 1.3 and Note 1.2, it is easy to verify that

$$R(X_1, X_2) = R(X_1, X_2, 0) + R(X_1, X_2, 1) + R(X_1, X_2, 2)$$

and

$$R(X_1) = R(X_1, 0) + R(X_1, 1) + R(X_1, 2).$$

SEPARABILITY AND NONINTERACTION

DEFINITION 1.4. An *n*-ary restriction $R(X_1, \ldots, X_n)$ is *separable* iff it can be expressed as the Cartesian product of unary restrictions

$$R(X_1, \ldots, X_n) = R(X_1) \times \cdots \times R(X_n) \qquad (1.25)$$

or, equivalently, as the intersection of cylindrical extensions [see Part I, Eq. (3.62)]

$$R(X_1, \ldots, X_n) = \bar{R}(X_1) \cap \cdots \cap \bar{R}(X_n). \qquad (1.26)$$

It should be noted that, if $R(X_1, \ldots, X_n)$ is normal, then so are its marginal restrictions (see Part I, Proposition 3.3). It follows, then, that the $R(X_i)$ in (1.25) are marginal restrictions induced by $R(X_1, \ldots, X_n)$. For, (1.25) implies that

$$\mu_{R(X_1, \ldots, X_n)}(u_1, \ldots, u_n) = \mu_{R(X_1)}(u_1) \wedge \cdots \wedge \mu_{R(X_n)}(u_n), \quad (1.27)$$

and hence by Eq. (3.57) of Part I,

$$P_i R(X_1, \ldots, X_n) = R(X_i), \qquad i = 1, \ldots, n. \qquad (1.28)$$

Unless stated to the contrary, we shall assume henceforth that $R(X_1, \ldots, X_n)$ is normal.

EXAMPLE 1.7. The relation matrix of the restriction shown below can be expressed as the max-min dyadic product of a column vector (a unary relation) and a row vector (a unary relation). This implies that the restriction in question is separable:

$$
\begin{bmatrix}
0.3 & 0.8 & 0.8 & 0.1 \\
0.3 & 0.8 & 1 & 0.1 \\
0.2 & 0.2 & 0.2 & 0.1 \\
0.3 & 0.6 & 0.6 & 0.1
\end{bmatrix}
=
\begin{bmatrix}
0.8 \\
1 \\
0.2 \\
0.6
\end{bmatrix}
\begin{bmatrix} 0.3 & 0.8 & 1 & 0.1 \end{bmatrix}
$$

EXAMPLE 1.8. The restrictions defined in Definition 1.2 and Example 1.3 are not separable.

An immediate consequence of separability is the following

PROPOSITION 4.2. *If $R(X_1, \ldots, X_n)$ is separable, so is every marginal restriction induced by $R(X_1, \ldots, X_n)$.*

Also, in consequence of (1.25), we can assert the

PROPOSITION 4.3. *The separable restriction $R(X_1) \times \cdots \times R(X_n)$ is the largest restriction with marginal restrictions $R(X_1), \ldots, R(X_n)$.*

The concept of separability is closely related to that of *noninteraction* of fuzzy variables. More specifically:

DEFINITION 1.5. The fuzzy variables $X_1, \ldots, X_n$ are said to be *noninteractive* iff the restriction $R(X_1, \ldots, X_n)$ is separable.

It will be recalled that, in the case of nonfuzzy variables, the justification for characterizing $X_1, \ldots, X_n$ as noninteractive is that if [see Part I, Eq. (2.18)]

$$R(X_1, \ldots, X_n) = R(X_1) \times \cdots \times R(X_n), \tag{1.29}$$

then the *n*-ary assignment equation

$$(x_1, \ldots, x_n) = (u_1, \ldots, u_n) : R(X_1, \ldots, X_n) \tag{1.30}$$

can be decomposed into a sequence of *n* unary assignment equations

$$x_1 = u_1 : R(X_1),$$
$$\cdot \qquad \cdot \tag{1.31}$$
$$x_n = u_n : R(X_n).$$

In the case of fuzzy variables, a basic consequence of noninteraction—from which Eq. (2.19) of Part I follows as a special case—is expressed by

PROPOSITION 4.4. *If the fuzzy variables $X_1, \ldots, X_n$ are noninteractive, then the n-ary assignment equation* (1.30) *can be decomposed into a sequence of n unary assignment equations* (1.31), *with the understanding that if $c(u_1, \ldots, u_n)$ is the compatibility of $(u_1, \ldots, u_n)$ with $R(X_1, \ldots, X_n)$, and if $c_i(u_i)$, $i = 1, \ldots, n$, is the compatibility of u_i with $R(X_i)$, then*

$$c(u_1, \ldots, u_n) = c_1(u_1) \wedge \cdots \wedge c_n(u_n). \tag{1.32}$$

Proof. By the definitions of compatibility, noninteraction and separability, we have at once

$$c(u_1, \ldots, u_n) = \mu_{R(X_1, \ldots, X_n)}(u_1, \ldots, u_n)$$
$$= \mu_{R(X_1)}(u_1) \wedge \cdots \wedge \mu_{R(X_n)}(u_n)$$
$$= c_1(u_1) \wedge \cdots \wedge c_n(u_n). \quad \text{Q.E.D.} \tag{1.33}$$

COMMENT 1.4. Pursuing the valise analogy further (see Comment 1.3), noninteractive fuzzy variables $X_1, \ldots, X_n$ may be likened to *n separate* soft valises with name-tags $X_1, \ldots, X_n$. The restriction associated with valise X_i

is characterized by the compatibility function $c(u_i)$. Then the overall compatibility function for the valises $X_1, \ldots. X_n$ is given by (1.32) (Fig. 4).

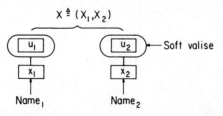

Fig. 4. Valise analogy for noninteractive fuzzy variables.

COMMENT 1.5. In terms of the base variables of $X_1, \ldots, X_n$ (see Definition 1.1), noninteraction implies that there are no constraints which jointly involve $u_1, \ldots, u_n$, where u_i is the base variable for X_i, $i = 1, \ldots, n$. For example, if the u_i are constrained by

$$u_1 + \cdots + u_n = 1,$$

then $X_1, \ldots, X_n$ are *interactive*, i.e., are not noninteractive. (See Part I, Comment 3.5.)

If $X_1, \ldots, X_n$ are interactive, it is still possible to decompose an n-ary assignment equation into a sequence of n unary assignment equations. However, the restriction on u_i will, in general, depend on the values assigned to $u_1, \ldots, u_{i-1}$. Thus, the n assignment equations will have the following form [see also Part I, Eq. (2.21)]:

$$x_1 = u_1 : R(X_1),$$

$$x_2 = u_2 : R(X_2 \mid u_1),$$

$$x_3 = u_3 : R(X_3 \mid u_1, u_2), \qquad\qquad (1.34)$$

$$\cdot \qquad \cdot \qquad \cdot$$

$$x_n = u_n : R(X_n \mid u_1, \ldots, u_{n-1}),$$

where $R(X_i \mid u_1, \ldots, u_{i-1})$ denotes the restriction on u_i conditioned on $u_1, \ldots, u_{i-1}$ (see Definition 1.3).

EXAMPLE 1.9. Taking Example 1.4, assume that $u_1 = 1$, $u_2 = 2$ and $u_3 = 0$. Then

$$R(X_1) = 0.8/0 + 1/1 + 0.4/2,$$

$$R(X_2 \,|\, u_1 = 1) = 1/0 + 0.8/1 + 0.9/2, \tag{1.35}$$

$$R(X_3 \,|\, u_1 = 1, u_2 = 2) = 0.9/0,$$

so that

$$c_1(1) = 1,$$

$$c_2(2) = 0.9, \tag{1.36}$$

$$c_3(0) = 0.9.$$

As in the case of (1.31), the justification for (1.34) is provided by

PROPOSITION 4.5. *If* $X_1, \ldots, X_n$ *are interactive fuzzy variables subject to the restriction* $R(X_1, \ldots, X_n)$, *and* $c_i(u_i)$, $i = 1, \ldots, n$, *is the compatibility of* u_i *with the conditioned restriction* $R(X_i \,|\, u_1, \ldots, u_{i-1})$ *in* (1.34), *then*

$$c(u_1, \ldots, u_n) = c_1(u_1) \wedge \cdots \wedge c_n(u_n), \tag{1.37}$$

where $c(u_1, \ldots, u_n)$ *is the compatibility of* $(u_1, \ldots, u_n)$ *with* $R(X_1, \ldots, X_n)$.

Proof. By the definition of a conditioned restriction [see (1.20)], we have, for all i, $1 \leqslant i \leqslant n$,

$$\mu_{R(X_i \,|\, u_1, \ldots, u_{i-1})}(u_i) = \mu_{R(X_1, \ldots, X_i)}(u_1, \ldots, u_i). \tag{1.38}$$

On the other hand, the definition of a marginal restriction [see (1.16)] implies that, for all i and all $u_1, \ldots, u_i$, we have

$$\mu_{R(X_1, \ldots, X_i)}(u_1, \ldots, u_i) \geqslant \mu_{R(X_1, \ldots, X_{i+1})}(u_1, \ldots, u_{i+1}), \tag{1.39}$$

and hence that

$$\mu_{R(X_{i+1} \,|\, u_1, \ldots, u_i)}(u_{i+1}) \wedge \mu_{R(X_i \,|\, u_1, \ldots, u_{i-1})}(u_i) = \mu_{R(X_{i+1} \,|\, u_1, \ldots, u_i)}(u_{i+1}). \tag{1.40}$$

Combining (1.40) with the defining equation

$$c_i(u_i) = \mu_{R(X_i \,|\, u_1, \ldots, u_{i-1})}(u_i), \tag{1.41}$$

we derive

$$c(u_1, \ldots, u_n) = c_1(u_1) \wedge \cdots \wedge c_n(u_n). \qquad \text{Q.E.D.} \qquad (1.42)$$

This concludes our discussion of some of the properties of fuzzy variables which are relevant to the concept of a linguistic variable. In the following section, we shall formalize the concept of a linguistic variable and explore some of its implications.

2. THE CONCEPT OF A LINGUISTIC VARIABLE

In our informal discussion of the concept of a linguistic variable in Part I, Sec. 1, we have stated that a linguistic variable differs from a numerical variable in that its values are not numbers but words or sentences in a natural or artificial language. Since words, in general, are less precise than numbers, the concept of a linguistic variable serves the purpose of providing a means of approximate characterization of phenomena which are too complex or too ill-defined to be amenable to description in conventional quantitative terms. More specifically, the fuzzy sets which represent the restrictions associated with the values of a linguistic variable may be viewed as summaries of various subclasses of elements in a universe of discourse. This, of course, is analogous to the role played by words and sentences in a natural language. For example, the adjective *handsome* is a summary of a complex of characteristics of the appearance of an individual. It may also be viewed as a label for a fuzzy set which represents a restriction imposed by a fuzzy variable named *handsome*. From this point of view, then, the terms *very handsome, not handsome, extremely handsome, quite handsome*, etc., are names of fuzzy sets which result from operating on the fuzzy set named *handsome* with the modifiers named *very, not, extremely, quite*, etc. In effect, these fuzzy sets, together with the fuzzy set labeled *handsome*, play the role of values of the linguistic variable *Appearance*.

An important facet of the concept of a linguistic variable is that it is a variable of a higher order than a fuzzy variable, in the sense that a linguistic variable takes fuzzy variables as its values. For example, the values of a linguistic variable named *Age* might be: *young, not young, old, very old, not young and not old, quite old*, etc., each of which is the name of a fuzzy variable. If X is the name of such a fuzzy variable, the restriction imposed by X may be interpreted as the *meaning* of X. Thus, if the restriction imposed by the fuzzy variable named *old* is a fuzzy subset of $U = [0, 100]$ defined by

$$R(old) = \int_{50}^{100} \left[1 + \left(\frac{u-50}{5} \right)^{-2} \right]^{-1} / u, \qquad u \in U, \qquad (2.1)$$

then the fuzzy set represented by $R(old)$ may be taken to be the meaning of *old* (Fig. 5).

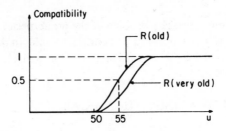

Fig. 5. Compatibility functions of *old* and *very old.*

Another important facet of the concept of a linguistic variable is that, in general, a linguistic variable is associated with two rules: (1) a *syntactic rule,* which may have the form of a grammar for generating the names of the values of the variable; and (2) a *semantic rule* which defines an algorithmic procedure for computing the meaning of each value. These rules constitute an essential part of the characterization of a *structured* linguistic variable.[1]

Since a linguistic variable is a variable of a higher order than a fuzzy variable, its characterization is necessarily more complex than that expressed by Definition 1.1. More specifically, we have

DEFINITION 2.1. A linguistic variable is characterized by a quintuple $(\mathscr{X}, T(\mathscr{X}), U, G, M)$ in which $\mathscr{X}$ is the name of the variable; $T(\mathscr{X})$ (or simply T) denotes the *term-set* of $\mathscr{X}$, that is, the set of names of *linguistic values* of $\mathscr{X}$, with each value being a fuzzy variable denoted generically by X and ranging over a universe of discourse U which is associated with the *base variable u; G* is a *syntactic rule* (which usually has the form of a grammar) for generating the names, X, of values of $\mathscr{X}$; and M is a *semantic rule* for associating with each X its *meaning,* $M(X)$, which is a fuzzy subset of U. A particular X, that is, a name generated by G, is called a *term.* A term consisting of a word or words which function as a unit (i.e., always occur together) is called an *atomic term.* A term which contains one or more atomic terms is a *composite term.* A concatenation of components of a composite term is a *subterm.* If $X_1, X_2, \ldots$ are terms in T, then T may be expressed as the union

$$T = X_1 + X_2 + \cdots . \tag{2.2}$$

---

[1]It is primarily the semantic rule that distinguishes a linguistic variable from the more conventional concept of a syntactic variable.

Where it is necessary to place in evidence that T is generated by a grammar G, T will be written as $T(G)$.

The meaning, $M(X)$, of a term X is defined to be the restriction, $R(X)$, on the base variable u which is imposed by the fuzzy variable named X. Thus

$$M(X) \overset{\Delta}{=} R(X), \tag{2.3}$$

with the understanding that $R(X)$—and hence $M(X)$—may be viewed as a fuzzy subset of U carrying the name X. The connection between $\mathscr{X}$, the linguistic value X and the base variable u is illustrated in Fig. 3 of Part I.

NOTE 2.1. In order to avoid a profusion of symbols, it is expedient to assign more than one meaning to some of the symbols occurring in Definition 2.1, relying on the context for disambiguation. Specifically:

(a) We shall frequently employ the symbol $\mathscr{X}$ to denote both the name of the variable and the generic name of its values. Likewise, X will be used to denote both the generic name of the values of the variable and the name of the variable itself.

(b) The same symbol will be used to denote a set and the name of that set. Thus, the symbols X, $M(X)$ and $R(X)$ will be used interchangeably, although strictly speaking X—as the name of $M(X)$ [or $R(X)$]—is distinct from $M(X)$. In other words, when we say that a term X (e.g. *young*) is a value of $\mathscr{X}$ (e.g., *Age*), it should be understood that the actual value is $M(X)$ and that X is merely the name of the value.

EXAMPLE 2.1. Consider a linguistic variable named *Age*, i.e., $\mathscr{X} = Age$, with $U = [0, 100]$. A linguistic value of *Age* might be named *old,* with *old* being an atomic term. Another value might be named *very old,* in which case *very old* is a composite term which contains *old* as an atomic component and has *very* and *old* as subterms. The value of *Age* named *more or less young* is a composite term which contains *young* as an atomic term and in which *more or less* is a subterm. The term-set associated with *Age* may be expressed as

$$T(Age) = old + very\ old + not\ old + more\ or\ less\ young + quite\ young$$
$$+ not\ very\ old\ and\ not\ very\ young + \cdots, \tag{2.4}$$

in which each term is the name of a fuzzy variable in the universe of discourse $U = [0, 100]$. The restriction imposed by a term, say $R(old)$, constitutes the meaning of *old*. Thus, if $R(old)$ is defined by (2.1), then the meaning of the linguistic value *old* is given by

$$M(old) = \int_{50}^{100} \left[1 + (\frac{u-50}{5})^{-2} \right]^{-1} /u, \tag{2.5}$$

or more simply (see Note 2.1),

$$old = \int_{50}^{100} \left[1 + (\frac{u-50}{5})^{-2}\right]^{-1} /u. \qquad (2.6)$$

Similarly, the meaning of a linguistic value such as *very old* may be expressed as (see Fig. 5)

$$M(\textit{very old}) = \textit{very old} = \int_{50}^{100} \left[1 + (\frac{u-50}{5})^{-2}\right]^{-2} /u. \qquad (2.7)$$

The assignment equation in the case of a linguistic variable assumes the form

$$X = \text{term in } T(\mathscr{X})$$

$$= \text{name generated by } G \qquad (2.8)$$

which implies that the meaning assigned to X is expressed by

$$M(X) = R(\text{term in } T(\mathscr{X}). \qquad (2.9)$$

In other words, the meaning of X is given by the application of the semantic rule M to the value assigned to X by the right-hand side of (2.8). Furthermore, as defined by (2.3), $M(X)$ is identical to the restriction imposed by X.

COMMENT 2.1. In accordance with Note 2.1(a), the assignment equation will usually be written as

$$\mathscr{X} = \text{name in } T(\mathscr{X}) \qquad (2.10)$$

rather than in the form (2.8). For example, if $\mathscr{X} = Age$, and *old* is a term in $T(\mathscr{X})$, we shall write

$$Age = old, \qquad (2.11)$$

with the understanding that *old* is a restriction on the values of u defined by (2.1), which is assigned by (2.11) to the linguistic variable named *Age*. It is important to note that the equality symbol in (2.10) does not represent a symmetric relation—as it does in the case of arithmetic equality. Thus, it would not be meaningful to write (2.11) as

$$old = Age$$

To illustrate the concept of a linguistic variable, we shall consider first a very elementary example in which $T(\mathscr{X})$ contains just a few terms and the syntactic and semantic rules are trivially simple.

EXAMPLE 2.2. Consider a linguistic variable named *Number* which is associated with the finite term-set

$$T(Number) = few + several + many, \qquad (2.12)$$

in which each term represents a restriction on the values of u in the universe of discourse

$$U = 1 + 2 + 3 + \cdots + 10. \qquad (2.13)$$

These restrictions are assumed to be fuzzy subsets of U which are defined as follows:

$$few = 0.4/1 + 0.8/2 + 1/3 + 0.4/4, \qquad (2.14)$$

$$several = 0.5/3 + 0.8/4 + 1/5 + 1/6 + 0.8/7 + 0.5/8, \qquad (2.15)$$

$$many = 0.4/6 + 0.6/7 + 0.8/8 + 0.9/9 + 1/10. \qquad (2.16)$$

Thus

$$R(few) = M(few) = 0.4/1 + 0.8/2 + 1/3 + 0.4/4, \qquad (2.17)$$

and likewise for the other terms in T. The implication of (2.17) is that *few* is the name of a fuzzy variable which is a value of the linguistic variable *Number*. The meaning of *few*—which is the same as the restriction imposed by *few*—is a fuzzy subset of U which is defined by the right-hand side of (2.17).

To assign a value such as *few* to the linguistic variable *Number*, we write

$$Number = few, \qquad (2.18)$$

with the understanding that what we actually assign to *Number* is a fuzzy variable named *few*.

EXAMPLE 2.3. In this case, we assume that we are dealing with a composite linguistic variable[2] named $(\mathscr{X}, \mathscr{Y})$ which is associated with the base variable (u, v) ranging over the universe of discourse $U \times V$, where

---

[2] Composite linguistic variables will be discussed in greater detail in Sec. 3 in connection with linguistic truth variables.

$$U \times V = (1 + 2 + 3 + 4) \times (1 + 2 + 3 + 4) \qquad (2.19)$$

$$= (1,1) + (1,2) + (1,3) + (1,4)$$

$$\cdots$$

$$\cdots$$

$$+ (4,1) + (4,2) + (4,3) + (4,4), \qquad (2.20)$$

with the understanding that

$$i \times j = (i,j), \qquad i,j = 1,2,3,4. \qquad (2.21)$$

Furthermore, we assume that the term-set of $(\mathscr{X}, \mathscr{Y})$ comprises just two terms:

$$T = approximately\ equal + more\ or\ less\ equal, \qquad (2.22)$$

where *approximately equal* and *more or less equal* are names of binary fuzzy relations defined by the relation matrices

$$approximately\ equal = \begin{bmatrix} 1 & 0.6 & 0.4 & 0.2 \\ 0.6 & 1 & 0.6 & 0.4 \\ 0.4 & 0.6 & 1 & 0.6 \\ 0.2 & 0.4 & 0.6 & 1 \end{bmatrix} \qquad (2.23)$$

and

$$more\ or\ less\ equal = \begin{bmatrix} 1 & 0.8 & 0.6 & 0.4 \\ 0.8 & 1 & 0.8 & 0.6 \\ 0.6 & 0.8 & 1 & 0.8 \\ 0.4 & 0.6 & 0.8 & 1 \end{bmatrix} \qquad (2.24)$$

In these relation matrices, the (i,j)th entry represents the compatibility of the pair (i,j) with the restriction in question. For example, the $(2,3)$ entry in *approximately equal*—which is 0.6—is the compatibility of the ordered pair $(2,3)$ with the binary restriction named *approximately equal*.

To assign a value, say *approximately equal*, to $(\mathscr{X}, \mathscr{Y})$, we write

$$(\mathscr{X},\mathscr{Y}) = approximately\ equal, \qquad\qquad (2.25)$$

where, as in (2.18), it is understood that what we assign to $(\mathscr{X},\mathscr{Y})$ is a binary fuzzy relation named *approximately equal,* which is a binary restriction on the values of (u,v) in the universe of discourse (2.20).

COMMENT 2.2. In terms of the valise analogy (see Comment 1.2), a linguistic variable as defined by Definition 2.1 may be likened to a hard valise into which we can put soft valises, as illustrated in Fig. 6. A soft valise corresponds to a fuzzy variable which is assigned as a linguistic value to $\mathscr{X}$, with X playing the role of the name-tag of the soft valise.

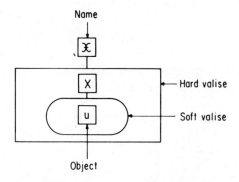

Fig. 6. Valise analogy for a linguistic variable.

STRUCTURED LINGUISTIC VARIABLES

In both of the above examples the term-set contains only a small number of terms, so that it is practicable to list the elements of $T(\mathscr{X})$ and set up a direct association between each element and its meaning. In the more general case, however, the number of elements in $T(\mathscr{X})$ may be infinite, necessitating the use of an algorithm, rather than a table look-up procedure, for generating the elements of $T(\mathscr{X})$ as well as for computing their meaning.

A linguistic variable $\mathscr{X}$ will be said to be *structured* if its term-set, $T(\mathscr{X})$, and the function, M, which associates a meaning with each term in the term-set, can be characterized algorithmically. In this sense, the syntactic and semantic rules associated with a structured linguistic variable may be viewed as algorithmic procedures for generating the elements of $T(\mathscr{X})$ and computing the meaning of each term in $T(\mathscr{X})$, respectively. Unless stated to the contrary, we shall assume henceforth that the linguistic variables we deal with are structured.

EXAMPLE 2.4. As a very simple illustration of the role played by the syntactic and semantic rules in the case of a structured linguistic variable, we shall consider a variable named *Age* whose terms are exemplified by: *old, very old, very very old, very very very old*, etc. Thus, the term set of *Age* can be written as

$$T(Age) = old + very\ old + very\ very\ old + \cdots . \qquad (2.26)$$

In this simple case, it is clear by inspection that every term in $T(Age)$ is of the form *old* or *very very . . . very old*. To deduce this rule in a more general way, we proceed as follows.

Let xy denote the concatenation of character strings x and y, e.g., $x = very$, $y = old$, $xy = very\ old$. If A and B are sets of strings, e.g.,

$$A = x_1 + x_2 + \cdots , \qquad (2.27)$$

$$B = y_1 + y_2 + \cdots , \qquad (2.28)$$

where x_i and y_j are character strings, then the concatenation of A and B is denoted by AB and is defined as the set of strings

$$AB = (x_1 + x_2 + \cdots)(y_1 + y_2 + \cdots)$$

$$= \sum_{i,j} x_i y_j \qquad (2.29)$$

For example, if $A = very$ and $B = old + very\ old$, then

$$very\ (old + very\ old) = very\ old + very\ very\ old. \qquad (2.30)$$

Using this notation, the given expression for $T(Age)$, or simply T, may be taken to be the solution of the equation[3]

$$T = old + very\ T, \qquad (2.31)$$

which, in words, means that every term in T is of the form *old* or *very* followed by some term in T.

Equation (2.31) can be solved by iteration, using the recursion equation

---

[3] As is well known in the theory of regular expressions (see [32]), the solution of (2.31) can be expressed as

$$T = (\lambda + very + very^2 + \cdots)\ old,$$

where λ is the null string. This expression for T is equivalent to that of (2.34).

$$T^{i+1} = old + very \ T^i, \qquad i = 0, 1, 2, \ldots, \tag{2.32}$$

with the initial value of T^i being the empty set θ. Thus

$$T^0 = \theta,$$

$$T^1 = old,$$

$$T^2 = old + very \ old, \tag{2.33}$$

$$T^3 = old + very \ old + very \ very \ old,$$

$$\ldots,$$

and the solution of (2.31) is given by

$$T = T^\infty = old + very \ old + very \ very \ old + very \ very \ very \ old + \cdots . \tag{2.34}$$

For the example under consideration, the syntactic rule, then, is expressed by (2.31) and its solution (2.34). Equivalently, the syntactic rule can be characterized by the production system

$$T \rightarrow old, \tag{2.35}$$

$$T \rightarrow very \ T, \tag{2.36}$$

for which (2.31) plays the role of an algebraic representation.[4] In this case, a term in T can be generated through a standard derivation procedure ([36], [37]) involving a successive application of the rewriting rules (2.35) and (2.36) starting with the symbol T. Thus, if T is rewritten as *very* T and then T in *very* T is rewritten as *old*, we obtain the term *very old*. In a similar fashion, the term *very very very old* can be obtained from T by the derivation chain

$$T \rightarrow very \ T \rightarrow very \ very \ T \rightarrow very \ very \ very \ T \rightarrow very \ very \ very \ old. \tag{2.37}$$

Turning to the semantic rule for *Age*, we note that to compute the meaning of a term such as *very . . . very old* we need to know the meaning of *old* and the meaning of *very*. The term *old* plays the role of a *primary term*, that is, a term whose meaning must be specified as an initial datum in order to provide

---

[4] A discussion of the algebraic representation of context-free grammars may be found in [33], [34] and [35]. Algebraic treatment of fuzzy languages is discussed in [6] and [58].

a basis for the computation of the meaning of composite terms in T. As for the term *very,* it acts as a *linguistic hedge,* that is, as a modifier of the meaning of its operand. If—as very simple approximation—we assume that *very* acts as a concentrator [see Part I, Eq. (3.40)], then

$$very\ old = \text{CON}\ (old)$$

$$= old^2. \tag{2.38}$$

Consequently, the semantic rule for *Age* may be expressed as

$$M(very \ldots very\ old) = old^{2n}, \tag{2.39}$$

where n is the number of occurrences of *very* in the term *very . . . very old* and $M(very \ldots very\ old)$ is the meaning of *very . . . very old.* Furthermore, if the primary term *old* is defined as

$$old = \int_{50}^{100} \left[1 + (\frac{u-50}{5})^{-2}\right]^{-1} /u, \tag{2.40}$$

then

$$M(very \ldots very\ old) = \int_{50}^{100} \left[1 + (\frac{u-50}{5})^{-2}\right]^{-2n} /u, \quad n = 1,2,\ldots. \tag{2.41}$$

This equation provides an explicit semantic rule for the computation of the meaning of composite terms generated by (2.31), from the knowledge of the meaning of the primary term *old* and the hedge *very.*

BOOLEAN LINGUISTIC VARIABLES

The linguistic variable considered in Example 2.4 is a special case of what might be called a *Boolean linguistic variable.* Typically, such a variable involves a finite number of primary terms, a finite number of hedges, the connectives *and* and *or,* and the negation *not.* For example, the term-set of a Boolean linguistic variable *Age* might be

$$T(Age) = young + old + not\ young + not\ old + very\ young + very\ very\ young$$
$$+ not\ very\ young\ and\ not\ very\ old + quite\ young + more\ or\ less\ old$$
$$+ extremely\ old + \cdots. \tag{2.42}$$

More formally, a Boolean linguistic variable may be defined recursively as follows.

DEFINITION 2.2. A *Boolean linguistic variable* is a linguistic variable whose terms, X, are Boolean expressions in variables of the form X_p, hX_p, X or hX, where h is a linguistic hedge, X_p is a primary term and hX is the name of a fuzzy set resulting from acting with h on X.

As an illustration, in the case of the linguistic variable *Age* whose term-set is defined by (2.42), the term *not very young and not very old* is of this form, with $h \triangleq$ *very*, $X_p \triangleq$ *young* and $X_p \triangleq$ *old*. Similarly, in the case of the term *very very young*, $h \triangleq$ *very very* and $X_p \triangleq$ *young*.

Boolean linguistic variables are particularly convenient to deal with because much of our experience in the manipulation and evaluation of Boolean expressions is transferable to variables of this type. To illustrate this point, we shall consider a simple example which involves two primary terms and a single hedge.

EXAMPLE 2.5. Let *Age* be a Boolean linguistic variable with the term-set

$$T(Age) = young + not\ young + old + not\ old + very\ young$$
$$+ not\ young\ and\ not\ old + young\ or\ old + young\ or\ (not$$
$$very\ young\ and\ not\ very\ old) + \cdots . \qquad (2.43)$$

If we identify *and* with intersection, *or* with union, *not* with complementation and *very* with concentration [see (2.38)], the meaning of a typical value of *Age* can be written down by inspection. For example,

$$M(not\ young) = \neg young,$$

$$M(not\ very\ young) = \neg(young^2),$$

$$M(not\ very\ young\ and\ not\ very\ old) = \neg(young^2) \cap \neg(old^2),$$

$$(2.44)$$

$$M(young\ or\ old) = young \cup old.$$

In effect, these equations express the meaning of a composite term as a function of the meanings of its constituent primary terms. Thus, if *young* and *old* are defined as

$$young = \int_0^{25} 1/u + \int_{25}^{100} \left[1 + \left(\frac{u-25}{5}\right)^2 \right]^{-1} /u, \qquad (2.45)$$

$$old = \int_{50}^{100} \left[1 + \left(\frac{u-50}{5}\right)^{-2} \right]^{-1} /u, \qquad (2.46)$$

then (see Fig. 7)

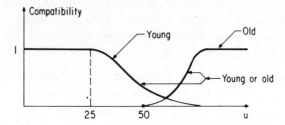

Fig. 7. Compatibility function for *young or old*.

$$M(\textit{young or old}) = \int_{0}^{25} 1/u + \int_{25}^{50} \left[1 + (\frac{u-25}{5})^2\right]^{-1} /u$$

$$+ \int_{50}^{100} \left[1 + (\frac{u-25}{5})^2\right]^{-1} \vee \left[1 + (\frac{u-50}{5})^{-2}\right]^{-1} /u. \tag{2.47}$$

The linguistic variable considered in the above example involves just one type of hedge, namely, *very*. More generally, a Boolean linguistic variable may involve a finite number of hedges, as in (2.42). The procedure for computing the meaning of a composite term remains the same, however, once the operations corresponding to the hedges are defined.

The question of what constitutes an appropriate representation for a particular hedge, e.g., *more or less* or *quite* or *essentially*, is by no means a simple one.[5] To illustrate the point, in some contexts the effect of the hedge *more or less* may be approximated by [see Part I, Eq. (3.41)]

$$M(\textit{more or less } X) = \text{DIL}(X) = X^{0.5}. \tag{2.48}$$

For example, if X = *old*, and *old* is defined by (2.46), then

$$\textit{more or less old} = \int_{50}^{100} \left[1 + (\frac{u-50}{5})^{-2}\right]^{-0.5} /u. \tag{2.49}$$

---

[5] A more detailed discussion of linguistic hedges from a fuzzy-set-theoretic point of view may be found in [27] and [38]. The idea of treating various types of linguistic hedges as operators on fuzzy sets originated in the course of the author's collaboration with Professor G. Lakoff.

In many instances, however, *more or less* acts as a fuzzifier in the sense of Part I, Eq. (3.48), rather than as a dilator. As an illustration, suppose that the meaning of a primary term *recent* is specified as

$$recent = 1/1974 + 0.8/1973 + 0.7/1972, \qquad (2.50)$$

and that *more or less recent* is defined as the result of acting with a fuzzifier F on *recent*, i.e.,

$$more \ or \ less \ recent = F(recent; K) \qquad (2.51)$$

where the kernel K of F is defined by

$$K(1974) = 1/1974 + 0.9/1973,$$
$$K(1973) = 1/1973 + 0.9/1972, \qquad (2.52)$$
$$K(1972) = 1/1972 + 0.8/1971.$$

On substituting the values of K into (3.48) of Part I, we obtain the meaning of *more or less recent*, i.e.,

$$more \ or \ less \ recent = 1/1974 + 0.9/1973 + 0.72/1972 + 0.56/1971. \quad (2.53)$$

On the other hand, if the hedge *more or less* were assumed to be a dilator, then we would have

$$more \ or \ less \ recent = (1/1974 + 0.8/1973 + 0.7/1972)^{0.5}$$
$$= 1/1974 + 0.9/1973 + 0.84/1972 \qquad (2.54)$$

which differs from (2.53) mainly in the absence of the term $0.56/1971$. Thus, if this term were of importance in the definition of *more or less recent*, then the approximation to *more or less* by a dilator would not be a good one.

In Example 2.5, we have deduced the semantic rule by inspection, taking advantage of our familiarity with the evaluation of Boolean expressions. To illustrate a more general technique, we shall consider the same linguistic variable as in Example 2.10, but use a method [39] which is an adaptation of the approach employed by Knuth in [40] to define the semantics of context-free languages.

EXAMPLE 2.6. It can readily be verified that the term-set of Example 2.5 is generated by a context-free grammar $G = (V_T, V_N, T, P)$ in which the nonterminals (syntactic categories) are denoted by T, A, B, C, D, i.e.,

$$V_N = T + A + B + C + D + E, \qquad (2.55)$$

while the set of terminals (components of terms in T) is expressed by

$$V_T = young + old + very + not + and + or + (+), \qquad (2.56)$$

and the production system, P, is given by

$$
\begin{array}{ll}
T \to A & C \to D, \\
T \to T \ or \ A, & C \to E, \\
A \to B, & D \to very \ D, \\
A \to A \ and \ B, & E \to very \ E, \qquad (2.57) \\
B \to C, & D \to young, \\
B \to not \ C, & E \to old, \\
C \to (T).
\end{array}
$$

The production system, P, can also be represented in an algebraic form as the set of equations (see Footnote 3)

$$
\begin{array}{l}
T = A + T \ or \ A, \\
A = B + A \ and \ B, \\
B = C + not \ C, \\
C = (T) + D + E, \qquad (2.58) \\
D = very \ D + young, \\
E = very \ E + old.
\end{array}
$$

The solution of this set of equations for T yields the term set T as expressed by (2.43). Similarly, the solutions for A, B, C, D and E yield sets of terms which constitute the syntactic categories denoted by A, B, C, D and E, respectively. The solution of (2.58) can be obtained iteratively, as in (2.32), by using the recursion equation

$$(T, A, B, C, D, E)^{i+1} = f((T, A, B, C, D, E)^i), \qquad i = 0, 1, 2, \ldots, \quad (2.59)$$

with

$$(T, A, B, C, D, E)^0 = (\theta, \ldots, \theta)$$

where (T, A, B, C, D, E) is a sextuple whose components are the nonterminals in (2.58); f is the mapping defined by the system of equations (2.58); θ is the empty set; and $(T, A, B, C, D, E)^i$ is the ith iterate of (T, A, B, C, D, E). The solution of (2.58), which is the fixed point of f, is given by $(T, A, B, C, D, E)^\infty$. However, it is true for all i that

$$(T, A, B, C, D, E)^i \subset (T, A, B, C, D, E), \qquad (2.60)$$

which means that every component in the sextuple on the left of (2.60) is a subset of the corresponding component on the right of (2.60). The implication of (2.60), then, is that we generate more and more terms in each of the syntactic categories T, A, B, C, D, E as we iterate (2.59) on i.

In a more conventional fashion, a term in T, say *not very young and not very old*, is generated by G through a succession of substitutions (derivations) involving the productions in P, with each derivation chain starting with T and terminating on a term generated by G. For example, the derivation chain for the term *not very young and not very old* is (see also Example 2.4),

$T \rightarrow A \rightarrow A$ *and* $B \rightarrow B$ *and* $B \rightarrow not\ C$ *and* $B \rightarrow not\ D$ *and* $B \rightarrow not\ very\ D$
and $B \rightarrow not\ very\ young$ *and* $B \rightarrow not\ very\ young\ and\ not\ C \rightarrow not\ very$
young and not $E \rightarrow not\ very\ young\ and\ not\ very\ E \rightarrow not\ very\ young$
and not very old. $\hspace{5cm}$ (2.61)

This derivation chain can be deduced from the syntax (parse) tree shown in Fig. 8, which exhibits the phrase structure of the term *not very young and not very old* in terms of the syntactic categories T, A, B, C, D, E. In effect, this procedure for generating the terms in T by the use of the grammar G constitutes the syntactic rule for the variable *Age*.

The semantic rule for *Age* is *induced* by the syntactic rule described above in the sense that the meaning of a term in T is determined, in part, by its syntax tree. Specifically, each production in (2.57) is associated with a relation between the fuzzy sets labeled by the corresponding terminal and nonterminal symbols. The resulting dual system of productions and associated equations has the appearance shown below, with the subscripts L and R serving to differentiate

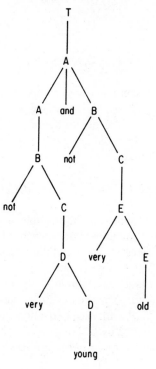

Fig. 8. Syntax tree for *not very young and not very old.*

between the symbols on the left- and right-hand sides of a production ($+ \triangleq$ union):

$$T \rightarrow A \qquad \Rightarrow T_L = A_R, \qquad\qquad (2.62)$$

$$T \rightarrow T \text{ or } A \quad \Rightarrow T_L = T_R + A_R, \qquad\qquad (2.63)$$

$$A \rightarrow B \qquad \Rightarrow A_L = B_R, \qquad\qquad (2.64)$$

$$A \rightarrow A \text{ and } B \quad \Rightarrow A_L = A_R \cap B_R, \qquad\qquad (2.65)$$

$$B \rightarrow C \qquad \Rightarrow B_L = C_R, \qquad\qquad (2.66)$$

$$B \rightarrow \text{not } C \qquad \Rightarrow B_L = \neg C_R. \qquad\qquad (2.67)$$

$$C \rightarrow (T) \qquad \Rightarrow C_L = T_R, \qquad\qquad (2.68)$$

$$C \to D \qquad \Rightarrow C_L = D_R, \qquad\qquad (2.69)$$

$$C \to E \qquad \Rightarrow C_L = E_R, \qquad\qquad (2.70)$$

$$D \to very\ D \qquad \Rightarrow D_L = (D_R)^2, \qquad\qquad (2.71)$$

$$E \to very\ E \qquad \Rightarrow E_L = (E_R)^2, \qquad\qquad (2.72)$$

$$D \to young \qquad \Rightarrow D_L = young, \qquad\qquad (2.73)$$

$$E \to old \qquad \Rightarrow E_L = old. \qquad\qquad (2.74)$$

This dual system is employed in the following manner to compute the meaning of a composite term in T.

1. The term in question, e.g., *not very young and not very old,* is parsed by the use of an appropriate parsing algorithm for G [37], yielding a syntax tree such as shown in Fig. 8. The leaves of this syntax tree are (a) primary terms whose meaning is specified *a priori;* (b) names of modifiers (i.e., hedges, connectives, negation, etc.); and (c) markers such as parentheses which serve as aids to parsing.

2. Starting from the bottom, the primary terms are assigned their meaning and, using the equations of (2.62), the meaning of nonterminals connected to the leaves is computed. Then the subtrees which have these nonterminals as their roots are deleted, leaving the nonterminals in question as the leaves of the pruned tree. This process is repeated until the meaning of the term associated with the root of the syntax tree is computed.

In applying this procedure to the syntax tree shown in Fig. 9, we first assign to *young* and *old* the meanings expressed by (2.45) and (2.46). Then, using (2.73) and (2.74), we find

$$D_7 = young \qquad\qquad (2.75)$$

and

$$E_{11} = old. \qquad\qquad (2.76)$$

Next, using (2.71) and (2.72), we obtain

$$D_6 = D_7^2 = young^2 \qquad\qquad (2.77)$$

and

$$E_{10} = E_{11}^2 = old^2 \qquad\qquad (2.78)$$

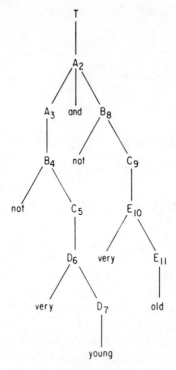

Fig. 9. Computation of the meaning
of *not very young and not very old.*

Continuing in this manner, we obtain

$$C_5 = D_6 = young^2 , \qquad (2.79)$$

$$C_9 = E_{10} = old^2 , \qquad (2.80)$$

$$B_4 = \neg C_5 = \neg(young^2), \qquad (2.81)$$

$$B_8 = \neg C_9 = \neg(old^2), \qquad (2.82)$$

$$A_3 = B_4 = \neg(young^2), \qquad (2.83)$$

$$A_2 = A_3 \cap B_8 = \neg(young^2) \cap \neg(old^2), \qquad (2.84)$$

and hence

$$not\ very\ young\ and\ not\ very\ old = \neg(young^2) \cap \neg(old^2),$$

which agrees with the expression which we had obtained previously by inspection [see (2.44)].

The basic idea behind the procedure described above is to relate the meaning of a composite term to that of its constituent primary terms by means of a system of equations which are determined by the grammar which generates the terms in T. In the case of the Boolean linguistic variable of Example 2.5, this can be done by inspection. More generally, the nature of the hedges in the linguistic variable and its grammar G might be such as to make the computation of the meaning of its values a nontrivial problem.

GRAPHICAL REPRESENTATION OF A LINGUISTIC VARIABLE

A linguistic variable may be represented in a graphical form which is similar to that of an object in the Vienna definition language [41, 42, 43]. Specifically, a variable $\mathscr{X}$ is represented as a fan (see Fig. 10) whose root is labeled $\mathscr{X}$ and whose edges are labeled with the names of the values of $\mathscr{X}$, i.e., $X_1, X_2, \ldots$. The object attached to the edge labeled X_i is the meaning of X_i. For example, in the case of the variable named *Age,* the edges might be labeled *young, old, not young,* etc., and the meaning of each such label can be represented as the graph of the membership function of the fuzzy set which is the meaning of the label in question (Fig. 11). It is important to note that, in the case of a structured linguistic variable, both the labels of the edges and the objects attached to them are generated algorithmically by the syntactic and semantic rules which are associated with the variable.

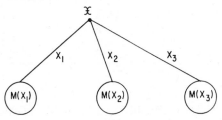

Fig. 10. Representation of a linguistic variable as a Vienna definition language object.

More generally, the graph of a linguistic variable may have the form of a tree rather than a single fan (see Fig. 12). In the case of a tree, it is understood that the name of a value of the variable is the concatenation of the names associated with an upward path from the leaf to the root. For example, in the tree of Fig. 12, the composite name associated with the path leading from node 3 to the root is *very tall · quite fat · extremely intelligent.*

This concludes our discussion of some of the basic aspects of the concept of a linguistic variable. In the following section and Part III, we shall focus our attention on some of the applications of this concept.

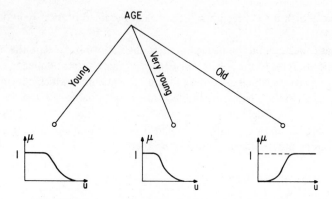

Fig. 11. Representation of the linguistic variable *Age* as a Vienna definition language object.

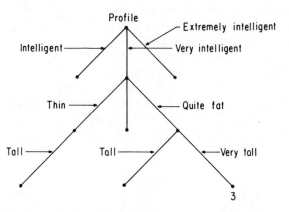

Fig. 12. Tree representation of the linguistic variable *Profile*.

3. LINGUISTIC TRUTH VARIABLES AND FUZZY LOGIC

In everyday discourse, we frequently characterize the degree of truth of a statement by expressions such as *very true, quite true, more or less true, essentially true, false, completely false,* etc. The similarity between these expressions and the values of a linguistic variable suggests that in situations in which the truth or falsity of an assertion is not well defined, it may be appropriate to treat *Truth* as a linguistic variable for which *true* and *false* are merely two of the primary terms in its term-set rather than a pair of extreme points in the universe of truth-values. Such a variable and its values will be called a *linguistic truth variable* and *linguistic truth-values,* respectively.

Treating truth as a linguistic variable leads to a *fuzzy linguistic logic,* or simply *fuzzy logic,* which is quite different from the conventional two-valued or even n-valued logic. This fuzzy logic provides a basis for what might be called *approximate reasoning,* that is, a mode of reasoning in which the truth-values and the rules of inference are fuzzy rather than precise. In many ways, approximate reasoning is akin to the reasoning used by humans in ill-defined or unquantifiable situations. Indeed, it may well be the case that much—perhaps most—of human reasoning is approximate rather than precise in nature.

In the sequel, the term *proposition* will be employed to denote statements of the form "*u* is *A*," where *u* is a name of an object and *A* is the name of a possibly fuzzy subset of a universe of discourse *U*, e.g., "John is young," "*X* is small," "apple is red," etc. If *A* is interpreted as a fuzzy predicate,[6] then the statement "*u* is *A*" may be paraphrased as "*u* has property *A*." Equivalently, "*u* is *A*" may be interpreted as an assignment equation in which a fuzzy set named *A* is assigned as a value to a linguistic variable which denotes an attribute of *u*, e.g.,

$$\text{John is young} \leftrightarrow Age \,(\text{John}) = young$$

$$X \text{ is small} \quad\leftrightarrow Magnitude \,(X) = small$$

$$\text{apple is red} \quad\leftrightarrow Color \,(\text{apple}) = red$$

A proposition such as "*u* is *A*" will be assumed to be associated with two fuzzy subsets: (i) The meaning of *A*, $M(A)$, which is a fuzzy subset of *U* named *A*; and (ii) the *truth-value* of "*u* is *A*," or simply *truth-value* of *A*, which is denoted by $v(A)$ and is defined to be a possibly fuzzy subset of a universe of truth-values, *V*. In the case of two-valued logic, $V = T + F$ ($T \triangleq$ true, $F \triangleq$ false). In what follows, unless stated to the contrary, it will be assumed that $V = [0, 1]$.

A truth-value which is a point in $[0, 1]$, e.g. $v(A) = 0.8$, will be referred to as a *numerical* truth-value. The numerical truth-values play the role of the values of the base variable for the linguistic variable *Truth*. The linguistic values of *Truth* will be referred to as *linguistic truth-values.* More specifically, we shall assume that *Truth* is the name of a Boolean linguistic variable in which the primary term is *true*, with *false* defined not as the negation of *true*,[7] but as its mirror image with respect to the point 0.5 in $[0, 1]$. Typically, the term-set of *Truth* will be assumed to be the following:

---

[6]More precisely, a fuzzy predicate may be viewed as the equivalent of the membership function of a fuzzy set. To simplify our terminology, both *A* and μ_A will be referred to as a fuzzy predicate.

[7]As will be seen later (3.11), the definition of *false* as the mirror image of *true* is a consequence of defining *false* as the truth-value of *not A* under the assumption that the truth-value of *A* is *true*.

$$T(Truth) = true + not\ true + very\ true + more\ or\ less\ true$$
$$+ very\ very\ true + essentially\ true + very\ (not\ true)$$
$$+ not\ very\ true + \cdots + false + not\ false + very\ false + \cdots$$
$$+ \cdots not\ very\ true\ and\ not\ very\ false + \cdots, \tag{3.1}$$

in which the terms are the names of the truth-values.

The meaning of the primary term *true* is assumed to be a fuzzy subset of the interval $V = [0, 1]$ characterized by a membership function of the form shown in Fig. 13. More precisely, *true* should be regarded as the name of a fuzzy variable whose restriction is the fuzzy set depicted in Fig. 13.

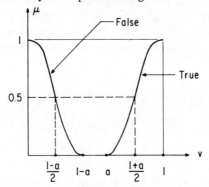

Fig. 13. Compatibility functions of linguistic truth-values *true* and *false*.

A possible approximation to the membership function of *true* is provided by the expression

$$\mu_{true}(v) = 0 \qquad \text{for } 0 \leqslant v \leqslant a$$
$$= 2\left(\frac{v-a}{1-a}\right)^2 \qquad \text{for } a \leqslant v \leqslant \frac{a+1}{2}$$
$$= 1 - 2\left(\frac{v-1}{1-a}\right)^2 \qquad \text{for } \frac{a+1}{2} \leqslant v \leqslant 1, \tag{3.2}$$

which has $v = (1+a)/2$ as its crossover point. (Note that the support of *true* is the interval $[a, 1]$.) Correspondingly, for *false*, we have (see Fig. 13)

$$\mu_{false}(v) = \mu_{true}(1-v), \qquad 0 \leqslant v \leqslant 1.$$

In some instances it is simpler to assume that *true* is a subset of the finite universe of truth-values

$$V = 0 + 0.1 + 0.2 + \cdots + 0.9 + 1 \qquad (3.3)$$

rather than of the unit interval $V = [0, 1]$. With this assumption, *true* may be defined as, say,

$$true = 0.5/0.7 + 0.7/0.8 + 0.9/0.9 + 1/1,$$

where the pair 0.5/0.7, for example, means that the compatibility of the truth-value 0.7 with *true* is 0.5.

In what follows, our main concern will be with relations of the general form

$$v(u \text{ is : linguistic value of a Boolean linguistic variable } \mathscr{X})$$
$$= \text{linguistic value of a Boolean linguistic truth variable } \mathscr{T} \qquad (3.4)$$

as in

$$v(\text{John is } tall \ and \ dark \ and \ handsome) = not \ very \ true \ and \ not \ very \ false,$$

where *tall and dark and handsome* is a linguistic value of a variable named $\mathscr{X} \triangleq$ *Appearance,* and *not very true and not very false* is that of a linguistic truth variable $\mathscr{T}$. In abbreviated form, (3.4) will usually be written as

$$v(X) = T,$$

where X is a linguistic value of $\mathscr{X}$ and T is that of $\mathscr{T}$.

Now suppose that X_1, X_2 and $X_1 * X_2$, where $*$ is a binary connective, are linguistic values of $\mathscr{X}$ with respective truth-values $v(X_1)$, $v(X_2)$ and $v(X_1 * X_2)$. A basic question that arises in this connection is whether or not it is possible to express $v(X_1 * X_2)$ as a function of $v(X_1)$ and $v(X_2)$, that is, write

$$v(X_1 * X_2) = v(X_1) *' v(X_2), \qquad (3.5)$$

where $*'$ is a binary connective associated with the linguistic truth variable $\mathscr{T}$ [8] It is this question that provides the motivation for the following discussion.

LOGICAL CONNECTIVES IN FUZZY LOGIC

To construct a basis for fuzzy logic it is necessary to extend the meaning of such logical operations as negation, disjunction, conjunction and implication to

---

[8] From an algebraic point of view, v may be regarded as a homomorphic mapping from $T(\mathscr{X})$, the term-set of $\mathscr{X}$, to $T(\mathscr{T})$, the term-set of $\mathscr{T}$, with $*'$ representing the operation in $T(\mathscr{T})$ induced by $*$.

operands which have linguistic rather than numerical truth-values. In other words, given propositions A and B, we have to be able to compute the truth-value of, say, A *and* B from the knowledge of the linguistic truth-values of A and B.

In considering this problem it is helpful to observe that, if A is a fuzzy subset of a universe of discourse U and $u \in U$, then the two statements

(a) The grade of membership of u in the fuzzy set A is $\mu_A(u)$.

(b) The truth-value of the fuzzy predicate A is $\mu_A(u)$.
                                                                        (3.6)

are equivalent. Thus, the question "What is the truth-value of A *and* B given the linguistic truth-values of A and B?" is similar to the question to which we had addressed ourselves in Part I, Sec. 3, namely, "What is the grade of membership of u in $A \cap B$ given the fuzzy grades of membership of u in A and B?"

To answer the latter question we made use of the extension principle. The same procedure will be followed to extend the meaning of *not, and, or* and *implies* to linguistic truth-values.

Specifically, if $v(A)$ is a point in $V = [0, 1]$ representing the truth-value of the proposition "u is A," (or simply A), where u is an element of a universe of discourse U, then truth-value of *not* A(or $\neg A$) is given by

$$v(not\ A) = 1 - v(A).\qquad(3.7)$$

Now suppose that $v(A)$ is not a point in $[0, 1]$ but a fuzzy subset of $[0, 1]$ expressed as

$$v(A) = \mu_1/v_1 + \cdots + \mu_n/v_n,\qquad(3.8)$$

where the v_i are points in $[0, 1]$ and the μ_i are their grades of membership in $v(A)$. Then, by applying the extension principle [Part I, Eq. (3.80)] to (3.7), we obtain the expression for $v(not\ A)$ as a fuzzy subset of $[0, 1]$, i.e.,

$$v(not\ A) = \mu_1/(1-v_1) + \cdots + \mu_n/(1-v_n).\qquad(3.9)$$

In particular, if the truth-value of A is *true*, i.e.,

$$v(A) = true,\qquad(3.10)$$

then the truth-value *false* may be defined as

$$false \overset{\Delta}{=} v(not\ A).\qquad(3.11)$$

For example, if

$$true = 0.5/7 + 0.7/0.8 + 0.9/0.9 + 1/1, \tag{3.12}$$

then the truth-value of *not A* is given by

$$false = v(not\ A) = 0.5/0.3 + 0.7/0.2 + 0.9/0.1 + 1/0.$$

COMMENT 3.1. It should be noted that if

$$true = \mu_1/v_1 + \cdots + \mu_n/v_n, \tag{3.13}$$

then by (3.33) of Part I,

$$not\ true = (1-\mu_1)/v_1 + \cdots + (1-\mu_n)/v_n. \tag{3.14}$$

By contrast, if

$$v(A) = true$$

$$= \mu_1/v_1 + \cdots + \mu_n/v_n, \tag{3.15}$$

then

$$false = v\ (not\ A)$$

$$= \mu_1/(1-v_1) + \cdots + \mu_n/(1-v_n). \tag{3.16}$$

The same applies to hedges. For example, by the definition of *very* [see (2.38)],

$$very\ true = \mu_1^2/v_1 + \cdots + \mu_n^2/v_n. \tag{3.17}$$

On the other hand, the truth-value of *very A* is expressed by

$$v(very\ A) = \mu_1/v_1^2 + \cdots + \mu_n/v_n^2. \tag{3.18}$$

Turning our attention to binary connectives, let $v(A)$ and $v(B)$ be the linguistic truth-values of propositions A and B, respectively. To simplify the notation, we shall adopt the convention of writing—as in the case where $v(A)$ and $v(B)$ are points in $[0, 1]$—

$$v(A) \wedge v(B) \quad \text{for} \quad v(A\ and\ B), \tag{3.19}$$

$$v(A) \vee v(B) \quad \text{for} \quad v(A \text{ or } B), \tag{3.20}$$

$$v(A) \Rightarrow v(B) \quad \text{for} \quad v(A \Rightarrow B), \tag{3.21}$$

and

$$\neg v(A) \qquad \text{for} \quad v(\text{not } A), \tag{3.22}$$

with the understanding that $\wedge$, $\vee$ and $\neg$ reduce to Min (conjunction), Max (disjunction) and 1- operations when $v(A)$ and $v(B)$ are points in $[0, 1]$.

Now if $v(A)$ and $v(B)$ are linguistic truth-values expressed as

$$v(A) = \alpha_1/v_1 + \cdots + \alpha_n/v_n, \tag{3.23}$$

$$v(B) = \beta_1/w_1 + \cdots + \beta_m/w_m, \tag{3.24}$$

where the v_i and w_j are points in $[0, 1]$ and the α_i and β_j are their respective grades of membership in A and B, then by applying the extension principle to $v(A \text{ and } B)$, we obtain

$$v(A \text{ and } B) = v(A) \wedge v(B)$$

$$= (\alpha_1/v_1 + \cdots + \alpha_n/v_n) \wedge (\beta_1/w_1 + \cdots + \beta_m/w_m)$$

$$= \sum_{i,j} (\alpha_i \wedge \beta_j)/(v_i \wedge w_j). \tag{3.25}$$

Thus, the truth-value of A *and* B is a fuzzy subset of $[0, 1]$ whose support comprises the points $v_i \wedge w_j$, $i = 1, \ldots, n$, $j = 1, \ldots, m$, with respective grades of membership $(\alpha_i \wedge \beta_j)$. Note that (3.25) is equivalent to the expression (3.107) of Part I for the membership function of the intersection of fuzzy sets having fuzzy membership functions.

EXAMPLE 3.2. Suppose that

$$v(A) = true$$

$$= 0.5/0.7 + 0.7/0.8 + 0.9/0.9 + 1/1 \tag{3.26}$$

and

$$v(B) = \textit{not true}$$

$$= 1/0 + 1/0.1 + 1/0.2 + 1/0.3 + 1/0.4 + 1/0.5 + 1/0.6$$

$$+ 0.5/0.7 + 0.3/0.8 + 0.1/0.9. \tag{3.27}$$

Then the use of (3.25) leads to

$$v(A \textit{ and } B) = \textit{true} \wedge \textit{not true}$$

$$= 1/(0 + 0.1 + 0.2 + 0.3 + 0.4 + 0.5 + 0.6) + 0.5/0.7$$

$$+ 0.3/0.8 + 0.1/0.9$$

$$= \textit{not true}. \tag{3.28}$$

In a similar fashion, for the truth-value of $A \textit{ or } B$, we obtain

$$v(A \textit{ or } B) = v(A) \vee v(B)$$

$$= (\alpha_1/v_1 + \cdots + \alpha_n/v_n) \vee (\beta_1/w_1 + \cdots + \beta_m/w_m)$$

$$= \sum_{i,j} (\alpha_i \wedge \beta_j)/(v_i \vee w_j). \tag{3.29}$$

The truth-value of $A \Rightarrow B$ depends on the manner in which the connective $\Rightarrow$ is defined for numerical truth-values. Thus, if we define [see Part III, Eq. (2.24)]

$$v(A \Rightarrow B) = \neg v(A) \vee v(A) \wedge v(B) \tag{3.30}$$

for the case where $v(A)$ and $v(B)$ are points in $[0, 1]$, then the application of the extension principle yields (see Part I, Comment 3.5)

$$v(A \Rightarrow B) = [(\alpha_1/v_1 + \cdots + \alpha_n/v_n) \Rightarrow (\beta_1/w_1 + \cdots + \beta_m/w_m)]$$

$$= \sum_{i,j} (\alpha_i \wedge \beta_j)/(1-v_i) \vee (v_i \wedge w_j) \tag{3.31}$$

for the case where $v(A)$ and $v(B)$ are fuzzy subsets of $[0, 1]$.

COMMENT 3.3. It is important to have a clear understanding of the difference between *and* in, say, *true and not true*, and $\wedge$ in *true $\wedge$ not true*. In the former, our concern is with the meaning of the term *true and not true*, and *and* is defined by the relation

$$M(\textit{true and not true}) = M(\textit{true}) \cap M(\textit{not true}), \qquad (3.32)$$

where M is the function mapping a term into its meaning (see Definition 2.1). By contrast, in the case of *true $\wedge$ not true* we are concerned with the truth-value of *true $\wedge$ not true*, which is derived from the equivalence [see (3.19)]

$$v(A \textit{ and } B) = v(A) \wedge v(B). \qquad (3.33)$$

Thus, in (3.32) $\cap$ is the operation of intersection of fuzzy sets, whereas in (3.33), $\wedge$ is that of conjunction. To illustrate the difference by a simple example, let $V = 0 + 0.1 + \cdots + 1$, and let P and Q be fuzzy subsets of V defined by

$$P = 0.5/0.3 + 0.8/0.7 + 0.6/1, \qquad (3.34)$$

$$Q = 0.1/0.3 + 0.6/0.7 + 1/1. \qquad (3.35)$$

Then

$$P \cap Q = 0.1/0.3 + 0.6/0.7 + 0.6/1, \qquad (3.36)$$

whereas

$$P \wedge Q = 0.5/0.3 + 0.8/0.7 + 0.6/1. \qquad (3.37)$$

Note that the same issue arises in the case of *not* and $\neg$, as pointed out in Comment 3.1.

COMMENT 3.4. It should be noted that in applying the extension principle [Part I, Eq. (3.96)] to the computation of $v(A \textit{ and } B)$, $v(A \textit{ or } B)$ and $v(A \Rightarrow B)$, we are tacitly assuming that $v(A)$ and $v(B)$ are noninteractive fuzzy variables in the sense of Part I, Comment 3.5. If $v(A)$ and $v(B)$ are interactive, then it is necessary to apply the extension principle as expressed by (3.97) of Part I rather than (3.96). It is of interest to observe that the issue of possible interaction between $v(A)$ and $v(B)$ arises even when $v(A)$ and $v(B)$ are points in [0, 1] rather than fuzzy variables.

COMMENT 3.5. By employing the extension principle to define the operations $\wedge, \vee, \neg$ and $\Rightarrow$ on linguistic truth-values, we are in effect treating fuzzy logic as an extension of multivalued logic. In the same sense, the classical three-valued logic may be viewed as an extension of two-valued logic [see Eqs. (3.64) *et seq.*].

The expressions for $v(not\ A)$, $v(A\ and\ B)$, $v(A\ or\ B)$ and $v(A \Rightarrow B)$ given above become more transparent if we first decompose $v(A)$ and $v(B)$ into level-sets and then apply the level-set form of the extension principle [see (3.86)] to the operations $\neg$, $\wedge, \vee$ and $\Rightarrow$. In this way, we are led to a simple graphical rule for computing the truth-values in question (see Fig. 14). Specifically, let the intervals $[a_1, a_2]$ and $[b_1, b_2]$ be the α-level sets for $v(A)$ and $v(B)$. Then, by using the extensions of the operations $\neg$, $\wedge$ and $\vee$ to intervals, namely [see Part I, Eq. (3.100)]

$$\neg [a_1, a_2] = [1 - a_2, 1 - a_1], \tag{3.38}$$

$$[a_1, a_2] \wedge [b_1, b_2] = [a_1 \wedge b_1, a_2 \wedge b_2], \tag{3.39}$$

$$[a_1, a_2] \vee [b_1, b_2] = [a_1 \vee b_1, a_2 \vee b_2], \tag{3.40}$$

we can find by inspection the α-level-sets for $v(not\ A)$, $v(A\ and\ B)$ and $v(A\ or\ B)$. Having found these level-sets, $v(not\ A)$, $v(A\ and\ B)$ and $v(A\ or\ B)$ can readily be determined by varying α from 0 to 1.

As a simple illustration, consider the determination of the conjunction of linguistic truth-values $v(A) \overset{\Delta}{=} true$ and $v(B) \overset{\Delta}{=} false,$ with the membership functions of *true* and *false* having the form shown in Fig. 15.

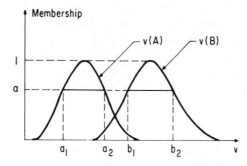

Fig. 14. Level-sets of truth-values of A and B.

We observe that, for all values of α,

$$[a_1, a_2] \wedge [b_1, b_2] = [b_1, b_2], \tag{3.41}$$

which implies that [see Part I, Eq. (3.118)]

$$[b_1, b_2] \leqslant [a_1, a_2]. \tag{3.42}$$

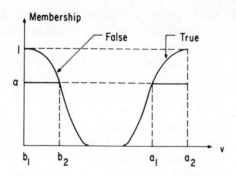

Fig. 15. Computation of the truth-value of the conjunction of *true* and *false.*

Consequently, merely on the basis of the form of the membership functions of *true* and *false,* we can conclude that

$$true \wedge false = false, \qquad (3.43)$$

which is consistent with (3.25).

TRUTH TABLES AND LINGUISTIC APPROXIMATION

In two-valued, three-valued and, more generally, n-valued logics the binary connectives $\wedge, \vee$ and $\Rightarrow$ are usually defined by a tabulation of the truth-values of *A and B, A or B* and $A \Rightarrow B$ in terms of the truth-values of A and B.

Since in a fuzzy logic the number of truth-values is, in general, infinite, $\wedge$, $\vee$ and $\Rightarrow$ cannot be defined by tabulation. However, it may be desirable to tabulate say, $\wedge$, for a finite set of truth-values of interest, e.g., *true, not true, false, very true, very (not true), more or less true,* etc. In such a table, for the entry in the ith row (say *not true*) and in the jth column (say *more or less true*), the (i,j)th entry would be

$$(i,j)\text{th entry} = i\text{th row label } (\overset{\Delta}{=} not\ true) \wedge j\text{th column label } (\overset{\Delta}{=} more\ or\ less\ true).$$
$$(3.44)$$

Given the definition of the primary term *true* and the definitions of the modifiers *not* and *more or less,* we can compute the right-hand side of (3.44), that is,

$$not\ true \wedge more\ or\ less\ true \qquad (3.45)$$

by using (3.25). However, the problem is that in most instances the result of the computation would be a fuzzy subset of the universe of truth-values which may not correspond to any of the truth-values in the term-set of *Truth*. Thus, if we wish to have a truth table in which the entries are linguistic, we must be content with an approximation to the exact truth-value of (*i*th row label $\wedge$ *j*th column label). Such an approximation will be referred to as a *linguistic* approximation. (See Part I, Fig. 5.)

As an illustration, suppose that the universe of truth-values is expressed as

$$V = 0 + 0.1 + 0.2 + \cdots + 1, \tag{3.46}$$

and that

$$true = 0.7/0.8 + 1/0.9 + 1/1, \tag{3.47}$$

$$more\ or\ less\ true = 0.5/0.6 + 0.7/0.7 + 1/0.8 + 1/0.9 + 1/1 \tag{3.48}$$

and

$$almost\ true = 0.6/0.8 + 1/0.9 + 0.6/1. \tag{3.49}$$

In the truth-table for $\vee$, assume that the *i*th row label is *more or less true* and the *j*th column label is *almost true*. Then, for the (i,j)th entry in the table, we have

$$
\begin{aligned}
more\ or\ less\ true \vee almost\ true &= (0.5/0.6 + 0.7/0.7 + 1/0.8 + 1/0.9 \\
&\quad + 1/1) \vee (0.6/0.8 + 1/0.9 + 0.6/1) \\
&= 0.6/0.8 + 1/0.9 + 1/1.
\end{aligned} \tag{3.50}
$$

Now, we observe that the right-hand side of (3.50) is approximately equal to *true* as defined by (3.47). Consequently, in the truth table for $\vee$, a linguistic approximation to the (i,j)th entry would be *true*.

THE TRUTH-VALUES UNKNOWN AND UNDEFINED

Among the truth-values that can be associated with the linguistic variable *Truth,* there are two that warrant special attention, namely, the empty set θ and the unit interval $[0, 1]$ —which correspond to the least and greatest elements (under set inclusion) of the lattice of fuzzy subsets of $[0, 1]$. The importance of these particular truth-values stems from their interpretability as the truth-values *undefined* and *unknown*,[9] respectively. For convenience we shall denote these truth-values by θ and ?, with the understanding that θ and ? are defined by

$$\theta \triangleq \int_0^1 0/v \tag{3.51}$$

and

$$? \triangleq V = \text{universe of truth-values}$$

$$= [0, 1]$$

$$= \int_0^1 1/w. \tag{3.52}$$

Interpreted as grades of membership, *undefined* and *unknown* enter also in the representation of fuzzy sets of type 1. For such sets, the grade of membership of a point u in U may have one of three possible forms: (i) a number in the interval $[0, 1]$; (ii) θ (*undefined*); and (iii) ? (*unknown*). As a simple example, let

$$U = a + b + c + d + e \tag{3.53}$$

and consider a fuzzy subset of U represented as

$$A = 0.1a + 0.9b + ?c + \theta d. \tag{3.54}$$

In this case, the grade of membership of c in A is *unknown* and that of d is *undefined*. More generally, we may have

$$A = 0.1\,a + 0.9\,b + 0.8\,?c + \theta d, \tag{3.55}$$

meaning that the grade of membership of c in A is *partially unknown*, with $0.8?c$ interpreted as

$$0.8?c \triangleq \left(\int_0^1 0.8/v \right)/c. \tag{3.56}$$

---

[9]The concept of *unknown* is related to that of *don't care* in the context of switching circuits [44]. Another related concept is that of quasi-truth-functionality [46].

It is important to have a clear understanding of the difference between 0 and θ. When we say that the grade of membership of a point u in A is θ, what we mean is that the membership function $\mu_A : U \to [0, 1]$ is undefined at u. For example, suppose that U is the set of real numbers and μ_A is a function defined on integers, with $\mu_A(u) = 1$ if u is an even integer and $\mu_A(u) = 0$ if u is an odd integer. Then the grade of membership of $u = 1.5$ in A is θ rather than 0. On the other hand, if μ_A were defined on real numbers and $\mu_A(u) = 1$ iff u is even, then the grade of membership of 1.5 in A would be 0.

Since we know how to compute the truth-values of A *and* B, A *or* B and *not* B given the linguistic truth-values of A and B, it is a simple matter to compute $v(A \text{ and } B)$, $v(A \text{ or } B)$ and $v(\text{not } B)$ when $v(B) = ?$. Thus, suppose that

$$v(A) = \int_0^1 \mu(v)/v \qquad (3.57)$$

and

$$v(B) = ? = \int_0^1 1/w. \qquad (3.58)$$

By applying the extension principle, as in (3.25), we obtain

$$v(A) \wedge ? = \int_0^1 \mu(v)/v \wedge \int_0^1 1/w$$

$$= \int_0^1 \int_0^1 \mu(v)/(v \wedge w), \qquad (3.59)$$

where

$$\int_0^1 \int_0^1 \overset{\Delta}{=} \int_{[0, 1] \times [0, 1]} , \qquad (3.60)$$

and which upon simplification reduces to

$$v(A) \wedge ? = \int_0^1 [\vee_{[w, 1]} \mu(v)]/w. \qquad (3.61)$$

In other words, the truth-value of A *and* B, where $v(B) = unknown$, is a fuzzy subset of $[0, 1]$ in which the grade of membership of a point w is given by the supremum of $\mu(v)$ (membership function of A) over the interval $[w, 1]$.

In a similar fashion, the truth-value of A *or* B is found to be expressed by

$$v(A \text{ or } B) = \int_0^1 \int_0^1 \mu(v)/(v \vee w)$$

$$= \int_0^1 [\vee_{[0, w]} \mu(v)] / w. \qquad (3.62)$$

It should be noted that both (3.61) and (3.62) can readily be obtained by the graphical procedure described earlier [see (3.38) *et seq.*]. An example illustrating its application is shown in Fig. 16.

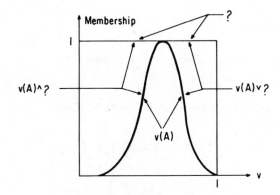

Fig. 16. Conjunction and disjunction of the truth-value of A with the truth-value *unknown* ($\triangleq$?).

Turning to the case where $v(B) = \theta$, we find

$$v(A) \wedge \theta = \int_0^1 \int_0^1 0/(v \wedge w)$$

$$= \int_0^1 0/w$$

$$= \theta \qquad (3.63)$$

and likewise for $v(A) \vee \theta$.

It is instructive to examine what happens to the above relations when we apply them to the special case of two-valued logic, that is, to the case where the universe V is of the form

$$V = 0 + 1, \qquad (3.64)$$

or, expressed more conventionally,

$$V = T + F, \qquad (3.65)$$

where T stands for *true* and F stands for *false*. Since ? is V, we can identify the truth-value *unknown* with *true or false,* that is,

$$? = T + F \qquad (3.66)$$

The resulting logic has four truth-values: θ, T, F and $T + F$ $(\stackrel{\triangle}{=} ?)$, and is an extension of two-valued logic in the sense of Comment 3.5.

Since the universe of truth-values has only two elements, it is expedient to derive the truth tables for $\vee$, $\wedge$ and $\Rightarrow$ in this four-valued logic directly rather than through specialization of the general formulae (3.25), (3.29) and (3.31). Thus, by applying the extension principle to $\wedge$, we find at once

$$T \wedge \theta = \theta, \qquad (3.67)$$

$$T \wedge (T + F) = T \wedge T + T \wedge F$$
$$= T + F, \qquad (3.68)$$

$$F \wedge (T + F) = F \wedge T + F \wedge F$$
$$= F + F$$
$$= F, \qquad (3.69)$$

$$(T + F) \wedge (T + F) = T \wedge T + T \wedge F + F \wedge T + F \wedge F$$
$$= T + F + F + F$$
$$= T + F, \qquad (3.70)$$

and consequently the extended truth-table for $\wedge$ has the form shown in Table 1.

TABLE 1

∧	θ	T	F	$T+F$
θ	θ	θ	θ	θ
T	θ	T	F	$T+F$
F	θ	F	F	F
$T+F$	θ	$T+F$	F	$T+F$

Upon suppression of the entry θ, this reads as shown in Table 2.

TABLE 2

∧	T	F	$T+F$
T	T	F	$T+F$
F	F	F	F
$T+F$	$T+F$	F	$T+F$

Similarly, for the operation ∨ we obtain Table 3.

TABLE 3

∨	T	F	$T+F$
T	T	T	T
F	T	F	$T+F$
$T+F$	T	$T+F$	$T+F$

These tables agree—as they should—with the corresponding truth tables for ∧ and ∨ in conventional three-valued logic [46].

The approach employed above provides some insight into the definition of ⇒ in two-valued logic—a somewhat controversial issue which motivated the development of modal logic [45, 47]. Specifically, instead of defining ⇒ in the conventional fashion, we may define ⇒ as a connective in three-valued logic by the partial truth table in Table 4,

TABLE 4

⇒	T	F	$T+F$
T	T	F	
F			T

which expresses the intuitively reasonable idea that if $A \Rightarrow B$ is *true* and A is *false*, then the truth-value of B is *unknown*. Now we can raise the question: How should the blank entries in the above table be filled in order to yield the entry T in the $(2,3)$ position in Table 4 upon the application of the extension principle? Thus, denoting the unknown entries in positions $(2, 1)$ and $(2, 2)$ by x and y, respectively, we must have

$$F \Rightarrow (T + F) = (F \Rightarrow T) + (F \Rightarrow F)$$

$$= x + y$$

$$= T, \tag{3.71}$$

which necessitates that

$$x = y = T. \tag{3.72}$$

In this way, we are led to the conventional definition of $\Rightarrow$ in two-valued logic, which is expressed by the truth table

$\Rightarrow$	T	F
T	T	F
F	T	T

As the above example demonstrates, the notion of the *unknown* truth-value in conjunction with the extension principle helps to clarify some of the concepts and relations in the conventional two-valued and three-valued logics. These logics may be viewed, of course, as degenerate cases of a fuzzy logic in which the truth-value *unknown* is the entire unit interval rather than the set $0 + 1$.

COMPOSITE TRUTH VARIABLES AND TRUTH-VALUE DISTRIBUTIONS

In the foregoing discussion, we have limited our attention to linguistic truth variables which are unary variables in the sense of Part I, Definition 2.1. In the following, we shall define the concept of a *composite* truth variable and dwell briefly on some of its implications.

Thus, let

$$\mathscr{T} \overset{\Delta}{=} (\mathscr{T}_1, \ldots, \mathscr{T}_n) \tag{3.73}$$

denote an *n-ary* composite linguistic truth variable in which each $\mathscr{T}_i$, $i = 1, \ldots, n$, is a unary linguistic truth variable associated with a term-set T_i, a universe of discourse V_i, and a base variable v_i (see Definition 2.1). For simplicity, we shall sometimes employ the symbol $\mathcal{T}_i$ in the dual role of (a) the name of the ith variable in (3.73), and (b) a generic name for the truth-values of $\mathcal{T}_i$. Furthermore, we shall assume that $T_1 = T_2 = \cdots = T_n$ and $V_1 = V_2 = \cdots = V_n = [0, 1]$.

Viewed as a composite variable whose component variables $\mathcal{T}_1, \ldots, \mathcal{T}_n$ take values in their respective universes $T_1, \ldots, T_n$, $\mathcal{T}$ is an n-ary nonfuzzy variable [see Part I, Eq. (2.3) et seq.]. Thus, the restriction $R(\mathcal{T})$ imposed by $\mathcal{T}$ is an n-ary nonfuzzy relation in $T_1 \times \cdots \times T_n$ which may be represented as an unordered list of ordered n-tuples of the form

$$
\begin{aligned}
R(\mathcal{T}) = &(\textit{true, very true, false, } \ldots, \textit{ quite true}) \\
&+ (\textit{quite true, true, very true, } \ldots, \textit{ very true}) \\
&+ (\textit{true, true, more or less true, } \ldots, \textit{ true}) \\
&+ \cdots
\end{aligned}
\tag{3.74}
$$

The n-tuples in $R(\mathcal{T})$ will be referred to as *truth-value assignment lists* since each such n-tuple may be interpreted as an assignment of truth-values to a list of propositions $A_1, \ldots, A_n$, with

$$
A \overset{\Delta}{=} (A_1, \ldots, A_n)
\tag{3.75}
$$

representing a composite proposition. For example, if

$$
A \overset{\Delta}{=} (\text{Scott is } \textit{tall}, \ \text{Pat is } \textit{dark-haired}, \ \text{Tina is } \textit{very pretty}),
$$

then a triple in $R(\mathcal{T})$ of the form (*very true, true, very true*) would represent the following truth-value assignments:

$$
v(\text{Scott is } \textit{tall}) = \textit{very true},
\tag{3.76}
$$

$$
v(\text{Pat is } \textit{dark-haired}) = \textit{true},
\tag{3.77}
$$

$$
v(\text{Tina is } \textit{very pretty}) = \textit{very true}.
\tag{3.78}
$$

Based on this interpretation of the n-tuples in $R(\mathcal{T})$, we shall frequently refer to $R(\mathcal{T})$ as a *truth-value distribution*. Correspondingly, the restriction $R(\mathcal{T}_{i_1}, \ldots, \mathcal{T}_{i_k})$ which is imposed by the k-ary variable $(\mathcal{T}_{i_1}, \ldots, \mathcal{T}_{i_k})$, where $q = (i_1, \ldots, i_k)$ is a subsequence of the index sequence $(1, \ldots, n)$, will be referred to as a *marginal truth-value distribution induced by* $R(\mathcal{T}_1, \ldots, \mathcal{T}_n)$ [see Part I, Eq. (2.8)]. Then, using the notation employed in Part I, Sec. 2 (see also Note 1.1 in this Part), the relation between $R(\mathcal{T}_{i_1}, \ldots, \mathcal{T}_{i_k})$ and $R(\mathcal{T}_1, \ldots, \mathcal{T}_n)$ may be expressed compactly as

$$R(\mathcal{T}_{(q)}) = P_q\, R(\mathcal{T}), \qquad\qquad (3.79)$$

where P_q denotes the operation of projection on the Cartesian product $T_{i_1} \times$

$\cdots \times T_{i_k}$

EXAMPLE 3.1. Suppose that $R(\mathcal{T})$ is expressed by

$$R(\mathcal{T}) \overset{\Delta}{=} R(\mathcal{T}_1, \mathcal{T}_2 \cdot \mathcal{T}_3)$$
$$= (true,\ quite\ true,\ very\ true)$$
$$+ (very\ true,\ true,\ very\ very\ true)$$
$$+ (true,\ false,\ quite\ true)$$
$$+ (false,\ false,\ very\ true). \qquad (3.80)$$

To obtain $R(\mathcal{T}_1, \mathcal{T}_2)$ we delete the $\mathcal{T}_3$ component in each triple, yielding

$$R(\mathcal{T}_1, \mathcal{T}_2) = (true,\ quite\ true)$$
$$+ (very\ true,\ true)$$
$$+ (true,\ false)$$
$$+ (false,\ false). \qquad (3.81)$$

Similarly, by deleting the $\mathcal{T}_2$ components in $R(\mathcal{T}_1 \cdot \mathcal{T}_2)$, we obtain

$$R(\mathcal{T}_1) = true + very\ true + false. \qquad (3.82)$$

If we view $\mathcal{T}$ as an n-ary nonfuzzy variable whose values are linguistic truth-values, the definition of noninteraction (Part I, Definition 2.2) assumes the following form in the case of linguistic truth variables.

DEFINITION 3.1. The components of an n-ary linguistic truth variable $\mathcal{T} = (\mathcal{T}_1, \ldots, \mathcal{T}_n)$ are λ-*noninteractive* (λ standing for linguistic) iff the truth-value distribution $R(\mathcal{T}_1, \ldots, \mathcal{T}_n)$ is separable in the sense that

$$R(\mathcal{T}_1, \ldots, \mathcal{T}_n) = R(\mathcal{T}_1) \times \cdots \times R(\mathcal{T}_n). \qquad (3.83)$$

The implication of this definition is that, if $\mathcal{T}_1, \ldots, \mathcal{T}_n$ are λ-noninteractive, then the assignment of specific linguistic truth-values to $\mathcal{T}_{i_1}, \ldots, \mathcal{T}_{i_k}$ does not affect the truth-values that can be assigned to the complementary components in $(\mathcal{T}_1, \ldots, \mathcal{T}_n)$, $\mathcal{T}_{j_i}, \ldots, \mathcal{T}_{j_m}$.

Before proceeding to illustrate the concept of λ-noninteraction by examples, we shall define another type of noninteraction which will be referred to as β-*noninteraction* (β standing for base variable).

DEFINITION 3.2. The components of an *n*-ary linguistic truth variable $\mathcal{T}$ $= (\mathcal{T}_1, \ldots . \mathcal{T}_n)$ are *β-noninteractive* iff their respective base variables $v_1, \ldots , v_n$ are noninteractive in the sense of Part I, Definition 2.2; that is, the v_i are not jointly constrained.

To illustrate the concepts of noninteraction defined above we shall consider a few simple examples.

EXAMPLE 3.2. For the truth-value distribution of Example 3.1, we have

$$R(\mathcal{T}_1) = true + very\ true + false,$$
$$R(\mathcal{T}_2) = quite\ true + true + false,$$
$$R(\mathcal{T}_3) = very\ true + very\ very\ true + quite\ true, \tag{3.84}$$

and thus

$$R(\mathcal{T}_1) \times R(\mathcal{T}_2) \times R(\mathcal{T}_3) = (true,\ quite\ true,\ very\ true)$$
$$+ (very\ true,\ quite\ true,\ very\ true)$$
$$\ldots$$
$$+ (false,\ false,\ quite\ true)$$
$$\neq R(\mathcal{T}_1\ \mathcal{T}_2, \mathcal{T}_3), \tag{3.85}$$

which implies that $R(\mathcal{T}_1, \mathcal{T}_2\ \mathcal{T}_3)$ is not separable and hence $\mathcal{T}_1 . \mathcal{T}_2, \mathcal{T}_3$ are λ-interactive.

EXAMPLE 3.3. Consider a composite proposition of the form $(A,\ not\ A)$ and assume for simplicity that $T(\mathcal{T}) = true + false$. In view of (3.11), if the truth-value of A is *true* then that of *not A* is *false*, and vice versa. Consequently, the truth-value distribution for the propositions in question must be of the form

$$R(\mathcal{T}_1, \mathcal{T}_2) = (true,\ false) + (false,\ true), \tag{3.86}$$

which induces

$$R(\mathcal{T}_1) = R(\mathcal{T}_2) = true + false. \tag{3.87}$$

Now

$$R(\mathcal{T}_1) \times R(\mathcal{T}_2) = (true + false) \times (true + false)$$
$$= (true,\ true) + (true,\ false)$$
$$+ (false,\ true) + (false,\ false), \tag{3.88}$$

and since

$$R(\mathcal{T}_1, \mathcal{T}_2) \neq R(\mathcal{T}_1) \times R(\mathcal{T}_2)$$

it follows that $\mathcal{T}_1$ and $\mathcal{T}_2$ are λ-interactive.

EXAMPLE 3.4. The above example can also be used as an illustration of β-interaction. Specifically, regardless of the truth-values assigned to A and *not* A, it follows from the definition of *not* [see Part I, Eq. (3.33)] that the base variables v_1 and v_2 are constrained by the equation

$$v_1 + v_2 = 1. \tag{3.89}$$

In other words, in the case of a composite proposition of the form $(A, not\ A)$, the sum of the numerical truth-values of A and *not* A must be unity.

REMARK 3.1. It should be noted that, in Example 3.4, β-interaction is a consequence of A_2 being related to A_1 by negation. In general, however, $\mathcal{T}_1$, $\ldots \mathcal{T}_n$ may be λ-interactive without being β-interactive.

A useful application of the concept of interaction relates to the truth-value *unknown* [see (3.52)]. Specifically, assuming for simplicity that $V = T + F$, suppose that

$$A_1 \triangleq \text{Pat lives in Berkeley,} \tag{3.90}$$

$$A_2 \triangleq \text{Pat lives in San Francisco,} \tag{3.91}$$

with the understanding that one and only one of these statements is true. This implies that, although the truth-values of A_1 and A_2 are *unknown* $(\triangleq ? = T + F)$, that is,

$$v(A_1) = T + F,$$
$$v(A_2) = T + F, \tag{3.92}$$

they are constrained by the relations

$$v(A_1) \vee v(A_2) = T, \tag{3.93}$$

$$v(A_1) \wedge v(A_2) = F. \tag{3.94}$$

Equivalently, the truth-value distribution associated with (3.90) and (3.91) may be regarded as the solution of the equations

$$v(A_1) \vee v(A_2) = T, \tag{3.95}$$

$$v(A_1) \wedge v(A_2) = F, \tag{3.96}$$

which is

$$R(\mathscr{T}_1, \mathscr{T}_2) = (T, F) + (F, T). \tag{3.97}$$

Note that (3.97) implies

$$v(A_1) = R(\mathscr{T}_1) = T + F \tag{3.98}$$

and

$$v(A_2) = R(\mathscr{T}_2) = T + F, \tag{3.99}$$

in agreement with (3.92). Note also that $\mathscr{T}_1$ and $\mathscr{T}_2$ are β-interactive in the sense of Definition 3.2, with $V = T + F$.

Now if A_1 and A_2 were changed to

$$A_1 \overset{\triangle}{=} \text{Pat lived in Berkeley,} \tag{3.100}$$

$$A_2 \overset{\triangle}{=} \text{Pat lived in San Francisco,} \tag{3.101}$$

with the possibility that both A_1 and A_2 could be true, then we would still have

$$v(A_1) = ? = T + F, \tag{3.102}$$

$$v(A_2) = ? = T + F, \tag{3.103}$$

but the constraint equation would become

$$v(A_1) \vee v(A_2) = T. \tag{3.104}$$

In this case, the truth-value distribution is the solution of (3.104), which is given by

$$R(\bar{\mathscr{T}}_1, \mathscr{T}_2) = (\textit{true, true}) + (\textit{true, false}) + (\textit{false, true}). \tag{3.105}$$

An important observation that should be made in connection with the above examples is that in some cases a truth-value distribution may be given in an implicit from, e.g., as a solution of a set of truth-value equations, rather than as an explicit list of ordered n-tuples of truth-values. In general, this will be the case where linguistic truth-values are assigned not to each A_i in $A = (A_1, \ldots, A_n)$, but to Boolean expressions involving two or more of the components of A.

Another point that should be noted is that truth-value distributions may be nested. As a simple illustration, in the case of a unary proposition we may have a nested sequence of assertions of the form

" " "Vera is *very very intelligent*" is very *true*" is *true*."   (3.106)

Restrictions induced by assertions of this type may be computed as follows.

Let the base variable in (3.106) be IQ, and let $R_0(\text{IQ})$ denote the restriction on the IQ of Vera. Then the proposition "Vera is *very very intelligent*" implies that

$$R_0(\text{IQ}) = very\ very\ intelligent.$$ (3.107)

Now, the proposition " "Vera is *very very intelligent*" is *very true*" implies that the grade of membership of Vera in the fuzzy set R_0 (IQ) is *very true* [see (3.6)]. Let $\mu_{very\ true}$ denote the membership function of *very true* [see (3.17)], and let μ_{R_0} denote that of R_0 (IQ). Regarding μ_{R_0} as a relation from the range of IQ to [0, 1], let $\mu_{R_0}^{-1}$ denote the inverse relation from [0, 1] to the range of IQ. This relation, then, induces a fuzzy set R_1 (IQ) expressed by

$$R_1\ (\text{IQ}) = \mu_{R_0}^{-1}\ (very\ true),$$ (3.108)

which can be computed by the use of the extension principle in the form given in Part I, Eq. (3.80). The fuzzy set R_1 (IQ) represents the restriction on IQ induced by the assertion " "Vera is *very very intelligent*" is *very true*."

Continuing the same argument, the restriction on IQ induced by the assertion " " "Vera is *very very intelligent*" is *very true*" is *true*" may be expressed as

$$R_2\ (\text{IQ}) = \mu_{R_1}^{-1}\ (true),$$ (3.109)

where $\mu_{R_1}^{-1}$ denotes the relation inverse to μ_{R_1}, which is the membership function of R_1 (IQ) given by (3.108). In this way, we can compute the restriction induced by a nested sequence of assertions such as that exemplified by (3.106).

The basic idea behind the technique sketched above is that an assertion of the form " "u is A" is T," where A is a fuzzy predicate and T is a linguistic truth-value, modifies the restriction associated with A in accordance with the expression

$$A' = \mu_A^{-1}\ (T),$$

where μ_A^{-1} is the inverse of the membership function of A, and A' is the restriction induced by the assertion " "u is A" is T."

REFERENCES

1. Sir William Thomson, *Popular Lectures and Addresses,* McMillan, London, 1891.
2. E. Feigenbaum, *Computers and Thought,* McGraw-Hill, New York, 1963.
3. M. Minsky and S. Papert, *Perceptrons: An Introduction to Computational Geometry,* M.I.T. Press, Cambridge, Mass., 1969.
4. M. Arbib, *The Metaphorical Brain,* Wiley-Interscience, New York, 1972.
5. A. Newell and H. Simon, *Human Problem Solving,* Prentice-Hall, Englewood Cliffs, N.J., 1972.
6. L. A. Zadeh, Fuzzy languages and their relation to human and machine intelligence, in *Proc. Int. Conf. on Man and Computer,* Bordeaux, France, S. Karger, Basel, 1972, pp. 130-165.
7. L. A. Zadeh, Outline of a new approach to the analysis of complex systems and decision processes, *IEEE Trans. Syst., Man and Cybern,* **SMC-3,** 28-44 (January 1973).
8. R. E. Bellman, R. E. Kalaba and L. A. Zadeh, Abstraction and pattern classification, *J. Math. Anal. Appl.* **13,** 1-7 (1966).
9. M. Black, Reasoning with loose concepts, *Dialogue* **2,** 1-12 (1963).
10. L. Wittgenstein, Logical form, *Proc. Aristotelian Soc.* **9,** 162-171 (1929).
11. M. Scriven, The logic of criteria, *J. Philos.* **56,** 857-868 (1959).
12. H. Khatchadourian, Vagueness, meaning and absurdity, *Am. Phil. Quart.* **2,** 119-129 (1965).
13. R. R. Verma, Vagueness and the principle of excluded middle, *Mind* **79,** 66-77 (1970).
14. J. A. Goguen, The Logic of Inexact Concepts, *Synthese* **19,** 325-373 (1969).
15. E. Adams, The logic of "Almost All", *J. Philos. Logic* **3,** 3–17 (1974).
16. K. Fine, Vagueness, truth and logic, Department of Philosophy, University of Edinburgh, 1973.
17. B.S. van Frassen, Presuppositions, supervaluations and free logic, in *The Logical Way of Doing Things* (K. Lambert, Ed.), Yale U. P., New Haven, Conn., 1969.
18. G. Lakoff, Linguistics and natural logic, in *Semantics of Natural Languages,* (D. David-son and G. Harman, Eds.), D. Reidel, Dordrecht, The Netherlands, 1971.
19. L. A. Zadeh, Shadows of fuzzy sets, *Probl. Transfer Inf.* **2,** 37-44 (1966).
20. A. Kaufmann, *Theory of Fuzzy Sets,* Masson, Paris, 1972.
21. J. Goguen, *L*-fuzzy sets, *J. Math. Anal. Appl.* **18,** 145-174 (1967).
22. J. G. Brown, A note on fuzzy sets, *Inf. Control* **18,** 32-39 (1971).
23. M. Mizumoto, J. Toyoda and K. Tanaka, General formulation of formal grammars, *Inf. Sci.* **4,** 87-100, 1972.
24. L. A. Zadeh, Similarity relations and fuzzy orderings, *Inf. Sci.* **3,** 177-200 (1971).
25. R. E. Bellman and L. A. Zadeh, Decision-making in a fuzzy environment, *Manage. Sci.* **17,** B-141-B-164 (1970).
26. R. E. Bellman and M. Giertz, On the analytic formalism of the theory of fuzzy sets, *Inf. Sci.* **5,** 149-156 (1973).
27. L. A. Zadeh, A fuzzy-set-theoretic interpretation of linguistic hedges, *J. Cybern.* **2,** 4-34 (1972).
28. E. F. Codd, Relational completeness of data base sublanguages, in *Courant Computer Science Symposia,* Vol. 6, Prentice-Hall, Englewood Cliffs, N.J., 1971.
29. L. A. Zadeh, Fuzzy sets, *Inf. Control* **8,** 338-353, 1965.
30. A. Thomasian, *The Structure of Probability Theory With Applications,* McGraw-Hill, New York, 1969.
31. R. E. Moore, *Interval Analysis,* Prentice-Hall, Englewood Cliffs, N.J., 1966.
32. J. A. Brzozowski, Regular expressions for linear sequential circuits, *IEEE Trans. Electron Comput.,* **EC-14,** 148-156 (1965).

33. E. Shamir, Algebraic, Rational and Context-Free Power Series in Noncommuting Variables, in M. Arbib's *Algebraic Theory of Machines, Languages and Semigroups,* Academic, New York, 1968, pp. 329-341.

34. A. Blikle, Equational languages, *Inf. Control* **21**, 134-147 (1972).

35. D. J. Rosenkrantz, Matrix equation and normal forms for context-free grammars, *J. Assoc. Comput. Mach.* **14**, 501-507, 1967.

36. J. E. Hopcroft and J. D. Ullman, *Formal Languages and Their Relations to Automata,* Addison-Wesley, Reading Mass., 1969.

37. A. V. Aho and J. D. Ullman, *The Theory of Parsing, Translation and Compiling,* Prentice-Hall, Englewood Cliffs, N.J., 1973.

38. G. Lakoff, Hedges: a study in meaning criteria and the logic of fuzzy concepts, in *Proc. 8th Reg. Meeting of Chic. Linguist. Soc.,* Univ. of Chicago Linguistics Dept., April 1972.

39. L. A. Zadeh, Quantitative fuzzy semantics, *Inf. Sci.* **3**, 159-176 (1971).

40. D. Knuth, Semantics of context-free languages, *Math. Syst. Theory,* **2**, 127-145 (1968).

41. P. Lucas et al., Method and notation for the formal definition of programming languages, Rept. TR 25.087, IBM Laboratory, Vienna, 1968.

42. P. Wegner, The Vienna definition language, *ACM Comput. Surv.* **4**, 5-63, (1972).

43. J. A. Lee, *Computer Semantics,* Van Nostrand-Reinhold, New York, 1972.

44. Z. Kohavi, *Switching and Finite Automata Theory,* McGraw-Hill, New York, 1970.

45. G. E. Hughes and M. J. Cresswell, *An Introduction to Modal Logic,* Methuen, London, 1968.

46. N. Rescher, *Many-Valued Logic,* McGraw-Hill, New York, 1969.

47. R. Barkan, A functional calculus of first order based on strict implication, *J. Symbol. Logic* **11**, 1-16 (1946).

48. L. A. Zadeh, Probability measures of fuzzy events, *J. Math. Anal. Appl.* **23**, 421-427 (1968).

49. A. DeLuca and S. Termini, A definition of non-probabilistic entropy in the setting of fuzzy set theory, *Inf. Control* **20**, 201-312 (1972).

50. J. Hintikka and P. Suppes (Eds.), *Aspects of Inductive Logic,* North-Holland, Amsterdam, 1966.

51. T. Winograd, *Understanding Natural Language,* Academic, New York, 1972.

52. A. DeLuca and S. Termini, Algebraic properties of fuzzy sets, *J. Math. Anal. Appl.* **40**, 373-386 (1972).

53. A. Rosenfeld, Fuzzy groups, *J. Math. Anal. Appl.* **35**, 512-517 (1971).

54. L. A. Zadeh, Fuzzy Algorithms, *Inf. Control* **12**, 94-102 (1968).

55. E. Santos, Fuzzy algorithms, *Inf. Control* **17**, 326-339 (1970).

56. S. K. Chang, On the execution of fuzzy programs using finite state machines, *IEEE Trans. Electron. Comput.,* **C-21**, 241-253 (1972).

57. S. S. L. Chang and L. A. Zadeh, Fuzzy mapping and control, *IEEE Trans. Syst., Man and Cybern.,* **SMC-2**, 30-34 (1972).

58. E. T. Lee, Fuzzy languages and their relation to automata, Dissertation, Dept. of Electr. Eng. and Comput. Sci., Univ. of Calif., Berkeley, 1972.

59. R. C. T. Lee, Fuzzy logic and the resolution principle, *J. Assoc. Comput. Mach.* **19**, 109-119 (1972).

60. K. M. Colby, S. Weber and F. D. Hilf, Artificial paranoia, *J. Artif. Intell.* **2**, 1-25 (1971).

Received November, 1973

The Concept of a Linguistic Variable and its Application to Approximate Reasoning—III*

L.A. ZADEH

Computer Sciences Division, Department of Electrical Engineering and Computer Sciences, and the Electronics Research Laboratory, University of California, Berkeley, California 94720

1. LINGUISTIC PROBABILITIES AND AVERAGES OVER FUZZY SETS

In the classical approach to probability theory, an *event, A,* is defined as a member of a σ-field, $\mathscr{A}$, of subsets of a sample space Ω. Thus, if P is a normed measure over a measurable space $(\Omega, \mathscr{A})$, the probability of A is defined as $P(A)$, the measure of A, and is a number in the interval $[0, 1]$.

There are many real-world problems in which one or more of the basic assumptions which are implicit in the above definition are violated. First, the event, A, is frequently ill-defined, as in the question, "What is the probability that it will be a *warm day* tomorrow?" In this instance, the event *warm day* is a *fuzzy event* in the sense that there is no sharp dividing line between its occurrence and nonoccurrence. As shown in [48], such an event may be characterized as a fuzzy subset, A, of the sample space Ω, with μ_A, the membership function of A, being a measurable function.

Second, even if A is a well-defined nonfuzzy event, its probability, $P(A)$, may be ill-defined. For example, in response to the question, "What is the probability that the Dow Jones average of stock prices will be higher in a month from now?" it would be patently unreasonable to give an unequivocal numerical answer, e.g., 0.7. In this instance, a vague response like "quite probable," would be much more commensurate with our lack of understanding of the dynamics of stock prices, and hence a more realistic—if less precise—characterization of the probability in question.

*This work was supported in part by the Navy Electronic Systems Command under Contract N00039-71-C-0255, the Army Research Office, Durham, N.C., under Grant DA-ARO-D-31-124-71-G174, and the National Science Foundation under Grant GK-10656X3. The writing of the paper was completed while the author was participating in a Joint Study Program with the Systems Research Department, IBM Research Laboratory, San Jose, California.

The limitations imposed by the assumption that A is well-defined may be removed, at least in part, by allowing A to be a fuzzy event, as was done in [48]. Another and perhaps more important step that can be taken to widen the applicability of probability theory to ill-defined problems is to allow P to be a linguistic variable in the sense defined in Part II, Sec. 3. In what follows, we shall outline a way in which this can be done and explore some of the elementary consequences of allowing P to be a linguistic variable.

LINGUISTIC PROBABILITIES

To simplify our exposition, we shall assume that the object of our concern is a variable, X, whose universe of discourse, U, is a finite set

$$U = u_1 + u_2 + \cdots + u_n. \tag{1.1}$$

Furthermore, we assume that the restriction imposed by X coincides with U. Thus, any point in U can be assigned as a value to X.

With each u_i, $i = 1, \ldots, n$, we associate a *linguistic probability*, $\mathscr{P}_i$, which is a Boolean linguistic variable in the sense of Part II, Definition 2.2, with $p_i, 0 \leqslant p_i \leqslant 1$, representing the base variable for $\mathscr{P}_i$. For concreteness, we shall assume that V, the universe of discourse associated with $\mathscr{P}_i$, is either the unit interval $[0, 1]$ or the finite set

$$V = 0 + 0.1 + \cdots + 0.9 + 1. \tag{1.2}$$

Using $\mathscr{P}$ as a generic name for the $\mathscr{P}_i$, the term-set for $\mathscr{P}$ will typically be the following.

$T(\mathscr{P}) =$ *likely + not likely + unlikely + very likely + more or less likely*
+ very unlikely + $\cdots$
+ probable + improbable + very probable + $\ldots$
+ neither very probable nor very improbable + $\cdots$
+ close to 0 + close to 0.1 + $\cdots$ + close to 1 + $\cdots$
+ very close to 0 + very close to 0.1 + $\cdots$, $\qquad$ (1.3)

in which *likely, probable* and *close to* play the role of primary terms.

The shape of the membership function of *likely* will be assumed to be like that of *true* [see Part II, Eq. (3.2)], with *not likely* and *unlikely* defined by

$$\mu_{not\ likely}\ (p) = 1 - \mu_{likely}\ (p), \tag{1.4}$$

and

$$\mu_{unlikely}\ (p) = \mu_{likely}\ (1-p), \tag{1.5}$$

where p is a generic name for the p_i.

EXAMPLE 1.1. A graphic example of the meaning attached to the terms *likely, not likely, very likely* and *unlikely* is shown in Fig. 1. In numerical terms, if the primary term *likely* is defined as

$$likely = 0.5/0.6 + 0.7/0.7 + 0.9/0.8 + 1/0.9 + 1/1, \tag{1.6}$$

then

$$not\ likely = 1/(0 + 0.1 + 0.2 + 0.3 + 0.4 + 0.5) + 0.5/0.6 + 0.3/0.7 + 0.1/0.8, \tag{1.7}$$

$$unlikely = 1/0 + 1/0.1 + 0.9/0.2 + 0.7/0.3 + 0.5/0.4 \tag{1.8}$$

and

$$very\ likely = 0.25/0.6 + 0.49/0.7 + 0.81/0.8 + 1/0.9 + 1/1. \tag{1.9}$$

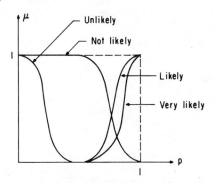

Fig. 1. Compatibility functions of *likely, not likely, unlikely* and *very likely*.

The term *probable* will be assumed to be more or less synonymous with *likely*. The term *close to* α, where α is a point in $[0, 1]$, will be abbreviated as $\vec{\alpha}$ or, alternatively, as "α",[1] suggesting that α is a "best example" of the fuzzy set "α". In this sense, then,

_____
[1] The symbol "α" will be employed in place of $\vec{\alpha}$ when the constraints imposed by type-setting dictate its use.

$$likely \overset{\Delta}{=} close\ to\ 1 \overset{\Delta}{=} \text{``1''}, \tag{1.10}$$

$$unlikely \overset{\Delta}{=} close\ to\ 0 \overset{\Delta}{=} \text{``0''}, \tag{1.11}$$

and

$$close\ to\ 0.8 \overset{\Delta}{=} \text{``0.8''} = 0.6/0.7 + 1/0.8 + 0.6/0.9, \tag{1.12}$$

from which it follows that

$$very\ close\ to\ 0.8 = very\ \text{``0.8''}$$
$$= (\text{``0.8''})^2 \qquad \text{[in the sense of Part II, Eq. (2.38)]}$$
$$= 0.36/0.7 + 1/0.8 + 0.36/0.9.$$

A particular term in $T(\mathscr{P})$ will be denoted by T_j, or T_{ji} in case a double subscript notation is needed. Thus, if $T_4 = very\ likely$, then T_{43} would indicate that *very likely* is assigned as a value to the linguistic variable $\mathscr{P}_3$.

The *n*-ary linguistic variable $(\mathscr{P}_1, \ldots, \mathscr{P}_n)$ constitutes a *linguistic probability assignment list* associated with X. A variable X which is associated with a linguistic probability assignment list will be referred to as a *linguistic random variable*. By analogy with linguistic truth-value distributions [see Part II, Eq. (3.74)], a collection of probability assignment lists will be referred to as a *linguistic probability distribution*.

The assignment of a probability-value T_j to P_i may be expressed as

$$P_i = T_j, \tag{1.13}$$

where P_i is used in a dual role as a generic name for the fuzzy variables which comprise $\mathscr{P}_i$. For example, we may write

$$P_3 = T_4$$

$$= very\ likely \tag{1.14}$$

in which case *very likely* will be identified as T_{43} (i.e., T_4 assigned to P_3).

An important characteristic of the linguistic probabilities $P_1, \ldots, P_n$ is that they are β-interactive in the sense of Part II, Definition 3.2. The interaction between the P_i is a consequence of the constraint ($+ \overset{\Delta}{=}$ arithmetic sum)

$$p_1 + p_2 + \cdots + p_n = 1, \tag{1.15}$$

in which the p_i are the base variables (i.e., numerical probabilities) associated with the P_i.

More concretely, let $R(p_1 + \cdots + p_n = 1)$ denote the nonfuzzy n-ary relation in $[0, 1] \times \cdots \times [0, 1]$ representing (1.15). Furthermore, let $R(P_i)$ denote the restriction on the values of p_i. Then the restriction imposed by the n-ary fuzzy variable $(P_1, \ldots, P_n)$ may be expressed as

$$R(P_1, \ldots, P_n) = R(P_1) \times \cdots \times R(P_n) \cap R(p_1 + \cdots + p_n = 1) \quad (1.16)$$

which implies that, apart from the constraint imposed by (1.15), the fuzzy variables $P_1, \ldots, P_n$ are noninteractive.

EXAMPLE 1.2. Suppose that

$$P_1 = likely$$
$$= 0.5/0.8 + 0.8/0.9 + 1/1 \quad (1.17)$$

and

$$P_2 = unlikely$$
$$= 1/0 + 0.8/0.1 + 0.5/0.2. \quad (1.18)$$

Then

$$R(P_1) \times R(P_2) = likely \times unlikely$$
$$= (0.5/0.8 + 0.8/0.9 + 1/1) \times (1/0 + 0.8/0.1 + 0.5/0.2)$$
$$= 0.5/(0.8,0) + 0.8/(0.9,0) + 1/(1,0)$$
$$+ 0.5/(0.8, 0.1) + 0.8/(0.9, 0.1) + 0.8/(1,0.1)$$
$$+ 0.5/(0.8,0.2) + 0.5/(0.9,0.2) + 0.5/(1,0.2). \quad (1.19)$$

As for $R(p_1 + \cdots + p_n = 1)$, it can be expressed as

$$R(p_1 + p_2 = 1) = \sum_k 1/(k, 1-k), \qquad k = 0,0.1, \ldots, 0.9, 1, \quad (1.20)$$

and forming the intersection of (1.19) and (1.20), we obtain

$$R(P_1, P_2) = 1/(1,0) + 0.8/(0.9,1) + 0.5/(0.8,0.2) \quad (1.21)$$

as the expression for the restriction imposed by (P_1, P_2). Obviously, $R(P_1, P_2)$ comprises those terms in $R(P_1) \times R(P_2)$ which satisfy the constraint (1.15).

REMARK 1.1. It should be observed that $R(P_1, P_2)$ as expressed by (1.21) is a normal restriction [see Part I, Eq. (3.23)]. This will be the case, more generally, when the P_i are of the form

$$P_i = \text{``}q_i\text{''}, \qquad i = 1, \ldots, n \tag{1.22}$$

and $q_1 + \cdots + q_n = 1$. Note that in Example 1.2, we have

$$P_1 = \text{``}1\text{''}, \tag{1.23}$$

$$P_2 = \text{``}0\text{''} \tag{1.24}$$

and

$$1 + 0 = 1. \tag{1.25}$$

COMPUTATION WITH LINGUISTIC PROBABILITIES

In many of the applications of probability theory, e.g., in the calculation of means, variances, etc., one encounters linear combinations of the form ($+ \overset{\Delta}{=}$ arithmetic sum)

$$z = a_1 p_1 + \cdots + a_n p_n, \tag{1.26}$$

where the a_i are real numbers and the p_i are probability-values in $[0, 1]$. Computation of the value of z given the a_i and the p_i presents no difficulties when the p_i are points in $[0, 1]$. It becomes, however, a nontrivial problem when the probabilities in question are linguistic in nature, that is, when

$$Z = a_1 P_1 + \cdots + a_n P_n, \tag{1.27}$$

where the P_i represent linguistic probabilities with names such as *likely, unlikely, very likely, close to* α, etc. Correspondingly, Z is not a real number—as it is in (1.26)—but a fuzzy subset of the real line $W \overset{\Delta}{=} (-\infty, \infty)$, with the membership function of Z being a function of those of the P_i.

Assuming that the fuzzy variables $P_1, \ldots, P_n$ are noninteractive [apart from the constraint expressed by (1.15)], the restriction imposed by $(P_1, \ldots, P_n)$ assumes the form [see (1.16)]

$$R(P_1, \ldots, P_n) = R(P_1) \times \cdots \times R(P_n) \cap R(p_1 + \cdots + p_n = 1). \tag{1.28}$$

Let $\mu(p_1, \ldots, p_n)$ be the membership function of $R(P_1, \ldots, P_n)$, and let $\mu_i(p_i)$ be that of $R(P_i)$, $i = 1, \ldots, n$. Then, by applying the extension principle

[Part I, Eq. (3.90)] to (1.26), we can express Z as a fuzzy set $(+ \overset{\Delta}{=}$ arithmetic sum)

$$Z = \int_W \mu(p_1, \ldots, p_n)/(a_1 p_1 + \cdots + a_n p_n), \qquad (1.29)$$

which in view of (1.28) may be written as

$$Z = \int_W \mu_1(p_1) \wedge \cdots \wedge \mu_n(p_n)/(a_1 p_1 + \cdots + a_n p_n) \qquad (1.30)$$

with the understanding that the p_i in (1.30) are subject to the constraint

$$p_1 + \cdots + p_n = 1. \qquad (1.31)$$

In this way, we can express a linear combination of linguistic probability-values . as a fuzzy subset of the real line.

The expression for Z may be cast into other forms which may be more convenient for computational purposes. Thus, let $\mu(z)$ denote the membership function of Z, with $z \in W$. Then (1.30) implies that

$$\mu(z) = \bigvee_{p_1, \ldots, p_n} \mu_1(p_1) \wedge \cdots \wedge \mu_n(p_n), \qquad (1.32)$$

subject to the constraints

$$z = a_1 p_1 + \cdots + a_n p_n, \qquad (1.33)$$

$$p_1 + \cdots + p_n = 1. \qquad (1.34)$$

In this form, the computation of Z reduces to the solution of a nonlinear programming problem with linear constraints. In more explicit terms, this problem may be expressed as: Maximize z subject to the constraints $(+ \overset{\Delta}{=}$ arithmetic sum)

$$\mu_1(p_1) \geqslant z,$$

$$\cdots$$

$$\mu_n(p_n) \geqslant z,$$

$$z = a_1 p_1 + \cdots + a_n p_n, \qquad (1.35)$$

$$p_1 + \cdots + p_n = 1.$$

EXAMPLE 1.3. As a very simple illustration, assume that

$$P_1 = likely \tag{1.36}$$

and

$$P_2 = unlikely, \tag{1.37}$$

where

$$likely = \int_0^1 \mu_{likely}(p)/p \tag{1.38}$$

and

$$unlikely = \neg\, likely \tag{1.39}$$

Thus [see (1.5)]

$$\mu_{unlikely}(p) = \mu_{likely}(1-p), \qquad 0 \leqslant p \leqslant 1. \tag{1.40}$$

Suppose that we wish to compute the expectation ($+ \triangleq$ arithmetic sum)

$$Z = a_1\, likely + a_2\, unlikely. \tag{1.41}$$

Using (1.32), we have

$$\mu(z) = \bigvee_{p_1, p_2} \mu_{likely}(p_1) \wedge \mu_{unlikely}(p_2), \tag{1.42}$$

subject to the constraints

$$z = a_1 p_1 + a_2 p_2, \tag{1.43}$$

$$p_1 + p_2 = 1.$$

Now in view of (1.40), if $p_1 + p_2 = 1$, then

$$\mu_{likely}(p_1) = \mu_{unlikely}(p_2), \tag{1.44}$$

and hence (1.42) reduces to

$$\mu(z) = \mu_{likely}(p_1), \tag{1.45}$$

$$z = a_1 p_1 + a_2(1 - p_1),$$

or, more explicitly,

$$\mu(z) = \mu_{likely}\left(\frac{z - a_2}{a_1 - a_2}\right). \tag{1.46}$$

This result implies that the fuzziness in our knowledge of the probability p_1 induces a corresponding fuzziness in the expectation of [see Fig. 2]

$$z = a_1 p_1 + a_2 p_2.$$

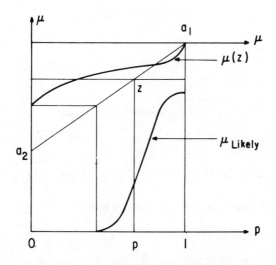

Fig. 2. Computation of the linguistic value of $a_1 p_1 + a_2 p_2$.

If the universe of probability-values is assumed to be $V = 0 + 0.1 + \cdots + 0.9 + 1$, then the expression for Z can be obtained more directly by using the extension principle in the form given in Part I, Eq. (3.97). As an illustration, assume that

$$P_1 = \text{"0.3"} = 0.8/0.2 + 1/0.3 + 0.6/0.4, \tag{1.47}$$

$$P_2 = \text{"0.7"} = 0.8/0.6 + 1/0.7 + 0.6/0.8, \tag{1.48}$$

and $(\oplus \overset{\Delta}{=} \text{arithmetic sum})$

$$Z = a_1 P_1 \oplus a_2 P_2, \tag{1.49}$$

where the symbol $\oplus$ is used to avoid confusion with the union.

On substituting (1.47) and (1.48) in (1.49), we obtain

$$Z = a_1(0.8/0.2 + 1/0.3 + 0.6/0.4) \oplus a_2(0.8/0.6 + 1/0.7 + 0.6/0.8)$$

$$= (0.8/0.2a_1 + 1/0.3a_1 + 0.6/0.4a_1) \oplus (0.8/0.6a_2 + 1/0.7a_2 + 0.6/0.8a_2). \tag{1.50}$$

In expanding the right-hand side of (1.50), we have to take into account the constraint $p_1 + p_2 = 1$, which means that a term of the form

$$\mu_1/p_1 a_1 \oplus \mu_2/p_2 a_2 \tag{1.51}$$

evaluates to

$$\mu_1/p_1 a_1 \oplus \mu_2/p_2 a_2 = \mu_1 \wedge \mu_2/(p_1 a_1 \oplus p_2 a_2) \qquad \text{if } p_1 + p_2 = 1$$

$$= 0 \qquad \text{otherwise.} \tag{1.52}$$

In this way, we obtain

$$Z = 1/(0.3a_1 \oplus 0.7a_2) + 0.6/(0.2a_1 \oplus 0.8a_2) + 0.6/(0.4a_1 \oplus 0.6a_2), \tag{1.53}$$

which expresses Z as a fuzzy subset of the real line $W = (-\infty, \infty)$.

AVERAGES OVER FUZZY SETS

Our point of departure in the foregoing discussion was the assumption that with each point u_i of a finite[2] universe of discourse U is associated a linguistic probability-value P_i which is a component of a linguistic probability distribution $(\mathscr{P}_1, \ldots, \mathscr{P}_n)$.

In this context, a fuzzy subset, A, of U plays the role of a *fuzzy event*. Let $\mu_A(u_i)$ be the grade of membership of u_i in A. Then, if the P_i are conventional numerical probabilities, p_i, $0 \leqslant p_i \leqslant 1$, then the probability of A, $P(A)$, is defined as (see [48]; $+ \overset{\Delta}{=}$ arithmetic sum)

---

[2]The assumption that U is a finite set is made solely for the purpose of simplifying our exposition. More generally, U can be a countable set or a continuum.

$$P(A) = \mu_A(u_1)p_1 + \cdots + \mu_A(u_n)p_n. \tag{1.54}$$

It is natural to extend this definition to linguistic probabilities by defining the linguistic probability[3] of A as

$$P(A) = \mu_A(u_1)P_1 + \cdots + \mu_A(u_n)P_n \tag{1.55}$$

with the understanding that the right-hand side of (1.55) is a linear form in the sense of (1.27). In connection with (1.55), it should be noted that the constraint

$$p_1 + \cdots + p_n = 1 \tag{1.56}$$

on the underlying probabilities, together with the fact that

$$0 \leqslant \mu_A(u_i) \leqslant 1, \qquad i = 1, \ldots, n,$$

insures that $P(A)$ is a fuzzy subset of $[0, 1]$.

EXAMPLE 1.4. As a very simple illustration, assume that

$$U = a + b + c, \tag{1.57}$$

$$A = 0.4a + b + 0.8c, \tag{1.58}$$

$$P_a = \text{"0.3"} = 0.6/0.2 + 1/0.3 + 0.6/0.4, \tag{1.59}$$

$$P_b = \text{"0.6"} = 0.6/0.5 + 1/0.6 + 0.6/0.7, \tag{1.60}$$

$$P_c = \text{"0.1"} = 0.6/0 + 1/0.1 + 0.6/0.2. \tag{1.61}$$

Then ($\oplus \overset{\Delta}{=}$ arithmetic sum)

$$P(A) = 0.4(0.6/0.2 + 1/0.3 + 0.6/0.4) \oplus (0.6/0.5 + 1/0.6 + 0.6/0.7)$$
$$\oplus 0.8(0.6/0 + 1/0.1 + 0.6/0.2), \tag{1.62}$$

subject to the constraint

$$p_1 + p_2 + p_3 = 1. \tag{1.63}$$

---

[3] It should be noted that the computation of the right-hand side of (1.55) defines $P(A)$ as a fuzzy subset of $[0, 1]$. In general, a linguistic approximation would be needed to express $P(A)$ as a linguistic probability-value.

Picking those terms in (1.62) which satisfy (1.63), we obtain

$$P(A) = 0.6/(0.4 \times 0.2 \oplus 0.6 \oplus 0.8 \times 0.2) \tag{1.64}$$

$$+ 0.6/(0.4 \times 0.2 \oplus 0.7 \oplus 0.8 \times 0.1)$$

$$+ 0.6/(0.4 \times 0.3 \oplus 0.5 \oplus 0.8 \times 0.2)$$

$$+ 1/(0.4 \times 0.3 \oplus 0.6 \oplus 0.8 \times 0.1)$$

$$+ 0.6/(0.4 \times 0.3 \oplus 0.7)$$

$$+ 0.6/(0.4 \times 0.4 \oplus 0.5 \oplus 0.8 \times 0.1)$$

$$+ 0.6/(0.4 \times 0.4 \oplus 0.6),$$

which reduces to

$$P(A) = 0.6/(0.84 + 0.86 + 0.78 + 0.82 + 0.74) + 1/0.8, \tag{1.65}$$

and which may be roughly approximated as

$$P(A) = \text{``0.8''}. \tag{1.66}$$

The linguistic probability of a fuzzy event as expressed by (1.55) may be viewed as a particular instance of a more general concept, namely, the *linguistic average* or, equivalently, the *linguistic expectation* of a function (defined on U) over a fuzzy subset of U. More specifically, let f be a real-valued function defined on U; let A be a fuzzy subset of U; and let $P_1, \ldots, P_n$ be the linguistic probabilities associated with $u_1, \ldots, u_n$, respectively. Then, the *linguistic average of f over A* is denoted by $Av(f;A)$ and is defined by ($+ \triangleq$ arithmetic sum)

$$Av(f;A) = f(u_1)\mu_A(u_1)P_1 + \cdots + f(u_n)\mu_A(u_n)P_n. \tag{1.67}$$

A concrete example of (1.67) is the following. Assume that individuals named $u_1, \ldots, u_n$ are chosen with linguistic probabilities $P_1, \ldots, P_n$, with P_i being a restriction on p_i, $i = 1, \ldots, n$. Suppose that u_i is fined an amount $f(u_i)$, which is scaled down in proportion to the grade of membership of u_i in a class A. Then, the linguistic average (expected) amount of the fine will be expressed by (1.67).

COMMENT 1.1. Note that (1.67) is basically a linear combination of the form (1.27) with

$$a_i = f(u_i)\mu_A(u_i). \tag{1.68}$$

Thus, to evaluate (1.67), we can employ the technique described earlier for the computation of linear forms in linguistic probabilities. In particular, it should be noted that, in the special case where $f(u_i) = 1$, the right-hand side of (1.67) becomes

$$\mu_A(u_1)P_1 + \cdots + \mu_A(u_n)P_n, \tag{1.69}$$

and $\text{Av}(f;A)$ reduces to $P(A)$.

In addition to subsuming the expression for $P(A)$, the expression for $\text{Av}(f;A)$ subsumes as special cases other types of averages which occur in various applications. Among them there are two that may be regarded as degenerate forms of (1.67) and which are encountered in many problems of practical interest. In what follows, we shall dwell briefly on these averages and, for convenience in exposition, will state their definitions in the form of answers to questions.

QUESTION 1.1. What is the number of elements in a given fuzzy set A? Clearly, this question is not well posed, since in the case of a fuzzy set the dividing line between membership and nonmembership is not sharp. Nevertheless, the concept of the *power* of a fuzzy set [49], which is defined as

$$|A| \overset{\Delta}{=} \sum_i \mu_A(u_i), \tag{1.70}$$

appears to be a natural generalization of that of the number of elements in A.

As an illustration of $|A|$, suppose that U is the universe of residents in a city, and A is the fuzzy set of the unemployed in that city. If $\mu_A(u_i)$ is interpreted as the grade of membership of an individual, u_i, in the class of the unemployed [e.g., $\mu_A(u_i) = 0.5$ if u_i is working half-time and is looking for a full-time job], then $|A|$ may be interpreted as the number of *full-time equivalent* unemployed.

QUESTION 1.2. Suppose that f is a real-valued function defined on U. What is the average value of f over a fuzzy subset, A, of U?

Using the same notation as in (1.67), let $\text{Av}(f;A)$ denote the average value of f over A. If A were nonfuzzy, $\text{Av}(f;A)$ would be expressed by

$$\text{Av}(f;A) = \frac{\sum_{u_i \in A} f(u_i)}{|A|}, \tag{1.71}$$

where $\Sigma_{u_i \in A}$ is the summation over those u_i which are in A, and $|A|$ is the number of the u_i which are in A. To extend (1.71) to fuzzy sets, we note that (1.71) may be rewritten as

$$\mathrm{Av}(f;A) = \frac{\Sigma_{u_i \in U} f(u_i) \mu_A(u_i)}{\Sigma_{u_i \in U} \mu_A(u_i)} , \qquad (1.72)$$

where μ_A is the characteristic function of A. Then, we adopt (1.72) as the definition of $\mathrm{Av}(f;A)$ for a *fuzzy* A by interpreting $\mu_A(u_i)$ as the grade of membership of u_i in A. In this way, we arrive at an expression for $\mathrm{Av}(f;A)$ which may be viewed as a special case of (1.67).

As an illustration of (1.72), suppose that U is the universe of residents in a city and A is the fuzzy subset of residents who are *young*. Furthermore, assume that $f(u_i)$ represents the income of u_i. Then, the average income of young residents in the city would be expressed by (1.72).

COMMENT 1.2. Since the expression for $|A|$ is a linear form in the $\mu_A(u_i)$, the power of a fuzzy set of type 2 (see Part I, Definition 3.1) can readily be computed by employing the technique which we had used earlier to compute $P(A)$. In the case of $\mathrm{Av}(f;A)$, however, we are dealing with a ratio of linear forms, and hence the computation of $\mathrm{Av}(f;A)$ for fuzzy sets of type 2 presents a more difficult problem.

In the foregoing discussion, our very limited objective was to indicate that the concept of a linguistic variable provides a basis for defining linguistic probabilities and, in conjunction with the extension principle, may be applied to the computation of linear forms in such probabilities. We shall not dwell further on this subject and, in what follows, will turn our attention to a basic rule of inference in fuzzy logic.

2. COMPOSITIONAL RULE OF INFERENCE AND APPROXIMATE REASONING

The basic rule of inference in traditional logic is the *modus ponens,* according to which we can infer the truth of a proposition B from the truth of A and the implication $A \Rightarrow B$. For example, if A is identified with "John is in a hospital," and B with "John is ill," then if it is true that "John is in a hospital," it is also true that "John is ill."

In much of human reasoning, however, *modus ponens* is employed in an approximate rather than exact form. Thus, typically, we know that A is true and that $A^* \Rightarrow B$, where A^* is, in some sense, an approximation to B. Then, from A and $A^* \Rightarrow B$ we may infer that B is approximately true.

In what follows, we shall outline a way of formalizing approximate reasoning based on the concepts introduced in the preceding sections. However, in a departure from traditional logic, our main tool will not be the *modus ponens*, but a so-called *compositional rule of inference* of which *modus ponens* forms a very special case.

COMPOSITIONAL RULE OF INFERENCE

The compositional rule of inference is merely a generalization of the following familiar procedure. Referring to Fig. 3, suppose that we have a curve $y = f(x)$ and are given $x = a$. Then from $y = f(x)$ and $x = a$, we can infer $y \triangleq b = f(a)$.

Next, let us generalize the above process by assuming that a is an interval and $f(x)$ is an interval-valued function such as shown in Fig. 4. In this instance, to find the interval $y \triangleq b$ which corresponds to the interval a, we first construct a cylindrical set, $\bar{a}$, with base a [see Part I, Eq. (3.58)] and find its intersection, I, with the interval-valued curve. Then we project the intersection on the OY axis, yielding the desired y as the interval b.

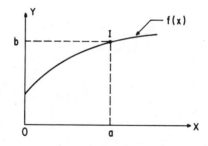

Fig. 3. Infering $y = b$ from $x = a$ and $y = f(x)$.

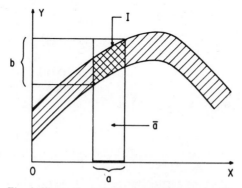

Fig. 4. Illustration of the compositional rule of inference in the case of interval-valued variables.

Going one step further in our chain of generalizations, assume that A is a fuzzy subset of the OX axis and F is a fuzzy relation from OX to OY. Again, forming a cylindrical fuzzy set $\bar{A}$ with base A and intersecting it with the fuzzy relation F (see Fig. 5), we obtain a fuzzy set $\bar{A} \cap F$ which is the analog of the point of intersection I in Fig. 3. Then, projecting this set on OY, we obtain y as a fuzzy subset of OY. In this way, from $y = f(x)$ and $x \triangleq A$ (fuzzy subset of OX), we infer y as a fuzzy subset, B, of OY.

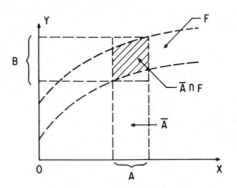

Fig. 5. Illustration of the compositional rule of inference for fuzzy variables.

More specifically, let μ_A, $\mu_{\bar{A}}$, μ_F and μ_B denote the membership functions of A, $\bar{A}$, F and B, respectively. Then, by the definition of $\bar{A}$ [see Part I, Eq. (3.58)]

$$\mu_{\bar{A}}(x, y) = \mu_A(x), \tag{2.1}$$

and consequently

$$\mu_{\bar{A} \cap F}(x,y) = \mu_{\bar{A}}(x,y) \wedge \mu_F(x,y)$$

$$= \mu_A(x) \wedge \mu_F(x,y). \tag{2.2}$$

Projecting $\bar{A} \cap F$ on the OY axis, we obtain from (2.2) and from Eq. (3.57) of Part I

$$\mu_B(y) = \vee_x \mu_A(x) \wedge \mu_F(x,y) \tag{2.3}$$

as the expression for the membership function of the projection (shadow) of $\bar{A} \cap F$ on OY. Comparing this expression with the definition of the composition of A and F [see Part I, Eq. (3.55)], we see that B may be represented as

$$B = A \circ F, \qquad (2.4)$$

where ∘ denotes the operation of composition. As stated in Part I, Sec. 3, this operation reduces to the max-min matrix product when A and F have finite supports.

EXAMPLE 2.1. Suppose that A and F are defined by

$$A = 0.2/1 + 1/2 + 0.3/3 \qquad (2.5)$$

and

$$F = 0.8/(1,1) + 0.9/(1,2) + 0.2/(1,3)$$
$$+ 0.6/(2,1) + 1/(2,2) + 0.4/(2,3)$$
$$+ 0.5/(3,1) + 0.8/(3,2) + 1/(3,3). \qquad (2.6)$$

Expressing A and F in terms of their relation matrices and forming the matrix product (2.4), we obtain

$$\underset{A}{[0.2 \quad 1 \quad 0.3]} \circ \overset{F}{\begin{bmatrix} 0.8 & 0.9 & 0.2 \\ 0.6 & 1 & 0.4 \\ 0.5 & 0.8 & 1 \end{bmatrix}} = \overset{B}{[0.6 \quad 1 \quad 0.4]} . \qquad (2.7)$$

The foregoing comments and examples serve to motivate the following rule of inference.

RULE 2.1. Let U and V be two universes of discourse with base variables u and v, respectively. Let $R(u)$, $R(u, v)$ and $R(v)$ denote restrictions on u, (u, v) and v, respectively, with the understanding that $R(u)$, $R(u, v)$ and $R(v)$ are fuzzy relations in U, $U \times V$ and V. Let A and F denote particular fuzzy subsets of U and $U \times V$. Then the *compositional rule of inference* asserts that the solution of the *relational* assignment equations

$$R(u) = A \qquad (2.8)$$

and

$$R(u, v) = F \qquad (2.9)$$

is given by

$$R(v) = A \circ F, \tag{2.10}$$

where $A \circ F$ is the composition of A and F. In this sense, we can *infer $R(v) = A \circ F$* from $R(u) = A$ and $R(u, v) = F$.

As a simple illustration of the use of this rule, assume that

$$U = V = 1 + 2 + 3 + 4, \tag{2.11}$$

$$A = small = 1/1 + 0.6/2 + 0.2/3 \tag{2.12}$$

and

$F = approximately\ equal$

$= 1/(1,1) + 1/(2,2) + 1/(3,3) + 1/(4,4)$

$+ 0.5/[(1,2) + (2,1) + (2,3) + (3,2) + (3,4) + (4,3)]. \tag{2.13}$

In other words, A is unary fuzzy relation in U named *small* and F is a binary fuzzy relation in $U \times V$ named *approximately equal*.

The relational assignment equations in this case read

$$R(u) = small, \tag{2.14}$$

$$R(u, v) = approximately\ equal \tag{2.15}$$

and hence

$R(v) = small \circ approximately\ equal$

$$= \begin{bmatrix} 1 & 0.6 & 0.2 & 0 \end{bmatrix} \circ \begin{bmatrix} 1 & 0.5 & 0 & 0 \\ 0.5 & 1 & 0.5 & 0 \\ 0 & 0.5 & 1 & 0.5 \\ 0 & 0 & 0.5 & 1 \end{bmatrix}$$

$$= \begin{bmatrix} 1 & 0.6 & 0.5 & 0.2 \end{bmatrix} \tag{2.16}$$

which may be approximated by the linguistic term

$$R(v) = more\ or\ less\ small \tag{2.17}$$

if *more or less* is defined as a fuzzifier [see Part I, Eq. (3.48)] , with

$$K(1) = 1/1 + 0.7/2,$$

$$K(2) = 1/2 + 0.7/3,$$

$$K(3) = 1/3 + 0.7/4,$$

$$K(4) = 1/4. \tag{2.18}$$

Note that the application of this fuzzifier to $R(u)$ yields

$$[1 \quad 0.7 \quad 0.42 \quad 0.14] \tag{2.19}$$

as an approximation to $[1 \quad 0.6 \quad 0.5 \quad 0.2]$.

In summary, then, by using the compositional rule of inference, we have infered from $R(u) = $ *small* and $R(u, v) = $ *approximately equal*

$$R(v) = [1 \quad 0.6 \quad 0.5 \quad 0.2] \qquad \text{exactly} \tag{2.20}$$

and

$$R(v) = \textit{more or less small} \qquad \text{as a linguistic approximation.} \tag{2.21}$$

Stated in English, this approximate inference may be expressed as

u is small	premiss
u and v are approximately equal	premiss
v is more or less small	approximate conclusion. (2.22)

The general idea behind the method sketched above is the following. Each fact or a premiss is translated into a relational assignment equation involving one or more restrictions on the base variables. These equations are solved for the desired restrictions by the use of the composition of fuzzy relations. The solutions to the equations then represent deductions from the given set of premisses.

MODUS PONENS *AS A SPECIAL CASE*
OF THE COMPOSITIONAL RULE OF INFERENCE

As we shall see in what follows, *modus ponens* may be viewed as a special case of the compositional rule of inference. To establish this connection, we shall first extend the notion of material implication from propositional variables to fuzzy sets.

In traditional logic, the material implication $\Rightarrow$ is defined as a logical connective for propositional variables. Thus, if A and B are propositional variables, the truth table for $A \Rightarrow B$ or, equivalently, IF A THEN B, is defined by Table 1 (see Part II, Table 2).

TABLE 1

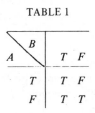

In much of human discourse, however, the expression IF A THEN B is used in situations in which A and B are fuzzy sets (or fuzzy predicates) rather than propositional variables. For example, in the case of the statement IF John is *ill* THEN John is *cranky*, which may be abbreviated as *ill* $\Rightarrow$ *cranky*, *ill* and *cranky* are, in effect, names of fuzzy sets. The same is true of the statement IF apple is *red* THEN apple is *ripe*, where *red* and *ripe* play the role of fuzzy sets.

To extend the notion of material implication to fuzzy sets, let U and V be two possibly different universes of discourse and let A, B and C be fuzzy subsets of U, V and V, respectively. First we shall define the meaning of the expression IF A THEN B ELSE C, and then we shall define IF A THEN B as a special case of IF A THEN B ELSE C.

DEFINITION 2.1. The expression IF A THEN B ELSE C is a binary fuzzy relation in $U \times V$ defined by

$$\text{IF } A \text{ THEN } B \text{ ELSE } C = A \times B + \neg A \times C. \tag{2.23}$$

That is, if A, B and C are unary fuzzy relations in U, V and V, then IF A THEN B ELSE C is a binary fuzzy relation in $U \times V$ which is the union of the Cartesian product of A and B [see Part I, Eq. (3.45)] and the Cartesian product of the negation of A and C.

Now IF A THEN B may be viewed as a special case of IF A THEN B ELSE C which results when C is allowed to be the entire universe V. Thus

$$\text{IF } A \text{ THEN } B \overset{\Delta}{=} \text{IF } A \text{ THEN } B \text{ ELSE } V$$

$$= A \times B + \neg A \times V. \tag{2.24}$$

In effect, this amounts to interpreting IF A THEN B as IF A THEN B ELSE *don't care.*[4]

It is helpful to observe that in terms of the relation matrices of A, B and C, (2.23) may be expressed as the sum of dyadic products involving A and B (and $\neg A$ and C) as column and row matrices, respectively. Thus,

$$\text{IF } A \text{ THEN } B \text{ ELSE } C = \begin{bmatrix} A \end{bmatrix} \begin{bmatrix} B \end{bmatrix} + \neg \begin{bmatrix} A \end{bmatrix} \begin{bmatrix} C \end{bmatrix}. \tag{2.25}$$

EXAMPLE 2.2. As a simple illustration of (2.23) and (2.24), assume that

$$U = V = 1 + 2 + 3, \tag{2.26}$$

$$A = small = 1/1 + 0.4/2, \tag{2.27}$$

$$B = large = 0.4/2 + 1/3, \tag{2.28}$$

$$C = not\ large = 1/1 + 0.6/2. \tag{2.29}$$

Then

$$\text{IF } A \text{ THEN } B \text{ ELSE } C = (1/1 + 0.4/2) \times (0.4/2 + 1/3) + (0.6/2 + 1/3) \times (1/1$$
$$+ 0.6/2)$$
$$= 0.4/(1,2) + 1/(1,3) + 0.6/(2,1) + 0.6/(2,2)$$
$$+ 0.4/(2,3) + 1/(3,1) + 0.6/(3,2). \tag{2.30}$$

which, represented as a relation matrix, reads

$$\text{IF } A \text{ THEN } B \text{ ELSE } C = \begin{bmatrix} 0 & 0.4 & 1 \\ 0.6 & 0.6 & 0.4 \\ 1 & 0.6 & 0 \end{bmatrix} \tag{2.31}$$

Similarly

$$\text{IF } A \text{ THEN } B = (1/1 + 0.4/2) \times (0.4/2 + 1/3) + (0.6/2 + 1/3) \times (1/1 + 1/2 + 1/3)$$
$$= 0.4/(1,2) + 1/(1,3) + 0.6/(2,1) + 0.6/(2,2)$$
$$+ 0.6/(2,3) + 1/(3,1) + 1/(3,2) + 1/(3,3),$$

---

[4] An alternative interpretation that is consistent with Lukasiewicz's definition of implication [46] is expressed by IF A THEN $B \triangleq \neg(A \times V) \oplus (U \times B)$, where the operation $\oplus$ (bounded-sum) is defined for fuzzy sets P, Q by $\mu_{P \oplus Q} \triangleq 1 \wedge (\mu_P + \mu_Q)$, with + denoting the arithmetic sum. More generally, IF A THEN B ELSE C $\triangleq [\neg(A \times V) \oplus (U \times B)] \cap [(A \times V) \oplus (U \times C)]$.

or equivalently

$$\text{IF } A \text{ THEN } B = \begin{bmatrix} 0 & 0.4 & 1 \\ 0.6 & 0.6 & 0.6 \\ 1 & 1 & 1 \end{bmatrix} \tag{2.32}$$

COMMENT 2.1. It should be noted that in defining IF A THEN B by (2.24) we are tacitly assuming that A and B are noninteractive in the sense that there is no joint constraint involving the base variables u and v. This would not be the case in the nonfuzzy statement IF $u \in A$ THEN $u \in B$, which may be expressed as IF $u \in A$ THEN $v \in B$, subject to the constraint $u = v$. Denoting this constraint by $R(u = v)$, the relation representing the statement in question would be

$$\text{IF } u \in A \text{ THEN } u \in B \overset{\Delta}{=} (A \times B + \neg A \times V) \cap [R(u = v)]. \tag{2.33}$$

REMARK 2.1. In defining $A \Rightarrow B$, we assumed that IF A THEN B is a special case of IF A THEN B ELSE C resulting from setting $C = V$. If we set C equal to θ (empty set) rather than V, the right-hand side of (2.23) reduces to the Cartesian product $A \times B$—which may be interpreted as A COUPLED WITH B (rather than A ENTAILS B). Thus, by definition,

$$A \text{ COUPLED WITH } B \overset{\Delta}{=} A \times B, \tag{2.34}$$

and hence

$$A \Rightarrow B \overset{\Delta}{=} A \text{ COUPLED WITH } B \text{ plus } \neg A \text{ COUPLED WITH } V. \tag{2.35}$$

More generally, an expression of the form

$$A_1 \times B_1 + \cdots + A_n \times B_n \tag{2.36}$$

would be expressed in words as

$$A_1 \text{ COUPLED WITH } B_1 \text{ plus} \ldots \text{plus } A_n \text{ COUPLED WITH } B_n. \tag{2.37}$$

It should be noted that expressions such as (2.37) may be employed to represent a fuzzy graph as a union of fuzzy points (see Fig. 6). For example, a fuzzy graph G may be represented as

$$G = \text{``}u_1\text{''} \times \text{``}v_1\text{''} + \text{``}u_2\text{''} \times \text{``}v_2\text{''} + \cdots + \text{``}u_n\text{''} \times \text{``}v_n\text{''}, \tag{2.38}$$

where the u_i and v_i are points in U and V, respectively, and "u_i" and "v_i", $i = 1, \ldots, n$, represent fuzzy sets named *close to u_i* and *close to v_i* [see (1.12)].

COMMENT 2.2. The connection between (2.24) and the conventional definition of material implication becomes clearer by noting that

$$\neg A \times B \subset \neg A \times V \tag{2.39}$$

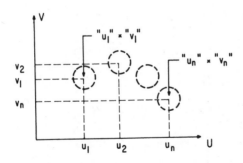

Fig. 6. Representation of a fuzzy graph as a union of fuzzy points.

and hence that (2.24) may be rewritten as

$$\text{IF } A \text{ THEN } B = A \times B + \neg A \times B + \neg A \times V$$

$$= (A + \neg A) \times B + \neg A \times V. \qquad (2.40)$$

Now, if A is a nonfuzzy subset of U, then

$$A + \neg A = U, \qquad (2.41)$$

and hence IF A THEN B reduces to

$$\text{IF } A \text{ THEN } B = U \times B + \neg A \times V, \qquad (2.42)$$

which is similar in form to the familiar expression for $A \Rightarrow B$ in the case of propositional variables, namely

$$A \Rightarrow B \equiv \neg A \vee B. \qquad (2.43)$$

Turning to the connection between *modus ponens* and the compositional rule of inference, we first define a *generalized modus ponens* as follows.

DEFINITION 2.2. Let A_1, A_2 and B be fuzzy subsets of U, U and V, respectively. Assume that A_1 is assigned to the restriction $R(u)$, and the relation $A_2 \Rightarrow B$ [defined by Eq. (3.24) of Part I] is assigned to the restriction $R(u, v)$. Thus

$$R(u) = A_1, \qquad (2.44)$$

$$R(u, v) = A_2 \Rightarrow B. \qquad (2.45)$$

As was shown earlier, these relational assignment equations may be solved for the restriction on v, yielding

$$R(v) = A_1 \circ (A_2 \Rightarrow B). \qquad (2.46)$$

An expression for this conclusion in the form

$$A_1 \qquad \text{premiss} \qquad (2.47)$$

$$A_2 \Rightarrow B \qquad \text{implication} \qquad (2.48)$$

$$\overline{A_1 \circ (A_2 \Rightarrow B)} \qquad \text{conclusion} \qquad (2.49)$$

constitutes the statement of the *generalized modus ponens.*[5]

COMMENT 2.3. The above statement differs from the traditional *modus ponens* in two respects: First, A_1, A_2 and B are allowed to be fuzzy sets, and second, A_1 need not be identical with A_2. To check on what happens when $A_1 = A_2 = A$ and A is nonfuzzy, we substitute the expression for $A_2 \Rightarrow B$ in (2.46), yielding

$$A \circ (A \Rightarrow B) = A \circ (A \times B + \neg A \times V)$$

$$= A_r A_c B_r + A_r (\neg A_c) V_r, \qquad (2.50)$$

where r and c stand for *row* and *column,* respectively; A_r and A_c denote the relation matrices for A expressed as a row matrix and a column matrix, respectively; and the matrix product is understood to be taken in the max-min sense.

Now, since A is nonfuzzy,

$$A_r (\neg A_c) = 0, \qquad (2.51)$$

and so long as A is normal [see Part I, Eq. (3.23)]

$$A_r A_c = 1. \qquad (2.52)$$

Consequently

$$A \circ (A \Rightarrow B) = B, \qquad (2.53)$$

which agrees with the conclusion yieled by *modus ponens.*

EXAMPLE 2.3. As a simple illustration of (2.49), assume that

$$U = V = 1 + 2 + 3, \qquad (2.54)$$

---

[5]The generalized *modus ponens* as defined here is unrelated to probabilistic rules of inference. A discussion of such rules and related issues may be found in [50].

$$A_2 = small = 1/1 + 0.4/2, \tag{2.55}$$

$$A_1 = more\ or\ less\ small = 1/1 + 0.4/2 + 0.2/3 \tag{2.56}$$

and

$$B = large = 0.4/2 + 1/3. \tag{2.57}$$

Then (see (2.32))

$$small \Rightarrow large = \begin{bmatrix} 0 & 0.4 & 1 \\ 0.6 & 0.6 & 0.6 \\ 1 & 1 & 1 \end{bmatrix} \tag{2.58}$$

and

$$more\ or\ less\ small \circ (small \Rightarrow large) = [1 \quad 0.4 \quad 0.2] \circ \begin{bmatrix} 0 & 0.4 & 1 \\ 0.6 & 0.6 & 0.6 \\ 1 & 1 & 1 \end{bmatrix}$$

$$= [0.4 \quad 0.4 \quad 1], \tag{2.59}$$

which may be roughly approximated as *more or less large.* Thus, in the case under consideration, the *generalized modus ponens* yields

u is *more or less small*	premiss
IF u is *small* THEN v is *large*	implication

$$\text{(2.60)}$$

v is *more or less large*	approximate conclusion

COMMENT 2.4. Because of the way in which $A \Rightarrow B$ is defined, namely,

$$A \Rightarrow B = A \times B + \neg A \times V,$$

the grade of membership of a point (u, v) will be high in $A \Rightarrow B$ if the grade of membership of u is low in A. This gives rise to an overlap between the terms $A \times B$ and $\neg A \times V$ when A is fuzzy, with the result that [see (2.50)], the inference drawn from A and $A \Rightarrow B$ is not B but[6]

---

[6] We assume that A is normal, so that $A_r A_c = 1$.

$$A \circ (A \Rightarrow B) = B + A \circ (\neg A \times V), \tag{2.61}$$

where the difference term $A \circ (\neg A \times V)$ represents the effect of the overlap.

To avoid this phenomenon it may be necessary to define $A \Rightarrow B$ in a way that differentiates between the numerical truth-values in $[0, 1]$ and the truth-value *unknown* [see Part II, Eq. (3.52)]. Also, it should be noted that for A COUPLED WITH B [see (2.34)], we do have

$$A \circ (A \text{ COUPLED WITH } B) = B \tag{2.62}$$

so long as A is a normal fuzzy set.

FUZZY THEOREMS

By a fuzzy theorem or an assertion we mean a statement, generally of the form IF A THEN B, whose truth-value is *true* in an approximate sense and which can be inferred from a set of axioms by the use of approximate reasoning, e.g., by repeated application of the generalized *modus ponens* or similar rules.

As an informal illustration of the concept of a fuzzy theorem, let us consider the theorem in elementary geometry which asserts that if M_1, M_2 and M_3 are the midpoints of the sides of a triangle (see Fig. 7), then the lines AM_1, BM_2 and CM_3 intersect at a point.

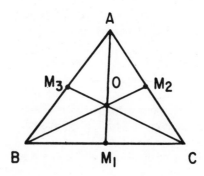

Fig. 7. An elementary theorem in geometry.

FUZZY THEOREM 2.1. *Let AB, BC and CA be approximate straight lines which form an approximate equilateral triangle with vertices A, B, C (see Fig. 8). Let M_1, M_2 and M_3 be approximate midpoints of the sides BC, CA and AB, respectively. Then the approximate straight lines AM_1, BM_2 and CM_3 form an approximate triangle $T_1 T_2 T_3$ which is more or less (more or less small) in relation to ABC.*

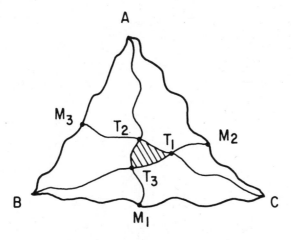

Fig. 8. A fuzzy theorem in geometry.

Before we can proceed to "prove" this fuzzy theorem, we must make more specific the sense in which the terms approximate straight line, approximate midpoint, etc. should be understood. To this end, let us agree that by an *approximate straight line AB* we mean a curve passing through A and B such that the distance of any point on the curve from the straight line AB is small in relation to the length of AB. With reference to Fig. 9, this implies that we are assigning a linguistic value *small* to the distance d, with the understanding that d is interpreted as a fuzzy variable.

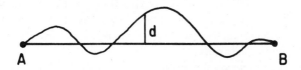

Fig. 9. Definition of *approximately straight line.*

Let $(AB)^0$ denote the straight line AB. Then, by an *approximate midpoint* of AB we mean a point on AB whose distance from M_1^0, the midpoint of $(AB)^0$, is *small.*

Turning to the statement of the fuzzy theorem, let O be the intersection of the straight lines $(AM_1^0)^0$ and $(BM_2^0)^0$ (Fig. 10). Since M_1 is assumed to be an approximate midpoint of BC, the distance of M_1 from M_1^0 is *small.* Consequently, the distance of any point on $(AM_1)^0$ from $(AM_1^0)^0$ is *small.* Furthermore, since the distance of any point on AM_1 from $(AM_1)^0$ is *small,* it follows that the distance of any point on AM_1 from $(AM_1^0)^0$ is *more or less small.*

The same argument applies to the distance of points on BM_2 from $(BM_2^0)^0$. Then, taking into consideration that the angle between $(AM_1)^0$ and $(BM_2)^0$ is approximately $120°$, the distance between an intersection of AM_1 and BM_2 and O is (*more or less*)^2 *small* [that is, *more or less* (*more or less small*)]. From this it follows that the distance of any vertex of the triangle $T_1 T_2 T_3$ from O is (*more or less*)^2 *small*. It is in this sense that the triangle $T_1 T_2 T_3$ is (*more or less*)^2 *small* in relation to ABC.

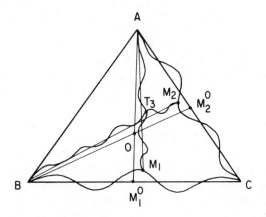

Fig. 10. Illustration of an approximate proof of the fuzzy theorem.

The reasoning used above is both approximate and qualitative in nature. It uses as its point of departure the fact that AM_1, BM_2 and CM_3 intersect at O, and employs what, in effect, are qualitative continuity arguments. Clearly, the "proof" would be longer and more involved if we had to start from the basic axioms of Euclidean geometry rather than the nonfuzzy theorem which served as our point of departure.

At this point, what we can say about fuzzy theorems is highly preliminary and incomplete in nature. Nonetheless, it appears to be an intriguing area for further study and eventually may prove to be of use in various types of ill-defined decision processes.

GRAPHICAL REPRESENTATION BY FUZZY FLOWCHARTS

As pointed out in [7], in the representation and execution of fuzzy algorithms it is frequently very convenient to employ flowcharts for the purpose of defining relations between variables and assigning values to them.

In what follows, we shall not concern ourselves with the many complex issues arising in the representation and execution of fuzzy algorithms. Thus, our limited objective is merely to clarify the role played by the decision boxes which are associated with fuzzy rather than nonfuzzy predicates by relating their function to the assignment of restrictions on base variables.

In the conventional flowchart, a decision box such as A in Fig. 11 represents a unary[7] predicate, $A(x)$. Thus, transfer from point 1 to point 2 signifies that $A(x)$ is *true*, while transfer from 1 to 3 signifies that $A(x)$ is *false*.

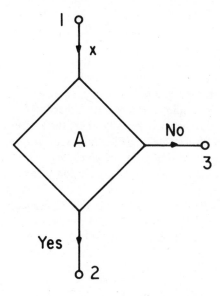

Fig. 11. A fuzzy decision box.

The concepts introduced in the preceding sections provide us with a basis for extending the notion of a decision box to fuzzy sets (or predicates). Specfically, with reference to Fig. 11, suppose that A is a fuzzy subset of U, and the question associated with the decision box is: "Is x A?" as in "Is x *small*?" where x is a generic name for the input variable. Flowcharts containing decision boxes of this type will be referred to as *fuzzy flowcharts*.

If the answer is simply YES, we assign A to the restriction on x. That is, we set

$$R(x) = A \tag{2.63}$$

and transfer x from 1 to 2.

---

[7]For simplicity, we shall not consider decision boxes having more than one input and two outputs.

On the other hand, if the answer is NO, we set

$$R(x) = \neg A \qquad\qquad (2.64)$$

and transfer x from 1 to 3.

As an illustration, if $A \overset{\Delta}{=} small$, then (2.63) would read

$$R(x) = small. \qquad\qquad (2.65)$$

If the answer is YES/μ, where $0 \leqslant \mu \leqslant 1$, then we transfer x to 2 with the conclusion that the grade of membership of x in A is μ. We also transfer x to 3 with the conclusion that the grade of membership of x in $\neg A$ is $1 - \mu$.

If the grade of membership, μ, is linguistic rather than numerical, we represent it as a linguistic truth-value. Typically, then, the answer would have the form YES/*true* or YES/*very true* or YES/*more or less true,* etc. As before, we conclude that the grade of membership of x in A is μ, where μ is a linguistic truth-value, and transfer x to 3 with the conclusion that the grade of membership of x in $\neg A$ is $1 - \mu$.

If we have a chain of decision boxes as in Fig. 12, a succession of YES answers would transfer x from 1 to $n + 1$ and would result in the assignment to $R(x)$ of the intersection of $A_1, \ldots, A_n$. Thus,

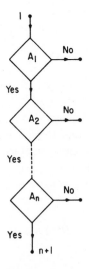

Fig. 12. A tandem combination of decision boxes.

$$R(x) = A_1 \cap \cdots \cap A_n, \qquad (2.66)$$

where $\cap$ denotes the intersection of fuzzy sets. (See also Fig. 13.)

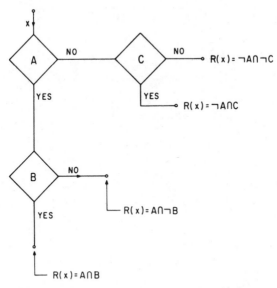

Fig. 13. Restrictions associated with various exits from a fuzzy flowchart.

As a simple illustration, suppose that $x =$ John, $A_1 = $ *tall* and $A_2 = $ *fat*. Then, if the response to the question "Is John *tall?*" is YES, and the response to "Is John *fat?*" is YES, the restriction imposed by John is expressed by

$$R(\text{John}) = tall \cap fat. \qquad (2.67)$$

It should be noted that "John" is actually the name of a binary linguistic variable with two components named *Height* and *Weight*. Thus (2.67) is equivalent to the assignment equations

$$Height = tall \qquad (2.68)$$

and

$$Weight = fat. \qquad (2.69)$$

As implied by (2.66), a tandem connection of decision boxes represents the intersection of the fuzzy sets (or, equivalently, the conjunction of the fuzzy

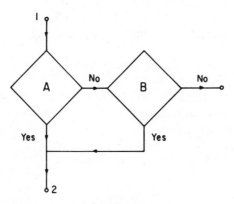

predicates) associated with them. In the case of nonfuzzy sets, their union may be realized by the scheme shown in Fig. 14. In this arrangement of decision boxes, it is clear that transfer from 1 to 2 implies that

$$R(u) = A + \neg A \cap B, \qquad (2.70)$$

and since

$$A \cap B \subset A, \qquad (2.71)$$

it follows that (2.70) may be rewritten as

$$R(u) = A + A \cap B + \neg A \cap B$$
$$= A + (A + \neg A) \cap B$$
$$= A + B, \qquad (2.72)$$

since

$$A + \neg A = U \qquad (2.73)$$

and

$$U \cap B = B. \qquad (2.74)$$

Fig. 14. A graphical representation of the disjunction of fuzzy predicates.

The same scheme would not yield the union of fuzzy sets, since the identity

$$A + \neg A = U \qquad (2.75)$$

does not hold exactly if A is fuzzy. Nevertheless, we can agree to interpret the arrangement of decision boxes in Fig. 14 as one that represents the union of A and B. In this way, we can remain on the familiar ground of flowcharts involving nonfuzzy decision boxes. The flowchart shown in Fig. 16 below illustrates the use of this convention in the definition of *Hippie*.

The conventions described above may be used to represent in a graphical form the assignment of a linguistic value to a linguistic variable. Of particular use in this connection is a tandem combination of decision boxes which represent a series of *bracketing* questions which are intended to narrow down the range of possible values of a variable. As an illustration, suppose that x = John and (see Fig. 15)

$$A_1 = tall,$$

$$A_2 = very\ tall,$$

$$A_3 = very\ very\ tall,$$

$$A_4 = extremely\ tall. \tag{2.76}$$

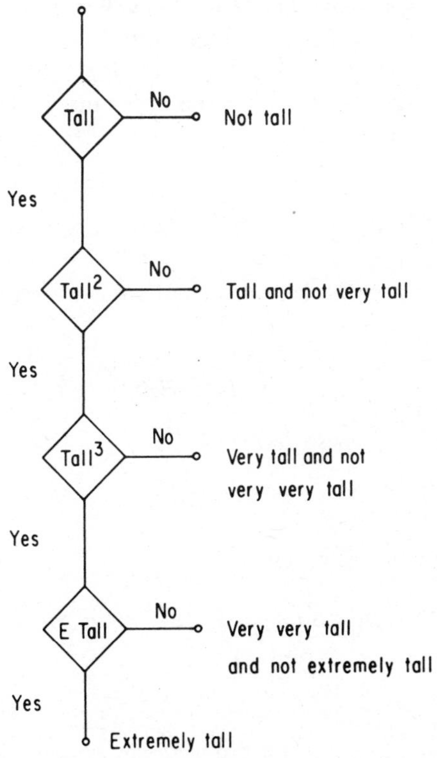

Fig. 15. Use of a tandem combination of decision boxes for purposes of bracketing.

If the answer to the first question is YES, we have

$$R(x) = tall. \qquad (2.77)$$

If the answer to the second question is YES and to the third question is NO, then

$$R(\text{John}) = very\ tall\ and\ not\ very\ very\ tall, \qquad (2.78)$$

which brackets the height of John between *very tall* and *not very very tall*.

By providing a mechanism—as in bracketing—for assigning linguistic values in stages rather than in one step, fuzzy flowcharts can be very helpful in the representation of algorithmic definitions of fuzzy concepts. The basic idea in this instance is to define a complex or a new fuzzy concept in terms of simpler or more familiar ones. Since a fuzzy concept may be viewed as a name for a fuzzy set, what is involved in this approach is, in effect, the decomposition of a fuzzy set into a combination of simpler fuzzy sets.

As an illustration, suppose that we wish to define the term *Hippie*, which may be viewed as a name of a fuzzy subset of the universe of humans. To this end, we employ the fuzzy flowchart[8] shown in Fig. 16. In essense, this flowchart defines the fuzzy set *Hippie* in terms of the fuzzy sets labeled *Long Hair*, *Bald*, *Shaved*, *Job* and *Drugs*. More specifically, it defines the fuzzy set *Hippie* as (+ $\overset{\Delta}{=}$ union)

$$Hippie = (Long\ Hair + Bald + Shaved) \cap Drugs \cap \neg Job \qquad (2.79)$$

Suppose that we pose the following questions and receive the indicated answers.

Does x have *Long Hair*	YES
Does x have a *Job*?	NO
Does x take *Drugs*?	YES

Then we assign to x the restriction

$$R(x) = Long\ Hair \cap \neg Job \cap Drugs,$$

---

[8]It should be understood, of course, that this highly oversimplified definition is used merely as an illustration and has no pretense at being accurate, complete or realistic.

and since it is contained in the right-hand side of (2.79), we conclude that x is a *Hippie*.

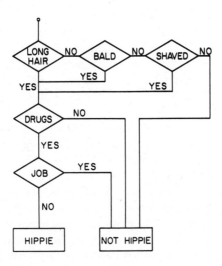

Fig. 16. Algorithmic definition of *Hippie* presented in the form of a fuzzy flowchart.

By modifying the fuzzy sets entering into the definition of *Hippie* through the use of hedges such as *very, more or less, extremely,* etc., and by allowing the answers to be of the form YES/μ or NO/μ, where μ is a numerical or linguistic truth-value, the definition of *Hippie* can be adjusted to fit more closely our conception of what we want to define. Furthermore, we may use a soft *and* (see Part I, Comment 3.1) to allow some trade-offs between the characteristics which define a hippie. And, finally, we may allow our decision boxes to have multiple inputs and multiple outputs. In this way, a concept such as *Hippie* can be defined as completely as one may desire in terms of a set of constituent concepts each of which, in turn, may be defined algorithmically. In essence, then, in employing a fuzzy flowchart to define a fuzzy concept such as *Hippie*, we are decomposing a statement of the general form

$$v(u \text{ is : linguistic value of a Boolean linguistic variable } \mathscr{X}) = $$
$$\text{linguistic value of a Boolean linguistic truth-variable } \mathscr{T} \quad (2.80)$$

into truth-value assignments of the same form, but involving simpler or more familiar variables on the left-hand side of (2.80).

CONCLUDING REMARKS

In this as well as in the preceding sections, our main concern has centered on the development of a conceptual framework for what may be called a *linguistic approach* to the analysis of complex or ill-defined systems and decision processes. The substantive differences between this approach and the conventional quantitative techniques of system analysis raise many issues and problems which are novel in nature and hence require a great deal of additional study and experimentation. This is true, in particular, of some of the basic aspects of the concept of a linguistic variable on which we have dwelt only briefly in our exposition, namely: linguistic approximation, representation of linguistic hedges, nonnumerical base variables, λ- and β-interaction, fuzzy theorems, linguistic probability distributions, fuzzy flowcharts and others.

Although the linguistic approach is orthogonal to what have become the prevailing attitudes in scientific research, it may well prove to be a step in the right direction, that is, in the direction of lesser preoccupation with exact quantitative analyses and greater acceptance of the pervasiveness of imprecision in much of human thinking and perception. It is our belief that, by accepting this reality rather than assuming that the opposite is the case, we are likely to make more real progress in the understanding of the behavior of humanistic systems than is possible within the confines of traditional methods.

REFERENCES

1. Sir William Thomson, *Popular Lectures and Addresses*, McMillan, London, 1891.
2. E. Feigenbaum, *Computers and Thought*, McGraw-Hill, New York, 1963.
3. M. Minsky and S. Papert, *Perceptrons: An Introduction to Computational Geometry*, M.I.T. Press, Cambridge, Mass., 1969.
4. M. Arbib, *The Metaphorical Brain*, Wiley-Interscience, New York, 1972.
5. A. Newell and H. Simon, *Human Problem Solving*, Prentice-Hall, Englewood Cliffs, N.J., 1972.
6. L. A. Zadeh, Fuzzy languages and their relation to human and machine intelligence, in *Proc. Int. Conf. on Man and Computer*, Bordeaux, France, S. Karger, Basel, 1972, pp. 130–165.
7. L. A. Zadeh, Outline of a new approach to the analysis of complex systems and decision processes, *IEEE Trans. Syst., Man and Cybern.* **SMC-3**, 28–44 (January 1973).
8. R. E. Bellman, R. E. Kalaba and L. A. Zadeh, Abstraction and pattern classification, *J. Math. Anal. Appl.* **13**, 1–7 (1966).
9. M. Black, Reasoning with loose concepts, *Dialogue* **2**, 1–12 (1963).
10. L. Wittgenstein, Logical form, *Proc. Aristotelian Soc.* **9**, 162–171 (1929).
11. M. Scriven, The logic of criteria, *J. Philos.* **56**, 857–868 (1959).
12. H. Khatchadourian, Vagueness, meaning and absurdity, *Am. Phil. Quart.* **2**, 119–129 (1965).
13. R. R. Verma, Vagueness and the principle of excluded middle, *Mind* **79**, 67–77 (1970).
14. J. A. Goguen, The Logic of Inexact Concepts, *Synthese* **19**, 325–373 (1969).
15. E. Adams, The logic of "Almost All", *J. Philos. Logic* **3**, 3-17 (1974).

16. K. Fine, Vagueness, truth and logic, Department of Philosophy, University of Edinburgh, 1973.
17. B.S. van Frassen, Presuppositions, supervaluations and free logic, in *The Logical Way of Doing Things* (K. Lambert, Ed.), Yale U. P., New Haven, Conn., 1969.
18. G. Lakoff, Linguistics and natural logic, in *Semantics of Natural Languages,* (D. Davidson and G. Harman, Eds.), D. Reidel, Dordrecht, The Netherlands, 1971.
19. L. A. Zadeh, Shadows of fuzzy sets, *Probl. Transfer Inf.* 2, 37–44 (1966).
20. A. Kaufmann, *Theory of Fuzzy Sets,* Masson, Paris, 1972.
21. J. Goguen, *L*-fuzzy sets, *J. Math. Anal. Appl.* 18, 145–174 (1967).
22. J. G. Brown, A note on fuzzy sets, *Inf. Control* 18, 32–39 (1971).
23. M. Mizumoto, J. Toyoda and K. Tanaka, General formulation of formal grammars, *Inf. Sci.* 4, 87–100, 1972.
24. L. A. Zadeh, Similarity relations and fuzzy orderings, *Inf. Sci.* 3, 177–200 (1971).
25. R. E. Bellman and L. A. Zadeh, Decision-making in a fuzzy environment, *Manage. Sci.* 17, B-141–B-164 (1970).
26. R. E. Bellman and M. Giertz, On the analytic formalism of the theory of fuzzy sets, *Inf. Sci.* 5, 149–156 (1973).
27. L. A. Zadeh, A fuzzy-set-theoretic interpretation of linguistic hedges, *J. Cybern.* 2, 4–34 (1972).
28. E. F. Codd, Relational completeness of data base sublanguages, in *Courant Computer Science Symposia,* Vol. 6, Prentice-Hall, Englewood Cliffs, N.J., 1971.
29. L. A. Zadeh, Fuzzy sets, *Inf. Control* 8, 338–353, 1965.
30. A. Thomasian, *The Structure of Probability Theory With Applications,* McGraw-Hill, New York, 1969.
31. R. E. Moore, *Interval Analysis,* Prentice-Hall, Englewood Cliffs, N.J., 1966.
32. J. A. Brzozowski, Regular expressions for linear sequential circuits, *IEEE Trans. Electron Comput.,* EC-14, 148–156 (1965).
33. E. Shamir, Algebraic, Rational and Context-Free Power Series in Noncommuting Variables, in M. Arbib's *Algebraic Theory of Machines, Languages and Semigroups,* Academic, New York, 1968, pp. 329–341.
34. A. Blikle, Equational languages, *Inf. Control* 21, 134–147 (1972).
35. D. J. Rosenkrantz, Matrix equation and normal forms for context-free grammars, *J. Assoc. Comput. Mach.* 14, 501–507, 1967.
36. J. E. Hopcroft and J. D. Ullman, *Formal Languages and Their Relations to Automata,* Addison-Wesley, Reading, Mass., 1969.
37. A. V. Aho and J. D. Ullman, *The Theory of Parsing, Translation and Compiling,* Prentice-Hall, Englewood Cliffs, N.J., 1973.
38. G. Lakoff, Hedges: a study in meaning criteria and the logic of fuzzy concepts, in *Proc. 8th Reg. Meeting of Chic. Linguist. Soc.,* Univ. of Chicago Linguistics Dept., April 1972.
39. L. A. Zadeh, Quantitative fuzzy semantics, *Inf. Sci.* 3, 159–176 (1971).
40. D. Knuth, Semantics of context-free languages, *Math. Syst. Theory,* 2, 127–145 (1968).
41. P. Lucas et al., Method and notation for the formal definition of programming languages, Rept. TR 25.087, IBM Laboratory, Vienna, 1968.
42. P. Wegner, The Vienna definition language, *ACM Comput. Surv.* 4, 5–63, (1972).
43. J. A. Lee, *Computer Semantics,* Van Nostrand–Reinhold, New York, 1972.
44. Z. Kohavi, *Switching and Finite Automata Theory,* McGraw-Hill, New York, 1970.
45. G. E. Hughes and M. J. Cresswell, *An Introduction to Modal Logic,* Methuen, London, 1968.
46. N. Rescher, *Many-Valued Logic*, McGraw-Hill, New York, 1969.

47. R. Barkan, A functional calculus of first order based on strict implication, *J. Symbol. Logic* **11**, 1–16 (1946).
48. L. A. Zadeh, Probability measures of fuzzy events, *J. Math. Anal. Appl.* **23**, 421–427 (1968).
49. A. DeLuca and S. Termini, A definition of non-probabilistic entropy in the setting of fuzzy set theory, *Inf. Control* **20**, 201–312 (1972).
50. J. Hintikka and P. Suppes (Eds.), *Aspects of Inductive Logic,* North-Holland, Amsterdam, 1966.
51. T. Winograd, *Understanding Natural Language,* Academic, New York, 1972.
52. A. DeLuca and S. Termini, Algebraic properties of fuzzy sets, *J. Math. Anal. Appl.* **40**, 373–386 (1972).
53. A. Rosenfeld, Fuzzy groups, *J. Math. Anal. Appl.* **35**, 512–517 (1971).
54. L. A. Zadeh, Fuzzy Algorithms, *Inf. Control* **12**, 94–102 (1968).
55. E. Santos, Fuzzy algorithms, *Inf. Control* **17**, 326–339 (1970).
56. S. K. Chang, On the execution of fuzzy programs using finite state machines, *IEEE Trans. Electron. Comput.,* C-21, 241–253 (1972).
57. S. S. L. Chang and L. A. Zadeh, Fuzzy mapping and control, *IEEE Trans. Syst., Man and Cybern.,* **SMC-2**, 30–34 (1972).
58. E. T. Lee, Fuzzy languages and their relation to automata, Dissertation, Dept. of Electr. Eng. and Comput. Sci., Univ. of Calif., Berkeley, 1972.
59. R. C. T. Lee, Fuzzy logic and the resolution principle, *J. Assoc. Comput. Mach.* **19**, 109–119 (1972).
60. K. M. Colby, S. Weber and F. D. Hilf, Artificial paranoia, *J. Artif. Intell.* **2**, 1–25 (1971).

Received November, 1973

A Theory of Approximate Reasoning

L.A. ZADEH

*Computer Science Division, University of California at Berkeley,
USA*

Summary

The theory of approximate reasoning outlined in this paper is concerned with the deduction of possibly imprecise conclusions from a set of imprecise premises.

The theory is based on a fuzzy logic, FL, in which the truth-values are linguistic, that is of the form *true, not true, very true, more or less true, false, not very false*, etc., and the rules of inference are approximate rather than exact. Furthermore, the premises are assumed to have the form of fuzzy propositions, for example, "(*X* is much smaller than *Y*) is quite true," "If *X* is small is possible then *Y* is very large is very likely," etc. By using the concept of a possibility — rather than probability — distribution, such propositions are translated into expressions in PRUF (Possibilistic Relational Universal Fuzzy), which is a meaning representation language for natural languages.

An expression in PRUF is a procedure for computing the possibility distribution which is induced by a proposition in a natural language. By applying the rules of inference in PRUF to such distributions, other distributions are obtained which upon retranslation and linguistic approximation yield the conclusions deduced from the fuzzy premises.

The principal rules of inference in fuzzy logic are the projection principle, the particularization/conjunction principle, and the entailment principle. The application of these rules to approximate reasoning is described and illustrated by examples.

1. INTRODUCTION

Informally, by *approximate* or, equivalently, **fuzzy reasoning** we mean the process or processes by which a possibly imprecise conclusion is deduced from a collection of imprecise premises. Such reasoning is, for the most part, qualitative rather than quantitative in nature, and almost all of it falls outside of the domain of applicability of classical logic. A thorough exposition of the foundations of fuzzy reasoning may be found in Gaines (1976a,b,c).

Approximate reasoning underlies the remarkable human ability to understand natural language, decipher sloppy handwriting, play games requiring
mental and/or physical skill and, more generally, make rational decisions in
complex and/or uncertain environments. In fact, it is the ability to reason in
qualitative, imprecise terms that distinguishes human intelligence from machine
intelligence. And yet, approximate reasoning has received little if any attention
within psychology, philosophy, logic, artificial intelligence and other branches
of cognitive sciences, largely because it is not consonant with the deeply
entrenched tradition of precise reasoning in science and contravenes the widely
held belief that precise, quantitative reasoning has the ability to solve the
extremely complex and ill-defined problems which pervade the analysis of
humanistic systems.

In earlier papers (Zadeh 1973, 1975a,b,c, 1976, 1977a,b), we have outlined
a conceptual framework for approximate reasoning based on the notions of
linguistic variable and fuzzy logic. In the present paper, a novel direction
involving the concept of a possibility distribution will be described (see also
Zadeh 1977). As will be seen in the sequel, the concept of a possibility distribution provides a natural basis for the representation of the meaning of
propositions expressed in a natural language, and thereby serves as a convenient
point of departure for the translation of imprecise premises into expressions
in a language PRUF to which the rules of inference associated with this language
can be applied.

Our exposition of approximate reasoning begins with a brief discussion
of the concept of a possibility distribution and its role in the translation of
fuzzy propositions expressed in a natural language. In Sec. 3, the concept
of a linguistic variable is introduced as a device for an approximate characterization of the values of variables and their interrelations. In Secs. 4 and 5, we
shall discuss some of the basic aspects of fuzzy logic — the logic that serves as
a foundation for approximate reasoning — and introduce the concepts of
semantic equivalence and semantic entailment. Finally, in Sec. 6, we formulate
the basic rules of inference in fuzzy logic and illustrate their application to
approximate reasoning by a number of simple examples.

2. THE CONCEPT OF A POSSIBILITY DISTRIBUTION

A basic assumption which underlies our approach to approximate reasoning is
that the imprecision which is instrinsic in natural languages is, in the main,
possibilistic rather than probabilistic in nature. The term **possibilistic** was coined
by B. R. Gaines and L. J. Kohout in their paper on possible automata (1975).

To illustrate the point, consider the proposition $p \triangleq X$ is an integer in the
interval [0,8]. The symbol $\triangleq$ stands for "is defined to be", or "denotes". Clearly,
such a proposition does not associate a unique integer with X; rather, it indicates
that any integer in the interval [0,8] could possibly be a value of X, and that
any integer not in the interval could not be a value of X.

This obvious observation suggests the following interpretation of p. The proposition "X is an integer in the interval $[0,8]$" induces a possibility distribution Π_X which associates with each integer n the possibility that n could be a value of X. Thus, for the proposition in question

$$Poss\{X = n\} = 1 \quad \text{for} \quad 0 \leqslant n \leqslant 8$$

and

$$Poss\{X = n\} = 0 \quad \text{for} \quad n < 0 \quad \text{or} \quad n > 8$$

where $Poss\{X = n\}$ is an abbreviation for "The possibility that X may assume the value n". Note that the possibility distribution induced by p is *uniform* in the sense that the possibility values are equal to unity for n in $[0,8]$ and zero elsewhere.

Next, consider the fuzzy proposition $q \triangleq X$ is a small integer, in which *small integer* is a fuzzy set defined by, say,

$$small\ integer = 1/0 + 1/1 + 0.8/2 + 0.6/3 + 0.4/4 + 0.2/5 \qquad (2.1)$$

in which + denotes the union rather than the arithmetic sum, and a singleton of the form $0.8/2$ signifies that the grade of membership of the integer 2 in the fuzzy set *small integer* is 0.8 (see A. Kaufmann (1975), C. V. Negoita and D. Ralescu (1975), and L. A. Zadeh, K. S. Fu, K. Tanaka and M. Shimura (1975)).

As an extension of our interpretation of the nonfuzzy proposition p, we shall interpret q as follows. The proposition $q \triangleq X$ is a small integer induces a possibility distribution Π_X which equates the possibility of X taking a value n to the grade of membership of n in the fuzzy set *small integer*. Thus

$$Poss\{X = 0\} = 1$$
$$Poss\{X = 2\} = 0.8$$
$$Poss\{X = 5\} = 0.2$$

and $\quad Poss\{X = 6\} = 0$.

More generally, we shall say that a fuzzy proposition of the form $p \triangleq X$ is F, where X is a variable taking values in a universe of discourse U, and F is a fuzzy subset of U, *induces a* **possibility distribution** Π_X *which is equal to F*, that is,

$$\Pi_X = F . \qquad (2.2)$$

Thus, in essence, the possibility distribution of X is a fuzzy set which serves to define the possibility that X could assume any specified value in U. Stated more concretely, if $u \in U$ and $\mu_F\colon U \to [0,1]$ is the membership function of F, then the possibility that $X = u$ given "X is F" is

$$Poss\{X = u \mid X \text{ is } F\} = \mu_F(u) , \quad u \in U . \qquad (2.3)$$

Since the concept of a possibility distribution coincides with that of a fuzzy set, possibility distributions may be manipulated by the rules governing the manipulation of fuzzy sets and, more particularly, fuzzy restrictions. A **fuzzy restriction** is a fuzzy set which serves as an elastic constraint on the values that may be assigned to a variable. A variable which is associated with a fuzzy restriction or, equivalently, with a possibility distribution, is a fuzzy variable. In what follows, we shall focus our attention only on those aspects of possibility distributions which are of relevance to approximate reasoning.

Possibility versus probability

What is the difference between possibility and probability? Intuitively, possibility relates to our perception of the degree of feasibility or ease of attainment, whereas probability is associated with the degree of belief, likelihood, frequency, or proportion. Thus, what is possible may not be probable, and what is improbable need not be impossible. A more concrete statement of this relation is embodied in the *possibility/probability consistency principle* (Zadeh, 1977a). More importantly, however, the distinction between possibility and probability manifests itself in the different rules which govern their combinations, especially under the union. More specifically, if A is a nonfuzzy subset of U, and Π_X is the possibility distribution induced by the proposition "X is F", then the possibility measure, $\Pi(A)$, of A is defined as

$$\Pi(A) \triangleq Poss\{X \in A\} \triangleq Sup_{u \in A}\ \mu_F(u)\ . \tag{2.4}$$

The possibility measure defined by (2.4) is a special case of the more general concept of a fuzzy measure defined by Sugeno (1974) and Terano and Sugeno (1975). More generally, if A is a fuzzy subset of U, then

$$\Pi(A) \triangleq Poss\{X \text{ is } A\} \triangleq Sup_u\ [\mu_F(u) \wedge \mu_A(u)] \tag{2.5}$$

where μ_A is the membership function of A and $\wedge \triangleq min$.

From the definition of possibility measure, it follows at once that, for arbitrary subsets A and B of U, the possibility measure of the union of A and B is given by

$$\Pi(A \cup B) = \Pi(A) \vee \Pi(B) \tag{2.6}$$

where $\vee \triangleq max$. Thus, the possibility measure does not have the basic additivity property of probability measure, namely,

$$P(A \cup B) = P(A) + P(B) \quad \text{if } A \text{ and } B \text{ are disjoint} \tag{2.7}$$

where $P(A)$ and $P(B)$ denote the probability measures of A and B, respectively.

Unlike probability, the concept of possibility in no way involves the notion of repeated experimentation. Thus, the concept of possibility is nonstatistical in character and, as such, is a natural concept to use when the imprecision or uncertainty in the phenomena under study are not susceptible of statistical analysis or characterization.

Possibility assignment equations

The reason why the concept of a possibility distribution plays such an important role in approximate reasoning relates to our assumption that a proposition in a natural language may be interpreted as an assignment of a fuzzy set to a possibility distribution. More specifically, if p is a proposition in a natural language, we shall say that p *translates* into a *possibility assignment equation*:

$$p \to \Pi_{(X_1, \ldots X_n)} = F \qquad (2.8)$$

where $X_1, \ldots X_n$ are variables which are explicit or implicit in p; $\Pi_{(X_1, \ldots X_n)}$ is the possibility distribution of the n-ary variable $X \triangleq (X_1, \ldots X_n)$; and F is a fuzzy relation, that is, a fuzzy subset of the cartesian product $U_1 \times \ldots \times U_n$, where U_i, $i = 1, \ldots n$, is the universe of discourse associated with X_i. In this context, the possibility assignment equation

$$\Pi_{(X_1, \ldots X_n)} = F \qquad (2.9)$$

will be referred to as the **translation** of p and, conversely, p will be said to be a **retranslation** of (2.9), in which case its relation to (2.9) will be represented as

$$p \leftarrow \Pi_{(X_1, \ldots X_n)} = F. \qquad (2.10)$$

In general, a proposition of the form $p \triangleq X$ is F, where X is the name of an object or a proposition, translates not into

$$p \to \Pi_X = F \qquad (2.11)$$

but into

$$p \to \Pi_{A(X)} = F \qquad (2.12)$$

where $A(X)$ is an implied attribute of X. For example,

$$\text{Joe is young} \to \Pi_{Age(Joe)} = young \qquad (2.13)$$

$$\text{Maria is blond} \to \Pi_{Colour(Hair(Maria))} = blond \qquad (2.14)$$

$$\text{Max is about as tall as Jim} \to$$
$$\Pi_{(Height(Max), Height(Jim))} = approximately\text{-}equal \qquad (2.15)$$

where *young*, *blond*, and *approximately equal* are specified fuzzy relations (unary and binary) in their respective universes of discourse. More concretely, if u is a numerical value of the age of Joe, then (2.13) implies that

$$Poss\{Age(Joe) = u\} = \mu_{young}(u). \qquad (2.16)$$

Similarly, if u is an identifying label for the colour of hair, then (2.14) implies that

$$Poss\{Colour(Hair(Maria)) = u\} = \mu_{blond}(u), \qquad (2.17)$$

while (2.15) signifies that

$$Poss\{Height(Max) = u, Height(Jim) = v\} =$$
$$\mu_{approximately\ equal}(u, v) \qquad (2.18)$$

where u and v are the generic values of the variables *Height(Max)* and *Height(Jim)*, respectively.

Projection and particularization

Among the operations that may be performed on a possibility distribution, there are two that are of particular relevance to approximate reasoning: projection and particularization.

Let $\Pi_{(X_1, \ldots X_n)}$ denote an n-ary possibility distribution which is a fuzzy realtion in $U_1 \times \ldots \times U_n$, with the possibility distribution function of $\Pi_{(X_1, \ldots X_n)}$ (that is, the membership function of $\Pi_{(X_1, \ldots X_n)}$) denoted by $\pi_{(X_1, \ldots X_n)}$ or, more simply, as π_X.

Let $s \triangleq (i_1, \ldots i_k)$ be a subsequence of the index sequence $(1, \ldots n)$ and let s' denote the complementary subsequence $s' \triangleq (j_1, \ldots j_m)$ (for example, for $n = 5$, $s = (1,3,4)$ and $s' = (2,5)$). In terms of such sequences, a k-tuple of the form $(A_{i_1}, \ldots A_{i_k})$ may be expressed in an abbreviated form as $A_{(s)}$. In particular, the variable $X_{(s)} = (X_{i_1}, \ldots X_{i_k})$ will be referred to as a k-ary **subvariable** of $X \triangleq (X_1, \ldots X_n)$, with $X_{(s')} = (X_{j_1}, \ldots X_{j_m})$ being a subvariable complementary to $X_{(s)}$.

The **projection** of $\Pi_{(X_1, \ldots X_n)}$ on $U_{(s)} \triangleq U_{i_1} \times \ldots \times U_{i_k}$ is a k-ary possibility distribution denoted by

$$\Pi_{X_{(s)}} \triangleq Proj_{U_{(s)}} \Pi_{(X_1, \ldots X_n)} \tag{2.19}$$

and defined by

$$\pi_{X_{(s)}}(u_{(s)}) \triangleq Sup_{u_{(s')}} \pi_X(u_1, \ldots u_n) \tag{2.20}$$

where $\pi_{X_{(s)}}$ is the possibility distribution function of $\Pi_{X_{(s)}}$. For example, for $n = 2$,

$$\pi_{X_1}(u_1) \triangleq Sup_{u_2} \pi_{(X_1,X_2)}(u_1,u_2)$$

is the expression for the possibility distribution function of the projection of $\Pi_{(X_1,X_2)}$ on U_1. By analogy with the concept of a marginal probability distribution, $\Pi_{X_{(s)}}$ will be referred to as a **marginal possibility distribution**. Note that our use of $\Pi_{X_{(s)}}$ in (2.19) to denote the projection of Π_X on $U_{(s)}$ anticipates (2.21).

The importance of the concept of a marginal possibility distribution derives from the fact that $\Pi_{X_{(s)}}$ may be regarded as the possibility distribution of the subvariable $X_{(s)}$. Thus, stated as the **projection principle** (in Sec. 6), the relation between $X_{(s)}$ and $\Pi_{X_{(s)}}$ may be expressed as follows.

From the possibility distribution, $\Pi_{(X_1, \ldots X_n)}$, of the variable $X \triangleq (X_1, \ldots X_n)$, the possibility distribution $\Pi_{X_{(s)}}$ of the subvariable $X_{(s)} \triangleq (X_{i_1}, \ldots X_{i_k})$ may be inferred by projecting $\Pi_{(X_1, \ldots X_n)}$ on $U_{(s)}$, that is,

$$\Pi_{X_{(s)}} = Proj_{U_{(s)}} \Pi_{(X_1, \ldots X_n)} . \tag{2.21}$$

As a simple illustration, assume that $n = 3$, $U_1 = U_2 = U_3 = a+b$ or, more conventionally, $\{a,b\}$ and $\Pi_{(X_1,X_2,X_3)}$ is expressed as a linear form

$$\Pi_{(X_1,X_2,X_3)} = 0.8\,aaa + 1\,aab + 0.6\,baa + 0.2\,bab + 0.5\,bbb \tag{2.22}$$

in which a term of the form $0.6\,baa$ signifies that

$$Poss\,\{X_1 = b, X_2 = a, X_3 = a\} = 0.6 \ . \tag{2.23}$$

To derive $\Pi_{(X_1, X_2)}$ from (2.22) it is sufficient to replace the value of X_3 in each term in (2.22) by the null string $\wedge$. This yields

$$\begin{aligned}
\Pi_{(X_1, X_2)} &= 0.8\,aa + 1\,aa + 0.6\,ba + 0.2\,ba + 0.5\,bb \\
&= 1\,aa + 0.6\,ba + 0.5\,bb
\end{aligned} \tag{2.24}$$

and similarly

$$\begin{aligned}
\Pi_{X_1} &= 1\,a + 0.6\,b + 0.5\,b \\
&= 1\,a + 0.6\,b \ .
\end{aligned} \tag{2.25}$$

Turning to the operation of particularization, let $\Pi_{(X_1, \ldots X_n)} = F$ denote the possibility distribution of $X = (X_1, \ldots X_n)$, and let $\Pi_{X_{(s)}} = G$ denote a specified possibility distribution (not necessarily the marginal distribution) of the subvariable $X_{(s)} = (X_{i_1}, \ldots X_{i_k})$.

Informally, by the **particularization** of $\Pi_{(X_1, \ldots X_n)}$ is meant the modification of $\Pi_{(X_1, \ldots X_n)}$ resulting from the stipulation that the possibility distribution of $\Pi_{X_{(s)}}$ is G. More specifically,

$$\Pi_{(X_1, \ldots X_n)} \, [\Pi_{X_{(s)}} = G] \triangleq F \cap \bar{G} \tag{2.26}$$

where the left-hand member places in evidence the X_i (that is, the attributes) which are particularized in $\Pi_{(X_1, \ldots X_n)}$, while the right-hand member defines the effect of particularization, with $\bar{G}$-denoting the cylindrical extension of G, that is, the cylindrical fuzzy set in $U_1 \times \ldots \times U_n$ whose projection on $U_{(s)}$ is G. Thus,

$$\begin{aligned}
\mu_{\bar{G}}(u_1, \ldots u_n) &\triangleq \mu_G(u_{i_1}, \ldots u_{i_k}) \ , \\
(u_1, \ldots u_n) &\in U_1 \times \ldots \times U_n \ .
\end{aligned} \tag{2.27}$$

As a simple illustration, consider the possibility distribution defined by (2.22) and assume that

$$\Pi_{(X_1, X_2)} = 0.4\,aa + 0.9\,ba + 0.1\,bb \ . \tag{2.28}$$

In this case,

$$\begin{aligned}
\bar{G} &= 0.4\,aaa + 0.4\,aab + 0.9\,baa + 0.9\,bab + 0.1\,bba + 0.1\,bbb \\
F \cap \bar{G} &= 0.4\,aaa + 0.4\,aab + 0.6\,baa + 0.2\,bab + 0.1\,bbb
\end{aligned}$$

and hence

$$\begin{aligned}
\Pi_{(X_1, X_2, X_3)} \, [\Pi_{(X_1, X_2)} = G] = \\
0.4\,aaa + 0.4\,aab + 0.6\,baa + 0.2\,bab + 0.1\,bbb
\end{aligned} \tag{2.29}$$

In general, some of the variables in a particularized possibility distribution (or a fuzzy relation) are assigned fixed values in their respective universes of discourse, while others are associated with possibility distributions. For example, in the case of a fuzzy relation which characterizes the fuzzy set of men who are tall, blond, and named Smith, the particularlized relation has the form

$$MAN[Name = Smith; \Pi_{Height} = TALL; \Pi_{Colour(Hair)} = BLOND]$$

(2.30)

(Note that the label of a relation is capitalized when it is desired to stress that it denotes a relation.) Similarly, the fuzzy set of men who have the above characteristics and, in addition, are approximately 30 years old, would be represented as

$$MAN[Name = Smith; \Pi_{Height} = TALL; \Pi_{Colour(Hair)} = BLOND;$$
$$\Pi_{Age} = APPROXIMATELY\ EQUAL\ [Age = 30]]\ .\qquad (2.31)$$

In this case, the possibility distribution which is associated with the variable *Age* is in itself a particularized possibility distribution.

It should be noted that the representations exemplified by (2.30) and (2.31) are somewhat similar in appearance to those that are commonly employed in semantic network and higher order predicate calculi representations of propositions in a natural language. Expositions of such representations may be found in Newell and Simon (1972), Miller and Johnson-Laird (1976), Bobrow and Collins (1975), Minsky (1975), and other books and papers listed in the bibliography. An essential difference, however, lies in the use of possibility distributions in (2.30) and (2.31) for the characterization of values of fuzzy variables, and in the concrete specification of the manner in which a possibility distribution is modified by particularization.

Meaning and information

Particularization as defined by (2.26) plays a particularly important role in PRUF — a language intended for the representation of the meaning of fuzzy propositions. A brief description of PRUF appears in Zadeh (1977b). A more detailed exposition of PRUF will be provided in a forthcoming paper.

Briefly, an expression, *P*, in PRUF is, in general, a procedure for computing a possibility distribution. More specifically, let *U* be a universe of discourse and let ℜ be a set of relations in *U*. Then, the pair

$$D \triangleq (U, \Re)$$

(2.32)

constitutes a **database**, with *P* defined on a subset of relations in ℜ. As defined here, the concept of a database is related to that of a possible world in modal logic (Hughes and Cresswell, 1968; Miller and Johnson-Laird, 1976).

If *p* is an expression in a natural language and *P* is its translation in PRUF, that is,

$$p \to P\ ,$$

then the procedure *P* may be viewed as defining the **meaning**, *M(p)*, of *p*, with the possibility distribution computed by *P* constituting the **information**, *I(p)*.

conveyed by p. (The procedure defined by an expression in PRUF and the possibility distribution which it yields are analogous to the intension and extension of a predicate in two-valued logic. (Cresswell, 1975.) When *meaning* is used loosely, no differentiation between $M(p)$ and $I(p)$ is made.)

As a simple illustration, consider the proposition

$$p \triangleq \text{John resides near Berkeley} \tag{2.33}$$

which in PRUF translates into

$$RESIDENCE\,[Subject = John;$$
$$\Pi_{Location} = Proj_{\mu \times City\,1}\ NEAR\,[City\,2 = Berkeley]] \tag{2.34}$$

where *NEAR* is a fuzzy relation with the frame $NEAR \| City\,1 | City\,2 | \mu |$ and the expression $Proj_{\mu \times City\,1}\ NEAR\,[City\,2 = Berkeley]$ represents the fuzzy set of cities which are near Berkeley. The frame of a fuzzy relation exhibits its name together with the names of its variables (that is, attributes) and μ — the grade of membership of each tuple in the relation.

The expression in PRUF represented by (2.34)) is, in effect, a procedure for computing the possibility distribution of the location of residence of John. Thus, given a relation *NEAR*, it will return a possibility distribution of the form ($\pi \triangleq$ possibility-value)

RESIDENCE	Subject	Location	π
	John	Oakland	1
	John	Palo Alto	0.6
	John	San Jose	0.2
	John	Orinda	0.8
	...	...	...

which may be regarded as the information conveyed by the proposition "John resides near Berkeley".

PRUF plays an essential role in approximate reasoning because it serves as a basis for translating the fuzzy premises expressed in a natural language into possibility assignment equations to which the rules of inference in approximate reasoning can be applied in a systematic fashion. In Sec. 4, we shall discuss in greater detail some of the basic translation rules in fuzzy logic which constitute a small subset of the translation rules in PRUF. This brief exposition of PRUF will suffice for our purposes in the present paper.

We turn next to the concept of a linguistic variable — a concept that plays a basic role in approximate reasoning, fuzzy logic, and the linguistic approach to systems analysis.

3. THE CONCEPT OF A LINGUISTIC VARIABLE

In describing the behaviour of humanistic — that is, human-centered — systems, we generally use words rather than numbers to characterize the values of variables as well as the relations between them. For example, the age of a person may be described as *very young*, intelligence as *quite high*, the relation with another person as *not very friendly*, and appearance as *quite attractive*.

Clearly, the use of words in place of numbers implies a lower degree of precision in the characterization of the values of a variable. In some instances, we elect to be imprecise because there is no need for a higher degree of precision. In most cases, however, the imprecision is forced upon by the fact that there are no units of measurement for the attributes of an object and no quantitative criteria for representing the values of such attributes as points on an anchored scale.

Viewed in this perspective, the concept of a linguistic variable may be regarded as a device for systematizing the use of words or sentences in a natural or synthetic language for the purpose of characterizing the values of variables and describing their interrelations. In this role, the concept of a linguistic variable serves a basic function in approximate reasoning both in the representation of values of variables and in the characterization of truth-values, probability-values, and possibility-values of fuzzy propositions.

In this section, we shall focus our attention only on those aspects of the concept of a linguistic variable which have a direct bearing on approximate reasoning. More detailed discussions of the concept of a linguistic variable and its applications may be found in Zadeh (1973, 1975c), Wenstop (1975, 1976), Mamdani and Assilian (1975), Procyk (1976), and other papers listed in the bibliography.

As a starting point for our discussion, it is convenient to consider a variable such as *Age*, which may be viewed both as a numerical variable ranging over, say, the interval [0,150], and as a linguistic variable which can take the values *young, not young, very young, not very young, quite young, old, not very young and not very old*, etc. Each of these values may be interpreted as a label of a fuzzy subset of the universe of discourse $U = [0,150]$, whose base variable, u, is the generic numerical value of *Age*.

Typically, the values of a linguistic variable such as *Age* are built up of one or more **primary terms** (the labels of *primary fuzzy sets* which play a role somewhat analogous to that of physical units in mechanistic systems), together with a collection of modifiers and connectives which allow a composite linguistic value to be generated from the primary terms. Usually, the number of such terms is two, with one being an antonym of the other. For example, in the case of *Age*, the primary terms are *young* and *old*.

A basic assumption underlying the concept of a linguistic variable is that the meaning of the primary terms is context-dependent, whereas the meaning of the modifiers and connectives is not. Furthermore, once the meaning of the primary terms is specified (or "calibrated") in a given context, the meaning of composite terms such as *not very young, not very young and not very old*, etc., may be computed by the application of a semantic rule.

Typically, the **term-set**, that is, the set of linguistic values of a linguistic variable, comprises the values generated from each of the primary terms together with the values generated from various combinations of the primary terms. For example, in the case of *Age*, a partial list of the linguistic values of *Age* is the following:

young	old	not young nor old
not young	not old	not very young and not very old
very young	very old	young or old
not very young	not very old	not young or not old
quite young	quite old	etc.
more or less young	more or less old	
extremely young	extremely old	
etc.	etc.	

What is important to observe is that most linguistic variables have the same basic structure as *Age*. For example, on replacing *young* with *tall* and *old* with *short*, we obtain the list of linguistic values of the linguistic variable *Height*. The same applies to the linguistic variables *Weight* (*heavy* and *light*), *Appearance* (*beautiful* and *ugly*), *Speed* (*fast* and *slow*), *Truth* (*true* and *false*), etc., with the words in parentheses representing the primary terms.

As is shown in Zadeh (1973, 1975c), a linguistic variable may be characterized by an attributed grammar (see Knuth 1968; Lewis *et al* 1974) which generates the term-set of the variable and provides a simple procedure for computing the meaning of a composite linguistic value in terms of the primary fuzzy sets which appear in its constituents.

As an illustration, consider the attributed grammar shown in which S, B, C, D, and E are nonterminals; not, and, a and b are terminals; a and b are the primary terms (and also the primary fuzzy sets); subscripted symbols are the fuzzy sets which are labelled by the corresponding nonterminals, with $L \triangleq$ left (that is, pertaining to the antecedent), $R \triangleq$ right (that is, pertaining to the consequent); and a production of the form

$$S \to S \text{ and } B \quad : \quad S_L = S_R \cap B_R \tag{3.1}$$

signifies that the fuzzy set which is the meaning of the antecedent, S, is the intersection of S_R, the fuzzy set which is the meaning of the consequent S, and B_R, the fuzzy set which is the meaning of the consequent B.

$$
\begin{aligned}
S &\to B & &: & S_L &= B_R \\
S &\to S \text{ and } B & &: & S_L &= S_R \cap B_R \\
B &\to C & &: & B_L &= C_R \\
B &\to \text{not } C & &: & B_L &= C_R' \ (\triangleq \text{complement of } C_R)
\end{aligned}
\tag{3.2}
$$

$$
\begin{array}{lll}
C \to S & : & C_L = S_R \\
C \to D & : & C_L = D_R \\
C \to E & : & C_L = E_R \\
D \to \text{very } D & : & D_L = D_R^2 \ (\triangleq \text{square of } D_R) \\
E \to \text{very } E & : & D_L = E_R^2 \ (\triangleq \text{square of } E_R) \\
D \to a & : & D_L = a \\
E \to b & : & E_L = b \ .
\end{array}
$$

The grammar in question generates the linguistic values exemplified by the list:

a	*b*	*a* and *b*
not *a*	not *b*	not *a* and *b*
very *a*	very *b*	not *a* and not *b*
not very *a*	not very *b*	not very *a* and not very *b*
not very very *a*	not very very *b*	etc.
etc.	etc.	

In general, to compute the meaning of a linguistic value, ℓ, generated by the grammar, the meaning of each node of the syntax tree of ℓ is computed — by the use of equations (3.2) — in terms of the meanings of its immediate descendants. In most cases, however, this can be done by inspection — which involves a straightforward application of the translation rules which will be formulated in Sec. 4. Thus, we readily obtain, for example:

$$\text{not very } a \to (a^2)' \tag{3.3}$$
$$\text{not very } a \text{ and not very } b \to (a^2)' \cap (b^2)'$$

where a' is the complement of a and a^2 is defined by

$$\mu_{a^2}(u) = (\mu_a(u))^2, \quad u \in U. \tag{3.4}$$

To characterize the primary fuzzy sets a and b, it is frequently convenient to employ standardized membership functions with adjustable parameters. One such function is the S-function, $S(u;\alpha,\beta,\gamma)$, defined by

$$
\begin{aligned}
S(u;\alpha,\beta,\gamma) &= 0 & \text{for } u \leqslant \alpha \\[4pt]
&= 2\left(\frac{u-\alpha}{\gamma-\alpha}\right)^2 & \text{for } \alpha \leqslant u \leqslant \beta \\[4pt]
&= 1 - 2\left(\frac{u-\gamma}{\gamma-\alpha}\right)^2 & \text{for } \beta \leqslant u \leqslant \gamma \\[4pt]
&= 1 & \text{for } u \geqslant \gamma
\end{aligned}
\tag{3.5}
$$

where the parameter $\beta \triangleq \frac{\alpha+\gamma}{2}$ is the crossover point, that is, the value of u at which $S(u;\alpha,\beta,\gamma) = 0.5$. For example, if $a \triangleq$ young and $b \triangleq$ old, we may have (see Fig. 1)

$$\mu_{young} = 1 - S(20,30,40) \tag{3.6}$$

and

$$\mu_{old} = S(40,55,70) \tag{3.7}$$

in which the argument u is suppressed for simplicity. Thus, in terms of (3.6), the translation of the proposition $p \triangleq$ Joe is young (see (2.13)), may be expressed more concretely as

$$\text{Joe is young} \rightarrow \pi_{Age(Joe)} = 1 - S(20,30,40) \tag{3.8}$$

where $\pi_{Age(Joe)}$ is the possibility distribution function of the linguistic variable $Age(Joe)$. Similarly,

$$\text{Joe is not very young} \rightarrow \pi_{Age(Joe)} = 1 - [1 - S(20,30,40)]^2. \tag{3.9}$$

An important aspect of the concept of a linguistic variable relates to the fact that, in general, the term-set of such a variable is not closed under the various operations that may be performed on fuzzy sets, for example, union, intersection, product, etc. For example, if ℓ is a linguistic value of a variable X, then, in general, ℓ^2 is not in the term-set of X.

The problem of finding a linguistic value of X whose meaning approximates to a given fuzzy subset of U is called the problem of **linguistic approximation** (Zadeh, 1975c, Wenstop, 1975; Procyk, 1976). We shall not discuss in the present paper the ways in which this nontrivial problem can be approached, but will assume that linguistic approximation is implicit in the retranslation of a possibility distribution (see (2.10)) into a proposition expressed in a natural language.

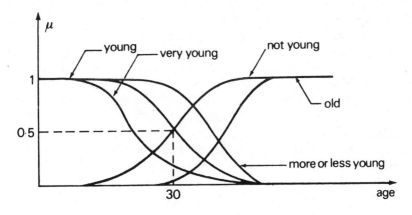

Fig. 1 – Graphical representation of linguistic values of Age.

4. FUZZY LOGIC (FL)

In a broad sense, fuzzy logic is the logic of approximate reasoning; that is, it bears the same relation to approximate reasoning that two-valued logic does to precise reasoning.

In this section, we shall focus our attention on a particular fuzzy logic, FL, whose truth-values are linguistic, that is, are expressible as the values of a linguistic variable *Truth* whose base variable takes values in the unit interval. In this sense, the base logic for FL is Lukasiewicz's L_{Aleph_1} logic whose truth-value set is the interval [0,1].

The principal constituents of FL are the following: (i) Translation rules, (ii) Valuation rules, and (iii) Inference rules.

By **translation rules** is meant a set of rules which yield the translation of a modified or composite proposition from the translations of its constituents. For example, if *p* and *q* are fuzzy propositions which translate into (see (2.8))

$$p \to \Pi_{(X_1, \ldots X_n)} = F \qquad (4.1)$$
and
$$q \to \Pi_{(Y_1, \ldots Y_m)} = G \qquad (4.2)$$

respectively, then the rule of conjunctive composition − which will be stated at a later point in this section − yields the translation of the composite proposition "*p* and *q*".

By **valuation rules** is meant the set of rules which yield the truth-value (or the probability-value or the possibility-value) of the modified or composite proposition from the specification of the truth-values (or probability-values or possibility-values) of its constituents. A typical example of the valuation rule is the conjunctive valuation rule which expresses the truth-value of the composite proposition "*p* and *q*" as a function of the truth-values of *p* and *q* − for example, *not very true* and *quite true*, respectively.

The principal rules of inference in FL are: (a) The projection principle, (b) The particularization/conjunction principle; and (c) The entailment principle. In combination, these rules lead to the **compositional rule of inference** which may be viewed as a generalization of the *modus ponens*.

In what follows, we shall discuss briefly only those aspects of fuzzy logic which are of direct relevance to approximate reasoning. A more detailed discussion of FL may be found in Zadeh (1975a) and Bellman and Zadeh (1976).

Translation rules

The translation rules in FL may be divided into several basic categories. Among these are:

Type I.   Rules pertaining to modification.
Type II.   Rules pertaining to composition.
Type III. Rules pertaining to quantification.
Type IV.   Rules pertaining to qualification.

Simple examples of propositions to which the rules in question apply are the following:

Type I.   *X* is very small.
           Therese is highly intelligent.
Type II.   *X* is small and *Y* is large.
           If *X* is small then *Y* is large.

Type III. Most Swedes are tall.

Many men are taller than most men.

Type IV. John is tall is very true.

John is tall is not very likely.

John is tall is quite possible.

In combination, the rules in question may be applied to the translation of more complex propositions exemplified by:

If ((X is small and Y is large) is very likely) then (Z is very large is not very likely).

((Many men are taller than most men) is very true) is quite possible.

Rules of Type I

A basic rule of Type I is the **modifier rule**, which may be stated as follows.

Let X be a variable taking values in $U = \{u\}$, let F be a fuzzy subset of U, and let p be a proposition of the form "X is F". If the translation of p is expressed by

$$X \text{ is } F \to \Pi_X = F \qquad (4.3)$$

then the translation of the modified proposition "X is mF", where m is a modifier such as *not, very, more or less*, etc., is given by

$$X \text{ is } mF \to \Pi_X = F^+ \qquad (4.4)$$

where F^+ is a modification of F induced by m. (More detailed discussions of various types of modifiers may be found in Zadeh (1972a, 1975c), Lakoff (1973a,b), Wenstop (1975), McVicar-Whelan (1975), Hersh and Caramazza (1976), and other papers listed in the bibliography.) More specifically,

If $m = not$, then $F^+ = F' \triangleq$ complement of F $\qquad (4.5)$

If $m = very$, then $F^+ = F^2$ $\qquad (4.6)$

where

$$F^2 = \int_U \mu_F^2(u)/u \qquad (4.7)$$

The "integral" representation of a fuzzy set in the form $F = \int_U \mu_F(u)/u$ signifies

that F is a union of the fuzzy singletons $\mu_F(u)/u$, $u \in U$, where μ_F is the membership function of F. Thus, (4.7) means that the membership function of F^2 is the square of that of F.

If $m = more\ or\ less$, then $F^+ = \sqrt{F}$ $\qquad (4.8)$

where

$$\sqrt{F} = \int_U \sqrt{\mu_F(u)}/u \qquad (4.9)$$

or,

$$F^+ = \int_U \mu_F(u) K(u) \tag{4.10}$$

where $K(u)$ is the **kernel** of more or less (Zadeh, 1972).

As a simple illustration, consider the proposition "X is small", where *small* is defined by

$$small = 1/0 + 1/1 + 0.8/2 + 0.6/3 + 0.4/4 + 0.2/5 . \tag{4.11}$$

Then

$$X \text{ is very small} \to \Pi_X = F^+ \tag{4.12}$$

where

$$F^+ = F^2 = 1/0 + 1/1 + 0.64/2 + 0.36/3 + 0.16/4 + 0.04/5 . \tag{4.13}$$

It is important to note that (4.6) and (4.8) should be regarded merely as standardized default definitions which may be replaced by other definitions whenever they do not fit the desired sense of the modifier m. Another point that should be noted is that X in (4.3) need not be a unary variable. Thus, (4.3) subsumes propositions of the form "X and Y are F", as in "X and Y are close", where *CLOSE* is a fuzzy binary relation in $U \times U$. Thus, if

then

$$X \text{ and } Y \text{ are close} \to \Pi_{(X,Y)} = CLOSE \tag{4.14}$$

$$X \text{ and } Y \text{ are very close} \to \Pi_{(X,Y)} = CLOSE^2 . \tag{4.15}$$

Rules of Type II

Compositional rules of Type II pertain to the translation of a proposition p which is a composite of propositions q and r. The most commonly employed modes of compositions are: conjunction, disjunction, and conditional composition (or implication). The translation rules for these modes of composition are as follows. (We are tacitly assuming that the compositions in question are noninteractive in the sense defined in Zadeh (1975c).)

Let X and Y be variables taking values in U and V, respectively, and let F and G be fuzzy subsets of U and V. If

$$X \text{ is } F \to \Pi_X = F \tag{4.16}$$

$$Y \text{ is } G \to \Pi_Y = G \tag{4.17}$$

then

(a) $X \text{ is } F \text{ and } Y \text{ is } G \to \Pi_{(X,Y)} = \bar{F} \cap \bar{G}$ $\tag{4.18}$
$$= F \times G$$

(b) $X \text{ is } F \text{ or } Y \text{ is } G \to \Pi_{(X,Y)} = \bar{F} + \bar{G} \tag{4.19}$

and (c_1) If X is F then Y is $G \to \Pi_{(X,Y)} = \bar{F}' \oplus \bar{G}$ $\tag{4.20}$

or (c_2) If X is F then Y is $G \to \Pi_{(X,Y)} = F \times G + F' \times V$ $\tag{4.21}$

where $\Pi_{(X,Y)}$ is the possibility distribution of the binary variable (X,Y), $\bar{F}$ and $\bar{G}$ are the cylindrical extensions of F and G, respectively, that is,

$$\bar{F} = F \times V \tag{4.22}$$

$$\bar{G} = U \times G ; \tag{4.23}$$

$F \times G$ is the Cartesian product of F and G, which may be expressed as $\bar{F} \cap \bar{G}$ and is defined by

$$\mu_{F \times G}(u,v) = \mu_F(u) \wedge \mu_G(v) , \quad u \in U, \ v \in V , \tag{4.24}$$

$+$ is the union, and $\oplus$ is the bounded sum, that is,

$$\mu_{\bar{F}' \oplus \bar{G}}(u,v) = 1 \wedge [1 - \mu_F(u) + \mu_G(v)] \tag{4.25}$$

where $+$ and $-$ denote the arithmetic sum and difference. Note that there are two interpretations of the conditional composition, (c_1) and (c_2). Of these, (c_1) is consistent with the definition of implication in L_{Aleph_1} logic, while (c_2) corresponds to the table

X	Y
F	G
F'	V

As a very simple illustration, assume that $U = V = 1 + 2 + 3$. (To be consistent with our notation for fuzzy sets, a finite nonfuzzy set $U = \{u_1, \ldots u_n\}$ may be expressed as $U = u_1 + \ldots + u_n$.)

$$F \triangleq \text{small} \triangleq 1/1 + 0.6/2 + 0.1/3 \tag{4.26}$$
$$G \triangleq \text{large} \triangleq 0.1/1 + 0.6/2 + 1/3$$

Then (4.18), (4.19), (4.20) and (4.21) yield

X is small and Y is large $\rightarrow \Pi_{(X,Y)} = 0.1/(1,1) + 0.6/(1,2) + 1/(1,3)$
$$+ 0.1/(2,1) + 0.6/(2,2) + 0.6/(2,3)$$
$$+ 0.1/(3,1) + 0.1/(3,2) + 0.1/(3,3)$$

X is small or Y is large $\rightarrow \Pi_{(X,Y)} = 1/(1,1) + 1/(1,2) + 1/(1,3)$
$$+ 0.6/(2,1) + 0.6/(2,2) + 1/(2,3)$$
$$+ 0.1/(3,1) + 0.6/(3,2) + 1/(3,3)$$

If X is small then Y is large $\rightarrow \Pi_{(X,Y)} = 0.1/(1,1) + 0.6/(1,2) + 1/(1,3)$
$$+ 0.5/(2,1) + 1/(2,2) + 1/(2,3)$$
$$+ 1/(3,1) + 1/(3,2) + 1/(3,3)$$

If X is small then Y is large $\rightarrow \Pi_{(X,Y)} = 0.1/(1,1) + 0.6/(1,2) + 1/(1,3)$
$$+ 0.4/(2,1) + 0.6/(2,2) + 0.6/(2,3)$$
$$+ 0.9/(3,1) + 0.9/(3,2) + 0.9/(3.3).$$

Rules of Type III

Quantificational rules of Type III apply to propositions of the general form

$$p \triangleq QX \text{ are } F \tag{4.27}$$

where Q is a fuzzy quantifier (*most, many, few, some, almost all*, etc.), X is a variable taking values in U, and F is a fuzzy subset of U. Simple examples of (4.27) are: "Most X's are small", "Some X's are small", "Many X's are very small". A somewhat less simple example is: "Most large X's are much smaller than α", where α is a specified number.

In general, a fuzzy quantifier is a fuzzy subset of the real line. However, when Q relates to a proportion, as is true of *most*, it may be represented as a fuzzy subset of the unit interval. Thus, the membership function of $Q \triangleq$ *most* may be represented as, say,

$$\mu_{most} = S(0.5, 0.7, 0.9) \tag{4.28}$$

where the S-function is defined by (3.5).

To be able to translate propositions of the form (4.27), it is necessary to define the cardinality of a fuzzy set, that is, the number (or the proportion) of elements of U which are in F. When U is a finite set $\{u_1, \ldots u_N\}$, a possible extension of the concept of cardinality of a nonfuzzy set — to which we shall refer as **fuzzy cardinality** — is the following. Let

$$F = \sum_{\alpha} \alpha F_{\alpha} \tag{4.29}$$

be the resolution (Zadeh, 1971) of F into its level-sets, that is,

$$F_{\alpha} \triangleq \{u \mid \mu_F(u) \geqslant \alpha\} \tag{4.30}$$

where αF_{α} is a fuzzy set defined by

$$\mu_{\alpha F_{\alpha}} = \alpha \mu_{F_{\alpha}} \tag{4.31}$$

and $\sum_{\alpha}$ denotes the union of the αF_{α} over $\alpha \in [0,1]$. Let $|F_{\alpha}|$ denote the cardinality of the nonfuzzy set F_{α}. Then, the fuzzy cardinality of F is denoted by $|F|_f$ and is defined to be the fuzzy subset of $\{0,1,2,\ldots\}$ expressed by

$$|F|_f = \sum_{\alpha} \alpha/|F_{\alpha}|. \tag{4.32}$$

As a simple example, consider the fuzzy subset *small* defined by (2.1). In this case,

$$
\begin{array}{ll}
F_1 = 0 + 1 & , \quad |F_1| = 2 \\
F_{0.8} = 0 + 1 + 2 & , \quad |F_{0.8}| = 3 \\
F_{0.6} = 0 + 1 + 2 + 3 & , \quad |F_{0.6}| = 4 \\
F_{0.4} = 0 + 1 + 2 + 3 + 4 & , \quad |F_{0.4}| = 5 \\
F_{0.2} = 0 + 1 + 2 + 3 + 4 + 5 & , \quad |F_{0.2}| = 6
\end{array}
$$

and

$$|F|_f = 1/2 + 0.8/3 + 0.6/4 + 0.4/5 + 0.2/6 . \tag{4.33}$$

Frequently, it is convenient or necessary to express the cardinality of a fuzzy set as a nonfuzzy real number (or an integer) rather than as a fuzzy number. In such cases, the concept of the **power** of a fuzzy set (DeLuca and Termini, 1972) may be used as a **numerical summary** of the fuzzy cardinality of a fuzzy set. Thus, the power of a fuzzy subset, F, of $U = \{u_1, \ldots u_N\}$ is defined by

$$|F| \triangleq \sum_{i=1}^{N} \mu_F(u_i) \qquad (4.34)$$

where $\mu_F(u_i)$ is the grade of membership of u_i in F, and $\sum$ denotes the arithmetic sum. For example, for the fuzzy set *small* defined by (2.1) we have

$$|F| = 1 + 1 + 0.8 + 0.6 + 0.4 + 0.2 = 4 .$$

For some applications, it is necessary to eliminate from the count those elements of F whose grade of membership falls below a specified threshold. This is equivalent to replacing F in (4.34) with $F \cap \Gamma$, where Γ is a fuzzy or nonfuzzy set which induces the desired threshold.

As N increases and U becomes a continuum, the concept of the power of F gives way to that of a **measure** of F (Zadeh, 1968; Sugeno, 1974), which may be regarded as a limiting form of the expression for the proportion of the elements of U which are in F. More specifically, if ρ is a density function defined on U, the measure in question is defined by

$$\|F\| \triangleq \int_U \rho(u)\mu_F(u)du \qquad (4.35)$$

where μ_F is the membership function of F. For example, if $\rho(u)du$ is the proportion of men whose height lies in the interval $[u, u+du]$, then the proportion of men who are tall is given by

$$\| tall \| = \int_0^\infty \rho(u)\mu_{tall}(u)du . \qquad (4.36)$$

Making use of the above definitions, the **quantifier rule** for propositions of the form "QX are F" may be stated as follows.

If $U = \{u_1, \ldots u_N\}$ and

$$X \text{ is } F \rightarrow \Pi_X = F \qquad (4.37)$$

then

$$QX \text{ are } F \rightarrow \Pi_{|F|} = Q \qquad (4.38)$$

and, if U is continuum,

$$QX \text{ are } F \rightarrow \Pi_{\|F\|} = Q \qquad (4.39)$$

which implies the more explicit rule

$$QX \text{ are } F \rightarrow \pi(\rho) = \mu_Q[\int_U \rho(u)\mu_F(u)du] \qquad (4.40)$$

where $\rho(u)du$ is the proportion of X's whose value lies in the interval $[u,u+du]$, $\pi(\rho)$ is the possibility of ρ, and μ_Q and μ_F are the membership functions of Q and F, respectively.

As a simple illustration, if *most* and *tall* are defined by (4.28) and $\mu_{tall} = S(160,170,180)$, respectively, then

Most men are tall $\rightarrow \pi(\rho) =$

$$S[\int_0^{200} \rho(u)S(u;160,170,180)du;0.5,0.7,0.9] \qquad (4.41)$$

where $\rho(u)du$ is the proportion of men whose height (in cm) is in the interval $[u,u+du]$. Thus, the proposition "Most men are tall" induces a possibility distribution of the height density function ρ which is expressed by the right-hand member of (4.41).

Rules of Type IV

Among the many ways in which a proposition, p, may be qualified there are three that are of particular relevance to approximate reasoning. These are: (a) by a linguistic truth-value, as in "p is very true", (b) by a linguistic probability-value, as in "p is highly probable"; and (c) by a linguistic possibility-value, as in "p is quite possible". Of these, we shall discuss only (a) in the sequel. Discussions of (b) and (c) may be found in Zadeh (1977).

As a preliminary to the formulation of translation rules pertaining to truth qualification, it is necessary to understand the role which a truth-value plays in modifying the meaning of proposition. Thus, in FL, the truth-value of a proposition, p, is defined as the compatibility of a reference proposition, r, with p. More specifically, let

$$p \triangleq X \text{ is } F$$

where F is a subset of U, and let r be a reference proposition of the special form

$$r \triangleq X \text{ is } u \qquad (4.42)$$

where u is an element of U. Then, the **compatibility** of r with p is defined as

$$Comp(X \text{ is } u/X \text{ is } F) \triangleq \mu_F(u) \qquad (4.43)$$

or, equivalently (in view of (2.3)),

$$Comp(X \text{ is } u/X \text{ is } F) \triangleq Poss\{X = u \mid X \text{ is } F\} . \qquad (4.44)$$

To extend (4.43) to the case where r is a fuzzy proposition of the form

$$r \triangleq X \text{ is } G , \quad G \subset U \qquad (4.45)$$

we apply the extension principle† to the evaluation of the expression $\mu_F(G)$, yielding

$$Comp\,\{X \text{ is } G/X \text{ is } F\} \triangleq \mu_F(G) \tag{4.46}$$

$$\triangleq \int_{[0,1]} \mu_G(u)/\mu_F(u)$$

in which the right-hand member is the union over the unit interval of the fuzzy singletons $\mu_G(u)/\mu_F(u)$. Thus, the compatibility of "X is G" with "X is F" is a fuzzy subset of $[0,1]$ defined by (4.46).

In FL, the **truth-value**, τ, of the proposition $p \triangleq X$ is F relative to the reference proposition $r \triangleq X$ is G is defined as the compatibility of r with p. Thus, by definition,

$$\tau \triangleq Tr\text{-}[X \text{ is } F/X \text{ is } G] \triangleq Comp\,\{X \text{ is } G/X \text{ is } F\} \tag{4.47}$$
$$= \mu_F(G)$$

$$= \int_{[0,1]} \mu_G(u)/\mu_F(u)$$

which implies that the truth-value, τ, of the proposition "X is F" relative to "X is G" is a fuzzy subset of the unit interval defined by (4.47). In this sense, then, a linguistic truth-value may be regarded as a linguistic approximation to the fuzzy subset, τ, represented by (4.47). (See Fig. 2.)

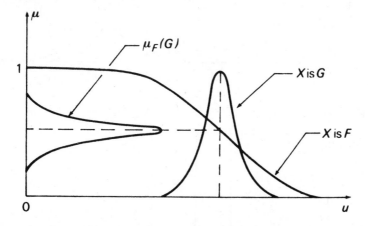

Fig. 2 — Graphical illustration of the concept of relative truth.

† The extension principle (Zadeh 1975c) serves to extend the definition of a mapping $f: U \to V$ to the set of fuzzy subsets of U. Thus, $f(F) \triangleq \int_U \mu_F(u)/f(u)$, where $f(F)$ and $f(u)$ are, respectively, the images of F and u in V.

A more explicit expression for τ which follows at once from (4.47) is the following. Let μ_τ denote the membership function of τ and let $v \in [0,1]$. Then

$$\mu_\tau(v) = Max_u \; \mu_G(u) \tag{4.48}$$

subject to

$$\mu_F(u) = v \; . \tag{4.49}$$

In particular, if μ_F is $1-1$, then (4.48) and (4.49) yield

$$\mu_\tau(v) = \mu_G(\mu_F^{-1}(v)) \; , \quad v \in [0,1] \; . \tag{4.50}$$

As a simple illustration, consider the propositions (see Fig. 3)

$$p \triangleq X \text{ is } F \tag{4.51}$$
$$r \triangleq X \text{ is } G \text{ where } G = [a,b] \; .$$

In this case, it follows from (4.50) that τ is the interval given by

$$\tau = [\mu_F(b), \mu_F(a)] \; .$$

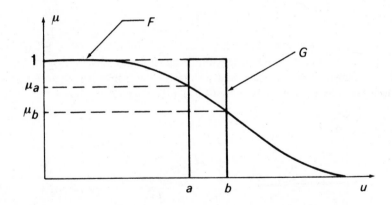

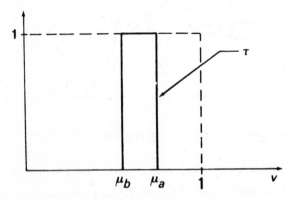

Fig. 3 – Interval-valued truth-value for an interval-valued reference proposition.

The definition of the truth-value of p as the compatibility of a reference proposition r with p provides us with a basis for the translation of truth-qualified propositions of the form "p is τ" when τ is a fuzzy subset of $[0,1]$. Specifically, from the relation

$$\tau = \mu_F(G) \tag{4.52}$$

which defines τ as the image of G under the mapping $\mu_F\colon U \to [0,1]$, it follows that the membership function of G may be expressed in terms of those of τ and μ_F by (see Fig. 4)

$$\mu_G(u) = \mu_\tau(\mu_F(u)) . \tag{4.53}$$

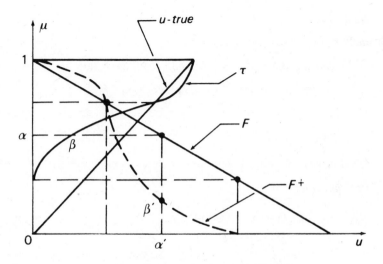

Fig. 4 – Effect of truth qualification on F. (β is mapped into β'.)

Now, if $r \triangleq X$ is G is the reference proposition for $p \triangleq X$ is F, we interpret the truth-qualified proposition

$$q \triangleq X \text{ is } F \text{ is } \tau \tag{4.45}$$

as the reference proposition $r \triangleq X$ is G. This leads us, then, to the following **rule for truth qualification**:

If

$$X \text{ is } F \to \Pi_X = F \tag{4.55}$$

then

$$X \text{ is } F \text{ is } \tau \to \Pi_X = F^+ \tag{4.56}$$

where

$$\mu_{F^+}(u) = \mu_\tau(\mu_F(u)) . \tag{4.57}$$

In particular, if τ is the **unitary** truth-value, that is

$$\tau \triangleq u\text{-}true \tag{4.58}$$

where

$$\mu_{u\text{-}true}(v) \triangleq v , \quad v \in [0,1] \tag{4.59}$$

then

$$X \text{ is } F \text{ is } u\text{-}true \to X \text{ is } F . \tag{4.60}$$

As an illustration of (4.56), consider the proposition

$$p \triangleq \text{Lucia is young is very true} \tag{4.61}$$

in which

$$\mu_{young} = 1 - S(25;35;45) \tag{4.62}$$
$$\mu_{true} = S(0.6,0.8,1.0)$$

and

$$\mu_{very\ true} = S^2(0.6,0.8,1.0)$$

On applying (4.56) to p, we obtain

$$p \to \pi_{Age(Lucia)}(u) = S^2[1 - S(u;25,35,45);0.6,0.8,1.0] \tag{4.63}$$

which may be roughly approximated by the proposition

$$p^+ \triangleq \text{Lucia is very young} . \tag{4.64}$$

Similarly, for the proposition

$$q \triangleq \text{Lucia is not young is very false} \tag{4.65}$$

where $false \triangleq ant\ true$, that is,

$$\mu_{false}(v) \triangleq \mu_{true}(1-v) , \quad v \in [0,1] \tag{4.66}$$
$$= 1 - S(v;0,0.2,0.4)$$

we obtain

$$q \to \pi_{Age(Lucia)} = \left[1 - S[S(u,25,35,45),0,0.2,0.4]\right]^2 \tag{4.67}$$

which, as can readily be verified, defines the same possibility distribution as (4.63).

The translation rules described above provide us with the necessary basis for the formulation of the rules of inference in FL and the related notions of semantic equivalence and semantic entailment. We turn to these issues in the following section.

5. SEMANTIC EQUIVALENCE AND SEMANTIC ENTAILMENT

In this section, we shall consider two related concepts in fuzzy logic that play an important role in approximate reasoning. These are the concepts of *semantic equivalence* and *semantic entailment*.

Informally, two propositions p and q are **semantically equivalent** if and only if the possibility distributions induced by p and q are equal. More specifically, if

$$p \to \Pi^p_{(X_1, \ldots X_n)} = F$$

and

$$q \to \Pi^q_{(X_1, \ldots X_n)} = G$$

where Π^p and Π^q are the possibility distributions induced by p and q, respectively, and $X_1, \ldots X_n$ are the variables that are implicit or explicit in p and q, then

$$p \leftrightarrow q \ iff \ \Pi^p_{(X_1, \ldots X_n)} = \Pi^q_{(X_1, \ldots X_n)} \tag{5.1}$$

where $\leftrightarrow$ denotes semantic equivalence.

When (5.1) holds for all fuzzy sets in p and q that have a context-dependent meaning, the semantic equivalence will be said to be **strong**.[†] For example, the semantic equivalence

$$\text{Adrienne is intelligent is true} \leftrightarrow \text{Adrienne is not intelligent is false} \quad (5.2)$$

holds for all definitions of *intelligent* and *true* (*false* $\triangleq$ antonym of true) and hence is a strong equivalence. On the other hand, the semantic equivalence

$$\text{Lucia is young is very true} \leftrightarrow \text{Lucia is very young} \tag{5.3}$$

is not a strong equivalence because it holds only for some particular definitions of *young* and *true*. (See (4.64) and (4.65) *et seq.*) Usually, a semantic equivalence which is not strong is approximate in nature, as is true of (5.3).

Generally, it is clear from the context whether a semantic equivalence is or is not strong. Where it is necessary to place in evidence that a semantic equivalence is strong, it will be denoted by $s\leftrightarrow$, while approximate semantic equivalence will be denoted by $a\leftrightarrow$.

The concept of **semantic entailment** is weaker than that of semantic equivalence in that p semantically entails q (or q is *semantically entailed* by p) if and only if $\Pi^p_{(X_1, \ldots X_n)} \subset \Pi^q_{(X_1, \ldots X_n)}$. Thus, in symbols,

$$p \mapsto q \ iff \ \Pi^p_{(X_1, \ldots X_n)} \subset \Pi^q_{(X_1, \ldots X_n)} \tag{5.4}$$

where $\mapsto$ denotes semantic entailment and $\Pi^p_{(X_1, \ldots X_n)}$ and $\Pi^q_{(X_1, \ldots X_n)}$ are the possibility distributions induced by p and q, respectively.

As in the case of semantic equivalence, semantic entailment is *strong* if (5.4) holds for all fuzzy sets in p and q that have a context-dependent meaning. As an illustration, the semantic entailment expressed by

$$X \text{ is very small} \mapsto X \text{ is small} \tag{5.5}$$

[†] The concept of strong semantic equivalence as defined here reduces to that of semantic equivalence in predicate logic (Lyndon, 1966) when p and q are nonfuzzy propositions.

is strong since it holds for all definitions of *small*. On the other hand, the validity of the semantic entailment expressed by

$$X \text{ is large} \mapsto X \text{ is not small} \tag{5.6}$$

depends on the way in which *large* and *small* are defined, and hence (5.6) is not an instance of strong semantic entailment.

In the case of propositions of the form $p \triangleq X$ is F and $q \triangleq X$ is G, it is evident that

$$X \text{ is } F \mapsto X \text{ is } G \text{ iff } F \subset G . \tag{5.7}$$

From this and the definition of conditional composition (4.20), it follows at once that

$$X \text{ is } F \mapsto X \text{ is } G \ \ iff \ \ \text{If } X \text{ is } F \text{ then } X \text{ is } G \rightarrow \Pi_X = U \tag{5.8}$$

or equivalently

$$X \text{ is } F \mapsto X \text{ is } G \ \ iff \ \ \text{If } X \text{ is } F \text{ then } X \text{ is } G \leftrightarrow X \text{ is } U \tag{5.9}$$

where Π_X is the possibility distribution of X and U is the universe of discourse associated with X. Similarly, from the definition of conjunctive composition, it follows that

$$X \text{ is } F \mapsto X \text{ is } G \ \ iff \ \ X \text{ is } F \text{ and } X \text{ is } G \rightarrow \Pi_X = F \tag{5.10}$$

or equivalently

$$X \text{ is } F \mapsto X \text{ is } G \ \ iff \ X \text{ is } F \text{ and } X \text{ is } G \leftrightarrow X \text{ is } F . \tag{5.11}$$

An intuitively appealing interpretation of (5.11) is that p semantically entails q if the information conveyed by "p and q" is the same as the information conveyed by p alone.

As a preliminary to applying the concepts of semantic equivalence and semantic entailment to approximate reasoning — which we shall do in Sec. 6 — it will be helpful to formulate several rules pertaining to the transformation of a given proposition, p, into other propositions that have the same meaning as p, that is, are strongly semantically equivalent to p.

A general rule governing such transformations may be stated informally as follows.

If m is a modifier and p is a proposition, than mp is semantically equivalent to the proposition which results from applying m to the possibility distribution which is induced by p.

Thus, on applying this rule to the case where $m \triangleq$ not and making use of the translation rules (4.5), (4.56) and (4.40), we arrive at the following specific rules governing the negation of a proposition:

$$(a) \ \ not(X \text{ is } F) \leftrightarrow X \text{ is not } F \tag{5.12}$$

for example,

$$not(X \text{ is small}) \leftrightarrow X \text{ is not small} ; \tag{5.13}$$

(b) $not(X$ is F is $\tau) \leftrightarrow X$ is F is not τ (5.14)

for example,

$not(X$ is small is very true$) \leftrightarrow X$ is small is not very true ;   (5.15)

(c) $not(QX$ are $F) \leftrightarrow (not\ Q)X$ are F (5.16)

for example,

$not($many men are tall$) \leftrightarrow ($not many$)$men are tall .   (5.17)

Similarly, for $m \triangleq$ very, we obtain

(a) $very(X$ is $F) \leftrightarrow X$ is very F (5.18)

(b) $very(X$ is F is $\tau) \leftrightarrow X$ is F is very τ (5.19)

(c) $very(QX$ are $F) \leftrightarrow ($very $Q)X$ are F (5.20)

In addition, from the translation formulas (4.5), (4.40), and (4.56), it follows at once that

X is F is $\tau \leftrightarrow X$ is not F is ant τ (5.21)

and

QX are $F \leftrightarrow ($ant $Q)X$ are not F (5.22)

where ant τ and ant Q denote the antonyms of τ and Q, respectively. (See (4.66).) Similarly, for $m =$ very, we have

X is F is $\tau \leftrightarrow X$ is very F is $^2\tau$ (5.23)

where the "left-square" operation on τ is defined by

$$^2\tau = \int_0^1 \mu_\tau(\nu)/\nu^2 , \quad \nu \in [0,1]$$ (5.24)

or equivalently

$$\mu_{2_\tau}(\nu) = \mu_\tau(\sqrt{\nu})$$ (5.25)

where μ_τ is the membership function of τ. However, as will be seen later, when F is modified to *very* F in "QX are F", we can assert only the semantic entailment — rather than the semantic equivalence — expressed by

QX are $F \to (^2Q)F$ are very F (5.26)

where

$$^2Q = \int_0^1 \mu_Q(\nu)/\nu^2$$ (5.27)

or equivalently

$$\mu_{2_Q}(\nu) = \mu_Q(\sqrt{\nu}) .$$ (5.28)

It should be noted in closing that the negation rule expressed by (5.16) appears to differ in form from the familiar negation rule in predicate calculus (Lyndon 1966), which, when F is interpreted as a nonfuzzy predicate, may be expressed as

$$not(\text{all } X \text{ are } F) \leftrightarrow \text{some } X \text{ are not } F. \tag{5.29}$$

However, by the use of (5.22) it is easy to show that the right-hand member of (5.29) is semantically equivalent to that of (5.16). Specifically, from (5.22) it follows that

$$(\text{not all})X \text{ are } F \leftrightarrow (\text{ant}(\text{not all})X)\text{are not } F$$

and if *some* is defined as

$$some \triangleq ant(\text{not all}) \tag{5.30}$$

then

$$(\text{not all})X \text{ are } F \leftrightarrow \text{some } X \text{ are not } F \tag{5.31}$$

in agreement with (5.29).

Remark

It should be observed that most of the definitions made in this and the preceding sections — especially in regard to the semantic equivalence and semantic entailment of fuzzy propositions — are nonfuzzy and, for the most part, quite precise. What should be understood, however, is that all such definitions may be fuzzified, if necessary, by the use of the following general convention.

Let U be a universe of discourse, with u denoting a generic element of U. A concept, C, in U is a subset, A, of U (or U^n, $n > 1$) which is defined by a predicate P such that $P(u)$ is true if $u \in A$, that is, u is an instance of C, and false otherwise. Assume that $P(u)$ is of the form $P(f(u))$, where $P(f(u))$ is true if $f(u) = 0$ and false if $f(u) > 0$. Then A — and hence the concept C which is associated with it — may be fuzzified by defining the grade of membership of u in A as a monotone function of $f(u)$ which assumes the value unity when $f(u) = 0$. (The definition of such a function is, in general, application-dependent rather than universal in nature.) In this sense, any definition which has the format stated above may be viewed as providing a mechanism for a fuzzification of the concept which it serves to define.

As an illustration of this convention, consider the concept of semantic equivalence as defined by (5.1). In this case, the concept of semantic equivalence may be fuzzified by defining the degree to which p and q are semantically equivalent as a monotone function of the "distance" between Π^p and Π^q, with the distance function defined in a way that reflects the specific nature of the domain of application. It should be understood, of course, that the concept in question may also be fuzzified in other ways which do not stem directly from its nonfuzzy definition.

6. RULES OF INFERENCE AND APPROXIMATE REASONING

As in any other logic, the rules of inference in FL govern the deduction of a proposition, q, from a set of premises $\{p_1, \ldots p_n\}$. However, in FL both the premises and the conclusion are allowed to be fuzzy propositions. Furthermore, because of the use of linguistic approximation in the process of retranslation, the final conclusion drawn from the premises $p_1, \ldots p_n$ is, in general, an approximate rather than exact consequence of $p_1, \ldots p_n$.

The principal rules of inference in FL are the following.

1. Projection principle

Let p be a fuzzy proposition whose translation is expressed as

$$p \to \Pi_{(X_1, \ldots X_n)} = F .$$

Let $X_{(s)}$ denote a subvariable of the variable $X \triangleq (X_1, \ldots X_n)$, that is,

$$X_{(s)} = (X_{i_1}, \ldots X_{i_k}) \tag{6.1}$$

where the index sequence $s \triangleq (i_1, \ldots i_k)$ is a subsequence of the sequence $(1, \ldots n)$.

Let $\Pi_{X_{(s)}}$ denote the marginal possibility distribution of $X_{(s)}$; that is,

$$\Pi_{X_{(s)}} = Proj_{U_{(s)}} F \tag{6.2}$$

where $U_i, i = 1, \ldots n$ is the universe of discourse associated with X_i;

$$U_{(s)} = U_{i_1} \times \ldots \times U_{i_k} , \tag{6.3}$$

and the projection of F on $U_{(s)}$ is defined by the possibility distribution function

$$\pi_{X_{(s)}}(u_{i_1} \ldots u_{i_k}) = Sup_{u_{j_1}, \ldots u_{j_m}} \mu_F(u_1, \ldots u_n) \tag{6.4}$$

where $s' \triangleq (j_1, \ldots j_m)$ is the index subsequence which is complementary to s, and μ_F is the membership function of F.

Let q be a retranslation of the possibility assignment equation

$$\Pi_{X_{(s)}} = Proj_{U_{(s)}} F . \tag{6.5}$$

Then, the projection principle asserts that q may be inferred from p. In a schematic form, this assertion may be expressed more transparently as

$$p \to \Pi_{(X_1, \ldots X_n)} = F \tag{6.6}$$
$$\downarrow$$
$$q \leftarrow \Pi_{X_{(s)}} = Proj_{U_{(s)}} F .$$

The statement of the projection principle assumes a particularly simple form for $n = 2$. In this case, writing X, Y, U, V for X_1, X_2, U_1, U_2 respectively, we have

$$p \to \Pi_{(X, Y)} = F \tag{6.7}$$

$$q \leftarrow \Pi_X = Proj_U F \tag{6.8}$$

and likewise for the projection on V.

A special case of (6.6) obtains when $\Pi_{(X,Y)}$ is the Cartesian product of normal fuzzy sets. Thus, if

$$p \to \Pi_{(X,Y)} = G \times H \tag{6.9}$$

then from p we can infer q and r, where

$$q \leftarrow \Pi_X = G \tag{6.10}$$

and

$$r \leftarrow \Pi_Y = H . \tag{6.11}$$

As a simple illustration, if

$$p \triangleq \text{John is tall and fat}$$

then from p we can infer

$$q \triangleq \text{John is tall}$$

and

$$r \triangleq \text{John is fat} .$$

2. Particularization/conjunction principle

Let p be a fuzzy proposition whose translation is expressed as

$$p \to \Pi_{(X_1, \ldots X_n)} = F , \quad F \subset U_1 \times \ldots \times U_n . \tag{6.12}$$

Then from p we can infer r, where r is a retranslation of a particularization of $\Pi_{(X_1, \ldots X_n)}$, that is,

$$r \leftarrow \Pi_{(X_1, \ldots X_n)}[\Pi_{X_{(s)}} = G] = F \cap \bar{G} \tag{6.13}$$

where $X_{(s)}$ is a subvariable of X, $\bar{G}$ is a cylindrical extension of G, $G \subset U$, and $\Pi_{(X_1, \ldots X_n)}[\Pi_{X_{(s)}} = G]$ denotes an n-ary possibility distribution which result from particularizing $X_{(s)}$ to G. Equivalently, the **particularization principle** may be expressed in the schematic form

$$p \to \Pi_{(X_1, \ldots X_n)} = F \tag{6.14}$$

$$q \to \Pi_{(X_{i_1}, \ldots X_{i_k})} = G$$

$$\overline{}$$

$$r \leftarrow \Pi_{(X_1, \ldots X_n)} = F \cap \bar{G} .$$

For the special case of $n = 2$, the particularization principle may be stated more simply as:

From

$$p \triangleq (X,Y) \text{ is } F \tag{6.15}$$

and

$$q \triangleq X \text{ is } G$$

we can infer

$$r \triangleq (X,Y) \text{ is } F \cap \bar{G} . \tag{6.16}$$

Thus, for example, from

$$p \triangleq X \text{ and } Y \text{ are approximately equal} \tag{6.17}$$

and

$$q \triangleq X \text{ is small}$$

we can infer (without the application of linguistic approximation)

$$r \triangleq X \text{ and } Y \text{ are } (approximately\ equal \cap (small \times V)). \tag{6.18}$$

As stated above, the particularization principle may be viewed as a special case of a somewhat more general principle which will be referred to as the **conjunction principle**. Specifically, assume that

$$p \to \Pi^p_{(Y_1, \ldots Y_k, X_{k+1}, \ldots X_n)} = F \tag{6.19}$$

$$q \to \Pi^q_{(Y_1, \ldots Y_k, Z_{k+1}, \ldots Z_m)} = G \tag{6.20}$$

where $Y_1, \ldots Y_k$ are variables which appear in both Π^p and Π^q, and U_i, V_j and W_ϱ are the universes of discourse associated with X_i, Y_j and Z_ϱ; let S be the smallest Cartesian product of the U_i, V_j, and W_ϱ which contains the Cartesian products $V_1 \times \ldots \times V_k \times U_{k+1} \times \ldots \times U_n$ and $V_1 \times \ldots \times V_k \times W_{k+1} \times \ldots \times W_m$; and let $\bar{F}$ and $\bar{G}$ be, respectively, the cylindrical extensions of F and G in S. Then, from p and q we can infer r, where (in schematic form)

$$p \to \Pi^p_{(Y,X)} = F \tag{6.21}$$

$$q \to \Pi^q_{(Y,Z)} = G$$

$$\overline{\qquad\qquad\qquad\qquad}$$

$$r \leftarrow \Pi_{(X,Y,Z)} = \bar{F} \cap \bar{G}$$

and $Y \triangleq (Y_1, \ldots Y_k)$, $X \triangleq (X_{k+1}, \ldots X_n)$ and $Z \triangleq (Z_{k+1}, \ldots Z_m)$.

A particular but important case of (6.21) which we shall use at a later point results when $n = 3$, and $k = 1$. For this case, (6.21) may be expressed as

$$p \to \Pi^p_{(X,Y)} = F \tag{6.22}$$

$$q \leftarrow \Pi^q_{(Y,Z)} = G$$

$$\overline{\qquad\qquad\qquad\qquad}$$

$$r \leftarrow \Pi_{(X,Y,Z)} = (F \times W) \cap (U \times G).$$

Although the particularization principle is subsumed by the conjunction principle, it is simpler that the latter, is employed more frequently, and has a somewhat greater intuitive appeal. For this reason, we use the designation "particularization/conjunction principle" to refer to the principle which, in most applications, is the particularization principle and, in some, the conjunction principle. It should be noted that, in predicate logic (Lyndon 1966), this principle implies the generalization rule.

3. Entailment principle

Stated informally, the **entailment principle** asserts that from any fuzzy proposition p we can infer a fuzzy proposition q if the possibility distribution induced by p is contained in the possibility distribution induced by q. Thus, schematically, we have

$$p \rightarrow \Pi_{(X_1, \ldots X_n)} = F \qquad (6.23)$$
$$\downarrow$$
$$q \leftarrow \Pi_{(X_1, \ldots X_n)} = G \supset F.$$

For example, from $p \triangleq X$ is very large we can infer $q \triangleq X$ is large.

The compositional rule of inference. In general, the inference principles stated above are used in sequence or in combination. A combination that is particularly effective involves an application of the particularization/conjunction principle followed by that of the projection principle. This combination will be referred to as the **compositional rule of inference** (Zadeh 1973). As will be seen later, the compositional rule of inference includes as a special case a generalization of the *modus ponens*.

For our purposes, it will be convenient to state the compositional rule of inference in the following schematic form

$$p \rightarrow \Pi_{(X,Y)} = F \qquad (6.24)$$
$$q \rightarrow \Pi_{(Y,Z)} = G$$
$$-------$$
$$r \leftarrow \Pi_{(X,Y)} = F \circ G$$

where X, Y and Z take values in U, V and W, respectively; F is a fuzzy subset of $U \times V$, G is a fuzzy subset of $V \times W$, and $F \circ G$ is the composition of F and G defined by

$$\mu_{F \circ G}(u,w) = Sup_v [\mu_F(u,v) \wedge \mu_G(v,w)] \qquad (6.25)$$

where $u \in U$, $v \in V$, $w \in W$ and μ_F and μ_G are the membership functions of F and G, respectively; and the dashed line signifies that, because of the use of linguistic approximation in retranslation, r is, in general, an approximate rather than exact consequence of p and q. It should be noted that the compositional rule of inference is analogous to the rule which yields the probability distribution of Y from the probability distribution of X and the conditional probability distribution of Y given X.

It is easy to demonstrate that the compositional rule of inference may be regarded as a result of applying the particularization/conjunction principle followed by the application of the projection principle. Specifically, on applying (6.21) to (6.24), we obtain

$$p \rightarrow \Pi_{(X,Y)} = F \tag{6.26}$$
$$q \rightarrow \Pi_{(Y,Z)} = G$$

$$s \leftarrow \Pi_{(X,Y,Z)} = (F \times W) \cap (U \times G)$$

where

$$\mu_{(F \times W) \cap (U \times G)}(u,v,w) = \mu_F(u,v) \wedge \mu_G(v,w) . \tag{6.27}$$

Next, on applying the projection principle to s and projecting $\Pi_{(X,Y,Z)}$ on $U \times W$, we have

$$Proj_{U \times W}[(F \times W) \cap (U \times G)] = \int_{U \times W} Sup_v[\mu_F(u,v) \wedge \mu_G(v,w)]/(u,w) \tag{6.28}$$

which upon comparison with (6.25) shows that the resulting proposition may be expressed − in agreement with (6.24) − as

$$r \leftarrow \Pi_{(X,Z)} = F \circ G . \tag{6.29}$$

An important special case of the compositional rule of inference obtains when p and q are of the form $p \triangleq X$ is F, $q \triangleq$ If X is G then Y is H. For this case, (4.20) and (6.24) yield the **compositional modus ponens**:

$$p \rightarrow \Pi_X = F \tag{6.30}$$
$$q \rightarrow \Pi_{(X,Y)} = \bar{G}' \oplus \bar{H}$$
$$\overline{}$$
$$r \leftarrow F \circ (\bar{G}' \oplus \bar{H})$$

which may be regarded as a generalization of the classical *modus ponens*, with the latter corresponding to the special case of (6.30) in which F, G and H are nonfuzzy and $F = G$. For this case, (6.30) reduces to

$$p \rightarrow \Pi_{(X} = F \tag{6.31}$$
$$q \rightarrow \Pi_{(X,Y)} = \bar{F}' \oplus \bar{H}$$

$$r \leftarrow \Pi_Y = F \circ (\bar{F}' \oplus \bar{H})$$

and since

$$F \circ (\bar{F}' \oplus \bar{H}) = H$$

it follows that

$$r \leftrightarrow Y \text{ is } H$$

which means that from $p \triangleq X$ is F and $q \triangleq$ If X is F then Y is H we can infer $r \triangleq Y$ is H, in agreement with the statement of the *modus ponens*.

The rules of inference presented in the foregoing discussion provide us with a basis for employing approximate reasoning for the purpose of question-answering and inference from fuzzy propositions. We shall illustrate the use of the methods based on these rules by applying them to several typical problems.

Semantic equivalence. As a simple example, assume that from the premise

$$p \triangleq \text{Ellen is not very tall}$$

we wish to deduce the answer to the question "Is Ellen tall $?\tau$", where the symbol $?\tau$ signifies that the answer to the question is expected to be of the form

$$q \triangleq \text{Ellen is tall is } \tau$$

where τ is a linguistic truth-value.

To obtain the answer to the question, we shall require that p and q be semantically equivalent, implying that the possibility distribution induced by p is equal to that induced by q.

Thus, by using the translation rules (4.5), (4.6) and (4.56), we obtain

$$\text{Ellen is not very tall} \rightarrow \pi_{Height(Ellen)}(u) = 1 - \mu_{tall}^2(u) \tag{6.32}$$

$$\text{Ellen is tall is } \tau \rightarrow \pi_{Height(Ellen)}(u) = \mu_\tau[\mu_{tall}(u)] \tag{6.33}$$

where μ_{tall}, the membership function of *tall*, is assumed to be given. From (6.32) and (6.33), then, it follows that the desired membership function μ_τ satisfies the identity

$$1 - \mu_{tall}^2(u) \equiv \mu_\tau[\mu_{tall}(u)] , \quad u \in [0,200] \tag{6.34}$$

from which we can conclude at once that μ_τ is given by (see Fig. 5)

$$\mu_\tau(v) = 1 - v^2 \tag{6.35}$$

to which a rough linguistic approximation may be expressed as

$$\tau \approx \text{not very true} . \tag{6.36}$$

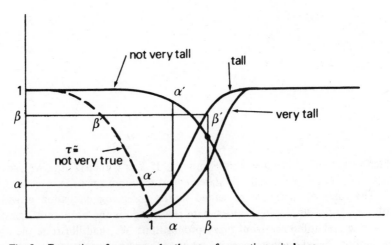

Fig. 5 – Extraction of an answer by the use of semantic equivalence.

It is instructive to obtain the same result by a succesive use of the rules governing the application of negation, truth qualification, and modification (by very). Thus, we can assert that

> John is not very tall
>> ↔ John is not very tall is *u-true*   (by (4.60))
>
> John is not very tall is *u-true*
>> ↔ John is very tall is *ant(u-true)*   (by (5.21))
>
> John is very tall is *ant(u-true)*
>> ↔ John is tall is $^{1/2}(ant(u\text{-}true))$ (by (5.23))

which implies that

$$\tau = {}^{1/2}(ant(u\text{-}true)) \tag{6.37}$$

that is, τ is the "left-square root" of $(ant(u\text{-}true))$, and since

$$\mu_{u\text{-}true}(v) = v \tag{6.38}$$

we have

$$\mu_\tau = 1 - v^2 \tag{6.39}$$

in agreement with (6.35).

Semantic entailment. Assume that we wish to deduce from the premise

> $p \triangleq$ Most Swedes are tall

the answer to the question "How many Swedes are very tall?"

Translating p by the use of (4.40), we have

$$\text{Most Swedes are tall} \rightarrow \pi_p(\rho) = \mu_{most}[\int_0^{200} \rho(u)\mu_{tall}(u)du] \tag{6.40}$$

where $\rho(u)du$ is the proportion of Swedes whose height is in the interval $[u, u+du]$ and π_p is the possibility distribution function of p. (Note that height is expressed in centimetres.)

Now, by (4.30) the proportion of Swedes who are very tall is given by

$$\gamma = \int_0^{200} \rho(u)\mu_{tall}^2(u)du . \tag{6.41}$$

Thus, our problem is to find the possibility distribution of γ from the knowledge of the possibility distribution of ρ — which is given by the right-hand member of (6.40). In a variational formulation (which follows from (4.48)), this problem may be expressed as

$$\pi(\gamma) = Max_\rho \, \mu_{most}[\int_0^{200} \rho(u)\mu_{tall}(u)du] \tag{6.42}$$

subject to

$$\gamma = \int_0^{200} \rho(u)\mu_{tall}^2(u)du \ .$$

(6.43)

The maximizing ρ for this problem is of the form (Bellman and Zadeh 1976).

$$\rho(u) = \delta(u - \alpha)$$

(6.44)

where δ is a δ-function and α is a point in the interval $[0,200]$. The δ-function density implies that all elements of the population have the same value of the attribute in question. Thus, from (6.43) we have

$$\gamma = \mu_{tall}^2(\alpha)$$

(6.45)

and hence

$$\pi(\gamma) = \mu_{most}(\mu_{tall}(\alpha))$$
$$= \mu_{most}(\sqrt{\gamma})$$

(6.47)

or equivalently (see (5.30))

$$\pi(\gamma) = \mu_{2most}(\gamma)$$

(6.47)

and hence the desired answer to the question "How many Swedes are very tall?" is (see Fig. 6)

$$q \triangleq {}^2most \text{ Swedes are very tall.}$$

(6.48)

To verify that p semantically entails q, we note that

$$q \rightarrow \pi_q(\rho) = \mu_{2most}[\int_0^{200} \rho(u)\mu_{tall}^2(u)du]$$

(6.49)

$$= \mu_{most}\left[\sqrt{\int_0^{200} \rho(u)\mu_{tall}^2(u)du}\right]$$

But, by Schwarz's inequality

$$\int_0^{200} \rho(u)\mu_{tall}(u)du \leqslant \sqrt{\int_0^{200} \rho(u)\mu_{tall}^2(u)du} \ ,$$

(6.50)

and since μ_{most} is a monotone function, it follows that

$$\pi_p(\rho) \leqslant \pi_q(\rho) \text{ for all } \rho \text{ and } \mu_{tall}$$

which implies that p semantically entails q, strongly.

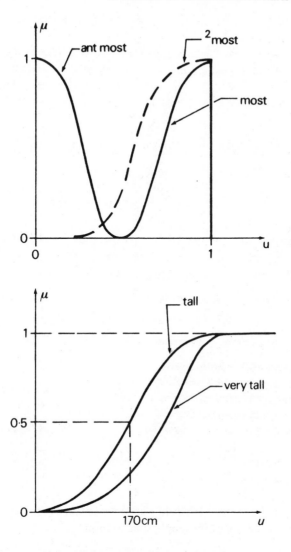

Fig. 6 — Representation of *most, tall* and their modifications.

Particularization and projection principles. An illustration of the application of the particularlization and projection principles is provided by the solution to the following simple problem.

Suppose that the premises are

$p \triangleq$ John is very big
$q \triangleq$ John is very tall

where big is a given fuzzy subset of $U \times V$ (that is, values of *Height* (in cms) $\times$

values of *Weight* (in kg)) and tall is a given fuzzy subset of *U*. The question is: "What is John's weight?"

Let us assume that the answer to the question is to be of the form $r \triangleq$ John is w, where w is a linguistic value of the weight of John (*heavy, very heavy, not very heavy*, etc.). Then, by employing the translation rule (4.6), the particularization principle, and the projection principle, we arrive at the retranslation relation

$$r \leftarrow Proj_{\mu \times Weight} BIG^2 \, [\Pi_{Height} = TALL] \qquad (6.51)$$

which expresses the answer to the posed question.

In more concrete terms, assume that the (incompletely tabulated) tables defining the fuzzy sets *BIG*, *TALL*, and *HEAVY* are of the form

BIG Height	Weight	μ
165	60	0.5
170	60	0.6
175	60	0.7
170	65	0.75
175	65	0.8
180	65	0.85
170	70	0.8
175	70	0.85
180	70	0.9
170	75	0.85
175	75	0.9
180	75	0.95
180	80	1

TALL Height	μ
165	0.7
170	0.8
175	0.9
180	1
185	1

HEAVY Weight	μ
60	0.7
65	0.8
70	0.9
75	0.95
80	1
85	1

On substituting these tables in (6.51) we obtain for the attribute *Weight* a possibility distribution of the (approximate) form

$$\Pi_{Weight} = 0.5/60 + 0.7/65 + 0.8/70 + 0.9/75 + 1/80 \qquad (6.52)$$

which upon retranslation (and linguistic approximation) yields the answer

$r \triangleq$ John is very heavy .

As an additional illustration, consider the following premises

$p \triangleq$ Romy lives near a small city
$q \triangleq$ Arnold lives near Romy

from which we wish to deduce an answer to the question "Where does Arnold live?"

Assume that the relations entering in p and q have the frames shown below.

$$NEARp \parallel City1 \mid City2 \mid \mu \mid \qquad NEARq \parallel City1 \mid City2 \mid \mu \mid$$
$$SMALL\ CITY \parallel City \mid \mu \mid$$

in which $NEARp$ and $NEARq$ refer to the relations $NEAR$ in p and q, respectively. In terms of these relations, the translations of p and q may be expressed as

$$p \rightarrow \Pi_{Location(Residence(Romy))} \tag{6.53}$$
$$= Proj_{\mu \times City1} NEARp\,[\Pi_{City2} = SMALL\ CITY]$$

$$q \rightarrow \Pi_{(Location(Residence(Romy)),Location(Residence(Arnold))} \tag{6.54}$$
$$= NEARq\ .$$

On substituting (6.53) in (6.54) and projecting on the attribute $Location$ $(Residence(Arnold))$, we obtain

$$r \leftarrow Proj_{\mu \times City2} NEARq\,[\Pi_{City1} = Proj_{\mu \times City1} NEARp\,[\Pi_{City2} = \tag{6.55}$$
$$SMALL\ CITY\,]]$$

as an expression for the answer to the posed question.

Compositional rule of inference. The compositional rule of inference is particularly convenient to use when the variables involved in the premises range over finite sets or can be approximated by variables ranging over such sets.

As a simple illustration, consider the premises

$$p \triangleq X \text{ is small}$$
$$q \triangleq X \text{ and } Y \text{ are approximately equal}$$

in which X and Y range over the set $U = 1 + 2 + 3 + 4$, and *small* and *approximately equal* are defined by

$$small = 1/1 + 0.6/2 + 0.2/3$$
$$approximately\ equal = 1/[(1,1) + (2,2) + (3,3) + (4,4)]$$
$$+ 0.5/[(1,2) + (2,1) + (2,3) + (3,2)$$
$$+ (3,4) + (4,3)]\ .$$

In terms of these sets, the translations of p and q may be expressed as

$$p \rightarrow \Pi_X = small \tag{6.56}$$
$$q \rightarrow \Pi_{(X,Y)} = approximately\ equal$$

and thus from p and q we may infer r, where

$$r \leftarrow \Pi_X \circ \Pi_{(X,Y)} = small \circ approximately\ equal\ . \tag{6.57}$$

The composition of *small* and *approximately equal* can readily be performed by computing the max-min product of the relation matrices corresponding to *small* and *approximately equal*. Thus, we obtain

$$[1 \quad 0.6 \quad 0.2 \quad 0] \circ \begin{bmatrix} 1 & 0.5 & 0 & 0 \\ 0.5 & 1 & 0.5 & 0 \\ 0 & 0.5 & 1 & 0.5 \\ 0 & 0 & 0.5 & 1 \end{bmatrix} = [1 \quad 0.6 \quad 0.5 \quad 0.2]$$

which implies that

$$\Pi_Y = \Pi_X \circ \Pi_{(X,Y)} = 1/1 + 0.6/2 + 0.5/3 + 0.2/4$$

and which upon retranslation yields the linguistic approximation

$$r \triangleq Y \text{ is more or less small} \leftarrow \Pi_Y = 1/1 + 0.6/2 + 0.5/3 + 0.2/4 \ .$$

Thus, from $p \triangleq X$ is small and $q \triangleq X$ and Y are approximately equal, we can infer, approximately, that $r \triangleq Y$ is more or less small.

As a simple illustration of the compositional *modus ponens*, assume that, as in Bellman and Zadeh (1976),

$$U = V = 1 + 2 + 3 + 4$$
$$F = 0.2/2 + 0.6/3 + 1/4$$
$$G = 0.6/2 + \quad 1/3 + 0.5/4$$
$$H = \quad 1/2 + 0.6/3 + 0.2/4$$

with

$$p \triangleq X \text{ is } F \to \Pi_X = F \tag{6.58}$$
$$q \triangleq \text{If } X \text{ is } G \text{ then } Y \text{ is } H \to \Pi_{(X,Y)} = \bar{G}' \oplus \bar{H}$$

and

$$r \leftarrow \Pi_Y = F \circ (\bar{G}' \oplus \bar{H}) \ .$$

In this case,

$$\bar{G}' \oplus \bar{H} = \begin{bmatrix} 1 & 1 & 1 & 1 \\ 0.4 & 1 & 1 & 0.6 \\ 0 & 1 & 0.6 & 0.6 \\ 0.5 & 1 & 1 & 0.7 \end{bmatrix}$$

and

$$F \circ (\bar{G}' \oplus \bar{H}) = [0 \quad 0.2 \quad 0.6 \quad 1] \circ \begin{bmatrix} 1 & 1 & 1 & 1 \\ 0.4 & 1 & 1 & 0.6 \\ 0 & 1 & 0.6 & 0.6 \\ 0.5 & 1 & 1 & 0.7 \end{bmatrix}$$

$$= [0.5 \quad 1 \quad 1 \quad 0.7] \ .$$

Thus, from p and q we can infer that

$$r \triangleq Y \text{ is } 0.5/1 + 1/2 + 1/3 + 0.7/4 \ .$$

The above example is intended merely to illustrate the computations involved in the application of the compositional *modus ponens* when X and Y range over finite sets. Detailed discussions of practical applications of the compositional rule of inference in the design of so-called fuzzy logic controllers may be found in the papers by Mamdani and Assilian (1975), Mamdani (1976a,b), Kickert and van Nauta Lemke (1976), Rutherford and Bloore (1975), and others given in the References and Bibliography.

7. CONCLUDING REMARK

The theory of approximate reasoning outlined in this paper may be viewed as an attempt at an accommodation with the pervasive imprecision of the real world.

Based as it is on fuzzy logic, approximate reasoning lacks the depth and universality of precise reasoning. And yet it may well prove to be more effective than precise reasoning in coming to grips with the complexity and ill-definedness of humanistic systems, and thus may contribute to the conception and development of intelligent systems which could approach the remarkable ability of the human mind to make rational decisions in the face of uncertainty and imprecision.

Acknowledgements

This research was supported by the National Science Foundation Grant MC76-06693, the Naval Electronic Systems Command Contract N000 39-77-C-0022, and the U.S. Army Research Office Grant DAHCO4-75-G0056.

REFERENCES AND BIBLIOGRAPHY

Adams, E. W. and Levine, H. P. (1975). On the uncertainties transmitted from premises to conclusions in deductive inferences, *Synthese*, 30, 429–460.

Aizerman, M. A. (1976). Fuzzy sets, fuzzy proofs and certain unsolved problems in the theory of automatic control, *Avtomatika i Telemehanika*, 171–177.

Bar-Hillel, Y. (1964). *Language and Information*. Reading, Mass.: Addison-Wesley.

Bellman, R. E. and Zadeh, L. A. (1976). Local and fuzzy logics, Electronics Research Laboratory *Memorandum M-584*, University of California, Berkeley. Also in *Modern Uses of Multiple-Valued Logics* (ed. D. Epstein). Dordrecht: D. Reidel.

Bezdek, J. C. (1974). Numerical taxonomy with fuzzy sets, *Journal of Mathematical Biology*, 1, 57–71.

Bezdek, J. C. and Dunn, J. C. (1975). Optimal fuzzy partitions: a heuristic for estimating the parameters in a mixture of normal distributions, *IEEE Transactions on Computers*, C-24, 835–838.

Black, M. (1963). Reasoning with loose concepts, *Dialogue*, 2, 1–12.

Bobrow, D. and Collins, A. (eds.) (1975). *Representation and Understanding*. New York: Academic Press.

Chang, S. K. and Ke, J. S. (1976), Database skeleton and its application to fuzzy query translation, *Research Memorandum*. Chicago: Department of Information Engineering, University of Illinois, Chicago Circle.

Chang, S. S. L. (1972). Fuzzy mathematics, man and his environment, *IEEE Transactions on Systems, Man and Cybernetics*, SMC-2, 92–93.

Cresswell, M. J. (1973), *Logics and Languages*. London: Methuen.

Damerau, F. J. (1975). On fuzzy adjectives, *Memorandum RC 5340*, Yorktown Heights, NY IBM Research Laboratory.

Davidson, D. (1967). Truth and meaning, *Synthese*, 17, 304–323.

DeLuca, A. and Termini, S. (1972). A definition of a nonprobabilistic entropy in the settin of fuzzy sets theory, *Information and Control*, 20, 301–312.

Dimitrov, V. D. (1975). Efficient governing humanistic systems by fuzzy instructions, *Thir International Congress of General Systems and Cybernetics*, Bucharest.

Dreyfuss, G. R., Kochen, M., Robinson, J. and Badre, A. N. (1975). On the psycholinguist reality of fuzzy sets. *Functionalism* (eds Grossman, R. E., San, L. J. and Vance, T. J. Chicago: University of Chicago Press.

Evans, G. and McDowell, J. (1976). *Truth and Meaning*. Oxford: Clarendon Press.

Fellinger, W. L. (1974). *Specifications for a fuzzy systems modelling language*. Ph.D. disse tation. Corvallis: Oregon State University.

Fine, K. (1975), Vagueness, truth and logic, *Synthese*, 30, 265–300.

Fodor, J. A. and Katz, J. J. (eds) (1964). *The Structure of Language*. Englewood Clifl N.J.: Prentice–Hall.

Gaines, B. R. (1975). Stochastic and fuzzy logics, *Electronics Letters*, 11, 444–445.

Gaines, B. R. (1976a), General fuzzy logics, *Proceedings of the Third European Meeting c Cybernetics and Systems Research*, Vienna.

Gaines, B. R. (1976b). Fuzzy reasoning and the logic of uncertainty, *Proceedings of the 6t International Symposium on Multiple-Valued Logic*, Utah, IEEE76CH 1111-4C, 175 188.

Gaines, B. R. (1976c). Foundations of fuzzy reasoning, *International Journal of Mar Machine Studies*, 6, 623–668.

Gaines, B. R. and Kohout, L. J. (1975). Possible automata, *Proceedings Internationi Symposium on Multiple-Valued Logic*, Bloomington, Indiana, 183–196.

Gaines, B. R. and Kohout, L. J. (1977). The fuzzy decade: a bibliography of fuzzy systen and closely related topics, *International Journal of Man-Machine Studies*, 9, 1–68.

Giles, R. (1976). Lukasiewicz logic and fuzzy set theory, *International Journal of Mai Machine Studies*, 8, 313–327.

Goguen, J. A. (1969). The logic of inexact concepts, *Synthese*, 19, 325–373.

Goguen, J. A. (1974). Concept representation in natural and artificial languages: axiom extension and applications for fuzzy sets, *International Journal of Man-Machin Studies*, 6, 513–561.

Gottinger, H. W. (1973). Toward a fuzzy reasoning in the behavioral science, *Cybernetic 2*, 113–135.

Haack, S. (1974). *Deviant Logic*. Cambridge: Cambridge University Press.

Hamacher, H. (1976). On logical connectives of fuzzy statements and their affiliated trut functions, *Proceedings of the Third European Meeting on Cybernetics and System Research*, Vienna.

Harris, J. I. (1974a). Fuzzy implication – comments on a paper by Zadeh, *DOAE Researc Working Paper*. Byfleet, Surrey: Ministry of Defence.

Harris, J. I. (1974b). Fuzzy sets: how to be imprecise precisely, *DOAE Research Workin Paper*. Byfleet, Surrey: Ministry of Defence.

Hendrix, G. G., Thompson, C. W. and Slocum, J. (1973). Language processing via canonic verbs and semantic models, *Proceedings of the Third Joint International Conference o Artificial Intelligence*, pp. 262-269. Menlo Park, Ca.: Stanford Research Institute.

Hersh, H. M. and Caramazza, A. (1976). A fuzzy set approach to modifiers and vagueness i natural language, *Journal of Experimental Psychology*, 105, 254–276.

Hughes, G. E. and Cresswell, M. J. (1968). *An Introduction to Modal Logic*. London Methuen.

Inagaki, Y. and Fukumura, F. (1975). On the description of fuzzy meaning of context-free language. *Fuzzy Sets and Their Applications to Cognitive and Decision Processes*, pp. 301-328 (eds Zadeh, L. A., Fu, K. S., Tanaka, K. and Shimura, M.) New York: Academic Press.

Jouault, J. P. and Luan, P. M. (1975). Application des concepts flous a la programmation en languages quasi-naturals. *Research Memorandum*. Paris: Inst. Inf. d'Enterprise, C.N.A.M.

Kandel, A. (1974). On the minimization of incompletely specified fuzzy functions, *Information and Control*, 26, 141-153.

Kaufmann, A. (1973). *Introduction to the Theory of Fuzzy Subsets*, Vol. 1. *Elements of Basic Theory*. Paris: Masson and Co.; (1975), Vol. 2. *Applications to Linguistics, Logic and Semantics*. Paris: Masson and Co., (1975), Vol. 3. *Applications to Classification and Pattern Recognition, Automata and Systems, and Choice of Criteria*. Paris: Masson and Co. Also English translation of Vol. 1, New York: Academic Press (1975).

Kickert, W. J. M. and van Nauta Lemke, H. R. (1976). Application of a fuzzy controller in a warm water plant, *Automatica*, 12, 301-308.

Kling, R. (1974). Fuzzy PLANNER: reasoning with inexact concepts in a procedural problem-solving language, *Journal of Cybernetics*, 4, 105-122.

Knuth, D. E. (1968), Semantics of context-free languages, *Mathematical Systems Theory*, 2, 127-145.

Kochen, M. and Badre, A. N. (1974). On the precision of adjectives which denote fuzzy sets, *Journal of Cybernetics*, 4, 49-59.

Lakoff, G. (1973a). Hedges: a study in meaning criteria and the logic of fuzzy concepts, *Journal of Philosophical Logic*, 2, 458-508. Also paper presented at 8th Regional Meeting of Chicago Linguistic Society (1972) and *Contemporary Research in Philosophical Logic and Linguistic Semantics*, pp. 221-271 (eds Hockney, D., Harper, W. and Freed, B.). The Netherlands: D. Reidel.

Lakoff, G. (1973b). Fuzzy grammar and the performance/competence terminology game, *Proceedings of Meeting of Chicago Linguistics Society*, 271-291.

Lee, E. T. and Chang, C. L. (1971). Some properties of fuzzy logic, *Information and Control*, 19, 417-431.

LeFaivre, R. A. (1974a). FUZZY: a programming language for fuzzy problem solving, *Technical Report 202*. Madison: Department of Computer Science, University of Wisconsin.

LeFaivre, R. A. (1974b). The representation of fuzzy knowledge, *Journal of Cybernetics*, 4, 57-66.

Lewis, P. M., Rosenkrantz, D. J. and Stearns, R. E. (1974). Attributed translations, *Journal of Computer and System Science*, 9, 279-307.

Lyndon, R. C. (1966). *Notes on Logic*. New York: D. Van Nostrand.

Machina, K. F. (1972). Vague predicates, *American Philosophical Quarterly*, 9, 225-233.

MacVicar-Whelan, P. J. (1975). Fuzzy sets, the concept of height and the hedge very, *Technical Memo 1*, Allendale, Michigan: Physics Department, Grand Valley State College.

Mamdani, E. H. (1976a). Application of fuzzy logic to approximate reasoning using linguistic synthesis, *Proceedings of 6th International Symposium on Multiple-Valued Logic*, Utah, IEEE76 CH1111-4C, 196-202.

Mamdani, E. H. (1976b). Advances in the linguistic synthesis of fuzzy controllers, *International Journal of Man-Machine Studies*, 8, 669-678.

Mamdani, E. H. and Assilian, S. (1975). An experiment in linguistic synthesis with a fuzzy logic controller, *International Journal of Man-Machine Studies*, 7, 1-13.

Marinos, P. N. (1969). Fuzzy logic and its application to switching systems, *IEEE Transactions on Electronic Computers*, C-18, 343-348.

Martin, W. A. (1973). Translation of English into MAPL using Winograd's syntax, state transition networks, and a semantic case grammar, *M.I.T. APG Internal Memo 11*.

Miller, G. A. and Johnson-Laird, P. N. (1976), *Language and Perception*. Cambridge, Mass.: Harvard University Press.

Minsky, M. (1975). A framework for representing knowledge. In *The Psychology of Computer Vision* (ed. Winston, P. H.). New York: McGraw Hill.

Moisil, G. C. (1975). Lectures on the logic of fuzzy reasoning, *Scientific Editions*, Bucarest.

Montague, R. (1974). *Formal Philosophy (Selected Papers)* (ed Thomason, R. H.). New Haven: Yale University Press.

Moore, J. and Newell, A. (1973). How can MERLIN understand? Pittsburgh: Department of Computer Science, Carnegie-Mellon University.

Nahmias, S. (1976). Fuzzy variables. *Technical Report 33*. Pittsburgh: Department of Industrial Engineering, Systems Management Engineering and Operations Research, University of Pittsburgh. Presented at ORSA/TIMS Meeting, Miami.

Nalimov, V. V. (1974). *Probabilistic Model of Language*. Moscow: Moscow State University.

Negoita, C. V. and Ralescu, D. A. (1975). *Applications of Fuzzy Sets to Systems Analysis*. Basel, Stuttgart: Birkhauser Verlag.

Newell, A. and Simon, H. A. (1972). *Human Problem Solving*. Englewood Cliffs, N. J.: Prentice-Hall.

Nguyen, H. T. (1976). A note on the extension principle for fuzzy sets, *Electronics Research Laboratory Memorandum M-611*. Berkeley: University of California.

Noguchi, K., Umano, M., Mizumoto, M. and Tanaka, K. (1976). Implementation of fuzzy artificial intelligence language FLOU, *Technical Report on Automation and Language of IECE*.

Norman, D. A. and Rumelhart, D. E. (eds) (1975). *Explorations in Cognition*. San Francisco: W. H. Freeman Co.

Ostegaard, J. J. (1976). Fuzzy logic control of a heat exchanger process, *Publication No. 7601*. Lyngby: Technical University of Denmark.

Pal, S. K. and Majumdar, D. D. (1977). Fuzzy sets and decision-making approaches in vowel and speaker recognition, *IEEE Transactions of Systems, Man, and Cybernetics*, SMC-7, 625-629.

Procyk, T. J. (1976). Linguistic representation of fuzzy variables, *Fuzzy Logic Working Group*. London: Queen Mary College.

Putman, H. (1976). *Meaning and Truth*. Sherman Lectures. London: University College.

Rescher, N. (1969). *Many-Valued Logic*. New York: McGraw-Hill.

Rieger, B. (1976). Fuzzy structural semantics, *Proceedings of 3rd European Meeting on Cybernetics and Systems Research*, Vienna.

Rieger, C. (1976). An organization of knowledge for problem solving and language comprehension, *Artificial Intelligence*, 7, 89-127.

Rutherford, G. and Bloore, G. C. (1975). The implication of fuzzy algorithms for control. Manchester: Control Systems Center, University of Manchester.

Sanchez, E. (1974). Fuzzy relations. Marseille: Faculty of Medicine, University of Marseille.

Sanford, D. H. (1975). Borderline logic, *American Philosophical Quarterly*, 12, 29-39.

Schank, R. C. (1973). Identification of conceptualizations underlying natural language. *Computer Models of Thought and Language*, pp. 187-247 (eds Schank, R and Colby. M). Englewood Cliffs, N. J.: Prentice-Hall.

Schotch, P. K. (1975). Fuzzy modal logic, *Proceedings of International Symposium on Multiple-Valued Logic*, pp. 176-182. Bloomington: University of Indiana.

Shimura, M. (1975). An approach to pattern recognition and associative memories using fuzzy logic. *Fuzzy Sets and Their Applications to Cognitive and Decision Processes*, pp. 449-476 (eds Zadeh, L. A., Fu, K. S., Tanaka, K. and Shimura, M.). New York: Academic Press.

Shortliffe, E. H. and Buchanan, B. G. (1975). A model of inexact reasoning in medicine, *Mathematical Biosciences*, 23, 351-379.

Simmons, R. F. (1973). Semantic networks, their computation and use for understanding English sentences. *Computer Models of Thought and Language*, pp. 63–113 (eds Schank, R. and Colby, K.). Englewood Cliffs, N. J.: Prentice-Hall.

Simon, H. A. and Siklossy, L. (1972). *Representation and Meaning: Experiments with Information Processing Systems.* Englewood Cliffs, N. J.: Prentice-Hall.

Siy, P. and Chen, C. S. (1974). Fuzzy logic for handwritten numerical character recognition, *IEEE Transactions on Systems, Man and Cybernetics*, SMC-4, 570–575.

Stalnaker, R. (1970). Probability and conditionals, *Philosophical Science*, 37, 64–80.

Sugeno, M. (1974). *Theory of fuzzy integrals and its applications*, Ph.D. Dissertation, Tokyo: Tokyo Institute of Technology.

Suppes, P. (1974a). The axiomatic method in the empirical sciences, *Proceedings of Tarski Symposium.* Providence, Rhode Island: American Mathematical Society.

Suppes, P. (1974b). Probabilistic metaphysics, *Filofiska Studier nr. 22.* Uppsala: Uppsala University.

Tarski, A. (1956). *Logic, Semantics, Metamathematics.* Oxford: Clarendon Press.

Terano, T. and Sugeno, M. (1975). Conditional fuzzy measures and their applications. *Fuzzy Sets and Their Applications to Cognitive and Decision Processes*, pp. 151–170 (eds Zadeh, L. A., Fu, K. S., Tanaka, K. and Shimura, M.). New York: Academic Press.

Uragami, M., Mizumoto, M. and Tanaka, K. (1976). Fuzzy robot controls, *Journal of Cybernetics*, 6, 39–64.

Wason, P. C. and Johnson-Laird, P. H. (1972). *Psychology of Reasoning Structure and Content.* Cambridge, Mass.: Harvard University Press.

Wechsler, H. (1975). Applications of fuzzy logic to medical diagnosis, *Proceedings of International Symposium on Multiple-Valued Logic*, pp. 162–174. Bloomington: University of Indiana.

Wenstop, F. (1975). *Application of linguistic variables in the analysis of organizations.* Ph.D. Dissertation. Berkeley: School of Business Administration, University of California.

Wenstop, F. (1976). Deductive verbal models of organizations, *International Journal of Man-Machine Studies*, 8, 293–311.

Zadeh, L. A. (1968). Probability measures of fuzzy events. *Journal of Mathematical Analysis and Applications*, 23, 421–427.

Zadeh, L. A. (1971). Similarity relations and fuzzy orderings, *Information Sciences*, 3, 177–200.

Zadeh, L. A. (1972a), A fuzzy-set-theoretic interpretation of linguistic hedges, *Journal of Cybernetics*, 2, 4–34.

Zadeh, L. A. (1972b). Fuzzy languages and their relation to human and machine intelligence, *Proceedings of International Conference on Man and Computer*, Bordeaux, France, pp. 130–165. Basel: S. Karger.

Zadeh, L. A. (1973). Outline of a new approach to the analysis of complex systems and decision processes, *IEEE Transactions on Systems, Man and Cybernetics*, SMC-3, 28–44.

Zadeh, L. A. (1975a). Fuzzy logic and approximate reasoning (In memory of Grigore Moisil), *Synthese*, 30, 407–428.

Zadeh, L. A. (1975b). Calculus of fuzzy restrictions. *Fuzzy Sets and Their Applications to Cognitive and Decision Processes*, pp. 1–39 (eds. Zadeh, L. A., Fu, K. S., Tanaka, K. Shimura, M.). New York: Academic Press.

Zadeh, L. A. (1975c). The concept of a linguistic variable and its application to approximate reasoning, Part I, *Information Sciences*, 8, 199–249; Part II, *Information Sciences*, 8, 301–357, Part III, *Information Sciences*, 9, 43–80.

Zadeh, L. A. (1976). A fuzzy-algorithmic approach to the definition of complex or imprecise concepts, *International Journal of Man-Machine Studies*, 8, 249–291.

Zadeh, L. A. (1977a). Fuzzy sets as a basis for a theory of possibility, *Electronics Research Laboratory Memorandum M77/12.* Berkeley: University of California. To appear in *Fuzzy Sets and Systems.*

Zadeh, L. A. (1977b), PRUF--a language for the representation of meaning in natural languages, *Proceedings of 5th International Joint Conference on Artificial Intelligence,* 918. Pittsburgh: Department of Computer Science, Carnegie-Mellon University.

Zadeh, L. A., Fu, K. S., Tanaka, K. and Shimura, M. (1975). *Fuzzy Sets and Their Application to Cognitive and Decision Processes.* New York: Academic Press.

Zimmermann, H. J. (1974). Optimization in fuzzy environments, Institute for Operations Research. Aachen: Technical University of Aachen.

The Role of Fuzzy Logic in the Management of Uncertainty in Expert Systems

L.A. ZADEH*

Computer Science Division, Department of EECS and ERL, University of California, Berkeley, CA 94720, USA

To Professor Elie Sanchez

Received July 1983

Management of uncertainty is an intrinsically important issue in the design of expert systems because much of the information in the knowledge base of a typical expert system is imprecise, incomplete or not totally reliable.

In the existing expert systems, uncertainty is dealt with through a combination of predicate logic and probability-based methods. A serious shortcoming of these methods is that they are not capable of coming to grips with the pervasive fuzziness of information in the knowledge base, and, as a result, are mostly ad hoc in nature. An alternative approach to the management of uncertainty which is suggested in this paper is based on the use of fuzzy logic, which is the logic underlying approximate or, equivalently, fuzzy reasoning. A feature of fuzzy logic which is of particular importance to the management of uncertainty in expert systems is that it provides a systematic framework for dealing with fuzzy quantifiers, e.g., *most, many, few, not very many, almost all, infrequently, about 0.8*, etc. In this way, fuzzy logic subsumes both predicate logic and probability theory, and makes it possible to deal with different types of uncertainty within a single conceptual framework.

In fuzzy logic, the deduction of a conclusion from a set of premises is reduced, in general, to the solution of a nonlinear program through the application of projection and extension principles. This approach to deduction leads to various basic syllogisms which may be used as rules of combination of evidence in expert systems. Among syllogisms of this type which are discussed in this paper are the intersection/product syllogism, the generalized modus ponens, the consequent conjunction syllogism, and the major-premise reversibility rule.

Keywords: Expert systems, Knowledge representation, Fuzzy logic, Fuzzy sets.

1. Introduction

An expert system, as its name implies, is an information system which provides the user with a facility for posing and obtaining answers to questions relating to the information stored in its knowledge base. Typically, such systems possess a nontrivial inferential capability and, in particular, have the capability to infer from premises which are imprecise, incomplete or not totally reliable.

* Research supported in part by NESC-N0039-83-C-0243.

Since the knowledge base of an expert system is a repository of human knowledge, and since much of human knowledge is imprecise in nature, it is usually the case that the knowledge base of an expert system is a collection of rules and facts which, for the most part, are neither totally certain nor totally consistent. Now, as a general principle, the uncertainty of information in the knowledge base of any question-answering system induces some uncertainty in the validity of its conclusions. Hence, to serve a useful purpose, the answer to a question must be associated – explicitly or, at least, implicitly – with an assessment of its reliability. For this reason, a basic issue in the design of expert systems is how to equip them with a computational capability to analyze the transmission of uncertainty from the premises to the conclusion and associate the conclusion with what is commonly called a *certainty factor*.

In the existing expert systems, the computation of certainty factors is carried out through a combination of methods which are based on, or, at least, not far removed from, two-valued logic and probability theory. However, it is widely recognized at this juncture that such methods have serious shortcomings and, for the most part, are hard to rationalize. In particular, what is open to question is the universally made assumption that if each premise is associated with a numerical certainty factor then the certainty factor of the conclusion is a number which may be expressed as a function of the certainty factors of the premises. As will be seen in the sequel, this assumption is, in general, invalid. It regains its validity, however, if the certainty factors are represented as fuzzy rather than crisp numbers.

More generally, a point of view which is articulated in the present paper is that the conventional approaches to the management of uncertainty in expert systems are intrinsically inadequate because they fail to come to grips with the fact that much of the uncertainty in such systems is possibilistic rather than probabilistic in nature. As an alternative, it is suggested that a fuzzy-logic-based computational framework be employed to deal with both possibilistic and probabilistic uncertainty within a single conceptual system. In this system, test-score semantics – which is the meaning-representational component of fuzzy logic – forms the basis for the representation of knowledge, while the inferential component of fuzzy logic is employed to deduce answers to questions and, when necessary, associate each answer with a probability which is represented as a fuzzy quantifier.

The employment of fuzzy logic as a framework for the management of uncertainty in expert systems makes it possible to consider a number of issues which cannot be dealth with effectively or correctly by conventional techniques. The more important of these issues are the following.

(1) The fuzziness of antecedents and/or consequents in rules of the form

(a) If X is A then Y is B,

(b) If X is A then Y is B with $CF = \alpha$,

where the antecedent, X is A, and the consequent, X is B, are fuzzy propositions, and α is a numerical value of the certainty factor, CF. For example,

If X is small then Y is large with $CF = 0.8$,

in which the antecedent "*X is small*" and the consequent "*Y is large*" are fuzzy propositions because the denotations of the predicates *small* and *large* are fuzzy subsets *SMALL* and *LARGE* of the real line.

(2) Partial match between the antecedent of a rule and a fact supplied by the user.

Since the number of rules in an expert system is usually relatively small (i.e., of the order of two hundred), there are likely to be many cases in which a fact such as "*X is A^**" may not match exactly the antecedent of any rule of the form "If X is A then Y is B with CF = α". The conventional rule-based systems usually avoid this issue or treat it in an ad hoc manner because partial matching does not lend itself to analysis within the confines of two-valued logic. By contrast, the gradation of truth and membership in fuzzy logic provides a natural way of dealing with partial matching through the use of the compositional rule of inference and interpolation.

(3) The presence of fuzzy quantifiers in the antecedent and/or the consequent of a rule.

In many cases, the antecedent and/or the consequent of a rule contain implicit or explicit fuzzy quantifiers such as *most, many, few, many more, usually, much of,* etc. As an illustration, consider the disposition[1]

$d \triangleq$ *Students are young,*

which may be interpreted as the proposition

$p \triangleq$ *Most students are young,*

which in turn may be expressed as a rule, or, equivalently, as the conditional proposition

$r \triangleq$ *If x is a student then it is likely that x is young,*

in which the fuzzy probability *likely* has the same denotation, expressed as a fuzzy subset of the unit interval, as the fuzzy quantifier *most*.

In the following two sections, we shall consider first the problems which arise when the antecedent and consequent components of a rule are fuzzy propositions, but no certainty factors are involved. In Section 4, then, we shall consider the more general case of rules in which the certainty factor is represented as a fuzzy quantifier.

Our exposition of the role of fuzzy logic in the management of uncertainty in expert systems is not intended to be definitive and/or complete. Rather, its much more limited objective is to suggest that fuzzy logic provides a natural conceptual framework for the analysis and design of expert systems, and point to some of the basic problem-areas which are in need of exploration.

---

[1] As defined in [89], a disposition is a proposition with implicit fuzzy quantifiers. Many of the rules in a typical expert system are, in effect, dispositions.

2. The effect of fuzziness in facts and rules

Typically, the knowledge base of an expert system consists of (a) a collection of propositions which represent the *facts*; and (b) a collection of conditional propositions which constitute the *rules*. For example, the facts may be:[2]

(a) Carol is a graduate student.
(b) Berkeley's population is *over 100 000*.
(c) San Francisco is a *foggy* city.
(d) John *has duodenal ulcer* (CF = 0.3).

In these examples, (a) is a nonfuzzy proposition since the class of graduate students is a crisp set; (b) is a fuzzy proposition because of an implicit understanding that *over 100 000* means *over 100 000 but not much over 100 000*; (c) is a fuzzy proposition because *foggy city* is a fuzzy predicate; and (d) is a fuzzy proposition because the predicate *has a duodenal ulcer* is a fuzzy predicate in the sense that having a duodenal ulcer is a matter of degree. Furthermore, although the certainty factor is stated to be equal to 0.3, it should be interpreted as a fuzzy number which is *approximately equal* to 0.3.

A typical rule in MYCIN [61] is exemplified by

Rule 0–47

If: (1) The site of the culture is blood, and
(2) The identity of the organism *is not known with certainty*, and
(3) The stain of the organism is gramneg, and
(4) The morphology of the organism is rod, and
(5) The patient has been *seriously burned*,
then: There is *weakly suggestive* evidence that the identity of the organism is pseudomonas.

Another typical rule in MYCIN reads:

If: (1) The route of the administration of the penicillin is oral, and
(2) There is a gastrointestinal factor which *may interfere* with the *absorption* of the penicillin,
then: There is *suggestive evidence* (0.6) that the route of administration of the penicillin is not *adequate*.

Typical rules in PROSPECTOR [20] are exemplified by:

(a) **If:**   *Abundant* quartz sulfide *veinlets* with *no apparent alteration halos*,
(b) **then:** (LS, LN) *alteration favorable* for the *potassic stage*.
(a) **If:**   Volcanic rocks in the region are *contemporaneous* with the *intrusive system* (*coeval volcanic rocks*),
(b) **then:** (LS, LN) the level of erosion is *favorable* for a *porphyry copper deposit*.

---

[2] In these and the following examples, the italicized constituents are fuzzy predicates. A proposition is defined to be fuzzy if its contains fuzzy predicates.

In these rules, LS and LN are real numbers representing likelihood ratios. Informally, if the ratio LS which is associated with a hypothesis H and evidence E is greater than unity, then the odds on H are increased by the presence of E. On the other hand, if LN is greater than unity, then the odds on H are increased by the absence of E. In consequence of the definitions of LS and LN, they cannot be simultaneously greater than unity.

The point we wish to make through these examples is that most of the facts and rules in expert systems contain fuzzy predicates and thus are fuzzy propositions. This is particularly true of the heuristic rules which are encoded as production rules in what are, in effect, fuzzy algorithms [72, 57, 74]. An example of such heuristic rules in the case of a rule-based system for playing poker is provided by the following excerpt [61, 66]:

"If your hand is *excellent* then bet *low* if the opponent *tends* to be a *conservative* player and has just bet *low*. Bet *high* if the opponent is not *conservative*, is not *easily bluffed*, and has just made a *sizable* bet. Call if the pot is *extremely large*, and the opponent has just made a *sizable* bet."

In the existing expert systems, the fuzziness of the knowledge base is ignored because neither predicate logic nor probability-based methods provide a systematic basis for dealing with it. As a consequence, fuzzy facts and rules are generallly manipulated as if they were nonfuzzy, leading to conclusions whose validity is open to question.

As a simple illustration of this point, consider the fact

John has duodenal ulcer (CF = 0.3).

Since *has duodenal ulcer* is a fuzzy predicate, so that John may have it to a degree, the meaning of the certainty factor becomes ambiguous. More specifically, does CF = 0.3 mean that (a) John had duodenal ulcer to the degree 0.3; or (b) that the probability of the fuzzy event "John has duodenal ulcer" is 0.3? Note that in order to make the latter interpretation meaningful, it is necessary to be able to define the probability of a fuzzy event. This can be done in fuzzy logic [71, 30] but not in classical probability theory.

As another illustration, consider a rule of the general form:

If X is A then Y is B with probability β,

where X and Y are variables, A and B are fuzzy predicates and $β$ is a fuzzy probability expressed as a fuzzy number, e.g., *about 0.8*, or as a linguistic probability, e.g., *very likely*. For example,

If Hans has a new red Porsche then it is likely that his wife is young,

in which case $X \triangleq$ make of Hans' car; $A \triangleq$ fuzzy set of new red Porsches; $Y \triangleq$ age of Hans' wife; $B \triangleq$ fuzzy subset *YOUNG* of the Age scale; and $β \triangleq LIKELY$, which is a fuzzy subset of the unit interval.

Expressed as a conditional probability, the rule in question may be written as

$$\Pr\{Y \text{ is } B \mid X \text{ is } A\} \text{ is } β. \qquad (2.1)$$

In the existing expert systems, such a rule would be treated as an ordinary conditional probability, from which it would follow that

$$\Pr\{Y \text{ is not } B \mid X \text{ is } A\} \text{ is } 1 - \beta. \tag{2.2}$$

However, as shown in [88], this conclusion is, in general, incorrect if A is a fuzzy set. The correct conclusion is weaker than (2.2), namely,

$$\Pr\{Y \text{ is not } B \mid X \text{ is } A\} + \Pr\{Y \text{ is } B \mid X \text{ is } A\} \geqslant 1, \tag{2.3}$$

with the understanding that the probabilities in question may be fuzzy numbers.

In short, an assumption which is treated as a truism in expert systems is that

$$P(H \mid E) \equiv 1 - P(\neg H \mid E), \tag{2.4}$$

where $P(H \mid E)$ is the conditional probability of a hypothesis H given an evidence E, and $\neg H$ is the negation of H. Our discussion shows that, in general, this assumption is not valid when E is a fuzzy proposition, as is frequently the case in most knowledge bases.

As was alluded to earlier, perhaps the most serious deficiency of the existing expert systems relates to the ways in which (a) the traditional rules of inference are applied to fuzzy rules; and (b) the computation of certainty factors is carried out when two or more rules are combined through conjunction, disjunction or chaining. In the case of chaining, in particular, the standard inference rule – *modus ponens* – loses much of its validity and must be replaced by the more general *compositional rule of inference* [75]. Furthermore, as shown in [88], the transitivity of implication, which forms the basis for both forward and backward chaining [7] in most expert systems, is a *brittle* property which must be applied with great caution. We shall discuss these issues in greater detail in the following sections.

3. Inference in fuzzy logic

Fuzzy logic provides a natural framework for the management of uncertainty in expert systems because – in contrast to traditional logical systems – it main purpose is to provide a systematic basis for representing and inferring from imprecise rather than precise knowledge. In effect, in fuzzy logic everything is allowed to be – but need not be – a matter of degree.

The greater expressive power of fuzzy logic derives from the fact that it contains as special cases the traditional two-valued as well as multi-valued logics. The main features of fuzzy logic which are of relevance to the management of uncertainty in expert systems are the following.

(a) In two-valued logical systems a proposition, p, is either true or false. In multi-valued logical systems, a proposition may be true, or have an intermediate truth-value which may be an element of a finite or infinite truth-value set T. In fuzzy logic, the truth-values are allowed to range over the fuzzy subsets of T. For example, if T is the unit interval, then a truth-value in fuzzy logic, e.g., *very true*, may be interpreted as a fuzzy subset of the unit interval which defines the possibility distribution associated with the truth-value in question. In this sense, a fuzzy truth-value may be viewed as an imprecise characterization of an intermediate truth-value.

(b) The predicates in two-valued logic are constrained to be crisp in the sense that the denotation of a predicate must be a nonfuzzy subset of the universe of discourse. In fuzzy logic, the predicates may be crisp, e.g., *mortal, even, father of,* etc. or, more generally, fuzzy, e.g., *ill, tired, large, tall, much heavier, friend of,* etc.

(c) Two-valued as well as multi-valued logics allow only two quantifiers: *all* and *some.* By contrast, fuzzy logic allows, in addition, the use of fuzzy quantifiers exemplified by *most, many, several, few, much of, frequently, occasionally, about ten,* etc. Such quantifiers may be interpreted as fuzzy numbers which provide an imprecise characterization of the cardinality of one or more fuzzy or nonfuzzy sets. In this perspective, a fuzzy quantifier may be viewed as a second order fuzzy predicate. Based on this view, fuzzy quantifiers may be used to represent the meaning of propositions containing fuzzy probabilities and thereby make it possible to manipulate probabilities within fuzzy logic.

(d) Fuzzy logic provides a method for representing the meaning of both nonfuzzy and fuzzy predicate-modifiers (or, simply, modifiers) exemplified by *not, very, more or less, extremely, slightly, much, a little,* etc. This, in turn, leads to a system for computing with *linguistic variables* [48, 77], that is, variables whose values are words or sentences in a natural or synthetic language. For example, *Age* is a linguistic variable when its values are assumed to be: *young, old, very young, not very old,* etc., with each value interpreted as a possibility distribution over the real line.

(e) In two-valued logical systems, a proposition, p, may be qualified, principally by associating with p a truth-value, *true* or *false*; a modal operator such as *possible* or *necessary*: and an intensional operator such as *know, believe,* etc. In fuzzy logic, the three principal modes of qualification are: (a) *truth qualification*, expressed as p is τ, in which τ is a fuzzy truth-value; *probability qualification*, expressed as p is λ, in which λ is a fuzzy probability; and *possibility qualification*, expressed as p is π, in which π is a fuzzy possibility, e.g., *quite possible, almost impossible,* etc. Furthermore, *knowing* and *believing* are assumed to be binary fuzzy predicates.

Types of propositions

Since the inference processes in the existing expert systems are based on two-valued logic and/or probability theory, the principal tools for inference are the modus ponens and/or its probabilistic analog – the Bayes rule and its variations. As was alluded to earlier, the validity of these inference processes is open to question since most of the information in the knowledge base of a typical expert system consists of a collection of fuzzy rather than nonfuzzy propositions.

To deal with such propositions, fuzzy logic draws, first, on test-score semantics [86] to represent their meaning; and, second, on a combination of the entailment and extension principles [83] to deduce the answer to a given question. In this way, the problem of inference from fuzzy propositions is reduced, in general, to the solution of a nonlinear program.

The basic ideas which underlie the deduction processes in fuzzy logic may be summarized as follows.

Assume that the knowledge base, KB, consists of a collection of propositions $\{p_1, \ldots, p_N\}$, some or all of which may be fuzzy in nature.

The propositions in a typical knowledge base may be divided into four principal categories.

(1) An unconditional, unqualified proposition.

Example: *Carol has a young daughter.*

Canonical form:[3] *X is F*, where *X* is a variable and *F* is a fuzzy predicate (i.e., a fuzzy subset of the domain of *X*).

(2) An unconditional, qualified proposition.

Example 1: *It is very likely that Carol has a young daughter*, or, equivalently, *Carol has a young daughter is very likely*, which exhibits more directly the proposition *Carol has a young daughter* and the qualifying fuzzy probability *very likely*.

Example 2: *Carol is very vivacious most of the time.* In this case, the qualifier *most of the time* plays the role of a fuzzy quantifier.

Canonical form 1: *X is F is λ*, where *X* is a variable, *F* is a fuzzy predicate and *λ* is a fuzzy probability.

Canonical form 2: *QU's are F's*, where *Q* is a fuzzy quantifier and *F* is a fuzzy subset of the universe of discourse, *U*.

(3) A conditional, unqualified proposition.

Example: *If X is a man then X is mortal.*

Canonical form: *If X is F then Y is G*, where *X* and *Y* are variables and *F* and *G* are fuzzy predicates.

(4) A conditional, qualified proposition.

Example 1: *If a car is old then it probabily is not very reliable* or, more precisely, *If X is an old car then (X is not very reliable is probable).*

Example 2: *If X is a young man then (Y is a young woman is likely)*, where *Y* denotes the girlfriend of *X* and *likely* is the qualifying fuzzy probability.

Example 3: *Most Swedes are blond.* This proposition may be expressed in the equivalent form *If X is a Swede then (X is blond is likely)* in which the fuzzy probability *likely* is equal, as a fuzzy number, to the fuzzy quantifier *most*.

Canonical form 1: *If X is F then Y is G is λ*, where *X* and *Y* are variables, *F* and *G* are fuzzy predicates and *λ* is a fuzzy probability.

Canonical form 2: *QF's are G's*, where *Q* is a fuzzy quantifier, and *F* and *G* are fuzzy predicates.

Most of the rules in the knowledge base of a typical expert system are of Types 3 and 4. In particular, many of the rules of Type 4 are *dispositions*, that is, propositions with implicit fuzzy quantifiers, e.g., *Snow is white, Small cars are unsafe, Young men like young women*, etc. Dispositions play an especially important role in the representation of – and inference from – commonsense knowledge [89].

---

[3] Canonical forms are discussed in greater detail at a later point in this section.

Canonical forms

As a preliminary to applying the rules of inference of fuzzy logic to propositions in KB, it is necessary to represent their meaning in a canonical form which places in evidence the constraints which are induced by each proposition. In fuzzy logic, this is accomplished through the use of test-score semantics [80, 86].

More specifically, the basic idea underlying test-score semantics is that a proposition, p, in a natural language may be viewed as a collection of *focal variables* $X_1, \ldots, X_n$ taking values in $U_1, \ldots, U_n$, respectively, which are constrained by a system of elastic (or fuzzy) constraints $F_1, \ldots, F_m$. In general, the variables $X_1, \ldots, X_n$ and the associated constraints $F_1, \ldots, F_m$ are implicit rather than explicit in p.

In more concrete terms, any unconditional proposition, p, may be represented in a *canonical form*, $cf(p)$,

$$p \rightarrow cf(p) \triangleq X \text{ is } F, \tag{3.1}$$

in which $X \triangleq (X_1, \ldots, X_n)$ is an n-ary focal variable whose components are $X_1, \ldots, X_n$; and F is a fuzzy relation in $U_1 \times \cdots \times U_n$ which represents an elastic constraint on X. Informally, what this means is that a proposition, p, may be viewed as a system of elastic constraints, and that the representation of the meaning of p is the process by which the implicit constraints in p are made explicit by expressing p in a form which places in evidence the constrained variable X and the elastic (or fuzzy) constraint F which is induced by p.

Let Π_X denote the possibility distribution of X, that is, the fuzzy set of possible values which X can take in $U \triangleq U_1 \times \cdots \times U_n$ [84]. Then, as shown in [86], the canonical proposition X *is* F may be interpreted as the *possibility assignment equation*

$$\Pi_X = F \tag{3.2}$$

or, more explicitly, as

$$\pi_X(u) \triangleq \text{Poss}\{X = u\}$$
$$= \mu_F(u), \tag{3.3}$$

where $u \triangleq (u_1, \ldots, u_n)$ is a generic point in $U = U_1 \times \cdots \times U_n$; $\pi_X : U \rightarrow [0, 1]$ is the *possibility distribution function* associated with Π_X; $\mu_F : U \rightarrow [0, 1]$ is the membership function of the fuzzy relation F; and $\text{Poss}\{X = u\}$ should be read as "the possibility that X may take u as its value". In this way, p may be *translated into* its canonical form (3.1) or, equivalently, its possibility assignment equation

$$p \rightarrow \Pi_X = F \tag{3.4}$$

in which $\Pi_{(X_1, \ldots, X_n)} = F$ is the possibility distribution induced by p.

If p is a conditional proposition, e.g., *If Roberta works near Washington then Roberta lives near Washington*, the canonical form of p may be expressed as

$$cf(p) \triangleq \text{If } X \text{ is } F \text{ then } Y \text{ is } G, \tag{3.5}$$

where the focal variables X and Y take values in U and V, respectively; F and G

are fuzzy subsets of U and V; and X is F and Y is G are the canonical forms of the antecedent and consequent components of p. Correspondingly, the possibility assignment equation associated with p becomes [83]

$$p \rightarrow \Pi_{(Y|X)} \text{ is } H, \tag{3.6}$$

in which $\Pi_{(Y|X)}$ denotes the conditional possibility distribution of Y given X and H is defined in terms of F and G by[4]

$$\mu_H(u, v) = 1 \wedge (1 - \mu_F(u) + \mu_G(v)), \tag{3.7}$$

where u and v are generic values of X and Y, respectively; $\mu_H : U \times V \rightarrow [0, 1]$ is the membership function of H; $\mu_F : U \rightarrow [0, 1]$ and $\mu_G : V \rightarrow [0, 1]$ are the membership functions of F and G; and $\wedge$ is the min operator. Thus, expressed in terms of the possibility distribution functions, (3.7) implies that

$$\pi_{(Y|X)}(u, v) \triangleq \text{Poss}\{Y = v \mid X = u\}$$
$$= 1 \wedge (1 - \mu_F(u) + \mu_G(v)). \tag{3.8}$$

In general, the translation of p into its canonical form requres the construction of (a) an *explanatory database*; (b) a test procedure which tests and scores the constraints induced by p; and (c) an aggregation function which combines the partial test scores into a single (or, more generally, a vector) test score τ which represents the compatibility of p with the explanatory database [86]. In the case of expert systems, however, the propositions in KB are usually simple enough to be amenable to translation by inspection. For example:

(a) *Carol has dark hair* $\rightarrow X$ is F,
where

$X \triangleq Color \ (Hair \ (Carol))$,

$F \triangleq DARK$

and *DARK* is a fuzzy subset of the set of colors of human hair.

(b) *John lives about two miles from Henry* $\rightarrow X$ is F,
where

$X \triangleq Distance \ (Location \ (Residence \ (John)), \ Location \ (Residence \ (Henry)))$

$F \triangleq ABOUT \ 2.$

(c) *Henry is much younger than George* $\rightarrow (X_1, X_2)$ is F,
where

$X_1 \triangleq Age \ (Henry)$,

$X_2 \triangleq Age \ (George)$,

$F \triangleq MUCH \ YOUNGER$

---

[4] Equation (3.7) expresses a particular definition of the conditional possibility distribution which is consistent with the definition of implication in Lukasiewicz's $L_{\text{Aleph }1}$ logic. A more detailed discussion of various forms of $\Pi_{(Y|X)}$ may be found in [5], [18] and [43].

and the fuzzy relation *MUCH YOUNGER* is a fuzzy subset of $[0, 120] \times [0, 120]$.

(d) *If Tong is blond then he is not Chinese* $\rightarrow$ *If X is F then Y is G,*

where

$X \triangleq Color\ (Hair\ (Tong)),$

$Y \triangleq Nationality\ (Tong),$

$F \triangleq BLOND,$

$G \triangleq NOT\ CHINESE.$

(e) *John has three sons* $\rightarrow$ *X is F,*

where

$X \triangleq Count\ (Sons\ (John)),$

$F \triangleq 3$

and $(Count\ (Sons\ (John))$ is the count of the number of elements in the set *Sons (John)*.

(f) *John has three young sons* $\rightarrow (X_1, X_2)$ *is F,*

where

$X_1 \triangleq Count\ (Sons\ (John)),$

$X_2 \triangleq Age\ (Oldest\ Son\ (John)),$

$F \triangleq (F_1, F_2),$

$F_1 \triangleq 3,$

$F_2 \triangleq YOUNG$

and *YOUNG* is a fuzzy subset of the interval $[0, 120]$.

A basic point which these simple examples are intended to illustrate is that either by inspection or, more generally, through the application of test-score semantics, a constitutent proposition, p, in KB may be expressed in a canonical form which places in evidence the variables constrained by p and the elastic or, equivalently, fuzzy constraints to which they are subjected.

By interpreting the canonical form of p as the possibility assignment equation, we are led to the possibility distribution, $\Pi^p_{(X_1,...,X_n)}$, which is induced by p. In this way, each constituent proposition in KB is converted into a possibility distribution which constrains the variables in KB. Then, through conjunction, we can construct the global possibility distribution, $\Pi_{(X_1,...,X_n)}$, which is induced by the totality of propositions in KB. As we shall see in the sequel, this is the point of departure for deduction in fuzzy logic.

Deduction

Assuming, as we have done already, that the knowledge base, KB, consists of a finite collection of propositions $\{p_1, \ldots, p_N\}$, let $\Pi^{p_i}_{(X_1,...,X_n)}$ or, simply, Π^i, denote the possibility distribution induced by p_j, $j = 1, \ldots, N$.[5] Then, under the assumption that the p_j are non-interactive [84], the possibility distribution function of the global possibility distribution may be expressed as[6]

$$\pi_{(X_1,\ldots,X_n)} = \pi^1_{(X_1,\ldots,X_n)} \wedge \cdots \wedge \pi^N_{(X_1,\ldots,X_n)} \tag{3.9}$$

where $\wedge \triangleq \min$, and

$$\pi_{(X_1,\ldots,X_n)}(u_1, \ldots, u_n) \triangleq \mathrm{Poss}\{X_1 = u_1, \ldots, X_n = u_n\}, \quad u_i \in U_i,$$
$$i = 1, \ldots, n. \tag{3.10}$$

Now suppose that we are interested in the possible values of a subset, $\{X_{i1}, \ldots, X_{ik}\}$, of the KB variables $\{X_1, \ldots, X_n\}$. In other words, we are interested in determining the *marginal possibility distribution* $\Pi_{(X_{i1},\ldots,X_{ik})}$ from the knowledge of the global possibility distribution $\Pi_{(X_1,\ldots,X_n)}$. As we shall see presently, the desired possibility distribution is given by the *projection* of $\Pi_{(X_1,\ldots,X_n)}$ on $U_{i1} \times \cdots \times U_{ik}$, which is written for simplicity as [80]

$$_{x_{i1}\times\cdots\times x_{ik}}\Pi_{(X_1,\ldots,X_n)} \triangleq \text{Proj. on } U_{i1} \times \cdots \times U_{ik} \text{ of } \Pi_{(X_1,\ldots,X_n)}. \tag{3.11}$$

For convenience, let $X_{(s)}$ denote the subvariable of the variable $X \triangleq (X_1, \ldots, X_n)$ which is the focus of our interest, i.e.,

$$X_{(s)} = (X_{i1}, \ldots, X_{ik}),$$

where the index sequence $s \triangleq (i1, \ldots, ik)$ is a subsequence of the sequence $(1, \ldots, n)$. Using the same notational device, any k-tuple of the form $(A_{i1}, \ldots, A_{ik})$ may be expressed more compactly as $A_{(s)}$.

Let $\Pi_{X(s)}$ be the *projection* of the global possibility distribution Π_X on $U_{(s)} \triangleq U_{i1} \times \cdots \times U_{ik}$. Then, by definition [83],

$$\pi_{X(s)}(u_{(s)}) \triangleq \sup_{u_{(s')}} \pi_X(u), \tag{3.12}$$

where $s' \triangleq (j1, \ldots, jm)$ is the index subsequence which is complementary to s. (E.g., if $n = 5$ and $s \triangleq (2, 3, 5)$ then $s' = (1, 4)$.)

Now the *entailment principle* [83] of fuzzy logic asserts that from any fuzzy proposition p we can infer a fuzzy proposition q if the possibility distribution induced by p is contained in that induced by q. This may be represented in the schematic form

$$p \rightarrow \Pi^p_X = F$$
$$\downarrow$$
$$q \leftarrow \Pi^q_X = G \supset F, \tag{3.13}$$

---

[5] Note that there is no loss of generality in assuming that the constituent propositions in KB have the same set of focal variables since the set $\{X_1, \ldots, X_n\}$ may be taken to be the union of the focal variables associated with each proposition.

[6] In the present paper, it will suffice four our purposes to use the standard connectives $\wedge \triangleq \min$ (conjunction) and $\vee \triangleq \max$ (disjunction). A more general treatment of connectives in fuzzy logic may be found in [18] and [32].

where the reverse arrow $\leftarrow$ signifies that q is a *retranslation* of the possibility assignment equation $\Pi_X^q = G$ if the latter is a translation of q. For simplicity, we shall say that Π_X^q may be *inferred* from Π_X^p if $\Pi_X^q \supset \Pi_X^p$, that is,

$$\pi_X^q(u) \geq \pi_X^p(u) \quad \text{for all } u \in U. \tag{3.14}$$

From the definition of $\pi_{X_{(s)}}(u_{(s)})$ as expressed by (3.12), it follows at once that

$$\pi_{X_{(s)}}(u_{(s)}) \geq \pi_X(u) \quad \text{for } u \in U, \tag{3.15}$$

and hence that $\Pi_{X_{(s)}}$ may be inferred from Π_X. As a consequence, $\Pi_{X_{(s)}}$ as defined by (3.12) represents the desired possibility distribution of the variable of interest, namely, $X_{(s)} \triangleq (X_{i1}, \ldots, X_{ik})$. Since $X_{(s)}$ is given by the projection of Π_X on $U_{(s)}$, we shall refer to the inference rule which yields $\Pi_{X_{(s)}}$ as the *P-rule*, with P standing for *projection*.

As a simple illustration, assume that the knowledge base contains three propositions p_1, p_2 and p_3 which induce, respectively, the possibility distributions $\Pi_{(X_2,X_3)}^1$, $\Pi_{(X_1)}^2$, and $\Pi_{(X_3,X_4)}^3$. Suppose that the variables of interest are X_2 and X_4. Then, the possibility distribution function of X_2 and X_4 is given by

$$\pi_{(X_2,X_4)}(u_2, u_4) = \sup_{u_1,u_3} (\pi_{(X_2,X_3)}^1(u_2, u_3) \wedge \pi_{(X_1)}^2(u_1) \wedge \pi_{(X_3,X_4)}^3(u_3, u_4)) \tag{3.16}$$

which reduces to

$$\pi_{(X_2,X_4)}(u_2, u_4) = \sup_{u_3} (\pi_{(X_2,X_3)}^1(u_2, u_3) \wedge \pi_{(X_3,X_4)}^3(u_3, u_4)) \tag{3.17}$$

if $\Pi_{(X_1)}^2$ is a *normal* possibility distribution, i.e.,

$$\sup_{u_1} \pi_{(X_1)}^2(u_1) = 1. \tag{3.18}$$

The right-hand member of (3.17) constitutes the *composition of* $\Pi_{(X_2,X_3)}^1$ and $\Pi_{(X_3,X_4)}^3$ with respect to X_3 [75].

In summary, given a knowledge base, KB, the possibility distribution, $\Pi_{X_{(s)}}$, of a specified subvariable, $X_{(s)}$, may be obtained by projecting the global possibility distribution Π_X on $U_{(s)}$. The resulting possibility distribution, $\Pi_{X_{(s)}}$, may be expressed as the composition with respect to $X_{(s')}$ of those constituent possibility distributions which contain variables which are linked directly or transitively to the variables in $X_{(s)}$.[7] For this reason, the P-rule may be described, more suggestively, as the *compositional rule of inference* [75, 83]. Expressed as a sequence of operations, the application of this rule to KB may be represented as the chain:

$$\{p_1, \ldots, p_N\} \xrightarrow{\text{translation}} \{\Pi^1, \ldots, \Pi^N\} \xrightarrow{\text{conjunction}}$$

$$\{\pi^1 \wedge \cdots \wedge \pi^N\} \xrightarrow{\text{projection}} \{\Pi_{X_{(s)}}\} \xrightarrow{\text{retranslation}} q, \tag{3.19}$$

---

[7] X_l and X_m, are linked directly if they appear in the same possibility distribution. X_l and X_m are linked transitively if there exists a chain of variables $X_{c_1}, \ldots, X_{c_r}$, such that (X_l, X_{c_1}) $\ldots$, (X_{c_r}, X_m) are linked directly.

where the *inferent* proposition q is a retranslation of $\Pi_{X_{(s)}}$, which is the marginal possibility distribution of the subvariable of $(X_1, \ldots, X_n)$, $X_{(s)}$, which is the target of the inference process.

Among the traditional rules of inference – which may be viewed as special cases of the compositional rule of inference – is the *modus ponens*. To establish this fact, we shall first derive from the compositional rule of inference a more general version of the modus ponens which in fuzzy logic is referred to as the *generalized modus ponens* [83, 35].

Consider a pair of propositions $\{p_1, p_2\}$ of the form

$$p_1 \triangleq \textit{If X is F then Y is G,}$$
$$p_2 \triangleq X \textit{ is } F^*, \tag{3.20}$$

in which F, F^* and G are fuzzy sets (or, equivalently, fuzzy predicates).

Applying (3.4), (3.6) and (3.7), the translations of p_1 and p_2 may be expressed as

$$p_1 \rightarrow \Pi_{(Y|X)} = H, \tag{3.21}$$
$$p_2 \rightarrow \Pi_X = F^*, \tag{3.22}$$

where $\mu_H(u, v)$ is given by (3.7).

By applying the compositional rule of inference to (3.21) and (3.22), the possibility distribution of Y is found to be given by

$$\Pi_Y = H \circ F^*, \tag{3.23}$$

where the right-hand member of (3.23) represents the composition of H and F^* with respect to X. More concretely,

$$\mu_Y(v) = \sup_u \left(\mu_H(u, v) \wedge \mu_{F^*}(u) \right)$$

$$= \sup_u \left(\mu_{F^*}(u) \wedge (1 - \mu_F(u) + \mu_G(v)) \right). \tag{3.24}$$

This conclusion may be stated in the form of the syllogism

$$\begin{array}{l} \textit{If x is F then Y is G} \\ \underline{X \textit{ is } F^*} \\ Y \textit{ is } H \circ F^* \end{array} \tag{3.25}$$

where $H \circ F^*$ is defined by (3.24). This syllogism expresses the generalized modus ponens.

The generalized modus ponens differs from its classical version in two respects. First, F^* is not required to be identical with F, as it is in the classical case. And second, the predicates F, G and F^* are not required to be crisp. It can readily be verified that when $F = F^*$ and the predicates are crisp, $H \circ F^*$ reduces to G

and (3.25) becomes[8]

> *If X is F then Y is G*
>
> *X is F* $\qquad\qquad\qquad\qquad\qquad\qquad\qquad\qquad$ (3.26)
> _____
>
> *Y is G*

The PE-*rule*

The compositional rule of inference makes it possible to deduce from KB the possibility distribution of a specified subvariable, $X_{(s)}$, of the KB variable $X \triangleq (X_1, \ldots, X_n)$. More generally, however, the target of the inference process is not $X_{(s)}$ but a specified function of $X_{(s)}$, say $f(X_{(s)})$. This may be viewed as a general formulation of the problem of finding an answer to a question which relates to the information resident in KB.

In fuzzy logic, the compositional rule of inference plays an essential role in the formulation and solution of this problem by making it possible to decompose the problem into two subproblems: (1) determination of the possibility distribution of $X_{(s)}$; and (2) determination of the possibility distribution of $f(X_{(s)})$ from the knowledge of that of $X_{(s)}$.

More specifically, assume that a question relates to the possibility distribution of a function $f(X_{(s)})$ whose argument is a subvariable of the KB variable $X \triangleq (X_1, \ldots, X_n)$, and that through the use of the compositional rule of inference we have determined the possibility distribution of $X_{(s)}$. Then, by the extension principle [83], the determination of the possibility distribution of $f(X_{(s)})$ is reduced to the solution of the following nonlinear program:

$$\pi_f(v) = \sup_{u_{(s)}} (\pi_{X_{(s)}}(u_{(s)})) \qquad\qquad (3.27)$$

subject to

$$f(u_{(s)}) = v, \qquad\qquad (3.28)$$

where v is a generic value of f, and

$$\pi_f(v) \triangleq \text{Poss}\{f = v\}. \qquad\qquad (3.29)$$

Deduction in fuzzy logic is based, in the main, on the solution of the nonlinear program expressed by (3.27). Summarizing what we have said so far, assume that we are interested in determining the value of an unknown variable q which may be expressed as a function of a set of variables $X_{(s)}$ which are constrained by a collection of propositions in a knowledge base KB. If $X_{(s)}$ is a proper subset of the variables $X \triangleq (X_1, \ldots, X_n)$ which are constrained by KB – as would usually be the case – then we first find the possibility distribution of $X_{(s)}$ by projecting the global possibility distribution Π_X on $U_{(s)}$. Then, we apply the

---

[8] Depending on the way in which $\Pi_{(Y|X)}$ is defined in terms of μ_F and μ_G, (3.26) may or may not hold when $F = F^*$ but F and G are not crisp. A more detailed discussion of this issue may be found in [22] and [43].

extension principle – as in (3.27) – to reduce the determination of the possibility distribution of q to the solution of a constrained maximization problem in which the function in question is treated as a constraint and the objective function is the possibility distribution function of $X_{(s)}$. For conveneience, we shall refer to this deduction process as the *PE-rule*, with P and E standing for *projection* and *extension*, respectively.

A simple illustration of the PE-rule is provided by the following problem. Suppose that the knowledge base contains the following propositions

$p_1 \triangleq$ *John lives about two miles from Henry,*

$p_2 \triangleq$ *Henry lives about three miles from Ed,*

$q \triangleq$ *How far away is John from Ed?*

The KB variables in this case are the coordinates of the residences of John, Henry and Ed. i.e., (X_J, Y_J), (X_H, Y_H) and (X_E, Y_E). Upon translation of p_1 and p_2, the possibility distribution functions which constrain these variables are found to be expressed by

$$\pi^1(X_J, Y_J, X_H, Y_H) = \mu_{ABOUT2}(((X_J - X_H)^2 + (Y_J - Y_H)^2)^{1/2}), \tag{3.30}$$

$$\pi^2(X_H, Y_H, X_E, Y_E) = \mu_{ABOUT3}(((X_H - X_E)^2 + (Y_H - Y_E)^2)^{1/2}), \tag{3.31}$$

where μ_{ABOUT2} and μ_{ABOUT3} are the membership functions of the fuzzy sets *ABOUT* 2 and *ABOUT* 3, respectively.

The function which characterizes the question q in this case is

$$f(X_J, Y_J, X_E, Y_E) = ((X_J - X_E)^2 + (Y_J - Y_E)^2)^{1/2}, \tag{3.32}$$

and hence the nonlinear program to which the solution of the problem reduces may be expressed as

$$\mu_f(d) = \sup_{X_J, Y_J, X_H, Y_H, X_E, Y_E} (\mu_{ABOUT2}(((X_J - X_H)^2 + (Y_J - Y_H)^2)^{1/2}))$$

$$\wedge (\mu_{ABOUT3}(((X_H - X_E)^2 + (Y_H - Y_E)^2)^{1/2})), \tag{3.33}$$

subject to

$$d = ((X_J - X_E)^2 + (Y_J - Y_E)^2)^{1/2},$$

where d denotes the distance of John from Ed.

This problem can readily be solved by employing the level-set technique in fuzzy mathematical programming [14, 46]. The solution yields d as a fuzzy interval (or, equivalently, a fuzzy number) which may be represented as

$$ABOUT\ 3 \ominus ABOUT\ 2 \leqslant d \leqslant ABOUT\ 3 \oplus ABOUT\ 2, \tag{3.34}$$

where $\oplus$ and $\ominus$ denote fuzzy addition and fuzzy subtraction, respectively [18].

A few observations concerning the solution are in order. First, the answer is similar in form to what it would be if the distances in p_1 and p_2 were specified as 2 miles and 3 miles instead of *about* 2 miles and *about* 3 miles, i.e.,

$$3 - 2 \leqslant d \leqslant 3 + 2. \tag{3.35}$$

However, whereas in the case of (3.34) we start with fuzzy distances, expressed as fuzzy numbers, and arrive at a fuzzy answer, likewise expressed as a fuzzy number, in the case of (3.35) we start with real numbers and end up with an interval-valued answer. Obviously this is so because the information in the knowledge base is incomplete in relation to the posed question. What is important to recognize is that this is a pervasive phenomenon in the case of expert systems, and is the reason – as was alluded to earlier – why the certainty factor of a conclusion should in general be an interval-valued or fuzzy number, rather than a real number, as it is usually assumed to be.

Second, if in the statement of the problem the distances were specified precisely, the upper and lower bounds in (3.35) would be given by the solutions of the following nonlinear programs

$$d_{max} = \sup_{X_J, Y_J, X_H, Y_H, X_E, Y_E} ((X_J - X_E)^2 + (Y_J - Y_E)^2)^{1/2}, \tag{3.36}$$

$$d_{min} = \inf_{X_J, Y_J, X_H, Y_H, X_E, Y_E} ((X_J - X_E)^2 + (Y_J - Y_E)^2)^{1/2}, \tag{3.37}$$

subject to

$$((X_J - X_H)^2 + (Y_J - Y_H)^2)^{1/2} = 2, \qquad ((X_H - X_E)^2 + (Y_H - Y_E)^2)^{1/2} = 3. \tag{3.38}$$

What is of interest to observe is that the roles of the constraints and objective functions in (3.36) and (3.37) are interchanged in relation to those in (3.33). In effect, the more general formulation expressed by (3.33) subsumes (3.36) and (3.37) and is dual to them.

Interpolation

As was pointed out earlier, an important problem which arises in the operation of any rule-based system is the following. Suppose that the user supplies a fact which in its canonical form may be expressed as X *is* F^*, where F is a fuzzy or nonfuzzy predicate. Furthermore, suppose that there is no conditional rule in KB whose antecedent matches F exactly. The question which arises is: Which rules should be executed and how should their results be combined?

The approach suggested by fuzzy logic involves the use of an interpolation technique which is based on the P-rule and is in the spirit of the generalized modus ponens [75].

Specifically, suppose that upon translation a group of propositions in KB may be expressed as a fuzzy relation of the form

R	X_1	X_2	$\cdots$	X_n	X_{n+1}	
	R_{11}	R_{12}	$\cdots$	R_{1n}	Z_1	(3.39)
	$\vdots$				$\vdots$	
	R_{m1}	R_{m2}	$\cdots$	R_{mn}	Z_m	

in which the entries are fuzzy sets; the input variables are $X_1, \ldots, X_n$, with domains $U_1, \ldots, U_n$; and the output variable is X_{n+1}, with domain U_{n+1}. The problem is: Given an input n-tuple $(R_1^*, \ldots, R_n^*)$, in which R_j^*, $j = 1, \ldots, n$, is a fuzzy subset of U_j, what is the value of X_{n+1} expressed as a fuzzy subset of U_{n+1}?

A fuzzy relation which is represented by a tableau of the form (3.39) may be defined in different ways. The definition which will be used here is that given in [78], which is in the spirit of the standard definition of a relation as a collection of tuples. Specifically,

$$R = R_{11} \times \cdots \times R_{1n} \times Z_1 + \cdots + R_{m1} \times \cdots \times R_{mn} \times Z_m, \qquad (3.40)$$

where $\times$ and $+$ denote the cartesian product and union, respectively.

Based on this interpretation of R, the desired value of X_{n+1} may be obtained as follows.

First, we compute for each pair (R_{ij}, R_j^*) the degree of consistency of the input R_j^* with the R_{ij} element of R, $i = 1, \ldots, m$, $j = 1, \ldots, n$. The degree of consistency, γ_{ij}, is defined as

$$\gamma_{ij} \triangleq \sup(R_{ij} \cap R_j^*)$$
$$= \sup_{u_j} (\mu_{R_{ij}}(u_j) \wedge \mu_{R_j^*}(u_j)), \qquad (3.41)$$

in which $\mu_{R_{ij}}$ and $\mu_{R_j^*}$ are the membership functions of R_{ij} and R_j^*, respectively, and u_j is a generic element of U_j.

Next, we compute the overall degree of consistency, γ_i, of the input n-tuple $(R_1^*, \ldots, R_n^*)$ with the ith row of R, $i = 1, \ldots, m$, by employing $\wedge$ (min) as the aggregation operator. Thus

$$\gamma_i = \gamma_{i1} \wedge \gamma_{i2} \wedge \cdots \wedge \gamma_{in}. \qquad (3.42)$$

As expressed by (3.42), γ_i may be interpreted as a conservative measure of agreement between the input n-tuple $(R_1^*, \ldots, R_n^*)$ and the ith row n-tuple $(R_{i1}, \ldots, R_{in})$. Then, employing γ_i as a weighting coefficient, the desired expression for X_{n+1} may be written as a 'linear' combination

$$X_{n+1} = \gamma_1 \wedge Z_1 + \cdots + \gamma_m \wedge Z_m \qquad (3.43)$$

in which $+$ denotes the union, and $\gamma_i \wedge Z_i$ is a fuzzy set defined by

$$\mu_{\gamma_i \wedge Z_i}(u_{i+1}) = \gamma_i \wedge \mu_{Z_i}(u_{i+1}), \quad i = 1, \ldots, m. \qquad (3.44)$$

It should be observed that if no row of R has a high degree of consistency with the input n-tuple, the value of the output variable X_{n+1} will be a subnormal fuzzy set, that is, its maximal grade of membership will be smaller than unity. Furthermore, the lower the degree of consistency, the higher the degree of subnormality. Thus, to achieve a high degree of normality of the output, it is necessary that at least one of the rows of R have a high degree of consistency with the input n-tuple.

4. Inference from quantified propositions

In the preceding section, we have restricted our discussion of inference in fuzzy logic to unqualified propositions. In the case of qualified propositions, the problem of inference becomes considerably more complex, and our discussion of it will touch upon only a few of the many issues which arise when some of the propositions in the knowledge base are associated with certainty factors.[9]

For simplicity, we shall restrict our attention to quantified propositions whose canonical form may be expressed as

$$q \triangleq QA\text{'s as } B\text{'s,} \tag{4.1}$$

e.g., *Most small cars are unsafe*, where A and B are fuzzy predicates and Q is a fuzzy quantifier such as *most, many, not very many, approximately 0.8, much more than a half*, etc. For convenience, A and B will be referred to as the *antecedent* and *consequent* of q. This is motivated by the observation that q may be interpreted as the conditional probability assignment

$$\text{Prob}\{B \mid A\} \text{ is } Q, \tag{4.2}$$

in which $\text{Prob}\{B \mid A\}$ denotes the conditional probability of the fuzzy event $\{X \text{ is } B\}$ given the fuzzy event $\{X \text{ is } A\}$ [71]. In both (4.1) and (4.2), Q plays the role of a fuzzy number which is a fuzzy subset of the unit interval.

Cardinality of fuzzy sets

To make the concept of a fuzzy quantifier meaningful, it is necessary to define a way of counting the number of elements in a fuzzy set or, equivalently, to determine its cardinality.

There are several ways in which this can be done [88]. For our purposes, it will suffice to employ the concept of a *sigma-count*, which is defined as follows.

Let F be a fuzzy subset of $U = \{u_1, \ldots, u_n\}$ expressed symbolically as

$$F = \mu_1/u_1 + \cdots + \mu_n/u_n = \sum_i \mu_i/u_i \tag{4.3}$$

or, more simply, as

$$F = \mu_1 u_1 + \cdots + \mu_n u_n, \tag{4.4}$$

in which the term μ_i/u_i, $i = 1, \ldots, n$, signifies that μ_i is the grade of membership of u_i in F, and the plus sign represents the union.[10]

The sigma-count of F is defined as the arithmetic sum of the μ_i, i.e.,

$$\sum Count(F) \triangleq \sum_{i=1}^{n} \mu_i, \tag{4.5}$$

---

[9] Some of the definitions and examples in this section are drawn from [88] and [89].

[10] In most cases, the context is sufficient to resolve the question of whether a plus sign should be interpreted as the union or the arithmetic sum.

with the understanding that the sum may be rounded, if need be, to the nearest integer. Furthermore, one may stipulate that the terms whose grade of membership falls below a specified threshold be excluded from the summation. The purpose of such an exclusion is to avoid a situation in which a large number of terms with low grades of membership become count-equivalent to a small number of terms with high membership.

The *relative sigma-count*, denoted by $\sum Count(F/G)$, may be interpreted as the proportion of elements of F which are in G. More explicitly,

$$\sum Count(F/G) = \frac{\sum Count(F \cap G)}{\sum Count(G)}, \tag{4.6}$$

where $F \cap G$, the intersection of F and G, is defined by

$$\mu_{F \cap G}(u_i) = \mu_F(u_i) \wedge \mu_G(u_i), \quad u_i \in U. \tag{4.7}$$

Thus, in terms of the membership functions of F and G, the relative sigma-count of F in G is given by

$$\sum Count(F/G) = \frac{\sum_i \mu_F(u_i) \wedge \mu_G(u_i)}{\sum_i \mu_G(u_i)}. \tag{4.8}$$

The concept of a relative sigma-count provides a basis for interpreting the meaning of propositions of the form QA's are B's, e.g., *Most young men are healthy*. More specifically, if the focal variable in the proposition in question is taken to be the proportion of B's in A's, then the corresponding translation rule may be expressed as

$$QA\text{'s are }B\text{'s} \rightarrow \sum Count(B/A) \text{ is } Q$$

or, equivalently, as

$$QA\text{'s are }B\text{'s} \rightarrow \Pi_X = Q \tag{4.9}$$

where

$$X = \frac{\sum_i \mu_A(u_i) \wedge \mu_B(u_i)}{\sum_i \mu_A(u_i)}. \tag{4.10}$$

The intersection/product syllogism

In fuzzy logic, a basic rule of inference for quantified propositions is the *intersection/product syllogism* which may be expressed as the schema [88]

$$\begin{array}{l} Q_1 A\text{'s are }B\text{'s} \\ \underline{Q_2(A \text{ and } B)\text{'s are }C\text{'s}} \\ (Q_1 \otimes Q_2) A\text{'s are } (B \text{ and } C)\text{'s} \end{array} \tag{4.11}$$

in which A, B and C are fuzzy predicates and $Q_1 \otimes Q_2$ is a fuzzy number which is the fuzzy product of the fuzzy numbers Q_1 and Q_2 [18]. For example, as a special case of (4.11), we may write

> *Most students are single*
>
> *A little more than a half of single students are male* (4.12)
> ───
> (*Most* ⊗ *A little more than a half*) *of students are single and male.*

Since the intersection of B and C is contained in C, the following corollary of (4.11) is its immediate consequence

> $Q_1 A$'s are B's
>
> $Q_2 (A$ *and* B)'s are C's (4.13)
> ─────────────────────────
> $\geq (Q_1 \otimes Q_2) A$'s are C's

where the fuzzy number $\geq (Q_1 \otimes Q_2)$ should be read as *at least* $(Q_1 \otimes Q_2)$, with the understanding that $\geq (Q_1 \otimes Q_2)$ represents the composition of the binary nonfuzzy relation $\geq$ with the unary fuzzy relation $(Q_1 \otimes Q_2)$. In particular, if the fuzzy quantifiers Q_1 and Q_2 are monotone increasing (e.g., when $Q_1 = Q_2 \triangleq most$), then as is stated in [88],

$$\geq (Q_1 \otimes Q_2) = Q_1 \otimes Q_2. \tag{4.14}$$

and (4.13) becomes

> $Q_1 A$'s are B's
>
> $Q_2 (A$ *and* B)'s are C's (4.15)
> ─────────────────────────
> $(Q_1 \otimes Q_2) A$'s are C's

The consequent conjunction syllogism

Another basic syllogism in fuzzy logic is the *consequent conjunction syllogism* [88] which may be expressed as the schema

> $Q_1 A$'s are B's
>
> $Q_2 A$'s are C's (4.16)
> ─────────────────
> QA's are $(B$ *and* C)'s

where

$$0 \vee (Q_1 \oplus Q_2 \ominus 1) \leq Q \leq Q_1 \wedge Q_2,$$

in which the operators $\wedge$, $\vee$, $\oplus$ and $\ominus$, and the inequality $\leq$ are the extensions of $\wedge$, $\vee$, $+$, $-$ and $\leq$, respectively, to fuzzy numbers [18].

The consequent conjunction syllogism plays the same role in fuzzy logic as the rule of combination of evidence for conjunctive hypotheses does in MYCIN [60] and PROSPECTOR [20]. However, the latter rules are ad hoc in nature whereas the consequent conjunction syllogism is not. Furthermore, the consequent conjunctive syllogism shows that the certainty-factor values used in the case of conjunctive hypotheses in MYCIN and PROSPECTOR correspond to the upper bound in (4.16).

As a simple illustration of the consequent conjunction syllogism, assume that

$p_1 \triangleq$ *Most Frenchmen are not tall,*

$p_2 \triangleq$ *Most Frenchmen are not short.* (4.17)

In this case, by the application of (4.16), we can infer that

Q Frenchmen are not tall and not short

where

$0 \oslash (2 \ most \ominus 1) \leq Q \leq most.$ (4.18)

In the above example, the variable of interest is the conjunction of the consequents *not tall* and *not short*. In a more general setting, the variable of interest may be a specified function of the variables constrained by the knowledge base. The following variation on (4.16) is inteneded to give an idea of how the value of the variable of interest may be inferred by an application of the PE-rule.

Infer from the propositions

$p_1 \triangleq$ *Most Frenchmen are not tall,*

$p_2 \triangleq$ *Most Frenchmen are not short* (4.19)

the answer to the question

$q \triangleq$ *What is the average height of a Frenchman?* (4.20)

Assume that propositions p_1 and p_2 refer to a population of Frenchmen, $\{Frenchman_1, \ldots, Frenchman_n\}$ with respective heights $X_1, \ldots, X_n$ which play the role of KB variables. On denoting the membership functions of the fuzzy predicates *tall* and *short* by μ_{TALL} and μ_{SHORT}, and that of the fuzzy quantifier *most* by μ_{MOST}, the possibility distributions induced by p_1 and p_2 may be expressed as

$$\pi^1_{(X_1,\ldots,X_n)}(h_1, \ldots, h_n) = \mu_{ANTMOST}\left(\frac{1}{n}\sum_i \mu_{TALL}(h_i)\right), \quad (4.21)$$

$$\pi^2_{(X_1,\ldots,X_n)}(h_1, \ldots, h_n) = \mu_{ANTMOST}\left(\frac{1}{n}\sum_i \mu_{SHORT}(h_i)\right), \quad (4.22)$$

where $h_1, \ldots, h_n$ are generic values of $X_1, \ldots, X_n$, and *ant* is an abbreviation for *antonym* [80], i.e.,

$$\mu_{ANTMOST}(u) \triangleq \mu_{MOST}(1-u), \quad u \in [0, 1]. \quad (4.23)$$

Now the variable of interest is the average height, Y, which, as a function of the KB variables $X_1, \ldots, X_n$ may be expressed as

$$Y = \frac{1}{n}(X_1 + \cdots + X_n). \quad (4.24)$$

Consequently, by applying the PE-rule (3.27), the determination of the possibility distribution of Y is reduced to the solution of the nonlinear program

$$\mu_Y(h) = \sup_{h_1,\ldots,h_n}\left(\mu_{ANTMOST}\left(\frac{1}{n}\sum_i \mu_{TALL}(h_i)\right) \wedge \mu_{ANTMOST}\left(\frac{1}{n}\sum_i \mu_{SHORT}(h_i)\right)\right)$$

(4.25)

subject to

$$h = \frac{1}{n} \sum_i h_i,$$

where h is the generic value of Y.

Alternatively, a simpler but less informative answer may be formulated by forming the intersection of the possibility distributions of Y which are induced separately by p_1 and p_2. More specifically, let $\Pi_{Y|p_1}$, $\Pi_{Y|p_2}$, $\Pi_{Y|p_1 \wedge p_2}$ be the possibility distributions of Y which are induced by p_1, p_2, and the conjunction of p_1 and p_2, respectively. Then it can readily be shown [88] that

$$\Pi_{Y|p_1} \cap \Pi_{Y|p_2} \supset \Pi_{Y|p_1 \wedge p_2}, \tag{4.26}$$

and hence we can invoke the entailment principle [83] to validate the intersection in question as the possibility distribution of Y. For the example under consideration, the desired possibility distribution is readily found to be given by

$$Poss\{Y = h\} = \mu_{ANTMOST}(\mu_{TALL}(h)) \wedge \mu_{ANTMOST}(\mu_{SHORT}(h)). \tag{4.27}$$

Chaining of propositions

An ordered pair, (p_1, p_2), of quantified propositions of the form

$$p_1 \triangleq Q_1 A\text{'s are } B\text{'s,}$$

$$p_2 \triangleq Q_2 B\text{'s are } C\text{'s,} \tag{4.28}$$

are said to form a *chain*. More generally, an *n-ary chain* may be represented as an ordered n-tuple

$$(Q_1 A_1\text{'s are } B_1\text{'s, } Q_2 A_2\text{'s are } B_2\text{'s, } \ldots, Q_n A_n\text{'s are } B_n\text{'s}), \tag{4.29}$$

in which $B_1 = A_2$, $B_2 = A_3, \ldots, B_{n-1} = A_n$.

Now assume that p_1 and p_2 appear as premises in the inference schema

$$Q_1 A\text{'s are } B\text{'s} \quad (\textit{major premise})$$

$$\underline{Q_2 B\text{'s are } C\text{'s} \quad (\textit{minor premise})} \tag{4.30}$$

$$?Q A\text{'s are } C\text{'s} \quad (\textit{conclusion})$$

in which $?Q$ is a fuzzy quantifier which is to be determined. This schema corresponds to the combining rules in MYCIN and PROSPECTOR in which uncertain evidence is combined with an uncertain rule.

When $Q_1 = Q_2 = all$, the transitivity of fuzzy set containment or, equivalently, the rule of property inheritance in AI, implies that $Q = all$, i.e.,

$$all\ A\text{'s are } B\text{'s}$$

$$\underline{all\ B\text{'s are } C\text{'s}} \tag{4.31}$$

$$all\ A\text{'s are } C\text{'s}$$

However, as shown in [88], property inheritance is a *brittle* property in the sense that even when Q_1 and Q_2 are arbitrarily close to *all*, there is nothing that can be said about Q, which means that, as a fuzzy number, $Q = [0, 1]$. Thus, to be able to constrain Q it is necessary to make restrictive assumptions about Q_1, Q_2, A and B.

As an illustration, assume that $B \subset A$ and hence that $A \cap B = B$. In this case, the intersection/product syllogism (4.11) yields

$$Q_1 A\text{'s are } B\text{'s}$$
$$\underline{Q_2 B\text{'s are } C\text{'s}} \qquad\qquad\qquad (4.32)$$
$$\geqslant (Q_1 \otimes Q_2) A\text{'s are } C\text{'s}$$

which implies that $Q = \geqslant (Q_1 \otimes Q_2)$. If, in addition, it is assumed that Q_1 and Q_2 are monotone increasing, so that $\geqslant (Q_1 \otimes Q_2) = Q_1 \otimes Q_2$, we obtain the *product chain rule* [89],

$$Q_1 A\text{'s are } B\text{'s}$$
$$\underline{Q_2 B\text{'s are } C\text{'s}} \qquad\qquad\qquad (4.33)$$
$$(Q_1 \otimes Q_2) A\text{'s are } C\text{'s}$$

In this case, the chain $(Q_1 A\text{'s are } B\text{'s}, Q_2 B\text{'s are } C\text{'s})$ will be said to be *product transitive*.[11]

As an illustration of (4.32), we can assert that

$$\text{most students are undergraduates}$$
$$\underline{\text{most undergraduates are young}}$$
$$\text{most}^2 \text{ students are young}$$

where most^2 represents the product of the fuzzy number *most* with itself.

Chaining under reversibility

An important chaining rule which is approximate in nature relates to the case where the major premise in the inference chain

$$Q_1 A\text{'s are } B\text{'s}$$
$$\underline{Q_2 B\text{'s are } C\text{'s}} \qquad\qquad\qquad (4.34)$$
$$Q A\text{'s are } C\text{'s}$$

is *reversible* in the sense that

$$Q_1 A\text{'s are } B\text{'s} \leftrightarrow Q_1 B\text{'s are } A\text{'s}, \qquad\qquad (4.35)$$

where $\leftrightarrow$ denotes approximate semantic equivalence [80]. For example,

$$\text{Most American cars are big} \leftrightarrow \text{Most big cars are American.} \qquad (4.36)$$

Under the assumption of reversibility, it is shown in [89] that the following

---

[11] More generally, an *n*-ary chain $(Q_1 A_1\text{'s are } B_1\text{'s}, \ldots, Q_n A_n\text{'s are } B_n\text{'s})$ will be said to be *product transitive* if from the premises which constitute the chain it may be inferred that $\geqslant (Q_1 \otimes \cdots \otimes Q_n) A_1\text{'s}$ are $B_n\text{'s}$.

syllogism holds in an approximate sense

$$Q_1 A\text{'s are } B\text{'s}$$
$$\underline{Q_2 B\text{'s are } C\text{'s}} \tag{4.37}$$
$$\geq (0 \oslash (Q_1 \oplus Q_2 \ominus 1)) A\text{'s are } C\text{'s}.$$

We shall refer to this syllogism as the *MPR-rule*, with MPR standing for *major premise reversibility*. The transitivity of fuzzy set containment which is implied by the MPR-rule will be referred to as the *BR-transitivity*, with B and R standing for Bezdek and Ruspini, respectively, who have employed this type of transitivity of fuzzy relations in their work on fuzzy clustering [11, 55].

Concluding remark

In the foregoing analysis, we have considered some of the representative problems which arise in the management of uncertainty in expert systems. Our analysis is intended to suggest that fuzzy logic provides a natural conceptual framework for knowledge representation and inference from knowledge bases which are imprecise, incomplete, or not totally reliable. Generally, the use of fuzzy logic reduces the problem of inference to that of solving a nonlinear program and leads to conclusions whose uncertainty is a cumulation of the uncertainties in the premises from which the conclusions are derived. As a consequence, the conclusions as well as the certainty factor (or their equivalents) are fuzzy sets which are characterized by their possibility distributions.

In our analysis, we have considered some – but by no means all – of the inference rules in fuzzy logic which are needed for the combination of evidence in expert systems. In particular, we have not considered the *antecedent conjunction syllogism* which is needed when the antecedents of fuzzy rules are combined conjunctively. In devising such syllogisms, it may be necessary to employ the theory of dispositions [89] to make use of the ultrafuzzy information about the antecedent and consequent predicates. Otherwise, little can be said, in general, about the value of the certainty factor of the conclusion.

References and related publications

[1] E.W. Adams and H.F. Levine, On the uncertainties transmitted from premises to conclusions in deductive inferences, Synthese 30 (1975) 429–460.
[2] K.P. Adlassing, A survey of medical diagnosis and fuzzy subsets, in: M.M. Gupta and E. Sanchez, Eds. Approximate Reasoning in Decision Analysis (North-Holland, Amsterdam, 1982) 203–217.
[3] J.F. Baldwin, A new approach to approximate reasoning using a fuzzy logic, Fuzzy Sets and Systems 2 (1979) 302–325.
[4] J.F. Baldwin and S.Q. Zhou, A fuzzy relational inference language, Report No. EM/FS132, University of Bristol (1982).
[5] W. Bandler and L. Kohout, Semantics of implication operators and fuzzy relational products, Internat. J. Man–Machine Studies 12 (1980) 89–116.
[6] W. Bandler, Representation and manipulation of knowledge in fuzzy expert systems, Proc.

Workshop on Fuzzy Sets and Knowledge-Based Systems, Queen Mary College, University of London (1983).

[7] A. Barr and E.W. Feigenbaum, The Handbook of Artificial Intelligence, Vol. 1, 2 and 3 (Kaufmann, Los Altos, 1982).

[8] J. Barwise and R. Cooper, Generalized quantifiers and natural language, Linguistics and Philosophy 4 (1981) 159–219.

[9] R.E. Bellman and L.A. Zadeh, Local and fuzzy logics, in: G. Epstein, Ed. Modern Uses of Multiple-Valued Logic (Reidel, Dordrecht, 1977) 103–165.

[10] M. Ben-Bassat, et al., Pattern-based interactive diagnosis of multiple disorders: The MEDAS system, IEEE Trans. Patern Anal. Machine Intell. (1980) 148–160.

[11] J. Bezdek, Pattern Recognition with Fuzzy Objective Function Algorithms. (Plenum Press, New York, 1981).

[12] R.J. Brachman and B.C. Smith, Special Issue on Knowledge Representation, SIGART 70 (1980).

[13] C. Carlsson, Tackling an MDCM-problem with the help of some results from fuzzy set theory, European Journal of Operational Research 10 (1982) 271–281.

[14] C. Carlsson, Solving ill-structured problems through well-structured fuzzy programming. in: J.P. Brans, Ed., Operational Research 81 (North-Holland, Amsterdam, 1981).

[15] S. Cushing, Quantifier Meanings – A Study in the Dimensions of Semantic Competence (North-Holland, Amsterdam, 1982).

[16] R. Davis and D.B. Lenat, Knowledge-Based Systems in Artificial Intelligence (McGraw-Hill, New York, 1982).

[17] A. DeLuca and S. Termini, A definition of non-probabilistic entropy in the setting of fuzzy sets theory, Inform. and Control 20 (1972) 301–312.

[18] D. Dubois and H. Prade, Fuzzy Sets and Systems: Theory and Applications (Academic Press, New York, 1980).

[19] R.O. Duda and E.H. Shortliffe, Expert systems research, Science 220 (1983) 261–268.

[20] R.O. Duda, P.E. Hart, K. Konolige and R. Reboh, A computer-based consultant for mineral exploration, Final Tech. Report, SRI International, Menlo Park, CA (1979).

[21] E.A. Feigenbaum, The art of artificial intelligence: Themes and case studies in knowledge engineering, Proc. IJCAI5 (1977).

[22] S. Fukami, M. Mizumoto and K. Tanaka, Some considerations on fuzzy conditional inference, Fuzzy Sets and Systems 4 (1980) 243–273.

[23] B.R. Gaines, Logical foundations for database systems, Internat. J. Man–Machine Studies 11 (1979) 481–500.

[24] J.A. Goguen, The logic of inexact concepts, Synthese 19 (1969) 325–373.

[25] S. Gottwald, Fuzzy-Mengen und ihre Anwendungen. Ein Überblick, Elektron. Informationsverarb. Kybernet. 17 (1981) 207–235.

[26] J. Higginbotham and R. May, Questions, quantifiers and crossing, The Liguistic Review 1 (1981) 41–80.

[27] J.K. Hintikka, Logic, Language-Games, and Information: Kantian Themes in the Philosophy of Logic (Oxford University Press, Oxford, 1973).

[28] D.N. Hoover, Probability logic, Annals Math. Logic 14 (1978) 287–313.

[29] M. Ishizuka, K.S. Fu and J.T.P. Yao, A rule-based inference with fuzzy set for structural damage assessment, in: M. Gupta and E. Sanchez, Ed., Fuzzy Information and Decision Processes (North-Holland, Amsterdam, 1982).

[30] E.P. Klement, W. Schwyhla and R. Lowen, Fuzzy probability measures, Fuzzy Sets and Systems 5 (1981) 83–108.

[31] E.P. Klement, Operations on fuzzy sets and fuzzy numbers related to triangular norms, Proc. 11th Conf. Multiple-Valued Logic, Univ. of Oklahoma, Norman (1981) 218–225.

[32] E.P. Klement, An axiomatic theory of operations on fuzzy sets, Institut fur Mathematik, Johannes Kepler Universitat Linz, Institutsbericht 159 (1981).

[33] C.A. Kulikowski, AI methods and systems for medical consultation, IEEE Trans. Pattern Anal. Machine Intell. (1980) 464–476.

[34] G. Lakoff, Hedges: A study in meaning criteria and the logic of fuzzy concepts, J. Phil. Logic 2 (1973) 458–508. Also in: D. Jockney, W. Harper and B. Freed, Eds., Contemporary Research in Philosophical Logic and Linguistic Semantics (Reidel, Dordrecht, 1973) 221–271.

[35] E.H. Mamdani and B.R. Gaines, Fuzzy Reasoning and its Applications (Academic Press, London, 1981).

[36] J. McCarthy, Circumscription: A non-monotonic inference rule, Artificial Intelligence 13 (1980) 27–40.

[37] J.D. McCawley, Everything that Linguists have Always Wanted to Know about Logic (University of Chicago Press, Chicago, 1981).

[38] D.V. McDermott and J. Doyle, Non-monotonic logic, I, Artificial Intelligence 13 (1980) 41–72.

[39] D.V. McDermott, Non-monotonic logic, II: non-monotonic modal theories, J. Assoc. Comp. Mach. 29 (1982) 33–57.

[40] R. Michalski, J.G. Carbonell and T.M. Mitchell, Machine Learning (Tioga Press, Palo Alto, 1983).

[41] G.A. Miller and P.N. Johnson-Laird, Language and Perception (Harvard University Press, Cambridge, 1976).

[42] M. Mizumoto, M. Umano and K. Tanaka, Implementation of a fuzzy-set-theoretic data-structure system, 3rd Int. Conf. Very Large Data Bases, Tokyo (1977).

[43] M. Mizumoto, S. Fukami and K. Tanaka, Fuzzy reasoning methods by Zadeh and Mamdani, and improved methods, Proc. Third Workshop on Fuzzy Reasoning, Queen Mary College, London (1979).

[44] G.C. Moisil, Lectures on the Logic of Fuzzy Reasoning (Scientific Editions, Bucarest, 1975).

[45] D. Nau, Expert computer systems, IEEE Computer 16 (1983) 63–85.

[46] C.V. Negoita and D. Ralescu, Applications of Fuzzy Sets to Systems Analysis (Birkhauser, Basel, 1975).

[47] K. Noguchi, M. Umano, M. Mizumoto and K. Tanaka, Implementation of fuzzy artificial intelligence language FLOU, Technical Report on Automation and Language of IECE, 1976.

[48] A.I. Orlov, Problems of Optimization and Fuzzy Variables (Znaniye, Moscow, 1980).

[49] P. Peterson, On the logic of *few*, *many* and *most*, Notre Dame J. of Formal Logic 20 (1979) 155–179.

[50] P.L. Peterson, Philosophy of language, Social Research 47 (1980) 749–774.

[51] R. Reiter and G. Criscuolo, Some representational issues in default reasoning, Computers and Mathematics 9 (1983) 15–28.

[52] N. Rescher, Plausible Reasoning (Van Gorcum, Amsterdam, 1976).

[53] E. Rich, Artificial Intelligence (McGraw-Hill, New York, 1983).

[54] C. Rieger, Conceptual memory, in: R. Schank, ed., Conceptual Information Processing (North-Holland, Amsterdam, 1975).

[55] E. Ruspini, Numerical methods for fuzzy clustering, Inform. Sci. 2 (1970) 319–350.

[56] E. Sanchez, Medical diagnosis and composite fuzzy relations, in: M.M. Gupta, R.K. Ragade and R. Yager, Eds., Advances in Fuzzy Set Theory and Applications (North-Holland, Amsterdam, 1979) 437–444.

[57] E.S. Santos, Fuzzy algorithms, Inform. and Control 18 (1970) 326–339.

[58] I. Scheffler, A Philosophical Inquiry into Ambiguity, Vagueness and Metaphor in Language (Routledge & Kegan Paul, London, 1981).

[59] L.K. Schubert, R.G. Goebel and N. Cercone, The structure and organization of a semantic net for comprehension and inference, in: N.V. Findler, Ed., Associative Networks (Academic Press, New York, 1979) 122–178.

[60] E.H. Shortliffe and B. Buchanan, A model of inexact reasoning in medicine, Mathematical Biosciences 23 (1975) 351–379.

[61] E.H. Shortliffe, Computer-Based Medical Consultations: MYCIN (American Elsevier, New York, 1976).

[62] E.H. Shortliffe, B.G. Buchanan and E.A. Feigenbaum, Knowledge engineering for medical decision making: A review of computer-based clinical decision aids, Proc. IEEE 67 (1979) 1207–1224.

[63] M. Sugeno and T. Tagaki, Multi-dimensional fuzzy reasoning, Fuzzy Sets and Systems 9 (1983) 313–325.

[64] P. Szolovits and S.G. Pauker, Categorical and probabilistic reasoning in medical diagnosis, Artificial Intelligence 11 (1978) 115–144.

[65] K. Tanaka and M. Mizumoto, Fuzzy programs and their execution, in: L.A. Zadeh, K.S. Fu, K. Tanaka and M. Shimura, Eds., Fuzzy Sets and Their Applications to Cognitive and Decision Processes (Academic Press, New York, 1975) 41–76.

[66] D.A. Waterman, Generalization learning techniques for automating the learning of heuristics, Artificial Intelligence 1 (1970) 121–170.

[67] S.M. Weiss and C.A. Kulikowski, EXPERT: A system for developing consultation models, Proc. IJCAI6 (1979).

[68] M. Wygralak, Fuzzy inclusion and fuzzy equality of two fuzzy subsets, fuzzy operations for fuzzy subsets, Fuzzy Sets and Systems 10 (1983) 157–168.

[69] R.R. Yager, Quantified propositions in a linguistic logic, in: E.P. Klement, Ed., Proceedings of the 2nd International Seminar on Fuzzy set Theory, Johannes Kepler University, Linz, Austria (1980).

[70] R.R. Yager, Some procedures for selecting fuzzy-set-theoretic operators, Internat. J. General Systems 8 (1982) 115–124.

[71] L.A. Zadeh, Probability measures of fuzzy events, J. Math. Anal. Appl. 23 (1968) 421–427.

[72] L.A. Zadeh, Fuzzy Algorithms, Inform. and Control 12 (1968) 94–102.

[73] L.A. Zadeh, Similarity relations and fuzzy orderings, Inform. Sci. 3 (1971) 177–200.

[74] L.A. Zadeh, On fuzzy algorithms, ERL Memorandum M-235 (February 1972).

[75] L.A. Zadeh, Outline of a new approach to the analysis of complex systems and decision processes, IEEE Trans. Systems Man Cybernet. 3 (1973) 28–44.

[76] L.A. Zadeh, Fuzzy logic and approximate reasoning (in memory of Grigore Moisil), Synthese 30 (1975) 407–428.

[77] L.A. Zadeh, The concept of a linguistic variable and its application to approximate reasoning, Inform. Sci. 8 (1975) 199–249; 301–357; 9, 43–80.

[78] L.A. Zadeh, A fuzzy algorithmic approach to the definition of complex or inprecise concepts, Internat. J. Man–Machine Studies 8 (1976) 249–291.

[79] L.A. Zadeh, Fuzzy sets and their application to pattern classification and clustering analysis, in: J. van Ryzin, Ed., Classification and Clustering (Academic Press, New York, 1977) 251–299.

[80] L.V. Zadeh, PRUF – a meaning representation language for natural languages, Internat. J. Man–Machine Studies 10 (1978) 395–460.

[81] L.A. Zadeh, On the validity of Dempster's rule of combination of evidence, Electronics Research Laboratory Memorandum M79/24, University of California, Berkeley (1979).

[82] L.A. Zadeh, Fuzzy sets and information granularity, in: M. Gupta, R. Ragade and R. Yager, Eds., Advances in Fuzzy Set Theory and Applications (North-Holland, Amsterdam, 1979) 3–18.

[83] L.A. Zadeh, A theory of approximate reasoning, Electronics Research Laboratory Memorandum M77/58, University of California, Berkeley (1977). Also in: J.E. Hayes, D. Michie and L.I. Kulich, Eds., Machine Intelligence 9 (Wiley, New York, 1979) 149–194.

[84] L.A. Zadeh, Fuzzy sets as a basis for a theory of possibility, Fuzzy sets and Systems 1 (1978) 3–28.

[85] L.A. Zadeh, Possibility theory and soft data analysis. Electronic Res. Laboratory Memorandum M79/59, University of California, Berkeley (1979). Also in: L. Cobb and R.M. Thrall, Eds., Mathematical Frontiers of the Social and Policy Sciences (Westview Press, Boulder, 1981) 69–129.

[86] L.A. Zadeh, Test-score semantics for natural languages and meaning-representation via PRUF, Tech. Note 247, AI Center, SRI International, Menlo Park CA (1981). Also in: B.B. Rieger, Ed., Empirical Semantics (Brockmeyer, Bochum, 1981) 281–349.

[87] L.A. Zadeh, Fuzzy probabilities and their role in decision analysis, Proc. 4th MIT/ONR Workshop on Command, Control and Communications, M.I.T. (1981) 159–179.

[88] L.A. Zadeh, A computational approach to fuzzy quantifiers in natural languages, Computers and Mathematics 9 (1983) 149–184.

[89] L.A. Zadeh, A theory of commonsense knowledge, ERL Memorandum No. UCB/ERL M83/26,

University of California, Berkeley, California (1983). Also in: H. Skala, Ed., Issues of Vagueness (Reidel, Dordrecht, 1983), to appear.

[90] A. Zimmer, Some experiments concerning the fuzzy meaning of logical quantifiers, in: L. Troncoli, Ed., General Surveys of Systems Methodology (Society for General Systems Research, Louisville, 1982) 435–441.

[91] H.-J. Zimmermann and P. Zysno, Latent connectives in human decision making, Fuzzy Sets and Systems 4 (1980) 37–52.

[92] H.-J. Zimmermann, Fuzzy programming and linear programming with several objective functions, Fuzzy Sets and Systems 1 (1978) 45–55.

Syllogistic Reasoning in Fuzzy Logic and its Application to Usuality and Reasoning with Dispositions

LOFTI A. ZADEH, FELLOW, IEEE

Abstract—A fuzzy syllogism in fuzzy logic is defined to be an inference schema in which the major premise, the minor premise and the conclusion are propositions containing fuzzy quantifiers. A basic fuzzy syllogism in fuzzy logic is the intersection/product syllogism. Several other basic syllogisms are developed that may be employed as rules of combination of evidence in expert systems. Among these is the consequent conjunction syllogism. Furthermore, we show that syllogistic reasoning in fuzzy logic provides a basis for reasoning with dispositions; that is, with propositions that are preponderantly but not necessarily always true. It is also shown that the concept of dispositionality is closely related to the notion of usuality and serves as a basis for what might be called a theory of usuality—a theory which may eventually provide a computational framework for commonsense reasoning.

I. INTRODUCTION

FUZZY logic may be viewed as a generalization of multivalued logic in that it provides a wider range of tools for dealing with uncertainty and imprecision in knowledge representation, inference, and decision analysis. In particular, fuzzy logic allows the use of a) fuzzy predicates exemplified by *small*, *young*, *nice*, etc; b) fuzzy quantifiers exemplified by *most*, *several*, *many*, *few*, *many more*, etc; c) fuzzy truth values exemplified by *quite true*, *very true*, *mostly false*, etc. d) fuzzy probabilities exem-

Manuscript received August 14, 1984; revised July 7, 1985. This research was supported by NSF grant IST-8320416, NASA grant NCC2-275, and DARPA contract N0039-84-C-0089.

The author is with the Division of Computer Science, University of California, Berkeley, CA 94720, USA.

plified by *likely*, *unlikely*, *not very likely* etc; e) fuzzy possibilities exemplified by *quite possible*, *almost impossible*, etc; and f) predicate modifiers exemplified by *very*, *more or less*, *quite*, *extremely*, etc.

What matters most about fuzzy logic is its ability to deal with fuzzy quantifiers as fuzzy numbers which may be manipulated through the use of fuzzy arithmetic. This ability depends in an essential way on the existence—within fuzzy logic—of the concept of cardinality or, more generally, the concept of measure of a fuzzy set. Thus if one accepts the classical view of Kolmogoroff that probability theory is a branch of measure theory, then, more generally, the theory of fuzzy probabilities may be subsumed within fuzzy logic. This aspect of fuzzy logic makes it particularly well-suited for the management of uncertainty in expert systems (Zadeh [48]). More specifically, by employing a single framework for the analysis of both probabilistic and possibilistic uncertainties, fuzzy logic provides a systematic basis for inference from premises which are imprecise, incomplete or not totally reliable. In this way it becomes possible, as is shown in this paper, to derive a set of rules for combining evidence through conjunction, disjunction, and chaining. In effect, such rules may be viewed as instances of syllogistic reasoning in fuzzy logic. However, unlike the rules employed in most of the existing expert systems, they are not *ad hoc* in nature.

Our concern in this paper is with fuzzy syllogisms of the general form

$$
\begin{array}{c}
p(Q_1) \\
\underline{q(Q_2)} \\
r(Q)
\end{array}
\qquad (1.1)
$$

in which the first premise $p(Q_1)$ is a fuzzy proposition containing a fuzzy quantifier Q_1; the second premise $q(Q_2)$ is a fuzzy proposition containing a fuzzy quantifier Q_2; and the conclusion $r(Q)$ is a fuzzy proposition containing a fuzzy quantifier Q. For example, the *intersection/product syllogism* may be expressed as

$$
\begin{array}{c}
Q_1 A\text{'s are } B\text{'s} \\
\underline{Q_2 (A \text{ and } B)\text{'s are } C\text{'s}} \\
Q A\text{'s are } (B \text{ and } C)\text{'s}
\end{array}
\qquad (1.2)
$$

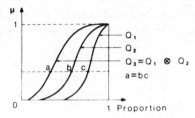

Fig. 1.   Multiplication of fuzzy quantifiers.

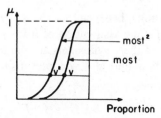

Fig. 2.   Representation of *most* and *most²*.

where A, B, and C are labels of fuzzy sets, and the fuzzy quantifier Q is given by the product of the fuzzy quantifiers Q_1 and Q_2, i.e.

$$Q = Q_1 \otimes Q_2 \qquad (1.3)$$

where $\otimes$ denotes the product in fuzzy arithmetic (Kaufmann and Gupta [21]) (Fig. 1).[1] It should be noted that (1.3) may be viewed as an analog of the basic probabilistic identity (Jaynes [19])

$$p(B, C/A) = p(B/A)p(C/A, B). \qquad (1.4)$$

A concrete example of the intersection/product syllogism is the following

most students are young

most young students are single $\qquad (1.5)$

most² students are young and single

where *most²* denotes the product of the fuzzy quantifier *most* with itself (Fig. 2).

---

[1] More generally, a circle around an arithmetic operator represents its extension to fuzzy operands.

For concreteness, we shall restrict our attention to syllogisms in which p, q, and r are propositions of the form

$$p \triangleq Q_1 A\text{'s are } B\text{'s} \qquad (1.6)$$

$$q \triangleq Q_2 C\text{'s are } D\text{'s} \qquad (1.7)$$

$$r \triangleq Q_3 E\text{'s are } F\text{'s} \qquad (1.8)$$

in which A, B, C, D, E, and F are interrelated fuzzy predicates.

The interrelations between A, B, C, D, E, and F provide a basis for a classification of fuzzy syllogisms. The more important of these syllogisms are the following ($\wedge \triangleq$ denotes conjunction, and $\vee \triangleq$ denotes disjunction).

Intersection/product syllogism:

$$C = A \wedge B, \quad E = A, \quad F = C \wedge D. \quad (1.9)$$

Chaining syllogism:

$$C = B, \quad E = A, \quad F = D. \quad (1.10)$$

Consequent conjunction syllogism:

$$A = C = E, \quad F = B \wedge D. \quad (1.11)$$

Consequent disjunction syllogism:

$$A = C = E, \quad F = B \vee D. \quad (1.12)$$

Antecedent conjunction syllogism:

$$B = D = F, \quad E = A \wedge C. \quad (1.13)$$

Antecedent disjunction syllogism:

$$B = D = F, \quad E = A \vee C. \quad (1.14)$$

In the context of expert systems, these and related syllogisms provide a set of inference rules for combining evidence through conjunction, disjunction and chaining [48].

An important application of syllogistic reasoning in fuzzy logic relates to what may be regarded as reasoning with *dispositions* [50]. A disposition, as its name suggests, is a proposition which is preponderantly but not necessarily always true. To capture this intuitive meaning of a disposition, we define a disposition as a proposition with implicit extremal fuzzy quantifiers, e.g., *most, almost all, almost always, usually, rarely, few, small fraction*. etc. This definition should be regarded as a *dispositional definition* in the sense that it may not be true in all cases.

Examples of commonplace statements of fact which may be viewed as dispositions are *overeating causes obesity, snow is white, glue is sticky, icy roads are slippery*, etc. An example of what appears to be a plausible conclusion drawn from dispositional premises is the following

icy roads are slippery

<u>*slippery roads are dangerous*</u>   (1.15)

icy roads are dangerous.

As will be seen in Section III, syllogistic reasoning with dispositions provides a basis for a formalization of the type of commonsense reasoning exemplified by (1.15).

The importance of the concept of a disposition stems from the fact that what is commonly regarded as commonsense knowledge may be viewed as a collection of dispositions [49]. For example, the dispositions

birds can fly

small cars are unsafe

professors are not rich

students are young

where there is smoke there is fire

Swedes are taller than Italians

Swedes are blond

it takes a little over one hour to drive
* from Berkeley to Stanford*

a TV set weighs about 50 pounds

may be regarded as a part of commonsense knowledge. The last two examples typify a particularly important type of disposition: *a dispositional valuation*, that is, a disposition which characterizes the usual value of a variable through the implicit fuzzy quantifier usually. Thus upon *explicitation*, that is, making explicit the implicit fuzzy quantifiers, the examples in question become

usually it takes a little over one hour to drive
* from Berkeley to Stanford*

usually a TV set weighs about 50 pounds.

Dispositional valuations have the general form

$$(usually)(X \text{ is } F)   (1.16)$$

where ($usually$) is an implicit fuzzy quantifier; X is a variable which is constrained by the disposition; and F is its fuzzy value. As an illustration, expressed in this form, the last example reads[2]

$(usually)(weight(TV) \text{ is } \text{ABOUT.FIFTY.POUNDS})$.

---

[2] Upper case symbols and periods serve to represent compound labels of fuzzy sets.

A basic syllogism which governs inference from dispositional valuations is the *dispositional modus ponens*

$$(\textit{usually})(X \text{ is } F)$$
$$\underline{\text{if } X \text{ is } F \text{ then } (\textit{usually})(Y \text{ is } G)} \qquad (1.17)$$
$$(\textit{usually}^2)(Y \text{ is } G),$$

where F and G are fuzzy predicates and $usually^2$ is the product of the fuzzy quantifier *usually* with itself.

To a first approximation, the fuzzy quantifier *usually* may be represented as a fuzzy number of the same form as *most* (Fig. 3). More generally, however, *usually* connotes a dependence on the assumption of "normality." More specifically, let Z denote what might be called a conditioning variable whose normal (or regular) value is R, where R in general is a fuzzy set which is the complement of a set of exceptions. In terms of the conditioning variable, the dispositional valuation $(\textit{usually})(X \text{ is } F)$ may be expressed as

$$(\textit{usually})(X \text{ is } F) \leftrightarrow \text{if } (Z \text{ is } R) \quad \text{then } (\textit{most } X\text{'s are } F).$$
$$(1.18)$$

For example

($usually$)($duration$($trip$($Berkeley$, $Stanford$))) is
LITTLE.OVER.ONE. HOUR) $\leftrightarrow$ if (($time$($trip$))) is
NOT.RUSH.HOUR) then ($most durations$($trip$
($Berkeley$, $Stanford$) are LITTLE.OVER.ONE.HOUR.

What these examples are intended to suggest is that the concept of a disposition plays an essential role in the representation and inference from commonsense knowledge. A more detailed discussion of this role of dispositions will be presented in Section III.

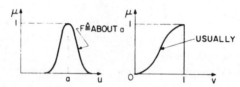

Fig. 3.   Representation of ($usually$)(X is ABOUT a).

II. Fuzzy Quantifiers, Compositionality, Robustness, and Usuality

As was stated in the introduction, the concept of a fuzzy quantifier is related in an essential way to the concept of cardinality—or, more, generally, the concept of measure—of fuzzy sets. More specifically, a fuzzy quantifier may be viewed as a fuzzy characterization of the absolute or relative cardinality of a collection of fuzzy sets. In this sense, then, a fuzzy quantifier is a second-order fuzzy predicate.

The cardinality of a fuzzy set may be defined in a variety of ways [47]. For simplicity, we shall employ the *sigma-count* for this purpose, which is defined as follows [6], [45].

Let A be a finite fuzzy subset of the universe of discourse U, with A expressed as

$$A = \mu_1/u_1 + \cdots + \mu_n/u_n \qquad (2.1)$$

where μ_1/u_1, $i = 1, \cdots, n$, signifies that μ_i is the grade of membership of u_i in A and $+$ denotes the union. Then, the sigma-count of A is defined as the real number

$$\Sigma\text{count}(A) = \Sigma_i \mu_i \qquad (2.2)$$

with the understanding that the sum may be rounded to the nearest integer, if necessary. Furthermore, one may stipulate that the terms whose grade of membership fall below a specified threshold be excluded from the summation. The purpose of such an exclusion is to avoid a situation in which a large number of terms with low grades of membership become count-equivalent to a small number of terms with high membership.

The relative sigma-count, denoted by $\Sigma\text{count}\,(B/A)$, may be interpreted as the proportion of elements of B in A. More explicitly,

$$\Sigma\text{count}(B/A) = \frac{\Sigma\text{count}(B \cap A)}{\Sigma\text{count}(A)} \qquad (2.3)$$

where $B \cap A$, the intersection of B and A, is defined by

$$\mu_{B \cap A}(u) = \mu_B(u) \wedge \mu_A(u), \qquad u \in U. \qquad (2.4)$$

Thus, in terms of the membership functions of B and A, the relative sigma-count of B in A is given by

$$\Sigma \, \text{count}(B/A) = \frac{\Sigma_i \mu_B(u_i) \wedge \mu_A(u_i)}{\Sigma_i \mu_A(u_i)}. \qquad (2.5)$$

The concept of a relative sigma-count provides a basis for interpreting the meaning of propositions of the form $p \triangleq QA$'s are B's, e.g., *most young men are healthy*. More specifically, the fuzzy quantifier Q in the proposition QA's are B's may be regarded as a fuzzy characterization of the relative sigma-count of B in A, which entails that the proposition in question may be translated as

$$QA\text{'s are }B\text{'s} \rightarrow \Sigma\text{count}(B/A) \text{ is } Q. \qquad (2.6)$$

The right-hand member of (2.6) implies that Q, viewed as a fuzzy number, defines the possibility distribution of $\Sigma \, \text{count}(B/A)$. This may be expressed as the *possibility assignment equation* [45]

$$\Pi_X = Q \qquad (2.7)$$

in which the variable X is the sigma-count in question and Π_X is its possibility distribution.

As was stated earlier, a fuzzy quantifier is a second-order fuzzy predicate. The interpretation expressed by (2.6) and (2.7) shows that the evaluation of a fuzzy quantifier may be reduced to that of a first order predicate if Q is interpreted as a fuzzy subset of the real line. Thus let us consider again the proposition $p \triangleq QA$'s are B's, in which A and B are fuzzy sets in their respective universes of discourse, U and V; and Q, regarded as a second-order fuzzy predicate, is assumed to be characterized by its membership function $\mu_Q(X, Y)$, with X and Y ranging over the fuzzy subsets of U and V. Then based on (2.6) and (2.7), we can define $\mu_Q(X, Y)$ through the equality.

$$\mu_Q(X, Y) = \mu_Q(\Sigma\text{count}(X/Y)) \qquad (2.8)$$

in the right-hand member of which Q is a unary first-order fuzzy predicate whose denotation is a fuzzy subset of the unit interval. Consequently, in the proposition QA's are B's, Q may be interpreted as a) a second-order fuzzy predicate defined on $U^* \times V^*$, where U^* and V^* are the fuzzy power sets of U and V or b) a first-order fuzzy predicate defined on the unit interval $[0, 1]$.

It is useful to classify fuzzy quantifiers into quantifiers of the first kind, second kind, third kind, etc., depending on the arity of the second-order fuzzy predicate which the quantifier represents. Thus, Q is a fuzzy quantifier of the first kind if it provides a fuzzy characterization of the cardinality of a fuzzy set; Q is of the second kind if it provides a fuzzy characterization of the relative cardinality of two fuzzy sets; and Q is of the third kind if it serves the same role in relation to three fuzzy sets. For example, the fuzzy quantifier labeled *several* is of the first kind; *most* is of the second kind; and *many more* in *there are many more A's in B's than A's in C's* is of the third kind. It should be noted that, in terms of this classification, the certainty factors employed in such expert systems as MYCIN [38] and PROSPECTOR [9] are fuzzy quantifiers of the third kind.

The concept of a fuzzy quantifier gives rise to a number of other basic concepts relating to syllogistic reasoning, among which are the concepts of *compositionality* and *robustness*.

Specifically, consider a fuzzy syllogism of the general form (1.1), i.e.

$$p(Q_1)$$
$$\underline{q(Q_2)} \qquad\qquad (2.9)$$
$$r(Q).$$

We shall say that the syllogism is strongly compositional if a) Q may be expressed as a function of Q_1 and Q_2 independent of the denotations of the predicates which enter into p and q, excluding the trivial case, where Q is the unit interval, and b) if Q_1 and Q_2 are numerical quantifiers, so is Q. Furthermore, we shall say that the syllogism is weakly compositional if only a) is satisfied, in which case if Q_1 and Q_2 are numerical quantifiers, Q may be interval-valued. As will be seen in the sequel, in order to achieve strict compositionality, it is necessary, in general, to make some restrictive assumptions concerning the predicates in p and q. For example, the syllogism

$$Q_1 A\text{'s are } B\text{'s}$$
$$\underline{Q_2 B\text{'s are } C\text{'s}} \qquad\qquad (2.10)$$
$$(Q_1 \otimes Q_2) A\text{'s are } (B \text{ and } C)\text{'s}$$

is strictly compositional if $B \subset A$.

Turning to the concept of robustness, suppose that we start with a nonfuzzy syllogism of the form

$$p(\text{all})$$

$$\frac{q(\text{all})}{r(\text{all})} \qquad (2.11)$$

an example of which is

all A's are B's

$$\frac{\text{all } B \text{'s are } C \text{'s}}{\text{all } A \text{'s are } C \text{'s.}} \qquad (2.12)$$

The original syllogism is *robust* if small perturbations in the quantifiers in p and q result in a small perturbation in the quantifier in r. For example, the syllogism represented by (2.12) is robust if its validity is preserved when a) the quantifier *all* in p and q is replaced by *almost all*, and b) the quantifier *all* in r is replaced by *almost almost all*. (In more concrete terms, this is equivalent to replacing *all* in p and q by the fuzzy number $1 \ominus \epsilon$, where ϵ is a small fuzzy number and b) replacing *all* in r by the fuzzy number $1 \ominus \epsilon$.) More generally, a syllogism is selectively robust if the above holds for perturbations in either the first or the second premise, but not necessarily in both. For example, it may be shown that the syllogism expressed by (2.12) is selectively robust with respect to perturbations in the first premise but not in the second premise. In fact, the syllogism in question is brittle with respect to perturbations in the second premise in the sense that the slightest perturbation in the quantifier *all* in q requires the replacement of the quantifier *all* in r by the vacuous quantifier *none-to-all*.

As was stated earlier, the concept of dispositionality is closely related to that of usuality. In particular, a dispositional valuation expressed as X is F may be interpreted as (*usually*)(X is F) or equivalently as $U(X) = F$, where *usually* is an implicit fuzzy quantifier, and $U(X)$ denotes the usual value of X. For example, *glue is sticky* may be interpreted as (*usually*)(*glue is sticky*); *lamb is more expensive than beef* may be interpreted as (*usually*)(*lamb is more expensive than beef*); and *most students are undergraduates* may be interpreted in some contexts as (*usually*)(*most students are undergraduates*).

To concretize the meaning of the fuzzy quantifier *usually*, assume that X takes a sequence of values $X_1, \cdots, X_n$ in a universe of discourse U. Then, *usually*, in its unconditioned sense, may be defined by

$$usually \ (X \text{ is } F) \triangleq most \ X \text{'s are } F \qquad (2.13)$$

where F plays the role of a usual value $U(X)$ of X. An immediate consequence of this definition is that a usual value of X is not unique. Rather, given a fuzzy value F, we can compute the degree τ to which it satisfies the definition (2.13) by employing the formula

$$\tau = \mu_{\text{MOST}}\left[\frac{1}{n}(\mu_F(X_1) + \cdots + \mu_F(X_n))\right] \quad (2.14)$$

where μ_{MOST} is the membership function of the fuzzy quantifier most; μ_F is the membership function of F; and the argument of μ_{MOST} is the relative sigma-count of the number of times X satisfies the proposition X is F.

More generally, when *usually* is defined in its conditioned sense via (1.18), the degree of compatibility of F with the definition of a usual value may be expressed as

$$\tau = \mu_{\text{MOST}}(\Sigma\text{count}(F/R))$$

in which the relative sigma-count is given by

$$\Sigma\text{count}(F/R) = \frac{\Sigma_i\mu_F(X_i) \wedge \mu_R(X_i)}{\Sigma_i\mu_R(X_i)}, \quad i = 1,\ldots,n.$$

The importance of the concept of usuality stems from the fact that it underlies almost all of human decisionmaking from the time of awakening in the morning till retirement at night. Thus in deciding on when to get up, we take into consideration how long it usually takes to dress, have breakfast, drive to work, etc. In fact, we could not function at all without a knowledge of the usual values of the variables which enter into our daily decisionmaking and an ability to employ them in decision analysis. It is of interest to observe that the usual values are usually fuzzy rather than crisp. How to manipulate the usual values in the context of fuzzy syllogisms will be discussed briefly in the following section.

III. Fuzzy Syllogisms and Reasoning with Dispositions

As was stated earlier, one of the basic syllogisms in fuzzy logic is the intersection/product syllogism expressed by (1.2).

In what follows, we shall employ this syllogism as a starting point for the derivation of other syllogisms which are of relevance to the important problem of combination of evidence in expert systems.

A derivative syllogism of this type is the multiplicative chaining syllogism

$$Q_1 A\text{'s are } B\text{'s}$$
$$\underline{Q_2 B\text{'s are } C\text{'s}}$$
$$\geqslant (Q_1 \otimes Q_2) A\text{'s are } C\text{'s} \tag{3.1}$$

in which $\geqslant (Q_1 \otimes Q_2)$ should be read as at least $Q_1 \otimes Q_2$. This syllogism is a special case of the intersection/product syllogism which results when $B \subset A$, i.e.

$$\mu_B(u_i) < \mu_A(u_i), \qquad u_i \in U, \; i = 1, \cdots . \tag{3.2}$$

For, in this case $A \cap B \equiv B$, and since $B \cap C$ is contained in C, it follows that

$$(Q_1 \otimes Q_2) A\text{'s are } (B \text{ and } C)\text{'s} \Rightarrow$$
$$\geqslant (Q_1 \otimes Q_2) A\text{'s are } C\text{'s}. \tag{3.3}$$

(It is of interest to note that if Q in the proposition QA's are B's is interpreted as the degree to which A is contained in B, then the multiplicative chaining syllogism shows that, under the assumption $B \subset A$, the fuzzy relation of fuzzy-set containment is *product transitive* [43], [48].

If, in addition to assuming that $\dot{B} \subset A$, we assume that Q_1 and Q_2 are monotone increasing [3], [47], i.e.

$$\geqslant Q_1 = Q_1$$
$$\geqslant Q_2 = Q_2, \tag{3.4}$$

which is true of the fuzzy quantifier *most*, then

$$\geqslant (Q_1 \otimes Q_2) = Q_1 \otimes Q_2 \tag{3.5}$$

and the multiplicative chaining syllogism becomes (Fig. 5)

$$Q_1 A\text{'s are } B\text{'s}$$
$$\underline{Q_2 B\text{'s are } C\text{'s}}$$
$$(Q_1 \otimes Q_2) A\text{'s are } C\text{'s}. \tag{3.6}$$

As an illustration, we shall consider an example in which the containment relation $B \subset A$ holds approximately, as in the proposition

$$p \triangleq most\ American\ cars\ are\ big. \tag{3.7}$$

Then, if

$$q \triangleq most\ big\ cars\ are\ expensive \tag{3.8}$$

we may conclude, by employing (3.6), that

$$r \triangleq most^2\ American\ cars\ are\ expensive$$

with the understanding that $most^2$ is the product of the fuzzy number *most* with itself [47].

It can readily be shown by examples that if no assumptions are made regarding A, B, and C, then the chaining inference schema

$$Q_1 A \text{'s are } B \text{'s}$$
$$\underline{Q_2 B \text{'s are } C \text{'s}} \qquad (3.9)$$
$$QA \text{'s are } C \text{'s.}$$

is not weakly compositional, which is equivalent to saying that, in general, Q is the vacuous quantifier *none-to-all*. However, if we assume, as done above, that $B \subset A$, then it follows from the intersection/product syllogism that (3.6) becomes weakly compositional, with

$$Q = \geqslant (Q_1 \otimes Q_2) \qquad (3.10)$$

and, furthermore, that (3.6) becomes strongly compositional if Q_1 and Q_2 are monotone increasing.

Another important observation relates to the robustness of the multiplicative chaining syllogism. Specifically, if we assume that

$$Q_1 = 1 \ominus \epsilon_1$$
$$Q_2 = 1 \ominus \epsilon_2$$

where ϵ_1 and ϵ_2 are small fuzzy numbers, then it can readily be verified that, approximately

$$Q_1 \otimes Q_2 \cong 1 \ominus \epsilon_1 \ominus \epsilon_2 \qquad (3.11)$$

which establishes that the multiplicative chaining syllogism is robust. However, in the absence of the assumption $B \subset A$, the inference schema (3.9) is robust only with respect to perturbations in Q_1. To demonstrate this, assume that $Q_1 = $ *almost all* and $Q_2 = $ *all*. Then, from the intersection/product syllogism it follows that $Q = \geqslant$ (*almost all*). On the other hand, if we assume that $Q_1 = $ *all* and $Q_2 = $ *almost all*, then $Q = $ *none-to-all*. Thus, as was stated earlier, the inference schema (3.9) is brittle with respect to perturbations in the second premise.

The MPR Chaining Syllogism

In the preceding discussion, we have shown that the assumption $B \subset A$ leads to a weakly compositional multiplicative chaining syllogism. Another type of assumption which also leads to a weakly compositional syllogism is that of major premise reversibility (MPR).[3] This assumption may be expressed as the semantic equivalence

$$Q_1 A \text{'s are } B \text{'s} \leftrightarrow Q_1 B \text{'s are } A \text{'s} \qquad (3.12)$$

which, in most cases, will hold approximately rather than exactly. For example,

most American cars are big ↔ *most big cars are American.*

It can be shown (Zadeh [49]) that under the assumption of reversibility the following chaining syllogism holds in an approximate sense

$$Q_1 A \text{'s are } B \text{'s}$$
$$\underline{Q_2 B \text{'s are } C \text{'s}}$$
$$\geqslant \left(0 \boxed{\vee} \left((Q_1 \oplus Q_2 \ominus 1) \right) \right) A \text{'s are } C \text{'s}.$$

$$(3.13)$$

We shall refer to this syllogism as the *MPR chaining syllogism*. It follows at once from (3.13) that the MPR chaining syllogism is weakly compositional and robust. A concrete instance of this syllogism is provided by the following example

most American cars are big
most big cars are heavy   (3.14)
$0 \boxed{\vee} (2 \text{ most} \ominus 1)$ *American cars are heavy.*

The Consequent Conjunction Syllogism

The consequent conjunction syllogism is an example of a basic syllogism which is not a derivative of the intersection/product syllogism. Its statement may be expressed as follows:

$$Q_1 A \text{'s are } B \text{'s}$$
$$\underline{Q_2 A \text{'s are } C \text{'s}} \qquad (3.15)$$
$$Q A \text{'s are } (B \text{ and } C) \text{'s}$$

where

$$0 \boxed{\vee} (Q_1 \oplus Q_2 \ominus 1) < Q < Q_1 \boxed{\wedge} Q_2. \qquad (3.16)$$

---

[3] It should be noted that, in the classical theory of syllogisms, the premise in question would be referred to as the minor premise.

From (3.16), it follows at once that the syllogism is weakly compositional and robust.

An illustration of (3.15) is provided by the example

most students are young

most students are single

_____

Q students are single and young

where

$$2\text{most} \ominus 1 \leqslant Q \leqslant \text{most}. \qquad (3.17)$$

This expression for Q follows from (3.16) by noting that

$$\text{most} \textcircled{\wedge} \text{most} = \text{most}$$

and

$$0 \textcircled{\vee} (2\text{most} \ominus 1) = 2\text{most} \ominus 1.$$

The importance of the consequent conjunction syllogism stems from the fact that it provides a formal basis for combining rules in an expert system through a conjunctive combination of hypotheses [38]. However, unlike such rules in MYCIN and PROSPECTOR, the consequent conjunction syllogism is weakly rather than strongly compositional. Since the combining rules in MYCIN and PROSPECTOR are *ad hoc* in nature, whereas the consequent conjunction syllogism is not, the validity of strong compositionality in MYCIN and PROSPECTOR is in need of justification.

The Antecedent Conjunction Syllogism

An issue which plays an important role in the management of uncertainty in expert systems relates to the question of how to combine rules which have the same consequent but different antecedents.

Expressed as an inference schema in fuzzy logic, the question may be stated as

$$Q_1 A\text{'s are } C\text{'s}$$

$$Q_2 B\text{'s are } C\text{'s}$$

_____

$$Q(A \text{ and } B)\text{'s are } C\text{'s} \qquad (3.18)$$

in which Q is the quantifier to be determined as a function of Q_1 and Q_2.

It can readily be shown by examples that, in the absence of any assumptions about A, B, C, Q_1, and Q_2, what can be said about Q is that it is the vacuous quantifier *none-to-all*. Thus, to be able to say more, it is necessary to make some restrictive assumptions which are satisfied, at least approximately, in typical situations.

The commonly made assumption in the case of expert systems [9], is that the items of evidence are conditionally independent given the hypothesis. Expressed in terms of the relative sigma-counts of A, B, and C, this assumption may be written as

$$\Sigma\text{count}(A \cap B/C) = \Sigma\text{count}(A/C)\Sigma\text{count}(B/C).$$
$$(3.19)$$

Using this equality, it is easy to show that

$$\Sigma\text{count}(C/A \cap B) = K\Sigma\text{count}(C/A)\Sigma\text{count}(C/B)$$
$$(3.20)$$

where the factor K is given by

$$K = \frac{\Sigma\text{count}(A)\ \Sigma\text{count}(B)}{\Sigma\text{count}(A \cap B)\Sigma\text{count}(C)}.\qquad (3.21)$$

The presence of this factor has the effect of making the inference schema (3.18) noncompositional. One way of getting around the problem is to employ—instead of the sigma-count—a count defined by

$$\rho\Sigma\text{count}(B) = \frac{\Sigma\text{count}(B)}{\Sigma\text{count}(\neg B)}\qquad (3.22)$$

$$\rho\Sigma\text{count}(B/A) = \frac{\Sigma\text{count}(B/A)}{\Sigma\text{count}(\neg B/A)}\qquad (3.23)$$

in which $\neg B$ denotes the negation of B (or, equivalently, the complement of B, if B is interpreted as a fuzzy set which represents the denotation of the predicate B). These counts will be referred to as *ρsigma-counts* (with ρ standing for ratio) and correspond to the odds which are employed in PROSPECTOR. Thus, expressed in words, we have

$$\rho\Sigma\text{count}(B) \triangleq \text{ratio of } B\text{'s } to \text{ non-}B\text{'s}\qquad (3.24)$$

$\rho\Sigma\text{count}(B/A) \triangleq$ ratio of B's *to* non-B's among A's.

$$(3.25)$$

In terms of ρsigma-counts, it can readily be shown that the assumption expressed by (3.19) entails the equality

$$\rho\Sigma\text{count}(C/A \cap B) = \rho\Sigma\text{count}(C/A)\rho\Sigma\text{count}(C/B)$$

$$\cdot \rho\Sigma\text{count}(\neg C). \quad (3.26)$$

This equality, then, leads to what will be referred to as the *antecedent conjunction syllogism*

> ratio of C's to non-C's among A's is R_1
> ratio of C's *to* non-C's among B's is R_2
> _____
> ratio of C's *to* non-C's
> among $(A$ and $B)$'s is $R_1 \otimes R_2 \otimes R_3$

$$(3.27)$$

where

$$R_3 \triangleq \text{ratio of } C\text{'s to non-}C\text{'s}.$$

It should be noted that this syllogism may be viewed as the fuzzy logic analog of the likelihood ratio combining rule in PROSPECTOR [9].

In the foregoing discussion, we have focused our attention on some of the basic syllogisms in fuzzy logic which may be employed as rules of combination of evidence in expert systems. Another important function which these and related syllogisms may serve is that of providing a basis for reasoning with dispositions, that is, with propositions in which there are implicit fuzzy quantifiers.

The basic idea underlying this application of fuzzy syllogisms is the following. Suppose that we are given two dispositions

icy roads are slippery
slippery roads are dangerous.

$$(3.28)$$

Can we infer from these dispositions what appears to be a plausible conclusion, namely

icy roads are dangerous?

As a first step, we have to restore the suppressed fuzzy quantifiers in the premises. For simplicity, assume that the desired restoration may be accomplished by prefixing the dispositions in question with the fuzzy quantifier most, i.e.

icy roads are slippery → *most icy roads are slippery*

slippery roads are dangerous

 → *most slippery roads are dangerous*.

Next, if we assume that the proposition *most slippery roads are dangerous* satisfies the major premise reversibility condition, i.e.,

most icy roads are slippery

 ↔ *most slippery roads are icy*

then by applying the MPR chaining syllogism (3.13) we have

most icy roads are slippery

<u>*most slippery roads are dangerous*,</u>   (3.29)

(2most ⊖ 1) *icy roads are dangerous*.

Finally, on suppressing the fuzzy quantifiers in (3.29), we are led to the chain of dispositions

icy roads are slippery

<u>*slippery roads are dangerous*</u>   (3.30)

icy roads are dangerous

which answers in the affirmative the question posed in (3.28), with the understanding that the implicit fuzzy quantifier in the conclusion of (3.28) is *2most ⊖ 1* rather than *most*.

In a more general setting, reasoning with dispositions may be viewed as a part of what might be called a *theory of usuality*. In earlier sections, we have already defined the concept of a usual value and stated a rule of inference which was referred to as the *dispositional modus ponens*. In what follows, we shall sketch a few additional rules of this nature which involve the concept of usuality. A more detailed account of the theory of usuality and its applications to commonsense reasoning will be presented in a forthcoming paper.

The rules described below have one basic characteristic in common, namely, they are dispositional variants of categorical (ie., nondispositional) rules of inference in fuzzy logic. In other words, such rules may be viewed as the result of replacing one or more categorical premises, say $p_1, \cdots, p_k$, with dispositional premises expressed as (*usually*)p_1, (*usually*)$p_2, \cdots$, (*usually*)p_k.

As a simple illustration, a basic inference rule in fuzzy logic is the *entailment principle*, which ı   be expressed as

$$X \text{ is } A$$
$$\frac{A \subset B}{X \text{ is } B} \qquad (3.31)$$

where A and B are fuzzy predicates. In plain words, this rule states that any assertion about X of the form X is A entails any less specific assertion X is B.

A dispositional variant of the entailment principle may be expressed as

$$(usually)(X \text{ is } A)$$
$$\frac{A \subset B}{(usually)(X \text{ is } B).} \qquad (3.32)$$

Another variant is

$$X \text{ is } A$$
$$\frac{(usually)(A \subset B)}{(usually)(X \text{ is } B).} \qquad (3.33)$$

Still another variant is

$$(usually)(X \text{ is } A)$$
$$\frac{(usually)(A \subset B)}{(usually^2)(X \text{ is } B).} \qquad (3.34)$$

Note that $usually^2$ is less specific than $usually$ (Fig. 4).

A special case of the entailment principle which plays an important role in fuzzy logic is the *projection rule*, which may be stated as follows. Let X and Y be variables ranging over U and V, respectively, and let R be a fuzzy relation in $U \times V$, which is characterized by its membership function $\mu_R(u, v)$, $u \in U$, $v \in V$. Then, in symbols, the projection rule may be expressed as

$$\frac{(X, Y) \text{ is } R}{X \text{ is } \text{proj}_U R} \qquad (3.35)$$

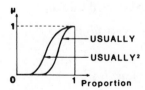

Fig. 4.   Representation of *usually* and *usually*^{2}.

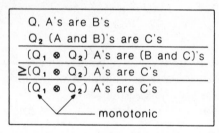

Fig. 5.   Representation of the intersection/product syllogism and its corollaries.

where the first premise signifies that X and Y are R-related, and $\text{proj}_U R$ denotes the projection of R on U. The membership function of $\text{proj}_U R$ is given by

$$\mu_{\text{Proj}_U} R(u) = \sup_v \mu_R(u, v). \qquad (3.36)$$

Since R is contained in $\text{proj}_U R$, it follows from (3.32) that

$$\frac{(usually)((X, Y) \text{ is } R)}{(usually)(X \text{ is Proj}_U R)}. \qquad (3.37)$$

This inference rule may be regarded as a dispositional version of the projection rule.

The dispositional inference rule expressed by (3.37) may be applied to the derivation of a dispositional form of the compositional rule of inference in fuzzy logic [45]

$$\begin{array}{c} X \text{ is } A \\ \underline{(X, Y) \text{ is } R} \\ Y \text{ is } A \cdot R \end{array} \qquad (3.38)$$

where $A \cdot R$ denotes the composition of A and R, which is defined by

$$\mu_{A \cdot R}(v) = \sup_u (\mu_A(u) \wedge \mu_R(u, v)). \qquad (3.39)$$

Specifically, assume that the premises in (3.38) are expressed as the dispositional valuations

$$(usually)(X \text{ is } A) \qquad (3.40)$$

and

$$(usually)((X, Y) \text{ is } R) \qquad (3.41)$$

in which the fuzzy quantifier usually is the same in the two premises in the sense that the conjunction of the premises in question may be expressed as

$$(usually)((X \text{ is } A) \land ((X, Y) \text{ is } R))$$

or, equivalently, as

$$(usually)(X, Y) \text{ is } A \land R) \qquad (3.42)$$

where the conjunction of A and R is defined by

$$\mu_{A \land R}(u, v) = \mu_A(u) \land \mu_R(u, v). \qquad (3.43)$$

Then, on applying the dispositional projection rule to (3.41), we have

$$\frac{(usually)((X \text{ is } A) \land ((X, Y) \text{ is } R))}{(usually)(Y \text{ is } \text{proj}_V(A \land R)} \qquad (3.44)$$

and since

$$A \cdot R = \text{proj}_V(A \land R)$$

we can assert that

$$\frac{(usually)(X \text{ is } A)}{(usually)((X, Y) \text{ is } R)} \qquad (3.45)$$
$$(usually)(Y \text{ is } A \cdot R),$$

which expresses a dispositional version of the compositional rule of inference. It should be noted that the validity of this version depends on the assumption that the two premises (3.40) and (3.41) may be combined conjunctively as in (3.42). Without this assumption, the fuzzy quantifier in the conclusion would be less specific than *usually*, e.g., it might be *usually*^{2}.

An important way in which usuality enters into syllogistic reasoning is the following.

Consider the chaining inference schema

$$\frac{Q_1 A \text{'s are } B \text{'s}}{Q_2 B \text{'s are } C \text{'s}} \qquad (3.46)$$
$$QA \text{'s are } C \text{'s}$$

in which no restrictions are placed on A, B, C. Then, as was pointed out earlier (see (3.9)), Q is the vacuous quantifier *none-to-all*. However, under the assumption that $B \subset A$, (3.46) leads to the **multiplicative chaining syllogism** (see (3.1))

$$Q_1 A \text{'s are } B\text{'s}$$
$$Q_2 B \text{'s are } C\text{'s}$$
$$\frac{B \subset A}{\geqslant (Q_1 \otimes Q_2) A \text{'s are } C\text{'s.}} \qquad (3.47)$$

Now assume that the categorical assumption $B \subset A$ is replaced by the dispositional assumption

$$(usually)(B \subset A). \qquad (3.48)$$

With this assumption, (3.45) becomes what we shall refer to as the multiplicative chaining *hypersyllogism*

$$Q_1 A \text{'s are } B\text{'s}$$
$$Q_2 B \text{'s are } C\text{'s}$$
$$\frac{(usually)(B \subset A)}{(usually)(\geqslant (Q_1 \otimes Q_2) A \text{'s are } C\text{'s})} \qquad (3.49)$$

with the understanding that the prefix *hyper* signifies that the fuzzy quantifier *usually* in application to the containment relation $B \subset A$ is a higher-order fuzzy quantifier.

CONCLUSION

Syllogistic reasoning in fuzzy logic is a basic mode of inference which provides a) a basis for a formalization of commonsense reasoning and b) a non *ad hoc* computational framework for combining evidence in expert systems. In the latter application, the specificity of conclusions may be sharpened by utilizing higher-order dispositional assumptions regarding the predicates that enter into the premises. In this way, we are led to the concept of *hypersyllogistic reasoning*—a mode of reasoning which may eventually play an important role in the management of uncertainty in knowledge-based systems.

ACKNOWLEDGMENT

This work is dedicated to the memory of King-Sun Fu.

REFERENCES

[1] E. W. Adams and H. F. Levine, "On the uncertainties transmitted from premises to conclusions in deductive inferences," *Synthese*, vol. 30, pp. 429–460, 1975.

[2] J. F. Baldwin and S. Q. Zhou, "A fuzzy relational inference language," rep. no. EM/FS132, University of Bristol, England, 1982.

[3] J. Barwise and R. Cooper, "Generalized quantifiers and natural language," *Linguistics and Philosophy*, vol. 4, pp. 159–219, 1981.

[4] R. E. Bellman and L. A. Zadeh, "Local and fuzzy logics," in *Modern Uses of Multiple-Valued Logic*, G. Epstein, Ed. Dordrecht: Reidel, 1977, pp. 103–165.

[5] S. Cushing, *Quantifier Meanings—A Study in the Dimensions of Semantic Competence*. Amsterdam: North-Holland, 1982.

[6] A. DeLuca and S. Termini, "A definition of non-probabilistic entropy in the setting of fuzzy sets theory," *Info. Control*, vol. 20, pp. 301–312, 1972.

[7] D. Dubois and H. Prade, *Fuzzy Sets and Systems: Theory and Applications*. New York: Academic, 1980.

[8] _____, *Theorie des Possibilites*. Paris: Masson, 1985.

[9] R. O. Duda, P. E. Hart, K. Konolige, and R. Reboh, "A computer-based consultant for mineral exploration," tech. rep., SRI International, Menlo Park, CA, 1979.

[10] P. S. Ekberg and L. Lopes, "Tests of a natural reasoning model for syllogistic reasoning," *Goteborg Psychological Reports*, University of Goteborg, Sweden, vol. 9, no. 3, 1979.

[11] J. M. Francioni, A. Kandel, and C. M. Eastman, "Imprecise decision tables and their optimization," in *Approximate Reasoning in Decision Analysis*, M. Gupta, and E. Sanchez, Eds. Amsterdam: North-Holland, pp. 89–96, 1982.

[12] B. R. Gaines, "Precise past—fuzzy future," *Int. J. of Man-Machine Studies*, vol. 19, pp. 117–134.

[13] J. A. Goguen, "The logic of inexact concepts," *Synthese*, vol. 19, pp. 325–373, 1969.

[14] I. Hacking, *Logic of Statistical Inference*. Cambridge: Cambridge University, 1965.

[15] C. G. Hempel, "Maximal specificity and lawlikeness," *Philosophy of Science*, vol. 35, pp. 116–133, 1968.

[16] J. Hintikka, "Statistics, induction and lawlikeness," *Synthese*, vol. 20, pp. 72–83, 1969.

[17] D. N. Hoover, "Probability logic," *Annals Math. Logic*, vol. 14, pp. 287–313, 1978.

[18] M. Ishizuka, K. S. Fu, and J. T. P. Yao, "A rule-based inference with fuzzy set for structural damage assessment," in *Fuzzy Information and Decision Processes*, M. Gupta, and E. Sanchez, Eds. Amsterdam: North-Holland, 1982.

[19] E. T. Jaynes, "New engineering applications of information theory," in *Proc. First Symp. Eng. Applications of Random Function Theory and Probability*, J. L. Bogdanoff and F. Kozin, Eds. New York: Wiley, pp. 163–203, 1963.

[20] R. C. Jeffrey, *The Logic of Decision*. New York: McGraw-Hill, 1965.

[21] A. Kaufmann and M. Gupta, *Introduction to Fuzzy Arithmetic*. New York: Van Nostrand, 1985.

[22] E. P. Klement, W. Schwyhla, and R. Lowen, "Fuzzy probability measures," *Fuzzy Sets and Systems*, vol. 5, pp. 83–108.

[23] H. E. Kyburg, *Probability and Inductive Logic*. New York: McMillan, 1970.

[24] E. H. Mamdani and B. R. Gaines, *Fuzzy Reasoning and its Applications*. London: Academic, 1981.

[25] J. D. McCawley, *Everything that Linguists have Always Wanted to Know about Logic*. Chicago: University of Chicago, 1981.

[26] J. McCarthy, "Circumscription: A non-monotonic inference rule," *Artificial Intelligence*, vol. 13, pp. 27–40, 1980.

[27] D. V. McDermott and J. Doyle, "Non-monotonic logic—Part I," *Artificial Intelligence*, vol. 13, pp. 41–72, 1980.

[28] D. V. McDermott, "Non-monotonic logic—Part II: non-monotonic modal theories," *J. Assoc. Comp. Mach.* vol. 29, pp. 33–57, 1982.

[29] D. H. Mellor, *Matter of Chance*.   Cambridge: Cambridge University, 1971.

[30] M. Mizumoto, S. Fukame, and K. Tanaka, "Fuzzy reasoning methods by Zadeh and Mamdani, and improved methods," *Proc. Third Workshop on Fuzzy Reasoning*, Queen Mary College, London, 1978.

[31] G. C. Moisil, *Lectures on the Logic of Fuzzy Reasoning*.   Bucarest: Scientific Editions, 1975.

[32] R. C. Moore and J. R. Hobbs, Eds., *Formal Theories of the Commonsense World*.   Harwood, NJ: Ablex, 1984.

[33] C. V. Negoita, *Expert Systems and Fuzzy Systems*.   Menlo Park: Benjamin/Cummings, 1985.

[34] N. Nilsson, *Probabilistic Logic*, SRI tech. note 321, Menlo Park, CA, 1984.

[35] P. Peterson, "On the logic of few, many and most," *Notre Dame J. Formal Logic*, vol. 20, pp. 155–179, 1979.

[36] R. Reiter and G. Criscuolo, "Some representational issues in default reasoning," *Computers and Mathematics*, vol. 9, pp. 15–28, 1983.

[37] N. Rescher, *Plausible Reasoning*.   Amsterdam: Van Gorcum, 1976.

[38] E. H. Shortliffe and B. Buchanan, "A model of inexact reasoning in medicine," *Mathematical Biosciences*, vol. 23, pp. 351–379, 1975.

[39] G. Soula and E. Sanchez, "Soft deduction rules in medical diagnostic processes," in *Approximate Reasoning in Decision Analysis*, M. Gupta, and E. Sanchez, Eds.   Amsterdam: North-Holland, 1982, pp. 77–88.

[40] M. Sugeno, "Fuzzy measures and fuzzy integrals: A survey," in *Fuzzy Automata and Decision Processes*, M. Gupta, G. N. Saridis, and B. R. Gaines, Eds.   Amsterdam: North-Holland, 1977, pp. 89–102.

[41] P. Szolovits and S. G. Pauker, "Categorical and probabilistic reasoning in medical diagnosis," *Artificial Intelligence*, vol. 11, pp. 115–144, 1978.

[42] R. R. Yager, "Quantified propositions in a linguistic logic," in *Proc. 2nd Int. Seminar Fuzzy Set Theory*, E. P. Klement, Ed. · Linz, Austria: Johannes Kepler University, 1980.

[43] L. A. Zadeh, "Similarity relations and fuzzy orderings," *Info. Sci.*, vol. 3, pp. 177–200, 1976.

[44] L. A. Zadeh, "Fuzzy logic and approximate reasoning (in memory of Grigore Moisil)," *Synthese*, vol. 30, pp. 407–428, 1975.

[45] L. A. Zadeh, "A theory of approximate reasoning," tech. Memo. no. M77/58, University of California, Berkeley, 1977. See also *Machine Intelligence*, (vol. 9), J. E. Hayes, D. Michie, and L. I. Kulich, Eds. New York: John Wiley, 1979, pp. 149–194.

[46] L. A. Zadeh, "Test-score semantics for natural languages and meaning-representation via PRUF," tech. memo. no. 247, AI Center, SRI International, Menlo Park, CA, 1981. See also *Empirical Semantics*, B. B. Rieger, Ed.   Bochum: Brockmeyer, 1981, pp. 281–349.

[47] L. A. Zadeh, "A computational approach to fuzzy quantifiers in natural languages," *Computers and Mathematics*, vol. 9, pp. 149–184, 1983.

[48] L. A. Zadeh, "The role of fuzzy logic in the management of uncertainty in expert systems," *Fuzzy Sets and Systems*, vol. 11, pp. 199–227, 1983.

[49] L. A. Zadeh, "A theory of commonsense knowledge," in *Issues of Vagueness*, H. J. Skala, S. Termini, and E. Trillas, Eds.   Dordrecht: Reidel, 1984, pp. 257–296.

[50] L. A. Zadeh, "A computational theory of dispositions," in *Proc. 1984 Int. Conf. Computational Linguistics*, 1984, pp. 312–318.

A Fuzzy-Set-Theoretic Interpretation of Linguistic Hedges

L.A. ZADEH†

*Department of Electrical Engineering and Computer Sciences
and the Electronics Research Laboratory, University of
California, Berkeley*

Abstract

A basic idea suggested in this paper is that a linguistic hedge such as *very, more or less, much, essentially, slightly,* etc. may be viewed as an operator which acts on the fuzzy set representing the meaning of its operand. For example, in the case of the composite term *very tall man,* the operator *very* acts on the fuzzy meaning of the term *tall man.*

To represent a hedge as an operator, it is convenient to define several elementary operations on fuzzy sets from which more complicated operations may be built up by combination or composition. In this way, an approximate representation for a hedge can be expressed in terms of such operations as complementation, intersection, concentration, dilation, contrast intensification, fuzzification, accentuation, etc.

Two categories of hedges are considered. In the case of hedges of Type I, e.g., *very, much, more or less, slightly,* etc., the hedge can be approximated by an operator acting on a single fuzzy set. In the case of hedges of Type II, e.g., *technically, essentially, practically,* etc., the effect of the hedge is more complicated, requiring a description of the manner in which the components of its operand are modified. If, in addition, the characterization of a hedge requires a consideration of a metric or proximity relation in the space of its operand, then the hedge is said to be of Type IP or IIP, depending on whether it falls into category I or II.

The approach is illustrated by constructing operator representations for several relatively simple hedges such as *very, more or less, much, slightly, essentially,* etc. More complicated hedges whose effect is strongly context-dependent, require the use of a fuzzy-algorithmic mode of characterization which is more qualitative in nature than the approach described in the present paper.

1. Introduction

Roughly speaking, a fuzzy set is a class with unsharp boundaries, that is, a class in which the transition from membership to non-membership is gradual rather than abrupt. In this sense, the class of tall men is a fuzzy set, as are the classes of beautiful women, young men, red flowers, small cars, etc.

Fuzziness plays an essential role in human cognition because most of the classes encountered in the real world are fuzzy—some only slightly and some markedly so. The pervasiveness of fuzziness in human thought processes suggests that much of the logic behind human reasoning is not the traditional two-valued or even multi-valued logic, but a logic with fuzzy truths, fuzzy connectives and fuzzy rules of inference. Indeed, it may be argued that it is the ability of the human brain to manipulate fuzzy concepts that

†Requests for reprints should be sent to Prof. Lotfi A. Zadeh, Dept. of Electrical Engineering and Computer Sciences and the Electronics Research Laboratory, University of California, Berkeley, California 94720.

distinguishes human intelligence from machine intelligence. And yet, despite its funda-mental importance, fuzziness has not been accorded much attention in the scientific literature,[1] partly because it is antithetic to the deeply entrenched traditions of scientific thinking based on Aristotelian logic and oriented toward exact quantitative analysis, and partly because fuzziness is susceptible of confusion with randomness. In fact, fuzziness and randomness are distinct phenomena which require different modes of treatment and mathematical analysis.

The theory of fuzzy sets[2] represents an attempt at constructing a conceptual framework for a systematic treatment of fuzziness in both quantitative and qualitative ways. A basic concept in this theory is that of a fuzzy subset A of a universe of discourse U, with A characterized by a membership function $\mu_A(x)$ which associates with each point x in U its "grade of membership" in A. Usually, but not necessarily, $\mu_A(x)$ is assumed to range in the interval [0, 1], so that the grade of membership is a number between 0 and 1, with 0 and 1 corresponding to non-membership and full membership, respectively. For example, if U is the set of integers from 0 to 100, then the grade of membership of a man who is 23 years old in the class of young men might be specified to be 0.9. In general, the grades of membership are subjective, in the sense that their specification is a matter of definition rather than objective experimentation or analysis. It should be noted that, although $\mu_A(x)$ may be interpreted as the truth-value of the statement "x belongs to A," it is more natural to view it simply as a grade of membership because the statement "x belongs to A" is not meaningful when A is a fuzzy set.

The pervasive fuzziness of the semantics and—to a lesser extent—the syntax of natural languages suggests that some aspects of linguistic theory might be amenable to analysis by techniques derived from the theory of fuzzy sets. A few preliminary steps in this direction were taken in [5], [6], and [7] with the aim of constructing a framework for a quantitative approach to fuzzy semantics and fuzzy syntax. In what follows, the focus of attention will be on a more specific application, namely, the construction of a fuzzy-set-theoretic interpretation of hedges, e.g., "very," "somewhat," "quite," "much," "more or less," "sort of," "essentially," etc., as operators acting on fuzzy subsets of the universe of discourse. Our main concern, however, will be with the basic aspects of this interpretation rather than with detailed analyses of particular hedges.

The possibility of defining hedges as operators acting on fuzzy sets provides a basis for a better understanding of their role in natural languages. More important, it suggests a way of constructing a system of both natural and artificial hedges which could be used to devise algorithmic languages for the description of the behavior of complex systems. Such languages might find significant applications in psychology, sociology, political science, physiology, economics, management science, information retrieval and other fields in which system behavior is frequently too complex or too ill-defined to admit of analysis in conventional mathematical terms.

2. Fuzzy Sets and Languages—Notation and Terminology

Let U be a universe of discourse, i.e., a collection of objects denoted generically by y. For our purposes, it will be convenient to regard a language, L, as a correspondence between a

---

[1] Discussions of vagueness and related questions from a philosophical point of view may be found in [11]–[15].

[2] Topics in the theory of fuzzy sets which are relevant to the subject of the present paper are discussed in [3]–[10].

set of terms, T, and the universe of discourse, U.[3] This correspondence will be assumed to be defined by a *naming relation,* N, which associates with each term x in T and each object y in U the degree, $\mu_N(x, y)$, to which x applies to y.[4] $\mu_N(x, y)$ will be assumed to be a number in the interval [0, 1], so that N is, in effect, a fuzzy relation from T to U.[5]

A term may be atomic, e.g., x = *red,* x = *barn,* x = *tall,* or composite, in which case it is a concatenation of atomic terms, e.g., x = *red barn,* x = *tall man,* x = *very beautiful woman,* x = *tall and dark,* x = *not very sweet or sour,* etc. In either case, when x is chosen to be a particular term in T, say x = *red,* the function μ_N (*red,* y) defines a fuzzy subset of U whose membership function $\mu_{red}(y)$ is given by

$$\mu_{red}(y) = \mu_N(red, y). \tag{1}$$

This fuzzy subset, denoted by M(*red*), is defined to be the *meaning*[6] of the term *red.* Equivalently, the term *red* may be viewed as a label for a fuzzy subset of U which "comprises" (in a fuzzy sense) those elements of U whose color is *red.*

In short, the *meaning,* M(x), of a term x is a fuzzy subset of U characterized by the membership function

$$\mu_{M(x)}(y) = \mu_N(x, y), \quad x \in T, \quad y \in U \tag{2}$$

where $\mu_N(x, y)$ is the membership function of the naming relation N. To illustrate, suppose that U is the set of integers from 0 to 100 representing ages. Then, the meaning of the term *young* may be specified to be a fuzzy subset M(*young*) of U whose membership function is expressed by[7]

$$\mu_{young}(y) = 1 \text{ for } y \leq 25$$

$$= \left[1 + \left(\frac{y - 25}{5}\right)^2\right]^{-1} \text{ for } y > 25 \tag{3}$$

and similarly

$$\mu_{old}(y) = 0 \text{ for } y < 50$$

$$= \left[1 + \left(\frac{y - 50}{5}\right)^{-2}\right]^{-1} \text{ for } y \geq 50. \tag{4}$$

The plots of the membership functions of M(*young*) and M(*old*) are shown in Fig. 1.

[3] A more detailed discussion may be found in [6] and [7]. For simplicity, we assume that T is a non-fuzzy set.

[4] It should be noted that $\mu_N(x, y)$ may not be defined for all $x \in T$ and $y \in U$. For example, the degree to which the term *jealous* applies to an inanimate object such as *chair* may be assumed to be undefined rather than zero.

[5] A fuzzy relation, R, from a set X to a set Y is a fuzzy subset of the cartesian product X × Y. (X × Y is the collection of ordered pairs (x, y), $x \in X$, $y \in Y$). R is characterized by a bivariate membership function $\mu_R(x, y)$. For example, if X = {Tom, Dick} and Y = {John, Jim}, then the fuzzy relation of *resemblance* between members of X and Y may be defined by a membership function $\mu_R(x, y)$ whose values might be μ_R(Tom, John) = 0.8, μ_R(Tom, Jim) = 0.6, μ_R(Dick, John) = 0.2 and μ_R(Dick, Jim) = 0.9. For additional discussion of fuzzy relations see [3], [8] and [9].

[6] It will be understood throughout that the meaning of a term depends–to a greater or lesser extent–on the context in which it is used.

[7] The definitions of *young* and *old* as expressed by Eqs. (3) and (4) should be viewed merely as illustrations of Eq. (2) rather than as accurate representations of the consensus regarding the meaning of these terms.

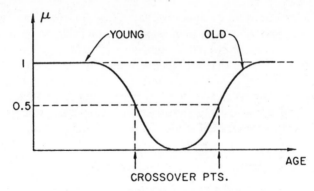

FIG. 1. Membership functions of *young* and *old*.

A *crossover* point in a fuzzy set is a point whose grade of membership in the set is 0.5. For example, the crossover point in M(*young*) is y = 30, while in M(*old*) it is y = 55.

The *support* of a fuzzy set A in U is defined to be the set of points in U for which $\mu_A(y)$ is positive. When the support of A is a finite set, it is frequently convenient to represent A as a linear combination of the elements of the support of A. Thus, if y_i, i = 1, . . . , n, is an element of the support of A and μ_i is its grade of membership in A, then A will be expressed as

$$A = \mu_1 y_1 + \mu_2 y_2 + \cdots + \mu_n y_n \tag{5}$$

or more compactly

$$A = \sum_{i=1}^{n} \mu_i y_i. \tag{6}$$

In cases where it is necessary to use a separator symbol to differentiate between μ_i and y_i, A will be written as

$$A = \mu_1/y_1 + \mu_2/y_2 + \cdots + \mu_n/y_n \tag{7}$$

where the separator / serves to identify the μ_i and y_i components of the string $\mu_i y_i$.

A *fuzzy singleton* is a fuzzy set whose support is a single point in U. If S is a fuzzy singleton, we write

$$S = \mu/y \tag{8}$$

where μ is the grade of membership of y in S. A singleton in the usual (non-fuzzy) sense will be denoted by 1/y.

The representation of a fuzzy set A in the form of Eq. (7) may be viewed as a decomposition of A into its constituent fuzzy singletons. The plus sign in this representation plays the role of the union (see Eq. (38)) of these singletons.

In most instances, the distinction between a term x and its meaning M(x) is implied by the context. Consequently, our notation can be simplified without incurring a significant risk of confusion by writing x for M(x). For example, if U is the set of ages from 0 to 100, i.e.,

$$U = \sum_{i=0}^{100} i \tag{9}$$

then the fuzzy subset[8] of U corresponding to the term *middle-aged* may be expressed as

$$\begin{aligned}
\textit{middle-aged} = \ & 0.3/40 + 0.5/41 + 0.8/42 + 0.9/43 + 1/44 + 1/45 + 1/46 + 1/47 \\
& + 1/48 + 0.9/49 + 0.8/50 + 0.7/51 + 0.6/52 + 0.5/53 + 0.4/54 \qquad (10) \\
& + 0.3/55
\end{aligned}$$

with the understanding that the left-hand member of Eq. (10) stands for the fuzzy subset M(*middle-aged*).

Where U is a countable set, e.g., the set of positive integers, we write

$$U = \sum_{i=1}^{\infty} y_i \tag{11}$$

and

$$A = \sum_{i=1}^{\infty} \mu_i/y_i. \tag{12}$$

For example, if U is the set of positive integers, and A is a fuzzy subset of U labeled *small integer,* we can represent A as

$$\textit{small integer} = \sum_{i=1}^{\infty} \left[1 + \left(\frac{i-1}{10}\right)^2\right]^{-1} \Big/ i = 1/1 + 0.99/2 + 0.96/3 + \cdots \tag{13}$$

When the universe of discourse, U, is a continuum, it is convenient to represent U as an "integral"

$$U = \int_U 1/y \tag{14}$$

with a fuzzy subset, A, of U represented as

$$A = \int_U \mu_A(y)/y \tag{15}$$

where μ_A is the membership function of A. It should be emphasized that in Eqs. (14) and (15) the integral sign is not used in its conventional sense. Rather, Eqs. (14) and (15) are merely continuous counterparts of Eqs. (11) and (12).[9]

---

[8] A is a *subset* of B, written as $A \subset B$, if and only if $\mu_A(y) \leqslant \mu_B(y)$ for all y in U. For example, the fuzzy set A = 0.8/3 + 0.6/4 is a subset of B = 0.9/3 + 0.7/4 + 0.3/5.

[9] Like Eq. (7), Eq. (14) represents the set-theoretic union (see Eq. (38)) of an indexed collection of fuzzy singletons $\mu_A(y)/y$. In this sense, the integral representation of a fuzzy set remains valid when U is a countable set rather than a continuum.

As an illustration of the above notation, Eq. (3) can be rewritten more simply as

$$young = \int_0^{25} 1/y + \int_{25}^{100} \left[1 + \left(\frac{y-25}{5}\right)^2\right]^{-1}/y \qquad (16)$$

similarly

$$old = \int_{50}^{100} \left[1 + \left(\frac{y-50}{5}\right)^{-2}\right]^{-1}/y \qquad (17)$$

Note. If y is an ordered n-tuple with components $v_1, \ldots, v_n$, i.e., $y = (v_1, \ldots, v_n)$ $v_i \in V_i \triangleq$ domain[10] of v_i, then Eq. (15) may be written as

$$A = \int_U \mu_A(v_1, \ldots, v_n)/(v_1, \ldots, v_n) . \qquad (18)$$

Using this notation, the relation of resemblance cited earlier (see footnote 5), may be expressed as

$$R = 0.8/(Tom, John) + 0.6/(Tom, Jim) + 0.2/(Dick, John) + 0.9/(Dick, Jim) . \qquad (19)$$

It should be noted that the grade of membership in a fuzzy set may in itself be a fuzzy set. For example, suppose that the universe of discourse comprises persons named John, Tom, Dick and Harry, i.e.,

$$U = John + Tom + Dick + Harry \qquad (20)$$

and that I is the fuzzy subset of *intelligent* men in U. Furthermore, suppose that there are three fuzzy grades of membership labeled *high*, *medium* and *low*, which are defined as fuzzy subsets of the universe V,

$$V = 0 + 0.1 + 0.2 + 0.3 + 0.4 + 0.5 + 0.6 + 0.7 + 0.8 + 0.9 + 1 \qquad (21)$$

Thus,
$$high = 0.5/0.7 + 0.7/0.8 + 0.9/0.9 + 1/1 \qquad (22)$$
$$medium = 0.5/0.4 + 0.7/0.5 + 1/0.6 + 0.7/0.7 + 0.5/0.8 \qquad (23)$$
$$low = 0.5/0.2 + 0.7/0.3 + 1/0.4 + 0.7/0.5 + 0.5/0.6. \qquad (24)$$

Then, we may have

$$I \triangleq intelligent = medium/John + high/Tom + low/Dick + low/Harry. \qquad (25)$$

In this way, the meaning of a term can be expressed as a fuzzy subset of the universe of discourse, with the grades of membership being numbers in the interval [0, 1] or fuzzy subsets of this interval.

This concludes our brief summary of those aspects of the notation and terminology relating to fuzzy sets which we shall need in the following sections.

3. Operations on Fuzzy Sets

As defined in Section 2, if $x_1, x_2, \ldots, x_n$ are atomic terms, then their concatenation

$$x = x_1 x_2 \cdots x_n \qquad (26)$$

[10] The symbol $\triangleq$ stands for "equal by definition" or "is defined to be" or "denotes."

is a *composite* term. For example, if x_1 = *very*, x_2 = *tall* and x_3 = *man*, then $x_1 x_2 x_3$ is the composite term *very tall man*.

In quantitative fuzzy semantics [6], the meaning of term x is a fuzzy subset, M(x), of the universe of discourse (see Eq. (2)). From this point of view, one of the basic problems in semantics is that of devising an algorithm for the computation of the meaning of a composite term $x = x_1 \cdots x_n$ from the knowledge of the meaning of each of its atomic components, $x_i, i = 1, \ldots, n$.

In the present paper, our main concern is with a special case of this problem in which x is of the form x = hu, where h is a hedge, e.g., h = *highly*, and u is a term, e.g., u = *intelligent man*.

The point of view developed in this paper is that a hedge, h, may be interpreted as an operator, with operand u, which transforms a fuzzy subset M(u) of U into the subset M(hu). To characterize this operator, it is convenient to define several primitive operations on fuzzy sets from which more complicated operators such as hedges may be built up by composition.

We shall begin with the basic set-theoretic operations of complementation, intersection and union, and follow these with several more specialized operations: product, normalization, concentration, dilation, contrast intensification, convex combination, and fuzzification.

Complementation

Complementation is a unary operation in the sense that it transforms a fuzzy set in U into another fuzzy set in U. More specifically, the complement of a fuzzy set A is denoted by $\neg$A and is defined by the relation[11]

$$\mu_{\neg A}(y) = 1 - \mu_A(y) \ , \ y \in U . \tag{27}$$

Thus, if A is the class of *rich men* and $\mu_A(John) = 0.8$, then $\mu_{\neg A}(John) = 0.2$. Equivalently, if u is a term whose meaning is M(u), then the meaning of M(*not* u) is given by [12]

$$M(not \ u) = \neg M(u). \tag{28}$$

In short, *not*(negation) is an operator[13] which transforms M(u) into $\neg$M(u).

Intersection

Intersection is a binary operation in the sense that it transforms a pair of fuzzy sets in U into a fuzzy set in U. More specifically, the *intersection* of two fuzzy sets A and B is a fuzzy set denoted by A $\cap$ B and defined by

$$\mu_{A \cap B}(y) = \mu_A(y) \wedge \mu_B(y) , \quad y \in U \tag{29}$$

[11] Note that if $\mu_A(y)$ is undefined at y, then the same is true of $\mu_{\neg A}(y)$.

[12] It should be noted that, in a natural language, the negation *not* frequently corresponds to a relative complement, that is, to a complement relative to a subset, V, of the universe of discourse, with V implicitly defined by the operand of *not* and the context in which it appears. For example, in the sentence "Fifi is *not* a poodle," it might be understood that Fifi is a dog other than a poodle rather than any object in the universe of discourse which is not a poodle.

[13] Strictly speaking, $\neg$ acts on a fuzzy set whereas *not* (negation) acts on its label. Thus, when we write $\neg$ u it should be understood that u stands for M(u).

where, for any real a and b, $a \wedge b$ denotes $\text{Min}(a, b)$, that is,

$$a \wedge b = a \quad \text{if} \quad a \leq b \qquad (30)$$
$$= b \quad \text{if} \quad a > b.$$

To a first approximation,[14] the conjunctive connective *and* may be identified with the intersection of fuzzy sets. Thus,. if u and v are terms in T, then the meaning of the composite term x = u *and* v is given by[15]

$$M (u \; and \; v) = M(u) \cap M(v). \qquad (31)$$

For example, if the universe of discourse is the set $U = 1 + 2 + 3 + 4 + 5 + 6$ and the meanings of u and v are expressed in the notation of Eq. (12) as

$$u = 0.8/3 + 1/5 + 0.6/6 \qquad (32)$$

and

$$v = 0.7/3 + 1/4 + 0.5/6 \qquad (33)$$

then

$$u \; and \; v = 0.7/3 + 0.5/6. \qquad (34)$$

More generally, if

$$u = \int_U \mu_A(y)/y \qquad (35)$$

and

$$v = \int_U \mu_B(y)/y \qquad (36)$$

then

$$u \; and \; v = \int_U (\mu_A(y) \wedge \mu_B(y))/y. \qquad (37)$$

Union

Like the intersection, the union of fuzzy sets is a binary operation. More concretely, the *union* of two fuzzy sets A and B is a fuzzy set denoted $A \cup B$—or, more conveniently, $A + B$—and defined by

$$\mu_{A + B}(y) = \mu_A(y) \vee \mu_B(y), \qquad y \in U \qquad (38)$$

where $a \vee b$ denotes $\text{Max}(a, b)$, that is

$$a \vee b = a \quad \text{if} \quad a \geq b \qquad (39)$$
$$= b \quad \text{if} \quad a < b.$$

Dual to the correspondence between the conjunctive connective *and* and $\cap$ is the correspondence between the disjunctive connective *or* and +.

Thus

$$M (u \; or \; v) = M(u) + M(v) \qquad (40)$$

---

[14] This qualification reflects the fact that in a natural language the meaning of *and* is somewhat context-dependent and is not always expressed by Eq. (31).

[15] It should be noted that, from an algebraic point of view, Eq. (31) may be regarded as the definition of a homomorphism from the set of labels of fuzzy subsets of U to the set of fuzzy subsets of U, with the corresponding operations being *and* and $\cap$.

or equivalently

$$u \ or \ v = \int_U (\mu_A(y) \lor \mu_B(y))/y \tag{41}$$

where u and v are defined by Eqs. (35) and (36).

As an illustration, for the terms defined by Eqs. (32) and (33), Eq. (41) yields

$$u \ or \ v = 0.8/3 + 1/4 + 1/5 + 0.6/6 . \tag{42}$$

It can readily be shown [3] that the union and intersection of fuzzy sets are associative as well as distributive operations. Furthermore, they satisfy the De Morgan identity

$$\neg(A \cap B) = \neg A + \neg B \tag{43}$$

which in terms of *and, or* and *not* may be stated as

$$not \ (u \ and \ v) = not \ u \ or \ not \ v. \tag{44}$$

Since $\neg \neg A = A$, Eq. (44) entails

$$not \ (u \ or \ v) = not \ u \ and \ not \ v. \tag{45}$$

Product

The *product* of two fuzzy sets A and B is denoted by AB and is defined by

$$\mu_{AB}(y) = \mu_A(y)\mu_B(y), \qquad y \in U . \tag{46}$$

Thus, if
$$A = 0.8/2 + 0.9/5 \tag{47}$$

and
$$B = 0.6/2 + 0.8/3 + 0.6/5 \tag{48}$$

then
$$AB = 0.48/2 + 0.54/5 . \tag{49}$$

More generally, using the integral representation of A and B, we can write

$$AB = \int_U \mu_A(y)\mu_B(y)/y . \tag{50}$$

For example, if
$$A = \int_0^\infty (1 + y^2)^{-1}/y \tag{51}$$

and
$$B = \int_0^\infty (1 + y^{-2})^{-1}/y \tag{52}$$

then
$$AB = \int_0^\infty y^2(1 + y^2)^{-2}/y . \tag{53}$$

An immediate extension of Eq. (50) leads to the following definition of the expression A^α, where α is any real number:

$$A^\alpha = \int_U [\mu_A(y)]^\alpha/y . \tag{54}$$

For example, if $\alpha = 2$ and A is expressed by Eq. (47), then
$$A^2 = 0.64/2 + 0.81/5 . \tag{55}$$

Similarly, if α is a non-negative real number, then the expression αA is defined by

$$\alpha A = \int_U \alpha \mu_A(y)/y .$$

(56)

For example, if A is expressed by Eq. (47) and $\alpha = 0.5$, then

$$0.5A = 0.4/2 + 0.45/5 .$$

(57)

A useful property of the product is that it distributes over the union and the intersection. Thus,

$$A(B + C) = AB + AC$$

(58)

and

$$A(B \cap C) = AB \cap AC .$$

(59)

Comment. It should be noted that, unlike the union and intersection, the product does not correspond to a commonly used connective. However, in some contexts the meaning of *and* may be more closely approximated by the product than by conjunction.

Comment. The product of A and B as defined above differs from the *cartesian* (or *direct product*) of A and B. Thus, if A is a fuzzy subset of a universe of discourse U and B is a fuzzy subset of a possibly different universe of discourse V, then (see Eq. (18))

$$A \times B = \int_{U \times V} \mu_A(u) \wedge \mu_B(v)/(u, v)$$

(60)

where $U \times V$ denotes the cartesian product of the non-fuzzy sets U and V, and $u \in U$, $v \in V$. Note that when A and B are non-fuzzy, Eq. (60) reduces to the conventional definition of the cartesian product of non-fuzzy sets.

The need for differentiation between the product in the sense of Eq. (50) and the cartesian product becomes particularly important when $A = B$, since A^2 is commonly used to denote $A \times A$ (when A is non-fuzzy), whereas Eq. (54) implies that $A^2 = AA$. In what follows, we shall adhere to the latter interpretation except where an explicit statement to the contrary is made.

Normalization

Let $\bar{\mu}_A$ denote the supremum of a membership function μ_A over the universe of discourse, i.e.,

$$\bar{\mu}_A = \underset{U}{\mathrm{Sup}}\ \mu_A(y).$$

(61)

A fuzzy set A is said to be *normal* if $\bar{\mu}_A = 1$; otherwise, A is *sub-normal.* For example, the set

$$A = 1/\mathrm{John} + 0.8/\mathrm{Jim} + 0.6/\mathrm{Tom}$$

(62)

is normal, while

$$A = 0.6/\mathrm{John} + 0.8/\mathrm{Jim} + 0.6/\mathrm{Tom}$$

(63)

is subnormal.

A subnormal fuzzy set A can be normalized by dividing μ_A by $\bar{\mu}_A$. Using the notation of Eq. (56), the operation of normalization may be expressed as

$$\text{NORM}(A) = (\bar{\mu}_A)^{-1} A, \quad \bar{\mu}_A \neq 0. \tag{64}$$

Thus, for the fuzzy set defined by Eq. (63), the normalization of A results in

$$\text{NORM}(A) = 0.75/\text{John} + 1/\text{Jim} + 0.75/\text{Tom}. \tag{65}$$

Concentration

Like complementation, *concentration* is a unary operation. As its name implies, the result of applying a concentrator to a fuzzy set A is a fuzzy subset of A such that the reduction in the magnitude of the grade of membership of y in A is relatively small for those y which have a high grade of membership in A and relatively large for the y with low membership. Thus, if we denote the result of applying a concentrator to A by CON(A), then the relation between the membership function of A and that of CON(A) will typically have the appearance shown in Fig. 2.

To be more specific, we shall assume that the operation of concentration has the effect of squaring the membership function of A. Thus,

$$\mu_{\text{CON}(A)}(y) = \mu_A^2(y), \quad y \in U \tag{66}$$

or, using the definition of A^2 (see Eq. (54))

$$\text{CON}(A) = A^2. \tag{67}$$

For example, if the meaning of the term *few* is defined by

$$few = 1/1 + 1/2 + 0.8/3 + 0.6/4 \tag{68}$$

then
$$\text{CON}(few) = 1/1 + 1/2 + 0.64/3 + 0.36/4. \tag{69}$$

It should be noted that concentration distributes over the union, intersection and product. Thus

$$\text{CON}(A + B) = \text{CON}(A) + \text{CON}(B) \tag{70}$$

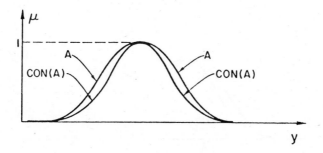

FIG. 2. The effect of concentration on a fuzzy set A.

$$CON(A \cap B) = CON(A) \cap CON(B) \tag{71}$$

and $$CON(AB) = CON(A) \, CON(B). \tag{72}$$

Equation (70) follows from the identities

$$(A + B)^2 = A^2 + AB + BA + B^2 \tag{73}$$

and $$AB + BA \subset A^2 + B^2 \tag{74}$$

which together imply that

$$(A + B)^2 = A^2 + B^2. \tag{75}$$

Equation (71) follows similarly, with + replaced by ∩.

The operation of concentration can be composed with itself. Thus

$$CON^2(A) = A^4 \tag{76}$$

and more generally

$$CON^\alpha(A) = A^{2\alpha} \tag{77}$$

where α is any integer $\geqslant 2$.

Dilation

The effect of *dilation* is the opposite of that of concentration. Thus, the result of applying a dilator to a fuzzy set A is a fuzzy set DIL(A) whose membership function is related to that of A as shown in Fig. 3.

More specifically, DIL(A) is defined by

$$DIL(A) = A^{0.5} \tag{78}$$

which implies that

$$\mu_{DIL(A)}(y) = \sqrt{\mu_A(y)}, \quad y \in U. \tag{79}$$

For example, if *few* is defined by Eq. (68), then

$$DIL \, (few) = 1/1 + 1/2 + 0.9/3 + 0.78/4.$$

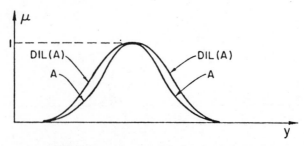

FIG. 3. The effect of dilation on the operand A.

Contrast Intensification

The operation of concentration has the effect of diminishing the value of $\mu_A(y)$ for every y (except where $\mu_A(y) = 1$), with the larger values of $\mu_A(y)$ diminished proportionately less than the smaller values.

The operation of *contrast intensification,* or simply *intensification,* differs from that of concentration in that it increases the values of $\mu_A(y)$ which are above 0.5 and diminishes those which are below this threshold. Thus, if the result of applying a contrast intensifier INT to a fuzzy set A is denoted by INT(A), we have

$$\mu_{INT(A)}(y) \geq \mu_A(y) \quad \text{for} \quad \mu_A(y) \geq 0.5 \tag{80}$$

and

$$\mu_{INT(A)}(y) \leq \mu_A(y) \quad \text{for} \quad \mu_A(y) \leq 0.5. \tag{81}$$

A simple concrete expression for an operator of this type is the following

$$\mu_{INT(A)}(y) = 2\mu_A^2(y) \quad \text{for} \quad 0 \leq \mu_A(y) \leq 0.5$$

$$\mu_{INT(A)}(y) = 1 - 2[1 - \mu_A(y)]^2 \quad \text{for} \quad 0.5 \leq \mu_A(y) \leq 1. \tag{82}$$

The effect of applying this intensifier to a fuzzy set A is shown in Fig. 4.

As in the case of concentration, intensification distributes over the union, intersection and product. Thus

$$\text{INT}(A + B) = \text{INT}(A) + \text{INT}(B) \tag{83}$$

$$\text{INT}(A \cap B) = \text{INT}(A) \cap \text{INT}(B) \tag{84}$$

and

$$\text{INT}(AB) = \text{INT}(A)\,\text{INT}(B). \tag{85}$$

Note. The function defined by Eq. (82) is of use also in the representation of membership functions of fuzzy sets (see Eq. (170)). For this purpose, it is convenient to define a function S from the real line to [0, 1] by the equations

$$
\begin{aligned}
S(u) &\triangleq 0 &&\text{for} \quad u < 0 \\
&\triangleq 2u^2 &&\text{for} \quad 0 \leq u \leq 0.5 \\
&\triangleq 1 - 2(1 - u)^2 &&\text{for} \quad 0.5 < u \leq 1 \\
&\triangleq 1 &&\text{for} \quad u > 1.
\end{aligned}
\tag{86}
$$

This function will be referred to as S-*function* to stress its resemblance to an S (see Fig. 5).

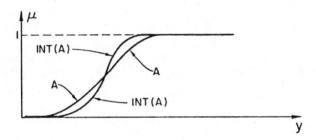

FIG. 4. The effect of intensification on the operand A.

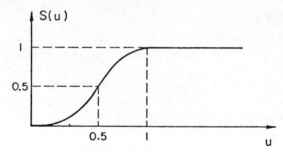

FIG. 5. The form of S-function.

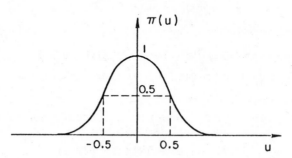

FIG. 6. The form of π-function.

Putting two S-functions back to back we obtain a *pulse-function,* $\pi(u)$, which has the appearance shown in Fig. 6. The defining equations for $\pi(u)$ are:

$$
\begin{aligned}
\pi(u) &\triangleq 0 & \text{for} \quad & u < -1 \\
&\triangleq 2(u + 1)^2 & \text{for} \quad & -1 \le u < -0.5 \\
&\triangleq 1 - 2u^2 & \text{for} \quad & -0.5 \le u \le 0.5 \\
&\triangleq 2(u - 1)^2 & \text{for} \quad & 0.5 < u \le 1 \\
&\triangleq 0 & \text{for} \quad & u > 1 .
\end{aligned}
\tag{87}
$$

Convex Combination

The *convex combination* [3] is an n-ary operation which combines a set of n fuzzy sets $A_1, \ldots, A_n$ into a single fuzzy set A.

The fuzzy set A is a *weighted* combination of $A_1, \ldots, A_n$ in the sense that the membership function of A is related to those of $A_1, \ldots, A_n$ by the expression

$$
\mu_A(y) = w_1(y)\mu_{A_1}(y) + \cdots + w_n(y)\mu_{A_n}(y)
\tag{88}
$$

where the weights $w_1(y), \ldots, w_n(y), 0 \le w_i(y) \le 1, i = 1, \ldots, n$, are such that

$$
w_1(y) + \cdots + w_n(y) = 1 \quad \text{for all } y \text{ in } U.
\tag{89}
$$

For example, for the sets expressed by Eqs. (32) and (33), the convex combination of u and v with constant weights $w_1 = 0.8$ and $w_2 = 0.2$ is given by

$$A = (0.8 \times 0.8 + 0.2 \times 0.7)/3 + 0.8/5 + 0.2/4 + (0.8 \times 0.6 + 0.2 \times 0.5)/6 \qquad (90)$$

or

$$A = 0.78/3 + 0.2/4 + 0.8/5 + 0.58/6 .$$

Fuzzification

The operation of contrast intensification has the effect of transforming a fuzzy set A into a fuzzy set A* which approximates to—and is less fuzzy than—A. As its name implies, the operation of fuzzification has the opposite effect. Thus, its main function is to provide a means of transforming a fuzzy (or non-fuzzy) set A into an approximating set $\underset{\sim}{A}$ which is more fuzzy than A.

The wavy bar $\sim$ plays the role of a *fuzzifier*. Thus, if U is the set of real numbers, then $\underset{\sim}{3}$ represents the fuzzy set of real numbers which are approximately equal to 3. Similarly, if $=$ denotes the relation of equality, then $\underset{\sim}{=}$ represents approximate equality; and if $>$ denotes *greater than*, then $\underset{\sim}{>}$ might be interpreted as *more or less greater than.*

There are many ways in which fuzzification can be accomplished. In what follows, we shall sketch two approaches which are particularly relevant to the definition of such linguistic hedges as *more or less.*

At the base of these approaches is a process of *point fuzzification* which transforms a singleton set $1/u$ in U into a fuzzy set $\underset{\sim}{u}$ which is concentrated around u. To place in evidence the dependence of $\underset{\sim}{u}$ on u, $\underset{\sim}{u}$ will be written as $\underset{\sim}{u} = K(u)$. Unless stated to the contrary, the grade of membership of u in K(u) will be assumed equal to 1.

The fuzzy set K(u) will be referred to as the *kernel* of the fuzzification. Usually, K(u) will be taken to be a *fuzzy interval*,[16] that is, a fuzzy set whose membership function $\mu_{K(u)}(y)$ is a non-increasing function of the distance between u and y. For example, suppose that the universe of discourse is defined by

$$U = 1 + 2 + 3 + 4 \qquad (91)$$

and the singleton set $1/2$ is transformed into the fuzzy set $0.6/1 + 1/2 + 0.8/3 + 0.3/4$, i.e.,

$$1/2 \rightarrow 0.6/1 + 1/2 + 0.8/3 + 0.3/4 . \qquad (92)$$

Then, in this case

$$K(2) = 0.6 /1 + 1/2 + 0.8/3 + 0.3/4 . \qquad (93)$$

Now consider a fuzzy set represented by

$$A = \mu_1/y_1 + \cdots + \mu_n/y_n \qquad (94)$$

---

[16] If U is an Euclidean n-space, then a fuzzy interval in U is a convex fuzzy subset of U. (Convex fuzzy sets are defined in [3].)

where μ_i is the grade of membership of y_i in A. If we postulate that fuzzification is a linear transformation, then Eq. (94) implies that

$$A = \mu_i/y_i + \cdots + \mu_n y_n \, . \tag{95}$$

At this point, it is natural to consider two special cases. In Case I, we hold μ_i constant and set

$$\mu_i/y_i = \mu_i/y_i \, , \quad i = 1, \ldots, n \, . \tag{96}$$

On the other hand, in Case II, we hold y_i constant and set

$$\mu_i/y_i = \mu_i/y_i \, . \tag{97}$$

Thus, in Case I we fuzzify each point in the support of A, while in Case II we fuzzify the grade of membership.

In Case I, suppose that the transformation which takes $1/y_i$ into y_i is characterized by the kernel $K(y_i)$. Then the fuzzification which takes A into $\underset{\sim}{A}$ is denoted by SF(A; K) (SF standing for *support fuzzification* or s-fuzzification, for short) and is defined by

$$\underset{\sim}{A} \triangleq SF(A; K) = \mu_1 K(y_1) + \cdots + \mu_n K(y_n) \tag{98}$$

where $\mu_i K(y_i)$ should be interpreted as a fuzzy set which is the product of a scalar constant μ_i and a fuzzy set $K(y_i)$ (see Eq. (56)), and + stands for the union of fuzzy sets.

As a simple illustration, suppose that

$$U = 1 + 2 + 3 + 4 \tag{99}$$

$$A = 0.8/1 + 0.6/2 \tag{100}$$

$$K(1) = 1/1 + 0.4/2 \tag{101}$$

and

$$K(2) = 1/2 + 0.4/1 + 0.4/3 \, . \tag{102}$$

Then

$$\begin{aligned} SF(A; K) &= 0.8(1/1 + 0.4/2) + 0.6(1/2 + 0.4/1 + 0.4/3) \\ &= 0.8/1 + 0.32/2 + 0.6/2 + 0.24/1 + 0.24/3 \\ &= 0.8/1 + 0.6/2 + 0.24/3 \, . \end{aligned} \tag{103}$$

Comment. Note that the effect of s-fuzzification is somewhat similar to but not quite the same as that of dilation. One important difference is that a dilation of a non-fuzzy set yields the same non-fuzzy set, whereas an s-fuzzification of a non-fuzzy set will, in general, yield a fuzzy set. However, in the case of the fuzzy set defined by

$$A = 1/1 + 0.8/2 + 0.6/3 \tag{104}$$

we have (see Eq. (78))

$$DIL(A) = 1/1 + 0.9/2 + \cdot 0.78/3 \tag{105}$$

and the equality

$$DIL(A) = SF(A; K) \tag{106}$$

can be realized with the kernel set

$$K(1) = 1/1 + 0.9/2$$
$$K(2) = 1/2 + 0.87/3. \tag{107}$$

Equation (98) defines the effect of s-fuzzification when A has a finite support. More generally, assume that A is defined by (see Eq. (15))

$$A = \int_U \mu_A(y)/y. \tag{108}$$

Then, SF(A; K) is given by

$$SF(A; K) = \int_U \mu_A(y) K(y), \quad y \in U \tag{109}$$

where, as in Eq. (98), $\mu_A(y)K(y)$ represents the product of the scalar constant $\mu_A(y)$, and K/y, $\int_U$ should be interpreted as the union of the family of fuzzy sets $\mu_A(y)K(y)$, $y \in U$.

Comment: It is of interest to observe that Eq. (109) is analogous to the integral representation of a linear operator. Note that if A is a singleton set $1/y$, then

$$SF(A; K) = K(y). \tag{110}$$

Thus, a singleton set is analogous to a delta-function and K(y) plays the role of impulse response.

As a simple illustration of Eq. (109), assume that A is a non-fuzzy set whose membership function is shown in Fig. 7a, and K is a fuzzy set whose membership function is depicted in Fig. 7b. Then, SF(A; K) is a fuzzy set whose membership function has the form shown in Fig. 7c.

It should be noted that s-fuzzification can be employed in an indirect manner to effect a translation of a fuzzy or non-fuzzy set within its universe. As an illustration, suppose that we wish to transform the non-fuzzy set A whose membership function is shown in Fig. 8a into the non-fuzzy set B whose membership function is a translate of μ_A (see Fig. 8b), i.e.,

$$\mu_B(y) = \mu_A(y - a). \tag{111}$$

Let K(y) be the interval $[y - a, y + a]$. Then, we can express B as

$$B = \neg SF(\neg A; K) \tag{112}$$

where $\neg$ is the operation of complementation (see Eq. (27)).

Turning to Case II, suppose that the point fuzzification which transforms μ_i into $\underset{\sim}{\mu_i}$ is defined by the kernel $K(\mu_i)$. Then, denoting by GF(A; K) (GF standing for *grade fuzzification* or *g-fuzzification,* for short) the fuzzification which transforms A into $\underset{\sim}{A}$, where

$$\underset{\sim}{A} = \underset{\sim}{\mu_1}/y_1 + \cdots + \underset{\sim}{\mu_n}/y_n \tag{113}$$

we can write

$$GF(A; K) \triangleq \underset{\sim}{A} = K(\mu_1)/y_1 + \cdots + K(\mu_n)/y_n. \tag{114}$$

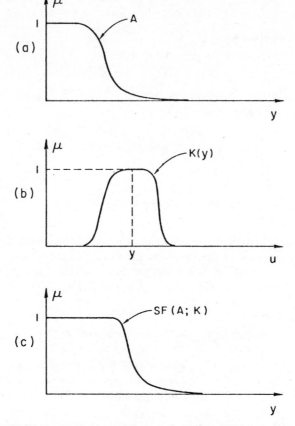

FIG. 7. The effect of fuzzification of a fuzzy set A with kernel set K.

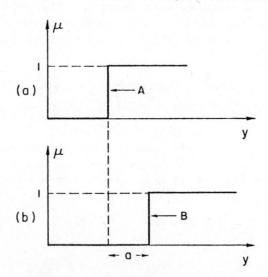

FIG. 8. Translation of A by means of fuzzification.

For example, if

$$U = 1 + 2 + 3 + 4 \qquad (115)$$

and

$$A = 0.8/1 + 0.6/2 \qquad (116)$$

then

$$GF(A;K) = \underset{\sim}{0.8}/1 + \underset{\sim}{0.6}/2 \qquad (117)$$

where $\underset{\sim}{0.8}$ and $\underset{\sim}{0.6}$ are fuzzy grades of membership defined as, say,

$$
\begin{aligned}
K(0.8) &\triangleq \underset{\sim}{0.8} = 1/0.8 + 0.6/0.7 + 0.6/0.9 \\
K(0.6) &\triangleq \underset{\sim}{0.6} = 1/0.6 + 0.6/0.5 + 0.6/0.7.
\end{aligned} \qquad (118)
$$

More generally, if

$$A = \int_U \mu_A(y)/y \qquad (119)$$

then a g-fuzzification of A yields the fuzzy set

$$GF(A;K) = \int_U \underset{\sim}{\mu_A(y)}/y \qquad (120)$$

in which the grade of membership of y in GF(A; K) is the fuzzy set $\mu_A(y) = K(y)$.

The various operations on fuzzy sets which we have defined in this section—especially complementation, intersection, product, concentration, convex combination and fuzzification—provide a basis for the characterization of hedges in terms of compositions or combinations of these operations. We turn to this subject in the following section.

4. Representation of Hedges as Operators

Before we can analyze the representation of hedges as operators acting on fuzzy subsets of the universe of discourse, it will be necessary to consider a basic question in the theory of fuzzy sets which relates to the representation of a fuzzy set in terms of other fuzzy sets.

We have already encountered several special instances of such representations in the preceding section. More generally, if $A_1, \ldots, A_n$ are fuzzy subsets of U with membership functions $\mu_{A_1}, \ldots, \mu_{A_n}$ respectively, then a fuzzy set A in U has $A_1, \ldots, A_n$ as its *components* if the membership function of A is expressible as some (nontrivial) function of $\mu_{A_1}, \ldots, \mu_{A_n}$. Thus, in symbols

$$\mu_A = f(\mu_{A_1}, \ldots, \mu_{A_n}). \qquad (121)$$

For example, if A is the intersection of A_1 and A_2, then by Eq. (29)

$$\mu_A = \mu_{A_1} \wedge \mu_{A_2} \qquad (122)$$

where for simplicity we have omitted the argument y.

Now if A represents the meaning of a fairly complex concept, then, in general, it is expedient to "resolve" A into a set of simpler components $A_1, \ldots, A_n$, so that the membership function of A becomes a function of those of $A_1, \ldots, A_n$. For example, if A is the class of *big men*, then, roughly, its components might be taken to be

$$A_1 = \text{class of } tall\ men \qquad (123)$$

$$A_2 = \text{class of } \textit{heavy men} \tag{124}$$

and in terms of these components A may be defined by the equation[17]

$$\mu_A(y) = 0.6\,\mu_{A_1}(y) + 0.4\,\mu_{A_2}(y). \tag{125}$$

For example, if

$$\mu_{A_1}(\text{John}) = 0.8 \tag{126}$$

and

$$\mu_{A_2}(\text{John}) = 0.5 \tag{127}$$

then

$$\mu_A(\text{John}) = 0.6 \times 0.8 + 0.4 \times 0.5 = 0.68. \tag{128}$$

Our motivation for considering the resolution of a fuzzy set into simpler components has to do with the fact that the representation of certain hedges such as *basically, technically, literally,* etc., involves in an essential way their effect on the components of the fuzzy sets on which they operate. On the other hand, the effect of simpler hedges such as *very, more or less, slightly,* etc., can be described without resort to the resolution of a set into its components. This suggests that hedges be divided into two somewhat fuzzy categories, which may be defined informally as follows.

Type I. Hedges in this category can be represented as operators acting on a fuzzy set. Typical hedges in this category are: *very, more or less, much, slightly, highly.*

Type II. Hedges in this category require a description of how they act on the components of the operand. Typical hedges in this category are: *essentially, technically, actually, strictly, in a sense, practically, virtually, regular,* etc.

Each of the above categories includes a subcategory of hedges—denoted by IP and IIP, respectively—whose effect is influenced by the notion of proximity of ordering in the domain of the operand. Such hedges—of which *slightly* is a simple example—are generally more context-dependent than those hedges of Types I and II which do not require the notion of proximity for their characterization.

In what follows, we shall examine in greater detail some of the basic aspects of the representation of hedges of Type I and II. As illustrations of our approach, we shall construct approximate operator representations for several typical hedges of Type I and II. It should be emphasized, however, that these representations are intended mainly to illustrate the approach rather than to provide accurate definitions of the hedges in question. Furthermore, it must be underscored that our analysis and its conclusions are tentative in nature and may require modification in later work.

5. Representation of Hedges of Type I

It will be convenient to begin our discussion by considering a relatively simple and yet very basic hedge, namely, *very.*

Very

Let A be a fuzzy set in U representing the meaning of a term such as

$$x = \textit{old men}. \tag{129}$$

We assume that A is characterized by a membership function of the form shown in Fig. 9.

---

[17] This definition is used merely as an illustration and is not intended to be an accurate representation of the concept of *big man.*

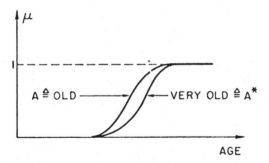

FIG. 9. The effect of the hedge *very*.

Now consider the term

$$x^* = very\ x = very\ old\ men \tag{130}$$

and let A* be the fuzzy set representing the meaning of x*.

The crux of our idea is to view a hedge such as *very* as an operator which transforms the fuzzy set A($\triangleq$ meaning of x) into the fuzzy set A*($\triangleq$ meaning of x*). If we accept this point of view, then the question that arises is: How can the operator *very* be defined?

Given the richness and complexity of natural languages, it is clear that questions of this kind do not admit of simple and definitive answers. Nevertheless, it is useful to attempt to concretize the meaning of a hedge such as *very* even if the postulated meaning does not have universal validity and is merely a fixed approximation to a variety of shades of meaning which *very* can assume in different contexts. It is in this perspective that the definitions of *very* and other hedges which are formulated in the sequel should be viewed.

Specifically, we assume that if the meaning of a term x is a fuzzy set[18] A, then the meaning of *very* x, A*, is given by

$$A^* = CON(A) \tag{131}$$

or, more explicitly (see Eq. (67))

$$A^* = A^2. \tag{132}$$

Using the same symbol to denote a term and its meaning (as in Eq. (10)), the definition of *very* may be expressed more compactly as

$$very\ x = x^2. \tag{133}$$

Thus, if

$$x = \mu_1/y_1 + \cdots + \mu_n/y_n, \quad y_i \in U, \quad i = 1, \ldots, n \tag{134}$$

then

$$very\ x = \mu_1^2/y_1 + \cdots + \mu_n^2/y_n \tag{135}$$

and, more generally, if

$$x = \int_U \mu(y)/y \tag{136}$$

---

[18] It should be observed that the operand of *very* must be a term with a fuzzy meaning, i.e., A must be a fuzzy set. Thus, strictly speaking, it is not correct to say *very rectangular* or *very pregnant*, since both *rectangular* and *pregnant* have non-fuzzy meaning.

then
$$very \; x \; = \; \int_U \mu^2(y)/y.$$
(137)

For example, if (see Eq. (17))

$$x \; = \; old \; men \; = \; \int_{50}^{100} \left[1 + \left(\frac{y - 50}{5} \right)^{-2} \right]^{-1} /y$$
(138)

then
$$very \; old \; men \; = \; \int_{50}^{100} \left[1 + \left(\frac{y - 50}{5} \right)^{-2} \right]^{-2} /y$$
(139)

which implies that if, say, the grade of membership of John in the class of *old men* is 0.8, then his grade of membership in the class of *very old men* is 0.64. (See Fig. 9 for illustration.)

Viewed as an operator, *very* can be composed with itself. Thus

$$very \; very \; old \; men \; = \; (old \; men)^4.$$
(140)

Using the definitions of *not* and *and*, we can compute the meaning of such composite terms as w = *not very old,* z = *not young and not very old,* etc., in which the term *men* is implicit. Thus

$$not \; very \; old \; = \; \neg \, old^2$$
(141)

and
$$not \; young \; and \; not \; very \; old \; = \; \neg \, young \; \cap \; \neg \, old^2.$$
(142)

For example, suppose that the grade of membership of David in the class of *old men* is 0.6 and in the class of *young men* is 0.1. Then his grade of membership in the class of *men* who are *not young and not very old* can be computed from Eq. (142) to be

$$\mu_z \; (David) \; = \; (1 - 0.1) \wedge (1 - 0.6^2)$$
$$= \; 0.64.$$
(143)

Comment. It should be noted that, basically, *very* modifies the adverb or the adjective which follows it. Consequently, it is not grammatical to write

$$x \; = \; very \; not \; exact$$
(144)

but if *not exact* is replaced by the single term *inexact,* then

$$x \; = \; very \; inexact$$
(145)

becomes meaningful and we can assert that

$$very \; inexact \; = \; inexact^2$$
$$= \; (not \; exact)^2.$$
(146)

On the other hand

$$not\ very\ exact\ =\ \neg(very\ exact)$$
$$=\ \neg(exact^2)$$

(147)

which is different from Eq. (146).

Comment. Under the definition of *very* expressed by Eq. (133), if the grade of membership of y (y ∈ U) is a fuzzy set labeled x is unity, then the same is true of *very* x, that is,

$$\mu_x(y)\ =\ 1 \quad \text{implies} \quad \mu_{very\ x}(y)\ =\ 1 .$$

(148)

For example, if the grade of membership of John in the class of *old men* is 1, then the same is true of the grade of membership of John in the class of *very old men.* Is this in accord with our intuition?

This basic question does not appear to have a clear-cut answer on purely intuitive grounds. It is easy to show, however, that Eq. (148) can be deduced as a consequence of the following two assumptions:

(a) *very* distributes over the union (e.g., *very* (*tall or fat*) = *very tall or very fat*)
(b) *very* x = x if x is non-fuzzy. (E.g., *very square = square*).

Specifically, suppose that there is a non-empty set of points in U whose grade of membership in x is unity. Let this non-fuzzy set be denoted by x_{nf}. Then x can be represented as the union of two disjoint sets: the non-fuzzy set x_{nf} and a possibly fuzzy set x_f whose support comprises those points in U whose grade of membership in x is less than unity. Thus

$$x\ =\ x_{nf}\ +\ x_f .$$

(149)

Now, by assumption (a)

$$very\ x\ =\ very\,(x_{nf}\ +\ x_f)$$
$$=\ very\ x_{nf}\ +\ very\ x_f$$

(150)

and by assumption (b)

$$very\ x_{nf}\ =\ x_{nf} .$$

(151)

Consequently

$$very\ x\ =\ x_{nf}\ +\ very\ x_f .$$

(152)

Now, if the grade of membership of y_1 in x is unity, then by the definition of x_{nf}, y_1 belongs to x_{nf}. But Eq. (152) implies that the grade of membership of y_1 in *very* x is greater than or equal to the grade of membership of y_1 in x_{nf}. This establishes that the grade of membership of y_1 in *very* x is unity and hence demonstrates that Eq. (148) is a consequence of (a) and (b).

As a part of a system of hedges which could be used to characterize the behavior of complex systems, it is convenient to have a hedge whose effect is milder than that of *very*. To this end, it is helpful to introduce two artificial hedges *plus* and *minus* which are defined below.

Plus and Minus

The artificial hedges *plus* and *minus* are, respectively, instances of what might be called *accentuators* and *deaccentuators* whose function is to increase the range of shades of meaning of various hedges by providing milder degrees of concentration and dilation than those associated with the operations CON and DIL (see Eqs. (67) and (78)). Thus, as operators acting on a fuzzy set labeled x, *plus* and *minus* are defined as follows

$$plus \ x \ = \ x^{1.25} \tag{153}$$

and
$$minus \ x \ = \ x^{0.75}. \tag{154}$$

The numerical values of the exponents in Eqs. (153) and (154) are chosen in such a way as to entail the approximate identity[19]

$$plus \ plus \ x \ = \ minus \ very \ x. \tag{155}$$

To illustrate the effects of *plus* and *minus*, plots of the membership functions of *old man*, *plus old man, minus very old man* and *very old man* are shown in Fig. 10.

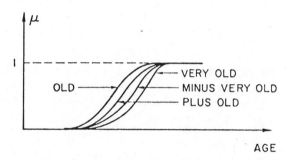

FIG. 10. The effect of the hedges *plus, minus* and *very.*

The main use for the artificial hedges *plus* and *minus* is likely to lie in applications—e.g., the description of fuzzy algorithms—in which the standards of precision of meaning are higher than in ordinary discourse. In addition, *plus* and *minus* can be used to define a natural hedge whose meaning differs slightly from that of some other natural hedge. For example, the hedge *highly* could be defined as

$$highly \ = \ plus \ very \tag{156}$$

or possibly as

$$highly \ = \ minus \ very \ very. \tag{157}$$

Thus, using Eq. (156) we would have the equality

$$highly \ intelligent \ = \ plus \ very \ intelligent \tag{158}$$

whereas Eq. (157) would entail

$$highly \ intelligent \ = \ minus \ very \ very \ intelligent. \tag{159}$$

Much

The hedge *much* is used mostly in conjunction with relations, e.g., *much greater, much more beautiful, much less likely,* etc.

---

[19] In order that Eq. (155) holds precisely, *plus* x and *minus* x should be defined as $plus \ x \triangleq x^{1+\alpha}$ and $minus \ x \triangleq x^{1-\alpha}$, where $\alpha = \sqrt{5} - 2$.

Consider the effect of *much* on the relation *greater than* defined in the universe of positive real numbers, $U = (0, \infty)$.

The meaning of the relation *greater than* or, more simply, $>$, is a non-fuzzy set in $U \times U$ which may be represented as

$$> = \int_D 1/(v, w) \tag{160}$$

where $v, w \in U$ and D is the set of all points in $U \times U$ at which $v > w$.

The hedge *much* acts as an s-fuzzifier (see Eq. (98)) in the sense that it transforms the non-fuzzy set $>$ into the fuzzy set $\gg$ ($\gg \triangleq$ *much greater than*). For simplicity, suppose that $\gg$ is defined as

$$\gg = \int_D (1 + (v - w)^{-2})^{-1}/(v, w) . \tag{161}$$

Then $\gg$ can be expressed as the result of acting with an s-fuzzifier on $>$, i.e.,

$$\gg = \neg SF(\neg >; K) \tag{162}$$

in which the kernel K is defined by

$$K(v, w) = \int_{D^*} (1 + (r - v - s + w)^2)^{-1}/(r, s) \tag{163}$$

where $r, s \in U$, and $D^* = \{(r,s)|r - v \geqslant s - w\}$.

The hedge *much* can be composed with itself by using the rule of composition of fuzzy relations (see [8]). Thus, if we denote *much much greater than* by $\ggg$, then $\ggg$ may be expressed as

$$\ggg = \int_D \bigvee_z ((1 + (v - z)^{-2})^{-1} \wedge (1 + (z - w)^{-2})^{-1})/(v, w) \tag{164}$$

where $\overset{\vee}{z}$ denotes the supremum over z of the parenthesized expression.

Observation. It should be noted that the term *greater than* is sometimes used in a somewhat fuzzy sense rather than in the non-fuzzy sense of Eq. (160). For example, *greater than* may be defined as a fuzzy relation expressed by

$$greater\ than = \int_D \left(0.8 + 0.2\left(1 + \left(\frac{v}{2w}\right)^{-2}\right)^{-1}\right)/(v, w) . \tag{165}$$

In such cases, *much greater than* may be interpreted as CON(*greater than*) or, alternatively, as the composition of *greater than* with itself.

More or Less

In such common uses of the hedge *more or less* as *more or less intelligent, more or less rectangular, more or less sweet, more or less* plays the role of a fuzzifier—usually an s-fuzzifier. When the operand of *more or less* is fuzzy, as in *more or less intelligent*, it may be possible to achieve the same effect by deaccentuation or dilation.

As an illustration, suppose that the term *recent* is defined by

$$recent = 1/1972 + 0.8/1971 + 0.7/1970 . \qquad (166)$$

Then, we may define the composite term *more* or *less recent* by the expression

$$more\ or\ less\ recent = SF\ (recent; K) \qquad (167)$$

in which the kernel set K is defined by

$$K(1972) = 1/1972 + 0.9/1971$$
$$K(1971) = 1/1971 + 0.9/1970 \qquad (168)$$
$$K(1970) = 1/1970 + 0.8/1969 .$$

Thus, on substituting Eq. (168) in Eq. (167), we obtain

$$more\ or\ less\ recent = 1/1972 + 0.9/1971 + 0.72/1970 + 0.56/1969 . \qquad (169)$$

Note that Eq. (169) cannot be obtained from Eq. (166) by dilation because the singleton 1/1969 is absent in Eq. (166).

As another example, suppose that the term *tall* (applying to *man*) is defined in terms of the S-function (see Eq. (86)) as follows

$$tall = \int_{62}^{\infty} S\left(\frac{h - 62}{12}\right)/h \qquad (170)$$

where h denotes the height in inches. (Note that, as defined by Eq. (170), the membership function of *tall*, $\mu_{tall}(h)$, is zero for $h < 62$; is equal to 0.5 at $h = 68$; and is equal to 1 for $h > 74$.).

Now, if we define *more or less tall* by

$$more\ or\ less\ tall = \neg SF\ (\neg tall; K) \qquad (171)$$

where the kernel set K(h) is the interval $[h - 2, h + 2]$, then Eq. (171) yields

$$more\ or\ less\ tall = \int_{60}^{\infty} S\left(\frac{h - 60}{12}\right)/h \qquad (172$$

which is equivalent to a translation (see Fig. 11) of the membership function of *tall*, i.e.,

$$\mu_{more\ or\ less\ tall}(h) = \mu_{tall}(h + 2), \quad h > 0 . \qquad (173)$$

Slightly

Basically, *slightly* is a hedge of Type IP, that is, its effect is dependent on the definition of proximity or ordering in the domain of its operand. There are cases, however, in which the meaning of *slightly* can be defined in terms of hedges of Type I, usually under the tacit assumption that the domain of the operand is a linearly ordered set.

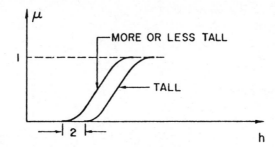

FIG. 11. The effect of the hedge *more or less.*

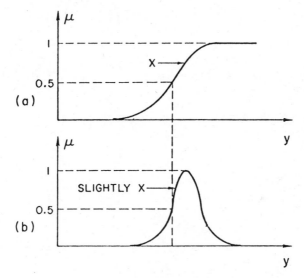

FIG. 12. The effect of the hedge *slightly.*

In such cases, various shades of the meaning of *slightly* might be defined by the following expressions

$$s\textit{lightly } x \;=\; \text{NORM} (x \text{ and not very } x) \tag{174}$$

$$s\textit{lightly } x \;=\; \text{INT} (\text{NORM} (\textit{plus } x \text{ and not very } x)) \tag{175}$$

$$s\textit{lightly } x \;=\; \text{INT} (\text{NORM} (\textit{plus } x \text{ and not plus very } x)) \tag{176}$$

in which the operation of normalization (see Eq. (64)) is needed because the fuzzy sets within the parentheses are, in general, subnormal, and the function of INT (see Eq. (82)) is to intensify the contrast between those elements that have a high grade of membership in its operand and those whose grade of membership is low. Essentially, Eqs. (174), (175), and (176) represent various degrees of approximation to an operator which transforms a fuzzy set x such as shown in Fig. 12a into a fuzzy set of the form shown in Fig. 12b.

As an illustration of the use of Eq. (174), suppose that x (e.g., x = *tall men*) is defined by

$$x = \int_0^\infty \left(1 + \left(\frac{y}{a}\right)^{-2}\right)^{-1} /y \tag{177}$$

where a is the crossover point of x, i.e., the value of y at which $\mu_x(y) = 0.5$. Then

$$very\ x = \int_0^\infty \left(1 + \left(\frac{y}{a}\right)^{-2}\right)^{-2} /y \tag{178}$$

$$not\ very\ x = \int_0^\infty \left(1 - \left(1 + \left(\frac{y}{a}\right)^{-2}\right)^{-2}\right)/y \tag{179}$$

and $\quad x\ and\ not\ very\ x = \int_0^\infty \left(1 + \left(\frac{y}{a}\right)^{-2}\right)^{-1} \wedge \left(1 - \left(1 + \left(\frac{y}{a}\right)^{-2}\right)^{-2}\right) /y . \tag{180}$

The value of y at which the membership function of x *and not very* x attains its maximum value is easily computed to be

$$y_{max} = a\sqrt{\frac{2}{\sqrt{5} - 1}} \eqsim 1.3 \tag{181}$$

and the value of the maximum is

$$\mu_{max} = \frac{2}{\sqrt{5} + 1} \eqsim 0.6. \tag{182}$$

Thus

$$NORM\,(x\ and\ not\ very\ x) \eqsim \int_0^{1.3a} 1.6\left(1 + \left(\frac{y}{a}\right)^{-2}\right)^{-1} /y$$

$$+ \int_{1.3a}^\infty 1.6\left(1 - \left(1 + \left(\frac{y}{a}\right)^{-2}\right)^{-2}\right)/y . \tag{183}$$

The membership function described by the right-hand side of Eq. (183) does not provide a good approximation to the desired shape of *slightly* x as depicted in Fig. 12b. More important, none of the expressions for *slightly* x given above is suitable for operation on terms having non-fuzzy meaning since in the case of such terms *plus* x = *very* x = x and hence *slightly* x yields an empty set.

To illustrate this point, suppose that x = *positive number,* with the requirement that the membership function of *slightly positive number* should have the form shown in Fig. 13. Then, using the definition of the pulse-function $\pi(u)$ (see Eq. (87)), we can approximate to *slightly* x by

$$slightly\ x = SF\,(x\,;K) \tag{184}$$

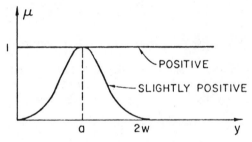

FIG. 13. The effect of *slightly* achieved by fuzzification.

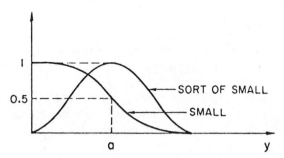

FIG. 14. The effect of the hedge *sort of*.

where

$$K(y) = \int_0^\infty \pi\left(\frac{u - a}{w}\right)/u \quad \text{for} \ \ y = a, \quad u \in U \tag{185}$$

$$= 0 \quad \text{for} \ \ y \neq a \ .$$

Note that the values of the parameters a and w are determined by the context in which the term *slightly positive number* is used.[20] In general, the context-dependence of the meaning of *slightly* x makes it difficult to construct a definition for the hedge *slightly* which has a broad validity.

Sort Of

Sort of is a member of a family of hedges which have the effect of reducing the grade of membership of those objects which are in the "center" of a class x and increasing those which are on its periphery. Figure 14 shows in graphical terms the relation between x and *sort of* x when x is a fuzzy set of the form

$$x = \int_0^\infty \left(1 + \left(\frac{y}{a}\right)^2\right)^{-1}/y \tag{186}$$

in which the parameter a is the crossover value of y.

---

[20] The parameter a defines the location of the peak of the pulse while w defines its width, i.e., the separation between the crossover points. Note that for $y \neq a$, the grade of membership of y in K(y) is not unity (see the definition of the kernel set).

For the fuzzy set defined by Eq. (186), *sort of* x can be approximated to by the following expression

$$\textit{sort of } x = \text{NORM}[\neg\text{CON}^2(x) \cap \text{DIL}(x)] \qquad (187$$

in which the term $\neg\,\text{CON}^2(x)$ serves to reduce the grade membership of those points which are close to zero, while $\text{DIL}(x)$ (see Eq. (78)) increases the grade of membership of points which are remote from zero. Thus, if x is interpreted as *small,* then Eq. (187) might read

$$\textit{sort of small} = \text{NORM}\,(\textit{more or less small but not very very small}) \qquad (188$$

where DIL is interpreted as *more or less* and *but* plays the role of *and.*

6. Representation of Hedges of Type II

As was stated in Section 4, the distinguishing feature of hedges of Type II is that their characterization as operators involves a description of the manner in which they affect the components of the operand.

For this reason, the characterization of hedges of Type II is a considerably more complex problem than that of hedges of Type I. Thus, in general, the definition of a hedge of Type II has to be formulated as a fuzzy algorithm involving hedges of Type I. Definitions of this kind will be discussed in a subsequent paper.

In what follows, we shall merely touch upon a relatively simple case in which the effect of a hedge can be described—to a first approximation—as a modification in the weighting coefficients of a convex combination.

Specifically, consider the hedge *essentially* and assume that its operand is a term x whose components are denoted by $x_1, \ldots, x_n$. For example, x = *decent,* with the components of x assumed to be $x_1 = \textit{kind},\ x_2 = \textit{honest},\ x_3 = \textit{polite}$ and $x_4 = \textit{attractive}$. We assume further, that the fuzzy set x is a convex combination of its components (see Eq. (88)), that is,

$$\mu = w_1\mu_1 + w_2\mu_2 + w_3\mu_3 + w_4\mu_4 \qquad (189)$$

where the w_i, i = 1, ..., n are non-negative weights whose sum is unity, μ is the grade of membership of an individual y in x, and μ_i, i = 1, ..., 4 is the grade of membership of y in x_i. For concreteness, suppose that $w_1 = 0.4,\ w_2 = 0.3,\ w_3 = 0.2$ and $w_4 = 0.1$. Then, if $\mu_1(\text{John}) = 0.9, \mu_2(\text{John}) = 0.8, \mu_3(\text{John}) = 0.9$ and $\mu_4(\text{John}) = 0.2$, we have

$$\begin{aligned} \mu\,(\text{John}) &= 0.4 \times 0.9 + 0.3 \times 0.8 + 0.2 \times 0.9 + 0.1 \times 0.2 \\ &= 0.8. \end{aligned} \qquad (190)$$

Now the magnitude of w_i is a measure of the importance of the attribute labeled x_i. Intuitively, the hedge *essentially* has the effect of increasing the weights of the important attributes and diminishing those that are relatively unimportant. To achieve this effect, let us first normalize the weights w_i so that if w_1 is the largest weight, then after normalization we have

$$w_1' = 1$$

$$w_2' = \frac{w_2}{w_1} = 0.75$$

$$w_3' = \frac{w_3}{w_1} = 0.5$$ (191)

and $$w_4' = \frac{w_4}{w_1} = 0.25$$

On squaring the normalized weights, forming their sum and dividing each normalized weight by the sum, we obtain (approximately)

$$w_1^* = \frac{1}{1.87} = 0.53$$

$$w_2^* = \frac{0.56}{1.87} = 0.3$$

$$w_3^* = \frac{0.25}{1.87} = 0.13$$ (192)

$$w_4^* = \frac{0.06}{1.87} = 0.04 .$$

In terms of these weights, the effect of *essentially* on *decent* can be approximately characterized by

$$\mu^* = 0.53\,\mu_1 + 0.3\,\mu_2 + 0.13\,\mu_3 + 0.04\,\mu_4$$ (193)

where μ^* denotes the membership function of *essentially decent*. Thus, in the case of John, we have

$$\mu^*(\text{John}) = 0.53 \times 0.9 + 0.3 \times 0.8 + 0.13 \times 0.9 + 0.04 \times 0.2$$
$$= 0.91$$ (194)

which is higher than $\mu(\text{John})$ since the low grade of membership of John in the fuzzy set *attractive* is weighed less heavily in *essentially decent* than it is in *decent*.

In short, in the very approximate characterization of the hedge *essentially* given above, the effect of *essentially* is described as a change in the coefficients of the convex combination, with the "important" components of x increased in magnitude in relation to the less important ones. The same general approach can be used to characterize the effect of hedges such as *technically, regular,* etc., provided the membership function of x can be expressed as a convex combination of its components. When this is not the case, the characterization of the effect of the hedge becomes a more complicated problem which, in general, requires a fuzzy-algorithmic approach to its solution.

Concluding Remarks

In the foregoing discussion, we have concerned ourselves with relatively simple hedges whose effect can be described in terms of combinations or compositions of the elementary operations of complementation, concentration, intersection, dilation, accentuation, contrast intensification and convex combination.

We have not considered hedges which are context-dependent to a degree where it is not feasible to represent them, even very approximately, in terms of a set of elementary operations such as those defined in this paper. Such hedges require the use of a fuzzy-algorithmic mode of characterization which is more qualitative than the approach described here. Examples of such hedges and their fuzzy-algorithmic definitions will be presented in subsequent papers.

References

[1] G. Lakoff, "Linguistics and natural logic," in *Semantics of Natural Languages,* D. Davidson and G. Harman (Eds.). Dondrecht, the Netherlands: D. Reidel, 1971.

[2] G. Lakoff, "Hedges: A study in meaning criteria and the logic of fuzzy concepts," *Proc. 8th Regional Meeting of Chicago Linguistic Society,* University of Chicago Linguistics Dept., April 1972.

[3] L. A. Zadeh, "Fuzzy sets," *Information and Control,* vol. 8, pp. 338–353, 1965.

[4] L. A. Zadeh, "Fuzzy algorithms," *Ibid.,* vol. 12, pp. 94–102, 1968.

[5] E. T. Lee and L. A. Zadeh, "Note on fuzzy languages," *Information Sciences,* vol. 1, pp. 421–434, 1969.

[6] L. A. Zadeh, "Quantitative fuzzy semantics," *Ibid.,* vol. 3, pp. 159–176, 1971.

[7] L. A. Zadeh, "Fuzzy languages and their relation to human and machine intelligence," *Proc. of Conference on Man and Computer,* Bordeaux, France, 1970. Memorandum M-302, 1971, Electronics Research Laboratory, University of California, Berkeley, Calif. 94720.

[8] L. A. Zadeh, "Similarity relations and fuzzy orderings," *Information Sciences,* vol. 3, pp. 177–200, 1971.

[9] S. Tamura, S. Higuchi, and K. Tanaka, "Pattern classification based on fuzzy relations," *Trans. on Systems, Man and Cybernetics,* vol. 1, pp. 61–66, 1971.

[10] J. A. Goguen, "The logic of inexact concepts," *Syntheses,* vol. 19, pp. 325–373, 1969.

[11] L. Wittgenstein, "Logical form," *Proc. of the Aristotelian Society,* vol. 9, pp. 162–171, 1929.

[12] M. Black, "Reasoning with loose concepts," *Dialogue,* vol. 2, pp. 1–12, 1963.

[13] M. Scriven, "The logic of criteria," *Jour. of Philosophy,* vol. 56, pp. 857–868, 1959.

[14] H. Khatchadourian, "Vagueness, meaning and absurdity," *American Philosophical Quarterly,* vol. 2, pp. 119–129, 1965.

[15] R. R. Verma, "Vagueness and the principle of excluded middle," *Mind,* vol. 79, pp. 67–77, 1970.

Research sponsored by the Army Research Office–Durham, Grant DA-ARO-D-31-124-71-G174, and the Joint Services Electronics Program, Contract F44620-71-C-0087. This work was inspired and influenced by many discussions with Professor G. Lakoff concerning the meaning of hedges and their interpretation in terms of fuzzy sets. Lakoff's analysis of hedges and related problems may be found in [1] and [2].

Received June 30, 1972

PRUF—a Meaning Representation Language for Natural Languages†

L.A. ZADEH

*Computer Science Division, Department of Electrical Engineering
and Computer Sciences and the Electronics Research Laboratory, University of California, Berkeley, CA 94720, U.S.A.*

(*Received 18 October 1977*)

PRUF—an acronym for Possibilistic Relational Universal Fuzzy—is a meaning representation language for natural languages which departs from the conventional approaches to the theory of meaning in several important respects.

First, a basic assumption underlying PRUF is that the imprecision that is intrinsic in natural languages is, for the most part, possibilistic rather than probabilistic in nature. Thus, a proposition such as "Richard is tall" translates in PRUF into a possibility distribution of the variable Height (Richard), which associates with each value of the variable a number in the interval [0,1] representing the possibility that Height (Richard) could assume the value in question. More generally, a proposition, p, translates into a procedure, P, which returns a possibility distribution, Π^p, with P and Π^p representing, respectively, the *meaning* of p and the *information* conveyed by p. In this sense, the concept of a possibility distribution replaces that of truth as a foundation for the representation of meaning in natural languages.

Second, the logic underlying PRUF is not a two-valued or multivalued logic, but a fuzzy logic, FL, in which the truth-values are linguistic, that is, are of the form *true, not true, very true, more or less true, not very true*, etc., with each such truth-value representing a fuzzy subset of the unit interval. The truth-value of a proposition is defined as its compatibility with a reference proposition, so that given two propositions p and r, one can compute the truth of p relative to r.

Third, the quantifiers in PRUF—like the truth-values—are allowed to be linguistic, i.e. may be expressed as *most, many, few, some, not very many, almost all*, etc. Based on the concept of the cardinality of a fuzzy set, such quantifiers are given a concrete interpretation which makes it possible to translate into PRUF propositions exemplified by "Many tall men are much taller than most men," "All tall women are blonde is not very true," etc.

The translation rules in PRUF are of four basic types: Type I—pertaining to modification; Type II—pertaining to composition; Type III—pertaining to quantification; and Type IV—pertaining to qualification and, in particular, to truth qualification, probability qualification and possibility qualification.

The concepts of semantic equivalence and semantic entailment in PRUF provide a basis for question-answering and inference from fuzzy premises. In addition to serving as a foundation for approximate reasoning, PRUF may be employed as a language for the representation of imprecise knowledge and as a means of precisiation of fuzzy propositions expressed in a natural language.

†To Professor I. M. Gel'fand, who had suggested—a decade ago—the application of the theory of fuzzy sets to natural languages.

1. Introduction

In a decade or so from now—when the performance of natural language understanding and question-answering systems will certainly be much more impressive than it is today—it may well be hard to comprehend why linguists, philosophers, logicians and cognitive scientists have been so reluctant to come to grips with the reality of the pervasive imprecision of natural languages and have persisted so long in trying to fit their theories of syntax, semantics and knowledge representation into the rigid conceptual mold of two-valued logic.†

A fact that puts this issue into a sharper perspective is that almost any sentence drawn at random from a text in a natural language is likely to contain one or more words that have a fuzzy‡ denotation—that is, are labels of classes in which the transition from membership to non-membership is gradual rather than abrupt. This is true, for example, of the italicized words in the simple propositions "John is *tall*," "May has *dark* hair," and "May is *much younger* than John," as well as in the somewhat more complex and yet commonplace propositions exemplified by: "*Most* Frenchmen are not *blond*," "It is *very true* that *many* Swedes are *tall*," "It is *quite possible* than *many wealthy* Americans have *high* blood pressure," and "It is *probably quite true* that *most* X's are *much larger* than *most* Y's."

The numerous meaning representation, knowledge representation and query representation languages which have been described in the literature§—prominent among which are semantic networks, predicate calculi, relation algebra, Montague grammar, conceptual dependency graphs, logical networks, AIMDS, ALPHA, CONVERSE, DEDUCE, DILOS, HAM-RPM, HANSA, ILL, KRL, KRS, LIFER, LSP, LUNAR, MAPL, MEANINGEX, MERLIN, OWL, PHILQAI, PLANES, QUEL, REL, REQUEST, SAM, SCHOLAR, SEQUEL, SQUARE and TORUS—are not oriented toward the representation of fuzzy propositions, that is, propositions containing labels of fuzzy sets and hence have no facilities for semantic—as opposed to syntactic—inference from fuzzy premises.‖ However, facilities for the representation and execution of fuzzy instructions are available in the programming languages FUZZY (LeFaivre, 1974), FLOU and FSTDS (Noguchi, Umano, Mizumoto & Tanaka, 1976, 1977) and in the system modelling language of Fellinger (1974).

To clarify this remark, it should be noted that, although a fuzzy proposition such as "Herb is tall," may be—and frequently is—represented in predicate notation as Tall (Herb), such a representation presupposes that Tall is a predicate which partitions a

†An incisive discussion of this and related issues may be found in Gaines (1976b).

‡Although the terms *fuzzy* and *vague* are frequently used interchangeably in the literature, there is, in fact, a significant difference between them. Specifically, a proposition, p, is *fuzzy* if it contains words which are labels of fuzzy sets; and p is *vague* if it is both fuzzy and insufficiently specific for a particular purpose. For example, "Bob will be back in a few minutes" is fuzzy, while "Bob will be back sometime" is vague if it is insufficiently informative as a basis for a decision. Thus, the vagueness of a proposition is a decision-dependent characteristic whereas its fuzziness is not.

§A list of representative papers and books dealing with the subject of meaning representation languages and related issues is presented in the appended bibliography.

‖Semantic inference differs from syntactic inference in that it involves the meaning of premises while syntactic inference involves only their surface structure.

collection of individuals, U, into two disjoint classes: those for which Tall(Herb) is true and those for which Tall(Herb) is false. One could, of course, interpret Tall as a predicate in a multivalued logic—in which case the extension of Tall would be a fuzzy subset of U—but even such more general representations cannot cope with quantified or qualified propositions of the form "Most tall men are fat," "It is very true that X is much larger than Y," "It is quite possible that if X is large then it is very likely that Y is small," etc.

In earlier papers (Zadeh, 1973, 1975a, b, c, 1976a, b; Bellman & Zadeh, 1976), we have argued that traditional logical systems are intrinsically unsuited for the manipulation of fuzzy knowledge—which is the type of knowledge that underlies natural languages as well as most of human reasoning—and have proposed a fuzzy logic, FL, as a model for approximate reasoning. In this logic, the truth-values are *linguistic*, i.e. of the form *true, not true, very true, not very true, more or less true, not very true and not very false* etc., with each truth-value representing a fuzzy subset of the unit interval. In effect, the fuzziness of the truth-values of FL provides a mechanism for the association of imprecise truth-values with imprecise propositions expressed in a natural language, and thereby endows FL with a capability for modeling the type of qualitative reasoning which humans employ in uncertain and/or fuzzy environments.

More recently, the introduction of the concept of a possibility distribution (Zadeh, 1977a, b) has clarified the role of the concept of a fuzzy restriction† in approximate reasoning, and has provided a basis for the development of a meaning representation language named PRUF (an acronym for *Possibilistic‡ Relational Universal Fuzzy*) in which—in a significant departure from tradition—it is the concept of a possibility distribution, as opposed to truth, that plays the primary role.

The conceptual structure of PRUF is based on the premise that, in sharp contrast to formal and programming languages, natural languages are intrinsically incapable of precise characterization on either the syntactic or semantic level. In the first place, the pressure for brevity of discourse tends to make natural languages *maximally ambiguous* in the sense that the level of ambiguity in human communication is usually near the limit of what is disambiguable through the use of an external body of knowledge which is shared by the parties in discourse.

Second, a significant fraction of sentences in a natural language cannot be characterized as strictly grammatical or ungrammatical. As is well known, the problem of partial grammaticality is accentuated in the case of sentences which are partially nonsensical in the real world but not necessarily in an imaginary world. Thus, a realistic grammar for a natural language should associate with each sentence its degree of grammaticality— rather than merely generate the sentences which are completely grammatical. The issue of partial grammaticality has the effect of greatly complicating the problem of automatic translation from a natural language into a meaning representation language—

---

†A *fuzzy restriction* is a fuzzy set which serves as an elastic constraint on the values that may be assigned to a variable. A variable which is associated with a fuzzy restriction or, equivalently, with a possibility distribution, is a *fuzzy variable*.

‡The term "possibilistic" was coined by Gaines & Kohout (1975). The concept of a possibility distribution is distinct from that of possibility in modal logic and related areas (Hughes & Cresswell, 1968; N. Rescher, 1975).

which is an important aspect of Montague-type grammars (Montague, 1974; Partee, 1976*b*).

Third, as was alluded to already, a word in a natural language is usually a summary of a complex, multifaceted concept which is incapable of precise characterization. For this reason, the denotation of a word is generally a fuzzy—rather than non-fuzzy—subset of a universe of discourse. For example, if U is a collection of individuals, the denotation of the term *young man* in U is a fuzzy subset of U which is characterized by a membership function $\mu_{young\ man} : U \rightarrow [0,1]$, which associates with each individual *u* in U the degree—on the scale from 0 to 1—to which *u* is a young man. When necessary or expedient, this degree or, equivalently, the grade of membership, $\mu_{young}(u)$, may be expressed in linguistic terms such as *high, not high, very high, not very high, low, more or less low*, etc., with each such term representing a fuzzy subset of the unit interval. In this case, the denotation of *young man* is a fuzzy set of Type 2, i.e. a fuzzy set with a fuzzy membership function.†

In essence, PRUF bears the same relation to FL that predicate calculus does to two-valued logic. Thus, it serves to translate a set of premises expressed in a natural language into expressions in PRUF to which the rules of inference in FL (or PRUF) may be applied, yielding other expressions in PRUF which upon retranslation become the conclusions inferred from the original premises. More generally, PRUF may be used as a basis for question-answering systems in which the knowledge-base contains imprecise data, i.e. propositions expressed in a natural or synthetic language which translate into a collection of possibility and/or probability distributions of a set of variables.

Typically, a simple proposition such as "John is young," translates in PRUF into what will be referred to as a *possibility assignment equation* of the form

$$\Pi_{Age(John)} = YOUNG \tag{1.1}$$

in which YOUNG—the denotation of young—is a fuzzy subset of the interval [0,100], and $\Pi_{Age\ (John)}$ is the possibility distribution of the variable Age(John). What (1.1) implies is that, if on the scale from 0 to 1, the degree to which a numerical age, say 30, is compatible with YOUNG is 0·7, then the possibility that John's age is 30 is also equal to 0·7. Equivalently, (1.1) may be expressed as

$$JOHN[\Pi_{Age} = YOUNG] \tag{1.2}$$

in which JOHN is the name of a relation which characterizes John and Age is an attribute of John which is particularized by the assignment of the fuzzy set YOUNG to its possibility distribution.

In general, an expression in PRUF may be viewed as a procedure which acts on a set of possibly fuzzy relations in a database and computes the possibility distribution of a set of variables. Thus, if *p* is a proposition in a natural language which translates into an expression P in PRUF, and Π^p is the possibility distribution returned by P, then P

---

†Expositions of the relevant aspects of the theory of fuzzy sets may be found in the books and papers noted in the bibliography, especially Kaufmann (1975), Negoita & Ralescu (1975), and Zadeh, Fu, Tanaka & Shimura (1975).

may be interpreted as the *meaning of p* while Π^p is the *information conveyed* by p.‡ The significance of these notions will be discussed in greater detail in section 3.

The main constituents of PRUF are (a) a collection of translation rules, and (b) a set of rules of inference.§ For the present, at least, the translation rules in PRUF are human-use oriented in that they do not provide a system for an automatic translation from a natural language into PRUF. However, by subordinating the objective of automatic translation to that of achieving a greater power of expressiveness, PRUF provides a system for the translation of a far larger subset of a natural language than is possible with the systems based on two-valued logic. Eventually, it may be possible to achieve the goal of machine translation into PRUF of a fairly wide variety of expressions in a natural language. It is not likely, however, that this goal could be attained through the employment of algorithms of the conventional type in translation programs. Rather, it is probable that recourse would have to be made to the use of fuzzy logic for the representation of imprecise contextual knowledge as well as for the characterization and execution of fuzzy instructions in translation algorithms.

At present, PRUF is still in its initial stages of development and hence our exposition of it in the present paper is informal in nature, with no pretense at definiteness or completeness. Thus, our limited aim in what follows is to explain the principal ideas underlying PRUF; to describe a set of basic translation rules which can serve as a point of departure for the development of other, more specialized, rules; and to illustrate the use of translation rules by relatively simple examples. We shall not consider the translation of imperative propositions nor the issues relating to the implementation of interactive connectives, reserving these and other important topics for subsequent papers.

In the following sections, our exposition of PRUF begins with an outline of some of the basic properties of the concept of a possibility distribution and its role in the representation of the meaning of fuzzy propositions. In section 3, we consider a number of basic concepts underlying PRUF, among them those of possibility assignment equation, fuzzy set descriptor, proposition, question, database, meaning, information, semantic equivalence, semantic entailment and definition.

Section 4 is devoted to the formalization of translation rules of Type I (modification), Type II (composition) and Type III (quantification). In addition, a translation rule for relations is derived as a corollary of rules of Type II, and a rule for forming the negation of a fuzzy proposition is formulated.

The concept of truth is defined in section 5 as a measure of the compatibility of two fuzzy propositions, one of which acts as a reference proposition for the other. Based on this conception of truth, a translation rule for truth-qualified propositions is developed in section 6. In addition, translation rules for probability-qualified and possibility-qualified propositions are established, and the concept of semantic equivalence is employed to derive several meaning-perserving transformations of fuzzy propositions. Finally, in section 7 a number of examples illustrating the application of various translation rules—both singly and in combination—are presented.

‡In effect, P and Π^p are the counterparts of the concepts of *intension* and *extension* in language theories based on two-valued logic (Cresswell, 1973; Linsky, 1971; Miller & Johnson-Laird, 1976).

§The rules of inference in PRUF and their application to approximate reasoning are described in a companion paper (Zadeh, 1977*b*).

2. The concept of a possibility distribution and its role in PRUF

A basic assumption underlying PRUF is that the imprecision that is intrinsic in natural languages is, in the main, possibilistic rather than probabilistic in nature.

As will be seen presently, the rationale for this assumption rests on the fact that most of the words in a natural language have fuzzy rather than non-fuzzy denotation. A conspicuous exception to this assertion are the terms in mathematics. Even in mathematics, however, there are concepts that are fuzzy, e.g. the concept of a sparse matrix, stiff differential equation, approximate equality, etc. More significantly, almost all mathematical concepts become fuzzy as soon as one leaves the idealized universe of mathematical constructs and comes in contact with the reality of pervasive ill-definedness, irreducible uncertainty and finiteness of computational resources.

To understand the relation between fuzziness and possibility,† it is convenient to consider initially a simple non-fuzzy proposition such as‡

$$p \triangleq X \text{ is an integer in the interval } [0,5].$$

Clearly, what this proposition asserts is that (a) it is possible for any integer in the interval [0,5] to be a value of X, and (b) it is not possible for any integer outside of this interval to be a value of X.

For our purposes, it is expedient to reword this assertion in a form that admits of extension to fuzzy propositions. More specifically, in the absence of any information regarding X other than that conveyed by p, we shall assert that: p induces a *possibility distribution* Π_x which associates with each integer u in [0,5] the possibility that u could be a value of X. Thus,

$$\text{Poss}\{X = u\} = 1 \text{ for } 0 \leq u \leq 5$$

and

$$\text{Poss}\{X = u\} = 0 \text{ for } u < 0 \text{ or } u > 5$$

where $\text{Poss}\{X = u\}$ is an abbreviation for "The possibility that X may assume the value u." For the proposition in question the possibility distribution Π_X is *uniform* in the sense that the possibility-values are equal to unity for u in [0,5] and zero elsewhere.

Next, let us consider a proposition q which may be viewed as a fuzzified version of p, namely,

$$q \triangleq X \text{ is a small integer}$$

where "small integer" is the label of a fuzzy set defined by, say,§

$$\text{SMALL INTEGER} = 1/0 + 1/1 + 0.8/2 + 0.6/3 + 0.4/4 + 0.2/5 \tag{2.1}$$

---

†A more detailed account of this and other issues related to the concept of a possibility distribution may be found in Zadeh (1977a).

‡The symbol $\triangleq$ stands for "is defined to be" or "denotes."

§To differentiate between a label and its denotation, we express the latter in upper case symbols. To simplify the notation, this convention will not be adhered to strictly where the distinction can be inferred from the context.

in which + denotes the union rather than the arithmetic sum and a fuzzy singleton of the form $0.6/3$ signifies that the grade of membership of the integer 3 in the fuzzy set SMALL INTEGER—or, equivalently, the *compatibility* of 3 with SMALL INTEGER— is 0.6.

At this juncture, we can make use of the simple idea behind our interpretation of p to formulate what might be called the *possibility postulate*—a postulate which may be used as a basis for a possibilistic intepretation of fuzzy propositions. In application to q, it may be stated as follows.

Possibility postulate. In the absence of any information regarding X other than that conveyed by the proposition $q \triangleq$ X is a small integer, q induces a possibility distribution Π_X which equates the possibility of X taking a value u to the grade of membership of u in the fuzzy set SMALL INTEGER. Thus

$$\text{Poss}\{X=0\}=\text{Poss}\{X=1\}=1$$
$$\text{Poss}\{X=2\}=0.8$$
$$\text{Poss}\{X=3\}=0.6$$
$$\text{Poss}\{X=4\}=0.4$$
$$\text{Poss}\{X=5\}=0.2$$

and

$$\text{Poss}\{X=u\}=0 \text{ for } u<0 \text{ or } u>5.$$

More generally, the postulate asserts that if X is a variable which takes values in U and F is a fuzzy subset of U, then the proposition

$$q \triangleq \text{X is F} \tag{2.2}$$

induces a possibility distribution Π_X *which is equal to* F, i.e.

$$\Pi_X = F \tag{2.3}$$

implying that

$$\text{Poss}\{X=u\}=\mu_F(u), \quad u \in U \tag{2.4}$$

where $\mu_F: U \to [0,1]$ is the membership function of F, and $\mu_F(u)$ is the grade of membership of u in F.

In essence, then, the possibility distribution of X is a fuzzy set which serves to define the possibility that X could assume any specified value u in U. The function $\pi_X: U \to [0,1]$ which is equal to μ_F and which associates with each $u \in U$ the possibility that X could take u as its value is called the *possibility distribution function* associated with X. In this connection, it is important to note that the possibility distribution defined by (2.3) depends on the definition of F and hence is purely subjective in nature.

We shall refer to (2.3) as the *possibility assignment equation* because it signifies that the proposition "X is F" translates into the assignment of a fuzzy set F to the possibility distribution of X. More generally, the possibility assignment equation corresponding to a proposition of the form "N is F," where F is a fuzzy subset of a universe of discourse U, and N is the name of (a) a variable, (b) a fuzzy set, (c) a proposition, or (d) an object, may be expressed as

$$\Pi_{X(N)} = F \tag{2.5}$$

or, more simply,

$$\Pi_X = F \tag{2.6}$$

where X is either N itself (when N is a variable) or a variable that is explicit or implicit in N, with X taking values in U. For example, in the case of the proposition "Nora is young," $N \triangleq$ Nora, X = Age(Nora), U = [0,100] and

$$\text{Nora is young} \rightarrow \Pi_{\text{Age(Nora)}} = \text{YOUNG} \tag{2.7}$$

where the symbol $\rightarrow$ stands for "translates into."

Since the concept of a possibility distribution is closely related to that of a fuzzy set,§ possibility distributions may be manipulated by the rules applying to such sets. In what follows, we shall discuss briefly some of the basic rules of this kind, focusing our attention only on those aspects of possibility distributions which are of direct relevance to PRUF.

POSSIBILITY *VS.* PROBABILITY

Intuitively, possibility relates to our perception of the degree of feasibility or ease of attainment, whereas probability is associated with the degree of likelihood, belief, frequency or proportion. All possibilities are subjective, as are most probabilities.† In general, probabilistic information is not as readily available as possibilistic information and is more difficult to manipulate.

Mathematically, the distinction between probability and possibility manifests itself in the different rules which govern their combinations, especially under the union. Thus, if A is a non-fuzzy subset of U, and Π_X is the possibility distribution induced by the proposition "N is F," then the *possibility measure*,‡ $\Pi(A)$, of A is defined as the supremum of μ_F over A, i.e.

$$\Pi(A) \triangleq \text{Poss}\{X \in A\} = \text{Sup}_{u \in A}\mu_F(u) \tag{2.8}$$

and, more generally, if A is a fuzzy subset of U,

$$\Pi(A) = \text{Poss}\{X \text{ is } A\} = \text{Sup}_u(\mu_A(u) \wedge \mu_F(u)) \tag{2.9}$$

where μ_A is the membership function of A and $\wedge \triangleq$ min.

§Strictly speaking, the concept of a possibility distribution is coextensive with that of a fuzzy restriction rather than a fuzzy set (Zadeh, 1973, 1975*b*).

†There are eminent authorities in probability theory (DeFinetti, 1974) who maintain that all probabilities are subjective.

‡The possibility measure defined by (2.8) is a special case of the more general concept of a fuzzy measure defined by Sugeno (1974) and Terano & Sugeno (1975).

From the definition of $\Pi(A)$, it follows at once that the possibility measure of the union of two arbitrary subsets of U is given by

$$\Pi(A \cup B) = \Pi(A)\Pi \vee (B) \qquad (2.10)$$

where $\vee \triangleq \max$. Thus, possibility measure does not have the basic additivity property of probability measure, namely,

$$P(A \cup B) = P(A) + P(B) \text{ if A and B are disjoint}$$

where $P(A)$ and $P(B)$ are the probability measures of A and B, respectively, and $+$ is the arithmetic sum.

An essential aspect of the concept of possibility is that it does not involve the notion of repeated or replicated experimentation and hence is nonstatistical in nature. Indeed, the importance of the concept of possibility stems from the fact that much—perhaps most—of human decision-making is based on information that is possibilistic rather than probabilistic in nature.§

POSSIBILITY DISTRIBUTIONS *VS.* FUZZY SETS

Although there is a close connection between the concept of a possibility distribution and that of a fuzzy set, there is also a significant difference between the two that must be clearly understood.

To illustrate the point by a simple example which involves a non-fuzzy set and a uniform possibility distribution, consider a variable labeled Sister(Dedre) to which we assign a set, as in

$$\text{Sister(Dedre)} = \text{Sue} + \text{Jane} + \text{Lorraine} \qquad (2.11)$$

or a possibility distribution, as in

$$\Pi_{\text{Sister(Dedre)}} = \text{Sue} + \text{Jane} + \text{Lorraine} \qquad (2.12)$$

where $+$ denotes the union. Now, the meaning of (2.11) is that Sue, Jane and Lorraine are sisters of Dedre. By contrast, the meaning of (2.12) is that the sister of Dedre is Sue *or* Jane *or* Lorraine, where *or* is the exclusive or. In effect, (2.12) signifies that there is uncertainty in our knowledge of who is the sister of Dedre, with the possibility that it is Sue being unity, and likewise for Jane and Lorraine. In the case of (2.11), on the other hand, we are certain that Sue, Jane and Lorraine are all sisters of Dedre. Thus, the set {Sue, Jane, Lorraine} plays the role of a possibility distribution in (2.12) but not in (2.11).

---

§In many realistic decision processes it is impracticable or impossible to obtain objective probabilistic information in the quantitative form that is needed for the application of statistical decision theory. Thus, the probabilities that are actually used in much of human decision-making are (a) subjective, and (b) linguistic (in the sense defined in Zadeh, 1975). Characterization of linguistic probabilities is related to the issue of probability qualification, which is discussed in section 6. A more detailed discussion of linguistic probabilities may be found in Nguyen (1976*a*).

Usually, it is clear from the context whether or not a fuzzy (or non-fuzzy) set should be interpreted as a possibility distribution. A difficulty arises, however, when a relation contains a possibility distribution, as is exemplified by the relation RESIDENT whose tableau is shown in Table 2.1.

TABLE 2.1

Resident	Subject	Location
	Jack	New Rochelle
	Jack	White Plains
	Ralph	New Rochelle
	Ralph	Tarrytown

In this case, the rows above the dotted line represent a relation in the sense that Jack resides both in New Rochelle and in White Plains. On the other hand, the rows below the dotted line represent a possibility distribution associated with the location of residence of Ralph, meaning that Ralph resides either in New Rochelle *or* in Tarrytown, but not both. It should be noted parenthetically that there is no provision for dealing with this kind of ambiguity in the conventional representations of relational models of data because the concept of a possibility distribution and the related issue of data uncertainty have not been an object of concern in the analysis of database management systems.

REPRESENTATION BY STANDARD FUNCTIONS

In the manipulation of possibility distributions, it is convenient to be able to express the membership function of a fuzzy subset of the real line as a standard function whose parameters may be adjusted to fit a given membership function in an approximate fashion. A standard function of this type is the S-function, which is a piecewise quadratic function defined by the equations:

$$
\begin{aligned}
S(u;\alpha,\beta,\gamma) &= 0 && \text{for } u \le \alpha \\
&= 2\left(\frac{u-\alpha}{\gamma-\alpha}\right)^2 && \alpha \le u \le \beta \\
&= 1-2\left(\frac{u-\gamma}{\gamma-\alpha}\right)^2 && \text{for } \beta \le u \le \gamma \\
&= 1 && \text{for } u \ge \gamma
\end{aligned}
\tag{2.13}
$$

in which the parameter $\beta \triangleq (\alpha+\gamma)/2$ is the *crossover* point, i.e. the value of u at which $S(u;\alpha,\beta,\gamma)=0\cdot 5$. Other types of standard functions which are advantageous when the arithmetic operations of addition, multiplication and division have to be performed on fuzzy numbers, are (i) piecewise linear (triangular) functions, and (ii) exponential (bell-shaped) functions. A discussion of these functions and their applications may be found in Nahmias (1976) and Mizumoto & Tanaka (1976).

There are two special types of possibility distributions which will be encountered in later sections. One is the *unity* possibility distribution, which is denoted by I and is defined by

$$\pi_I(u) = 1 \text{ for } u \in U \tag{2.14}$$

where π_I is the possibility distribution function of I. The other, which is defined on the unit interval, is the *unitary* possibility distribution (or the *unitary fuzzy set* or the *unitor*, for short), which is denoted by $\perp$ and is defined by

$$\pi_\perp(v) = v \text{ for } v \in [0,1]. \tag{2.15}$$

In the particular case where a truth-value in FL is the unitary fuzzy set, it will be referred to as the *unitary* truth-value. On denoting this truth-value by *u*-true, we have

$$\mu_{u\text{-true}}(v) \triangleq v, \ v \in [0,1]. \tag{2.16}$$

PROJECTION AND MARGINAL POSSIBILITY DISTRIBUTIONS

The possibility distributions with which we shall be concerned in the following sections are, in general, *n*-ary distributions denoted by $\Pi_{(X_1, \ldots, X_n)}$, where $X_1, \ldots, X_n$ are variables—or, equivalently, *attributes*—taking values in their respective universes of discourse $U_1, \ldots, U_n.$† As a simple example in which $n=2$, consider the proposition "John is a big man," in which BIG MAN is a fuzzy relation F defined by Table 2.2, with the variables Height and Weight expressed in centimeters and kilograms, respectively.

The relation in question may also be expressed as a linear form

$$\text{BIG MAN} = 0.5/(165,60) + 0.6/(170,60) + \ldots + 1/(180,80) + \ldots \tag{2.17}$$

TABLE 2.2

BIG MAN	Height	Weight	μ
	165	60	0·5
	170	60	0·6
	175	60	0·7
	170	65	0·75
	—	—	—
	180	70	0·9
	175	75	0·9
	180	75	0·95
	180	80	1
	185	75	1

†When it is necessary to place in evidence that X takes values in U (i.e. the domain of X is U), we shall express the domain of X as U(X) or, where no confusion can arise, as X (see (2.23)).

in which a term such as 0·6/(170,60) signifies that the grade of membership of the pair (170,60) in the relation BIG MAN—or, equivalently, its compatibility with the relation BIG MAN—is 0·6.

The possibility postulate implies that the proposition "John is a big man" induces a binary possibility distribution $\Pi_{(\text{Weight(John), Height(John)})}$ whose tableau is identical with Table 2.2 except that the label of the last column is changed from μ to π in order to signify that the compatibility-values in that column assume the role of possibility-values. What this means is that, by inducing the possibility distribution $\Pi_{(\text{Height(John),Weight(John)})}$, the proposition "John is a big man" implies that the possibility that John's height and weight are, say, 170 cm and 60 kg, respectively, is 0·6.

It should be noted that, in general, the entries in a relation F need not be numbers, as they are in Table 2.2. Thus, the entries may be pointers to—or identifiers of—physical or abstract objects. For example, the u's in the relation CUP shown in Table 2.3:

TABLE 2.3

CUP	Identifier	μ
	u_1	0·8
	u_2	0·9
	u_3	1·0
	u_4	0·2

may be pictures of cups of various forms. In this case, given the relation CUP, the proposition "X is a cup" induces a possibility distribution Π_X such that $\text{Poss}\{X=u_1\}=$ 0·8 and likewise for other rows in the table.

In the translation of expressions in a natural language into PRUF, there are two operations on possibility distributions (or fuzzy relations) that play a particularly important role: *projection* and *particularization*.

Specifically, let $X \triangleq (X_1, \ldots, X_n)$ be a fuzzy variable which is associated with a possibility distribution $\Pi_{(X_1, \ldots, X_n)}$ or, more simply, Π_X, with the understanding that Π_X is an n-ary fuzzy relation in the Cartesian product, $U = U_1 \times \ldots \times U_n$, of the universes of discourse associated with $X_1, \ldots, X_n$. We assume that Π_X is characterized by its possibility distribution function—or, equivalently, membership function—$\pi_{(X_1, \ldots, X_n)}$ (or π_X, for short).

A variable of the form

$$X_{(s)} \triangleq (X_{i_1}, \ldots, X_{i_k}), \qquad (2.18)$$

where $s \triangleq (i_1, \ldots, i_k)$ is a subsequence of the index sequence $(1, \ldots, n)$, constitutes a *subvariable* of $X \triangleq (X_1, \ldots, X_n)$. By analogy with the concept of a marginal probability distribution, the *marginal possibility distribution* associated with $X_{(s)}$ is defined by

$$\Pi_{X_{(s)}} = \text{Proj}_{U_{(s)}} \Pi_{(X_1, \ldots, X_n)}, \qquad (2.19)$$

where $U_{(s)} \triangleq U_{i_1} \times \cdots \times U_{i_k}$, and the operation of projection is defined—in terms of possibility distribution functions—by

$$\pi_{X_{(s)}}(u_{(s)}) = \text{Sup}_{u_{(s')}} \, \pi_X(u_1, \ldots, u_n), \qquad (2.20)$$

where $u_{(s)} \triangleq (u_{i_1}, \ldots, u_{i_k})$ and $u_{(s')} \triangleq (u_{j_1}, \ldots, u_{j_l})$, with s' denoting the index sequence complementary to s (e.g. if $n=5$ and $s=(2,3)$, then $(s')=(1,4,5)$). For example, for $n=2$ and $s=(2)$, (2.20) yields

$$\pi_{X_2}(u_2) = \text{Sup}_{u_1} \, \pi_{(X_1, X_2)}(u_1, u_2) \qquad (2.21)$$

as the expression for the marginal possibility distribution function of X_2.

The operation of projection is very easy to perform when Π_X is expressed as a linear form. As an illustration, assume that $U_1 = U_2 = a + b$, or, more conventionally, $\{a,b\}$, and

$$\Pi_{(X_1, X_2)} = 0 \cdot 8aa + 0 \cdot 6ab + 0 \cdot 4ba + 0 \cdot 2bb \qquad (2.22)$$

in which a term of the form $0 \cdot 6ab$ signifies that

$$\text{Poss}\{X_1 = a, \, X_2 = b\} = 0 \cdot 6.$$

To obtain the projection of Π_X on, say, U_2 it is sufficient to replace the value of X_1 in each term in (2.22) by the null string Λ. Thus†

$$\begin{aligned}
\text{Proj}_{U_2} \, \Pi_{(X_1, X_2)} &= 0 \cdot 8a + 0 \cdot 6b + 0 \cdot 4a + 0 \cdot 2b \\
&= 0 \cdot 8a + 0 \cdot 6b.
\end{aligned}$$

To simplify the notation, it is convenient—as is done in SQUARE (Boyce *et al.*, 1974)— to omit the word Proj in (2.19) and interpret $U_{(s)}$ as $X_{i_1} \times \cdots \times X_{i_k}$ (see footnote on p. 404). Thus,

$$\text{Proj}_{U_{(s)}} \Pi_{(X_1, \ldots, X_n)} \triangleq_{U_{(s)}} \Pi_{(X_1, \ldots, X_n)} \triangleq_{X_{i_1} \times \cdots \times X_{i_k}} \Pi_{(X_1, \ldots, X_n)}. \qquad (2.23)$$

In some applications, it is convenient to have at one's disposal not only the operation of projection, as defined by (2.20), but also its *dual, conjunctive projection*,‡ which is defined by (2.20) with Sup replaced by Inf. It is easy to verify that the latter can be expressed in terms of the former as

$$\overline{\text{Proj}_{U_{(s)}}} \Pi_{(X_1, \ldots, X_n)} = (\text{Proj}_{U_{(s)}} \Pi'_{(X_1, \ldots, X_n)})' \qquad (2.24)$$

---

†If r and s are two tuples and α and β are their respective possibilities, then $\alpha r + \beta r = (\alpha \vee \beta) r$. Additional details may be found in Zadeh (1977a).

‡A more detailed discussion of conjunctive projections may be found in Zadeh (1966). It should be noted that the concept of a conjunctive projection is related to that of a conjunctive mapping in SQUARE (Boyce *et al.*, 1974) and to universal quantification in multivalued logic (Rescher, 1969).

in which $\overline{\text{Proj}}$ stands for conjunctive projection and $'$ denotes the complement, where the complement of a fuzzy set F in U is a fuzzy set F$'$ defined by

$$\mu_{F'}(u) = 1 - \mu_F(u), \quad u \in U. \tag{2.25}$$

PARTICULARIZATION

Informally, by the *particularization* of a fuzzy relation or a possibility distribution which is associated with a variable $X \triangleq (X_1, \ldots, X_n)$, is meant the effect of specification of the possibility distributions of one or more subvariables of X. In the theory of non-fuzzy relations, the resulting relation is commonly referred to as a *restriction* of the original relation; and, in the particular case where the values of some of the constituent variables are specified, the degenerate restriction becomes a *section* of the original relation.

Particularization in PRUF may be viewed as the result of forming the conjunction of a proposition of the form "X is F," where X is an *n*-ary variable, $X \triangleq (X_1, \ldots, X_n)$, with particularizing propositions of the form "$X_{(s)}$ is G," where $X_{(s)}$ is a subvariable of X, and F and G are fuzzy subsets of $U \triangleq U_1 \times \ldots \times U_n$ and $U_{(s)} = U_{i_1} \times \ldots \times U_{i_k}$, respectively.

More specifically, let $\Pi_X \triangleq \Pi_{(X_1, \ldots, X_n)} = F$ and $\Pi_{X_{(s)}} \triangleq \Pi_{(X_{i_1}, \ldots, X_{i_k})} = G$ be the possibility distributions induced by the propositions "X is F" and "$X_{(s)}$ is G," respectively. By definition, the *particularization of* Π_X *by* $X_{(s)} = G$ (or, equivalently, of F by G) is denoted by $\Pi_X[\Pi_{X_{(s)}} = G]$ (or $F[\Pi_{X_{(s)}} = G]$) and is defined as the intersection$\ddagger$ of F and G, i.e.

$$\Pi_X[\Pi_{X_{(s)}} = G] = F \cap \overline{G}, \tag{2.26}$$

where $\overline{G}$ is the cylindrical extension of G, i.e. the cylindrical fuzzy set in $U_1 \times \ldots \times U_n$ whose projection on $U_{(s)}$ is G and whose membership function is expressed by

$$\mu_{\overline{G}}(u_1, \ldots, u_n) \triangleq \mu_G(u_{i_1}, \ldots, u_{i_k}), \quad (u_1, \ldots, u_n) \in U_1 \times \ldots \times U_n. \tag{2.27}$$

As a simple illustration, assume that $U_1 = U_2 = U_3 = a + b$,

$$\Pi_{(X_1, X_2, X_3)} = 0 \cdot 8aab + 0 \cdot 6baa + 0 \cdot 1bab + 1bbb \tag{2.28}$$

and

$$\Pi_{(X_1, X_2)} = G = 0 \cdot 5aa + 0 \cdot 2ba + 0 \cdot 3bb.$$

In this case

$$\overline{G} = 0 \cdot 5aaa + 0 \cdot 5aab + 0 \cdot 2baa + 0 \cdot 2bab + 0 \cdot 3bba + 0 \cdot 3bbb$$
$$F \cap \overline{G} = 0 \cdot 5aab + 0 \cdot 2baa + 0 \cdot 1bab + 0 \cdot 3bbb$$

$\ddagger$If A and B are fuzzy subsets of U, their *intersection* is defined by $\mu_{A \cap B}(u) = \mu_A(u) \wedge \mu_B(u)$, $u \in U$. Thus, $\mu_{F \cap \overline{G}}(u_1, \ldots, u_n) = \mu_F(u_1, \ldots, u_n) \wedge \mu_G(u_{i_1}, \ldots, u_{i_k})$. Dually, the *union* of A and B is denoted as A+B (or $A \cup B$) and is defined by $\mu_{A + B}(u) \triangleq \mu_A(u) \vee \mu_B(u)$. ($\vee \triangleq$ max and $\wedge \triangleq$ min.)

and hence

$$\Pi_{(x_1, x_2, x_3)}[\Pi_{(x_1, x_2)} = G] = 0 \cdot 5aab + 0 \cdot 2baa + 0 \cdot 1bab + 0 \cdot 3bbb.$$

As will be seen in section 4, the right-hand member of (2.26) represents the possibility distribution induced by the conjunction of "X is F" and "$X_{(s)}$ is G," that is, the proposition "X is F and $X_{(s)}$ is G". It is for this reason that the particularized possibility distribution $\Pi_X[\Pi_{X(s)} = G]$ may be viewed as the possibility distribution induced by the proposition "X is F and $X_{(s)}$ is G".

In cases in which more than one subvariable is particularized, e.g. the particularizing propositions are "$X_{(s)}$ is G," and "$X_{(r)}$ is H," the particularized possibility distribution will be expressed as

$$\Pi_X[\Pi_{X(s)} = G; \Pi_{X(r)} = H]. \tag{2.29}$$

Furthermore, particularization may be *nested*, as in

$$\Pi_X[\Pi_{X(s)} = G[\Pi_{Y(t)} = J]] \tag{2.30}$$

where the particularizing relation G is, in turn, particularized by the proposition "$Y_{(t)}$ is J," where $Y_{(t)}$ is a subvariable of the variable associated with G.

It is of interest to observe that, as its name implies, particularization involves an imposition of a restriction on the values that may be assumed by a variable. However, by dualizing the definition of particularization as expressed by (2.26), that is, by replacing the intersection with the union the opposite effect is achieved, with the resulting possibility distribution corresponding to the disjunction of "X is F" and "$X_{(s)}$ is G". We shall not make an explicit use of the dual of particularization in the present paper.

As a simple illustration of particularization, consider the proposition $p \triangleq$ John is big, where BIG is defined by Table 2.2, and assume that the particularizing proposition is $q \triangleq$ John is tall, in which TALL is defined by Table 2.4.

The assertion "John is big" may be expressed equivalently as "Size(John) is big," which is of the form "X is F," with $X \triangleq$ Size(John) and $F \triangleq$ BIG. Similarly, "John is tall" may be expressed as "Height(John) is tall," or, equivalently, Y is G, where $Y \triangleq$ Height (John) and $G \triangleq$ TALL.

TABLE 2.4

TALL	Height	μ
	165	0·6
	170	0·7
	175	0·8
	180	0·9
	185	1

TABLE 2.5

BIG(Π_{Height}=TALL)	Height	Weight	μ
	165	60	0·5
	170	60	0·6
	175	60	0·7
	170	65	0·7
	—	—	—
	180	70	0·9
	175	75	0·8
	180	75	0·9
	180	80	0·9
	185	75	1

Using (2.26), the tableau of the particularized relation BIG[Π_{Height}=TALL] is readily found to be given by Table 2.5.

The value of μ for a typical row in this table, say for (Height=180, Weight=75), is obtained by computing the minimum of the values of μ for the corresponding rows in BIG and TALL (i.e. (180,75) in BIG and (180) in TALL). As is pointed out in section 4, this mode of combination of μ's corresponds to *non-interactive* conjunction, which is assumed to be a standard default definition of conjunction in PRUF. However, PRUF allows any definition of conjunction which is specified by the user to be employed in place of the standard definition.

As an additional example, consider the particularized possibility distribution (see (2.23))

$$\text{PROFESSOR[Name}=\text{Simon; Sex}=\text{Male;}$$
$$\Pi_{\text{Age}}=_{\mu \times \text{Age2}}\text{APPROXIMATELY[Age1}=45]] \qquad (2.31)$$

which describes a subset of a set of professors whose name is Simon, who are male and who are approximately 45 years old. In this case, the possibility distribution of the variable Age is a particularized relation APPROXIMATELY in which the first variable, Age1, is set equal to 45, and which is projected on the Cartesian product of U(μ) and U(Age2), yielding the fuzzy set of values of Age which are approximately equal to 45.

It should be noted that some of the attributes in (2.31) (e.g. Name) are assigned single values, while others—whose values are uncertain—are associated with possibility distributions. As will be seen in the following sections, this is typical of the particularized possibility distributions arising in the translation of expressions in a natural language into PRUF.

Expressions of the form (2.31) are similar in appearance to the commonly employed semantic network, query language and predicate calculus representations of propositions in a natural language. An essential difference, however, lies in the use of possibility distributions in (2.31) for the characterization of the values of fuzzy variables and in the concrete specification of the manner in which possibility distributions and fuzzy relations are modified by particularization and other operations which will be described in sections 4 and 6.

3. Basic concepts underlying translation into PRUF

The concept of a possibility distribution provides a natural point of departure for the formalization of many other concepts which underlie the translation of expressions in a natural language into PRUF. We shall present a brief exposition of several such concepts in this section, without aiming at the construction of an embracing formal framework.

In speaking somewhat vaguely of expressions in a natural language, what we have in mind is a variety of syntactic, semantic and pragmatic forms exemplified by sentences, propositions, phrases, clauses, questions, commands, exclamations, etc. In what follows, we shall restrict our attention to expressions which are (a) fuzzy propositions (or assertions); (b) fuzzy questions; and (c) what will be referred to as *fuzzy set descriptors* or simply *descriptors*.

PROPOSITIONS

Basically, a fuzzy proposition may be regarded as an expression which translates into a possibility assignment equation in PRUF. This is analogous to characterizing a non-fuzzy proposition as an expression which translates into a well-formed formula (or, equivalently, a closed sentence) in predicate calculus.

The types of fuzzy propositions to which our analysis will apply are exemplified by the following. (Italics place in evidence the words that have fuzzy denotation.)

Ronald is *more or less young*	(3.1)
Miriam was *very rich*	(3.2)
Harry *loves* Ann	(3.3)
X is *much smaller* than Y	(3.4)
X and Y are *approximately equal*	(3.5)
If X is *large* then Y is *small*	(3.6)
Most Swedes are *blond*	(3.7)
Many men are *much taller* than *most men*	(3.8)
Most Swedes are *tall* is *not very true*	(3.9)
The man in the *dark* suit is *walking slowly toward* the door	(3.10)
Susanna gave *several expensive* presents to each of her *close friends*	(3.11)
If X is *much greater* than Y then (Z is *small* is *very probable*)	(3.12)
If X is *much greater* than Y then (Z is *small* is *quite possible*)	(3.13)

In these examples, propositions (3.9), (3.12) and (3.13) are, respectively, truth qualified, probability qualified and possibility qualified; propositions (3.7), (3.8), (3.9) and (3.11) contain fuzzy quantifiers; and proposition (3.10) contains a fuzzy relative clause.

FUZZY SET DESCRIPTORS

Informally, a fuzzy set descriptor or simply a descriptor is an expression which is a label of a fuzzy set or a characterization of a fuzzy set in terms of other fuzzy sets. Simple examples of fuzzy set descriptors in English are

Very tall man	(3.14)
Tall man wearing a brown hat	(3.15)

The dishes on the table	(3.16)
Small integer	(3.17)
Numbers which are much larger than 10	(3.18)
Most	(3.19)
All	(3.20)
Several	(3.21)
Many tall women	(3.22)
Above the table	(3.23)
Much taller than	(3.24)

A descriptor differs from a proposition in that it translates, in general, into a fuzzy relation rather than a possibility distribution or a possibility assignment equation. In this connection, it should be noted that a non-fuzzy descriptor (i.e. a description of a non-fuzzy set) would, in general, translate into an open sentence (i.e. a formula with free variables) in predicate calculus. However, while the distinction between open and closed sentences is sharply drawn in predicate calculus, the distinction between fuzzy propositions and fuzzy set descriptors is somewhat blurred in PRUF.

QUESTIONS

For the purposes of translation into PRUF, a question will be assumed to be expressed in the form B is ?A, where B is the body of the question —e.g., How tall is Vera—and A indicates the form of an admissible answer, which might be (a) a possibility distribution or, as a special case, an element of a universe of discourse; (b) a truth-value; (c) a probability-value; and (d) a possibility-value. To differentiate between these cases, A will be expressed as Π in (a) and, more particularly, as α when a numerical value of an attribute is desired; as τ in (b); as λ in (c); and as ω in (d).

To simplify the treatment of questions, we shall employ the artifice of translating into PRUF not the question itself but rather the answer to it, which, in general, would have the form of a fuzzy proposition. As an illustration,

How tall is Tom?$\Pi\to$Tom is ?Π	(3.25)
How tall is Tom ?$\alpha\to$Tom is ?α tall	(3.26)
Where does Tom live$\to$Tom lives in ?α	(3.27)
Is it true that Fran is blonde$\to$Fran is blonde is ?τ	(3.28)
Is it likely that X is small$\to$X is small is ?λ	(3.29)
Is it possible that (Jan is tall is false)$\to$(Jan is tall is false) is ?ω	(3.30)

In this way, the translation of questions stated in a natural language may be carried out by the application of translation rules for fuzzy propositions, thus making it unnecessary to have separate rules for questions.

POSSIBILITY ASSIGNMENT EQUATIONS

The concept of a possibility assignment equation and its role in the translation of propositions in a natural language into PRUF have been discussed briefly in section 2.

In what follows, we shall focus our attention on several additional aspects of this concept which relate to the translation rules which will be formulated in sections 4 and 6.

As was stated earlier, a proposition of the form $p \triangleq N$ is F in which N is the name of (a) a variable, (b) a fuzzy set, (c) a proposition, or (d) an object, and F is a fuzzy subset of a universe of discourse U, translates, in general, into a possibility assignment equation of the form

$$\Pi_{X(N)} = F \tag{3.31}$$

or, more simply,

$$\Pi_X = F \tag{3.32}$$

where X is a variable taking values in U, with X being either N itself (when N is a variable) or a variable that is explicit or implicit in N.

To place in evidence that (3.32) is a translation of "N is F," we write

$$p \triangleq N \text{ is } F \rightarrow \Pi_X = F \tag{3.33}$$

and, conversely,

$$p \triangleq N \text{ is } F \leftarrow \Pi_X = F \tag{3.34}$$

with the left-hand member of (3.34) referred to as a *retranslation* of its right-hand member.

In general, the variable X is an *n*-ary variable which may be expressed as $X \triangleq (X_1, \ldots, X_n)$, with $X_1, \ldots, X_n$ varying over $U_1, \ldots, U_n$, respectively. In some instances, the identification of the X_i and F is quite straightforward; in others, it may be a highly non-trivial task requiring a great deal of contextual knowledge.† For this reason, the identification of the X_i is difficult to formulate as a mechanical process. However, as is usually the case in translation processes, the problem can be greatly simplified by a decomposition of *p* into simpler constituent expressions, translating each expression separately, and then combining the results. The translation rules formulated in sections 4 and 6 are intended to serve this purpose.

In general, a constituent variable, X_i, has a nested structure of the form

$$X_i = \text{Attribute name(Part name(Part name} \ldots \text{(N)))} \tag{3.35}$$

which is similar to the structure of selectors in the Vienna Definition Language (Lucas *et al.*, 1968; Wegner, 1972). As a simple illustration,

$$\text{Myrna is blonde} \rightarrow \Pi_{\text{Color(Hair(Myrna))}} = \text{BLONDE} \tag{3.36}$$

where Color(Hair(Myrna)) is a nested variable of the form (3.35) and BLONDE is the fuzzy denotation of blonde in the universe of discourse which is associated with the proposition in question.

---

†In one form or another, this problem arises in all meaning representation languages. However, it is a much more difficult problem in machine–oriented languages than in PRUF, because in PRUF the task of identifying the X_i is assumed to be performed by a human.

A problem that arises in some cases relates to the lack of an appropriate attribute name. For example, to express the translation of "Manuel is kind," in the form (3.33), we need a designation in English for the attribute which takes "kind" as a value. When such a name is not available in a language, it will be denoted by the symbol A, with a subscript if necessary, to indicate that "kind" is a value of A. However, what is really needed in cases like this is a possibly algorithmic definition of the concept represented by A which decomposes it into simpler concepts for which appropriate names are available.

In the foregoing examples, N represents the name of an object, e.g. the name of a person. More generally, N may be a descriptor, which is usually expressed as a relative clause, as in

$$\text{The man standing near the door is tall.} \tag{3.37}$$

N may also be a proposition, as in

$$\text{Lucia is tall is false.} \tag{3.38}$$

In (3.37), $N \triangleq$ The man standing in the door, while in (3.38), $N \triangleq$ Lucia is tall and X(N) is the truth-value of the proposition "Lucia is tall".

An important point concerning propositions of the form "N is F" which can be clarified at this juncture, is that "N is F" should be regarded not as a restricted class of propositions, but as a canonical form for all propositions which admit of translation into a possibility assignment equation†. Thus, if p is any proposition such that

$$p \rightarrow \Pi_X = F \tag{3.39}$$

then upon retranslation it may be expressed as "X is F," which is of the form "N is F".

As an illustration, the proposition "Paul was rich," may be translated as

$$\text{Paul was rich} \rightarrow \Pi_{(\text{Wealth}(\text{Paul}), \text{Time})} = \text{RICH} \times \text{PAST} \tag{3.40}$$

where (Wealth(Paul),Time) is a binary variable whose first component is the wealth of Paul (expressed as net worth) and the second component is the time at which net worth is assessed; RICH is a fuzzy subset of U(Wealth); PAST is a fuzzy subset of the time-interval extending from the present into the past; and RICH×PAST is the Cartesian product‡ of RICH and PAST.

Similarly, the proposition "X and Y are approximately equal," where X and Y are real numbers, may be translated as

---

†This is equivalent to saying that "N is F" is a canonical form for all propositions which can be expressed in the form "N is F" through the application of a meaning-preserving transformation. Such transformations will be defined later in this section in connection with the concept of semantic equivalence.

‡If A and B are fuzzy subsets of U and V, respectively, their *Cartesian product* is defined by $\mu_{A \times B}$ $(u,v) \triangleq \mu_A(u) \wedge \mu_B(v)$, $u \in U$, $v \in V$.

X and Y are approximately equal $\rightarrow \Pi_{(X, Y)} =$ APPROXIMATELY EQUAL   (3.41)

where APPROXIMATELY EQUAL is a fuzzy relation in R^2. Upon retranslation, (3.41) yields the equivalent proposition

$$(X, Y) \text{ is approximately equal} \tag{3.42}$$

which, though ungrammatical, is in canonical form.

A related issue which concerns the form of possibility assignment equations is that, in general, such equations may be expressed equivalently in the form of possibility distributions. More specifically, if we have

$$N \text{ is } F \rightarrow \Pi_X = F \tag{3.43}$$

then the possibility assignment equation in (3.43) may be expressed as a possibility distribution (labeled N) of the variable X(N), with the tableau of N having the form:

TABLE 3.1

N	X(N)	π
	u_1	π_1
	u_2	π_2
	.	.
	u_n	π_n

where the π_i are the possibility-values of the u_i.

As a simple illustration, in the translation

$$\text{Brian is tall} \rightarrow \Pi_{\text{Height(Brian)}} = \text{TALL} \tag{3.44}$$

where TALL is a fuzzy set defined by, say,

$$\text{TALL} = 0{\cdot}5/160 + 0{\cdot}6/165 + 0{\cdot}7/170 + 0{\cdot}8/175 + 0{\cdot}9/180 + 1/185 \tag{3.45}$$

the possibility assignment equation may be replaced by the possibility distribution

BRIAN	Height	π
	160	0·5
	165	0·6
	170	0·7
	175	0·8
	180	0·9
	185	1·0

which in turn may be expressed as the particularized possibility distribution

$$\text{BRIAN}[\Pi_{\text{Height}} = \text{TALL}] \tag{3.46}$$

on the understanding that, initially,

$$\Pi_{\text{BRIAN}} = I; \tag{3.47}$$

that is, BRIAN is a unity possibility distribution with

$$\pi_{\text{BRIAN}}(u) = 1 \text{ for } u \in U. \tag{3.48}$$

It is this equivalence between (3.44) and (3.46) that forms the basis for the statement made in section 1 regarding the equivalence of (1.1) and (1.2).

DEFINITION

All natural languages provide a mechanism for defining a concept in terms of other concepts and, more particularly, for designating a complex descriptor by a single label. Consequently, it is essential to have a facility for this purpose in every meaning representation language, including PRUF.†

A somewhat subtle issue that arises in this connection in PRUF relates to the need for normalizing‡ the translation of the definiens into PRUF. As an illustration of this point, suppose that the descriptor *middle-aged* is defined as

$$\text{middle-aged} \triangleq \text{not young and not old.} \tag{3.49}$$

Now, as will be seen in section 5, the translation of the right-hand member of (3.49) is expressed by

$$\text{not young and not old} \rightarrow \text{YOUNG}' \cap \text{OLD}' \tag{3.50}$$

where YOUNG and OLD are the translations of young and old, respectively, and ' denotes the complement. Consequently, for some definitions of YOUNG and OLD the definition of *middle-aged* by (3.49) would result in a subnormal fuzzy set, which would imply that there does not exist any individual who is middle-aged to the degree 1.

While this may be in accord with one's intuition in some cases, it may be counterintuitive in others. Thus, to clarify the intent of the definition, it is necessary to indicate whether or not the definiens is to be normalized.§ For this purpose, the notation

†Concept definition plays a particularly important role in conceptual dependency graphs (Schank, 1973), in which a small number of primitive concepts are used as basic building blocks for more complex concepts.

‡A fuzzy set F is *normal* if and only if $\text{Sup}_u \, \mu_F(u) = 1$. If F is *subnormal*, it may be normalized by dividing μ_F by $\text{Sup}_u \, \mu_F(u)$. Thus, the membership function of normalized F, Norm(F), is given by $\mu_{\text{Norm}(F)}(u) \triangleq \mu_F(u)/\text{Sup}_u \, \mu_F(u)$.

§The need for normalization was suggested by some examples brought to the author's attention by P. Kay (U.C., Berkeley) and W. Kempton (U.T., San Antonio). (See Kay 1975).)

$$\text{definiendum} \triangleq \text{Norm(definiens)} \tag{3.51}$$

e.g.

$$\text{middle-aged} \triangleq \text{Norm(not young and not old)} \tag{3.52}$$

may be employed to indicate that the translation of the definiens ought to be normalized.

EXPRESSIONS IN PRUF

Expressions in PRUF are not rigidly defined, as they are in formal, programming and machine-oriented meaning representation languages. Typically, an expression in PRUF may assume the following forms.

(a) A label of a fuzzy relation or a possibility distribution. Examples: CUP, BIG MAN, APPROXIMATELY EQUAL.

(b) A particularized fuzzy relation or a possibility distribution. Examples:

$$\text{CUP}[\Pi_{\text{Color}} = \text{RED}; \text{Weight} = 35 \text{ gr}] \tag{3.53}$$

$$\text{CAR}[\text{Make} = \text{Ford}; \Pi_{\text{Size(Trunk)}} = \text{BIG};$$
$$\Pi_{\text{Weight}} = {}_{\mu \times \text{Weight2}}\text{APPROXIMATELY}[\text{Weight1} = 1500 \text{ kg}]].$$

(c) A possibility assignment equation. Examples:

$$\Pi_{\text{Height(Valentina)}} = \text{TALL} \tag{3.54}$$

$$\Pi_X = \text{CUP}[\Pi_{\text{Color}} = \text{RED}; \text{Weight} = 35 \text{ gr}].$$

(d) A definition. Examples:

$$F \triangleq H + G[\Pi_{X(s)} = K] \tag{3.55}$$

where + denotes the union, H is a fuzzy relation and $G[\Pi_{X(s)} = K]$ is a particularized fuzzy relation

$$F \triangleq \text{HOUSE}[\Pi_{\text{Color}} = \text{GREY}; \Pi_{\text{Price}} = \text{HIGH}] \tag{3.56}$$

which defines a fuzzy set of houses which are grey in color and high-priced.

(e) A procedure—expressed in a natural, algorithmic or programming language—for computing a fuzzy relation or a possibility distribution. Examples: examples (t), (u) and (v) in section 7.

In general, a fuzzy set descriptor will translate into an expression of the form (a), (b) or (d), while a fuzzy proposition will usually translate into (b), (c) or (d). In all these cases, an expression in PRUF may be viewed as a procedure which—given a set of relations in a database—returns a fuzzy relation, a possibility distribution or a possibility assignment equation.†

---

†It should be noted that an expression in PRUF may also be interpreted as a probability—rather than possibility—manipulating procedure. Because of the need for normalization, operations on probability distributions are, in general, more complex than the corresponding operations on possibility distributions.

DATABASE, MEANING AND INFORMATION

By a *relational database* or, simply, a *database* in the context of PRUF is meant a collection, $\mathscr{D}$, of fuzzy, time-varying relations which may be characterized in various ways, e.g. by tables, predicates, recognition algorithms, generation algorithms, etc. A simple self-explanatory example of a database, $\mathscr{D}$, consisting of fixed (i.e. time-invariant) relations POPULATION, YOUNG and RESEMBLANCE is shown in Table 3.2. What is implicit in this representation is that each of the variables (i.e. attributes) which appear as column headings, is associated with a specified universe of discourse (i.e. a domain). For example, the universe of discourse associated with the variable Name in POPULATION is given by

$$U(\text{Name}) = \text{Codd} + \text{King} + \text{Chen} + \text{Chang}. \tag{3.57}$$

In general, two variables which have the same name but appear in different tables may be associated with different universes of discourse.

The relations YOUNG and RESEMBLANCE in Table 3.2 are *purely extensional†* in the sense that YOUNG and RESEMBLANCE are defined directly

TABLE 3.2

POPULATION	Name		RESEMBLANCE	Name1	Name2	μ
	Codd			Codd	King	0·8
	King			Codd	Chen	0·6
	Chen			Codd	Chang	0·6
	Chang			King	Chen	0·5
				. . .	. . .	. . .
				Chang	Chen	0·8

YOUNG	Name	μ
	Codd	0·7
	King	0·9
	Chen	0·8
	Chang	0·9

as fuzzy subsets of POPULATION and not through a procedure which would allow the computation of YOUNG and RESEMBLANCE for any given POPULATION. To illustrate the point, if POPULATION and YOUNG were defined as shown in Table

†In the theories of language based on two-valued logic (Linsky, 1971; Quine, 1970*a*; Cresswell, 1973) the dividing line between *extensional* and *intensional* is sharply drawn. This is not the case in PRUF—in which there are levels of intensionality (or, equivalently, levels of procedural generality), with pure extensionality constituting one extreme. This issue will be discussed in greater detail in a forthcoming paper.

TABLE 3.3

POPULATION	Name	Age		YOUNG	Age	μ
	Codd	45			30	0·8
	King	31			31	0·75
	Chen	42			32	0·70
	Chang	33			33	0·60
					...	...
					42	0·4
					45	0·3

3.3, then it would be possible to compute the fuzzy subset YOUNG of any given POPULATION by employing the procedure expressed by

$$YOUNG =_{\mu \times Name} POPULATION[\Pi_{Age} = YOUNG] \qquad (3.58)$$

where YOUNG in the right-hand member is a fuzzy subset of U(Age), and μ is implicit in POPULATION.

Since an expression in PRUF is a procedure, it involves, in general, not the relations in the database but only their *frames*.† In addition, an expression in PRUF may involve the names of universes of discourse and/or their Cartesian products; the names of some of the relation elements; and possibly the values of some attributes of the relations in the database (e.g. the number of rows).

As an illustration, the *frame of the database* shown in Table 3.2 (i.e. the collection of frames of its constituent relations) is comprised of:

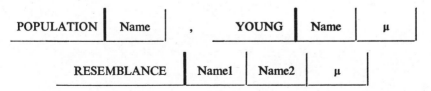

Correspondingly, an expression in PRUF such as

$$_{\mu \times Name1} RESEMBLANCE[Name2 = King] \qquad (3.59)$$

represents a procedure which returns the fuzzy subset of POPULATION comprising names of individuals who resemble King.

Ultimately, each of the symbols or names in a database is assumed to be defined ostensively (Lyons, 1968) or, equivalently, by exemplification; that is, by pointing or otherwise focussing on a real or abstract object and indicating the degree—on the scale from 0 to 1—to which it is compatible with the symbol in question. In this sense, then,

---

†By the *frame* of a relation is meant its name and column headings (i.e. the names of variables or, equivalently, attributes). The rest of the relation (i.e. the table without column headings) is its *body*.

a database may viewed as an interface with an external world which might be real or abstract or a combination of the two.‡

In general, the correspondence between a database and an external world is difficult to formalize because the universe of discourse associated with an external world comprises not just a model of that world, say M, but also the set of fuzzy subsets of M, the set of fuzzy subsets of fuzzy subsets of M, etc. To illustrate this point, it is relatively easy to define by exemplification the denotation of *red*, which is a fuzzy subset of M; much more difficult to define the concept of *color*, which is a subset of $\mathscr{P}(M)$, the set of fuzzy subsets of M; and much much more difficult to define the concept of *attribute*, which is a subset of $\mathscr{P}(\mathscr{P}(M))$ (Zadeh, 1971b).

Viewed in this perspective, the issues related to the correspondence between a database and an external world are similar to those which arise in pattern recognition and are even harder to formulate and resolve within a formal framework. As a direct consequence of this difficulty, a complete formalization of the concept of *meaning* does not appear to be an attainable goal in the foreseeable future.

In the context of PRUF, the concept of meaning is defined in a somewhat restricted way, as follows:

Let *e* be an expression in a natural language and let E be its translation into PRUF, i.e.

$$e \to E \tag{3.60}$$

and, more particularly,

$$p \to P \tag{3.61}$$

if *e* is a proposition; and

$$d \to D \tag{3.62}$$

if *e* is a descriptor. To illustrate:

$$\text{cup} \to \text{CUP} \tag{3.63}$$

$$\text{red cup} \to \text{CUP}[\Pi_{\text{Color}} = \text{RED}] \tag{3.64}$$

$$\text{George is young} \to \Pi_{\text{Age(George)}} = \text{YOUNG} \tag{3.65}$$

or, equivalently,

$$\text{George is young} \to \text{GEORGE}[\Pi_{\text{Age}} = \text{YOUNG}]. \tag{3.66}$$

Stated informally, the procedure, E, may be viewed as the *meaning* of *e* in the sense that, if $e \triangleq d$, then for any given database $\mathscr{D}$ on which D is defined, D computes (or returns) a fuzzy relation F^d which is a fuzzy denotation (or extension) of *d* in its universe of discourse (which may be different from $\mathscr{D}$). Similarly, if $e \triangleq p$, then P is a procedure which, for any given database $\mathscr{D}$ on which P is defined, computes a possibility distribution Π^p. This distribution, then, may be regarded as the *information* conveyed by *p*.†

‡In this sense, the concept of a database is related to that of a possible world in possible world semantics and modal logic (Kripke, 1963; Hughes & Cresswell, 1968; Partee, 1976a).

†It should be noted that a non-probabilistic measure of information was introduced by Kampe de Feriet & Forte (1967, 1977). In the present paper, however, our concern is with the information itself, which is represented by a possibility distribution, rather than with its measure, which is a real number.

In particular, if Π^p is the possibility distribution of a variable X and $X_{(s)}$ is a subvariable of X, then the *information conveyed by p about* $X_{(s)}$ is given by the projection of Π^p on $U_{(s)}$. When it is necessary to indicate that Π^p is the result of acting with P on a particular database $\mathscr{D}$, Π^p will be referred to as the possibility distribution induced by *p* (or the information conveyed by *p*) *in application to* $\mathscr{D}$.

As an illustration, consider the proposition

$$p \triangleq \text{Mike recently lived near Boston} \qquad (3.67)$$

which in PRUF translates into

$$\text{RESIDENCE[Subject = Mike; } \Pi_{\text{Time}} = \text{RECENT PAST;}$$
$$\Pi_{\text{Location}} = {}_{\mu \times \text{City1}} \text{ NEAR[City2 = Boston]]} \qquad (3.68)$$

where NEAR is a fuzzy relation with the frame

NEAR	City1	City2	μ

,

RECENT PAST is a fuzzy relation with the frame

RECENT PAST	Time	μ

(in which Time is expressed in years counting from the present to the past), and $_{\mu \times \text{City1}}$NEAR[City2 = Boston] is the fuzzy set of cities which are near Boston. Given a database, $\mathscr{D}$, (3.68) would return a possibility distribution such as shown (in a partially tabulated form) in Table 3.4, in which

TABLE 3.4

RESIDENCE	Subject	Location	Time	π
	Mike	Cambridge	1	1
	Mike	Cambridge	2	0·8
	Mike	Cambridge	3	0·6
	Mike	Wayland	1	0·9
	Mike	Wayland	2	0·8
	...	...	...	...

the third row, for example, signifies that the possibility that Mike lived in Cambridge 3 years ago is 0·6. In this example, (3.68) constitutes the meaning of *p*, while the possibility distribution whose tableau is given by Table 3.4 is the information conveyed by *p*.

In addition to representing the meaning of an expression, *e*, in a natural language, the corresponding expression, E, in PRUF may be viewed as its *deep structure*—not in the technical sense employed in the literature of linguistics (Chomsky, 1965, 1971)—but in the sense of being dependent not on the surface structure of e but on its meaning. This implies that the form of E is independent of the natural language in which e is expressed, thus providing the basis for referring to PRUF as a universal language. The same can be said, of course, of most of the meaning representation languages that have been described in the literature.

Another characteristic of PRUF that is worthy of mention is that it is an *intentional*† language in the sense that an expression in PRUF is supposed to convey the intended rather than the literal meaning of the corresponding expression in a natural language. For example, if the proposition $p \triangleq$ John is no genius is intended to mean that $q \triangleq$ John is dumb, then the translation of p into PRUF would be that of q rather than p itself. As an example illustrating a somewhat different point, consider the proposition

$$p \triangleq \text{Alla has red hair.} \tag{3.69}$$

In PRUF, its translation could be expressed in one of two ways:

(a) $$\text{Alla has red hair} \rightarrow \Pi_{\text{Color(Hair(Alla))}} = \varphi \tag{3.70}$$

where φ is an identifier of the color that is commonly referred to as *red* in the case of hair; or

(b) $$\text{Alla has red hair} \rightarrow \Pi_{\text{Color(Hair(Alla))}} = \text{RED}^f \tag{3.71}$$

in which the superscript f (standing for *footnote*) points to a non-standard definition of RED which must be used in (3.71). The same convention is employed, more generally, whenever a non-standard definition of any entity in an expression in PRUF must be employed.

SEMANTIC EQUIVALENCE AND SEMANTIC ENTAILMENT

The concepts of semantic equivalence and semantic entailment are two closely related concepts in PRUF which play an important role in fuzzy logic and approximate reasoning.

Informally, let p and q be a pair of expressions in a natural language and let Π^p and Π^q be the possibility distributions (or the fuzzy relations) induced by p and q in application to a database $\mathscr{D}$. Then, we shall say that p and q are *semantically equivalent*, expressed as

$$p \leftrightarrow q, \tag{3.72}$$

if and only if $\Pi^p = \Pi^q$. Furthermore, if (3.72) holds for all databases,† the semantic equivalence between p and q is said to be *strong*.‡ Thus, the definition of strong semantic

---

†A thorough discussion of the concept of intentionality may be found in Grice (1968) and Searle (1971).

†Generally, "all databases" should be interpreted as all databases which are related in a specified way to a reference database. This is analogous to the role of the alternativeness relation in possible world semantics (Hughes & Cresswell, 1968).

‡The concept of strong semantic equivalence as defined here reduces to that of semantic equivalence in predicate logic (see Lyndon, 1966) when p and q are non-fuzzy propositions.

equivalence implies that p and q have the same meaning if and only if they are strongly semantically equivalent. In this sense, then, any transformation which maps p into q is *meaning-preserving*.

To illustrate, as will be seen in section 6, the propositions

$$p \triangleq \text{Jeanne is tall is true} \qquad (3.73)$$

and

$$q \triangleq \text{Jeanne is not tall is false} \qquad (3.74)$$

in which *false* is the antonym of *true*, i.e.

$$\mu_{FALSE}(v) = \mu_{TRUE}(1-v), \quad v \in [0,1] \qquad (3.75)$$

are semantically equivalent no matter how TALL and TRUE are defined. Consequently, p and q are strongly semantically equivalent and hence have the same meaning. On the other hand, the propositions

$$p \triangleq \text{Jeanne is tall is very true} \qquad (3.76)$$

and

$$q \triangleq \text{Jeanne is very tall} \qquad (3.77)$$

can be shown to be semantically equivalent when TRUE is the unitary fuzzy set (see (2.15)), that is

$$\mu_{TRUE}(v) = v, \quad v \in [0,1]$$

but not when TRUE is an arbitrary fuzzy subset of [0,1]. Consequently, p and q are not strongly semantically equivalent.

Usually, it is clear from the context whether a semantic equivalence is or is not strong. When it is necessary to place in evidence that a semantic equivalence is strong, it will be denoted by $s\leftrightarrow$. Correspondingly, if the equality between Π^p and Π^q is approximate in nature, the approximate semantic equivalence between p and q will be expressed as $p\; a\leftrightarrow q$.

While the concept of semantic equivalence relates to the equality of possibility distributions (or fuzzy relations), that of *semantic entailment* relates to inclusion.† More specifically, on denoting the assertion "p semantically entails q (or q is semantically entailed by p)," by $p \mapsto q$, we have

$$p \mapsto q \text{ iff } \Pi^p \subset \Pi^q \qquad (3.78)$$

where Π^p and Π^q are the possibility distributions induced by the propositions p and q, respectively.

As in the case of semantic equivalence, semantic entailment is *strong* if the relation $\mapsto$ holds for all databases. For example, as will be seen in section 4, the possibility distribution induced by the proposition "Gary is very tall" is contained in that induced by

---

†If A and B are fuzzy subsets of U, then $A \subset B$ iff $\mu_A(u) \le \mu_B(u)$, $u \in U$.

"Gary is tall" no matter how TALL is defined. Consequently, we can assert that

$$\text{Gary is very tall } s\!\mapsto \text{Gary is tall} \qquad (3.79)$$

where $s\!\mapsto$ denotes strong semantic entailment. On the other hand, the validity of the semantic entailment

$$\text{Gary is very tall}\mapsto \text{Gary is not short} \qquad (3.80)$$

depends on the definitions of *tall* and *short*, and hence (3.80) does not represent strong semantic entailment.

As was stated earlier, the concepts of semantic equivalence and semantic entailment play an important role in fuzzy logic and approximate reasoning (Zadeh, 1977b). In the present paper, we shall make use of the concept of semantic equivalence in sections 4 and 6 to derive several useful meaning-preserving transformations.

4. Translation rules of Types I, II and III

To facilitate the translation of expressions in a natural language into PRUF, it is desirable to have a stock of translation rules which may be applied singly or in combination to yield an expression, E, in PRUF, which is a translation of a given expression, *e*, in a natural language.

The translation rules which apply to descriptors may readily be deduced from the corresponding rules for propositions. Consequently, we shall restrict our attention in the sequel to the translation of propositions.

The translation rules for propositions may be divided into several basic categories, the more important of which are the following:

Type I.   Rules pertaining to modification.
Type II.   Rules pertaining to composition.
Type III.   Rules pertaining to quantification.
Type IV.   Rules pertaining to qualification.

Simple examples of propositions to which the rules in question apply are the following:

Type I.   X is very small   (4.1)
            X is much larger than Y   (4.2)
            Eleanor was very upset   (4.3)
            The man with the blond hair is very tall   (4.4)

Type II.   X is small and Y is large (conjunctive composition)   (4.5)
            X is small or Y is large (disjunctive composition)   (4.6)
            If X is small then Y is large   (4.7)
               (conditional composition)
            If X is small then Y is large else Y is very large   (4.8)
               (conditional and conjunctive composition)

Type III.	Most Swedes are tall	(4.9)
	Many men are much taller than most men	(4.10)
	Most tall men are very intelligent	(4.11)

Type IV.	Abe is young is not very true	(4.12)
	(truth qualification)	
	Abe is young is quite probable	(4.13)
	(probability qualification)	
	Abe is young is almost impossible	(4.14)
	(possibility qualification)	

Rules of Types I, II and III will be discussed in this section. Rules of Type IV will be discussed in section 6, following an exposition of the concepts of consistency, compatibility and truth in section 5.

Translation rules in PRUF are generally expressed in a conditional format exemplified by

If $\qquad\qquad p \to P$ (4.15)

then $\qquad\qquad p^+ \to P^+$

where p^+ and P^+ are modifications of p and P, respectively. In effect, a rule expressed in this form states that if in a certain context p translates into P, then in the *same context* a specified modification of p, p^+, translates into a specified modification of P, P^+. In this way, the rule makes it explicit that the translation of a modified proposition, p^+, depends on the translation of p. The simpler notation employed in (4.28) conveys the same information, but does so less explicitly.

RULES OF TYPE I

A basic rule of Type I is the *modifier rule*, which may be stated as follows.

If the proposition

$$p \triangleq N \text{ is } F \qquad\qquad (4.16)$$

translates into the possibility assignment equation (see (3.31))

$$\Pi_{(x_1, \ldots, x_n)} = F \qquad\qquad (4.17)$$

then the translation of the modified proposition

$$p^+ \triangleq N \text{ is } mF \qquad\qquad (4.18)$$

where m is a modifier such as *not, very, more or less, quite, extremely,* etc., is given by

$$N \text{ is } mF \to \Pi_{(x_1, \ldots, x_n)} = F^+ \qquad\qquad (4.19)$$

where F^+ is a modification of F induced by m. In particular:

(a)   if $m \triangleq$ not, then $F^+ = F' \triangleq$ complement of F;   (4.20)

(b)   if $m \triangleq$ very, then $F^+ = F^2$, where†   (4.21)

$$F^2 = \int_U \mu_F^2(u)/u;$$ (4.22)

(c)   if $m \triangleq$ more or less, then $F^+ = \sqrt{F}$ where   (4.23)

$$\sqrt{F} = \int_U \sqrt{\mu_F(u)}/u;$$ (4.24)

or, alternatively,

$$F^+ = \int_U \mu_F(u)K(u)$$ (4.25)

where $K(u)$ is the kernel of *more or less*.‡

As a simple illustration of (4.21), let p be the proposition "Lisa is young," where YOUNG is a fuzzy subset of the interval [0,100] whose membership function is expressed in terms of the S-function (2.13) as (omitting the arguments of μ and S):

$$\mu_{YOUNG} = 1 - S(25,35,45).$$ (4.26)

Then, the translation of "Lisa is very young" is given by

$$\text{Lisa is very young} \rightarrow \Pi_{Age(Lisa)} = YOUNG^2$$ (4.27)

where

$$\mu_{YOUNG^2} = (1 - S(25,35,45))^2.$$

Note that we can bypass the conditional format of the translation rule (4.16) and assert directly that

$$\text{Lisa is very young} \rightarrow \Pi_{Age(Lisa)} = YOUNG^2$$ (4.28)

on the understanding that YOUNG is the denotation of *young* in the context in which the proposition "Lisa is very young" is asserted. As was stated earlier, the conditional format serves the purpose of making this understanding more explicit.

In some cases, a modifier such as *very* may be implicit rather than explicit in a proposition. Consider, for example, the proposition

---

†The "integral" representation of a fuzzy set in the form $F = \int_U \mu_F(u)/u$ signifies that F is a union of the fuzzy singletons $\mu_F(u)/u$, $u \in U$, where μ_F is the membership function of F. Thus, (4.22) means that the membership function of F^2 is the square of that of F.

‡More detailed discussions of various types of modifiers may be found in Zadeh (1972a, 1975c), Lakoff (1973a,b), Wenstop (1975, 1976), Mizumoto *et al.* (1977), Hersh & Caramazza (1976), and other papers listed in the bibliography. It is important to note that (4.21) and (4.23) should be regarded merely as standardized default definitions which may be replaced, if necessary, by the user-supplied definitions.

$$p \triangleq \text{Vera and Pat are close friends.} \qquad (4.29)$$

As an approximation, p may be assumed to be semantically equivalent to

$$q = \text{Vera and Pat are friends}^2 \qquad (4.30)$$

so that (using (4.22)) the translation of p may be expressed as (see (7.21))

$$\pi(\text{FRIENDS}) = \mu\text{FRIENDS}^2[\text{Name1} = \text{Vera}; \text{ Name2} = \text{Pat}] \qquad (4.31)$$

where $\pi(\text{FRIENDS})$ is the possibility of the relation FRIENDS in $\mathcal{D}$. Thus, what (4.31) implies is that the relation FRIENDS in $\mathcal{D}$ is such that

$$\Pi_X = \perp^2 \qquad (4.32)$$

where

$$X \triangleq \mu\text{FRIENDS}[\text{Name1} = \text{Vera}; \text{ Name2} = \text{Pat}] \qquad (4.33)$$

and $\perp$ is the unitor defined by (2.15).

RULES OF TYPE II

Translation rules of Type II pertain to the translation of propositions of the form

$$p = q * r \qquad (4.34)$$

where $*$ denotes an operation of composition, e.g. conjunction (and), disjunction (or), implication (if . . . then), etc.

Under the assumption that the operation of composition is non-interactive (Bellman & Zadeh, 1976),† the rules in question may be stated as follows.
  If

$$q \triangleq M \text{ is } F \rightarrow \Pi_{(X_1, \ldots, X_m)} = F$$

and

$$r \triangleq N \text{ is } G \rightarrow \Pi_{(Y_1, \ldots, Y_n)} = G \qquad (4.35)$$

then

(a) M is F and N is $G \rightarrow \Pi_{(X_1, \ldots, X_m, Y_1, \ldots, Y_n)} = \bar{F} \cap \bar{G}$ (4.36)
$$= F \times G$$

(b) M is F or N is $G \rightarrow \Pi_{(X_1, \ldots, X_m, Y_1, \ldots, Y_n)} = \bar{F} + \bar{G}$ (4.37)

and

(c_1) If M is F then N is $G \rightarrow \Pi_{(X_1, \ldots, X_m, Y_1, \ldots, Y_n)} = \bar{F}' \oplus \bar{G}$ (4.38)

or

(c_2) If M is F then N is $G \rightarrow \Pi_{(X_1, \ldots, X_m, Y_1, \ldots, Y_n)} = F \times G + F' \times V$ (4.39)

---

†Informally, a binary operation $*$ on real numbers u, v is *non-interactive* if an increase in the value of u (or v) cannot be compensated by a decrease in the value of v (or u). It should be understood that the non-interactive definitions of *and* and *or* in (4.36) and (4.37) may be replaced, if necessary, by user-supplied interactive definitions.

where F and G are fuzzy subsets of $U \triangleq U_1 \times \ldots \times U_n$ and $V = V_1 \times \ldots \times V_n$, respectively; $\overline{F}'$ and $\overline{G}$ are the cylindrical extensions of F' and G, i.e.

$$\overline{F}' = F' \times V \tag{4.40}$$

$$\overline{G} = U \times G; \tag{4.41}$$

$F \times G$ is the Cartesian product of F and G which may be expressed as $\overline{F} \cap \overline{G}$ and is defined by

$$\mu_{F \times G}(u,v) = \mu_F(u) \wedge \mu_G(v), \quad u \in U, \quad v \in V; \tag{4.42}$$

$+$ is the union and $\oplus$ is the bounded-sum, i.e.

$$\mu_{\overline{F}' \oplus \overline{G}}(u,v) = 1 \wedge (1 - \mu_F(u) + \mu_G(v)) \tag{4.43}$$

where $u \triangleq (u_1, \ldots, u_n)$, $v \triangleq (v_1, \ldots, v_n)$, $\wedge \triangleq \min$, $+ \triangleq$ arithmetic sum and $- \triangleq$ arithmetic difference.† Note that there are two distinct rules for the conditional composition, (c_1) and (c_2). Of these, (c_1) is consistent with the definition of implication in Łukasiewicz's $Ł_{Aleph_1}$ logic (Rescher, 1969), while (c_2)—in consequence of (4.53)—corresponds to the relation expressed by the table:

TABLE 4.1

M	N
F	G
F'	V

As a very simple illustration, assume, as in Zadeh (1977b), that $U = V = 1 + 2 + 3$, $M \triangleq X$, $N \triangleq Y$,

$$F \triangleq SMALL \triangleq 1/1 + 0.6/2 + 0.1/3 \tag{4.44}$$

and

$$G \triangleq LARGE \triangleq 0.1/1 + 0.6/2 + 1/3. \tag{4.45}$$

Then (4.36), (4.37), (4.38) and (4.39) yield

X is small and Y is large $\rightarrow$ (4.46)
$$\Pi_{(X, Y)} = 0.1/(1,1) + 0.6/(1,2) + 1/(1,3) + 0.1/(2,1) + 0.6/(2,2) + 0.6/(2,3)$$
$$+ 0.1/(3,1) + 0.1/(3,2) + 0.1/(3,3)$$

---

†If the variables $X \triangleq (X_1, \ldots, X_n)$ and $Y \triangleq (Y_1, \ldots, Y_n)$ have a subvariable, say Z, in common, i.e. $X \triangleq (S,Z)$ and $Y \triangleq (T,Z)$, then F and G should be interpreted as cylindrical extensions of F and G in $U(S) \times U(T) \times U(Z)$ rather than in $U(X) \times U(Y)$, where $U(S)$, $U(T)$ and $U(Z)$ denote, respectively, the universes in which S, T and Z take their values. Additionally, the possibility distributions in (7.38) and (7.39) should be interpreted, in a strict sense, as conditional distributions.

X is small or Y is large $\rightarrow$ (4.47)
$$\Pi_{(X, Y)} = 1/(1,1) + 1/(1,2) + 1/(1,3) + 0 \cdot 6/(2,1) + 0 \cdot 6/(2,2) + 1/(2,3) + 0 \cdot 1/(3,1) \\ + 0 \cdot 6/(3,2) + 1/(3,3)$$

If X is small then Y is large $\rightarrow$ (4.48)
$$\Pi_{(X, Y)} = 0 \cdot 1/(1,1) + 0 \cdot 6/(1,2) + 1/(1,3) + 0 \cdot 5/(2,1) + 1/(2,2) + 1/(2,3) \\ + 1/(3,1) + 1/(3,2) + 1/(3,3)$$

If X is small then Y is large $\rightarrow$ (4.49)
$$\Pi_{(X, Y)} = 0 \cdot 1/(1,1) + 0 \cdot 6/(1,2) + 1/(1,3) + 0 \cdot 4/(2,1) + 0 \cdot 6/(2,2) + 0 \cdot 6/(2,3) \\ + 0 \cdot 9/(3,1) + 0 \cdot 9/(3,2) + 0 \cdot 9/(3,3)$$

The rules stated above may be employed in combination, yielding a variety of corollary rules which are of use in the translation of more complex forms of composite propositions and descriptors. Among the basic rules of this type are the following.

(d) If M is F then N is G else N is H (4.50)

$$\rightarrow \Pi_{(X_1, \ldots, X_m, Y_1, \ldots, Y_n)} = (\overline{F}' \oplus \overline{G}) \cap (\overline{F} \oplus \overline{H})$$

where $F \subset U \triangleq U_1 \times \ldots \times U_m$ and $G, H \subset V \triangleq V_1 \times \ldots \times V_n$. This rule follows from the semantic equivalence:

If M is F then N is G else N is H (4.51)
$$\leftrightarrow (\text{If M is F then N is G}) \text{ and } (\text{If M is not F then N is H})$$

and the application of (a) and (c_1).

(e) *Translation rule for relations*
Consider a relation, R, whose tableau is of the form shown in Table 4.2.

TABLE 4.2

R	X_1	X_2	...	X_n
	F_{11}	F_{12}	...	F_{1n}
	...	...	...	...
	F_{m1}	...	...	F_{mn}

in which the F_{ij} are fuzzy subsets of the U_j, respectively. On interpreting R as

$$R = X_1 \text{ is } F_{11} \text{ and } X_2 \text{ is } F_{12} \text{ and } \ldots \text{ and } X_n \text{ is } F_{1n} \text{ or} \qquad (4.52)$$
$$X_1 \text{ is } F_{21} \text{ and } X_2 \text{ is } F_{22} \text{ and } \ldots \text{ and } X_n \text{ is } F_{2n} \text{ or } \ldots \text{ or}$$
$$X_1 \text{ is } F_{m1} \text{ and } X_2 \text{ is } F_{m2} \text{ and } \ldots \text{ and } X_n \text{ is } F_{mn}$$

it follows from (a) and (b) that

$$R \rightarrow F_{11} \times \ldots \times F_{1n} + \ldots + F_{m1} \times \ldots \times F_{mn} \qquad (4.53)$$

which will be referred to as the *tableau rule*. This rule plays an important role in applications to pattern recognition, decision analysis, medical diagnosis and related areas, in which binary relations are employed to describe the features of a class of objects (Zadeh, 1976*a,b*).

As a simple illustration, consider the relation defined by Table 4.3

TABLE 4.3

X	Y
small very small not small	large not very large very small

in which X and Y are real-valued variables and

$$\text{small} \rightarrow \text{SMALL}$$
$$\text{large} \rightarrow \text{LARGE}$$

where SMALL and LARGE are specified fuzzy subsets of the real line.
First, by the application of (4.20) and (4.21), we have

$$\text{very small} \rightarrow \text{SMALL}^2 \qquad (4.54)$$
$$\text{not small} \rightarrow \text{SMALL}' \qquad (4.55)$$
$$\text{not very large} \rightarrow (\text{LARGE}^2)'. \qquad (4.56)$$

Then, on applying (4.53), we obtain

$$R \rightarrow \text{SMALL} \times \text{LARGE} + (\text{SMALL}^2) \times (\text{LARGE}^2)' + \text{SMALL}' \times \text{SMALL}^2$$

which is the desired translation of the relation in question.

LINGUISTIC VARIABLES

The modifier rule in combination with the translation rules for conjunctive and disjunctive compositions provides a simple method for the translation of linguistic values of so-called *linguistic variables* (Zadeh, 1973, 1975c).

Informally, a *linguistic variable* is a variable whose *linguistic values* are words or sentences in a natural or synthetic language, with each such value being a label of a fuzzy subset of a universe of discourse. For example, a variable such as Age may be viewed both as a numerical variable ranging over, say, the interval [0,150], and as a linguistic variable which can take the values *young, not young, very young, not very young, quite young, old, not very young and not very old*, etc. Each of these values may be interpreted as a label of a fuzzy subset of the universe of discourse $U = [0,150]$, whose base variable, u, is the generic numerical value of Age.

Typically, the values of a linguistic variable such as Age are built up of one or more *primary terms* (which are the labels of *primary fuzzy sets*†), together with a collection of modifiers which allow a composite linguistic value to be generated from the primary terms through the use of conjunctions and disjunctions. Usually the number of primary terms is two, with one being an antonym of the other. For example, in the case of Age, the primary terms are *young* and *old*, with *old* being the antonym of *young*.

Using the translation rules (4.20), (4.21), (4.36) and (4.37) in combination, the linguistic values of a linguistic variable such as Age may be translated by inspection. To illustrate, suppose that the primary terms *young* and *old* are defined by

$$\mu_{\text{YOUNG}} = 1 - S(20,30,40) \tag{4.57}$$

and

$$\mu_{\text{OLD}} = S(40,55,70). \tag{4.58}$$

Then

$$\text{not very young} \rightarrow (\text{YOUNG}^2)' \tag{4.59}$$

and

$$\text{not very young and not very old} \rightarrow (\text{YOUNG}^2)' \cap (\text{OLD}^2)' \tag{4.60}$$

and thus

$$\text{John is not very young} \rightarrow \Pi_{\text{Age(John)}} = (\text{YOUNG}^2)' \tag{4.61}$$

where

$$\mu_{(\text{YOUNG}^2)'} = 1 - (1 - S(20,30,40))^2. \tag{4.62}$$

The problem of finding a linguistic value of Age whose meaning approximates to a given fuzzy subset of U is an instance of the problem of *linguistic approximation* (Zadeh, 1975c; Wenstop, 1975; Procyk, 1976). We shall not discuss in the present paper the ways in which this non-trivial problem can be approached, but will assume that linguistic approximation is implicit in the retranslation of a possibility distribution into a proposition expressed in a natural language.

---

†Such sets play a role which is somewhat analogous to that of physical units.

RULES OF TYPE III

Translation rules of Type III pertain to the translation of propositions of the general form

$$p \triangleq QN \text{ are } F \tag{4.63}$$

where N is the descriptor of a possibly fuzzy set, Q is a fuzzy quantifier (e.g. *most, many, few, some, almost all,* etc.) and F is a fuzzy subset of U. Simple examples of (4.63) are:

$$\text{Most Swedes are tall} \tag{4.64}$$

$$\text{Many tall men are fat} \tag{4.65}$$

$$\text{Some men are much taller than most men.} \tag{4.66}$$

In general, a fuzzy quantifier is a fuzzy subset of the set of integers, the unit interval or the real line. For example, we may have

$$\text{SEVERAL} \triangleq 0\cdot2/3 + 0\cdot6/4 + 1/5 + 1/6 + 0\cdot6/7 + 0\cdot2/8 \tag{4.67}$$

$$\text{MOST} \triangleq \int_0^1 S(u;0\cdot5,0\cdot7,0\cdot9)/u \tag{4.68}$$

(which means that MOST is a fuzzy subset of the unit interval whose membership function is given by $S(0\cdot5,0\cdot7,0\cdot9)$) and -

$$\text{LARGE NUMBER} \triangleq \int_0^\infty (1+(\tfrac{u}{100})^{-2})^{-1}/u. \tag{4.69}$$

In order to be able to translate propositions of the form (4.63), it is necessary to define the *cardinality* of a fuzzy set, i.e. the number (or the proportion) of elements of U which are in F. Strictly speaking, the cardinality of a fuzzy set should be a fuzzy number, which could be defined as in Zadeh (1977b). It is simpler, however, to deal with the *power* of a fuzzy set (DeLuca & Termini, 1972), which in the case of a fuzzy set with a finite support† is defined by‡

$$|F| \triangleq \sum_i \mu_F(u_i), \quad u_i \in \text{Support of F} \tag{4.70}$$

where $\mu_F(u_i)$, $i=1,\ldots,$ N, is the grade of membership of u_i in F and $\sum$ denotes the arithmetic sum. For example, for the fuzzy set SMALL defined by

†The *support* of a fuzzy subset F of U is the set of all points in U at which $\mu_F(u) > 0$.

‡For some applications, it is necessary to eliminate from the count those elements of F whose grade of membership falls below a specified threshold. This is equivalent to replacing F in (4.70) with $F \cap \Gamma$, where Γ is a fuzzy or non-fuzzy set which induces the desired threshold.

$$\text{SMALL} \triangleq 1/0 + 1/1 + 0{\cdot}8/2 + 0{\cdot}6/3 + 0{\cdot}4/4 + 0{\cdot}2/5 \qquad (4.71)$$

we have

$$|F| = 1 + 1 + 0{\cdot}8 + 0{\cdot}6 + 0{\cdot}4 + 0{\cdot}2 = 4.$$

In the sequel, we shall usually employ the more explicit notation Count(F) to represent the power of F, with the understanding that F should be treated as a bag§ rather than a set. Furthermore, the notation Prop(F/G) will be used to represent the "proportion" of F in G, i.e.

$$\text{Prop}\{F/G\} \triangleq \frac{\text{Count}(F \cap G)}{\text{Count}(G)} \qquad (4.72)$$

and more explicitly

$$\text{Prop}\{F/G\} = \frac{\sum_i (\mu_F(u_i) \wedge \mu_G(u_i))}{\sum_j \mu_G(u_j)} \qquad (4.73)$$

where the summation ranges over the values of i for which $u_i \in$ Support of $F \cap$ Support of G. In particular, if $G \triangleq U \triangleq$ finite non-fuzzy set, then (4.73) becomes

$$\text{Prop}\{F/U\} = \frac{1}{N} \sum_{i=1}^{N} \mu_F(u_i) \qquad (4.74)$$

where N is the cardinality of U. For convenience, the number Prop$\{F/U\}$ will be referred to as the *relative cardinality* of F expressed as

$$\text{Prop}(F) \triangleq \text{Prop}\{F/U\} = \frac{1}{N} \sum_{i=1}^{N} \mu_F(u_i). \qquad (4.75)$$

As N increases and U becomes a continuum, the expression for the power of F tends in the limit to that of the *additive measure* of F (Zadeh, 1968; Sugeno, 1974), which may be regarded as a continuous analog of the proportion of the elements of U which are "in" F. More specifically, if $\rho(u)$ is a density function defined on U, the measure in question is defined by†

$$\text{Prop}(F) = \int_U \rho(u)\mu_F(u)du. \qquad (4.76)$$

For example, if $\rho(u)du$ is the proportion of Swedes whose height lies in the interval $[u, u+du]$, then the proportion of tall Swedes is given by

$$\text{Prop(tall Swedes)} = \int_0^{200} \rho(u)\mu_{\text{TALL}}(u)du \qquad (4.77)$$

§The elements of a bag need not be distinct. For example, the collection of integers $\{2,3,5,3,5\}$ is a bag if $\{2,3,5,3,5\} \neq \{2,3,5\}$.

†We employ the notation Prop(F) even in the continuous case to make clearer the intuitive meaning of measure.

where μ_{TALL} is the membership function of *tall* and height is assumed to be measured in centimeters.

In a similar fashion, the expression for Prop{F/G} tends in the limit to that of the *relative measure* of F in G, which is defined by

$$\text{Prop(F/G)} \triangleq \frac{\displaystyle\int_{U\times V} \rho(u,v)(\mu_F(u)\wedge\mu_G(v))dudv}{\displaystyle\int_V \rho(v)\mu_G(v)dv} \tag{4.78}$$

where $\rho(u,v)$ is a density function defined on $U\times V$ and

$$\rho(v) = \int_U \rho(u,v)du. \tag{4.79}$$

For example, if $F \triangleq$ TALL MEN and $G \triangleq$ FAT MEN, (4.78) becomes

$$\text{Prop\{TALL MEN/FAT MEN\}} = \frac{\displaystyle\int_{[0,200]\times[0,100]} \rho(u,v)\mu_{TALL}(u)\wedge\mu_{FAT}(v)dudv}{\displaystyle\int_{[0,100]} \rho(v)\mu_{FAT}(v)dv} \tag{4.80}$$

where $\rho(u,v)dudv$ is the proportion of men whose height lies in the interval $[u,u+du]$ and whose weight lies in the interval $[v,v+dv]$.

The above definitions provide the basis for the *quantifier rule* for the translation of propositions of the form "QN are F". More specifically, assuming for simplicity that N is a descriptor of a non-fuzzy set, the rule in question may be stated as follows.

If $U = \{u_1, \ldots, u_N\}$ and

$$N \text{ is } F \rightarrow \Pi_X = F \tag{4.81}$$

then

$$QN \text{ are } F \rightarrow \Pi_{Count(F)} = Q \tag{4.82}$$

and, if U is a continuum,

$$QN \text{ are } F \rightarrow \Pi_{Prop(F)} = Q \tag{4.83}$$

which implies the more explicit rule

$$QN \text{ are } F \rightarrow \pi(\rho) = \mu_Q(\int_U \rho(u)\mu_F(u)du) \tag{4.84}$$

where $\rho(u)du$ is the proportion of X's whose value lies in the interval $[u,u+du]$, $\pi(\rho)$ is the possibility of ρ, and μ_Q and μ_F are the membership functions of Q and F, respectively.

As a simple illustration, if MOST and TALL are defined by (4.68) and $\mu_{TALL} =$ S(160,170,180), respectively, then

$$\text{Most men are tall}\rightarrow\pi(\rho)=S(\int_{0}^{200}\rho(u)S(u;160,1970,180)du;0{\cdot}5,0{\cdot}7,0{\cdot}9) \tag{4.85}$$

where $\rho(u)du$ is the proportion of men whose height (in cm) is in the interval $[u,u+du]$. Thus, the proposition "Most men are tall" induces a possibility distribution of the height density function ρ which is expressed by the right-hand member of (4.85).

MODIFIER RULE FOR PROPOSITIONS

The modifier rule which was stated earlier in this section (4.16) provides a basis for the formulation of a more general modifier rule which applies to propositions and which leads to a rule for transforming the negation of a proposition into a semantically equivalent form in which the negation has a smaller scope.

The *modifier rule for propositions* may be stated as follows.

If a proposition p translates into a procedure P, i.e.

$$p\rightarrow P \tag{4.86}$$

and P returns a possibility distribution Π^p in application to a database $\mathscr{D}$, then mp, where m is a modifier, is semantically equivalent to a retranslation of mP, i.e.

$$mp\leftrightarrow q \tag{4.87}$$

where

$$q\leftarrow mP. \tag{4.88}$$

In (4.88), mP is understood to be a procedure which returns (in application to $\mathscr{D}$):

$$(\Pi^p)' \text{ if } m\triangleq\text{not} \tag{4.89}$$

$$(\Pi^p)^2 \text{ if } m\triangleq\text{very} \tag{4.90}$$

and

$$(\Pi^p)^{0{\cdot}5} \text{ if } m\triangleq\text{more or less.} \tag{4.91}$$

For simplicity, the possibility distribution defined by (4.89), (4.90) and (4.91) will be denoted as $m\Pi^p$.

On applying this rule to a proposition of the form $p\triangleq N$ is F and making use of the translation rules (4.20), (4.21), (4.23), (4.36), (4.37) and (4.87), we obtain the following general forms of (strong) semantic equivalence:

(a)
$$m(\text{N is F})\leftrightarrow\text{N is } m\text{F} \tag{4.92}$$

and, in particular,

$$\text{not}(\text{N is F})\leftrightarrow\text{N is not F} \tag{4.93}$$

$$\text{very}(\text{N is F})\leftrightarrow\text{N is very F} \tag{4.94}$$

$$\text{more or less(N is F)} \leftrightarrow \text{N is more or less F.} \tag{4.95}$$

(b) $\qquad m\text{(M is F and N is G)} \leftrightarrow \text{(X,Y) is } m\text{(F} \times \text{G)} \tag{4.96}$

and, in particular (in virtue of (4.20), (4.36) and 4.37)),

$$\text{not(M is F and N is G)} \leftrightarrow \text{(X,Y) is (F} \times \text{G)}' \tag{4.97}$$
$$\leftrightarrow \text{(X,Y) is } \overline{\text{F}}' + \overline{\text{G}}' \tag{4.98}$$
$$\leftrightarrow \text{M is not F or N is not G} \tag{4.99}$$
$$\text{very(M is F and N is G)} \leftrightarrow \text{M is very F and N is very G} \tag{4.100}$$
$$\text{more or less(M is F and N is G)} \leftrightarrow \text{M is more or less F and N is}$$
$$\text{more or less G} \tag{4.101}$$

and dually for disjunctive composition.

(c) $\qquad m\text{(QN are F)} \leftrightarrow (m\text{Q})\text{N are F} \tag{4.102}$

and, in particular,

$$\text{not(QN are F)} \leftrightarrow \text{(not Q)N are F} \tag{4.103}$$

which may be regarded as a generalization of the standard negation rules in predicate calculus, viz.

$$\neg\,(\forall x)F(x) \leftrightarrow (\exists x)\neg\,F(x), \tag{4.104}$$

$$\neg\,(\exists x)F(x) \leftrightarrow (\forall x)\neg\,F(x). \tag{4.105}$$

To see the connection between (4.104), say, and (4.102), we first note that, in consequence of (4.84), we can assert the semantic equivalence

$$\text{QN are F} \leftrightarrow \text{ant Q are not F} \tag{4.106}$$

where ant Q, the antonym of Q, is defined by

$$\mu_{\text{ant Q}}(v) = \mu_Q(1-v), \quad v \in [0,1]. \tag{4.107}$$

Thus, on combining (4.103) and (4.106), we have

$$\text{not(QN are F)} \leftrightarrow \text{(ant(not Q))N are not F} \tag{4.108}$$

which for $Q \triangleq$ all gives

$$\text{not(all N are F)} \leftrightarrow \text{(ant(not all))N are not F.} \tag{4.109}$$

Then, the right-hand member of (4.109) may be expressed as

$$\text{not(all N are F)} \leftrightarrow \text{some N are not F} \tag{4.110}$$

if we assume that

$$\text{some} \triangleq \text{ant(not all).} \tag{4.111}$$

In a similar fashion, the modifier rule for propositions may be employed to derive the negation rules for qualified propositions of the form $q \triangleq p$ is γ, where γ is a truth-value, a probability-value, or a possibility-value. Rules of this type will be formulated in section 5.

5. Consistency, compatibility and truth

Our aim in this section is to lay the groundwork for the translation of truth-qualified propositions of the form "p is τ," where τ is a linguistic truth-value. To this end, we shall have to introduce two related concepts—consistency and compatibility—in terms of which the *relative* truth of a proposition p with respect to a reference proposition r may be defined.

The concept of truth has traditionally been accorded a central place in logic and philosophy of language. In recent years, it has also come to play a primary role in the theory of meaning—especially in Montague grammar and possible world semantics.

By contrast, it is the concept of a possibility distribution rather than truth that serves as a basis for the definition of meaning as well as other primary concepts in fuzzy logic and PRUF. Thus, as we shall see in the sequel, the concept of truth in PRUF serves in the main as a mechanism for assessing the consistency or compatibility of a pair of propositions rather than—as in classical logic—as an indicator of the correspondence between a proposition and "reality".

CONSISTENCY AND COMPATIBILITY

Let p and q be two propositions of the form $p \triangleq N$ is F and $q \triangleq N$ is G, which translate, respectively, into

$$p \triangleq N \text{ is F} \rightarrow \Pi^p_{(X_1, \ldots, X_n)} = \text{F} \tag{5.1}$$

and

$$q \triangleq N \text{ is G} \rightarrow \Pi^q_{(X_1, \ldots, X_n)} = \text{G} \tag{5.2}$$

where $(X_1, \ldots, X_n)$ takes values in U. Intuitive considerations suggest that the *consistency* of p with q (or vice versa) be defined as the possibility that "N is F" given that "N is G" (or vice versa). Thus, making use of (2.9), we have

$$\text{Cons}\{N \text{ is F}, N \text{ is G}\} \triangleq \text{Poss}\{N \text{ is F} \mid N \text{ is G}\} \tag{5.3}$$
$$= \sup_{u \in U} (\mu_F(u) \wedge \mu_G(u))$$

where $u \triangleq (u_1, \ldots, u_n)$ denotes the generic value of $(X_1, \ldots, X_n)$, and μ_F and μ_G are the membership functions of F and G, respectively.

As a simple illustration, assume that

$$p \triangleq N \text{ is a small integer} \tag{5.4}$$

$$q \triangleq N \text{ is not a small integer} \tag{5.5}$$

where

$$\text{SMALL INTEGER} \triangleq 1/0 + 1/1 + 0 \cdot 8/2 + 0 \cdot 6/3 + 0 \cdot 4/4 + 0 \cdot 2/5. \tag{5.6}$$

In this case, $\text{Cons}\{p,q\} = 0 \cdot 4$.

As a less simple example, consider the propositions

$$p \triangleq \text{Most men are tall} \tag{5.7}$$

and

$$q \triangleq \text{Most men are short} \tag{5.8}$$

which translate into (see (4.84))

$$\pi^p(\rho) = \mu_{\text{MOST}}\left(\int_0^{200} \rho(u)\mu_{\text{TALL}}(u)du \right) \tag{5.9}$$

and

$$\pi^q(\rho) = \mu_{\text{MOST}}\left(\int_0^{200} \rho(u)\mu_{\text{SHORT}}(u)du \right). \tag{5.10}$$

In this case, assuming that μ_{MOST} is a monotone function, we have

$$\text{Cons}\{p,q\} = \mu_{\text{MOST}}\left(\text{Sup}_\rho \left((\int_0^{200} \rho(u)\mu_{\text{TALL}}(u)du) \wedge (\int_0^{200} \rho(u)\mu_{\text{SHORT}}(u)du) \right) \right). \tag{5.11}$$

If q is assumed to be a *reference* proposition, which we shall denote by r, then the *truth of p relative to r* could be defined as the consistency of p with r. It appears to be more appropriate, however, to define the truth of p relative to r through the concept of *compatibility* rather than consistency. More specifically, assume that the reference proposition r is of the form

$$r \triangleq N \text{ is } u \tag{5.12}$$

where u is an element of U. Then, by definition,

$$\text{Comp}\{N \text{ is } u / N \text{ is } F\} \triangleq \mu_F(u) \tag{5.13}$$

which coincides with the definition of $\text{Poss}\{X \text{ is } u \mid N \text{ is } F\}$ (see (2.4)) as well as with the definition of the consistency of "N is u" with "N is F". However, when the reference proposition is of the form $r \triangleq N$ is G, the definitions of compatibility and consistency cease to coincide. More specifically, by employing the extension principle,† (5.13) becomes

$$\text{Comp}\{N \text{ is } G/N \text{ is } F\} = \mu_F(G) \tag{5.14}$$

$$= \int_{[0,1]} \mu_G(u)/\mu_F(u)$$

in which the right-hand member is the union over the unit interval of the fuzzy singletons $\mu_G(u)/\mu_F(u)$. Thus, (5.14) signifies that the compatibility of "N is G" with "N is F" is a fuzzy subset of $[0,1]$ defined by (5.14).

The concept of compatibility as defined by (5.14) provides the basis for the following definition of Truth.

Truth. Let p be a proposition of the form "N is F," and let r be a reference proposition, $r \triangleq N$ is G, where F and G are subsets of U. Then, the *truth*, τ, of p *relative to* r is defined as the compatibility of r with p, i.e.

$$\tau \triangleq \text{Tr}\{N \text{ is } F/N \text{ is } G\} \triangleq \text{Comp}\{N \text{ is } G/N \text{ is } F\} \tag{5.15}$$
$$\triangleq \mu_F(G)$$
$$\triangleq \int_{[0,1]} \mu_G(u)/\mu_F(u).$$

It should be noted that τ, as defined by (5.15), is a fuzzy subset of the unit interval, implying that a linguistic truth-value may be regarded as a linguistic approximation to the subset defined by (5.15).

A more explicit expression for τ which follows at once from (5.15) is:

$$\mu_\tau(v) = \text{Max}_u \mu_G(u), \quad v \in [0,1] \tag{5.16}$$

subject to

$$\mu_F(u) = v.$$

Thus, if μ_F is 1–1, then the membership function of τ may be expressed in terms of those of F and G as

$$\mu_\tau(v) = \mu_G(\mu_F^{-1}(v)). \tag{5.17}$$

Another immediate consequence of (5.15) is that the truth-value of p relative to itself is given by

$$\mu_\tau(v) = v$$

rather than unity. Thus, in virtue of (2.15), we have

$$\text{Tr}\{N \text{ is } F/N \text{ is } F\} = \perp \tag{5.18}$$
$$= u\text{-true}.$$

---

†The extension principle (Zadeh, 1975c) serves to extend the definition of a mapping $f : U \to V$ to the set of fuzzy subsets of U. Thus, $f(F) \triangleq \int_U \mu_F(u)/f(u)$, where $f(F)$ and $f(u)$ are, respectively, the images of F and u in V.

As an illustration of (5.15), assume that

$$p \triangleq N \text{ is not small} \tag{5.19}$$

and

$$r \triangleq N \text{ is small} \tag{5.20}$$

where SMALL is defined by (5.6). Then, (5.15) yields

$$\tau = 1/0 + 0 \cdot 8/0 \cdot 2 + 0 \cdot 6/0 \cdot 4 + 0 \cdot 4/0 \cdot 6 + 0 \cdot 2/0 \cdot 8 \tag{5.21}$$

which may be regarded as a discretized version of the antonym of u-true (see (4.107)). Thus,

$$\text{Tr}\{N \text{ is not small}/N \text{ is small}\} = \text{ant } u\text{-true} \tag{5.22}$$

which, as will be seen later, is a special case of the strong semantic equivalence

$$\text{Tr}\{N \text{ is F}/N \text{ is not F}\} = \text{ant } u\text{-true}. \tag{5.23}$$

As can be seen from the foregoing discussion, in our definition of the truth-value of a proposition p, τ serves as a measure of the compatibility of p with a reference proposition r. To use this definition as a basis for the translation of truth-qualified propositions, we adopt the following postulate.

Postulate. A truth-qualified proposition of the form "p is τ" is semantically equivalent to the reference proposition, r, relative to which

$$\text{Tr}\{p/r\} = \tau. \tag{5.24}$$

We shall use this postulate in the following section to establish translation rules for truth-qualified propositions.

6. Translation rules of Type IV

Our concern in this section is with the translation of qualified propositions of the form $q \triangleq p$ is γ, where γ might be a truth-value, a probability-value, a possibility-value or, more generally, the value of some specified propositional function, i.e. a function from the space of propositions (or n-tuples of propositions) to the set of fuzzy subsets of the unit interval.

Typically, a translation rule of Type IV may be viewed as an answer to the following question: Suppose that a proposition p induces a possibility distribution Π^p. What, then, is the possibility distribution induced by the qualified proposition $q \triangleq p$ is γ, where γ is a specified truth-value, probability-value or possibility-value?

In what follows, we shall state the translation rules pertaining to (a) truth qualification; (b) probability qualification; and (c) possibility qualification. These are the principal modes of qualification which are of more or less universal use in natural languages.

RULE FOR TRUTH QUALIFICATION

Let p be a proposition of the form

$$p \triangleq N \text{ is F} \tag{6.1}$$

and let q be a truth-qualified version of p expressed as

$$q \triangleq N \text{ is F is } \tau \tag{6.2}$$

where τ is a linguistic truth-value. As was stated in section 5, q is semantically equivalent to the reference proposition r, i.e.

$$N \text{ is F is } \tau \leftrightarrow N \text{ is G} \tag{6.3}$$

where F, G and τ are related by

$$\tau = \mu_F(G). \tag{6.4}$$

Equation (6.4) states that τ is the image of G under the mapping $\mu_F \colon U \to [0,1]$. Consequently (Zadeh, 1965), the expression for the membership function of G in terms of those of τ and F is given by

$$\mu_G(u) = \mu_\tau(\mu_F(u)). \tag{6.5}$$

Using this result, the rule for truth qualification may be stated as follows.
If

$$N \text{ is F} \to \Pi_X = F \tag{6.6}$$

then

$$N \text{ is F is } \tau \to \Pi_X = F^+ \tag{6.7}$$

where

$$\mu_{F^+}(u) = \mu_\tau(\mu_F(u)). \tag{6.8}$$

In particular, if τ is the unitary truth-value, that is,

$$\tau = u\text{-true} \tag{6.9}$$

where

$$\mu_{u\text{-true}}(v) = v, \quad v \in [0,1] \tag{6.10}$$

then

$$N \text{ is F is } u\text{-true} \to N \text{ is F.} \tag{6.11}$$

As an illustration of (6.5), if

$$q \triangleq N \text{ is small is very true} \tag{6.12}$$

where

$$\mu_{\text{SMALL}} = 1 - S(5,10,15), \quad u \in [0, \infty) \tag{6.13}$$

and

$$\mu_{\text{TRUE}} = S(0 \cdot 6, 0 \cdot 8, 1 \cdot 0) \tag{6.14}$$

then

$$q \to \pi_X(u) = S^2(1 - S(u; 5,10,15); 0 \cdot 6, 0 \cdot 8, 1 \cdot 0). \tag{6.15}$$

RULE FOR PROBABILITY QUALIFICATION

Let p be a proposition of the form (6.1) and let q be a probability-qualified version of p expressed as

$$q \triangleq N \text{ is F is } \lambda \tag{6.16}$$

where λ is a linguistic probability-value such as probable, very probable, not very probable, or, equivalently, likely, very likely, not very likely, etc.

We shall assume that q is semantically equivalent to the proposition

$$\text{Prob}\{N \text{ is F}\} \text{ is } \lambda \tag{6.17}$$

in which $p \triangleq N$ is F is interpreted as a fuzzy event (Zadeh, 1968). More specifically, let $p(u)du$ be the probability that $X \in [u, u+du]$, where $X \triangleq X(N)$. Then

$$\text{Prob}\{N \text{ is F}\} = \int_U p(u)\mu_F(u)du \tag{6.18}$$

and hence (6.17) implies that

$$\Pi_{\int_U p(u)\mu_F(u)du} = \lambda. \tag{6.19}$$

Equation (6.19) provides the basis for the following statement of the rule for probability qualification.

If

$$N \text{ is F} \rightarrow \Pi_X = F \tag{6.20}$$

then

$$N \text{ is F is } \lambda \rightarrow \Pi_{\int_U p(u)\mu_F(u)du} = \lambda \tag{6.21}$$

or more explicitly

$$\pi(p(\cdot)) = \mu_\lambda \left(\int_U p(u)\mu_F(u)du \right) \tag{6.22}$$

where $\pi(p(\cdot))$ is the possibility of the probability density function $p(\cdot)$.

As an illustration of (6.22), assume that

$$q \triangleq N \text{ is small is likely} \tag{6.23}$$

where LIKELY is defined by

$$\mu_{\text{LIKELY}} = (0\cdot7, 0\cdot8, 0\cdot9) \tag{6.24}$$

and SMALL is given by (6.13). Then

$$N \text{ is small is likely} \rightarrow \pi(p(\cdot)) = \int_0^\infty p(u)(1 - S(u;5,10,15))du. \tag{6.25}$$

Note that in this case the proposition in question induces a possibility distribution of the probability density of $X \triangleq N$.

Our concern here is with the translation of possibility-qualified propositions of the form

$$q \triangleq N \text{ is F is } \omega \tag{6.26}$$

where ω is a linguistic possibility-value such as *quite possible, very possible, almost impossible*, etc., with each such value representing a fuzzy subset of the unit interval.

By analogy with our interpretation of probability-qualified propositions, q may be interpreted as

$$N \text{ is F is } \omega \leftrightarrow \text{Poss}\{X \text{ is F}\} \text{ is } \omega \tag{6.27}$$

which implies that

$$\Pi_{\text{Poss}\{ X \text{ is } F\}} = \omega. \tag{6.28}$$

Now suppose that we wish to find a fuzzy set G such that

$$N \text{ is F is } \omega \leftrightarrow N \text{ is G}. \tag{6.29}$$

Then, from the definition of possibility measure (2.9), we have

$$\text{Poss}\{N \text{ is F} \mid N \text{ is G}\} = \text{Sup}_{u}(\mu_F(u) \wedge \mu_G(u)) \tag{6.30}$$

and hence

$$N \text{ is F is } \omega \to \pi(\mu_G(\cdot)) = \mu_\omega\left(\text{Sup}_{u}(\mu_F(u) \wedge \mu_G(u))\right) \tag{6.31}$$

where μ_ω is the membership function of ω. Note that (6.31) is analogous to the translation rule for probability-qualified propositions (6.22).†

Although the interpretation expressed by (6.31) is consistent with (6.22), it is of interest to consider alternative interpretations which are not in the spirit of (6.28). One such interpretation which may be employed as a basis for possibility qualification is the following.

Assume that $\omega \triangleq 1$-possible (i.e. $\mu_\omega(v) = 1$ for $v = 1$ and $\mu_\omega(v) = 0$ for $v \in [0,1)$), and let

$$p \triangleq N \text{ is F} \to \Pi_X = F. \tag{6.32}$$

Then

$$q \triangleq N \text{ is F is 1-possible} \to \Pi_X = G \tag{6.33}$$

---

†A more detailed discussion of this issue may be found in Zadeh (1977*a*).

where G is a fuzzy set of Type 2† which has an interval-valued **membership function** defined by

$$\mu_G(u) = [\mu_F(u), 1], \quad u \in U \tag{6.34}$$

with the understanding that (6.34) implies that Poss$\{X = u\}$ may be any number in the interval $[\mu_F(u), 1]$.

More generally, if $\omega \triangleq \alpha$-possible (i.e. $\mu_\omega(v) = \alpha$ for $v = 1$ and $\mu_\omega(v) = 0$ for $v \in [0,1)$), then

$$N \text{ is } F \text{ is } \alpha\text{-possible} \rightarrow \Pi_X = G \tag{6.35}$$

where G is a fuzzy set of Type 2 defined by

$$\mu_G(u) = [\alpha \wedge \mu_G(u), \ \alpha \oplus (1 - \mu_F(u))], \quad u \in U \tag{6.36}$$

and $\oplus$ denotes the bounded sum (see (4.43)). The rules expressed by (6.33) and (6.35) should be regarded as provisional in nature, since further experience in the use of possibility distributions may suggest other, more appropriate, interpretations of the concept of possibility qualification.

MODIFIER RULES FOR QUALIFIED PROPOSITIONS

As in the case of translation rules of Types I, II and III, the modifier rule for propositions may be applied to translation rules of Type IV to yield, among others, the negation rule for qualified propositions. In what follows, we shall restrict our attention to the application of this rule to truth-qualified propositions.

Specifically, on applying the modifier rule for propositions to (6.7), we obtain the following general form of strong semantic equivalence

$$m(N \text{ is } F \text{ is } \tau) \leftrightarrow N \text{ is } F \text{ is } m\tau \tag{6.37}$$

which implies that

$$\text{not}(N \text{ is } F \text{ is } \tau) \leftrightarrow N \text{ is } F \text{ is not } \tau \tag{6.38}$$

$$\text{very}(N \text{ is } F \text{ is } \tau) \leftrightarrow N \text{ is } F \text{ is very } \tau \tag{6.39}$$

and

$$\text{more or less}(N \text{ is } F \text{ is } \tau) \leftrightarrow N \text{ is } F \text{ is more or less } \tau. \tag{6.40}$$

On the other hand, from (6.7) it also follows that

$$N \text{ is not } F \text{ is } \tau \leftrightarrow N \text{ is } F \text{ is ant } \tau \tag{6.41}$$

where ant τ is the antonym of τ. Thus, for example,

$$\text{false} \triangleq \text{ant true} \tag{6.42}$$

---

†A fuzzy set F is of Type 2 if, for each $u \in U$, $\mu_F(u)$ is a fuzzy subset of Type 1, i.e. $\mu_{\mu_F(u)} : [0,1] \rightarrow [0,1]$.

i.e.

$$\mu_{FALSE}(v) = \mu_{TRUE}(1-v), \quad v \in [0,1] \tag{6.43}$$

where FALSE and TRUE are the fuzzy denotations of false and true, respectively. Similarly, from (6.7) it follows that

$$N \text{ is very } F \text{ is } \tau \leftrightarrow N \text{ is } F \text{ is } {}^{0 \cdot 5}\tau \tag{6.44}$$

where the "left square-root" of τ is defined by

$$\mu_{0 \cdot 5\tau}(v) \triangleq \mu_\tau(v^2), \quad v \in [0,1] \tag{6.45}$$

and, more generally, for a "left-exponent" α,

$$\mu_{\alpha_\tau}(v) \triangleq u_\tau(v^{1/\alpha}), \quad v \in [0,1]. \tag{6.46}$$

On applying these rules in combination to a proposition such as "Barbara is not very rich," we are led to the following chain of semantically equivalent propositions:

Barbara is not very rich	(6.47)
Barbara is not very rich is u-true	(6.48)
Barbara is very rich is ant u-true	(6.49)
Barbara is rich is ${}^{0 \cdot 5}$(ant u-true)	(6.50)

where

$$\mu_{0 \cdot 5(\text{ant } u\text{-true})}(v) = 1 - v^2. \tag{6.51}$$

If *true* is assumed to be approximately semantically equivalent to *u-true*, the last proposition in the chain may be approximated by

$$\text{Barbara is rich is not very true.} \tag{6.52}$$

Thus, if we know that "Barbara is not very rich," then by using the chain of reasoning represented by (6.48), (6.49), (6.50) and (6.52), we can assert that an approximate answer to the question "Is Barbara rich?τ" is "not very true".

This example provides a very simple illustration of a combined use of the concepts of semantic equivalence and truth qualification for the purpose of deduction of an approximate answer to a given question, given a knowledge base consisting of a collection of fuzzy propositions. Additional illustrations relating to the application of PRUF to approximate reasoning may be found in Zadeh (1977*b*).

7. Examples of translation into PRUF

As was stated earlier, the translation rules formulated in the preceding sections are intended to serve as an aid to a human user in the translation of propositions (or descriptors) expressed in a natural language into PRUF. The use of the rules in question is illustrated by the following examples, with the understanding that, in general, in the translation of an expression, *e*, in a natural language into an expression, E, in PRUF, E is a procedure whose form depends on the frame of the database and hence is not unique.

For convenience of the reader, the notation employed in the examples is summarized below.

In a translation $e \rightarrow E$, if *w* is a word in *e* then its correspondent, W, in E is the name of a relation in $\mathscr{D}$ (the database).

$$F \triangleq \text{fuzzy relation with membership function } \mu_F$$
$$\Pi_X \triangleq \text{possibility distribution of the variable } X$$
$$\pi_X \triangleq \text{possibility distribution function of } \Pi_X \text{ (or X) ((2.2 et seq.)}$$
$$F[\Pi_{X(s)} = G] \triangleq \text{fuzzy relation F which is particularized by the proposition}$$
$$\text{``}X_{(s)} \text{ is G,'' where } X_{(s)} \text{ is a subvariable of the variable, X,}$$
$$\text{associated with F (2.26)}$$
$$x_{l_1} \times \cdots \times x_{l_k} \quad F \triangleq \text{Proj F on } U_{l_1} \times \ldots \times U_{l_k}, \ U_{l_k} \triangleq U_{l_k}(X_{l_k}) \text{ (2.23)}$$
$$F^2 \triangleq \text{square of F (4.21)}$$
$$\sqrt{F} \triangleq \text{square root of F (4.24)}$$
$$+ \triangleq \text{union or arithmetic sum}$$
$$\vee \triangleq \text{max}$$
$$\wedge \triangleq \text{min}$$
$$' \triangleq \text{complement (2.25)}$$
$$\cap \triangleq \text{intersection (footnote on p. 407)}$$
$$\times \triangleq \text{Cartesian product (footnote on p. 413)}$$
$$\oplus \triangleq \text{bounded sum (4.43)}$$
$$\perp \triangleq \text{unitor (2.15)}$$
$$\text{Count(F)} \triangleq \text{Cardinality (power) of F (4.70)}$$
$$\text{Prop(F)} \triangleq \text{Count(F)/Cardinality of universe of discourse (4.75)}$$
$$\text{Prop(F/G)} \triangleq \text{Count(F} \cap \text{G)/Count(G) (4.73)}$$
$$\text{Name}_i \triangleq \text{Name of } i\text{th object in a population}$$
$$\text{Support(F)} \triangleq \text{set of all points } u \text{ in U for which } \mu_F(u) > 0$$
$$U(X) \triangleq \text{universe of discourse associated with X}$$

Example (a)

$$\text{Ed is 30 years old} \rightarrow \text{Age(Ed)} = 30 \tag{7.1}$$

$$\text{Ed is young} \rightarrow \Pi_{\text{Age(Ed)}} = \text{YOUNG} \tag{7.2}$$

$$\text{Ed is not very young} \rightarrow \Pi_{\text{Age(Ed)}} = (\text{YOUNG}^2)', \tag{7.3}$$

where the frame of YOUNG is

YOUNG	Age	μ

Alternatively,

$$Ed\ is\ young \rightarrow ED[\Pi_{Age} = YOUNG]. \tag{7.4}$$

Example (b)

$$Sally\ is\ very\ intelligent \rightarrow \Pi_X = \perp^2, \tag{7.5}$$

where

$$X \underset{\mu}{\triangleq} INTELLIGENT[Name = Sally] \tag{7.6}$$

(that is, X is the degree of intelligence of Sally in the table

INTELLIGENT	Name	μ

).

Note that (7.5) implies that

$$\pi(X) = X^2, \quad X \in [0,1]. \tag{7.7}$$

Example (c)

$$Edith\ is\ tall\ and\ blonde \rightarrow \Pi_{(Height(Edith),\,Color(Hair(Edith)))} = TALL \times BLONDE. \tag{7.8}$$

Alternatively,

$$Edith\ is\ tall\ and\ blonde \rightarrow EDITH[\Pi_{Height} = TALL; \Pi_{Color(Hair)} = BLONDE]. \tag{7.9}$$

Example (d)

$$A\ man\ is\ tall \rightarrow \Pi_{Height(X)} = TALL \tag{7.10}$$

where X is the name of the tallest man in the relation

POPULATION	Name	Height

.

Equivalently,

$$A\ man\ is\ tall \rightarrow \Pi_{Height(Name_1)} = TALL \tag{7.11}$$
$$\text{or } \Pi_{Height(Name_2)} = TALL$$
$$\cdots\cdots\cdots\cdots\cdots$$
$$\text{or } \Pi_{Height(Name_N)} = TALL$$

Example (e)

$$All\ men\ are\ tall \rightarrow \Pi_{Height(X)} = TALL \tag{7.12}$$

where X is the name of the shortest man in the relation

POPULATION	Name	Height

.

Equivalently,

$$All\ men\ are\ tall \rightarrow \Pi_{Height(Name_1)} = TALL \tag{7.13}$$
$$\cdots\cdots\cdots\cdots\cdots$$
$$\Pi_{Height(Name_N)} = TALL$$

Example (f)

$$Most\ men\ are\ tall. \tag{7.14}$$

Case 1. The frame of $\mathscr{D}$ is comprised of

POPULATION	Name	μ

MOST	ρ	μ

where μ_i in POPULATION is the degree to which Name$_i$ is TALL, and μ_j in MOST is the degree to which ρ_j is compatible with MOST. Then

$$Most\ men\ are\ tall \rightarrow \Pi_{\text{Prop(TALL)}} = \text{MOST} \qquad (7.15)$$

where

$$\text{Prop(TALL)} = \frac{\sum_\mu \underset{i}{\text{POPULATION[Name}=\text{Name}_i]}}{\text{Count(POPULATION)}}. \qquad (7.16)$$

Case 2. The frame of $\mathscr{D}$ is comprised of

POPULATION	Name	Height

TALL	Height	μ

MOST	ρ	μ

In this case, the translation is still expressed by (7.15), but with Prop(TALL) given by

$$\text{Prop(TALL)} = \frac{\sum_\mu \underset{i}{\text{TALL[Height}=\,_{\text{Height}}\text{POPULATION[Name}=\text{Name}_i]]}}{\text{Count(POPULATION)}}. \qquad (7.17)$$

Example (g)

$$Three\ tall\ men \rightarrow \mu(X) = \text{Min}_i\ \mu_i\ \text{for Name}_i \in \text{Support}(X)\ \text{and} \qquad (7.18)$$
$$\text{Count(Support}(X)) = 3$$
$$= 0 \qquad \text{otherwise}$$

where X is a fuzzy subset of

POPULATION	Name	μ

and μ_i is the degree to which Name$_i$ is tall. The left-hand member of (7.18) is a descriptor, while the right-hand member defines the membership function of a fuzzy subset of the fuzzy power set of $_{\text{Name}}$POPULATION (i.e. the set of all fuzzy subsets of the names of individuals in POPULATION).

More generally,

$$\text{Several tall men} \rightarrow \mu(X) = \text{Min}_i \mu_i \wedge \mu_{\text{SEVERAL}}(\text{Count}(\text{Support}(X))) \qquad (7.19)$$

where, as in (7.18), Min_i is taken over all i such that $\text{Name}_i \in \text{Support}(X)$.

Example (h)

$$\text{Expensive red car with big trunk} \rightarrow \qquad (7.20)$$
$$\text{CAR}[\Pi_{\text{Price}} = \text{EXPENSIVE}; \ \Pi_{\text{Color}} = \text{RED}; \ \Pi_{\text{Size(Trunk)}} = \text{BIG}].$$

Example (i)

$$\text{John loves Pat} \rightarrow \Pi_X = \perp \qquad (7.21)$$

where

$$X \underset{\mu}{\triangleq} \text{LOVES}(\text{Name1} = \text{John}; \ \text{Name2} = \text{Pat}), \qquad (7.22)$$

with the right-hand member of (7.21) implying that

$$\pi(X) = X. \qquad (7.23)$$

It should be noted that in the special case where LOVES is a non-fuzzy relation, (7.21) reduces to the conventional predicate representation LOVES(John,Pat).

Example (j)

$$\text{John loves someone} \rightarrow \Pi_X = \perp \qquad (7.24)$$

where

$$\mu_i \underset{\mu}{\triangleq} \text{LOVES}[\text{Name1} = \text{John}; \ \text{Name2} = \text{Name}_i] \qquad (7.25)$$
$$\triangleq \text{degree to which John loves Name}_i$$

and

$$X \triangleq \text{Max}_i \ \mu_i. \qquad (7.26)$$

Note that when LOVES is a non-fuzzy relation, (7.24) reduces to $(\exists y)\text{LOVES}(\text{John},y)$.

Example (k)

$$\text{John loves everyone} \rightarrow \Pi_X = \perp \qquad (7.27)$$

where

$$X \triangleq \text{Min}_i \ \mu_i \qquad (7.28)$$

and μ_i is expressed by (7.25).

Example (l)

$$\text{Someone loves someone} \rightarrow \Pi_X = \perp \qquad (7.29)$$

where

$$X = \text{Max}_{i,j} \ \mu_{ij} \qquad (7.30)$$

and μ_{ij} is expressed by

$$\mu_{ij} \underset{\mu}{\triangleq} \text{LOVES}[\text{Name1} = \text{Name}_i; \ \text{Name2} = \text{Name}_j]. \qquad (7.31)$$

Example (m)

$$\text{Someone loves everyone} \rightarrow \Pi_X = \bot \qquad (7.32)$$

where

$$X \triangleq \text{Max}_i \text{ Min}_j \mu_{ij} \qquad (7.33)$$

and μ_{ij} is given by (7.31).

Example (n)

$$\text{Jill has many friends} \rightarrow \Pi_X = \text{MANY} \qquad (7.34)$$

where

$$X \triangleq \text{Count}(_{\mu \times \text{Name2}} \text{FRIENDS}(\text{Name1} = \text{Jill})). \qquad (7.35)$$

Note that the argument of Count is the fuzzy set of friends of Jill.

Example (o)

$$\text{The man near the door is young} \rightarrow \Pi_{\text{Age}(N)} = \text{YOUNG} \qquad (7.36)$$

where

$$N = \text{MAN} \cap_{\mu \times \text{Object1}} \text{NEAR}[\text{Object2} = \text{DOOR}]. \qquad (7.37)$$

Implicit in (7.37) is the assumption that the descriptor "The man near the door" identifies a man uniquely. The frame of MAN is

MAN	Name

.

Example (p)

$$\text{Kent was walking slowly toward the door} \rightarrow \text{WALKING}[\text{Name} = \text{Kent}; \qquad (7.38)$$
$$\Pi_{\text{Speed}} = \text{SLOW}; \ \Pi_{\text{Time}} = \text{PAST}; \ \Pi_{\text{Direction}} = \text{TOWARD}(\text{Object} = \text{DOOR})].$$

Example (q)

$$\text{Herta is not very tall is very true} \rightarrow \qquad (7.39)$$
$$\pi_{\text{Height}(\text{Herta})}(u) = \mu^2_{\text{TRUE}}(1 - \mu^2_{\text{TALL}}(u)), \quad u \in [0,200]$$

where the frames are

TRUE	v	μ

TALL	Height	μ

$v \in [0,1]$.

Example (r)

$$\text{Carole is very intelligent is very likely.} \qquad (7.40)$$

Let

$$v \triangleq_{\mu} \text{INTELLIGENT}[\text{Name} = \text{Carole}] \qquad (7.41)$$

i.e. v is the degree to which Carole is intelligent, and the frame of INTELLIGENT is assumed to be

INTELLIGENT	Name	μ

Then (see (7.5))

$$Carole\ is\ very\ intelligent \rightarrow \Pi_v = \perp^2 \qquad (7.42)$$

in which the right-hand member is equivalent to

$$\pi(v) = v^2. \qquad (7.43)$$

Next, let

$$X = \int_0^1 p(v)v^2 dv \qquad (7.44)$$

where $p(v)dv$ is the probability that Carole's degree of intelligence falls in the interval $[v, v+dv]$. Then, using the translation rule for probability qualification, we obtain

$$(Carole\ is\ very\ intelligent)\ is\ very\ likely \rightarrow \Pi_X = LIKELY^2 \qquad (7.45)$$

in which the right-hand member is equivalent to

$$\pi(p(\cdot)) = \mu^2_{LIKELY}\left(\int_0^1 p(v)v^2 dv\right) \qquad (7.46)$$

and the frame of LIKELY is

LIKELY	p	μ

$p \in [0,1]$. Expressed in this form, the translation defines a possibility distribution of the probability density function $p(\cdot)$.

Example (s)

$$X\ is\ small\ is\ very\ true\ is\ likely \rightarrow$$

$$\pi(p(\cdot)) = \mu_{LIKELY}\left(\int_0^\infty p(u)\mu^2_{TRUE}(\mu_{SMALL}(u))du\right) \qquad (7.47)$$

where $U \triangleq [0,\infty)$ and $p(u)du \triangleq Prob\{X \in [u, u+du]\}$. As in the previous example, (7.47) defines a possibility distribution of the probability density function of X.

Example (*t*)

$$\text{Men who are much taller than most men} \to F \qquad (7.48)$$

where the fuzzy subset F of POPULATION is computed by the following procedure. (For simplicity, the procedure is stated in plain English.)

Assume that the frame of $\mathcal{D}$ is comprised of:

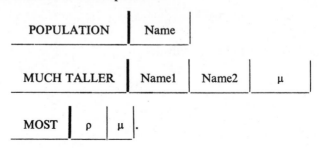

1. Compute
   $F_i \triangleq _{\mu \times \text{Name2}}\text{MUCH TALLER}[\text{Name1} = \text{Name}_i]$
   $\triangleq$ fuzzy set of men in relation to whom Name$_i$ is much taller.

2. Compute the relative cardinality of F_i, i.e.

$$\text{Prop}(F_i) = \frac{\text{Count}(F_i)}{\text{Count}(\text{POPULATION})}. \qquad (7.49)$$

3. Compute
   $\delta_i \triangleq \mu_{\text{MOST}}(\text{Prop}(F_i)) \qquad (7.50)$
   $\triangleq$ degree to which Name$_i$ is much taller than most men.

4. The fuzzy set of men who are much taller than most men is given by
   $$F = \delta_1/\text{Name}_1 + \ldots + \delta_N/\text{Name}_N \qquad (7.51)$$

where + denotes the union and Name$_1$, . . ., Name$_N$ are the elements of U(Name) in POPULATION. Alternatively, assume that the frame of $\mathcal{D}$ is comprised of:

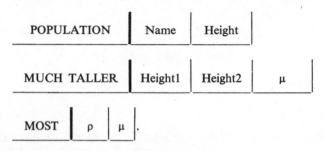

In this case, the procedure assumes the following form.

1. Compute
$$h_i \triangleq \text{Height}(\text{Name}_i) = {}_{\text{Height}}\text{POPULATION}[\text{Name} = \text{Name}_i]. \tag{7.52}$$

2. Compute
$$\gamma_{ij} = \text{MUCH TALLER}[\text{Height1} = h_i; \text{Height2} = h_j] \tag{7.53}$$
$\triangleq$ degree to which Name$_i$ is much taller than Name$_j$.

3. Compute the fuzzy set
$$F_i = \gamma_{i1}/\text{Name}_1 + \ldots + \gamma_{iN}/\text{Name}_N \tag{7.54}$$
$\triangleq$ fuzzy set of men in relation to whom Name$_i$ is much taller.

4. Same as Step 3 in previous procedure.

5. Same as Step 4 in previous procedure.

Example (u)
$$\text{Many men are much taller than most men} \rightarrow \pi(\text{POPULATION}) = \mu_G \tag{7.55}$$

where μ_G is computed by the following procedure.

Assume that the frame of $\mathscr{D}$ is comprised of

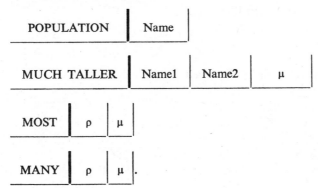

1. Compute F as in Example (t).
2. Compute
$$\gamma = \text{Prop}(F)$$
$\triangleq$ Proportion of men who are much taller than most men.
3. The possibility of the relation POPULATION is given by

$$\pi(\text{POPULATION}) = {}_\mu\text{MANY}[\rho = \gamma] \tag{7.56}$$

in which the right-hand member defines μ_G.

Example (v)
$$\text{Beth gave several big apples to each of her close friends} \rightarrow \pi(\text{GAVE}) = \mu_G. \tag{7.57}$$

The following procedure computes μ_G on the assumption that the frame of $\mathscr{D}$ is comprised of:

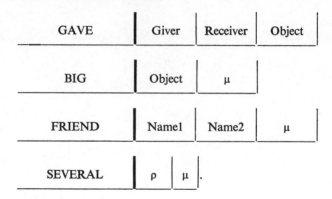

GAVE	Giver	Receiver	Object

BIG	Object	μ	

FRIEND	Name1	Name2	μ

SEVERAL	ρ	μ	

.

1. Compute
$$G_i \triangleq {}_{\text{Object}}\text{GAVE[Giver} = \text{Beth; Receiver} = \text{Name}_i]$$ (7.58)
   $\triangleq$ Set of objects received from Beth by Name_i.

2. Compute
$$H = \text{BIG[Object} = \text{APPLE]}$$ (7.59)
   $\triangleq$ fuzzy set of big apples.

3. Compute
$$K = G_i \cap H$$ (7.60)
   $\triangleq$ fuzzy set of big apples received from Beth by Name_i.

4. Compute
$$\gamma_i = {}_\mu \text{SEVERAL}[\rho = \text{Count}(K)]$$ (7.61)
   $\triangleq$ degree to which Name_i received several big apples from Beth.

5. Compute
$$\delta_i = {}_\mu \text{FRIEND}^2[\text{Name1} = \text{Beth; Name2} = \text{Name}_i]$$ (7.62)
   $\triangle$ degree to which Name_i is a close friend of Beth.

6. Compute
$$\sigma_i \triangleq 1 \wedge (1 - \delta_i + \gamma_i)$$ (7.63)
   $\triangleq$ degree to which (If Name_i is a close friend of Beth then Name_i received several big apples from Beth).

7. Compute
$$\pi(\text{GAVE}) \triangleq \text{Min}_i \; \sigma_i$$ (7.64)
   $\triangleq$ degree to which all close friends of Beth received from her several big apples.

It should be noted that when the translation of a proposition, p, into PRUF requires the execution of a procedure, P, which cannot be expressed as a relatively simple expression in PRUF—as is true of Examples (t), (u) and (v)—the relationship between p and P ceases to be transparent. A higher degree of transparency in cases of this type may be achieved through the introduction into PRUF of higher-level constructions relating to quantification, qualification, particularization and definition. This and other issues concerning the translation of more complex propositions than those considered here will be treated in subsequent papers.

8. Concluding remarks

In essence, PRUF may be regarded as a relation-manipulating language which serves the purposes of (a) precisiation of expressions in a natural language; (b) exhibiting their logical structure; and (c) providing a system for the characterization of the meaning of a proposition by a procedure which acts on a collection of fuzzy relations in a database and returns a possibility distribution.

By serving these purposes, PRUF provides a basis for a formalization of approximate reasoning. More specifically, through the use of PRUF, a set of imprecise premises expressed in a natural or synthetic language may be translated into possibility distributions to which the rules of inference in FL (or PRUF) may be applied, yielding other possibility distributions which upon retranslation lead to approximate consequents of the original premises. In this respect, PRUF plays the same role in relation to fuzzy premises and fuzzy conclusions that predicate calculus does in relation to non-fuzzy premises and non-fuzzy conclusions.

An important aspect of PRUF is a concomitant of its break with the long-standing tradition in logic, linguistics and philosophy of language—the tradition of employing the concept of truth as a foundation for theories of meaning. By adopting instead the concept of a possibility distribution as its point of departure, PRUF permits a uniform treatment of truth-qualification, probability-qualification and possibility-qualification of fuzzy propositions and thereby clarifies the roles played by the concepts of truth, probability and possibility not only in logic and language theory, but also in information analysis, decision analysis and related application areas.

As was stated in the Introduction, our exposition of PRUF in the preceding sections is neither definitive nor complete. There are many issues that remain to be explored, the most complex of which is that of automatic translation from a natural language into PRUF. However, to view this issue in a proper perspective, it must be recognized that the existing systems for automatic translation from a small subset of a natural language into a meaning representation language (and especially, a query language) have very narrow versatility since they are limited in their use to highly restricted domains of semantic discourse and human concept comprehension.

Although PRUF is still in its initial stages of development, its somewhat unconventional conceptual framework puts into a different perspective many of the long-standing issues in language theory and knowledge representation, expecially those pertaining to vagueness, uncertainty and inference from fuzzy propositions. By so doing, PRUF points a way toward the conception of question-answering systems having the capability to act on imprecise, incomplete or unreliable information which is resident in a database. To implement such systems, however, we shall need (a) a better system of linguistic modifiers than those that are available in natural languages, and (b) special-purpose hardware that is oriented toward the storage and manipulation of fuzzy rather than non-fuzzy data.

I wish to thank Barbara Cerny, Christian Freksa and Lucia Vaina for reading the manuscript and offering helpful criticisms. I also wish to acknowledge stimulating discussions with Zoltan Domotor, Brian Gaines, Ellen Hisdal, Hung Nguyen, Paul Kay, Elie Sanchez, Pat Suppes and

Hans Zimmermann on various issues related to those considered in this paper. This research was supported by National Science Foundation Grant MCS77–07568 and Naval Electronic Systems Command Contract N00039–77–C–0022.

References

AIZERMAN, M. A. (1976). Fuzzy sets, fuzzy proofs and certain unsolved problems in the theory of automatic control. *Avtomatika i Telemehanika*, 171–177.

ANDERSON, J. & BOWER, G. (1973). *Human Associative Memory*. Washington, D.C.: Winston.

BALLMER, T. T. (1976). Fuzzy punctuation or the continuum of grammaticality. *Electronics Research Laboratory Memo M-590*. University of California, Berkeley.

BAR-HILLEL, Y. (1964). *Language and Information*. Reading, Mass.: Addison-Wesley.

BELLINGER, D. (1972). *Degree Words*. The Hague: Mouton.

BELLMAN, R. E. & GIERTZ, M. (1973). On the analytic formalism of the theory of fuzzy sets. *Information Sciences*, **5**, 149–156.

BELLMAN, R. E. & ZADEH, L. A. (1976). Local and fuzzy logics. *Electronics Research Laboratory Memorandum M-584*. University of California, Berkeley. In EPSTEIN, G., Ed., *Modern Uses of Multiple-Valued Logic*. Dordrecht: D. Reidel, pp. 103–165.

BEZDEK, J. D. & DUNN, J. C. (1975). Optimal fuzzy partitions: a heuristic for estimating the parameters in a mixture of normal distributions. *IEEE Transactions on Computers*, **C-24**, 835–838.

BIERWISCH, M. (1970). Semantics. In LYONS, J., Ed., *New Horizons in Linguistics*. Baltimore: Penguin.

BISS, K., CHIEN, K. & STAHL, F. (1971). R2—a natural language question–answering system. *Proceedings of the Fall Joint Computer Conference*, **38**, 303–308.

BLACK, M. (1963). Reasoning with loose concepts. *Dialogue*, **2**, 1–12.

BOBROW, D. G. (1977). A panel on knowledge representation. *Proceedings of the Fifth International Conference on Artificial Intelligence*, M.I.T., Cambridge, Mass., pp. 983–992.

BOBROW, D. & COLLINS, A., Eds (1975). *Representation and Understanding*. New York: Academic Press.

BOBROW, D. G. & WINOGRAD, T. (1977). An overview of KRL, a knowledge representation language. *Cognitive Science*, **1**, 3–46.

BOYCE, R. F., CHAMBERLIN, D. D., KING, W. F. III & HAMMER, M. M. (1974). Specifying queries as relational expressions. In KLIMBIE, J. W. & KOFFEMAN, K. L., Eds, *Data Base Management*. Amsterdam: North-Holland, pp. 169–176.

BRACCHI, G., FEDELI, A. & PAOLINI, P. (1972). A multilevel relational model for data base management systems. In KLIMBIE, J. W. & KOFFEMAN, K. L., Eds, *Data Base Management*. Amsterdam: North-Holland, pp. 211–223.

BRACHMAN, R. J. (1977). What's in a concept: structural foundations for semantic networks. *International Journal of Man–Machine Studies*, **9**, 127–152.

BRIABRIN, V. M. & POSPELOV, D. A. (1977). DILOS—Dialog system for information retrieval, computation and logical inference. In MICHIE, D., Ed., *Machine Intelligence*, **19**. New York: American Elsevier.

BRIABRIN, V. M. & SENIN, G. V. (1977). Natural language processing within a restricted context. *Proceedings of the International Workshop on Natural Language for Interactions with Data Bases*, IIASA, Vienna.

CARBONELL, J. R. & COLLINS, A. M. (1973). Natural semantics in artificial intelligence. *Proceedings of the Third Joint Conference on Artificial Intelligence*, Stanford University, Stanford, pp. 344–351.

CARLSTROM, I. F. (1975). Truth and entailment for a vague quantifier. *Synthese*, **30**, 461–495.

CARNAP, R. (1937). *The Logical Syntax of Language*. New York: Harcourt, Brace & World.

CHAMBERLIN, D. D. & BOYCE, R. F. (1974). SEQUEL: a structured English query language. *Proceedings of the ACM SIGMOD Workshop on Data Description, Access and Control*, pp. 249–264.

CHANG, C. L. (1975). Interpretation and execution of fuzzy programs. In ZADEH, L. A. *et al.* (1975), op. cit., pp. 191–218.

CHANG, C. L. (1976). DEDUCE—a deductive query language for relational data bases. In CHEN, C. H., Ed., *Pattern Recognition and Artificial Intelligence*. New York: Academic Press, pp. 108–134.

CHANG, S. K. & KE, J. S. (1976). Database skeleton and its application to fuzzy query translation. Dept of Information Engineering, University of Illinois, Chicago, Illinois.

CHANG, S. S. L. (1972). Fuzzy mathematics, man and his environment. *IEEE Transactions on Systems, Man and Cybernetics*, SMC-2, 92–93.

CHARNIAK, E. (1972). Toward a model of children's story comprehension. *AITR-266*. Artificial Intelligence Laboratory, M.I.T., Cambridge, Mass.

CHARNIAK, E. (1973). Context and the reference problem. In RUSTIN, R. (1973), op. cit.

CHARNIAK, E. (1975). Organization and inference in a frame-like system of common sense knowledge. In SCHANK, R. & NASH-WEBBER, B. L., Eds, *Theoretical Issues in Natural Language Processing*. Cambridge.

CHOMSKY, N. (1957). *Syntactic Structures*. The Hague: Mouton.

CHOMSKY, N. (1965). *Aspects of the Theory of Syntax*. Cambridge, Mass.: M.I.T. Press.

CHOMSKY, N. (1971). Deep structure, surface structure, and semantic interpretation. In STEINBERG, D. D. & JAKOBOVITS, L. A., Eds, *Semantics: An Interdisciplinary Reader in Philosophy, Linguistics and Psychology*. Cambridge: Cambridge University Press.

CLIFF, N. (1959). Adverbs as multipliers. *Psychology Review*, 66, 27–44.

CODD, E. F. (1971). A data base sublanguage founded on the relational calculus. *Proceedings of the ACM SIGFIDET Workshop on Data Description, Access and Control*, pp. 35–68.

CODD, E. F. (1974). Seven steps to rendezvous with the casual user. In KLIMBIE, J. W. & KOFFEMAN, K. L., Eds, *Data Base Managemment.* Amsterdam: North-Holland, pp. 179–199.

CRESSWELL, M. J. (1973). *Logics and Languages*. London: Methuen.

DAMERAU, F. J. (1975). On fuzzy adjectives. *Memorandum RC 5340*. IBM Research Laboratory, Yorktown Heights, N.Y.

DAVIDSON, D. (1964). The method of extension and intension. In SCHILPP, Ed., *The Philosophy of Rudolf Carnap*. La Salle, Illinois: Open Court, pp. 311–350.

DAVIDSON, D. (1967). Truth and meaning. *Synthese*, 17, 304–323.

DEFINETTI, B. (1974). *Probability Theory*. New York: Wiley.

DELUCA, A. & TERMINI, S. (1972a). A definition of a non-probabilistic entropy in the setting of fuzzy sets theory. *Information and Control*, 20, 301–312.

DELUCA, A. & TERMINI, S. (1972b). Algebraic properties of fuzzy sets. *Journal of Mathematical Analysis and Applications*, 40, 373–386.

DIMITROV, V. D. (1975). Efficient governing hymanistic systems by fuzzy instructions. *Third International Congress of General Systems and Cybernetics*, Bucarest.

DREYFUSS, G. R., KOCHEN, M., ROBINSON, J. & BADRE, A. N. (1975). On the psycholinguistic reality of fuzzy sets. In GROSSMAN, R. E., SAN, L. J. & VANCE, T. J., Eds, *Functionalism*. Chicago, Illinois: University of Chicago Press.

DUDA, R. O., HART, P. E., NILSSON, N. & SUTHERLAND, G. L. (1977). Semantic network representations in rule-based inference systems. *Technical Note 136*. Artificial Intelligence Center, Stanford Research Institute, Menlo Park, Calif. In WATERMAN, D. A. & HAYES-ROTH, F., Eds, *Pattern-directed Inference Systems*. New York: Academic Press, to appear.

EVANS, G. & MCDOWELL, J. (1976). *Truth and Meaning*. Oxford: Clarendon Press.

FELLINGER, W. L. (1974). Specifications for a fuzzy systems modelling language. Ph.D. thesis. Oregon State University, Corvallis.

FILLMORE, C. J. (1968). The case for case. In BACH, E. & HARMS, R. T., Eds, *Universals in Linguistic Theory*. New York: Holt, Rinehart & Winston.

FINE, K. (1975). Vagueness, truth and logic. *Synthese*, 30, 265–300.

FODOR, J. A. (1975). *The Language of Thought*. New York: Crowell.

FODOR, J. A. & KATZ, J. J., Eds (1964). *The Structure of Language: Readings in the Philosophy of Language*. Englewood Cliffs, N.J.: Prentice-Hall.

FREDERIKSEN, C. (1975). Representing logical and semantic structure of knowledge acquired from discourse. *Cognitive Psychology*, **7**, 371–458.

GAINES, B. R. (1976a). General fuzzy logics. *Proceedings of the Third European Meeting on Cybernetics and Systems Research*, Vienna.

GAINES, B. R. (1976b). Foundations of fuzzy reasoning. *International Journal of Man–Machine Studies*, **6**, 623–668.

GAINES, B. R. & KOHOUT, L. J. (1975). Possible automata. *Proceedings of International Symposium on Multiple-Valued Logic*, Bloomington, Indiana, pp. 183–196.

GAINES, B. R. & KOHOUT, L. J. (1977). The fuzzy decade: a bibliography of fuzzy systems and closely related topics. *International Journal of Man–Machine Studies*, **9**, 1–68.

GILES, R. (1976). Łukasiewicz logic and fuzzy set theory. *International Journal of Man–Machine Studies*, **8**, 313–327.

GOGUEN, J. A. (1969). The logic of inexact concepts. *Synthese*, **19**, 325–373.

GOGUEN, J. A. (1974). Concept representation in natural and artificial languages: axioms, extension and applications for fuzzy sets. *International Journal of Man–Machine Studies*, **6**, 513–561.

GOLDSTEIN, I. & PAPERT, S. (1977). Artificial intelligence, language and the study of knowledge. *Cognitive Science*, **1**, 84–123.

GOTTINGER, H. W. (1973). Toward a fuzzy reasoning in the behavioral science. *Cybernetica*, **2**, 113–135.

GREENWOOD, D. (1957). *Truth and Meaning*. New York: Philosophical Library.

GRICE, H. P. (1968). Utterer's meaning, sentence-meaning and word-meaning. *Foundations of Language*, **4**, 225–242.

HAACK, S. (1974). *Deviant Logic*. Cambridge: Cambridge University Press.

HAACK, S. (1976). The pragmatist theory of truth. *British Journal of the Philosophy of Science*, **27**, 231–249.

HAMACHER, H. (1976). On logical connectives of fuzzy statements and their affiliated truth functions. *Proceedings of the Third European Meeting on Cybernetics and Systems Research*, Vienna.

HARRIS, J. I. (1974a). Fuzzy implication—comments on a paper by Zadeh. DOAE Research Working Paper, Ministry of Defence, Byfleet, Surrey, U.K.

HARRIS, J. I. (1974b). Fuzzy sets: how to be imprecise precisely. DOAE Research Working Paper, Ministry of Defense, Byfleet, Surrey, U.K.

HELD, G. D. & STONEBRAKER, M. R. (1975). Storage structures and access methods in the relational data base management system INGRES. *Proceedings of ACM Pacific Meeting 1975*, pp. 26–33.

HELD, G. D., STONEBRAKER, M. R. & WONG, E. (1975). INGRES—a relational data base system. *Proceedings AFIPS-National Computer Conference*, **44**, 409–416.

HENDRIX, G. G. (1975). Expanding the utility of semantic networks through partitioning. *Advance Papers of the Fourth International Joint Conference on Artificial Intelligence Tbilisi, 1975*, pp. 115–121.

HENDRIX, G. G. (1977). The LIFER manual. *Technical Note 138*. Stanford Research Institute, Menlo Park, California.

HENDRIX, G. G., THOMPSON, C. W. & SLOCUM, J. (1973). Language processing via canonical verbs and semantics models. *Proceedings of the Third Joint International Conference on Artificial Intelligence*, Stanford, pp. 262–269.

HERSH, H. M. & CARAMAZZA, A. (1976). A fuzzy set approach to modifiers and vagueness in natural language. *Journal of Experimental Psychology*, **105**, 254–276.

HINTIKKA, J. (1967). Individuals, possible worlds and epistemic logic, *Nous*, **1**, 33–62.

HUGHES, G. E. & CRESSWELL, M. J. (1968). *An Introduction to Modal Logic*. London: Methuen.

INAGAKI, Y. & FUKUMURA, F. (1975). On the description of fuzzy meaning of context-free language. In ZADEH, L. A. *et al.* (1975), op. cit., pp. 301–328.

JACOBSON, R., Ed. (1961). *On the Structure of Language and Its Mathematical Aspects.* Providence, R.I.: American Mathematical Society.

JOSHI, A. K. & ROSENSCHEIN, S. J. (1976). Some problems of inferencing: relation of inferencing to decomposition of predicates. *Proceedings of International Conference on Computational Linguistics*, Ottawa, Canada.

JOUAULT, J. P. & LUAN, P. M. (1975). Application des concepts flous a la programmation en languages quasi-naturels. Inst. Inf. d'Entreprise, C.N.A.M., Paris.

KAMPÉ DE FERIET, J. (1977). Mesure de l'information fournie par un évenement. In PICARD, C. F., Ed., *Structures de l'Information.* Paris: Centre National de la Recherche Scientifique, University of Pierre and Marie Curie, pp. 1–30.

KAMPÉ DE FERIET, J. & FORTE, B. (1967). Information et probabilité. *Comptes Rendus, Academy of Sciences (Paris)*, **265A**, 142–146, 350–353.

KATZ, J. J. (1964). Analyticity and contradiction in natural language. In FODOR, J. A. & KATZ, J. J. (1964), op. cit.

KATZ, J. J. (1966). *The Philosophy of Language.* New York: Harper & Row.

KATZ, J. J. (1967). Recent issues in semantic theory. *Foundations of Language*, **3**, 124–194.

KAUFMANN, A. (1973). *Introduction to the Theory of Fuzzy Subsets, Vol. 1, Elements of Basic Theory.* Paris: Masson and Co. English translation. New York: Academic Press.

KAUFMANN, A. (1975a). *Introduction to the Theory of Fuzzy Subsets, Vol. 2. Applications to Linguistics, Logic and Semantics.* Paris: Masson and Co.

KAUFMANN, A. (1975b). *Introduction to the Theory of Fuzzy Subsets, Vol. 3. Applications to Classification and Pattern Recognition, Automata and Systems and Choice of Criteria.* Paris: Masson and Co.

KAY, P. (1975). A model-theoretic approach to folk taxonomy. *Social Science Information*, **14**, 151–166.

KELLOGG, C. H., BURGER, J., DILLER, T. & FOGT, K. (1971). The CONVERSE natural language data management system: current status and plans. *Proceedings of ACM Symposium on Information Storage and Retrieval*, University of Maryland, College Park, pp. 33–46.

KHATCHADOURIAN, H. (1965). Vagueness, meaning and absurdity. *American Philosophical Quarterly*, **2**, 119–129.

KLEIN, S. (1973). Automatic inference of semantic deep structure rules in generative semantic grammars. *Technical Report No. 180.* Computer Science Department, University of Wisconsin, Madison.

KLING, R. (1974). Fuzzy PLANNER: reasoning with inexact concepts in a procedural problem-solving language. *Journal of Cybernetics*, **4**, 105–122.

KNEALE, W. C. (1972). Propositions and truth in natural languages. *Mind*, **81**, 225–243.

KNUTH, D. E. (1968). Semantics of context-free languages. *Mathematical Systems Theory*, **2**, 127–145.

KOCHEN, M. & BADRE, A. N. (1974). On the precision of adjectives which denote fuzzy sets. *Journal of Cybernetics*, **4**, 49–59.

KRIPKE, S. (1963). Semantical analysis of modal logic I. *Zeitshrift für Mathematische Logik und Grundlagen der Mathematik*, **9**, 67–96.

KRIPKE, S. (1971). Naming and necessity. In DAVIDSON, D. & HARMAN, G., Eds, *Semantics of Natural Languages.* Dordrecht, Holland: D. Reidel.

KUHNS, J. L. (1967). Answering questions by computer. *Memorandum RM-5428-PR.* Rand Corporation, Santa Monica, California.

LABOV, W. (1973). The boundaries of words and their meanings. In BAILEY, C.-J. N. & SHUY, R. W., Eds, *New Ways of Analyzing Variation in English, Vol. 1.* Washington: Georgetown University Press.

LAKOFF, G. (1971). Linguistics and natural logic. In DAVIDSON, D. & HARDMAN, G., Eds, *Semantics of Natural Languages*. Dordrecht, Holland: D. Reidel.

LAKOFF, G. (1973*a*). Hedges: a study in meaning criteria and the logic of fuzzy concepts. *Journal of Philosophical Logic*, **2**, 458–508. In HOCKNEY, D., HARPER, W. & FREED, B., Eds, *Contemporary Research in Philosophical Logic and Linguistic Semantics*. Dordrecht, Holland: D. Riedel, pp. 221–271.

LAKOFF, G. (1973*b*). Fuzzy grammar and the performance/competence terminology game. *Proceedings of Meeting of Chicago Linguistics Society*, pp. 271–291.

LAMBERT, K. & VAN FRAASSEN, B. C. (1970). Meaning relations, possible objects and possible worlds. *Philosophical Problems in Logic*, 1–19.

LEE, E. T. (1972). Fuzzy languages and their relation to automata. *Ph.D. thesis*. Department of Electrical Engineering and Computer Sciences, University of California, Berkeley.

LEE, E. T. & CHANG, C. L. (1971). Some properties of fuzzy logic. *Information and Control*, **19**, 417–431.

LEE, E. T. & ZADEH, L. A. (1969). Note on fuzzy languages. *Information Sciences*, **1**, 421–434.

LEFAIVRE, R. A. (1974*a*). FUZZY: a programming language for fuzzy problem solving. *Technical Report 202*. Department of Computer Science, University of Wisconsin, Madison.

LEFAIVRE, R. A. (1974*b*). The representation of fuzzy knowledge. *Journal of Cybernetics*, **4**, 57–66.

LEHNERT, W. (1977). Human and computational question answering. *Cognitive Science*, **1**, 47–73.

LEWIS, D. K. (1970). General semantics. *Synthese*, **22**, 18–67.

LEWIS, P. M., ROSENKRANTZ, D. J. & STEARNS, R. E. (1974). Attributed translations. *Journal of Computer and System Sciences*, **9**, 279–307.

LINSKY, L. (1971). *Reference and Modality*. London: Oxford University Press.

LUCAS, P. *et al.* (1968). Method and notation for the formal definition of programming languages. *Report TR 25.087*. IBM Laboratory, Vienna.

LYNDON, R. C. (1966). *Notes on Logic*. New York: D. Van Nostrand.

LYONS, J. (1968). *Introduction to Theoretical Linguistics*. Cambridge: Cambridge University Press.

MACHINA, K. F. (1972). Vague predicates. *American Philosophical Quarterly*, **9**, 225–233.

MAMDANI, E. H. (1976). Advances in the linguistic synthesis of fuzzy controllers. *International Journal of Man–Machine Studies*, **8**, 669–678.

MAMDANI, E. H. & ASSILIAN, S. (1975). An experiment in linguistic synthesis with a fuzzy logic controller. *International Journal of Man–Machine Studies*, **7**, 1–13.

MARINOS, P. N. (1969). Fuzzy logic and its application to switching systems. *IEEE Transactions on Electronic Computers*, **EC-18**, 343–348.

MARTIN, W. A. (1973). Translation of English into MAPL using Winograd's syntax, state transition networks, and a semantic case grammar. *M.I.T. APG Internal Memo 11*. Project MAC, M.I.T., Cambridge, Mass.

MCCARTHY, J. & HAYES, P. (1969). Some philosophical problems from the standpoint of artificial intelligence. In MICHIE, D. & MELTZER, B., Eds, *Machine Intelligence 4*. Edinburgh University Press, pp. 463–502.

MCDERMOTT, D. V. (1974). Assimilation of new information by a natural language-understanding system. *AI TR-291*. M.I.T.

MESEGUER, J. & SOLS, I. (1975). Fuzzy semantics in higher order logic and universal algebra. University of Zaragoza, Spain.

MILLER, G. A. & JOHNSON-LAIRD, P. N. (1976). *Language and Perception*. Cambridge: Harvard University Press.

MINSKY, M. (1975). A framework for representing knowledge. In WINSTON, P., Ed., *The Psychology of Computer Vision*. New York: McGraw-Hill.

MISHELEVICH, D. J. (1971). MEANINGEX—a computer-based semantic parse approach to the analysis of meaning. *Proceedings of Fall Joint Computer Conference*, **39**, 271–280.

MIZUMOTO, M., UMANO, M. & TANAKA, K. (1977). Implementation of a fuzzy-set-theoretic data structure system. *Third International Conference on Very Large Data Bases*, Tokyo, Japan, 6–8 October 1977. *ACM Transactions on Data Base Systems*, to appear.

MIZUMOTO, M. & TANAKA, K. (1976). Algebraic properties of fuzzy numbers. *Proceedings of the International Conference on Cybernetics and Society*, Washington, D.C., pp. 559–563.

MOISIL, G. C. (1975). Lectures on the logic of fuzzy reasoning. *Scientific Editions*. Bucarest.

MONTAGUE, R. (1974). *Formal Philosophy (Selected Papers)*. New Haven: Yale University Press.

MONTGOMERY, C. A. (1972). Is natural language an unnatural query language? *Proceedings of ACM National Conference*, New York, pp. 1075–1078.

MOORE, J. & NEWELL, A. (1973). How can MERLIN understand? In GREGG, L., Ed., *Knowledge and Cognition*. Hillsdale, N.J.: Lawrence Erlbaum Associates.

MYLOUPOULOS, J., BORGIDA, A., COHEN, P., ROUSSOPOULOS, N., TSOTSOS, J. & WONG, H. (1976). TORUS: a step towards bridging the gap between data bases and the casual user. *Information Systems*, **2**, 49–64.

MYLOUPOULOS, J., SCHUSTER, S. A. & TSICHRITZIS, D. C. (1975). A multi-level relational system. *Proceedings of AFIPS-National Computer Conference*, **44**, 403–408.

NAGAO, M. & TSUJII, J.-I. (1977). Programs for natural language processing. *Information Processing Society of Japan*, **18**(1), 63–75.

NAHMIAS, S. (1976). Fuzzy variables. *Technical Report 33*. Department of Industrial Engineering, Systems Management Engineering and Operations Research, University of Pittsburgh.

NALIMOV, V. V. (1974). *Probabilistic Model of Language*. Moscow: Moscow State University.

NASH-WEBBER, B. (1975). The role of semantics in automatic speech understanding. In BOBROW, D. G. & COLLINS, A. M. (1975), op. cit., pp. 351–383.

NEGOITA, C. V. & RALESCU, D. A. (1975). *Applications of Fuzzy Sets to Systems Analysis*. Basel, Stuttgart: Birkhauser Verlag.

NEWELL, A. & SIMON, H. A. (1972). *Human Problem Solving*. Englewood Cliffs, N.J.: Prentice-Hall.

NGUYEN, H. T. (1976a). On fuzziness and linguistic probabilities. *Memorandum M-595*. Electronics Research Laboratory, University of California, Berkeley.

NGUYEN, H. T. (1976b). A note on the extension principle for fuzzy sets. *Memorandum M-611*. Electronics Research Laboratory, University of California, Berkeley.

NOGUCHI, K., UMANO, M., MIZUMOTO, M. & TANAKA, K. (1976). Implementation of fuzzy artificial intelligence language FLOU. *Technical Report on Automation and Language of IECE*.

NORMAN, D. A., RUMELHART, D. E. & LNR RESEARCH GROUP. *Explorations in Cognition*. San Francisco: W. H. Freeman.

PAL, S. K. & MAJUMDAR, D. D. (1977). Fuzzy sets and decision-making approaches in vowel and speaker recognition. *IEEE Transactions on Systems, Man and Cybernetics*, SMC-7, 625–629.

PARSONS, C. (1974). Informal axiomatization, formalization and the concept of truth. *Synthese*, **27**, 27–47.

PARTEE, B. (1976a). Possible world semantics and linguistic theory. Department of Linguistics, University of Massachusetts, Amherst. *Monist*, to appear.

PARTEE, B. (1976b). *Montague Grammar*. New York: Academic Press.

PARTEE, B. (1977). Montague grammar, mental representations and reality. *Proceedings of the Symposium on Philosophy and Grammar*, University of Uppsala, Uppsala, Sweden, to appear.

PETRICK, S. R. (1973). Semantic interpretation in the REQUEST system. *IBM Research Report RC4457*. IBM Research Center, Yorktown Heights, New York.

PROCYK, T. J. (1976). Linguistic representation of fuzzy variables. Fuzzy Logic Working Group, Queen Mary College, London, U.K.

PUTNAM, H. (1975). The meaning of "meaning." In GUNDERSON, K., Ed., *Language, Mind and Knowledge*. Minneapolis, Minn.: University of Minnesota Press.

PUTNAM, H. (1976). *Meaning and Truth*. Sherman Lectures, University College, London, U.K.

PYOTROVSKII, R. G., BEKTAYEV, K. B. & PYOTROVSKAYA, A. A. (1977). *Mathematical Linguistics*. Moscow: Higher Education Press.

QUILLIAN, M. R. (1968). Semantic memory. In MINSKY, M., Ed., *Semantic Information Processing*. Cambridge: M.I.T. Press.

QUINE, W. V. (1970a). *Philosophy of Logic*. Englewood Cliffs, N.J.: Prentice-Hall.

QUINE, W. V. (1970b). Methodological reflections on current linguistic theory. *Synthese*, **21**, 387–398.

RESCHER, N. (1969). *Many-Valued Logic*. New York: McGraw-Hill.

RESCHER, N. (1973). *The Coherence Theory of Truth*. Oxford: Oxford University Press.

RESCHER, N. (1975). *Theory of Possibility*. Pittsburgh: University of Pittsburgh Press.

RIEGER, B. (1976). Fuzzy structural semantics. *Proceedings of Third European Meeting on Cybernetics and Systems Research*, Vienna.

RIEGER, C. (1976). An organization of knowledge for problem solving and language comprehension. *Artificial Intelligence*, **7**, 89–127.

RÖDDER, W. (1975). On "and" and "or" connectives in fuzzy set theory. Institute for Operations Research, Technical University of Aachen.

ROSCH, E. (1973). On the internal structure of perceptual and semantic categories. In MOORE, T. M., Ed., *Cognitive Development and the Acquisition of Language*. New York: Academic Press.

ROSCH, E. (1975). Cognitive representations of semantic categories. *Journal of Experimental Psychology: General*, **104**, 192–233.

ROSS, J. R. (1970). A note on implicit comparatives. *Linguistic Inquiry*, **1**, 363–366.

ROUSSOPOULOS, N. D. (1976). A semantic network model of data bases. Department of Computer Science, University of Toronto, Canada.

RUSTIN, R., Ed. (1973). *Courant Computer Science Symposium 8: Natural Language Processing*. New York: Algorithmics Press.

SAGER, N. (1977). Natural language analysis and processing. In BELZER, J., HOLZMAN, A. G. & KENT, A., Eds. New York: Marcel Dekker, to appear.

SANCHEZ, E. (1974). Fuzzy relations. Faculty of Medicine, University of Marseille, France.

SANCHEZ, E. (1977). On possibility qualification in natural languages. *Electronics Research Laboratory Memorandum M77/28*. University of California, Berkeley.

SANDEWALL, E. (1970). Formal methods in the design of question-answering systems. *Report No. 28*. Uppsala University, Sweden.

SANFORD, D. H. (1975). Borderline logic. *American Philosophical Quarterly*, **12**, 29–39.

SANTOS, E. (1970). Fuzzy algorithms. *Information and Control*, **17**, 326–339.

SCHANK, R. C. (1973). Identification of conceptualizations underlying natural language. In SCHANK, R. & COLBY, K., Eds, *Computer Models of Thought and Language*. Englewood Cliffs, N.J.: Prentice-Hall.

SCHANK, R. C., Ed. (1975). *Conceptual Information Processing*. Amsterdam: North-Holland.

SCHOTCH, P. K. (1975). Fuzzy modal logic. *Proceedings of International Symposium on Multiple-Valued Logic*. University of Indiana, Bloomington, pp. 176–182.

SCHUBERT, L. K. (1972). Extending the expressive power of semantic networks. *Artificial Intelligence*, **2**, 163–198.

SEARLE, J., Ed. (1971). *The Philosophy of Language*. Oxford: Oxford University Press.

SHAUMJAN, S. K. (1965). *Structural Linguistics*. Moscow: Nauka.

SHIMURA, M. (1975). An approach to pattern recognition and associative memories using fuzzy logic. In ZADEH, L. A. *et al.* (1975), op. cit., pp. 449–476.

SHORTLIFFE, E. (1976). *MYCIN: Computer-based Medical Consultations*. New York: American Elsevier.

SHORTLIFFE, E. H. & BUCHANAN, B. G. (1975). A model of inexact reasoning in medicine. *Mathematical Biosciences*, **23**, 351–379.

SIMMONS, R. F. (1973). Semantic networks, their computation and use for understanding English sentences. In SCHANK, R. & COLBY, K., Eds, *Computer Models of Thought and Language*. Englewood Cliffs, N.J.: Prentice-Hall, pp. 63–113.

SIMON, H. A. (1973). The structure of ill structured problems. *Artificial Intelligence*, **4**, 181–201.

SIMON, H. A. & SIKLOSSY, L. (1972). *Representation and Meaning: Experiments with Information Processing Systems*. Englewood Cliffs, N.J.: Prentice-Hall.

SIY, P. & CHEN, C. S. (1974). Fuzzy logic for handwritten numerical character recognition. *IEEE Transactions on Systems, Man and Cybernetics*, **SMC-4**, 570–575.

SLOMAN, A. (1971). Interactions between philosophy and artificial intelligence: the role of intuition and non-logical reasoning in intelligence. *Artificial Intelligence*, **2**, 209–225.

SRIDHARAN, M. J. (1976). A frame-based system for reasoning about actions. *Technical Report CBM-TM-56*. Department of Computer Science, Rutgers University, New Brunswick, N.J.

STAAL, J. F. (1969). Formal logic and natural languages. *Foundations of Language*, **5**, 256–284.

STALNAKER, R. (1970). Probability and conditionals. *Philosophical Science*, **37**, 64–80.

STITCH, S. P. (1975). Logical form and natural language. *Philosophical Studies*, **28**, 397–418.

STONEBRAKER, M. R., WONG E. & KREPS, P. (1976). The design and implementation of INGRES. *ACM Transactions on Database Systems*, **1**, 189–222.

SUGENO, M. (1974). Theory of fuzzy integrals and its applications. *Ph.D. thesis*. Tokyo Institute of Technology, Japan.

SUGENO, M. & TERANO, T. (1977). A model of learning based on fuzzy information. *Kybernetes*, **6**, 157–166.

SUPPES, P. (1974a). The axiomatic method in the empirical sciences. In HENKIN, J., Ed., *Proceedings of the Tarski Symposium*. Rhode Island: American Mathematical Society.

SUPPES, P. (1974b). Probabilistic metaphysics. *Filofiska Studier nr. 22*. Uppsala University, Sweden.

SUPPES, P. (1976). Elimination of quantifiers in the semantics of natural languages by use of extended relation algebras. *Revue Internationale de Philosophie*, **117–118**, 243–259.

SUSSMAN, G. (1973). *A Computational Model of Skill Acquisition*. Amsterdam: North-Holland.

TAMURA, S. & TANAKA, K. (1973). Learning of fuzzy formal language. *IEEE Transactions on Systems, Man and Cybernetics*, **SMC-3**, 98–102.

TARSKI, A. (1956). *Logic, Semantics, Metamathmatics*. Oxford: Clarendon Press.

TERANO, T. & SUGENO, M. (1975). Conditional fuzzy measures and their applications. In ZADEH, L. A. *et al.* (1975), op.cit., pp. 151–170.

THOMPSON, F. P., LOCKEMANN, P. C., DOSTERT, B. H. & DEVERILL, R. (1969). REL: a rapidly extensible language system. *Proceedings of the 24th ACM National Conference*, New York, pp. 399–417.

THORNE, J. P., BRATLEY, P. & DEWAR, H. (1968). The syntactic analysis of English by machine. In MICHIE, D., Ed., *Machine Intelligence 3*. New York: American Elsevier.

ULLMANN, S. (1962). *Semantics: An Introduction to the Science of Meaning*. Oxford: Blackwell.

URAGAMI, M., MIZUMOTO, M. & TANAKA, K. (1976). Fuzzy robot controls. *Journal of Cybernetics*, **6**, 39–64.

VAN FRAASSEN, B. C. (1971). *Formal Semantics and Logic*. New York: Macmillan.

WALTZ, D. L. (1977). Natural language interfaces. *SIGART Newsletter*, **61**, 16–65.

WASON, P. C. & JOHNSON-LAIRD, P. N. (1972). *Psychology of Reasoning: Structure and Content*. Cambridge, Mass.: Harvard University Press.

WECHSLER, H. (1975). Applications of fuzzy logic to medical diagnosis. *Proceedings of International Symposium on Multiple-valued Logic*, University of Indiana, Bloomington, pp. 162–174.

WEGNER, P. (1972). The Vienna definition language. *ACM Computing Surveys*, **4**, 5–63.

WENSTØP, F. (1975). Application of linguistic variables in the analysis of organizations. *Ph.D. thesis*. School of Business Administration, University of California, Berkeley.

WENSTØP, F. (1976). Deductive verbal models of organizations. *International Journal of Man–Machine Studies*, **8**, 293–311.

WHEELER, S. C. (1975). Reference and vagueness. *Synthese*, **30**, 367–380.

WILKS, Y. (1974). Natural language understanding systems within the AI paradigm. *SAIL Memo AIM-237*. Stanford University.

WINOGRAD, T. (1972). *Understanding Natural Language*. New York: Academic Press.

WINSTON, P. (1975). Learning structural descriptions from examples. In WINSTON, P., Ed., *The Psychology of Computer Vision*. New York: McGraw-Hill.

WOODS, W. A. (1973). Progress in natural language understanding—an application to lunar geology. *Proceedings AFIPS-National Computer Conference*, **42**, 441–450.

WOODS, W. A. (1975). What is in a link: foundations for semantic networks. In BOBROW, D. B. & COLLINS, A. (1975), op. cit., pp. 35–82.

WOODS, W. A., KAPLAN, R. M. & NASH-WEBBER, B. (1972). The lunar sciences natural language information system. Cambridge, Mass.: Bolt, Beranek & Newman.

WRIGHT, C. (1975). On the coherence of vague predicates. *Synthese*, **30**, 325–365.

ZADEH, L. A. (1966). Shadows of fuzzy sets. *Probl. Transmission Inf.* (in Russian), **2**, 37–44.

ZADEH, L. A. (1968a). Probability measures of fuzzy events. *Journal of Mathematical Analysis and Applications*, **23**, 421–427.

ZADEH, L. A. (1968b). Fuzzy algorithms. *Information and Control*, **12**, 94–102.

ZADEH, L. A. (1971a). Similarity relations and fuzzy orderings. *Information Sciences*, **3**, 177–200.

ZADEH, L. A. (1971b). Quantitative fuzzy semantics. *Information Sciences*, **3**, 159–176.

ZADEH, L. A. (1972a). A fuzzy-set-theoretic interpretation of linguistic hedges. *Journal of Cybernetics*, **2**, 4–34.

ZADEH, L. A. (1972b). Fuzzy languages and their relation to human and machine intelligence. *Proceedings of International Conference on Man and Computer*, Bordeaux, France, pp. 130–165. Basel: S. Karger.

ZADEH, L. A. (1973). Outline of a new approach to the analysis of complex systems and decision processes. *IEEE Transactions on Systems, Man and Cybernetics*, **SMC-3**, 28–44.

ZADEH, L. A. (1975a). Fuzzy logic and approximate reasoning (in memory of Grigore Moisil). *Synthese*, **30**, 407–428.

ZADEH, L. A. (1975b). Calculus of fuzzy restrictions. In ZADEH, L. A. *et al.* (1975), op. cit., pp. 1–39.

ZADEH, L. A. (1975c). The concept of a linguistic variable and its application to approximate reasoning, Part I. *Information Sciences*, **8**, 199–249; Part II, *Information Sciences*, **8**, 301–357; Part III, *Information Sciences*, **9**, 43–80.

ZADEH, L. A. (1976a). A fuzzy-algorithmic approach to the definition of complex or imprecise concepts. *International Journal of Man–Machine Studies*, **8**, 249–291.

ZADEH, L. A. (1976b). Fuzzy sets and their application to pattern classification and cluster analysis. *Memorandum M-607*. Electronics Research Laboratory, University of California, Berkeley. In VAN RYZIN, J., Ed., *Classification and Clustering*. New York: Academic Press, pp. 251–299.

ZADEH, L. A. (1977a). Fuzzy sets as a basis for a theory of possibility. *Memorandum M77/12*. Electronics Research Laboratory, University of California, Berkeley. In *Fuzzy Sets and Systems*, **1**, 3–28.

ZADEH, L. A. (1977b). A theory of approximate reasoning. *Memorandum M77/58*. Electronics Research Laboratory, University of California, Berkeley.

ZADEH, L. A., FU, K. S., TANAKA, K. & SHIMURA,. M. (1975). *Fuzzy Sets and Their Application to Cognitive and Decision Processes*. New York: Academic Press.

ZIMMERMANN, H. J. (1974). Optimization in fuzzy environments. Institute for Operations Research, Technical University of Aachen.

ZIMMERMANN, H. J. (1978). Fuzzy programming and linear programming with several objective functions. *Fuzzy Sets and Systems*, **1**, 45–56.

A Computational Approach to Fuzzy Quantifiers in Natural Languages†

LOTFI A. ZADEH

Computer Science Division, University of California, Berkeley, CA 94720, U.S.A.

Abstract—The generic term *fuzzy quantifier* is employed in this paper to denote the collection of quantifiers in natural languages whose representative elements are: *several, most, much, not many, very many, not very many, few, quite a few, large number, small number, close to five, approximately ten, frequently,* etc. In our approach, such quantifiers are treated as fuzzy numbers which may be manipulated through the use of fuzzy arithmetic and, more generally, fuzzy logic.

A concept which plays an essential role in the treatment of fuzzy quantifiers is that of the cardinality of a fuzzy set. Through the use of this concept, the meaning of a proposition containing one or more fuzzy quantifiers may be represented as a system of elastic constraints whose domain is a collection of fuzzy relations in a relational database. This representation, then, provides a basis for inference from premises which contain fuzzy quantifiers. For example, from the propositions "Most U's are A's" and "Most A's are B's," it follows that "Most^2 U's are B's," where *most^2* is the fuzzy product of the fuzzy proportion *most* with itself.

The computational approach to fuzzy quantifiers which is described in this paper may be viewed as a derivative of fuzzy logic and test-score semantics. In this semantics, the meaning of a semantic entity is represented as a procedure which tests, scores and aggregates the elastic constraints which are induced by the entity in question.

1. INTRODUCTION

During the past two decades, the work of Montague and others[57, 67, 18] has contributed much to our understanding of the proper treatment of the quantifiers *all, some* and *any* when they occur singly or in combination in a proposition in a natural language.

Recently, Barwise and Cooper and others[7, 68] have described methods for dealing with so-called *generalized quantifiers* exemplified by *most, many,* etc. In a different approach which we have described in a series of papers starting in 1975[86–90, 92], the quantifiers in question—as well as other quantifiers with imprecise meaning such as *few, several, not very many,* etc.—are treated as fuzzy numbers and hence are referred to as *fuzzy quantifiers*. As an illustration, a fuzzy quantifier such as *most* in the proposition "Most big men are kind" is interpreted as a fuzzily defined proportion of the fuzzy set of kind men in the fuzzy set of big men. Then, the concept of the cardinality‡ of a fuzzy set is employed to compute the proportion in question and find the degree to which it is compatible with the meaning of *most*.

To Professor Kokichi Tanaka.

†Research supported in part by the NSF Grants IST-8018196 and MCS79–06543.

‡Informally, the cardinality of a fuzzy set F is a real or fuzzy number which serves as a count of the number of elements "in" F. A more precise definition of cardinality will be given in Section 2.

We shall employ the class labels "fuzzy quantifiers of the first kind" and "fuzzy quantifiers of the second kind" to refer to absolute and relative counts, respectively, with the understanding that a particular quantifier, e.g. *many*, may be employed in either sense, depending on the context. Common examples of quantifiers of the first kind are: *several, few, many, not very many, approximately five, close to ten, much larger than ten, a large number*, etc. while those of the second kind are: *most, many, a large fraction, often, once in a while, much of*, etc. Where needed, ratios of fuzzy quantifiers of the second kind will be referred to as *fuzzy quantifiers of the third kind*. Examples of quantifiers of this type are the likelihood ratios and certainty factors which are encountered in the analysis of evidence, hypothesis testing and expert systems [73, 24, 5].

An important aspect of fuzzy quantifiers is that their occurrence in human discourse is, for the most part, implicit rather than explicit. For example, when we assert that "Basketball players are very tall," what we usually mean is that "Almost all basketball players are very tall." Likewise, the proposition, "Lynne is never late," would normally be interpreted as "Lynne is late very rarely." Similarly, by "Overeating causes obesity," one may mean that "Most of those who overeat are obese," while "Heavy smoking causes lung cancer," might be interpreted as "The incidence of lung cancer among heavy smokers is much higher than among nonsmokers."

An interesting observation that relates to this issue is that *property inheritance*—which is exploited extensively in knowledge representation systems and high-level AI languages [5]—is a *brittle* property with respect to the replacement of the nonfuzzy quantifier *all* with the fuzzy quantifier *almost all*.† What this means is that if in the inference rule‡

$$p \stackrel{\Delta}{=} all\ A\text{'s are } B\text{'s}$$

$$q \stackrel{\Delta}{=} all\ B\text{'s are } C\text{'s}$$

$$r \stackrel{\Delta}{=} all\ A\text{'s are } C\text{'s}$$

the quantifier *all* in *p* and *q* is replaced by *almost all*, then the quantifier *all* in *r* should be replaced by *none-to-all*. Thus, a slight change in the quantifier *all* in the premises may result in a large change in the quantifier *all* in the conclusion.§

Another point which should be noted relates to the close connection between fuzzy quantifiers and fuzzy probabilities. Specifically, it can be shown [86, 87] that a proposition of the form $p \stackrel{\Delta}{=} Q$ A's are B's, where Q is a fuzzy quantifier (e.g. $p \stackrel{\Delta}{=}$ most doctors are not very tall), implies that the conditional probability of the event B given the event A is a fuzzy probability which is equal to Q. What can be shown, in fact, is that most statements involving fuzzy

---

†The brittleness of property inheritance is of relevance to nonmonotonic logic, default reasoning and exception handling.

‡The symbol $\stackrel{\Delta}{=}$ stands for "denotes" or "is defined to be."

§An example which relates to this phenomenon is: What is rare is expensive. A cheap apartment in Paris is rare. Therefore, a cheap apartment in Paris is expensive. This example was suggested to the author in a different connection by Prof. O. Botta of the University of Lyon.

probabilities may be replaced by semantically equivalent statements involving fuzzy quantifiers. This connection between fuzzy quantifiers and fuzzy probabilities plays an important role in expert systems and fuzzy temporal logic, but we shall not dwell on it in the present paper.

As was stated earlier, the main idea underlying our approach to fuzzy quantifiers is that the natural way of dealing with such quantifiers is to treat them as fuzzy numbers. However, this does not imply that the concept of a fuzzy quantifier is coextensive with that of a fuzzy number. Thus, in the proposition "Vickie is several years younger than Mary," the fuzzy number *several* does not play the role of a fuzzy quantifier, whereas in "Vickie has several good friends," it does. More generally, we shall view a fuzzy quantifier as a fuzzy number which provides a fuzzy characterization of the absolute or relative cardinality of one or more fuzzy or nonfuzzy sets. For example, in "Vickie has several credit cards," *several* is a fuzzy characterization of the cardinality of the nonfuzzy set of Vickie's credit cards; in "Vickie has several good friends," *several* is a fuzzy characterization of the cardinality of the fuzzy set of Vickie's good friends; and in "Most big men are kind," *most* is a fuzzy characterization of the relative cardinality of the fuzzy set of kind men in the fuzzy set of big men. There are propositions, however, in which the question of whether or not a constituent fuzzy number is a fuzzy quantifier does not have a clear cut answer.

A simple example may be of help at this point in providing an idea of how fuzzy quantifiers may be treated as fuzzy numbers. Specifically, consider the propositions

$$p \overset{\Delta}{=} 80\% \text{ of students are single}$$

$$q \overset{\Delta}{=} 60\% \text{ of single students are male}$$

$$r \overset{\Delta}{=} Q \text{ of students are single and male}$$

in which r represents the answer to the question "What percentage of students are single males?" given the premises expressed by p and q.

Clearly, the answer is: $80\% \times 60\% = 48\%$, and, more generally, we can assert that:

$$p \overset{\Delta}{=} Q_1 \text{ of } A\text{'s are } B\text{'s} \tag{1.1}$$

$$q \overset{\Delta}{=} Q_2 \text{ of } (A \text{ and } B)\text{'s are } C\text{'s}$$

$$r \overset{\Delta}{=} Q_1 Q_2 \text{ of } A\text{'s are } (B \text{ and } C)\text{'s}$$

where Q_1 and Q_2 are numerical percentages, and A, B and C are labels of nonfuzzy sets or, equivalently, names of their defining properties.

Now suppose that Q_1 and Q_2 are fuzzy quantifiers of the second kind, as in the following example:

$$p \overset{\Delta}{=} most \text{ students are single}$$

$$q \overset{\Delta}{=} a \text{ } little \text{ } more \text{ } than \text{ } a \text{ } half \text{ } of \text{ } single \text{ } students \text{ } are \text{ } male$$

$$r \overset{\Delta}{=} ? \text{ } Q \text{ } of \text{ } students \text{ } are \text{ } single \text{ } and \text{ } male$$

where the question mark indicates that the value of Q is to be inferred from p and q.

By interpreting the fuzzy quantifiers *most*, *a little more than a half*, and Q as fuzzy numbers which characterize, respectively, the proportions of single students among students, males among single students and single males among students, we can show that Q may be expressed as the product, in fuzzy arithmetic (see Appendix), of the fuzzy numbers *most* and *a little more than a half*. Thus, in symbols,

$$Q = most \otimes a \text{ } little \text{ } more \text{ } than \text{ } a \text{ } half \qquad (1.2)$$

and, more generally, for fuzzy Q's, A's, B's and C's, we can assert the syllogism:

$$p \overset{\Delta}{=} Q_1 \text{ } of \text{ } A\text{'s } are \text{ } B\text{'s} \qquad (1.3)$$

$$q \overset{\Delta}{=} Q_2 \text{ } of \text{ } (A \text{ } and \text{ } B)\text{'s } are \text{ } C\text{'s}$$

$$r \overset{\Delta}{=} Q_1 \otimes Q_2 \text{ } of \text{ } A\text{'s } are \text{ } (B \text{ } and \text{ } C)\text{'s},$$

which will be referred to as the *intersection/product* syllogism. A pictorial representation of (1.2) is shown in Fig. 1.

The point of this example is that the syllogism (or the *inference schema*) expressed by (1.1) generalizes simply and naturally to fuzzy quantifiers when they are treated as fuzzy numbers. Furthermore, through the use of *linguistic approximation* [87, 45]—which is analogous to rounding to an integer in ordinary arithmetic—the expression for Q may be approximated to by a fuzzy quantifier which is an element of a specified context-free language. For example, in the case of (1.2), such a quantifier may be expressed as *about a half*, or *more or less close to a half*, etc. depending on how the fuzzy numbers *most*, *a little more than a half*, and *close to a half* are defined through their respective possibility distributions (see Appendix).

In our discussion so far, we have tacitly assumed that a fuzzy quantifier is a fuzzy number of type 1, i.e. a fuzzy set whose membership function takes values in the unit interval. More generally, however, a fuzzy quantifier may be a fuzzy set of type 2 (or higher), in which case we shall refer to it as an *ultrafuzzy quantifier*. The membership functions of such quantifiers take values in the space of fuzzy sets of type 1, which implies that the compatibility of an ultrafuzzy quantifier with a real number is a fuzzy number of type 1. For example, the fuzzy quantifier *not so many* would be regarded as an ultrafuzzy quantifier if the compatibility of *not so many* with 5, say, would be specified in a particular context as *rather high*, where *rather high* is interpreted as a fuzzy number in the unit interval.

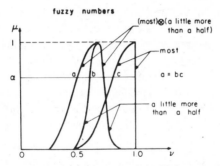

Fig. 1. The intersection/product syllogism with fuzzy quantifiers.

Although the rule of inference expressed by (1.3) remains valid for ultrafuzzy quantifiers if $\otimes$ is interpreted as the product of ultrafuzzy numbers (see Fig. 2), we shall restrict our attention in the present paper to fuzzy quantifiers of type 1, with the understanding that most of the inference schemas derived on this assumption can readily be generalized to fuzzy quantifiers of higher type.

As will be seen in the sequel, a convenient framework for the treatment of fuzzy quantifiers as fuzzy numbers is provided by a recently developed meaning-representation system for natural languages termed *test-score semantics* [93]. Test-score semantics represents a break with the traditional approaches to semantics in that it is based on the premise that almost everything that relates to natural languages is a matter of degree. The acceptance of this premise necessitates an abandonment of bivalent logical systems as a basis for the analysis of natural languages and suggests the adoption of fuzzy logic [86, 90, 8], as the basic conceptual framework for the representation of meaning, knowledge and strength of belief.

Viewed from the perspective of test-score semantics, a semantic entity such as a proposition, predicate, predicate-modifier, quantifier, qualifier, command, question, etc. may be regarded as a system of elastic constraints whose domain is a collection of fuzzy relations in a database—a database which describes a state of affairs [11] or a possible world [44] or, more generally, a set of objects or derived objects in a universe of discourse. The meaning of a semantic entity, then, is represented as a test which when applied to the database yields a collection of partial test scores. Upon aggregation, these test scores lead to an overall vector test score, τ, whose components are numbers in the unit interval, with τ serving as a measure of the compatibility of the semantic entity with the database. In this respect, test-score semantics

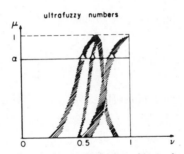

Fig. 2. The intersection/product syllogism with ultrafuzzy quantifiers.

subsumes both truth-conditional and possible-world semantics as limiting cases in which the partial and overall test scores are restricted to {pass, fail} or, equivalently, {true, false} or {1,0}.

In more specific terms, the process of meaning representation in test-score semantics involves three distinct phases. In Phase I, an *explanatory database frame* or EDF, for short, is constructed. EDF consists of a collection of relational frames, i.e. names of relations, names of attributes and attribute domains, whose meaning is assumed to be known. In consequence of this assumption, the choice of EDF is not unique and is strongly influenced by the knowledge profile of the addressee of the representation process as well as by the desideratum of explanatory effectiveness. For example, in the case of the proposition $p \overset{\Delta}{=}$ Over the past few years Nick earned far more than most of his close friends, the EDF might consist of the following relations:† *INCOME* [*Name*; *Amount*; *Year*], which lists the income of each individual identified by his/her name as a function of the variable *Year*; *FRIEND* [*Name*; μ], where μ is the degree to which *Name* is a friend of Nick; *FEW* [*Number*; μ], where μ is the degree to which *Number* is compatible with the fuzzy quantifier *FEW*; *MOST* [*Proportion*; μ], in which μ is the degree to which *Proportion* is compatible with the fuzzy quantifier *MOST*; and *FAR. MORE* [*Income 1*; *Income 2*; μ], where μ is the degree to which *Income 1* fits the fuzzy predicate *FAR. MORE* in relation to *Income 2*. Each of these relations is interpreted as an elastic constraint on the variables which are associated with it.

In Phase 2, a test procedure is constructed which acts on relations in the explanatory database and yields the test scores which represent the degrees to which the elastic constraints induced by the constituents of the semantic entity are satisfied. For example, in the case of *p*, the test procedure would yield the test scores for the constraints induced by the relations *FRIEND, FEW, MOST* and *FAR. MORE*.

In Phase 3, the partial test scores are aggregated into an overall test score, τ, which, in general, is a vector which serves as a measure of the compatibility of the semantic entity with an instantiation of EDF. As was stated earlier, the components of this vector are numbers in the unit interval, or, more generally, possibility/probability distributions over this interval. In particular, in the case of a proposition, *p*, for which the overall test score is a scalar, τ may be interpreted—in the spirit of truth-conditional semantics—as the degree of truth of the proposition with respect to the explanatory database ED (i.e. an instantiation of EDF). Equivalently, τ may be interpreted as the possibility of ED given *p*, in which case we may say that *p induces a possibility distribution*. More concretely, we shall say that *p translates into a possibility assignment equation* [88]:

$$p \to \Pi_{(X_1, \ldots X_n)} = F, \tag{1.4}$$

where F is a fuzzy subset of a universe of discourse $U, X_1, \ldots X_n$ are variables which are explicit or implicit in *p*, and $\Pi_{(X_1, \ldots X_n)}$ is their joint possibility distribution. For example, in the case of the proposition $p \overset{\Delta}{=}$ Danielle is tall, we have

$$Danielle\ is\ tall \to \Pi_{Height\,(Danielle)} = TALL, \tag{1.5}$$

---

†Generally, we follow the practice of writing the names of fuzzy subsets and fuzzy relations in uppercase symbols.

where *TALL* is a fuzzy subset of the real-line, *Height(Danielle)* is a variable which is implicit in p, and $\Pi_{Height(Danielle)}$ is the possibility distribution of the variable *Height(Danielle)*. Equation (1.5) implies that

$$Poss\{Height(Danielle) = u\} = \mu_{TALL}(u),$$

where u is a specified value of the variable *Height(Danielle)*, $\mu_{TALL}(u)$ is the grade of membership of u in the fuzzy set *TALL*, and $Poss\{X = u\}$ should be read as "the possibility that X is u." In effect, (1.5) signifies that the proposition "Danielle is tall," may be interpreted as an elastic constraint on the variable *Height(Danielle)*, with the elasticity of the constraint characterized by the unary relation *TALL* which is defined as a fuzzy subset of the real line.

The same basic idea may be applied to propositions containing one or more fuzzy quantifiers. As a simple illustration, let us consider the proposition

$$p \overset{\Delta}{=} \textit{Vickie has several credit cards,}$$

in which *several* is regarded as a fuzzy quantifier which induces an elastic constraint on the number of credit cards possessed by Vickie. In this case, X may be taken to be the count of Vickie's cards, and the possibility assignment equation becomes

$$\textit{Vickie has several credit cards} \rightarrow \Pi_{Count(Cards(Vickie))} = SEVERAL, \tag{1.6}$$

in which *SEVERAL* plays the role of a specified fuzzy subset of the integers 1, 2, ... 10. Thus, if the integer 4, say, is assumed to be compatible with the meaning of *several* to the degree 0.8, then (1.6) implies that, given p and the definition of *several*, the possibility that Vickie has four credit cards is expressed by

$$Poss\{Count(Cards(Vickie)) = 4\} = 0.8.$$

In the above example, the class of Vickie's credit cards is a nonfuzzy set and hence there is no problem in counting their number. By contrast, in the proposition

$$p \overset{\Delta}{=} \textit{Vickie has several close friends}$$

the class of close friends is a fuzzy set and thus we must first resolve the question of how to count the number of elements in a fuzzy set or, equivalently, how to determine its cardinality. This issue is addressed in the following section.

1. CARDINALITY OF FUZZY SETS

In the case of a crisp (nonfuzzy) subset, A, of a universe of discourse, U, the proposition "u is an element of A," is either true or false, and hence there is just one way in which the cardinality of A, i.e. the count of elements of A, may be defined. However, even though the count may be defined uniquely, there may be some uncertainty about its value if there is an uncertainty regarding the membership status of points of U in A.

By contrast, in the case of a fuzzy subset, F, of U, the proposition "u is an element of F," is generally true to degree, with the result that the concept of cardinality admits of a variety of definitions. Among them, some associate with a fuzzy set F a real number, in which case the cardinality of a fuzzy set is nonfuzzy. Others associate with F a fuzzy number, since it may be argued that the cardinality of a fuzzy set should be a fuzzy number. A brief discussion of these viewpoints is presented in the following. For simplicity, we shall restrict our attention to finite universes of discourse, in which case a fuzzy subset, F, of $U = \{u_1, \ldots u_n\}$ may be expressed symbolically as

$$F = \mu_1/u_1 + \ldots + \mu_n/u_n$$

or, more simply, as

$$F = \mu_1 u_1 + \ldots + \mu_n u_n,$$

in which the term μ_i/u_i, $i = 1, \ldots n$, signifies that μ_i is the grade of membership of u_i in F, and the plus sign represents the union.†

Nonfuzzy cardinality

A simple way of extending the concept of cardinality to fuzzy sets is to form the *sigma-count* [17, 85], which is the arithmetic sum of the grades of membership in F. Thus

$$\Sigma Count(F) \overset{\Delta}{=} \Sigma_i u_i, \quad i = 1, \ldots n, \tag{2.1}$$

with the understanding that the sum may be rounded, if need be, to the nearest integer. Furthermore, one may stipulate that the terms whose grade of membership falls below a specified threshold be excluded from the summation. The purpose of such an exclusion is to avoid a situation in which a large number of terms with low grades of membership become count-equivalent to a small number of terms with high membership.

As a simple illustration of the concept of sigma-count, assume that the fuzzy set of close friends of Teresa is expressed as

$$F = 1/Enrique + 0.8/Ramon + 0.7/Elie + 0.9/Sergei + 0.8/Ron.$$

In this case,

$$\Sigma Count(F) = 1 + 0.8 + 0.7 + 0.9 + 0.8$$

$$= 4.2.$$

A sigma-count may be *weighted*, in the sense that if $w = (w_1, \ldots w_n)$ is an n-tuple of nonnegative real numbers, then the *weighted sigma-count of F with respect to w* is defined by

$$\Sigma Count(F; w) \overset{\Delta}{=} \Sigma_i w_i \mu_i. \quad i = 1, \ldots n.$$

---

†For the most part we shall rely on the context to disambiguate the meaning of +.

This definition implies that $\Sigma Count(F;w)$ may be interpreted as the sigma-count of a fuzzy multiset† $'F$ in which the grade of membership and the multiplicity of u_i, $i = 1, \ldots n$, are, respectively, μ_i and w_i. The concept of a weighted sigma-count is closely related to that of the *measure* of a fuzzy set[83, 75, 40–42].

Whether weighted or not, the sigma-count of a fuzzy set is a real number. As was stated earlier, it may be argued that the cardinality of a fuzzy set should be a fuzzy number. If one accepts this argument, then a natural way of defining fuzzy cardinality is the following[90].

Fuzzy cardinality‡

In this case, the point of departure is a stratified representation of F in terms of its *level sets*[84], i.e.

$$F = \Sigma_\alpha \alpha F_\alpha,$$

in which the α-level-sets F_α are nonfuzzy sets defined by

$$F_\alpha \overset{\Delta}{=} \{u | \mu_F(u) \geq \alpha\}, \qquad 0 < \alpha \leq 1,$$

and

$$\mu_{\alpha F_\alpha}(u) = \alpha \mu_F(u), \qquad u \in U.$$

In terms of this representation, there are three fuzzy counts, *FCounts*, that may be associated with F. First, the *FGCount* is defined as the conjunctive fuzzy integer§[93]

$$FGCount(F) = 1/0 + \Sigma_\alpha \alpha / Count(F_\alpha), \qquad a > 0.$$

Second, the *FLCount* is defined as

$$FLCount(F) = (FGCount(F))' \ominus 1$$

where $'$ denotes the complement and $\ominus 1$ means that 1 is subtracted from the fuzzy number $FGCount(F)$. And finally, the *FECount*(F) is defined as the intersection of $FGCount(F)$ and $FLCount(F)$, i.e.

$$FECount(F) = FGCount(F) \cap FLCount(F).$$

---

†A fuzzy multiset, $'F$, may be represented as $'F = \Sigma_i \mu_i / m_i \times u_i$, in which m_i is the *multiplicity* of u_i and μ_i is the grade of membership of u_i in the fuzzy set $F = \Sigma_i \mu_i / u_i$. The multiplicity, m_i, is a nonnegative real number which is usually, but not necessarily, an integer. Thus, a fuzzy multiset may have identical elements, or elements which differ only in their grade of membership.

‡Although it is perhaps a more natural extension of the concept of cardinality than the sigma-count, fuzzy cardinality is a more complex concept and is more difficult to manipulate. The exposition of fuzzy cardinality in this section may be omitted on first reading.

§It should be noted that the membership function of a conjunctive fuzzy number is not a possibility distribution.

Equivalently—and more precisely—we may define the counts in question via the membership function of F, i.e.

$$\mu_{FECount(F)}(i) \overset{\Delta}{=} sup_\alpha\{\alpha|Count(F_\alpha) \geq i\}, \qquad i = 0, 1, \ldots n, \tag{2.2}$$

$$\mu_{FLCount(F)}(i) \overset{\Delta}{=} sup_\alpha\{\alpha|Count(F_\alpha) \geq n - i\} \tag{2.3}$$

$$\mu_{FECount(F)}(i) \overset{\Delta}{=} \mu_{FGCount(F)}(i) \wedge \mu_{FLCount(F)}(i), \tag{2.4}$$

where $\wedge$ stands for min in infix position.

As a simple illustration, consider the fuzzy set expressed as

$$F = 0.6/u_1 + 0.9/u_2 + 1/u_3 + 0.7/u_4 + 0.3/u_5. \tag{2.5}$$

In this case,

$$F_1 = u_3$$
$$F_{0.9} = u_2 + u_3$$
$$F_{0.7} = u_2 + u_3 + u_4$$
$$F_{0.6} = u_1 + u_2 + u_3 + u_4$$
$$F_{0.3} = u_1 + u_2 + u_3 + u_4 + u_5,$$

which implies that, in stratified form, F may be expressed as

$$F = 1(u_3) + 0.9(u_2 + u_3) + 0.7(u_2 + u_3 + u_4) + 0.6(u_1 + u_2 + u_3 + u_4) + 0.3(u_1 + u_2 + u_3 + u_4 + u_5),$$

and hence that

$$FGCount(F) = 1/0 + 1/1 + 0.9/2 + 0.7/3 + 0.6/4 + 0.3/5$$
$$FLCount(F) = 0.1/2 + 0.3/3 + 0.4/4 + 0.7/5 + 1/6 + \cdots \ominus 1$$
$$= 0.1 \cdot 1 + 0.3/2 + 0.4/3 + 0.7/4 + 1/5 + \cdots$$
$$FECount(F) = 0.1/1 + 0.3/2 + 0.4/3 + 0.6/4 + 0.3/5$$

while, by comparison,

$$\Sigma Count(F) = 0.6 + 0.9 + 1.0 + 0.7 + 0.3$$
$$= 3.5.$$

A useful interpretation of the defining relations (2.2)–(2.4) may be stated as follows:

(a) $\mu_{FGCount}(i)$ is the truth value of the proposition "F contains at least i elements."
(b) $\mu_{FLCount}(i)$ if the truth value of the proposition "F contains at most i elements."
(c) $\mu_{FECount}(i)$ is the truth value of the proposition "F contains i and only i elements."

From (a), it follows that $FGCount(F)$ may readily be obtained from F by first sorting F in the order of decreasing grades of membership and then replacing u_i with i and adding the term $1/0$. For example, for F defined by (2.5), we have

$$F\downarrow \ = 1/u_3 + 0.9/u_2 + 0.7/u_4 + 0.6/u_1 + 0.3/u_5 \tag{2.6}$$

$$NF\downarrow \ = 1/1 + 0.9/2 + 0.7/3 + 0.6/4 + 0.3/5$$

and

$$FGCount(F) = 1/0 + 1/1 + 0.9/2 + 0.7/3 + 0.6/4 + 0.3/5,$$

where $F\downarrow$ denotes F sorted in descending order, and $NF\downarrow$ is $F\downarrow$ with *ith* u replaced by *i*. An immediate consequence of this relation between $\Sigma Count(F)$ and $FGCount(F)$ is the identity

$$\Sigma Count(F) = \Sigma_i \mu_{FGCount}(i) - 1,$$

which shows that, as a real number, $\Sigma Count(F)$ may be regarded as a "summary" of the fuzzy number $FGCount(F)$.

Relative count

A type of count which plays an important role in meaning representation is that of *relative count* (or *relative cardinality*)[87]. Specifically, if F and G are fuzzy sets, then the relative sigma-count of F and G is defined as the ratio:

$$\Sigma Count(F/G) = \frac{\Sigma Count(F \cap G)}{\Sigma Count(G)}, \tag{2.7}$$

which represents the proportion of elements of F which are in G, with the intersection $F \cap G$ defined by

$$\mu_{F \cap G}(u) = \mu_F(u) \wedge \mu_G(u). \tag{2.8}$$

The corresponding definition for the $FGCount$ is

$$FGCount(F/G) = \Sigma_\alpha \alpha \left/ \frac{Count(F_\alpha \cap G_\alpha)}{Count(G_\alpha)} \right. , \tag{2.9}$$

where the F_α and G_α represent the α-sets of F and G, respectively. It should be noted that the right-hand member of (2.9) should be treated as a fuzzy multiset, which implies that terms of the form α_1/u and α_2/u should not be combined into a single term $(\alpha_1 \vee \alpha_2)/u$, as they would be in the case of a fuzzy set.

The $\Sigma Count$ and $FCounts$ of fuzzy sets have a number of basic properties of which only a few will be stated here. Specifically, if F and G are fuzzy sets, then from the identity

$$a \vee b + a \wedge b = a + b$$

which holds for any real numbers, it follows at once that

$$\Sigma Count(F \cap G) + \Sigma Count(F \cup G) = \Sigma Count(F) + \Sigma Count(G) \qquad (2.10)$$

since

$$\mu_{F \cap G}(u) = \mu_F(u) \wedge \mu_G(u), \qquad u \in U$$

and

$$\mu_{F \cup G}(u) = \mu_F(u) \vee \mu_G(u).$$

Thus, if F and G are disjoint (i.e. $F \cap G = \theta$), then

$$\Sigma Count(F \cup G) = \Sigma Count(F) + \Sigma Count(G) \qquad (2.11)$$

and, more generally,

$$\Sigma Count(F) \vee \Sigma Count(G) \leq \Sigma Count(F \cup G) \leq \Sigma Count(F) + \Sigma Count(G) \qquad (2.12)$$

and

$$0 \vee (\Sigma Count(F) + \Sigma Count(G) - Count(U)) \leq \Sigma Count(F \cap G) \leq \Sigma Count(F) \wedge \Sigma Count(G). \qquad (2.13)$$

These inequalities follow at once from (2.10) and

$$\Sigma Count(F \cap G) \leq \Sigma Count(F)$$
$$\Sigma Count(F \cap G) \leq \Sigma Count(G)$$
$$\Sigma Count(F \cup G) \leq \Sigma Count(U).$$

In the case of $FCounts$ and, more specifically, the $FGCount$, the identity corresponding to (2.10) reads [92, 93, 23],

$$FGCount(F \cap G) \oplus FGCount(F \cup G) = FGCount(F) \oplus FGCount(G), \qquad (2.14)$$

where $\oplus$ denotes the addition of fuzzy numbers, which is defined by (see Appendix)

$$\mu_{A \oplus B}(u) = sup_v(\mu_A(v) \wedge \mu_B(u - v)), \qquad u,v \in (-\infty, \infty), \qquad (2.15)$$

where A and B are fuzzy numbers, and μ_A and μ_B are their respective membership functions. A basic identity which holds for relative counts may be expressed as:

$$\Sigma Count(F \cap G) = \Sigma Count(G) \Sigma Count(F/G) \tag{2.16}$$

for sigma-counts, and as

$$FGCount(F \cap G) = FCount(G) \otimes FGCount(F/G) \tag{2.17}$$

for *FGCounts*, where $\otimes$ denotes the multiplication of fuzzy numbers, which is defined by (see Appendix)

$$\mu_{A \otimes B}(u) = sup_v \left(\mu_A(v) \quad \mu_B\left(\frac{u}{v}\right) \right), \qquad u, v \in (-\infty, \infty), \quad v \neq 0. \tag{2.18}$$

An inequality involving relative sigma-counts which is of relevance to the analysis of evidence in expert systems is the following:

$$\Sigma Count(F/G) + \Sigma Count(\neg F/G) \geq 1 \tag{2.19}$$

where $\neg F$ denotes the complement of F, i.e. $= 1$ *if G is nonfuzzy*,

$$\mu_{\neg F}(u) = 1 - \mu_F(u), \qquad u \in U. \tag{2.20}$$

Note that (2.19) implies that if the relative sigma-count $\Sigma Count(F/G)$ is identified with the conditional probability $Prob(F/G)$[94], then

$$Prob\,(\neg F/G) \geq 1 - Prob(F/G) \tag{2.21}$$

rather than

$$Prob\,(\neg F/G) = 1 - Prob(F/G), \tag{2.22}$$

which holds if G is nonfuzzy.

The inequality in question follows at once from

$$\Sigma Count(\neg F_i/G) = \frac{\Sigma_i (1 - \mu_F(u_i)) \wedge \mu_G(u_i)}{\Sigma_i \mu_G(u_i)} \tag{2.23}$$

$$\geq \frac{\Sigma_i (1 - \mu_F(u_i)) \mu_G(u_i)}{\Sigma_i \mu_G(u_i)}$$

$$\geq 1 - \frac{\Sigma_i \mu_F(u_i) \mu_G(u_i)}{\Sigma_i \mu_G(u_i)}$$

$$\geq 1 - \frac{\Sigma_i \mu_F(u_i) \wedge \mu_G(u_i)}{\Sigma_i \mu_G(u_i)}$$

since

$$\Sigma Count(F/G) = \frac{\Sigma_i \mu_F(u_i) \wedge \mu_G(u_i)}{\Sigma_i \mu_G(u_i)}.$$

This concludes our brief exposition of some of the basic aspects of the concept of cardinality of fuzzy sets. As was stated earlier, the concept of cardinality plays an essential role in representing the meaning of fuzzy quantifiers. In the following sections, this connection will be made more concrete and a basis for inference from propositions containing fuzzy quantifiers will be established.

3. FUZZY QUANTIFIERS AND CARDINALITY OF FUZZY SETS

As was stated earlier, a fuzzy quantifier may be viewed as a fuzzy characterization of absolute or relative cardinality. Thus, in the proposition $p \overset{\Delta}{=} Q$ A's *are* B's, where Q is a fuzzy quantifier and A and B are labels of fuzzy or nonfuzzy sets, Q may be interpreted as a fuzzy characterization of the relative cardinality of B and A. The fuzzy set A will be referred to as the *base set*.

When both A and B are nonfuzzy sets, the relative cardinality of B in A is a real number and Q is its possibility distribution. The same is true if A and/or B are fuzzy sets and the sigma-count is employed to define the relative cardinality. The situation becomes more complicated, however, if an *FCount* is employed for this purpose, since Q, then, is the possibility distribution of a conjunctive fuzzy number.

To encompass these cases, we shall assume that the following propositions are semantically equivalent [89]:

$$There\ are\ Q\ A\text{'s} \leftrightarrow Count(A)\ is\ Q \tag{3.1}$$

$$Q\ A\text{'s}\ are\ B\text{'s} \leftrightarrow Prop(B/A)\ is\ Q, \tag{3.2}$$

where the more specific term *Proportion* or *Prop*, for short, is used in place of *Count* in (3.2) to underscore that *Prop(B/A)* is the relative cardinality of B in A, with the understanding that both *Count* in (3.1) and *Prop* in (3.2) may be fuzzy or nonfuzzy counts. In the sequel, we shall assume for simplicity that, except where stated to the contrary, both absolute and relative cardinalities are defined via the sigma-count.

The right-hand members of (3.1) and (3.2) may be translated into possibility assignment equations (see 1.1). Thus we have

$$Count(A)\ is\ Q \rightarrow \Pi_{Count(A)} = Q \tag{3.3}$$

and

$$Prop(B/A)\ is\ Q \rightarrow \Pi_{Prop(B/A)} = Q, \tag{3.4}$$

in which $\Pi_{Count(A)}$ and $\Pi_{Prop(B/A)}$ represent the possibility distributions of *Count(A)* and *Prop(B/A)*, respectively. Furthermore, in view of (3.1) and (3.2), we have

$$There\ are\ Q\ A\text{'s} \rightarrow \Pi_{Count(A)} = Q \tag{3.5}$$

$$Q\ A\text{'s}\ are\ B\text{'s} \rightarrow \Pi_{Prop(B/A)} = Q. \tag{3.6}$$

These translation rules in combination with the results established in Section 2, provide a basis for deriving a variety of syllogisms for propositions containing fuzzy quantifiers, an instance of which is the *intersection/product* syllogism described by (1.3), namely,

$$Q_1\ A\text{'s}\ are\ B\text{'s} \tag{3.7}$$

$$\underline{Q_2(A\ and\ B)\text{'s}\ are\ C\text{'s}}$$

$$Q_1 \otimes Q_2\ A\text{'s}\ are\ (B\ and\ C)\text{'s}$$

in which Q_1, Q_2, A, B and C are assumed to be fuzzy, as in

$$most\ tall\ men\ are\ fat \tag{3.8}$$

$$\underline{many\ tall\ and\ fat\ men\ are\ bald}$$

$$most \otimes many\ tall\ men\ are\ fat\ and\ bald.$$

To establish the validity of syllogisms of this form, we shall rely, in the main, on the semantic· entailment principle [89, 90], and on a special case of this principle which will be referred to as the *quantifier extension principle.*

Stated in brief, the semantic entailment principle asserts that a proposition p entails proposition q, which we shall express as $p \rightarrow q$ or

$$\frac{p}{q},$$

if and only if the possibility distribution which is induced by p, $\Pi^p_{(X_1,\dots X_n)}$, is contained in the possibility distribution induced by q, $\Pi^q_{(X_1,\dots X_n)}$ (see (1.4)). Thus, stated in terms of the possibility distribution functions of Π^p and Π^q, we have [88]

$$\frac{p}{q}\ if\ and\ only\ if\ \pi^p_{(X_1,\dots X_n)} \leq \pi^q_{(X_1,\dots X_n)} \tag{3.9}$$

for all points in the domain of π^p and π^q.

Informally, (3.9) means that p entails q if and only if q is less specific than p. For example, the proposition $p \overset{\Delta}{=}$ Diana is 28 years old, entails the proposition $q \overset{\Delta}{=}$ Diana is in her late twenties, because p is less specific than q, which in turn is a consequence of the containment of the nonfuzzy set "28" in the fuzzy set "late twenties."

It should be noted that, in the context of test-score semantics, the inequality of possibilities in (3.9) may be expressed as a corresponding inequality of overall test scores. Thus, if τ^p and τ^q are the overall test scores associated with p and q, respectively, then

$$\frac{p}{q} \text{ if and } only \text{ if } \tau^p \leq \tau^q, \tag{3.10}$$

with the understanding that the tests yielding τ^p and τ^q are applied to the same explanatory database and that the inequality holds for all instantiations of EDF.

In our applications of the entailment principle, we shall be concerned, for the most part, with an entailment relation between a collection of propositions $p_1, \ldots p_n$ and a proposition q which is entailed by the collection. Under the assumption that the propositions which constitute the premises are noninteractive[89], the statement of the entailment principle (3.9) becomes:

$$p_1 \text{ if and } only \text{ if } \pi^{p_1} \wedge \cdots \wedge \pi^{p_n} \leq \pi^{q'} \tag{3.11}$$

$$\vdots$$

$$\frac{p_n}{q}$$

where $\pi^{p_1}, \ldots \pi^{p_n}, \pi^q$, are the possibility distribution functions induced by $p_1, \ldots p_n, q$, respectively, and likewise for (3.10).

We are now in a position to formulate an important special case of the entailment principle which will be referred to as the *quantifier extension principle*. This principle may also be viewed as an inference rule which is related to the transformational rule of inference described in[91].

Specifically, assume that each of the propositions $p_1, \ldots p_n$ is a fuzzy characterization of an absolute or relative cardinality which may be expressed as $p_i \overset{\Delta}{=} C_i$ is Q_i, $i = 1, \ldots n$, in which C_i is a count and Q_i is a fuzzy quantifier, e.g.

$$p_i \overset{\Delta}{=} \Sigma \, Count(B/A) \text{ is } Q_i$$

or, more concretely,

$$p_i \overset{\Delta}{=} most \, A\text{'s } are \, B\text{'s}.$$

Now, in general, a syllogism involving fuzzy quantifiers has the form of a collection of premises of the form $p_i \overset{\Delta}{=} C_i$ is Q_i, $i = 1, \ldots n$, followed by a conclusion of the same form, i.e. $q \overset{\Delta}{=} C$ is Q, where C is a count that is related to $C_i, \ldots C_n$, and Q is a fuzzy quantifier which is related to $Q_i, \ldots Q_n$. The quantifier extension principle makes these relations explicit, as represented in the following inference schema:

---

†The composition, RoS, of a binary relation R with a unary relation S is defined by $\mu_{RoS}(v) = \vee_u (\mu_R(v,u) \wedge \mu_S(u))$, $u \in U, v \in V$, where $\mu_R, \mu_S,$ and μ_{RoS} are the membership functions of R, S and RoS, respectively, and $\vee_u$ denotes the supremum over U. Where no confusion can result, the symbol o may be suppressed.

$$C_1 \text{ is } Q_1$$

$$\underline{C_n \text{ is } Q_n} \qquad\qquad (3.12)$$

$$C \text{ is } Q,$$

where Q is given by

$$If \ C = g(C_1, \dots C_n) \ then \ Q = g(Q_1, \dots Q_n),$$

in which g is a function which expresses the relation between C and the C_i, and the meaning of $Q = g(Q_1, \dots Q_n)$ is defined by the extension principle (see Appendix). A somewhat more general version of the quantifier extension principle which can also be readily deduced from the extension principle is the following:

$$C_1 \text{ is } Q_1 \qquad\qquad (3.13)$$

$$\underline{C_n \text{ is } Q_n}$$

$$C \text{ is } Q,$$

where Q is given by

$$If \ f(C_1, \dots C_n) \le C \le g(C_1, \dots C_n) \ then \ f(Q_1, \dots Q_n) \le Q \le g(Q_1, \dots Q_n).$$

As in (3.12), the meaning of the inequalities which bound Q is defined by the extension principle. In more concrete terms, these inequalities imply that Q is a fuzzy interval which may be expressed as

$$Q = (\ge f(Q_1, \dots Q_n)) \cap (\le g(Q_1, \dots Q_n)), \qquad\qquad (3.14)$$

where the fuzzy s-number $\ge f(Q_1, \dots Q_n)$ and the fuzzy z-number $\le g(Q_1, \dots Q_n)$ (see Appendix) should be read as "at least $f(Q_1, \dots Q_n)$" and "at most $g(Q_1, \dots Q_n)$," respectively, and are the compositions† of the binary relations $\ge$ and $\le$ with $f(Q_1, \dots Q_n)$ and $g(Q_1, \dots Q_n)$. In terms of (3.14), then, the relation between C and Q may be expressed as:

$$If \ f(C_1, \dots C_n) \le C \le g(C_1, \dots C_n) \ then \ Q = (\ge f(Q_1, \dots Q_n)) \cap (\le g(Q_1, \dots Q_n)). \quad (3.15)$$

An important special case of (3.12) and (3.15) is one where f and g are arithmetic or boolean expressions, as in

$$C = C_1 C_2 + C_3$$

and

$$C_1 + C_2 - 1 \leq C \leq C_1 \wedge C_2.$$

For these cases, the quantifier extension principle yields

$$Q = Q_1 \otimes Q_2 \oplus Q_3$$

and

$$Q = (\geq (Q_1 \oplus Q_2 \ominus 1)) \cap \leq (Q_1 \oslash Q_2),$$

where Q, Q_1, Q_2 and Q_3 are fuzzy numbers, and $\otimes$, $\oplus$ and $\oslash$ are the product, sum and min in fuzzy arithmetic.‡

We are now in a position to apply the quantifier extension principle to the derivation of the intersection/product syllogism expressed by (3.7). Specifically, we note that

$$Q_1 \ A\text{'s } are \ B\text{'s} \leftrightarrow Prop(B/A) \ is \ Q_1 \tag{3.16}$$

$$Q_2(A \ and \ B)\text{'s } are \ C\text{'s} \leftrightarrow Prop(C/A \cap B) \ is \ Q_2 \tag{3.17}$$

and

$$Q \ A\text{'s } are \ (B \ and \ C)\text{'s} \leftrightarrow Prop(B \cap C/A) \ is \ Q, \tag{3.18}$$

where

$$Prop(B/A) = \frac{\Sigma Count(B \cap A)}{\Sigma Count(A)} \tag{3.19}$$

$$Prop(C/A \cap B) = \frac{\Sigma Count(A \cap B \cap C)}{\Sigma Count(A \cap B)} \tag{3.20}$$

$$Prop(B \cap C/A) = \frac{\Sigma Count(A \cap B \cap C)}{\Sigma Count(A)}. \tag{3.21}$$

From (3.19) to (3.21), it follows that the relative counts $C_1 \overset{\Delta}{=} Prop(B/A)$, $C_2 \overset{\Delta}{=} Prop(C/A \cap B)$ and $C \overset{\Delta}{=} Prop(B \cap C/A)$ satisfy the identity

$$Prop(B \cap C/A) = Prop(B/A)Prop(C/A \cap B). \tag{3.22}$$

and hence

$$C = C_1 C_2. \tag{3.23}$$

‡Where typographical convenience is a significant consideration, a fuzzy version of an arithmetic operation * may be expressed more simply as (*).

On the other hand, from (3.16) to (3.18), we see that Q_1, Q_2 and Q are the respective possibility distributions of C_1, C_2 and C. Consequently, from the quantifier extension principle applied to arithmetic expressions, it follows that the fuzzy quantifier Q is the fuzzy product of the fuzzy quantifiers Q_1 and Q_2, i.e.

$$Q = Q_1 \otimes Q_2, \qquad (3.24)$$

which is what we wanted to establish.

As a corollary of (3.7), we can deduce at once the following syllogism:

$$Q_1 \text{ A's } are \text{ B's} \qquad (3.25)$$

$$\underline{Q_2(A \text{ and } B)\text{'s } are \text{ C's}}$$

$$(\geq (Q_1 \otimes Q_2)) \text{ A's } are \text{ C's,}$$

where the quantifier $(\geq (Q_1 \otimes Q_2))$, which represents the composition of the binary relation $\geq$ with the unary relation $Q_1 \otimes Q_2$, should be read as $at\ least(Q_1 \otimes Q_2)$. This syllogism is a consequence of (3.7) by virtue of the inequality

$$\Sigma Count(B \cap C) \leq \Sigma Count(C), \qquad (3.26)$$

which holds for all fuzzy or nonfuzzy B and C. For, if we rewrite (3.7) in terms of proportions,

$$Prop(B/A) \text{ is } Q_1 \qquad (3.27)$$

$$\underline{Prop(C/A \cap B) \text{ is } Q_2}$$

$$Prop(B \cap C/A) \text{ is } (Q_1 \otimes Q_2),$$

then from (3.26) it follows that

$$Prop(B \cap C/A) \text{ is } (Q_1 \otimes Q_2) \Rightarrow Prop(C/A) \text{ is } (\geq (Q_1 \otimes Q_2)). \qquad (3.28)$$

Thus, based on (3.28), the syllogism (3.7) and its corollary (3.25) may be represented compactly in the form:

$$Q_1 \text{ A's } are \text{ B's} \qquad (3.29)$$

$$\underline{Q_2(A \cap B)\text{'s } are \text{ C's}}$$

$$\underline{(Q_1 \otimes Q_2) \text{ A's } are \text{ (B and C)'s}}$$

$$(\geq (Q_1 \otimes Q_2)) \text{ A's } are \text{ C's.}$$

As an additional illustration of the quantifier extension principle, consider the inequality established in Section 2, namely,

$$O \vee (\Sigma Count(A) + \Sigma Count(B) - Count(U)) \leq \Sigma Count(A \cap B) \leq \Sigma Count(A) \wedge \Sigma Count(B). \tag{3.30}$$

Let Q, Q_1 and Q_2 be the fuzzy quantifiers which characterize $C \overset{\Delta}{=} \Sigma Count(A \cap B)$, $C_1 \overset{\Delta}{=} \Sigma Count(A)$, and $C_2 \overset{\Delta}{=} \Sigma Count(B)$, respectively. Then

$$O \oslash (Q_1 \oplus Q_2 \ominus 1) \leq Q \leq Q_1 \oslash Q_2. \tag{3.31}$$

where, as stated earlier, $\oplus$, $\otimes$, $\oslash$ and $\oslash$ are the operations of sum, product, min and max in fuzzy arithmetic. Consequently, as a special case of (3.31), we can assert that in the inference schema

$$most\ students\ are\ single \tag{3.32}$$

$$many\ students\ are\ male$$

$$Q\ students\ are\ single\ and\ male,$$

Q is a fuzzy interval given by

$$Q = (\geq (O \oslash (most \oplus many \ominus 1))) \cap (\leq (most \oslash many)). \tag{3.33}$$

Monotonicity

In the theory of generalized quantifiers [7], a generalized quantifier Q is said to be *monotonic* if a true proposition of the form $p \overset{\Delta}{=} Q$ A's *are* B's, where A and B are nonfuzzy sets, remains true when B is replaced by any superset (or any subset) of B. In this sense, *most* is a monotonic generalized quantifier under the assumption that B is replaced by a superset of B.

In the case of fuzzy quantifiers of the first or second kinds, a similar but more general definition which is valid for fuzzy sets may be formulated in terms of the membership function or, equivalently, the possibility distribution function of Q. More specifically:

A fuzzy quantifier Q is *monotone nondecreasing* (*nonincreasing*) if and only if the membership function of Q, μ_Q, is monotone nondecreasing (nonincreasing) over the domain of Q. From this definition, it follows at once that

$$Q\ is\ monotone\ nondecreasing \Leftrightarrow \geq Q = Q \tag{3.34}$$

$$Q\ is\ monotone\ nonincreasing \Leftrightarrow \leq Q = Q, \tag{3.35}$$

where, as stated earlier, $\geq Q$ and $\leq Q$ should be read as "at least Q" and "at most Q," respectively. Furthermore, from (2.7) it follows that, if $B \subset C$, then

$$Q \text{ is monotone nondecreasing} \Leftrightarrow \tag{3.36}$$

$$Prop(B|A) \text{ is } Q \Rightarrow Prop(C|A) \text{ is } Q$$

and

$$Q \text{ is monotone nonincreasing} \Leftrightarrow \tag{3.37}$$

$$Prop(C|A) \text{ is } Q \Leftrightarrow Prop(B|A) \text{ is } Q.$$

If Q is a fuzzy quantifier of the second kind, the *antonym* of Q, *ant* Q, is defined by [89]

$$\mu_{antQ}(u) = \mu_Q(1 - u), \quad u \in [0,1]. \tag{3.38}$$

Thus, if *few* is interpreted as the antonym of *most*, we have

$$\mu_{FEW}(u) = \mu_{MOST}(1 - u), \quad u \in [0,1]. \tag{3.39}$$

A graphic illustration of (3.39) is shown in Fig. 3.

An immediate consequence of (3.38) is the following:

If Q is monotone nondecreasing (e.g. most), then its antonym (e.g. few) is monotone nonincreasing.

We are now in a position to derive additional syllogisms for fuzzily-quantified propositions and, inter alia, establish the validity of the example given in the abstract, namely,

$$most \; U\text{'s are } A\text{'s} \tag{3.40}$$

$$\underline{most \; A\text{'s are } B\text{'s}}$$

$$most^2 \; U\text{'s are } B\text{'s,}$$

where by U's we mean the elements of the universe of discourse U, and *most* is assumed to be monotone nondecreasing.

Specifically, by identifying A in (3.25) with U in (3.40), B in (3.25) with A in (3.40), C in (3.25) with B in (3.40), and noting that

$$U \cap A = A,$$

we obtain as a special case of (3.25) the inference schema

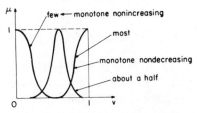

Fig. 3. The fuzzy quantifier *few* as an antonym of *most*.

$$most\ U\text{'s}\ are\ A\text{'s} \tag{3.41}$$

$$\underline{most\ A\text{'s}\ are\ B\text{'s}}$$

$$\geq 0\ (most\ \otimes\ most)\ U\text{'s}\ are\ B\text{'s}$$

$$most^2\ U\text{'s}\ are\ B\text{'s},$$

where $most^2$ denotes $most \otimes most$. More generally, for any monotone nondecreasing fuzzy quantifiers Q_1 and Q_2, we can assert that

$$Q_1\ U\text{'s}\ are\ A\text{'s} \tag{3.42}$$

$$\underline{Q_2\ A\text{'s}\ are\ B\text{'s}}$$

$$(Q_1 \otimes Q_2)\ U\text{'s}\ are\ B\text{'s}.$$

If one starts with a rule of inference in predicate calculus, a natural question which arises is: how does the rule in question generalize to fuzzy quantifiers? An elementary example of an answer to a question of this kind is the following inference schema:

$$\frac{Q_1\ A\text{'s}\ are\ B\text{'s}}{(\geq Q_2)\ A\text{'s}\ are\ B\text{'s}} \quad \text{if} \quad Q_2 \leq Q_1, \tag{3.43}$$

which is a generalization of the basic rule:

$$\frac{(\forall x)P(x)}{(\exists x)P(x)},$$

where P is a predicate. In (3.43), the inequality $Q_2 \leq Q_1$ signifies that, as a fuzzy number, Q_2 is less than or equal to the fuzzy number Q_1 (see Fig. 4).

To establish the validity of (3.43), we start with the inference rule

$$\frac{Q_1\ A\text{'s}\ are\ B\text{'s}}{Q_2\ A\text{'s}\ are\ B\text{'s}} \quad \text{if} \quad Q_1 \subset Q_2, \tag{3.44}$$

which is an immediate consequence of the entailment principle (3.9), since the conclusion in (3.44) is less specific than the premise. Then, (3.43) follows at once from (3.44) and the containment relation

$$Q_1 \leq Q_2 \Rightarrow Q_2 \subset (\geq Q_1) \tag{3.45}$$

which, in words, means that, if a fuzzy number Q_1 is less than or equal to Q_2, then, as a fuzzy set, Q_2 is contained in the fuzzy set which corresponds to the fuzzy number "at least Q_1."

In inferring from fuzzily-quantified propositions with negations, it is useful to have rules which concern the semantic equivalence or semantic entailment of such propositions. In what

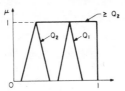

Fig. 4. The possibility distributions of Q_2 and Q_1, with $Q_2 \leq Q_1$.

follows, we shall derive a few basic rules of this type.

The first rule, which applies to fuzzy quantifiers of the first kind, and to fuzzy quantifiers of the second kind when the base set, A, is nonfuzzy, is the following:

$$Q \ A\text{'s are } B\text{'s} \leftrightarrow (antQ) \ A\text{'s are not } B\text{'s}, \tag{3.46}$$

where $antQ$ denotes the antonym of Q (see (3.38)). For example,

$$most \ men \ are \ tall \leftrightarrow (ant \ most) \ men \ are \ not \ tall, \tag{3.47}$$

and

$$most \ men \ are \ tall \leftrightarrow few \ men \ are \ not \ tall$$

if *few* is interpreted as the antonym of *most*.

To establish (3.46), we note that, in consequence of (3.6), we have

$$Q \ A\text{'s are } B\text{'s} \rightarrow \Pi_{\Sigma Count(B/A)} = Q. \tag{3.48}$$

The possibility assignment equation in (3.48) implies that the test score, τ_1, associated with the proposition "$Q \ A$'s are B's," is given by

$$\tau_1 = \mu_Q(\Sigma Count(B/A)). \tag{3.49}$$

where μ_Q is the membership function of Q.

Similarly, the test score associated with the proposition "$(antQ) \ A$'s *are not* B's," is given by

$$\tau_2 = \mu_{antQ}(\Sigma Count(\neg B/A)). \tag{3.50}$$

Thus, to demonstrate that the two propositions are semantically equivalent, it will suffice to show that $\tau_1 = \tau_2$.

To this end, we note that

$$\Sigma Count \, (\neg B/A) = \frac{\Sigma Count(A \cap (\neg B))}{\Sigma Count(A)} \tag{3.51}$$

$$= \frac{\Sigma_i \mu_A(u_i) \wedge (1 - \mu_B(u_i))}{\Sigma_i \mu_A(u_i)}$$

and, if A is nonfuzzy, the right-hand member of (3.51) may be written as:

$$\frac{\Sigma_i \mu_A(u_i) \wedge (1 - \mu_B(u_i))}{\Sigma_i \mu_A(u_i)} = 1 - \Sigma Count(B/A). \tag{3.52}$$

Now, from the definition of the antonym (3.38), it follows that

$$\mu_{antQ}(1 - \Sigma Count(B/A)) = \mu_Q(\Sigma Count(B/A)), \tag{3.53}$$

and hence that $\tau_1 = \tau_2$, which is what we had to establish.

In the more general case where A is fuzzy, the semantic equivalence (3.46) does not hold. Instead, the following semantic entailment may be asserted:

If Q is monotone nonincreasing, then

$$\frac{Q \ A\text{'s } are \ B\text{'s}}{(antQ) \ A\text{'s } are \ not \ B\text{'s}} \tag{3.54}$$

To validate (3.54), we note that in Section 2 we have established the inequality (see 2.19)

$$1 - \Sigma Count(\neg B/A) \leq \Sigma Count(B/A). \tag{3.55}$$

Now, if Q is monotone nonincreasing, then on application of μ_Q to both sides of (3.55) the inequality is reversed, yielding

$$\mu_Q(1 - \Sigma Count(\neg B/A)) \geq \mu_Q(\Sigma Count(B/A))$$

or, equivalently,

$$\mu_{antQ}(\Sigma Count(\neg B/A)) \geq \mu_Q(\Sigma Count(B/A)), \tag{3.56}$$

which establishes that the consequence in (3.54) is less specific than the premise and thus, by the entailment principle, is entailed by the premise.

In general, an application of the entailment principle for the purpose of demonstrating the validity of an inference rule reduces the computation of a fuzzy quantifier to the solution of a variational problem or, in discrete cases, to the solution of a nonlinear program. As an illustration, we shall consider the following inference schema

$$\frac{Q_1 \ A\text{'s } are \ B\text{'s}}{?Q \ A\text{'s } are \ (very \ B)\text{'s}}, \tag{3.57}$$

where $?Q$ is the quantifier to be computed; the base set A is nonfuzzy and the modifier *very* is an intensifier whose effect is assumed to be defined by [85]

$$very \ B = {}^2B, \tag{3.58}$$

†In earlier papers, the meanings of B^2 and ${}^2B$ were interchanged.

where the left exponent 2 signifies that the membership function of ^2B is the square of that of B.† Since A is nonfuzzy, we can assume, without loss of generality, that $A = U$.

With this assumption, the translation of the premise in (3.57) is given by

$$Q_1 \ U\text{'s are } B\text{'s} \to \Pi_{\Sigma Count(B/U)} = Q_1 \tag{3.59}$$

while that of the consequent is

$$Q \ U\text{'s are } ^2B\text{'s} \to \Pi_{\Sigma Count(^2B/U)} = Q. \tag{3.60}$$

Let $\mu_1, \ldots \mu_n$ be the grades of membership of the points $u_1, \ldots u_n$ in B. Then, (3.59) and (3.60) imply that the overall test scores for the premise and the consequent are, respectively,

$$\tau_1 = \mu_{Q_1}\left(\frac{1}{N}\Sigma_i\mu_i\right) \tag{3.61}$$

$$\tau_2 = \mu_Q\left(\frac{1}{N}\Sigma_i\mu_i^2\right), \tag{3.62}$$

where $N = \Sigma Count(U)$.

The problem we are faced with at this point is the following. The premise, $Q_1 \ U$'s *are* B's, defines via (3.61) a fuzzy set, P_1, in the unit cube $C^N = \{\mu_1, \ldots \mu_N\}$ such that the grade of membership of the point $\mu = (\mu_1, \ldots \mu_N)$ in P_1 is τ_1. The mapping $C^N \to [0,1]$ which is defined by the sigma-count

$$\Sigma Count(very \ B/U) = \frac{1}{N}\Sigma_1\mu_i^2, \tag{3.63}$$

induces the fuzzy set, Q, in $[0,1]$ whose membership function, μ_Q, is what we wish to determine. For this purpose, we can invoke the extension principle, which reduces the determination of μ_Q to the solution of the following nonlinear program:

$$\mu_Q(v) = \max_\mu\left(\mu_{Q_1}\left(\frac{1}{N}\Sigma_i\mu_i\right)\right), \quad v \in [0,1], \tag{3.64}$$

subject to the constraint

$$v = \frac{1}{N}\Sigma_i\mu_i^2.$$

As shown in [90], this nonlinear program has an explicit solution given by

$$\mu_Q(v) = \mu_{Q_1}(\sqrt{v}), \quad v \in [0,1],$$

which implies that

$$Q = Q_1^2 = Q_1 \otimes Q_1. \tag{3.65}$$

We are thus led to the inference schema

$$\frac{Q_1 \ A\text{'s are } B\text{'s}}{Q_1^2 \ A\text{'s are } (\textit{very } B)\text{'s}} \tag{3.66}$$

and, more generally, for any positive m and nonfuzzy A,

$$\frac{Q_1 \ A\text{'s are } B\text{'s}}{Q_1^m \ A\text{'s are } (^m B)\text{'s}}, \tag{3.67}$$

and

$$\frac{Q_1 \ A\text{'s are } {}^m B\text{'s}}{Q_1^{1/m} \ A\text{'s are } B\text{'s}} \tag{3.68}$$

where

$$\mu_{Q_1^m}(v) = \mu_{Q_1}(v^{1/m}), \quad v \in [0,1] \tag{3.69}$$

$$\mu_{Q^{1/m}}(v) = \mu_{Q_1}(v^m), \tag{3.70}$$

and

$$\mu_{m_B}(u) = (\mu_B(u))^m, \quad u \in U. \tag{3.71}$$

As a simple example, assume that the premise in (3.66) is the proposition "Most men over sixty are bald." Then, the inference schema represented by (3.66) yields the syllogism:

$$\frac{\textit{most men over sixty are bald}}{\textit{most}^2 \textit{ men over sixty are very bald}}. \tag{3.72}$$

It should be noted that an inference schema may be formed by a composition of two or more other inference schemas. For example, by combining (3.46) and (3.68), we are led to the following schema:

$$\frac{Q_1 A\text{'s are } (\textit{not very } B)\text{'s}}{(\textit{ant } Q_1)^{0.5} \ A\text{'s are } B\text{'s}}, \tag{3.73}$$

in which the base set A is assumed to be nonfuzzy. Thus, the syllogism

$$\frac{\textit{most Frenchmen are not very tall}}{(\textit{ant most})^{0.5} \textit{ Frenchmen are tall}} \tag{3.74}$$

may be viewed as an instance of this schema (see Fig. 5).

In the foregoing discussion, we have attempted to show how the treatment of fuzzy quantifiers as fuzzy numbers makes it possible to derive a wide variety of inference schemas for fuzzily-quantified propositions. These propositions were assumed to have a simple structure like "Q A's are B's," which made it unnecessary to employ the full power of test-score semantics for representing their meaning. We shall turn our attention to more complex propositions in the following section and will illustrate by examples the application of test-score semantics to the representation of meaning of various types of fuzzily-quantified semantic entities.

4. MEANING REPRESENTATION BY TEST-SCORE SEMANTICS

As was stated in the Introduction, the process of meaning representation in test-score semantics involves three distinct phases: Phase I, in which an explanatory database frame, EDF, is constructed; Phase II, in which the constraints induced by the semantic entity are tested and scored; and Phase III, in which the partial test scores are aggregated into an overall test score which is a real number in the interval [0,1] or, more generally, a vector of such numbers.

In what follows, the process is illustrated by several examples in which Phase I and Phase II are merged into a single test which yields the overall test score. This test represents the meaning of the semantic entity and may be viewed as a description of the process by which the meaning of the semantic entity is composed from the meanings of the constituent relations in EDF.

In some cases, the test which represents the meaning of a given semantic entity may be expressed in a higher level language of logical forms. The use of such forms is illustrated in Examples 4 and 5.

When a semantic entity contains one or more fuzzy quantifiers, its meaning is generally easier to represent through the use of $\Sigma Counts$ than $FCounts$. However, there may be cases in which a $\Sigma Count$ may be a less appropriate representation of cardinality than an $FGCount$ or an $FECount$. This is particularly true of cases in which the cardinality of a set is low, i.e. is a small fuzzy number like *several*, *few*, etc. Furthermore, what should be borne in mind is that $\Sigma Count$ is a summary of an $FGCount$ and hence is intrinsically less informative.

In some of the following examples, we employ alternative counts for purposes of comparison: In others, only one type of count, usually the $\Sigma Count$, is used.

Example 1

$$SE \overset{\Delta}{=} several \ balls \ most \ of \ which \ are \ large.$$

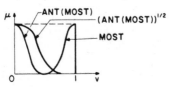

Fig. 5. The possibility distributions of the fuzzy quantifiers *most*, ant most and $(ant \ most)^{0.5}$.

For this semantic entity, we shall assume that *EDF* comprises the following relations:

$$EDF \stackrel{\Delta}{=} BALL\,[Identifier;\ Size] +$$

$$LARGE\,[Size;\ \mu] +$$

$$SEVERAL\,[Number;\ \mu] +$$

$$MOST\,[Proportion;\ \mu].$$

In this *EDF*, the first relation has n rows and is a list of the identifiers of balls and their respective sizes; in *LARGE*, μ is the degree to which a ball of size *Size* is large; in *SEVERAL*, μ is the degree to which *Number* fits the description *several*; and in *MOST*, μ is the degree to which *Proportion* fits the description *most*.

The test which yields the compatibility of *SE* with *ED* and thus defines the meaning of *SE* depends on the definition of fuzzy set cardinality. In particular, using the sigma-count, the test procedure may be stated as follows:

(1) Test the constraint induced by *SEVERAL*:

$$\tau_1 = {}_\mu SEVERAL[Number = n],$$

which means that the value of *Number* is set to *n* and the value of μ is read, yielding the test score τ_1 for the constraint in question.

(2) Find the size of each ball in *BALL*:

$$Size_i = {}_{Size}BALL[Identifier = Identifier_i],$$

$$i = 1,\ldots n.$$

(3) Test the constraint induced by *LARGE* for each ball in *BALL*:

$$\mu_{LB}(i) = {}_\mu LARGE[Size = Size_i].$$

(4) Find the sigma-count of large balls in *BALL*:

$$\Sigma Count(LB) = \Sigma_i \mu_{LB}(i).$$

(5) Find the proportion of large balls in *BALL*:

$$PLB = \frac{1}{n}\Sigma_i \mu_{LB}(i).$$

(6) Test the constraint induced by *MOST*:

$$\tau_2 = {}_\mu MOST[Proportion = PLB].$$

(7) Aggregate the partial test scores:

$$\tau = \tau_1 \wedge \tau_2,$$

where τ is the overall test score. The use of the min operator to aggregate τ_1 and τ_2 implies that we interpret the implicit conjunction in SE as the cartesian product of the conjuncts.

The use of fuzzy cardinality affects the way in which τ_2 is computed. Specifically, the employment of $FGCount$ leads to:

$$\tau_2 = \sup_i (FGCount(LB) \cap nMOST),$$

which expressed in terms of the membership functions of $FGCount$ (LB) and $MOST$ may be written as

$$\tau_2 = \sup_i \left(\mu_{FGCount(LB)}(i) \wedge \mu_{MOST}\left(\frac{1}{n}\right) \right).$$

The rest of the test procedure is unchanged.

Example 2

$$SE \overset{\Delta}{=} several\ large\ balls.$$

In this case, we assume that the EDF is the same as in Example 1, with $MOST$ deleted.

As is pointed out in [93], the semantic entity in question may be interpreted in different ways. In particular, using the so-called compartmentalized interpretation in which the constraints induced by $SMALL$ and $SEVERAL$ are tested separately, the test procedure employing the $FGCount$ may be stated as follows:

(1) Test the constraint induced by $SEVERAL$:

$$\tau_1 \overset{\Delta}{=} {}_\mu SEVERAL[Number = n].$$

(2) Find the size of the smallest ball:

$$SSB \overset{\Delta}{=} {}_{Size}\min{}_{Size}(BALL),$$

in which the right-hand member signifies that the smallest entry in the column *Size* of the relation $BALL$ is read and assigned to the variable SSB (Smallest Size Ball).

(3) Test the constraint induced by $LARGE$ by finding the degree to which the smallest ball is large:

$$\tau_2 \overset{\Delta}{=} {}_\mu LARGE[Size = SSB].$$

(4) Aggregate the test scores:

$$\tau = \tau_1 \wedge \tau_2.$$

Example 3

$p \overset{\Delta}{=}$ *Hans has many acquaintances* and *a few close friends most of whom are highly intelligent.*

Assume that the *EDF* comprises the following relations:

$$ACQUAINTANCE\,[Name\,1;\,Name\,2;\,\mu]+$$

$$FRIEND\,[Name\,1;\,Name\,2;\,\mu]+$$

$$INTELLIGENT\,[Name;\,\mu]+$$

$$MANY\,[Number;\,\mu]+$$

$$FEW\,[Number;\,\mu]+$$

$$MOST\,[Proportion;\,\mu].$$

In *ACQUAINTANCE*, μ is the degree to which *Name* 1 is an acquaintance of *Name* 2; in *FRIEND*, μ is the degree to which *Name* 1 is a friend of *Name* 2; in *INTELLIGENT*, μ is the degree to which *Name* is intelligent; *MANY* and *FEW* are fuzzy quantifiers of the first kind, and *MOST* is a fuzzy quantifier of the second kind.

The test procedure may be stated as follows:

(1) Find the fuzzy set of Hans' acquaintances:

$$HA \overset{\Delta}{=} {}_{Name\,1\times\mu}ACQUAINTANCE[Name\,2 = Hans],$$

which means that in each row in which *Name* 2 is Hans, we read *Name* 1 and μ and form the fuzzy set *HA*.

(2) Count the number of Hans' acquaintances:

$$CHA \overset{\Delta}{=} \Sigma\,Count(HA).$$

(3) Find the test score for the constraint induced by *MANY*:

$$\tau_1 = {}_{\mu}MANY[Name\,1 = CHA].$$

(4) Find the fuzzy set of friends of Hans:

$$FH \overset{\Delta}{=} {}_{Name\,1\times\mu}FRIEND[Name\,2 = Hans].$$

(5) Intensify *FH* to account for *close* [89]:

$$CFH \stackrel{\Delta}{=} {}^2FH.$$

(6) Determine the count of close friends of Hans:

$$CCFH \stackrel{\Delta}{=} \Sigma Count({}^2FH).$$

(7) Find the test score for the constraint induced by *FEW*:

$$\tau_2 \stackrel{\Delta}{=} {}_\mu FEW[Number = CCFH].$$

(8) Intensify *INTELLIGENT* to account for *highly*. (We assume that this is accomplished raising *INTELLIGENT* to the third power.)

$$HIGHLY.INTELLIGENT = {}^3INTELLIGENT.$$

(9) Find the fuzzy set of close friends of Hans who are highly intelligent:

$$CFH.HI \stackrel{\Delta}{=} CFH \cap {}^3INTELLIGENT.$$

(10) Determine the count of close friends of Hans who are highly intelligent:

$$CCFH.HI \stackrel{\Delta}{=} \Sigma Count(CFH \cap {}^3INTELLIGENT).$$

(11) Find the proportion of those who are highly intelligent among the close friends of ns:

$$\gamma \stackrel{\Delta}{=} \frac{\Sigma Count(CFH \cap {}^3INTELLIGENT)}{\Sigma Count(CFH)}.$$

(12) Find the test score for the constraint induced by *MOST*:

$$\tau_3 \stackrel{\Delta}{=} {}_\mu MOST[Proportion = \gamma].$$

(13) Aggregate the partial test scores:

$$\tau = \tau_1 \wedge \tau_2 \wedge \tau_3.$$

The test described above may be expressed more concisely as a logical form which is nantically equivalent to *p*. The logical form may be expressed as follows:

$$p \leftrightarrow Count(_{Name\ 1 \times \mu}ACQUAINTANCE[Name\ 2 = Hans])\ is\ MANY\ \wedge$$

$$Count(_{Name\ 1 \times \mu}{}^2FRIEND[Name\ 2 = Hans])\ is\ FEW\ \wedge$$

$$Prop(^3INTELLIGENT/_{Name\,1\times\mu}{}^2FRIEND[Name\,2 = Hans]) \text{ is MOST}$$

where $\wedge$ denotes the conjunction.

Example 4

Consider the proposition

$$p \overset{\Delta}{=} \text{ Over the past few years Nick earned far more than most of his close friends.}$$

In this case, we shall assume that *EDF* consists of the following relations:

$$EDF \overset{\Delta}{=} INCOME[Name;\ Amount;\ Year] +$$

$$FRIEND[Name;\ \mu] +$$

$$FEW[Number;\ \mu] +$$

$$FAR.MORE[Income\,1;\ Income\,2;\ \mu] +$$

$$MOST[Proportion;\ \mu].$$

Using the sigma-count, the test procedure may be described as follows:
(1) Find Nick's income in $Year_i$, $i = 1, 2, \ldots$ counting backward from present:

$$IN_i \overset{\Delta}{=} {}_{Amount}INCOME[Name = Nick;\ Year = Year_i].$$

(2) Test the constraint induced by *FEW*:

$$\mu_i \overset{\Delta}{=} {}_\mu FEW[Year = Year_i].$$

(3) Compute Nick's total income during the past few years:

$$TIN = \Sigma_i\mu_i IN_i,$$

in which the μ_i play the role of weighting coefficients.
(4) Compute the total income of each $Name_j$ (other than Nick) during the past several years:

$$TIName_j = \Sigma_i\mu_i IName_{ji},$$

where $IName_{ji}$ is the income of $Name_j$ in $Year_i$.
(5) Find the fuzzy set of individuals in relation to whom Nick earned far more. The grade of membership of $Name_j$ in this set is given by:

$$\mu_{FM}(Name_j) = {}_\mu FAR.MORE[Income\ 1 = TIN\,;\ Income\ 2 = TIName_j].$$

(6) Find the fuzzy set of close friends of Nick by intensifying the relation *FRIEND*:

$$CF = {}^2FRIEND,$$

which implies that

$$\mu_{CF}(Name_j) = ({}_\mu FRIEND[Name = Name_j])^2.$$

(7) Using the sigma-count, count the number of close friends of Nick:

$$\Sigma Count(CF) = \Sigma_j \mu^2_{FRIEND}(Name_j).$$

(8) Find the intersection of *FM* with *CF*. The grade of membership of *Name*$_j$ in the intersection is given by

$$\mu_{FM \cap CF}(Name_j) = \mu_{FM}(Name_j) \wedge \mu_{CF}(Name_j).$$

(9) Compute the sigma-count of *FM* ∩ *CF*:

$$\Sigma Count(FM \cap CF) = \Sigma_j \mu_{FM}(Name_j) \wedge \mu_{CF}(Name_j).$$

(10) Compute the proportion of individuals in *FM* who are in *CF*:

$$\rho \overset{\Delta}{=} \frac{\Sigma Count(FM \cap CF)}{\Sigma Count(CF)}.$$

(11) Test the constraint induced by *MOST*:

$$\tau = {}_\mu MOST[Proportion = \rho],$$

which expresses the overall test score and thus represents the desired compatibility of ρ with the explanatory database.

For the proposition under consideration, the logical form has a more complex structure than in Example 3. Specifically, we have

$$Prop((\Sigma_j \mu_j / Name_j) / {}^2 FRIEND[Name\ 2 = Nick])\ is\ MOST$$

where

$$\mu_j = {}_\mu FAR.MORE[Income\ 1 = TIN\,;\ Income\ 2 = TIName_j]$$

where *Name*$_j \neq$ *Nick* and

$$TIN = \Sigma_i \mu_{FEW}(i)_{Amount} INCOME[Name = Nick;\ Year = Year_i]$$

and

$$TIName_j = \Sigma_i \mu_{FEW}(i)_{Amount} INCOME[Name = Name_j;\ Year = Year_i].$$

Example 5

$$p \overset{\Delta}{=} \textit{They like each other.}$$

In this case there is an implicit fuzzy quantifier in p which reflects the understanding that not all members of the group referred to as *they* must necessarily like each other.

Since the fuzzy quantifier in p is implicit, it may be interpreted in many different ways. The test described below represents one such interpretation and involves, in effect, the use of an *FCount*.

Specifically, we associate with p the *EDF*

$$EDF \overset{\Delta}{=} THEY[Name] +$$

$$LIKE[Name\ 1;\ Name\ 2;\ \mu] +$$

$$ALMOST.ALL[Proportion;\ \mu],$$

in which *THEY* is the list of names of members of the group to which p refers; *LIKE* is a fuzzy relation in which μ is the degree to which *Name* 1 likes *Name* 2; and *ALMOST.ALL* is a fuzzy quantifier in which μ is the degree to which a numerical value of *Proportion* fits a subjective perception of the meaning of *almost all*.

Let μ_{ij} be the degree to which *Name$_i$* likes *Name$_j$*, $i \neq j$. If there are n names in *THEY*, then there are $(n^2 - n)\mu_{ij}$'s in *LIKE* with $i \neq j$. Denote the relation *LIKE* without its diagonal elements by *LIKE**.

The test procedure which yields the overall test score τ may be described as follows:
(1) Count the number of members in *THEY*:

$$n \overset{\Delta}{=} Count(THEY).$$

(2) Compute the *FGCount* of *LIKE**:

$$C \overset{\Delta}{=} FGCount(LIKE^*).$$

Note that in view of (2.6), C may be obtained by sorting the elements of *LIKE** in descending order, which yields $LIKE^* \downarrow$. Thus,

$$FGCount(LIKE^*) = NLIKE^* \downarrow.$$

(3) Compute the height (i.e. the maximum value) of the intersection of C and the fuzzy number $(n^2 - n)ALMOST.ALL$:

$$\tau = sup(FGCount(LIKE^*) \cap (n^2 - n)ALMOST.ALL).$$

The result, as shown in Fig. 6, is the overall test score.

The last two examples in this Section illustrate the application of test-score semantics to question-answering. The basic idea behind this application is the following.

Suppose that the answer to a question, q, is to be deduced from a knowledge base which consists of a collection of prcpositions:

$$KB = \{p_1, \ldots p_n\}. \tag{4.1}$$

Furthermore, assume that the p_i are noninteractive and that each p_i induces a possibility distribution, Π^i, which is characterized by its possibility distribution function, π^i, over a collection of base variables $X = \{X_1, \ldots X_m\}$. This implies (a) that p_i, $i = 1, \ldots, n$, translates into the possibility assignment equation

$$p_i \to \Pi^i_{(X_1, \ldots X_m)} = F_i,$$

where F_i is a fuzzy subset of U, the cartesian product of the domains of $X_1, \ldots X_m$, i.e.

$$U = U_1 \times \cdots \times U_m,$$

in which U_i is the domain of X_i; and (b) that the collection KB induces a combined possibility distribution Π whose possibility distribution function is given by

$$\pi_{(X_1, \ldots X_m)} = \pi^1_{(X_1, \ldots X_m)} \wedge \cdots \wedge \pi^n_{(X_1, \ldots X_m)}. \tag{4.2}$$

In test-score semantics, the translation of a question is a procedure which expresses the answer to the question as a function of the explanatory database. In terms of the framework described above, this means that the answer is expressed as a function of $(X_1, \ldots X_m)$, i.e.

$$ans(q) = f(X_1, \ldots X_m).$$

Thus, given the possibility distribution Π over U and the function f, we can obtain the possibility distribution of $ans(q)$ by using the extension principle. In more specific terms, this reduces to the solution of the nonlinear program:

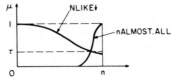

Fig. 6. Computation of the test score for "They like each other."

$$\mu_{ans(q)}(v) = \max_{(u_1, \ldots u_m)} \pi_{(X_1, \ldots X_m)}(u_1, \ldots u_m) \qquad (4.3)$$

subject to

$$v = f(u_1, \ldots, u_m),$$

where u_i denotes the generic value of X_i and $u_i \in U_i$, $i = 1, \ldots, m$. An example of such a program which we have encountered earlier is provided by (3.64).

In many cases, the nonlinear program (4.3) has special features which reduce it to a simpler problem which can be solved by elementary means. This is what happens in the following examples.

Example 6

$p_1 \overset{\Delta}{=}$ *There are about twenty graduate students in his class.*

$p_2 \overset{\Delta}{=}$ *There are a few more undergraduate students than graduate students in his class.*

$q \overset{\Delta}{=}$ *How many undergraduate students are there in his class?*

Let C_g, C_u and D denote, respectively, the number of graduate students, the number of undergraduate students, and the difference between the two counts, so that

$$C_u = C_g + D.$$

Applying the quantifier extension principle to this relation, we obtain

$$ans(q) = about\ 20 \oplus few,$$

where *ans(q)*, *about* 20 and *few* are fuzzy numbers which represent the possibility distributions of C_u, C_g and D, respectively. Using the addition rule for fuzzy numbers (see Appendix), the membership function of *ans(q)* may be expressed more explicitly as

$$\mu_{ans(q)}(v) = sup_u(\mu_{ABOUT20}(u) \wedge \mu_{FEW}(v - u)).$$

Example 7

$p \overset{\Delta}{=}$ *Brian is much taller than most of his close friends*

$q \overset{\Delta}{=}$ *How tall is Brian?*

Following the approach described earlier, we shall (a) determine the possibility distribution induced by p through the use of test-score semantics; (b) express the answer to q as a function defined on the domain of the possibility distribution; and (c) compute the possibility distribution of the answer.

To represent the meaning of p, we assume, as in Examples 4 and 5, that the *EDF* comprises the following relations:

$$EDF \overset{\Delta}{=} POPULATION[Name; Height] +$$

$$MUCH.TALLER[Height\ 1; Height\ 2; \mu] +$$

$$MOST[Proportion; \mu].$$

For this *EDF*, the test procedure may be described as follows:
(1) Determine the height of each *Name$_i$*, $i = 1, \ldots, n$, in *POPULATION*:

$$HN_i = {}_{Height}POPULATION[Name = Name_i],$$

and, in particular,

$$HB \overset{\Delta}{=} {}_{Height}POPULATION[Name = Brian].$$

(2) Determine the degree to which Brian is much taller than *Name$_i$*:

$$\mu_{BMT}(Name_i) \overset{\Delta}{=} {}_\mu MUCH.TALLER[Height\ 1 = HB; Height\ 2 = HN_i].$$

(3) Form the fuzzy set of members of *POPULATION* in relation to whom Brian is much taller:

$$BMT \overset{\Delta}{=} \Sigma_i \mu_{BMT}(Name_i)/Name_i, \quad Name_i \neq Brian.$$

(4) Determine the fuzzy set of close friend of Brian by intensifying *FRIEND*:

$$CFB = {}^2({}_{\mu \times Name\ 1}FRIEND[Name\ 2 = Brian])$$

which implies that

$$\mu_{CFB}(Name_i) = ({}_\mu FRIEND[Name\ 1 = Name_i: Name\ 2 = Brian])^2, \quad Name_i \neq Brian.$$

(5) Find the proportion of *BMT*'s in *CFB*'s:

$$\Sigma\,Count(BMT/CFB) = \frac{\Sigma_i \mu_{BMT}(Name_i) \wedge \mu_{CFB}(Name_i)}{\Sigma_i \mu_{CFB}(Name_i)}.$$

(6) Find the test score for the constraint induced by the fuzzy quantifier *most*:

$$\tau = {}_\mu MOST[Proportion = \Sigma\,Count(BMT/CFB)].$$

This test score represents the overall test score for the test which represents the meaning of p. Expressed as a logical form, the test may be represented more compactly as:

$$Prop(BMT/^2(_{\mu \times Name\,1}FRIEND[Name\,2 = Brian]) \text{ is } MOST,$$

where the fuzzy set BMT is defined in Steps 2 and 3.

To place in evidence the variables which are constrained by p, it is expedient to rewrite the expression for τ as follows:

$$\tau = \mu_{MOST}\left(\frac{\Sigma_i \mu_{MT}(h_B, h_i) \wedge \mu_{FB}^2(Name_i)}{\Sigma_i \mu_{FB}^2(Name_i)}\right),$$

in which h_B is the height of Brian; h_i is the height of $Name_i$; $\mu_{FB}(Name_i)$ is the degree to which $Name_i$ is Brian's friend; $\mu_{MT}(h_B, h_i)$ is the degree to which Brian is much taller than $Name_i$; and μ_{MOST} is the membership function of the quantifier *most*.

Now the variables $X_1, \ldots X_m$ are those entries in the relations in the explanatory database which are the arguments of τ, with the value of τ representing their joint possibility $\pi_{(X_1, \ldots X_m)}$. In the example under consideration, these variables are the values of $h_i \overset{\Delta}{=} Height(Name_i)$, $i = 1, \ldots n$; the values of $\mu_{MT}(h_B, h_i)$; the values of $\mu_{FB}(Name_i)$; and the values of $\mu_{MOST}(Pro\text{-}portion)$, where $Proportion$ is the value of the argument of μ_{MOST} in the expression for τ.

Since we are interested only in the height of Brian, it is convenient to let $X_1 \overset{\Delta}{=} h_B = Height(Brian)$. With this understanding, the possibility distribution function of $Height(Brian)$ given the values of $X_2, \ldots X_m$ may be expressed as

$$Poss\{Height(Brian) = u | X_2, \ldots X_m\} = \mu_{MOST}\left(\frac{\Sigma_i \mu_{MT}(u, h_i) \wedge \mu_{FB}^2(Name_i)}{\Sigma_i \mu_{FB}^2(Name_i)}\right),$$

where the range of the index i in Σ_i excludes $Name_i = Brian$. Correspondingly, the unconditional possibility distribution function of $Height\ (Brian)$ is given by the projection of the possibility distribution $\Pi_{(X_1, \ldots X_m)}$ on the domain of X_1. The expression for the projection is given by the supremum of the possibility distribution function of $(X_1, \ldots X_m)$ over all variables other than X_1 [89, 90]. Thus

$$Poss\{Height(Brian) = u\} = sup_{(X_2, \ldots X_m)}\mu_{MOST}\left(\frac{\Sigma_i \mu_{MOST}(u, h_i) \wedge \mu_{FB}^2(Name_i)}{\Sigma_i \mu_{FB}^2(Name_i)}\right).$$

Example 8

$p_1 \overset{\Delta}{=}$ *Most Frenchmen are not tall*

$p_2 \overset{\Delta}{=}$ *Most Frenchmen are not short*

$q \overset{\Delta}{=}$ *What is the average height of a Frenchman?*

Because of the simplicity of p_1 and p_2, the constraints induced by the premises may be found directly. Specifically, using (3.46), p_1 and p_2 may be replaced by the semantically equivalent premises

$$p_i' \overset{\Delta}{=} ant\ most\ Frenchmen\ are\ tall$$

$$p_i' \overset{\Delta}{=} ant\ most\ Frenchmen\ are\ short.$$

To formulate the constraints induced by these premises, let $h_1, \ldots h_n$ denote the heights of $Frenchman_1, \ldots, Frenchman_n$, respectively. Then, the test scores associated with the constraints in question may be expressed as

$$\tau_1 = \mu_{ANT\,MOST}\left(\frac{1}{n}\Sigma_i\mu_{TALL}(h_i)\right)$$

and

$$\tau_2 = \mu_{ANT\,MOST}\left(\frac{1}{n}\Sigma_i\mu_{SHORT}(h_i)\right),$$

where

$$\mu_{ANT\,MOST}(u) = \mu_{MOST}(1-u), \quad u \in [0,1],$$

and μ_{TALL} and μ_{SHORT} are the membership functions of *TALL* and *SHORT*, respectively. Correspondingly, the overall test score may be expressed as

$$\tau = \tau_1 \wedge \tau_2.$$

Now, the average height of a Frenchman and hence the answer to the question is given by

$$ans(q) = \frac{1}{n}\Sigma_i h_i.$$

Consequently, the possibility distribution of $ans(q)$ is given by the solution of the nonlinear program

$$\mu_{ans(q)}(h) = \max_{h_1, \ldots h_n}(\tau)$$

subject to

$$h = \frac{1}{n}\Sigma_i h_i.$$

Alternatively, a simpler but less informative answer may be formulated by forming the intersection of the possibility distributions of $ans(q)$ which are induced separately by p_1' and p_2'. More specifically, let $\Pi_{ans(q)|p_1'}$, $\Pi_{ans(q)|p_2'}$, $\Pi_{ans(q)|p_1' \wedge p_2'}$ be the possibility distributions of $ans(q)$ which are induced by p_1', p_2', and the conjunction of p_1' and p_2', respectively. Then, by using the minimax inequality[84], it can readily be shown that

$$\Pi_{ans(q)|p_1'} \cap \Pi_{ans(q)|p_2'} \supset \Pi_{ans(q)|p_1' \wedge p_2'}$$

and hence we can invoke the entailment principle to validate the intersection in question as the possibility distribution of $ans(q)$. For the example under consideration, the possibility distribution is readily found to be given by

$$Poss\{ans(q) = h\} = \mu_{ANT\,MOST}(\mu_{TALL}(h)) \wedge \mu_{ANT\,MOST}(\mu_{SHORT}(h)).$$

CONCLUDING REMARK

As was stated in the Introduction, the basic idea underlying our approach to fuzzy quantifiers is that such quantifiers may be interpreted as fuzzy numbers—a viewpoint which makes it possible to manipulate them through the use of fuzzy arithmetic and, more generally, fuzzy logic.

By applying test-score semantics to the translation of fuzzily-quantified possibilities, a method is provided for inference from knowledge bases which contain such propositions—as most real-world knowledge bases do. The examples presented in this section are intended to illustrate the translation and inference techniques which form the central part of our approach. There are many computational issues, however, which are not addressed by these examples. One such issue is the solution of nonlinear programs to which the problem of inference is reduced by the application of the extension principle. What is needed for this purpose are computationally efficient techniques which are capable of taking advantage of the tolerance for imprecision which is intrinsic in inference from natural language knowledge bases.

REFERENCES

1. E. W. Adams, The logic of "almost all." *J. Philos. Logic* **3**, 3–17 (1974).
2. E. W. Adams and H. F. Levine, On the uncertainties transmitted from premises to conclusions in deductive inferences. *Synthese* **30**, 429–460 (1975).
3. E. W. Adams and I. F. Carlstrom, Representing approximate ordering and equivalence relations. *J. Math. Psych.* **19**, 182–207 (1979).
4. E. W. Adams, Improbability transmissibility and marginal essentialness of premises in inferences involving indicative conditionals. *J. Philos. Logic* **10**, 149–177 (1981).
5. A. Barr and E. W. Feigenbaum, *The Handbook of Artificial Intelligence*, Vols. 1–3. Kaufmann, Los Altos (1982).
6. R. Bartsch and T. Vennemann, *Semantic Structures*. Attenaum Verlag, Frankfurt (1972).
7. J. Barwise and R. Cooper, Generalized quantifiers and natural language. *Linguistics and Philos.* **4**, 159–219 (1981).
8. R. E. Bellmann and L. A. Zadeh, Local and fuzzy logics. In *Modern Uses of Multiple-Valued Logic* (Edited by G. Epstein), pp. 103–165. Reidel, Dordrecht (1977).
9. N. Blanchard, Theories cardinales et ordinales des ensembles flou: les multiensembles. Thesis, University of Claude Bernard, Lyon (1981).
10. I. F. Carlstrom, Truth and entailment for a vague quantifier. *Synthese* **30**, 461–495 (1975).
11. R. Carnap, *Meaning and Necessity*. University of Chicago Press (1952).
12. W. S. Cooper, *Logical Linguistics*. Reidel, Dordrecht (1978).
13. M. J. Cresswell, *Logic and Languages*. Methuen, London (1973).
14. S. Cushing, The formal semantics of quantification. In *Indiana Univ. Linguistics Club* (1977).
15. S. Cushing, *Quantifier Meanings—A Study in the Dimensions of Semantic Competence*. North-Holland, Amsterdam (1982).
16. F. J. Damerau, On fuzzy adjectives. *Memo. RC 5340, IBM Research Laboratory*, Yorktown Heights, New York (1975).
17. A. DeLuca and S. Termini, A definition of non-probabilistic entropy in the setting of fuzzy sets theory. *Inform. Control* **20**, 301–312 (1972).
18. D. R. Dowty *et al.*, *Introduction to Montague Semantics*. Reidel, Dordrecht (1981).
19. D. Dubois, A new definition of fuzzy cardinality of finite fuzzy sets. *BUSEFAL* **8**, 65–67 (1981).
20. D. Dubois and H. Prade, *Fuzzy Sets and Systems: Theory and Applications*. Academic Press, New York (1980).
21. D. Dubois and H. Prade, Operations on fuzzy numbers. *Int. J. Systems Sci.* **9**, 613–626 (1978).

22. D. Dubois and H. Prade, Addition of interactive fuzzy numbers. *IEEE Trans. on Automatic Control* **26**, 926–936 (1981).
23. D. Dubois, Proprietes de la cardinalite floue d'un ensemble flou fini. *BUSEFAL* **8**, 11–12 (1981).
24. R. O. Duda, J. Gaschnig and P. E. Hart, Model design in the PROSPECTOR consultation system for mineral exploration. In *Expert Systems in the Micro-Electronic Age* (Edited by D. Michie), pp. 153–167. Edinburgh University Press, Edinburgh (1979).
25. H.-D. Ebbinghaus, Uber fur-fast-alle quantoren. *Archiv fur Math. Logik und Grundlageusforschung* **12**, 39–53 (1969).
26. B. R. Gaines and L. J. Kohout, The fuzzy decade: a bibliography of fuzzy systems and closely related topics. *Int. J. Man-Machine Studies* **9**, 1–68 (1977).
27. B. R. Gaines, Logical foundations for database systems. *Int. J. Man-Machine Studies* **11**, 481–500 (1979).
28. J. A. Goguen, The logic of inexact concepts. *Synthese* **19**, 325–373 (1969).
29. H. M. Hersh and A. Caramazza, A fuzzy set approach to modifiers and vagueness in natural language. *J. Exper. Psych.* **105**, 254–276 (1976).
30. R. Hilpinen, Approximate truth and truthlikeness. In *Proc. Conf. Formal Methods in the Methodology of the Empirical Sciences*, Warsaw, 1974 (Edited by M. Przelecki and R. Wojcicki). Reidel, Dordrecht (1976).
31. J. K. Hintikka, *Logic, Language-Games, and Information: Kantian Themes in the Philosophy of Logic.* Oxford University Press, Oxford (1973).
33. J. Hobbs, Making computational sense of Montague's intensional logic. *Artificial Intell.* **9**, 287–306 (1978).
34. T. R. Hofmann, Qualitative terms for quantity, In *Proc. 6th European Meeting on Cybernetics and Systems Research* (Edited by R. Trappl). North-Holland, Amsterdam (1982).
35. D. N. Hoover, Probability logic. *Annals Math. Logic* **14**, 287–313 (1978).
36. M. Ishizuka, K. S. Fu and J. T. P. Yao, A rule-based inference with fuzzy set for structural damage assessment. In *Fuzzy Information and Decision Processes* (Edited by M. Gupta and E. Sanchez). North-Holland, Amsterdam (1982).
37. P. N. Johnson-Laird, Procedural semantics. *Cognition* **5**, 189–214 (1977).
38. A. Kaufmann, La theorie des numbres hybrides. *BUSEFAL* **8**, 105–113 (1981).
39. E. L. Keenan, Quantifier structures in English. *Foundations of Language* **7**, 255–336 (1971).
40. E. P. Klement, W. Schwyhla and R. Lowen, Fuzzy probability measures. *Fuzzy Sets and Systems* **5**, 83–108 (1981).
41. E. P. Klement, Operations on fuzzy sets and fuzzy numbers related to triangular norms. *Proc. 11th Conf. Multiple-Valued Logic*, Univ. of Oklahoma, Norman, pp. 218–225, 1981.
42. E. P. Klement, An axiomatic theory of operations on fuzzy sets. Institut fur Mathematik, Johannes Kepler Universitat Linz, Institutsbericht 159 (1981).
43. G. Lakoff, Hedges: A study in meaning criteria and the logic of fuzzy concepts. *J. Phil. Logic* **2**, 458–508 (1973) Also in *Contemporary Research in Philosophical Logic and Linguistic Semantics* (Edited by D. Jockney, W. Harper and B. Freed), pp. 221–271. Reidel, Dordrecht (1973).
44. K. Lambert and B. C. van Fraassen, Meaning relations, possible objects and possible worlds. *Philos. Problems in Logic* 1–19 (1970).
45. E. H. Mamdani and B. R. Gaines, *Fuzzy Reasoning and its Applications.* Academic Press, London (1981).
46. J. McCarthy, Circumscription: A non-monotonic inference rule. *Artificial Intell.* **13**, 27–40 (1980).
47. J. D. McCawley, *Everything that Linguists have Always Wanted to Know about Logic.* University of Chicago Press, Chicago (1981).
48. D. V. McDermott and J. Doyle, Non-monotonic logic. I. *Artificial Intell.* **13**, 41–72 (1980).
49. D. V. McDermott, Non-monotonic logic II: non-monotonic modal theories. *J. Assoc. Comp. Mach.* **29**, 33–57 (1982).
50. J. S. Mill, *A System of Logic.* Harper, New York (1895).
51. D. Miller, Popper's qualitative theory of verisimilitude. *Brit. J. Philos. Sci.* **25**, 166–177 (1974).
52. G. A. Miller and P. N. Johnson-Laird, *Language and Perception.* Harvard University Press, Cambridge (1976).
53. M. Mizumoto, S. Fukame and K. Tanaka, Fuzzy reasoning Methods by Zadeh and Mamdani, and improved methods. *Proc. 3rd Workshop on Fuzzy Reasoning*, Queen Mary College, London, 1978.
54. M. Mizumoto and K. Tanaka, Some properties of fuzzy numbers. In *Advances in Fuzzy Set Theory and Applications* (Edited by M. M. Gupta, R. K. Ragade and R. R. Yager), pp. 153–164. North-Holland, Amsterdam (1979).
55. M. Mizumoto, M. Umano and K. Tanaka, Implementation of a fuzzy-set-theoretic data structure system. *3rd Int. Conf. Very Large Data Bases*, Tokyo, 1977.
56. G. C. Moisil, *Lectures on the Logic of Fuzzy Reasoning.* Scientific Editions, Bucarest (1975).
57. R. Montague, *Formal Philosophy.* In: *Selected Papers* (Edited by R. Thomason). Yale University Press, New Haven (1974).
58. R. E. Moore, *Interval Analysis.* Prentice-Hall, Englewood Cliffs (1966).
59. C. F. Morgenstern, The measure quantifier. *J. Symbolic Logic* **44**, 103–108 (1979).
60. A. Mostowski, On a generalization of quantifiers. *Fund. Math.* **44**, 17–36 (1957).
61. A. S. Naranyani, Methods of modeling incompleteness of data in knowledge bases. In *Knowledge Representation and Modeling of Processes of Understanding*, pp. 153–162. Academy of Sciences of U.S.S.R., Novosibirsk (1980).
62. H. T. Nguyen, Toward a calculus of the mathematical notion of possibility. In *Advances in Fuzzy Set Theory and*

Applications (Edited by M. M. Gupta, R. K. Ragade and R. R. Yager), pp. 235–246. North-Holland, Amsterdam (1979).
63. I. Niiniluoto and R. Tuomela, *Theoretical Concepts and Hypothetico-inductive Inference.* Reidel, Dordrecht (1973).
64. I. Niiniluoto, On the truthlikeness of generalizations. In *Basic Problems in Methodology and Linguistics* (Edited by J. Hintikka and R. Butts), pp. 121–147. Reidel, Dordrecht (1977).
65. K. Noguchi, M. Umano, M. Mizumoto and K. Tanaka, Implementation of fuzzy artificial intelligence language FLOU. *Tech. Rep.*, Automation and Language of IECE (1976).
66. A. I. Orlov, *Problems of Optimization and Fuzzy Variables.* Znaniye, Moscow (1980).
67. B. Partee, *Montague Grammar.* Academic Press, New York (1976).
68. P. Peterson, On the logic of *few, many* and *most. Notre Dame. J. Formal Logic* **20**, 155–179 (1979).
69. R. Reiter, A logic for default reasoning. *Artificial Intell.* **13**, 81–132 (1980).
70. N. Rescher, *Plausible Reasoning.* Van Gorcum, Amsterdam (1976).
71. L. K. Schubert, R. G. Goebel and N. Cercone, The structure and organization of a semantic net for comprehension and inference. In *Associative Networks* (Edited by N. V. Findler), pp. 122–178. Academic Press, New York (1979).
72. J. Searle (Ed.) *The Philosophy of Language.* Oxford University Press, Oxford (1971).
73. E. H. Shortliffe, *Computer-based Medical Consultations: MYCIN.* American Elsevier, New York (1976).
74. A. B. Slomson, Some problems in mathematical logic. Thesis, Oxford University (1967).
75. M. Sugeno, Fuzzy measures and fuzzy integrals: a survey, In: *Fuzzy Automata and Decision Processes* (Edited by M. M. Gupta, G. N. Saridis and B. R. Gaines), pp. 89–102. North-Holland, Amsterdam (1977).
76. P. Suppes, Elimination of quantifiers in the semantics of natural languages by use of extended relation algebras. *Revue Int. Philosphie*, 117–118, 243–259 (1976).
77. T. Terano and M. Sugeno, Conditional fuzzy measures and their applications. *Fuzzy Sets and Their Applications to Cognitive and Decision Processes* (Edited by L. A. Zadeh, K. S. Fu, K. Tanaka and M. Shimura), pp. 151–170. Academic Press, New York (1975).
78. K. Van Lehn, Determining the scope of English quantifiers, *Tech. Rep.* 483, AI Laboratory, M.I.T. (1978).
79. Y. Wilks, Philosophy of language. In *Computational Linguistics* (Edited by E. Charniak and Y. Wilks), pp. 205–233. North-Holland, Amsterdam (1976).
80. R. R. Yager, A note on probabilities of fuzzy events. *Inform. Sci.* **18**, 113–129 (1974).
81. R. R. Yager, Quantified propositions in a linguistic logic. In *Proc. 2nd Int. Seminar on Fuzzy Set Theory* (Edited by E. P. Klement). Johannes Kepler University, Linz, Austria (1980).
82. R. R. Yager, A foundation for a theory of possibility. *J. Cybernetics* **10**, 177–204 (1980).
83. L. A. Zadeh, Probability measures of fuzzy events. *J. Math. Anal. Appl.* **23**, 421–427 (1968).
84. L. A. Zadeh, Similarity relations and fuzzy orderings. *Inform. Sci.* **3**, 177–200 (1971).
85. L. A. Zadeh, Fuzzy languages and their relation to human and machine intelligence. *Proc. Int. Conf. on Man and Computer*, Bordeaux, France, pp. 130–165. Karger, Basel (1972).
86. L. A. Zadeh, Fuzzy logic and approximate reasoning (in memory of Grigore Moisil). *Synthese* **30**, 407–428 (1975).
87. L. A. Zadeh, The concept of a linguistic variable and its application to approximate reasoning. *Inform. Sci.* **8**, 199–249, 301–357, **9**, 43–80 (1975).
88. L. A. Zadeh, Fuzzy sets as a basis for a theory of possibility. *Fuzzy Sets and Systems* 3–28 (1978).
89. L. A. Zadeh, PRUF—a meaning representation language for natural languages. *Int. J. Man-Machine Studies* **10**, 395–460 (1978).
90. L. A. Zadeh, A theory of approximate reasoning. *Electronics Research Laboratory Mem. M77/58* University of California, Berkeley (1977); also in *Machine Intelligence* 9 (Edited by J. E. Hayes, D. Michie and L. I. Kulich), pp. 149–194. Wiley, New York (1979).
91. L. A. Zadeh, Inference in fuzzy logic. *Proc. 10th Inter. Symp. on Multiple-valued Logic*, Northwestern University, pp. 124–131, 1980.
92. L. A. Zadeh, Possibility theory and soft data analysis. *Electronics Res. Laboratory Mem. M79/59*, University of California, Berkeley (1979); also in *Mathematical Frontiers of the Social and Policy Sciences* (Edited by L. Cobb and R. M. Thrall), pp. 69–129. Westview Press, Boulder (1981).
93. L. A. Zadeh, Test-score semantics for natural languages and meaning-representation via PRUF. *Tech. Note* 247, AI Center, SRI International, Menlo Park, Calif. (1981); also in *Empirical Semantics* (Edited by B. B. Rieger), pp. 281–349. Brockmeyer, Bochum (1981).
94. L. A. Zadeh, Fuzzy probabilities and their role in decision analysis. *Proc. 4th MIT/ONR Workshop on Command, Control and Communications*, M.I.T. pp. 159–179, 1981.
95. A. Zimmer, Some experiments concerning the fuzzy meaning of logical quantifiers. In *General Surveys of Systems Methodology* (Edited by L. Troncoli), pp. 435–441. Society for General Systems Research, Louisville (1982).
96. H.-J. Zimmermann and P. Zysno, Latent connectives in human decision making. *Fuzzy Sets and Systems* **4**, 37–52 (1980).

APPENDIX

The extension principle

Let f be a function from U to V. The extension principle—as its name implies—serves to extend the domain of definition of f from U to the set of fuzzy subsets of U. In particular, if F is a finite fuzzy subset of U expressed as

$$F = \mu_1/u_1 + \cdots + \mu_n/u_n$$

then $f(F)$ is a finite fuzzy subset of V defined as

$$f(F) = f(\mu_1/u_1 + \cdots + \mu_n/u_n) \tag{A1}$$

$$= \mu_1/f(u_1) + \cdots + \mu_n/f(u_n).$$

Furthermore, if U is the cartesian product of $U_1, \ldots U_N$, so that $u = (u^1, \ldots u^N)$, $u^i \in U_i$, and we know only the projections of F on $U_1, \ldots U_N$, whose membership functions are, respectively, $\mu_{F1}, \ldots \mu_{FN}$, then

$$f(F) = \Sigma_u \mu_{F1}(u^1) \wedge \cdots \wedge \mu_{FN}(u^N)/f(u^1, \ldots u^N), \tag{A2}$$

with the understanding that, in replacing $\mu_F(u^1, \ldots u^N)$ with $\mu_F(u^1) \wedge \cdots \wedge \mu_{FN}(u^N)$, we are tacitly invoking the *principle of maximal possibility* [87]. This principle asserts that in the absence of complete information about a possibility distribution Π, we should equate Π to the maximal (i.e. least restrictive) possibility distribution which is consistent with the partial information about Π.

As a simple illustration of the extension principle, assume that $U = \{1, 2, \ldots 10\}$; f is the operation of squaring; and *SMALL* is a fuzzy subset of U defined by

$$SMALL = 1/1 + 1/2 + 0.8/3 + 0.6/4 + 0.4/5.$$

Then, it follows from (A2) that the *right square* of *SMALL* is given by

$$SMALL^2 \overset{\Delta}{=} 1/1 + 1/4 + 0.8/9 + 0.6/16 + 0.4/25.$$

On the other hand, the *left square* of *SMALL* is defined by

$$^2SMALL \overset{\Delta}{=} 1/1 + 1/2 + 0.64/3 + 0.36/4 + 0.16/5$$

and, more generally, for a subset F of U and any real m, we have

$$\mu_{m_F}(u) \overset{\Delta}{=} (\mu_F(u))^m, \quad u \in U. \tag{A3}$$

Fuzzy numbers†

By a fuzzy number, we mean a number which is characterized by a possibility distribution or is a fuzzy subset of real numbers. Simple examples of fuzzy numbers are fuzzy subsets of the real line labeled *small, approximately 8, very close to 5, more or less large, much larger than 6, several*, etc. In general, a fuzzy number is either a convex or a concave fuzzy subset of the real line. A special case of a fuzzy number is an interval. Viewed in this perspective, fuzzy arithmetic may be viewed as a generalization of interval arithmetic [58].

Fuzzy arithmetic is not intended to be used in situations in which a high degree of precision is required. To take advantage of this assumption, it is expedient to represent the possibility distribution associated with a fuzzy number in a standardized form which involves a small number of parameters—usually two—which can be adjusted to fit the given distribution. A system of standardized possibility distributions which suits this purpose in most cases of practical interest is the following [87].

(1) *π-numbers.* The possibility distribution of such numbers is bell-shaped and piecewise-quadratic. The distribution is characterized by two parameters: (a) the peak-point, i.e. the point at which $\pi = 1$, and (b) the bandwidth, β, which is defined as the distance between the cross-over points, i.e. the points at which $\pi = 0.5$. Thus, a fuzzy π-number, x, is expressed as (p, β), where p is the peak-point and β is the bandwidth; or, alternatively, as (p, β'), where β' is the normalized bandwidth, i.e. $\beta' = \beta/p$. As a function of u, $u \in (-\infty, \infty)$, the values of $\pi_x(u)$ are defined by the equations

†A more detailed exposition of the properties of fuzzy numbers may be found in Dubois and Prade (1980).

$$\pi_x(u) = 0 \text{ for } u \leq p - \beta \text{ and } u \geq p + \beta \tag{A4}$$

$$= \frac{2}{\beta^2}(u - p + \beta)^2 \text{ for } p - \beta \leq u \leq p - \frac{\beta}{2}$$

$$= 1 - \frac{2}{\beta^2}(u - p)^2 \text{ for } p - \frac{\beta}{2} \leq u \leq p + \frac{\beta}{2}$$

$$= \frac{2}{\beta^2}(u - p - \beta)^2 \text{ for } p + \frac{\beta}{2} \leq u \leq p + \beta.$$

(2) s-*numbers*. As its name implies, the possibility distribution of an s-number has the shape of an s. Thus, the equations defining an s-number, expressed as (p/β), are:

$$\pi_x(u) = 0 \text{ for } u \leq p - \beta \tag{A5}$$

$$= \frac{2}{\beta^2}(u - p + \beta)^2 \text{ for } p - \beta \leq u \leq p - \frac{\beta}{2}$$

$$= 1 - \frac{2}{\beta^2}(u - p)^2 \text{ for } p - \frac{\beta}{2} \leq u \leq p$$

$$= 1 \text{ for } u \geq p,$$

where β (the bandwidth) is the length of the transition interval from $\pi_x = 0$ to $\pi_x = 1$ and p is the left peak-point, i.e. the right end-point of the transition interval.

(3) z-*numbers*. A z-number is a mirror image of an s-number. Thus, the defining equations for a z-number, expressed as $(p \setminus \beta)$, are:

$$\pi_x(u) = 0 \text{ for } u \leq p - \beta \tag{A6}$$

$$= \frac{2}{\beta^2}(u - p + \beta)^2 \text{ for } p - \beta \leq u \leq p - \frac{\beta}{2}$$

$$= 1 - \frac{2}{\beta^2}(u - p)^2 \text{ for } p - \frac{\beta}{2} \leq u \leq p + \beta$$

$$= 0 \text{ for } u \geq p + \beta,$$

where p is the right peak-point and β is the bandwidth.

(4) s/z-*numbers*. An s/z-number has a flat-top possibility distribution which may be regarded as the intersection of the possibility distributions of an s-number and a z-number, with the understanding that the left peak-point of the s-number lies to the left of the right peak-point of the z-number. In some cases, however, it is expedient to disregard the latter restrictions and allow an s/z-number to have a sharp peak rather than a flat top. An s/z-number is represented as an ordered pair $(p_1/\beta_1; p_2 \setminus \beta_2)$ in which the first element is an s-number and the second element is a z-number.

(5) z\s-*numbers*. The possibility distribution of a $z \setminus s$ number is the complement of that of an s/z-number. Thus, whereas an s/z-number is a convex fuzzy subset of the real line, a $z \setminus s$-number is a concave fuzzy subset. Equivalently, the possibility distribution of a $z \setminus s$-number may be regarded as the union of the possibility distributions of a z-number and an s-number. A z\s'-number is represented as $(p_1 \setminus \beta_1; p_2/\beta_2)$.

Arithmetic operations on fuzzy numbers

Let * denote an arithmetic operation such as addition, subtraction, multiplication or division, and let $x*y$ be the result of applying * to the fuzzy numbers x and y.

By the use of the extension principle, it can readily be established that the possibility distribution function of $x*y$ may be expressed in terms of those of x and y by the relation

$$\pi_{x*y}(w) = \vee_{u,v}(\pi_x(u) \wedge \pi_y(v)), \tag{A7}$$

subject to the constraint

$$w = u*v, \quad u, v, w \in (-\infty, \infty)$$

where $\vee_{u,v}$ denotes the supremum over u, v, and $\wedge \stackrel{\Delta}{=} \min$.

As a special case of a general result established by Dubois and Prade[20] for so-called L-R numbers, it can readily be deduced from (A7) that if x and y are numbers of the same type (e.g. π-numbers), then so are $x + y$ and $x - y$. Furthermore, the characterizing parameters of $x + y$ and $x - y$ depend in a very simple and natural way on those of x and y. More specifically, if $x = (p, \beta)$ and $y = (q, \gamma)$, then

$$(p,\beta) + (q,\gamma) = (p + q, \beta + \gamma)$$

$$(p/\beta) + (q/\gamma) = (p + q/\beta + \gamma)$$

$$(p\setminus\beta) + (q\setminus\gamma) = (p + q\setminus\beta + \gamma)$$

$$(p_1/\beta_1; p_2\setminus\beta_2 + (q_1/\gamma_1; q_2\setminus\gamma_2)$$

$$= (p_1 + q_1/\beta_1 + \beta_2; p_2 + q_2\setminus\gamma_1 + \gamma_2)$$

$$(p,\beta) - (q,\gamma) = (p - q, \beta + \gamma)$$

and similarly for other types of numbers.

In the case of multiplication, it is true only as an approximation that if x and y are π-numbers then so is $x \times y$. However, the relation between the peak-points and normalized bandwidths which is stated below is exact:

$$(p,\beta') \times (q,\gamma') = (p \times q, \beta' + \gamma'). \tag{A8}$$

The operation of division, x/y, may be regarded as the composition of (a) forming the reciprocal of y, and (b) multiplying the result by x. In general, the operation $1/y$ does not preserve the type of y and hence the same applies to x/y. However, if y is a π-number whose peak point is much larger than 1 and whose normalized bandwidth is small, then $1/y$ is approximately a π-number defined by

$$1/(p, \beta') \cong (1/p, \beta') \tag{A9}$$

and consequently

$$(p, \beta')/(q, \gamma') \cong (p/q, \beta' + \gamma') \tag{A10}$$

As a simple example of operations on fuzzy numbers, suppose that x is a π-number (p,β) and y is a number which is much larger than x. The question is: What is the possibility distribution of y?

Assume that the relation $y \gg x$ is characterized by a conditional possibility distribution $\Pi_{(y|x)}$ (i.e. the conditional possibility distribution of y given x) which for real values of x is expressed as an s-number

$$\Pi_{(y|x)} = (q(x)/\gamma(x)) \tag{A11}$$

whose peak-point and bandwidth depend on x.

On applying the extension principle to the composition of the binary relation $\gg$ as defined by (A11) with the unary relation x, it is readily found that y is an s-number which is approximately characterized by

$$y = (q(p)/[q(p) - q(p - \beta)]). \tag{A12}$$

In this way, then, the possibility distribution of y may be expressed in terms of the possibility distribution of x and the conditional possibility distribution of y given x.

Because of the reproducibility property of possibility distributions, the computational effort involved in the manipulation of fuzzy numbers is generally not much greater than that required in interval arithmetic. The bounds on the results, however, are usually appreciably tighter because in the case of fuzzy numbers the possibility distribution functions are allowed to take intermediate values in the interval [0,1], and not just 0 or 1, as in the case of intervals.

A Theory of Commonsense Knowledge

L.A. ZADEH

ABSTRACT. The theory outlined in this paper is based on the idea that what is commonly called *commonsense knowledge* may be viewed as a collection of *dispositions*, that is, propositions with implied fuzzy quantifiers. Typical examples of dispositions are: *Icy roads are slippery. Tall men are not very agile. Overeating causes obesity. Bob loves women. What is rare is expensive*, etc. It is understood that, upon restoration of fuzzy quantifiers, a disposition is converted into a proposition with explicit fuzzy quantifiers, e.g., *Tall men are not very agile → Most tall men are not very agile.*

Since traditional logical systems provide no methods for representing the meaning of propositions containing fuzzy quantifiers, such systems are unsuitable for dealing with commonsense knowledge. It is suggested in this paper that an appropriate computational framework for dealing with common-sense knowledge is provided by *fuzzy logic*, which, as its name implies, is the logic underlying fuzzy (or approximate) reasoning. Such a framework, with an emphasis on the re-presentation of dispositions, is outlined and illustrated with examples.

1. INTRODUCTION

It is widely agreed at this juncture that one of the impor-tant problem-areas in AI relates to the representation of commonsense knowledge. In general, such knowledge may be re-garded as a collection of propositions exemplified by: *Snow is white. Icy roads are slippery. Most Frenchmen are not very tall. Virginia is very intelligent. If a car which is offered for sale is cheap and old then it probably is not in good shape. Heavy smoking causes lung cancer*, etc. Represen-tation of propositions of this type plays a particularly important role in the design of expert systems [3].

The conventional knowledge representation techniques based on the use of predicate calculus and related methods are not well-suited for the representation of commonsense knowledge because the predicates in propositions which represent commonsense knowledge do not, in general, have crisp denotations. For example, the proposition *Most Frenchmen are not very tall* cannot be represented as a well-formed formula in predicate calculus because the sets which constitute the denotations of the predicate *tall* and the quantifier *most* in their respective universes of discourse are fuzzy rather than crisp.

More generally, the inapplicability of predicate calculus and related logical systems to the representation of commonsense knowledge reflects the fact that such systems make no provision for dealing with uncertainty. Thus, in predicate logic, for example, a proposition is either true or false and no gradations of truth or membership are allowed. By contrast, in the case of commonsense knowledge, a typical proposition contains a multiplicity of sources of uncertainty. For example, in the case of the proposition *If a car which is offered for sale is cheap and much more than ten years old then it probably is not in good shape*, there are five sources of uncertainty: (i) the temporal uncertainty associated with the fuzzy predicate *much more than ten years old*; (ii) the uncertainty associated with the fuzzy predicate *cheap*; (iii) the uncertainty associated with the fuzzy predicate *not in good shape*; (iv) the probabilistic uncertainty associated with the event *The car is not in good shape*; and (v) the uncertainty associated with the fuzzy characterization of the probability of the event in question *as probable*.

The approach to the representation of commonsense knowledge which is described in this paper is based on the idea that propositions which characterize commonsense knowledge are, for the most part, *dispositions*, that is, propositions with implied fuzzy quantifiers. In this sense, the proposition *Tall men are not very agile* is a disposition which upon restoration is converted into the proposition *Most tall men are not very agile*. In this proposition, *most* is an explicit fuzzy quantifier which provides an approximate characterization of the proportion of *men who are not very agile* among *men who are tall* [63].

To deal with dispositions in a systematic fashion, we shall employ *fuzzy logic* - which is the logic underlying *approximate* or *fuzzy reasoning* [5, 19, 26, 55, 58]. Basically, fuzzy logic has two principal components. The first component is, in effect, a translation system for representing the meaning of propositions and other types of semantic entities. We shall employ the suggestive term *test-score semantics* to refer to this translation system because it involves an aggregation of the test scores of elastic constraints which are induced by the semantic entity whose meaning is represented.

The second component is an inferential system for arriving at an answer to a question which relates to the information which is resident in a knowledge base. In the present paper, the focus of our attention will be the problem of meaning representation in the context of commonsense knowledge, and our discussion of the inferential component will be limited in scope.[1]

2. MEANING REPRESENTATION IN TEST-SCORE SEMANTICS

Test score semantics is concerned with representation of the meaning of various types of semantic entities, e.g., propositions, predicates, commands, questions,modifiers, etc. However, knowledge, whether commonsense or not, may be viewed as a collection of propositions. For this reason, we shall restrict our discussion of test-score semantics to the representation of the meaning of propositions.[2]

In test-score semantics, as in PRUF [57], a proposition is regarded as a collection of elastic, or, equivalently, fuzzy constraints. For example, the proposition *Pat is tall* represents an elastic constraint on the height of Pat. Similarly, the proposition *Charlotte is blonde* represents an elastic constraint on the color of Charlotte's hair. And, the proposition *Most tall men are not very agile* represents an elastic constraint on the proportion of men who are not very agile among tall men.

In more concrete terms, representing the meaning of a proposition, p, through the use of test-score semantics involves the following steps.

1. Identification of the variables $X_1,...,X_n$ whose values are constrained by the proposition. Usually, these variables are implicit rather than explizit in p.
2. Identification of the constraints $C_1,...,C_m$ which are induced by p.
3. Characterization of each constraint, C_i, by describing a testing procedure which associates with C_i a test score τ_i representing the degree to which C_i is

satisfied. Usually τ_i is expressed as a number in the interval [0,1]. More generally, however, a test score may be a probability/possibility distribution over the unit interval.

4. Aggregation of the partial test scores $\tau_1, \ldots, \tau_m$ into a smaller number of test scores $\bar{\tau}_1, \ldots, \bar{\tau}_k$, which are represented as an *overall vector test score* $\tau = (\bar{\tau}_1, \ldots, \bar{\tau}_k)$. In most cases $k = 1$, so that the overall test scores is a scalar. We shall assume that this is the case unless an explicit statement to the contrary is made.

It is important to note that, in test-score semantics, the meaning of p is represented not by the overall test score τ but the procedure which leads to it. Viewed in this perspective, test-score semantics may be regarded as a generalization of truth-conditional, possible-world and model-theoretic semantics [8, 9, 28, 32]. However, by providing a computational framework for dealing with uncertainty - which the conventional semantical systems disregard - test-score semantics achieves a much higher level of expressive power and thus provides a basis for representing the meaning of a much wider variety of propositions in a natural language.

In test-score semantics, the testing of the constraints induced by p is performed on a collection of fuzzy relations which constitute an *explanatory data-base*, or *ED* for short. A basic assumption which is made about the explanatory database is that it is comprised of relations whose meaning is known to the addressee of the meaning-representation process. In an indirect way, then, the testing and aggregation procedures in test-score semantics may be viewed as a description of a process by which the meaning of p is composed from the meanings of the constituent relations in the explanatory database. It is this explanatory role of the relations in *ED* that motivates its description as an *explanatory database*.

As will be seen in the sequel, in describing the testing procedures we need not concern ourselves with the actual entries in the constituent relations. Thus, in general, the description of a test involves only the frames[3] of the constituent relations, that is, their names, their variables (or attributes) and the domain of each variable. When this is the case, the explanatory database will be referred to as the *explanatory database frame*, or *EDF* for short.

As a simple illustration, consider the proposition

p = *Debbie is a few years older than Dana.*   (2.1)

In this case, a suitable explanatory database frame may be represented as

EDF $\underset{=}{\Delta}$ *POPULATION* [*Name; Age*] + *FEW* [*Number; μ*].

which signifies that the explanatory database frame consists of two relations: (a) a nonfuzzy relation *POPULATION* [*Name; Age*],which lists names of individuals and their age; and (b) a fuzzy relation *FEW* [*Number; μ*], which associates with each value of Number the degree, μ, to which *Number* is compatible with the intended meaning of *few*. In general, the domain of each variable in the *EDF* is implicitly determined by p, and is not spelled-out explicitly unless it is necessary to do so to define the testing procedure.

As another example, consider the disposition[4]

p $\underset{=}{\Delta}$ *Snow ist white,*

which is frequently used in the literature to explain the basic ideas underlying truth-conditional semantics.[5]

To construct an *EDF* for this disposition, we first note that what is generally meant by *Snow is white* is *Usually snow ist white*, in which *usually* may be interpreted as a fuzzy quantifier. Consequently, on the assumption that the proposition

p' = *Usually snow is white*   (2.2)

is a restoration of p, a natural choice for the *EDF* would be

EDF $\underset{=}{\Delta}$ *WHITE* [*Sample; μ*] + *USUALLY* [*Proportion; μ*].(2.3)

In this *EDF*, the relation *WHITE* is a listing of samples of snow together with the degree, μ, to which each sample is white, while the relation *USUALLY* defines the degree to which a numerical value of *Proportion* is compatible with the intended meaning of *usually*.

In proposition (2.1), the constrained variable is the difference in the ages of Debbie and Dana. Thus,

X $\underset{=}{\Delta}$ *Age(Debbie) - Age(Dana).*   (2.4)

The elastic constraint which is induced by p is determined by the fuzzy relation *FEW*. More specifically, let Π_X denote

the possibility distribution of X, i.e., the fuzzy set of possible values of X. Then, the constraint on X may be expressed as the *possibility assignment equation*

$$\Pi_X = FEW, \tag{2.5}$$

which assigns the fuzzy relation *FEW* to the possibility distribution of X. This equation implies that

$$\pi_X(u) \triangleq Poss\ \{X=u\} = \mu_{FEW}(u), \tag{2.6}$$

where *Poss* $\{X = u\}$ is the possibility that X may take u as its value; $\mu_{FEW}(u)$ is the grade of membership of u in the

fuzzy relation *FEW*; and the function π_X (from the domain of

X to the unit interval) is the *possibility distribution function* associated with X.

For the example under consideration, what we have done so far may be summarized by stating that the proposition

$$p \triangleq Debbie\ is\ a\ few\ years\ older\ than\ Dana$$

may be expressed in a canonical form, namely,

$$X\ is\ FEW, \tag{2.7}$$

which places in evidence the implicit variable X which is constrained by p. The canonical proposition implies and is implied by the possibility assignment equation (2.5), which defines via (2.6) the possibility distribution of X and thus characterizes the elastic constraint on X which is induced by p.

The foregoing analysis may be viewed as an instantiation of the basic idea underlying PRUF, namely, that any proposition, p, in a natural language may be expressed in the canonical form

$$cf(p) \triangleq X\ is\ F, \tag{2.8}$$

where $X = (X_1, \ldots, X_n)$ is an n-ary *focal variable* whose

constituent variables $X_1, \ldots, X_n$ range over the universes
$U_1, \ldots, U_n$, respectively; F is an n-ary fuzzy relation in the
product space $U = U_1 \times \ldots \times U_n$ and $cf(p)$ is an abbreviation
for *canonical form of* p. The canonical form, in turn, may be
expressed more concretely as the possibility assignment
equation

$$\Pi_X = F. \tag{2.9}$$

which signifies that the possibility distribution of X is
given by F. Thus, we may say that p *translates* into the
possibility assignment equation (2.9), i.e.

$$p \to \Pi_X = F \tag{2.10}$$

in the sense that (2.9) exhibits the implicit variable which
is constrained by p and defines the elastic constraint which
is induced by p.

When we employ test-score semantics, the meaning of a
proposition, p, is represented as a test which associates
with each *ED* (i.e., an instantiation of *EDF*) an overall
test score τ which may be viewed as the *compatibility* of p
with *ED*. This compatibility may be interpreted in two
equivalent ways: (a) as the truth of p given *ED*; and (b) as
the possibility of *ED* given p. The latter interpretation
shows that the representation of the meaning of p as a test
is equivalent to representing the meaning of p by a possibility assignment equation.[6]

The connection between the two points of view will
become clearer in Section 4, where we shall discuss several
examples of propositions representing commonsense knowledge.
As a preliminary, we shall present in the following section
a brief exposition of some of the basic techniques which
will be needed in Section 4 to test the constituent relations
in the explanatory database and aggregate the partial test
scores.

3. TESTING AND TRANSLATION RULES

A typical relation in *EDF* may be expressed as
$R[X_1; \ldots; X_n; \mu]$, where R is the name of the relation;

the $X_i, i = 1, \ldots, n$, are the names of the variables (or, equivalently, the attributes of R), with U_i and u_i representing, respectively, the domain of X_i and its generic value; and μ is the grade of membership of a generic n-tuple $u = (u_1, \ldots, u_n)$ in R.

In the case of nonfuzzy relations, a basic operation on R which is a generalization of the familiar operation of looking up the value of a function for a given value of its argument, is the so-called *mapping operation* [11]. The counterpart of this operation for fuzzy relations is the operation of *transduction* [61].

Transduction may be viewed as a combination of two operations: (a) *particularization*[7], which constrains the values of a subset of variables of R; and *projection*, which reads the induced constraints on another subset of variables of R. The subsets in question may be viewed as the *input* and *output* variables, respectively.

To define particularization, it is helpful to view a fuzzy relation as an elastic constraint on n-tuples in $U_1 \times \ldots \times U_n$, with the μ-value for each row in R representing the degree (or the test score) with which the constraint is satisfied.

For concreteness, assume that the input variables are X_1, X_2 and X_3, and that the constraints on these variables are expressed as canonical propositions. For example

$$X_1 \; is \; F$$

and

$$(X_2, X_3) \; is \; G,$$

where F and G are fuzzy subsets of U_1, and $U_2 \times U_3$, respectively. Equivalently, the constraints in question may be expressed as

$$\Pi_{X_1} = F$$

and

$$\Pi_{(X_2, X_3)} = G,$$

where Π_{X_1} and $\Pi_{(X_2, X_3)}$ are the respective possibility distributions of X_1 and X_2, X_3. To place in evidence the input constraints, the particularized relation is written as

$$R^* \underset{=}{\Delta} R[X_1 \ is \ F; \ (X_2, X_3) \ is \ G] \qquad (3.1)$$

or, equivalently, as

$$R^* \underset{=}{\Delta} R[\Pi_{X_1} = F; \ \Pi_{(X_2, X_3)} = G]. \qquad (3.2)$$

As a concrete illustration, assume that R is a relation whose frame is expressed as

$$RICH \ [Name; \ Age; \ Height; \ Weight, \ Sex; \ \mu] \qquad (3.3)$$

in which Age, Height, Weight and Sex are attributes of Name, and μ is the degree to which Name is *rich*. In this case, the input constraints might be:

> Age is YOUNG
> (Height, Weight) is BIG
> Sex is MALE

and, correspondingly, the particularized relation reads

$$R^* \underset{=}{\Delta} RICH \ [Age \ is \ YOUNG; \ (Height, \ Weight) \ is \ BIG; \\ Sex \ is \ MALE]. \qquad (3.4)$$

To concretize the meaning of a particularized relation it is necessary to perform a *row test* on each row of R. Specifically, with reference to (3.1), let

$$\tau_t = (u_{1t}, \ldots, u_{nt}, \mu_t) \text{ be the } t^{th} \text{ row of } R, \text{ where}$$

$u_{1t}, \ldots, u_{nt}, \mu_t$ are the values of $X_1, \ldots, X_n, \mu$, respectively.

Furthermore, let μ_F and μ_G be the respective membership functions of F and G. Then, for τ_i, the test score for the

constraints on X_1 and (X_2,X_3) may be expressed as

$$\tau_{1t} = \mu_F(u_{1t})$$
$$\tau_{2t} = \mu_G(u_{2t},u_{3t}).$$

To aggregate the test scores with μ_t, we employ the min

operator $\wedge^8$, which leads to the overall test score for τ_t:

$$\tau_t = \tau_{1t} \wedge \tau_{2t} \wedge \mu_t. \tag{3.5}$$

Then, the particularized relation (3.1) is obtained by
replacing each μ_t in τ_t, $t = 1,2,\ldots$, by τ_t. An example

illustrating these steps in the computation of a particularized relation may be found in [61].

As was stated earlier, when a fuzzy relation R is particularized by constraining a set of input variables, we
may focus our attention on a subset of variables of R which
are designated as *output* variables and ask the question:
What are the induced constraints on the output variables? As
in the case of nonfuzzy relations, the answer is yielded by
projecting the particularized relation on the cartesian
product of the domains of output variables. Thus, for example,
if the input variables are X_2,X_3 and X_5, and the output

variables are X_1 and X_4, then the induced constraints on

X_1 and X_4 are determined by the projection, G, of the

particularized relation $R*$ $[(X_2,X_3,X_5)$ *is* $F]$ on $U_1 \times U_2$. The

relation which represents the projection is expressed as in
[27][9].

$$G \overset{\Delta}{=} {}_{X_1 \times X_2} R \; [(X_2,X_3,X_5) \; is \; F], \tag{3.6}$$

with the understanding that $X_1 \times X_2$ in (3.6) should be

interpreted as $U_1 \times U_2$. In more transparent terms, (3.6) may be restated as the *transduction:*

$$\text{If } (X_2, X_3, X_5) \text{ is } F, \text{ then } (X_1, X_2) \text{ is } G, \qquad (3.7)$$

where G is given by (3.6). Equivalently, (3.7) may be interpreted as the instruction:

$$\text{Read } (X_1, X_2) \text{ given that } (X_2, X_3, X_5) \text{ is } F. \qquad (3.8)$$

For example, the transduction represented by the expression

$$\underset{Name \times \mu}{RICH} [Age \text{ is } YOUNG; \text{ (Height, Weight) is } BIG; \\ Sex \text{ is } MALE]$$

may be interpreted as the fuzzy set of names of rich men who
are young and big. It may also be interpreted in an imperative
sense as the instruction: Read the name and grade of membership in the fuzzy set of rich men of all those who are young
and big.

Remark. When the constraint set which is associated with
an input variable, say X_1, is a singleton, say $\{a\}$, we write
simply

$$X = a$$

instead of X is a. For example,

$$\underset{Name \times \mu}{RICH} [Age = 25; \text{ Weight} = 136, \text{ Sex} = Male]$$

represents the fuzzy set of rich men whose age and weight
are equal to 25 and 136, respectively.

Composition of Elastic Constraints

In testing the constituent relations in *EDF*, it is helpful
to have a collection of standardized translation rules for
computing the test score of a combination of elastic constraints $C_1, \ldots, C_k$ from the knowledge of the test scores

of each constraint considered in isolation. For the most part,

such rules are *default* rules in the sense that they are intended to be used in the absence of alternative rules
supplied by the user.

For purposes of commonsense knowledge representation,
the principal rules of this type are the following.[10]

3.1. Rules pertaining to modification

If the test score for an elastic constraint C in a specified
context is τ, then in the same context the test score for

> *(a) not C is* $1 - \tau$ *(negation)*   (3.9)
>
> *(b) very C is* τ^2 *(concentration)*   (3.10)
>
> *(c) more or less C is* $\tau^{\frac{1}{2}}$ *(diffusion).*   (3.11)

3.2. Rules pertaining to composition

If the test scores for elastic constraints C_1 and C_2 in a

specified context are τ_1 and τ_2 respectively, then in the

same context the test score for

> *(a)* C_1 *and* C_2 *is* $\tau_1 \wedge \tau_2$ *(conjunction),where* $\wedge \triangleq min.$ (3.12)
>
> *(b)* C_1 *or* C_2 *is* $\tau_1 \vee \tau_2$ *(disjunction),where* $\vee \triangleq max.$ (3.13)
>
> *(c) If* C_1 *then* C_2 *is* $1 \wedge (1-\tau_1+\tau_2)$ *(implication).*   (3.14)

3.3. Rules pertaining to quantification

The rules in question apply to propositions of the general
form *Q A's and B's* where Q is a fuzzy quantifier, e.g., most,
many, several, few, etc. and A and B are fuzzy sets, e.g.,
tall men, intelligent men, etc. As was stated earlier, when
the fuzzy quantifiers in a proposition are implied rather
than explicit, their suppression may be placed in evidence
by referring to the proposition as a *disposition*. In this
sense, the proposition *Overeating causes obesity* is a
disposition which results from the suppression of the fuzzy
quantifier *Most* in the proposition *Most of those who overeat
are obese.*

To make the concept of a fuzzy quantifier meaningful, it is necessary to define a way of counting the number of elements in a fuzzy set or, equivalently, to determine its cardinality.

There are several ways in which this can be done [61]. For our purposes, it will suffice to employ the concept of a *sigma-count*, which is defined as follows.

Let F be a fuzzy subset of $U = \{u_1, \ldots, u_n\}$

expressed symbolically as

$$F = \mu_1/u_1 + \ldots + \mu_n/u_n = \Sigma_i \mu_i/u_i \tag{3.15}$$

or, more simply, as

$$F = \mu_1 u_1 + \ldots + \mu_n u_n, \tag{3.16}$$

in which the term μ_i/u_i, $i = 1, \ldots, n$, signifies that μ_i

is the grade of membership of u_i in F, and the plus sign represents the union.[11]

The sigma-count of F is defined as the arithmetic sum of the μ_i, i.e.

$$Count\ (F) \triangleq \Sigma_i \mu_i, \quad i = 1, \ldots, n, \tag{3.17}$$

with the understanding that the sum may be rounded, if need be, to the nearest integer. Furthermore, one may stipulate that the terms whose grade of membership falls below a specified threshold be excluded from the summation. The purpose of such an exclusion is to avoid a situation in which a large number of terms with low grades of membership become count-equivalent to a small number of terms with high membership.

The *relative sigma-count*, denoted by $\Sigma Count(F/G)$, may be interpreted as the proportion of elements of F which are in G. More explicitly,

$$\Sigma Count\ (F/G) = \frac{\Sigma Count\ (F \cap G)}{\Sigma Count\ (G)} \tag{3.18}$$

where $F \cap G$, the intersection of F and G, is defined by

$$F \cap G = \Sigma_i (\mu_B(u_i) \wedge \mu_G(u_i))/u_i, \quad i = 1, \ldots, n. \tag{3.19}$$

Thus, in terms of the membership functions of F and G, the relative sigma-count of F in G is given by

$$\Sigma Count \ (F/G) = \frac{\Sigma_i \mu_F (u_i) \wedge \mu_G(u_i)}{\Sigma_i \mu_G (u_i)} \tag{3.20}$$

The concept of a relative sigma-count provides a basis for interpreting the meaning of propositions of the form Q A's are B's, e.g., *Most young men are healthy*. More specifically, if the focal variable (i.e., the constrained variable) in the proposition in question is taken to be the proportion of B's in A's, then the corresponding translation rule may be expressed as

$$Q \ A\text{'s are } B\text{'s} \rightarrow \Sigma Count \ (B/A) \ is \ Q \tag{3.21}$$

or, equivalently, as

$$Q \ A\text{'s are } B\text{'s} \rightarrow \Pi_X = Q \tag{3.22}$$

where

$$X = \frac{\Sigma_i \mu_A (u_i) \wedge \mu_B(u_i)}{\Sigma_i \mu_A (u_i)} . \tag{3.23}$$

As will be seen in the following section, the quantification rule (3.21) together with the other rules described in this section provide a basic conceptual framework for the representation of commonsense knowledge. We shall illustrate the representation process through the medium of several examples in which the meaning of a disposition is represented as a test on a collection of fuzzy relations in an explanatory database.

4. REPRESENTATION OF DISPOSITIONS

To clarify the difference between the conventional approaches to meaning representation and that described in the present paper, shall consider as our first example the disposition

$$d \ \underline{\underline{\Delta}} \ Snow \ is \ white \ , \tag{4.1}$$

which, as was stated earlier, is frequently employed as an

illustration in introductory expositions of truth-conditional
semantics (see footnote 5).

The first step in the representation process involves a
restoration of the suppressed quantifiers in d. We shall
assume that the intended meaning of d is conveyed by the
proposition

$$p \underline{\underline{\Delta}} \; Usually \; snow \; is \; white, \qquad (4.2)$$

and, as an *EDF* for (4.2), we shall use (2.3), i.e.

$$EDF \; \underline{\underline{\Delta}} \; WHITE \; [Sample;\mu] \; + \; USUALLY \; [Proportion;\mu]. \quad (4.3)$$

Let $S_1, \ldots, S_m$ denote samples of snow and let

τ_i, $i = 1, \ldots, m$, denote the degree to which the color of

S_i matches white. Thus, τ_i may be interpreted as the test

score for the constraint on the color of S_i which is in-

duced by *WHITE*.

Using this notation, the steps in the testing procedure
may be described as follows:

1. Find the proportion of samples whose color is white:

$$\rho = \frac{\Sigma Count \; (WHITE)}{m} \qquad (4.4)$$

$$= \frac{\tau_1 + \ldots + \tau_m}{m}$$

2. Compute the degree to which ρ satisfies the constraint
induced by *USUALLY*:

$$\tau = {}_\mu USUALLY \; [Proportion = \rho] \qquad (4.5)$$

In (4.5), τ represents the overall test score and the
right-hand member signifies that the relation *USUALLY* is
particularized by setting Proportion equal to ρ and projecting the resulting relation on μ. The meaning of d, then, is
represented by the test procedure which leads to the value
of τ.

Equivalently, the meaning of d may be represented as a possibility assignment equation. Specifically, let X denote the focal variable (i.e., the constrained variable) in p. Then we can write

$$d \rightarrow \Pi_X = USUALLY \qquad (4.6)$$

where

$$X \underset{=}{\Delta} \frac{1}{m} \ \Sigma Count \ (WHITE).$$

Example 2.

To illustrate the use of translation rules relating to modification, we shall consider the disposition

$$d \underset{=}{\Delta} Frenchmen \ \ are \ not \ very \ tall. \qquad (4.7)$$

After restoration, the intended meaning of d is assumed to be represented by the proposition

$$p \underset{=}{\Delta} Most \ Frenchmen \ are \ not \ very \ tall. \qquad (4.8)$$

To represent the meaning of p, we shall employ an EDF whose constituent relations are:

$$(4.9)$$

$EDF \underset{=}{\Delta} POPULATION \ [Name; \ Height]+$
$\qquad TALL \ [Height; \ \mu]+$
$\qquad MOST \ [Proportion; \ \mu].$

The relation $POPULATION$ is a tabulation of $Height$ as a function of $Name$ for a representative group of Frenchmen. In $TALL, \mu$ is the degree to which a value of $Height$ fits the description $tall$; and in $MOST$, μ is the degree to which a numerical value of $Proposition$ fits the intended meaning of $most$.

The test procedure which represents the meaning of p involves the following steps:

1. Let $Name_i$ be the name of ith individual in $POPULATION$.

For each $Name_i$, $i = 1,\ldots,m$, find the height of $Name_i$:

$$Height \ (Name_i) \ \underset{=}{\Delta} \ _{Height}POPULATION \ [Name = Name_i].$$

2. For each $Name_i$, compute the test score for the constraint induced by $TALL$:
$$\tau_i = \ _{\mu}TALL[Height = Height \ (Name_i)].$$

3. Using the translation rules (3.9) and (3.10), compute the test score for the constraint induced by $NOT \ VERY \ TALL$:

$$\tau'_i = 1 - \tau_i^2.$$

4. Find the relative sigma-count of Frenchmen who are not very tall:

$$p \ \underset{=}{\Delta} \ \Sigma Count \ (NOT.VERY.TALL/POPULATION)$$

$$= \frac{\Sigma_i \tau'_i}{m} \ .$$

5. Compute the test score for the constraint induced by $MOST$:

$$\tau = \ _{\mu}MOST \ [Proportion = \rho]. \hspace{3em} (4.10)$$

The test score given by (4.10) represents the overall test score for d, and the test procedure which yields τ represents the meaning of d.

Example 3.

Consider the disposition

$$d \ \underset{=}{\Delta} \ Overeating \ causes \ obesity \hspace{3em} (4.11)$$

which after restoration is assumed to read[12]

$$p \ \underset{=}{\Delta} \ Most \ of \ those \ who \ overeat \ are \ obese. \hspace{2em} (4.12)$$

To represent the meaning of p, we shall employ an EDF whose constituent relations are:

$$EDF \triangleq POPULATION \;\; [Name; \; Overeat; \; Obese] \; + \qquad (4.13)$$
$$MOST \;\; [Proportion; \; \mu].$$

The relation *POPULATION* is a list of names of individuals, with the variables *Overeat* and *Obese* representing, respectively, the degrees to which *Name* overeats and is obese. In *MOST*, μ is the degree to which a numerical value of *Proportion* fits the intended meaning of *MOST*.

The test procedure which represents the meaning of d involves the following steps.

1. Let $Name_i$ be the name of ith individual in *POPULATION*.

   For each $Name_i$, $i = 1,...,m$, find the degrees to which $Name_i$ overeats and is obese:

$$\alpha_i \triangleq \mu_{OVEREAT}(Name_i) \triangleq {}_{Overeat}POPULATION[Name = Name_i]$$
$$(4.14)$$

and

$$\beta_i \triangleq \mu_{OBESE}(Name_i) \triangleq {}_{Obese} POPULATION \;\; [Name = Name_i].$$
$$(4.15)$$

2. Compute the relative sigma-count of *OBESE* in *OVEREAT*:

$$p \triangleq \Sigma Count(OBESE/OVEREAT) = \frac{\Sigma_i \alpha_i \wedge \beta_i}{\Sigma_i \alpha_i} . \qquad (4.16)$$

3. Compute the test score for the constraint induced by *MOST*:

$$\tau = {}_\mu MOST \; [Proportion = \rho]. \qquad (4.17)$$

This test score represents the compatibility of d with the explanatory database.

Example 4.

Consider the disposition

$$d \triangleq Heavy \; smoking \; causes \; lung \; cancer. \qquad (4.18)$$

Although it has the same form as (4.11), we shall interpret it differently. Specifically, the restored proposition will be assumed to be expressed as

$p \triangleq$ *The incidence of cases of lung cancer among* (4.19)
heavy smokers is much higher than among those
who are not heavy smokers.

The *EDF* for this proposition is assumed to have the following constituents:

$EDF \triangleq$ *POPULATION* [*Name; Heavy. Smoker; Lung.Cancer*]+ (4.20)
MUCH. HIGHER [*Proportion 1; Proportion 2; μ*].

In *POPULATION*, *Heavy. Smoker* represents the degree to which *Name* is a heavy smoker, and the variable *Lung. Cancer* is 1 or 0 depending on whether or not *Name* has lung cancer. In *MUCH. HIGHER*, μ is the degree to which *Proportion 1* is much higher than *Proportion 2*.

The steps in the test procedure may be summarized as follows:

1. For each *Name$_i$*, $i = 1,\ldots,m$, determine the degree to which *Name$_i$* is a heavy smoker:

$$\alpha_i \triangleq {}_{Heavy.Smoker} POPULATION \ [Name = Name_i]. \ (4.21)$$

Then, the degree to which *Name$_i$* is not a heavy smoker is

$$\beta_i = 1 - \alpha_i. \tag{4.22}$$

2. For each *Name$_i$*, determine if *Name$_i$* has lung cancer:

$$\lambda_i \triangleq {}_{Lung.Cancer} POPULATION \ [Name = Name_i]. \ (4.23)$$

3. Compute the relative sigma-counts of those who have lung cancer among (a) heavy smokers; and (b) not heavy smokers:

$$\rho_1 = \Sigma Count \ (LUNG. \ CANCER/HEAVY. \ SMOKER)$$

$$= \frac{\Sigma_i \lambda_i \wedge \alpha_i}{\Sigma_i \alpha_i}$$

$$\rho_2 = \Sigma Count \ (LUNG. \ CANCER/NOT. \ HEAVY. \ SMOKER)$$

$$= \frac{\Sigma_i \lambda_i \wedge (1 - \alpha_i)}{\Sigma_i 1 - \alpha_i}$$

4. Test the constraint induced by MUCH. HIGHER:

$$\tau = {}_\mu MUCH.HIGHER[Proportion \ 1 = \rho_1 ; Proportion \ 2 = \rho_2] \quad (4.24)$$

Example 5.

Consider the disposition

$$d \triangleq Small \ families \ are \ friendly \qquad (4.25)$$

which we shall interpret as the proposition

$$p \triangleq In \ most \ small \ families \ almost \ all \ of \ the \qquad (4.26)$$
$$members \ are \ friendly \ with \ one \ another.$$

It should be noted that the quantifier *most* in p is a
second order fuzzy quantifier in the sense that it represents a fuzzy count of fuzzy sets (i.e., *small families*).

The *EDF* for p is assumed to be expressed by

$$EDF \triangleq \quad POPULATION \ [Name; \ Family. \ Identifier] + \qquad (4.27)$$
$$SMALL \ [Number; \ \mu] +$$
$$FRIENDLY \ [Name \ 1; \ Name \ 2; \ \mu] +$$
$$MOST \ [Proportion; \ \mu] +$$
$$ALMOST. \ ALL \ [Proportion; \ \mu] \ .$$

The relation *POPULATION* is assumed to be partitioned (by
rows) into disjoint families $F_1, \ldots, F_k$. In *FRIENDLY*, μ is

the degree to which *Name 1* is friendly toward *Name 2*, *with
Name 1 ≠ Name 2.*
    The test procedure may be described as follows:

1. For each family, F_i, find the count of its members:

$$C_i \triangleq Count \ POPULATION \ [Family.Identifier = F_i]. \qquad (4.28)$$

2. For each family, test the constraint on C_i induced by SMALL:

$$\alpha_i \triangleq {}_\mu SMALL \ [Number = C_i].$$ (4.29)

3. For each family, compute the relative sigma-count of its members who are friendly with one another:

$$\beta_i = \frac{1}{(C_i^2 - C_i)} \ \Sigma_{jk} \ ({}_\mu FRIENDLY \ [Name \ 1 = Name_j;$$

$$Name \ 2 = Name_k])$$ (4.30)

where $Name_j$ and $Name_k$ range over the members of F_i and $Name_i \neq Name_j$. The normalizing factor $C_i^2 - C_i$ represents the total number of links between pairs of distinct individuals in F_i.

4. For each family, test the constraint on β_i which is induced by ALMOST. ALL:

$$\gamma_i = {}_\mu ALMOST. \ ALL \ [Proportion = \beta_i].$$ (4.31)

5. For each family, aggregate the test scores α_i and γ_i by using the min operator ($\wedge$):

$$\delta_i \triangleq \alpha_i \wedge \gamma_i.$$ (4.32)

6. Compute the relative sigma-count of small families in which almost all members are friendly with one another:

$$\rho = \frac{1}{k} \ (\delta_1 + \ldots + \delta_k)$$ (4.33)

7. Test the constraint on ρ induced by MOST:

$$\tau = {}_\mu MOST \ [Proportion = \rho].$$ (4.35)

The value of τ given by (4.35) represents the compatibility of d with the explanatory database.

The foregoing examples are intended to illustrate the basic idea underlying our approach to the representation of commonsense knowledge, namely, the conversion of a disposition into a proposition, and the construction of a test procedure which acts on the constituent relations in an explanatory database and yields its compatibility with the restored proposition.

5. INFERENCE FROM DISPOSITIONS[13]

A basic issue which will be addressed only briefly in the present paper is the following. Assuming that we have represented a collection of dispositions in the manner described above, how can an answer to a query be determined from the representations in question? In what follows, we shall consider a few problems of this type which are of relevance to the computation of certainty factors in expert systems [3, 15, 45, 49, 51].

Conjunction of consequents

Consider two dispositions d_1 and d_2 which upon restoration become propositions of the general form

$$d_1 \rightarrow p_1 \triangleq Q_1 \; A\text{'s are } B\text{'s} \tag{5.1}$$

$$d_2 \rightarrow p_2 \triangleq Q_2 \; A\text{'s are } C\text{'s}. \tag{5.2}$$

Now assume that p_1 and p_2 appear as premises in the inference schema[14]

$$Q_1 \; A\text{'s are } B\text{'s} \tag{5.3}$$

$$\underline{Q_2 \; A\text{'s are } C\text{'s}}$$

$$?\; Q \; A\text{'s are } (B \text{ and } C)\text{'s}.$$

in which $?Q$ is a fuzzy quantifier which is to be determined. We shall refer to this schema as the *conjunction of consequents*.

As stated in the following assertion, the fuzzy quantifier Q is bounded by two fuzzy numbers. More specifically, on interpreting Q_1 and Q_2 as fuzzy numbers, we can assert that

$$0 \ \pmb{\otimes} \ (Q_1 \ \oplus \ Q_2 \ \ominus \ 1) \ \leq \ Q \ \leq \ Q_1 \ \pmb{\otimes} \ Q_2, \qquad (5.4)$$

in which the operators $\pmb{\otimes}$, $\pmb{\oplus}$, $\oplus$, and $\ominus$, and the inequality $\leq$ are the extensions of $\wedge$, $\vee$, $+$, $-$ and $\leq$, respectively, to fuzzy numbers [14].

Proof. We shall consider the upper bound first. To this end it will suffice to show that

$$\Sigma Count \ (B \cap C/A) \ \leq \ \Sigma Count \ (B/A) \wedge \Sigma Count \ (C/A), \qquad (5.5)$$

since, in view of (3.21), the fuzzy quantifiers Q, Q_1 and

Q_2 may be regarded as fuzzy characterizations of the

corresponding sigma-counts.

For convenience, let α_i, β_i and δ_i, $i = 1,\ldots,n$, denote,

respectively, the grades of membership of μ_i in A, B and C.

Then, on using (3.18) – (3.20), we may write

$$\Sigma Count \ (B \cap C/A) \ = \ \frac{\Sigma Count((A \cap B) \cap (A \cap C))}{\Sigma Count \ (A)} \qquad (5.6)$$

$$= \ \frac{1}{\Sigma Count(A)} \ [\ \Sigma_i(\alpha_i \wedge \beta_i) \wedge (\alpha_i \wedge \delta_i)].$$

Now

$$\Sigma_i(\alpha_i \wedge \beta_i) \wedge (\alpha_i \wedge \delta_i) \leq \Sigma_i(\alpha_i \wedge \beta_i)$$

$$\Sigma_i(\alpha_i \wedge \beta_i) \wedge (\alpha_i \wedge \delta_i) \leq \Sigma_i(\alpha_i \wedge \delta_i)$$

and hence

$$\Sigma_i(\alpha_i \wedge \beta_i) \wedge (\alpha_i \wedge \delta_i) \leq \Sigma_i(\alpha_i \wedge \beta_i) \wedge \Sigma_i(\alpha_i \wedge \delta_i). \qquad (5.7)$$

Consequently,

$$\Sigma \text{Count}(B \cap C/A) \leq \frac{1}{\Sigma Count(A)} [\Sigma Count(A \cap B) \wedge \Sigma Count(A \cap C)]$$

$$\leq \Sigma Count(B/A) \wedge \Sigma Count(C/A).$$

which is what we set out to establish.

To deduce the lower bound, we note that for any real numbers a, b, we have

$$a \wedge b = a + b - a \vee b. \qquad (5.8)$$

Consequently,

$$\frac{1}{\Sigma Count(A)} [\Sigma_i(\alpha_i \wedge \beta_i) \wedge (\alpha_i \wedge \delta_i)] =$$

$$\frac{1}{\Sigma Count(A)} [\Sigma_i(\alpha_i \wedge \beta_i) + \Sigma_i(\alpha_i \wedge \delta_i) - \Sigma_i((\alpha_i \wedge \beta_i) \vee (\alpha_i \wedge \delta_i))],$$

and since

$$\alpha_i \geq (\alpha_i \wedge \beta_i) \vee (\alpha_i \wedge \delta_i)$$

it follows that

$$\Sigma Count(B \cap C/A) \geq \frac{1}{\Sigma Count(A)} [\Sigma_i \alpha_i \wedge \beta_i + \Sigma_i \alpha_i \wedge \delta_i - \Sigma_i \alpha_i]$$

or, equivalently,

$$\Sigma Count(B \cap C/A) \geq \Sigma Count(B/A) + (C/A) - 1, \qquad (5.9)$$

from which (5.4) follows by an application of the extension principle and the observation that the left-hand member of (5.5) must be non-negative [63].

In conclusion, the simple proof given above establishes the validity of the following inference schema, which, for convenience, will be referred to as the *consequent conjunction syllogism*:

$$Q_1 \ A\text{'s are } B\text{'s}$$

$$\underline{Q_2 \ A\text{'s are } C\text{'s}} \qquad\qquad (5.10)$$

$$Q \ A\text{'s are } (B \text{ and } C)\text{'s}.$$

where

$$0 \odot (Q_1 \oplus Q_2 \ominus 1) \le Q \le Q_1 \odot Q_2.$$

As an illustration, from

$$p_1 \triangleq \textit{Most Frenchmen are not tall}$$

$$p_2 \triangleq \textit{Most Frenchmen are not short} \qquad (5.11)$$

we can infer that

$$Q \textit{ Frenchmen are not tall and not short}$$

where

$$0 \odot (2 \textit{ most} \ominus 1) \le Q \le \textit{most}. \qquad (5.12)$$

In the above example, the variable of interest is the proportion of Frenchmen who are *not tall* and *not short*. In a more general setting, the variable of interest may be a specified function of the variables constrained by the knowledge base. The following variation on (5.11) is intended to give an idea of how the value of the variable of interest may be inferred by an application of the extension principle [56].

Example 6.

Infer from the propositions

$$p_1 \triangleq \textit{Most Frenchmen are not tall} \qquad (5.13)$$

$$p_2 \triangleq \textit{Most Frenchmen are not short}$$

the answer to the question

$$q \triangleq \textit{What is the average height of a Frenchman?} \quad (5.14)$$

Because of the simplicity of p_1 and p_2, the constraints induced by the premises may be found directly. Specifically, let $h_1, \ldots h_n$ denote the heights of $\textit{Frenchman}_1, \ldots, \textit{French-man}_n$, respectively. Then, the test scores associated with the constraints in question may be expressed as

$$\tau_1 = \mu_{ANT\ MOST}(\frac{1}{n}\ \Sigma_i \mu_{TALL}(h_i)) \tag{5.15}$$

and

$$\tau_2 = \mu_{ANT\ MOST}(\frac{1}{n}\ \Sigma_i \mu_{SHORT}(h_i)), \tag{5.16}$$

where ANT is an abbreviation for $antonym$, i.e.,

$$\mu_{ANT\ MOST}(u) = \mu_{MOST}(1-u),\quad u\varepsilon[0,1], \tag{5.17}$$

and μ_{TALL} and μ_{SHORT} are the membership functions of $TALL$

and $SHORT$, respectively. Correspondingly, the overall test score may be expressed as

$$\tau = \tau_1 \wedge \tau_2.$$

Now, the average height of a Frenchman and hence the answer to the question is given by

$$ans(q) = \frac{1}{n}\ \Sigma_i h_i. \tag{5.18}$$

Consequently, the possibility distribution of $ans(q)$ is given by the solution of the nonlinear program

$$\mu_{ans(q)}(h) = \max_{h_1,\ldots,h_n}(\tau) \tag{5.19}$$

subject to the constraint

$$h = \frac{1}{n}\ \Sigma_i h_i. \tag{5.20}$$

Alternatively, a simpler but less informative answer may be formulated by forming the intersection of the possibility distributions of $ans(q)$ which are induced separately by p_1 and p_2. More specifically, let

$\Pi_{ans(q)|p_1}$, $\Pi_{ans(q)|p_2}$, $\Pi_{ans(q)|p_1 \wedge p_2}$ be the possibility

distributions of $ans(q)$ which are induced by p_1, p_2, and

the conjunction of p_1 and p_2, respectively. Then, by using

the minimax inequality [54], it can readily be shown that

$$\Pi_{ans(q)|p_1} \cap \Pi_{ans(q)|p_2} \supset \Pi_{ans(q)|p_1 \wedge p_2} \qquad (5.21)$$

and hence we can invoke the entailment principle [58] to validate the intersection in question as the possibility distribution of $ans(q)$. For the example under consideration, the desired possibility distribution is readily found to be given by

$$Poss\{ans(q)=h\}=\mu_{ANT\ MOST}(\mu_{TALL}(h)) \wedge \mu_{ANT\ MOST}(\mu_{SHORT}(h)).$$

$$(5.22)$$

Chaining of dispositions

As in (5.1) and (5.2), consider two dispositions d_1 and d_2. which upon restoration become propositions of the general form

$$d_1 \to p_1 \triangleq Q_1 \; A\text{'s are } B\text{'s}$$

$$d_2 \to p_2 \triangleq Q_2 \; B\text{'s are } C\text{'s}.$$

An ordered pair, (p_1,p_2), of propositions of this form will be said to form a *chain*. More generally, an *n-ary chain* may be represented as an ordered n-tuple

$$(Q_1 A_1 \text{'s are } B_1 \text{'s}, Q_2 A_2 \text{'s are } B_2 \text{'s}, \ldots, Q_n A_n \text{ are } B_n \text{'s}), \quad (5.23)$$

in which $B_1 = A_2$, $B_2 = A_3, \ldots, B_{n-1} = A_n$.

Now assume that p_1 and p_2 appear as premises in the inference schema

Q_1 A's are B's	*(major premise)*	(5.24)
Q_2 B's are C's	*(minor premise)*	
$?Q$ A's are C's	*(conclusion)*	

in which $?Q$ is a fuzzy quantifier which is to be determined.

A basic rule of inference which is established in [63] and which has a direct bearing - as we shall see presently - on the determination of Q, is the *intersection/product syllogism*

$$Q_1 \text{ A's are B's} \tag{5.25}$$

$$\frac{Q_2 \text{ (A and B)'s are C's}}{(Q_1 \otimes Q_2 \text{ A's are (B and C)'s,}}$$

in which $Q_1 \otimes Q_2$ is a fuzzy number which is the fuzzy product of the fuzzy numbers Q_1 and Q_2. For example, as a special case of (5.25), we may write

$$\frac{\begin{array}{l}\textit{Most students are single} \hspace{2em} (5.26)\\ \textit{A little more than a half of single students are male}\end{array}}{\begin{array}{l}\textit{(Most} \otimes \textit{A little more than a half) of students are}\\ \textit{single and male.}\end{array}}$$

Since the intersection of B and C is contained in C, the following corollary of (5.25) is its immediate consequence

$$Q \text{ A's are B's} \tag{5.27}$$

$$\frac{Q_2 \text{ (A and B)'s are C's}}{\geq (Q_1 \otimes Q_2) \text{ A's are C's,}}$$

where the fuzzy number $\geq (Q_1 \otimes Q_2)$ should be read as *at least* $(Q_1 \otimes Q_2)$, with the understanding that $\geq (Q_1 \otimes Q_2)$ represents the composition of the binary nonfuzzy relation $\geq$ with the unary fuzzy relation $(Q_1 \otimes Q_2)$. In particular, if the fuzzy quantifiers Q_1 and Q_2 are monotone nondecreasing (e.g., when $Q_1 = Q_2 \triangleq most$), then as is stated in [63],

$$\geq (Q_1 \otimes Q_2) = Q_1 \otimes Q_2, \tag{5.28}$$

and (5.27) becomes

$$Q_1 \ A\text{'s are } B\text{'s} \qquad\qquad\qquad\qquad (5.29)$$

$$\frac{Q_2 \ (A \text{ and } B\text{'s are } C\text{'s}}{(Q_1 \circledast Q_2) \ A\text{'s are } C\text{'s}}$$

There is an important special case in which the premises in (5.29) form a chain. Specifically, if $B \subset A$, then

$$A \cap B = B$$

and (5.29) reduces to what will be referred to as the *product chain rule*, namely,

$$Q_1 \ A\text{'s are } B\text{'s} \qquad\qquad\qquad\qquad (5.30)$$

$$\frac{Q_2 B\text{'s are } C\text{'s}}{(Q_1 \circledast Q_2) \ A\text{'s are } C\text{'s}.}$$

In this case, the chain $(Q_1 \ A\text{'s are } B\text{'s}, \ Q_2 \ B\text{'s are } C\text{'s})$ will be said to be *product transitive.*[15]

As an illustration of (5.30), we can assert that

$$\frac{Most \ students \ are \ undergraduates}{Most \ undergraduates \ are \ young}$$

$$Most^2 \ students \ are \ young,$$

where $Most^2$ represents the product of the fuzzy number *Most* with itself.

Chaining under reversibility

An important chaining rule which is approximate in nature relates to the case where the major premise in the inference chain

$$Q_1 \ A\text{'s are } B\text{'s} \qquad\qquad\qquad\qquad (5.31)$$

$$\frac{Q_2 \ B\text{'s are } C\text{'s}}{Q \ A\text{'s are } C\text{'s}}$$

is *reversible* in the sense that

$$Q_1 \ A's \ are \ B's \leftrightarrows Q_1 \ B's \ are \ A's, \qquad (5.32)$$

where $\leftrightarrows$ denotes approximate semantic equivalence [57]. -
For example,

Most American cars are big $\leftrightarrows$ Most big cars are   (5.33)
American.

Under the assumption of reversibility, the following
syllogism holds in an approximate sense

$$Q \ A's \ are \ B's \qquad (5.34)$$

$$\underline{Q_2 \ B's \ are \ C's}$$

$$\geq (0 \odot (Q_1 \oplus Q_2 \ominus 1)) \ A's \ are \ C's.$$

We shall refer to this syllogism as the *R-rule*.
To demonstrate the approximate validity of this rule we
shall first establish the following lemma.

Lemma

$$If \ \Sigma Count(A) \ = \ \Sigma Count(B) \qquad (5.35)$$

and $$\Sigma Count(B/A) \ \geq \ q_1 \qquad (5.36)$$

$$\Sigma Count(C/B) \ \geq \ q_2 \qquad (5.37)$$

then

$$\Sigma Count(C/A) \ \geq \ (0 \lor (q_1 + q_2 - 1)). \qquad (5.38)$$

Proof.

We have

$$\Sigma Count(B/A) \ = \ \frac{\Sigma Count(A \cap B)}{\Sigma Count(A)}$$

and
$$\Sigma Count(C/B) = \frac{\Sigma Count(B \cap C)}{\Sigma Count(B)}$$

$$= \frac{\Sigma Count(B \cap C)}{\Sigma Count(A)}$$

in virtue of (5.35).

For simplicity, we shall denote $\mu_A(u_i)$, $\mu_B(u_i)$

and $\mu_C(u_i)$, $i = 1,\ldots,n$, by α_i, β_i, γ_i, respectively. Then,

$$\frac{\Sigma Count(A \cap B)}{\Sigma Count(A)} = \frac{\Sigma_i\, \alpha_i \wedge \beta_i}{\Sigma_i\, \alpha_i} \geq q_1 \qquad (5.39)$$

$$\frac{\Sigma Count(B \cap C)}{\Sigma Count(A)} = \frac{\Sigma_i\, \beta_i \wedge \gamma_i}{\Sigma_i\, \alpha_i} \geq q_2 \qquad (5.40)$$

and hence

$$\frac{\Sigma_i(\alpha_i \wedge \beta_i + \beta_i \wedge \gamma_i)}{\Sigma_i\, \alpha_i} \geq q_1 + q_2, \qquad (5.41)$$

and

$$\frac{\Sigma_i(\alpha_i \wedge \beta_i + \beta_i \wedge \gamma_i - \alpha_i)}{\Sigma_i\, \alpha_i} \geq q_1 + q_2 - 1. \qquad (5.42)$$

Since

$$\frac{\Sigma Count(C \cap A)}{\Sigma Count(A)} = \frac{\Sigma_i\, \alpha_i \wedge \gamma_i}{\Sigma_i\, \alpha_i}, \qquad (5.43)$$

it follows from (5.42) and (5.43) that, to establish (5.38), it will suffice to show that

$$\Sigma_i(\alpha_i \wedge \gamma_i) \geq \Sigma_i(\alpha_i \wedge \beta_i + \beta_i \wedge \gamma_i - \alpha_i), \qquad (5.44)$$

or, equivalently,

$$\Sigma_i[(\alpha_i - \alpha_i \wedge \beta_i) + (\beta_i - \beta_i \wedge \gamma_i)] \geq \Sigma_i(\alpha_i - \alpha_i \wedge \gamma_i), (5.45)$$

which is a consequence of the equality

$$\Sigma_i\, \alpha_i = \Sigma_i\, \beta_i, \qquad (5.46)$$

which in turn follows from (5.35).

Now, to establish (5.45) it is sufficient to show that, for each i, $i = 1,\ldots,n$, we have

$$(\alpha_i - \alpha_i \wedge \beta_i) + (\beta_i - \beta_i \wedge \gamma_i) \geq (\alpha_i - \alpha_i \wedge \gamma_i), \qquad (5.48)$$

in which the summands as well as the right-hand member are non-negative. To this end, we shall verify that (5.48) holds for all possible values of α_i, β_i and γ_i.

Case 1.

Consider all values of α_i, β_i and γ_i which satisfy the inequality $\alpha_i \leq \gamma_i$. In this case, the right-hand member of (5.48) is zero and thus the inequality is verified.

Case 2.

$\quad\quad$ $\alpha_i > \gamma_i$. In this case, we shall consider four subcases.

(i)$\quad$ $\alpha_i \leq \beta_i$, $\beta_i \leq \gamma_i$, which contradicts $\alpha_i > \gamma_i$.

(ii)$\quad$ $\alpha_i > \beta_i$, $\beta_i \leq \gamma_i$, which verifies the inequality.

(iii)$\quad$ $\alpha_i \leq \beta_i$, $\beta_i > \gamma_i$, which verifies the inequality.

(iv)$\quad$ $\gamma_i > \beta_i$, $\beta_i > \gamma_i$, which verifies the inequality.

This concludes the proof of the lemma.
$\quad$ Now, if condition (5.35), i.e.,

$$\Sigma_i \alpha_i = \Sigma_i \beta_i,$$

were not needed to prove the lemma, we could invoke the extension principle [56] to extend the inequality

$$\frac{\Sigma Count(C/A)}{\Sigma Count(A)} \geq 0 \vee (q_1 + q_2 - 1),$$

which holds for real numbers, to

$$Q \geq 0 \bullet (Q_1 \oplus Q_2 \ominus 1), \quad\quad\quad\quad (5.49)$$

which holds for the fuzzy quantifiers in (5.31). As it is, the assumption of reversibility, i.e.,

$$\frac{\Sigma Count(A \cap B)}{\Sigma Count(A)} \text{ is } Q_1$$

$$\frac{\Sigma Count(B \cap A)}{\Sigma Count(B)} \text{ is } Q_1$$

implies the equality

$$\Sigma Count(A) = \Sigma Count(B)\dot{}$$

only in an approximate sense. Consequently, as was stated
earlier, the *R-rule* (5.34) also holds only in an approximate
sense. The question of how this sense could be defined more
precisely presents a nontrivial problem which will not be
addressed in this paper.

Concluding Remark

The point of departure in this paper is the idea that commonsense knowledge may be regarded as a collection of dispositions. Based on this idea, the representation of commonsense
knowledge may be reduced, in most cases, to the representation of fuzzily-quantified propositions through the use of
testscore semantics. Then, the rules of inference of fuzzy
logic may be employed to deduce answers to questions which
relate to the information resident in a knowledge base.
    The computational framework for dealing with common-
sense knowledge which is provided by fuzzy logic is of relevance to the management of uncertainty in expert systems. The
advantage of employing fuzzy logic in this application-area
is that it provides a systematic framework for syllogistic
reasoning and thus puts on a firmer basis the derivation of
combining functions for uncertain evidence. The consequent
conjunction syllogism which we established in Section 5 is,
in effect, an example of such a combining function. What it
demonstrates, however, is that, in general, combining
functions cannot be expected to yield real-valued probabilities or certainty factors, as they do in MYCIN, PROSPECTOR
and other expert systems. Thus, in general, the value returned
by a combining function should be a fuzzy number or an
n-tuple of such numbers.

Computer Science Division, Department of EECS and ERL;
University of California, Berkeley.

ACKNOWLEDGMENT

This paper is dedicated to Professor Hans J. Zimmermann.

NOTES

1. A more detailed discussion of the inferential component of fuzzy logic may be found in [5, 58, 59]. Recent literature, [26], contains a number of papers dealing with fuzzy logic and its applications. Descriptions of implemented fuzzy-logic-based inferential systems may be found in [2], [22] and [36].

2. A more detailed exposition of test-score semantics may be found in [61].

3. In the literature of database management systems, some authors employ the term *schema* in the same sense as we employ *frame*. More commonly, however, the term schema is used in a narrower sense [11], to describe the frame of a relation together with the dependencies between the variables.

4. As was stated earlier, a *disposition* is a proposition with implied fuzzy quantifiers. (Note that this definition, too, is a disposition.) For example, the proposition *Small cars are unsafe* is a disposition, since it may be viewed as an abbreviation of the proposition *Most small cars are unsafe*, in which *most* is a fuzzy quantifier. In general, a disposition may be restored in more than one way.

5. In truth-conditional semantics, the truth condition for the proposition *Snow is white* is described as snow is white, which means, in plain terms, that the proposition *Snow is white* is true if and only if snow is white.

6. As is pointed out in [61], the translation of p into a possibility assignment equation is an instance of a *focused* translation. By contrast, representation of the meaning of p by a test on *EDF* is an instance of an *unfocused* translation. The two are equivalent in principle but differ in detail.

7. In the case of nonfuzzy relation, particularization is usually referred to as *selection* [11], or *restriction*.

8. Here and elsewhere in the paper the aggregation operation min ($\wedge$) is used as a default choice when no alternative (e.g., arithmetic mean, geometric mean, etc.) is specified.

9. If R is a fuzzy relation its projection on $U_1 \times U_2$ is

obtained by deleting from R all columns other than X_1 and X_2, and forming the union of the resulting tuples.

10. A more detailed discussion of such rules in the context of PRUF may be found in [57].
11. In most cases, the context is sufficient to resolve the question of whether a plus sign should be interpreted as the union or the arithmetic sum.
12. It should be understood that (4.12) is just one of many possible interpretations of (4.11), with no implication that it constitutes a prescriptive interpretation of causality. See [48] for a thorough discussion of relevant issues.
13. Inference from dispositions may be viewed as an alternative approach to default reasoning and non-monotinic logic [27, 29, 30, 40].
14. This schema has a bearing on the rule of combination of evidence for conjunctive hypotheses in MYCIN [46].
15. More generally, an n-ary chain
$$(Q_1 \ A_1\text{'s are } B_1\text{'s},\ldots,Q_n \ A_n\text{'s are } B_n\text{'s})$$ will be said to

be *product transitive* if from the premises which constitute the chain it may be inferred that
$$\geq (Q_1 \ \otimes \ldots \otimes \ Q_n) \ A_1\text{'s are } B_n\text{'s}.$$

REFERENCES

[1] Adams, E.W., H.F. Levine: 1975, 'On the uncertainties transmitted from premises to conclusions in deductive inferences', <u>Synthese</u> 30, 429-460.
[2] Baldwin, J.F., S.Q. Zhou: 1982, 'A fuzzy relational inference language', Report No. EM/FS 132, University of Bristol.
[3] Barr, A., E.W. Feigenbaum: 1982, 'The Handbook of Artificial Intelligence', vols. 1,2 and 3. Los Altos: Kaufmann.
[4] Barwise, J., R. Cooper: 1981, 'Generalized quantifiers and natural language', <u>Linguistics and Philosophy</u> 4, 159-219.
[5] Bellman, R.E., L.A. Zadeh: 1977, 'Local and fuzzy logics', in: Modern Uses of Multiple-Valued Logic, Epstein, G. (ed.), Dordrecht: Reidel, 103-165.
[6] Blanchard, N.: 1981, 'Theories cardinales et ordinales des ensembles flou: les multiensembles', Thesis, University of Claude Bernard, Lyon.
[7] Brachman, R.J., B.C. Smith: 1980, 'Special Issue on Knowledge Representation, SIGART 70.

[8] Cooper, W.S.: 1978, 'Logical Linguistics', Dordrecht: Reidel.

[9] Cresswell, M.J.: 1973, 'Logic and Languages', London: Methuen.

[10] Cushing, S.: 1982, 'Quantifier Meanings - A Study in the Dimensions of Semantic Competence', Amsterdam: North-Holland.

[11] Date, C.J.: 1977, 'An Introduction to Database Systems', Reading: Addison-Wesley.

[12] Davis, R., D.B. Lenat: 1982, 'Knowledge-Based Systems in Artificial Intelligence', New York: McGraw-Hill.

[13] DeLuca, A., S.A. Termini: 1972, 'A definition of non-probabilistic entropy in the setting of fuzzy sets theory', Information and Control 20, 301-312.

[14] Dubois, D., H. Prade: 1980, 'Fuzzy Sets and Systems: Theory and Applications', New York: Academic Press.

[15] Duda, R.O., P.E. Hart, K. Konolige, R. Reboh: 1979, 'A computer-based consultant for mineral exploration', Tech. Report, SRJ International, Menlo Park, CA.

[16] Feigenbaum, E.A.: 1977, 'The art of artificial intelligence: Themes and case studies in knowledge engineering', Proc. IJCAI5.

[17] Findler, N.V. (ed.): 1979, 'Associative Networks: Representation and Use of Knowledge by Computer', New York: Academic Press.

[18] Gaines, B.R.: 1979, 'Logical foundations for database systems', Int. J. Man-Machine Studies 11, 481-500.

[19] Goguen, J.A.: 1969, 'The logic of inexact concepts', Synthese 19, 325-373.

[20] Hofmann, T.R.: 1982, 'Qualitative terms for quantity', In Proc. 6th European Meeting on Cybernetics and Systems Research, R. Trappl, (ed.), Amsterdam: North-Holland.

[21] Hoover, D.N.: 1978, 'Probability logic', Annals Math. Lgoic 14, 287-313.

[22] Ishizuka, M., K.S. Fu, J.T.P. Yao: 1982, 'A rule-based inference with fuzzy set for structural damage assessment', Fuzzy Information and Decision Processes, M. Gupta, and E. Sanchez (eds.), Amsterdam: North-Holland.

[23] Keenan, E.L.: 1971, 'Quantifier structures in English', Foundations of Language 7, 255-336.

[24] Klement, E.P., W. Schwyhla, R. Lowen: 1981, 'Fuzzy probability measures', Fuzzy Sets and Systems 5, 83-108.

[25] Klement, E.P.: 1981, 'Operations on fuzzy sets and
     fuzzy numbers related to triangular norms',
     Proc. 11th Conf. Multiple-Valued Logic, Univ. of
     Oklahoma, Norman, 218-225.
[26] Mamdani, E.H., B.R. Gaines: 1981, 'Fuzzy Reasoning and
     its Applications', London: Academic Press.
[27] McCarthy, J.: 1980, 'Circumscription: A non-monotonic
     inference rule', Artificial Intelligence 13, 27-40.
[28] McCawley, J.D.: 1981, 'Everything that Linguists have
     Always Wanted to Know about Logic', Chicago: Univer-
     sity of Chicago Press.
[29] McDermott, D.V., J. Doyle: 1980, 'Non-monotonic logic I',
     Artificial Intelligence 13, 41-72.
[30] McDermott, D.V.: 1982, 'Non-monotonic logic, II: non-
     monotonic modal theories', J. Assoc. Comp. Mach. 29,
     33-57.
[31] Michalski, R., J.G. Carbonell, T.M. Mitchell: 1983,
     'Machine Learning', Palo Alto: Tioga Press.
[32] Miller, G.A., P.N. Johnson-Laird: 1976, 'Language and
     Perception', Cambridge: Harvard University Press.
[33] Mizumoto, M., M. Umano, K. Tanaka: 1977, 'Implementation
     of a fuzzy-set-theoretic data-structure system',
     3rd Int. Conf. Very Large Data Bases, Tokyo.
[34] Mizumoto, M., S. Fukame, K. Tanaka: 1978, 'Fuzzy rea-
     soning Methods by Zadeh and Mamdani, and improved
     methods', Proc. Third Workshop on Fuzzy Reasoning,
     Queen Mary College, London.
[35] Moisil, G.C.: 1975, 'Lectures on the Logic of Fuzzy
     Reasoning', Bucarest: Scientific Editions.
[36] Noguchi, K., M. Umano, M. Mizumoto, K. Tanaka: 1976,
     'Implementation of fuzzy artificial intelligence
     language FLOU', Technical Report on Automation and
     Language of IECE.
[37] Orlov, A.I.: 1980, 'Problems of Optimization and Fuzzy
     Variables', Moscow: Znaniye.
[38] Partee, B.: 1976, 'Montague Grammar', New York: Academic
     Press.
[39] Peterson, P.: 1979, 'On the logic of few, many and most',
     Notre Dame J. Formal Logic 20, 155-179.
[40] Reiter, R., G. Criscuolo: 1983, 'Some representational
     issues in default reasoning', Computers and Mathema-
     tics 9, 15-28.
[41] Rescher, N.: 1976, 'Plausible Reasoning', Amsterdam:
     Van Gorcum.

[42] Rich, E.: 1983, 'Artificial Intelligence', New York: McGraw-Hill.

[43] Rieger, C.: 1975, 'Conceptual memory', in Conceptual Information Processing, Schank, R. (ed.), Amsterdam: North-Holland.

[44] Schubert, L.K., R.G. Goebel, N. Cercone: 1979, 'The structure and organization of a semantic net for comprehension and inference', in: Associative Networks, Findler, N.V., (ed.), New York: Academic Press, 122-178.

[45] Shortliffe, E.H., B. Buchanan: 1975, 'A model of inexact reasoning in medicine', Mathematical Biosciences 23, 351-379.

[46] Shortliffe, E.H.: 1976, 'Computer-Based Medical Consultations: MYCIN', New York: American Elsevier.

[47] Sugeno, M.: 1977, 'Fuzzy measures and fuzzy integrals: a survey', in Fuzzy Automata and Decision Processes, Gupta, M.M., Saridis, G.N. and Gaines, B.R. (eds.) Amsterdam: North Holland, 89-102.

[48] Suppes, P.: 1970, 'A Probabilistic Theory of Causality', Amsterdam: North-Holland.

[49] Szolovits, P., S.G. Pauker: 1978, 'Categorical and probabilistic reasoning in medical diagnosis', Artificial Intelligence 11, 115-144.

[50] Terano, T., M. Sugeno: 1975, 'Conditional fuzzy measures and their applications', Fuzzy Sets and Their Applications to Cognitive and Decision Processes, Zadeh, L.A., Fu, K.S., Tanaka, K., Shimura, M. (eds.), New York: Academic Press, 151-170.

[51] Weiss, S.M., C.A. Kulikowski: 1979, 'EXPERT: A system for developing consultation models', Proc. IJAI6.

[52] Yager, R.R.: 1980, 'Quantified propositions in a linguistic logic', in: Proceedings of the 2nd International Seminar on Fuzzy Set Theory, Klement, E.P. (ed.), Johannes Kepler University, Linz, Austria.

[53] Zadeh, L.A.: 1968, 'Probability measures of fuzzy events', J. Math. Anal. Appl. 23, 421-427.

[54] Zadeh, L.A.; 1971, 'Similarity relations and fuzzy orderings', Inf. Sci. 3, 177-200.

[55] Zadeh, L.A.: 1975, 'Fuzzy logic and approximate reasoning' (in memory of Grigore Moisil), Synthese 30, 407-428.

[56] Zadeh, L.A.: 1975, 'The concept of a linguistic variable and its application to approximate reasoning', Inform. Sci. 8, 199-249, 301-357, 9, 43-80.

[57] Zadeh, L.A.: 1978, 'PRUF - a meaning representation language for natural languages', <u>Int. J. Man-Machine Studies</u> 10, 395-460.

[58] Zadeh, L.A.: 1979, 'A theory of approximate reasoning' Electronics Research Laboratory Memorandum M77/58 University of California, Berkeley. Also in: Machine Intelligence 9, Hayes, J.E., Michie, D., Kulich, L.I. (eds.), New York: Wiley 149-194.

[59] Zadeh, L.A.: 1980, 'Inference in fuzzy logic', Proc. 10th Inter. Symp. on Multiple-valued Logic, Northwestern University, 124-131.

[60] Zadeh, L.A.: 1980, 'Possibility theory and soft data analysis', Electronic Res. Laboratory Memorandum M79/59 University of California, Berkeley; also in Mathematical Frontiers of the Social and Policy Sciences, Cobb, L., and Thrall, R.M. (eds.), Boulder: Westview Press, 69-129 (1981).

[61] Zadeh, L.A.: 1981, 'Test-score semantics for natural languages and meaning-representation via PRUF', Techn. Note 247, AI Center, SRI International, Menlo Park, CA. Also in Empirical Semantics, Rieger, B.B., (ed.), Bochum: Brockmeyer, 281-349.

[62] Zadeh, L.A.: 1981, 'Fuzzy probabilities and their role in decision analysis', Proc. 4th MIT/ONR Workshop on Command, Control and Communications M.I.T., 159-179.

[63] Zadeh, L.A.: 1983, 'A computational approach to fuzzy quantifiers in natural languages', <u>Computers and Mathematics</u> 9, 149-184.

[64] Zimmer, A.: 1982, 'Some experiments concerning the fuzzy meaning of logical quantifiers', in General Surveys of Systems Methodology, Troncoli, L. (ed.), 435-441. Society for General Systems Research, Louisville.

[65] Zimmermann, H.-J., P. Zysno: 1980, 'Latent connectives in human decision making', <u>Fuzzy Sets and Systems</u> 4, 37-52.

Test-Score Semantics as a Basis for a Computational Approach to the Representation of Meaning

L.A. ZADEH

University of California, Berkeley, USA

Abstract

A basic idea which underlies test-score semantics is that a proposition in a natural language may be interpreted as a system of elastic constraints which is analogous to a non-linear program. Viewed in this perspective, meaning representation may be regarded as a process which (a) identifies the variables which are constrained, and (b) characterizes the constraints to which they are subjected. In test-score semantics, this is accomplished through the construction of a test procedure which tests, scores and aggregates the elastic constraints which are implicit in the proposition, yielding an overall test score which serves as a measure of compatibility between the proposition, on the one hand, and what is referred to as an explanatory database, on the other.

Test-score semantics provides a framework for the representation of the meaning of dispositions, that is, propositions which are preponderantly, but not necessarily always, true. Another important concept which is a part of test-score semantics is that of a canonical form, which may be viewed as a possibilistic analog of an assignment statement. The concepts of a disposition and canonical form play particularly important roles in the representation of—and reasoning with—commonsense knowledge.

Introduction

During the past decade, the emergence of natural language processing as one of the major areas of research in artificial intelligence has provided a strong stimulus for the development of computationally-oriented theories of

Correspondence: L. A. Zadeh, Computer Science Division, University of California, Berkeley, CA 94720. Research supported in part by NESC Contract N0039-83-C-0243 and NASA Grant NCC 2-275. To Professor Abe Mamdani, Dr Janet Efstathiou and Dr Richard Tong.

meaning, knowledge representation and approximate reasoning.

The traditional approaches to the theory of meaning—among which the best known are truth-conditional semantics, possible-world semantics, Montague semantics, procedural semantics and, in a broader sense, model-theoretic semantics—are based on two-valued logic and the concomitant assumption that the extension of a predicate is a crisply defined set in a universe of discourse. It may be argued, however, that, in the case of natural languages, the extension of a predicate is, in general, a fuzzy set in which the transition from membership to nonmembership is gradual rather than abrupt. This is true, for example, of such commonly used predicates as *tall man, loud music, attractive woman*; quantifiers (1) such as *most, several, few*; temporal quantifiers such as *frequently, once in a while, almost always*; and qualifiers such as *very likely, quite true, almost impossible*, etc. Indeed, it is evident that almost all concepts in or about natural languages are fuzzy in nature. Viewed in this perspective, it is hard to rationalise the almost exclusive use of two-valued logic in the traditional approaches to the semantics of natural languages.

In a departure from this tradition, we have described in a series of papers starting in 1971, an approach to the semantics of natural languages based on the theory of fuzzy sets and, more particularly, on possibility theory. During the past several years, this approach has evolved into a meaning-representation system termed *test-score semantics* (Zadeh, 1978b, 1981), which is computational in nature and which is based on the premise that almost everything that relates to natural languages is a matter of degree. Like almost all theories of meaning, test-score semantics is referential in nature in the sense that it deals with the correspondence between expressions in a language and their denotations in a universe, or family of universes, of discourse. However, unlike the traditional approaches to the theory of meaning, test-score semantics does not make use of the machinery of first-order or intensional logic, and is based, instead, on *fuzzy logic*—the logic of approximate or fuzzy reasoning. In what follows, we shall present an informal exposition of some of the basic ideas underlying test-score semantics and illustrate its application to the representation of the meaning of various types of semantic entities, among them propositions, predicates, dispositions and commands.

Composition of Meaning

The point of departure in test-score semantics is the assumption that the problem of meaning representation is that of composing the meaning of a given semantic entity, *s*, from a collection of fuzzy relations—termed the *composants* of *s*—whose meaning is assumed to be known by the addressee of the representation process.

As an illustration, assume that the semantic entity in question is the proposition $p \triangleq$ *Over the past few years Stan made a lot of money*. Furthermore, assume that the relations whose meaning is assumed to be known are:

(*a*)   *INCOME [Name; Amount; Year]*,

which is a collection of tuples, e.g., (*Ted, 150,000, 1982*), whose first element is the name of an individual; whose second element is the income of that individual; and whose third element is the year in which the income was made;

(*b*)   *FEW [Number; μ]*,

which serves to calibrate the meaning of the fuzzy number *few* by associating with each real value of the variable *Number*, the degree, μ, to which that number fits the intended meaning of *few*; and

(*c*)   *LOT.OF.MONEY[Amount; μ]*,

which calibrates the meaning of *lot of money* by associating with each value of *Amount*, the degree, μ, to which the value of *Amount* fits the intended meaning of *lot of money*.

We shall show at a later point how the meaning of *p* may be composed from the meaning of its composants. What is important to note at this juncture is that the choice of the relations in question reflects our assumption concerning the knowledge profile of the addressee of the representation process. In this sense, then, the representation of the meaning of *p* is strongly dependent on the choice of relations from which the meaning of *p* is composed.

In its strict interpretation—which is the interpretation underlying Montague semantics—Frege's principle of compositionality (Hintikka, 1982) (Janssen, 1978) requires that the meaning of a proposition be composed from the meanings of its constituents. In test-score semantics, the composants of *s* are, in general, implicit rather than explicit in *s*. Furthermore, their number is, in general, much smaller than the number of constituents in *s*. As will be seen later, by allowing the meaning of *s* to be composed from its composants rather than constituents, test-score semantics achieves a higher expressive power than is possible under a strict interpretation of Frege's principle of compositionality (Zadeh, 1983c).

Since the composants of *s* serve to explain the meaning of *s*, they constitute, collectively, a collection of fuzzy relations which is referred to as an *explanatory database*, or ED, for short. In general, the procedure which describes the composition of the meaning of *s* involves only the *frames* of the relations in ED, that is, their names, their attributes and the domains of their attributes. In test-score semantics, the collection of frames in ED is referred to as the *explanatory database frame*, or, more simply, as EDF. In effect, the role played by EDF in test-score semantics is analogous to that of a collection of possible worlds in possible-world, or, more generally, model-theoretic semantics (McCawley, 1981).

Testing and Scoring of Elastic Constraints

A basic idea which underlies test-score semantics—and motivates its name—is that any semantic entity—and, especially, a proposition—may be viewed as a system of implicity defined elastic or, equivalently, fuzzy constraints whose domain is the collection of fuzzy relations in the explanatory database. Viewed in this perspective, the meaning of a semantic entity, *s*, may be represented as a procedure which tests, scores and aggregates the elastic constraints which are induced by *s*. In more concrete terms, assume, for simplicity, that *s* is a proposition, *p*. Representation of the meaning of *p* through the use of test-score semantics involves, in general, the following steps.

1. Selection of an appropriate explanatory database, that is, a set of fuzzy relations which collectively constitute the composants of *p*.
2. Explicitation of the elastic constraints $C_1, ..., C_m$ which are induced by *p*. For example, in the case of the proposition $p \triangleq Susan\ is\ young$, the implicit constrained variable is the age of Susan and the associated elastic constraint is characterized by the fuzzy predicate *young*. Less obviously, in the case of the proposition $p \triangleq Most\ students\ are\ young$, the implicit constrained variable is the proportion of *young students* among *students*, and the associated

elastic constraint is characterized by the fuzzy quantifier *most*.

3. Characterization of each constraint C_i, $i = 1$, $\ldots, m$, by a test which yields the test score, τ_i, representing the degree to which the constraint is satisfied. Usually, the test score is represented as a number in the interval [0, 1]. More generally, however, a test score may be a fuzzy number or a probability/possibility distribution over the unit interval. The test scores in test-score semantics play a role which is somewhat analogous to that of truth values in truth-conditional semantics.

4. Aggregation of the partial test scores $\tau_1, \ldots, \tau_m$ into a smaller number of test scores $\bar{\tau}_1, \ldots, \bar{\tau}_k$, represented as an *overall vector test score* τ,

$$\tau = (\bar{\tau}_1, \ldots, \bar{\tau}_k). \tag{3.1}$$

In most cases $k = 1$, so that the overall test score is a scalar. The overall test score serves as a measure of the *compatibility* of p with the explanatory database, ED. Alternatively, τ may be interpreted as the *truth* of p given ED or, equivalently, as the *possibility* of ED given p. What is important to note is that the meaning of p is represented not by the overall test score τ, but by the test procedure which computes τ for any given ED.

As a simple illustration of the concept of a test procedure, consider the proposition $p \triangleq$ *Joan is young and attractive*. The EDF in this case will be assumed to consist of the following relations:

$$EDF \triangleq POPULATION \text{ [}Name;\ Age;\ \mu Attractive\text{]}$$

$$+ YOUNG \text{ [}Age;\ \mu\text{]}, \tag{3.2}$$

in which + should read as 'and'.

The relation labeled *POPULATION* consists of a collection of triples whose first element is the name of an individual; whose second element is the age of that individual; and whose third element is the degree to which the individual in question is attractive. The relation *YOUNG* is a collection of pairs whose first element is a value of the variable *Age* and whose second element is the degree to which that value of *Age* satisfies the elastic constraint characterized by the fuzzy predicate *young*. In effect, this relation serves to calibrate the meaning of the fuzzy predicate *young* in a particular context by representing its denotation as a fuzzy subset, *YOUNG*, of the interval [0, 100].

With this EDF, the test procedure which computes the overall test score may be described as follows:

1. Determine the age of Joan by reading the value of *Age* in *POPULATION*, with the variable *Name* bound to Joan. In symbols, this may be expressed as

$$Age\ (Joan) = {}_{Age}\,POPULATION\ [Name = Joan].$$

$$(3.3)$$

In this expression, we use the notation ${}_Y R[X = a]$ to signify that X is bound to a in R and the resulting relation is projected on Y, yielding the values of Y in the tuples in which $X = a$ (Zadeh, 1978b).

2. Test the elastic constraint induced by the fuzzy predicate *young*:

$$\tau_1 = {}_\mu\,YOUNG[Age = Age(Joan)]. \quad (3.4)$$

3. Determine the degree to which Joan is attractive:

$$\tau_2 = {}_{\mu Attractive}\,POPULATION[Name = Joan].$$

$$(3.5)$$

4. Compute the overall test score by aggregating the partial test scores τ_1 and τ_2. For this purpose, we shall use the min operator $\wedge$ as the aggregation operator, yielding

$$\tau = \tau_1 \wedge \tau_1, \quad\quad\quad (3.6)$$

which signifies that the overall test score is taken to be the smaller of the operands of $\wedge$. The overall test score, as expressed by (3.6), represents the compatibility of $p \triangleq$ *Joan is young and attractive* with the data resident in the explanatory database.

Remark. In the example under consideration, Joan has two attributes—*Youth* and *Attractiveness*—of which *Youth* is measurable in objective terms via *Age* while *Attractiveness* is subjective in nature. For this reason, the degree of attractiveness in the relation *POPULATION* is tabulated directly rather than through a measurable attribute. In so doing, we are implicitly taking advantage of the human ability to assign a (possibly fuzzy) grade of membership in a class without a conscious employment of quantitative criteria to arrive at its value.

As an additional illustration which relates to a semantic entity other than a proposition, assume that s is a second-order fuzzy predicate, namely,

$$s \triangleq many \ large \ balls. \qquad (3.7)$$

In this case, we shall employ the following EDF:

$$EDF \triangleq POPULATION[Identifier; Size]$$

$$+ LARGE[Size; \mu]$$

$$+ MANY[Number; \mu]. \qquad (3.8)$$

In this EDF, *POPULATION* is a collection of pairs in which the first element is a label which identifies a ball, i.e., an identifier, while the second element is the size of that ball; *LARGE* is a relation which calibrates the meaning of the fuzzy predicate *large*; and *MANY* is a relation which calibrates the fuzzy number *many* by associating with each numerical value of the variable *Number* the degree to which that value fits the intended meaning of *many*.

1. Assume that *POPULATION* consists of a collection of n balls, $b_1, ..., b_n$, of various sizes. Determine the size of each ball:

$$Size(b_i) = {}_{Size} POPULATION[Identifier = b_i],$$

$$i = 1, ..., n. \qquad (3.9)$$

2. For each ball, test the constraint induced by *large*:

$$\tau_i = {}_{\mu} LARGE[Size = Size(b_i)], i = 1, ..., n.$$

$$(3.10)$$

3. Count the number of large balls by adding the τ_i: (2)

$$\Sigma Count(LARGE.BALL) = \Sigma_i \tau_i. \qquad (3.11)$$

4. Test the constraint induced by *many*:

$$\tau = {}_{\mu} MANY[Number$$

$$= \Sigma Count(LARGE.BALL)], \qquad (3.12)$$

which represents the overall test score yielded by the test procedure.

The intent of these simple examples is to provide an idea of how the meaning of a semantic entity, s, may be represented by a test procedure which associates with each instance of an explanatory database, ED, the degree to which that ED is compatible with *s*. In the examples which we considered, the degree of compatibility is a single number. More generally, however, and particularly in the case of a proposition which is associated with presuppositions, the degree of compatibility may be vector-valued.

Aggregation and Composition

In test-score semantics, the manner in which the partial test scores are aggregated is left to the discretion of the constructor of the test procedure. It is helpful, however, to have a collection of standardized rules for dealing with the aggregation and combination of elastic constraints which are associated with conjunction, disjunction, implication, quantification and modification. These standardized rules should be regarded as default rules, that is, rules to be used when the context does not require special-purpose rules which may provide a better fit to the intended mode of aggregation. The basic standardized rules (3) in test-score semantics may be summarized as follows.

Modification Rules

If the test score for an elastic constraint *C* in a specified context is τ, then in the same context:

(*a*) the text score for *not C* is $1 - \tau$ (negation)

$$(4.1)$$

(*b*) the test score for *very C* is τ^2

(intensification)

$$(4.2)$$

(*c*) the test score for *more or less C* is $\tau^{1/2}$

(diffusion)

$$(4.3)$$

Composition Rules

If the test scores for elastic constraints C_1 and C_2 in a specified context are τ_1 and τ_2, respectively, then in the same context:

(*a*) the test score for C_1 *and* C_2 is $\tau_1 \wedge \tau_2$

(conjunction)

$$(4.4)$$

(b) the test score for C_1 *or* C_2 is $\tau_1 \vee \tau_2$
(disjunction)

(4.5)

(c) the test score for *if* C_1 *then* C_2 is $1 \wedge (1 - \tau_1 + \tau_2)$,
(implication)

(4.6)

where $\wedge \triangleq$ min and $\vee \triangleq$ max.

Quantification rules

These rules apply to semantic entities which contain fuzzy quantifiers, e.g., *several tall men, a few successes, it rained often, many more cats than dogs, most Frenchmen are not very tall*, etc. In such semantic entities, a fuzzy quantifier serves to provide a fuzzy characterization of an absolute or relative count of elements in one or more fuzzy sets.

Since membership in a fuzzy set is a matter of degree, it is not obvious how the count of elements in a fuzzy set may be defined. Among the various ways in which this can be done, the simplest is to employ the concept of the power (DeLuca and Termini, 1970) or, equivalently, the *sigma-count* of a fuzzy set (Zadeh 1975, 1983), which is defined as the arithmetic sum of the grades of membership (4). More specifically, assume that F is a fuzzy subset of a finite universe of discourse $U = \{u_1, \ldots, u_n\}$ which comprises the objects $u_1, \ldots, u_n$. For convenience, F may be represented symbolically as the linear form

$$F = \mu_1/u_1 + \mu_2/u_2 + \ldots + \mu_n/u_n, \qquad (4.7)$$

in which the term μ_i/u_i, $i = 1, \ldots, n$, signifies that μ_i, $0 \leq \mu_i \leq 1$, is the grade of membership of u_i in F, and the plus sign should be read as 'and'.

The *sigma-count* of F, denoted as $\Sigma Count(F)$, is defined as the arithmetic sum

$$\Sigma Count(F) \triangleq \Sigma_i \mu_i, i = 1, \ldots, n, \qquad (4.8)$$

with the understanding that the sum may be rounded if necessary to the nearest integer. Furthermore, one may stipulate that terms with grades of membership below a specified threshold be excluded from the summation in order to keep a large number of terms with low grades of membership from becoming count-equivalent to a small number of terms with high grades of membership.

In the case of a pair of fuzzy sets, (F, G), the *relative*

sigma-count, denoted by $\Sigma Count(F/G)$, may be interpreted as the proportion of elements of F which are in G. More explicitly,

$$\Sigma Count(F/G) = \frac{\Sigma Count(F \cap G)}{\Sigma Count(G)}, \qquad (4.9)$$

where $F \cap G$, the intersection of F and G, is defined by

$$\mu_{F \cap G}(u) = \mu_F(u) \wedge \mu_G(u), \, u \in U. \qquad (4.10)$$

Thus, in terms of the membership functions of F and G, the relative sigma-count of F and G may be expressed as

$$\Sigma Count(F/G) = \frac{\Sigma_i \mu_F(u_i) \wedge \mu_G(u_i)}{\Sigma_i \mu_G(u_i)}. \qquad (4.11)$$

The concept of a relative sigma-count provides a basis for computing the test score for the elastic constraint induced by the proposition

$$p \triangleq Q \ A's \ are \ B's, \qquad (4.12)$$

where A and B are fuzzy sets—or, equivalently, fuzzy predicates—and Q is a fuzzy quantifier such as *most, many, almost all, few*, etc. More specifically, if μ_Q is the membership function of Q, then

$$p \rightarrow \tau = \mu_Q(\Sigma Count(B/A)), \qquad (4.13)$$

where τ is the compatibility of p with an explanatory database whose constituents are the fuzzy relations A, B and Q, and the arrow $\rightarrow$ should be read as 'induces' or 'translates into'.

As an illustration of the rules discussed above, consider the proposition

$$p \triangleq Most \ Frenchmen \ are \ neither \ very \ tall \ nor \ very \ fat.$$

$$(4.14)$$

The EDF in this case will be assumed to have the following relations as constituents:

$$EDF \triangleq POPULATION.FRENCHMEN$$

$$[Name; \ Weight; \ Height]$$

$$+ \ FAT[Weight; \ Height; \ \mu]$$

$$+ \ TALL[Height; \ \mu]$$

$$+ \ MOST[Proportion; \ \mu].$$

In this EDF, the first relation is a listing of a representative group of Frenchmen, with *Weight* and *Height* being the attributes of *Name*; the second relation serves to calibrate the fuzzy predicate *fat* as a function of *Weight* and *Height*; the third relation calibrates the fuzzy predicate *tall*; and the last relation calibrates the fuzzy quantifier *most* as a function of *Proportion*. Correspondingly, the test procedure with computes the overall test score and thus represents the meaning of *p* assumes the following form:

1. Let $Name_i$ be the name of *i*th Frenchman in POPULATION. For each $Name_i$, $i = 1, \ldots, m$, find the weight and height of $Name_i$:

$$Weight\,(Name_i) = {}_{Weight}$$
$$POPULATION[Name = Name_i]. \qquad (4.15)$$

$$Height\,(Name_i) = {}_{Height}$$
$$POPULATION[Name = Name_i]. \qquad (4.16)$$

2. For each $Name_i$, compute the test scores for the constraints induced by *fat* and *tall*:

$$\alpha_i = {}_{\mu}FAT[Weight = Weight(Name_i)] \quad (4.17)$$

$$\beta_i = {}_{\mu}TALL[Height = Height(Name_i)]. \quad (4.18)$$

3. Intensify the constraints induced by *fat* and *tall* to account for the modifier *very*. Using (4.2), the corresponding test scores may be expressed as:

$$\alpha_i^* = \alpha_i^2, i = 1, \ldots, m \qquad (4.19)$$

$$\beta_i^* = \beta_i^2. \qquad (4.20)$$

4. Modify the constraints induced by *very fat* and *very tall* to account for the negation *not*:

$$\alpha_i' = 1 - \alpha_i^* = 1 - \alpha_i^2 \qquad (4.21)$$

$$\beta_i' = 1 - \beta_i^* = 1 - \beta_i^2. \qquad (4.22)$$

5. For each $Name_i$, aggregate conjunctively the test scores for the constraints induced by *not very fat* and *not very tall*:

$$\tau_i = (1 - \alpha_i^2) \wedge (1 - \beta_i^2), i = 1, \ldots, m.$$

$$(4.23)$$

The resulting test score represents the degree to which $Name_i$ is *not very fat and not very tall*.

6. By using the relative sigma-count, compute the proportion of Frenchmen who are *not very fat and not very tall*:

$$\rho = \frac{1}{m}\Sigma_i \tau_i$$

$$= \frac{1}{m}\Sigma_i(1 - \alpha_i^2) \wedge (1 - \beta_i^2). \qquad (4.24)$$

7. Compute the test score for the constraint induced by *most*:

$$\tau = {}_\mu MOST[Proportion = \rho]. \qquad (4.25)$$

This test score is the desired overall test score for the proposition under consideration, and the test procedure which computes τ represents the meaning of *p*.

Vector Test Scores

In the examples considered so far, the overall test score is a scalar. In the case of the following example (Zadeh, 1981):

$$p \triangleq By\ far\ the\ richest\ man\ in\ France\ is\ bald, (5) \quad (5.1)$$

the overall test score assumes the form of an ordered pair,

$$\tau \triangleq (\tau_0, \tau_p),$$

in which τ_0 is the test score associated with the fuzzy presupposition

$$p^* \triangleq There\ exists\ by\ far\ the\ richest\ man\ in\ France,$$

$$(5.2)$$

and τ_p is the degree to which the by far the richest man in France is bald. The presupposition expressed by (5.2) is fuzzy by virtue of the fuzziness of the predicate *by far the richest man in France*. What this implies is that the existence of *by far the richest man in France* is a matter of degree.

To represent the meaning of *p* (5.1), we choose the following EDF:

$$EDF \triangleq POPULATION[Name;\ Wealth;\ \mu Bald]$$

$$+ BY.FAR.RICHEST[Wealth1;\ Wealth2;\ \mu]. \quad (5.3)$$

In the first relation in (5.3), *Wealth* is interpreted as the net worth of *Name* and $\mu Bald$ is the degree to which *Name* is bald. In the second relation, *Wealth1* is the wealth of the richest man, *Wealth2* is the wealth of the second richest man, and μ is the degree to which *Wealth1* and *Wealth2* qualify the richest man in France (who is assumed to be unique) to be regarded as by far the richest man in France.

To compute the compatibility of p with the explanatory database, perform the following test.

1. Sort POPULATION in descending order of Wealth. Denote the result by POPULATION↓ and let $Name_i$ be the ith name in POPULATION↓.

2. Determine the degree to which the richest man in France is bald:

$$\tau_p = {}_{\mu Bald}POPULATION{\downarrow}[Name = Name_1].$$

(5.4)

3. Determine the wealth of the richest and second richest man in France:

$$w_1 = {}_{Wealth}POPULATION{\downarrow}[Name = Name_1]$$

$$w_2 = {}_{Wealth}POPULATION{\downarrow}[Name = Name_2].$$

4. Determine the degree to which the richest man in France is by far the richest man in France:

$$\tau_0 = {}_{\mu}BY.FAR.RICHEST$$

$$[Wealth1 = w_1; Wealth2 = w_2].$$

(5.5)

5. The overall test score is taken to be the ordered pair

$$\tau = (\tau_o, \tau_p).$$

(5.6)

Thus, instead of aggregating τ_1 and τ_2 into a single test score, we maintain their separate identities in the overall test score. We do this because the aggregated test score

$$\tau = \tau_o \wedge \tau_p$$

would be creating a misleading impression when τ_o is small, that is, when the test score for the constraint on the existence of *by far the richest man in France* is low.

More generally, the need for a vector test score arises when the degree of summarization which is implicit in a scalar test score may be excessive in relation to the purpose of meaning-representation in a particular context. As an illustration, consider the proposition

$$p \triangleq Berkeley\ has\ a\ temperate\ climate,$$

(5.7)

and suppose that the intended meaning of p is represented by the following three propositions:

$p_1 \triangleq$ *The average temperature in Berkeley is around* $70°$.

$p_2 \triangleq$ *The fraction of hot days is small.*

$p_3 \triangleq$ *The fraction of cold days is small.*

The EDF for p will be assumed to be the union of the EDF's for p_1, p_2 and p_3, which are:

$EDF_1 \triangleq AROUND[T_1; T_2; \mu]$.

$EDF_2 \triangleq HOT[T; \mu] + SMALL[Proportion; \mu]$.

$EDF_3 \triangleq COLD[T; \mu] + SMALL[Proportion; \mu]$.

In $AROUND$, μ is the degree to which a temperature, T_1, is *around* T_2; in HOT, μ is the degree to which T is *hot*; in $COLD$, μ is the degree to which T is *cold*; and in $SMALL$, μ is the degree to which *Proportion* is *small*. Note that all of these relations serve to calibrate the meanings of the predicates which they represent.

Now suppose that the overall test scores for p_1, p_2 and p_3 are τ_1, τ_2 and τ_3, respectively. We can compute the overall test score by aggregating these test scores, yielding

$$\tau = \tau_1 \wedge \tau_2 \wedge \tau_3. \qquad (5.8)$$

On the other hand, for some purposes, the single test score represented by (5.8) might be insufficiently informative. Then, a vector test score represented by the triple

$$\tau = (\tau_1, \tau_2, \tau_3) \qquad (5.9)$$

might be more appropriate.

Representation of the Meaning of Dispositions

A concept which plays an important role in natural languages is that of a *disposition*. Informally, a disposition is a proposition which is preponderantly, but not necessarily always, true. For example, *snow is white* is a disposition, as are the propositions *trees are green, small cars are unsafe, John is always drunk, overeating causes obesity*, etc. Technically, a disposition may be defined as a proposition with implicit fuzzy quantifiers (Zadeh, 1983bcd). This definition is a *dispositional definition* in the sense that it, too, is a disposition. It should be noted that what is usually called *commonsense knowledge* may be viewed as a collection of dispositions (Zadeh, 1983d).

Usually, a disposition may be derived from a dispositional proposition by a suppression of one or more fuzzy quantifiers. For example, by suppressing the fuzzy temporal quantifier *usually* in the dispositional proposition

$$dp \triangleq \text{Usually snow is white,} \qquad (6.1)$$

we obtain the disposition

$$d \triangleq \text{Snow is white.} \qquad (6.2)$$

Similarly, by suppressing the fuzzy quantifier *most of* in the dispositional proposition

$$dp \triangleq \text{Most of those who overeat are obese,} \qquad (6.3)$$

we arrive at the disposition

$$d \triangleq \text{Those who overeat are obese,} \qquad (6.4)$$

which, in an associational sense of causality (Suppes, 1970) may be expressed as

$$d \triangleq \text{Overeating causes obesity.} \qquad (6.5)$$

As a further illustration, by suppressing the fuzzy quantifiers *most* and *mostly* in the dispositional proposition

$$dp \triangleq \text{Most young men like mostly young women,}$$
$$(6.6)$$

we obtain the disposition

$$d \triangleq \text{Young men like young women.} \qquad (6.7)$$

Frequently, a proposition with nonfuzzy quantifiers is intended to be understood as a dispositional proposition. For example,

$$p \triangleq \text{John is always drunk,} \qquad (6.8)$$

would usually be understood as

$$dp \triangleq \text{John is frequently drunk,} \qquad (6.9)$$

in which the fuzzy temporal quantifier *frequently* conveys the intended meaning of *always*. (Note, however, that the fuzzy quantifier *frequently* in (6.9) cannot be suppressed without distorting the intended meaning of *p*.) The replacement of a nonfuzzy quantifier with a fuzzy quantifier is an instance of *fuzzification*.

More generally, the transformation of a proposition with implicit fuzzy quantifiers into one in which the fuzzy quantifiers are explicit may be viewed as an instance of *explicitation*. The process of explicitation constitutes the first step in representing the meaning of a disposition.

In general, explicitation is an interpretation-dependent process in the sense that the restoration of suppressed fuzzy quantifiers and/or the fuzzification of nonfuzzy quantifiers depends on the intended meaning of the disposition. As an illustration, consider the disposition

$$d \triangleq Overeating\ causes\ obesity. \qquad (6.10)$$

The intended meaning of this disposition may be conveyed by the restored dispositional proposition

$$dp \triangleq Most\ of\ those\ who\ overeat\ are\ obese.$$
$$(6.11)$$

On the other hand, the intended meaning of the disposition

$$d \triangleq Heavy\ smoking\ causes\ lung\ cancer, \qquad (6.12)$$

which is similar in form to (6.10), may be conveyed by the dispositional proposition

$dp \triangleq$ *The proportion of cases of lung cancer among heavy smokers is much higher than among nonsmokers.*

Similarly,

$$d \triangleq Young\ men\ like\ young\ women \qquad (6.13)$$

has a number of different interpretations, among them:

(a) $dp \triangleq Most\ young\ men\ like\ most\ young\ women,$

$$(6.14)$$

and, what appears to be much more reasonable:

(b) $dp \triangleq Most\ young\ men\ like\ mostly\ young\ women.$

$$(6.15)$$

At a later point in this section, we shall use (6.15) to illustrate the representation of the meaning of a disposition through the use of test-score semantics.

The concept of a dispositional proposition opens the door to the construction of a number of other concepts with a *dispositional* flavor, e.g., *dispositional predicate, dispositional containment* (in the sense of set inclusion), *dispositional command, dispositional preference relation,* etc. For example, *smokes* in *Virginia smokes cigarettes; like* in *Young men like young women; like* in *Frenchmen are like Spaniards; loves* in *Ann loves men* are dispositional predicates. *Keep under refrigeration, Avoid overeating,*

Stay away from bald men are dispositional commands; and *Gentlemen prefer blondes, Mike is a better tennis player than Claudine, Tokyo is much safer than New York,* are examples of dispositional preference or ordering relations.

As an illustration of how the meaning of a dispositional predicate may be defined, consider the proposition

$$d \triangleq Virginia\ smokes\ cigarettes, \qquad (6.16)$$

which is a disposition by virtue of containing the dispositional predicate *smokes*.

Assume that the intended meaning of *d* is conveyed by the proposition

$$p \triangleq On\ the\ average\ Virginia\ smokes\ at\ least\ a\ few$$
$$cigarettes\ a\ day,$$

$$(6.17)$$

in which *smokes* is used in its literal (nondispositional) sense. At this point, we can employ test-score semantics to represent the meaning of *p*. Specifically, let the following relations be the constituents of EDF:

$$EDF \triangleq RECORD[Day;\ Number]$$

$$+ FEW[Number;\ \mu], \qquad (6.18)$$

in which *RECORD* is a daily record of the number of cigarettes smoked by Virginia during a representative period, say a month; and *FEW* is a fuzzy relation which calibrates the meaning of the fuzzy number *few*.

The steps in the test procedure which represents the meaning of *p* may be described as follows:

1. Let Day_i denote the *i*th day in *RECORD*, $i = 1$, ..., 30. Determine the number of cigarettes smoked by Virginia on Day_i:

$$Number(Day_i) = {}_{Number}RECORD[Day = Day_i].$$

$$(6.19)$$

2. Compute the average number of cigarettes smoked by Virginia during the period under consideration:

$$\rho = \frac{1}{30} \Sigma_i Number(Day_i). \qquad (6.20)$$

3. Test the constraint induced by the fuzzy quantifier *at least a few*:

$$\tau = {}_\mu \geq FEW[Number = \rho], \qquad (6.21)$$

in which $\geq FEW$ represents the relation *at least a few*, and τ is the overall test score.

The ability to represent the meaning of a dispositional predicate is of use in representing the meaning of complex dispositions. This aspect of the representation of the meaning of dispositions is illustrated by the following example (Zadeh, 1983b):

$$d \triangleq Young \ men \ like \ young \ women, \qquad (6.22)$$

which will be interpreted as the proposition

$$p \triangleq Most \ young \ men \ like \ mostly \ young \ women.$$

$$(6.23)$$

Now, in d the predicate

$$D \triangleq likes \ young \ women$$

may be viewed as a dispositional predicate, so that (6.23) may be rewritten as

$$p \triangleq Most \ young \ men \ are \ D. \qquad (6.24)$$

In this way, representation of the meaning of d (6.22) may be accomplished in two steps: first, we construct a test procedure to represent the meaning of D, and second, a test procedure is constructed to represent the meaning of p (6.24).

We shall assume that the EDF for p consists of the following relations:

$$EDF \triangleq POPULATION[Name; Age; Sex]$$

$$+ LIKE[Name1; Name2; \mu]$$

$$+ YOUNG[Age; \mu]$$

$$+ MUCH.HIGHER[Proportion1; Proportion2; \mu]$$

$$+ MOST[Proportion; \mu].$$

In $LIKE$, μ is the degree to which $Name1$ likes $Name2$; and in $MUCH.HIGHER$, μ is the degree to which $Proportion1$ is much higher than $Proportion2$.

The meaning of D may be represented by the following test procedure:

1. Divide *POPULATION* into the population of males, *M.POPULATION*, and population of females, *F.POPULATION*:

$$M.POPULATION = {}_{Name, Age}POPULATION$$

$$[Sex = Male]$$

$$F.POPULATION = {}_{Name, Age}POPULATION$$

$$[Sex = Female], \qquad (6.26)$$

where ${}_{Name, Age}POPULATION$ denotes the projection of $POPULATION$ on the attributes $Name$ and Age.

2. For each $Name_j$, $j = 1, ..., K$, in $F.POPULA$-$TION$, find the age of $Name_j$:

$$A_j = {}_{Age}F.POPULATION[Name = Name_j].$$

$$(6.27)$$

3. For each $Name_j$, find the degree to which $Name_j$ is young:

$$\alpha_i = {}_\mu YOUNG[Age = A_j], \qquad (6.28)$$

where α_i may be interpreted as the grade of membership of $Name_j$ in the fuzzy set, YW, of young women.

4. For each $Name_i$, $i = 1, ..., L$, in $M.POPULA$-$TION$, find the age of $Name_i$:

$$B_i = {}_{Age}M.POPULATION[Name = Name_i].$$

$$(6.29)$$

5. For each $Name_i$, find the degree to which $Name_i$ is young:

$$\delta_i = {}_\mu YOUNG[Age = B_i], \qquad (6.30)$$

where δ_i may be interpreted as the grade of membership of $Name_i$ in the fuzzy set, YM, of young men.

6. For each $Name_j$, find the degree to which $Name_i$ likes $Name_j$:

$$\beta_{ij} = {}_\mu LIKE[Name1 = Name_i; Name2 = Name_j],$$

$$(6.31)$$

with the understanding that β_{ij} may be interpreted as the grade of membership of $Name_j$ in the fuzzy set, WL_i, of women whom $Name_i$ likes.

7. For each $Name_j$, find the degree to which $Name_i$ likes $Name_j$ and $Name_j$ is young:

$$\gamma_{ij} = \alpha_j \wedge \beta_{ij}. \qquad (6.32)$$

Note: As in previous examples, we employ the aggregation operator min ($\wedge$) to represent the meaning of conjunction. In effect, γ_{ij} is the grade of membership of $Name_j$ in the intersection of the fuzzy sets WL_i and YW.

8. Compute the relative sigma-count of young women among the women whom $Name_i$ likes:

$$\rho_i = \Sigma Count(YW/WL_i)$$

$$= \frac{\Sigma Count(YW \cap WL_i)}{\Sigma Count(WL_i)}$$

$$= \frac{\Sigma_j \gamma_{ij}}{\Sigma_j \beta_{ij}}$$

$$= \frac{\Sigma_j \alpha_j \wedge \beta_{ij}}{\Sigma_j \beta_{ij}}. \tag{6.33}$$

9. Compute the relative sigma-count of (not young) women among the women whom $Name_i$ likes:

$$\gamma_i = \frac{\Sigma_j(1 - \alpha_j) \wedge \beta_{ij}}{\Sigma_j \beta_{ij}}. \tag{6.34}$$

10. Compute the degree to which ρ_i and γ_i satisfy the constraint induced by $MUCH.HIGHER$:

$$\tau_i = {}_\mu MUCH.HIGHER[Proportion1 = \rho_i;$$

$$Proportion2 = \gamma_i]. \tag{6.35}$$

This test score represents the overall test score for the proposition '$Name_i$ is D', and may be interpreted as the grade of membership of $Name_i$ in the fuzzy set, D, of men who have property D.

We are now in a position to compute the overall test score for p. Thus, on continuing the test, we have

11. Compute the relative sigma-count of men who have property D among young men:

$$\rho = \Sigma Count(D/YM)$$

$$= \frac{\Sigma Count(D \cap YM)}{\Sigma Count(YM)}$$

$$= \frac{\Sigma_i \tau_i \wedge \delta_i}{\Sigma_i \delta_i}. \tag{6.36}$$

12. Test the constraint induced by $MOST$:

$$\tau = {}_\mu MOST[Proportion = \rho]. \tag{6.37}$$

The test score expressed by (6.37) represents the overall test score for the disposition $d \triangleq$ *Young men like young women.*

As the final example in this Section, we shall consider the dispositional command:

$$c \triangleq \textit{Stay away from bald men.} \qquad (6.38)$$

In general, to represent the meaning of a command, c, it is necessary to associate with c its *compliance criterion*, cc, which may then be viewed as the definition of c. In the case of c (6.28), the compliance criterion will be assumed to be represented by the proposition

$$cc \triangleq \textit{Staying away from most bald men.} \qquad (6.39)$$

To represent the meaning of cc (6.29), we shall employ the following EDF:

$$EDF \triangleq RECORD[Name; \mu Bald; Action]$$

$$+ MOST[Proportion; \mu]. \qquad (6.40)$$

In this EDF, the relation *RECORD* is the record of successive actions which constitute the execution of c over a representative period of time; $\mu Bald$ is the degree to which *Name* is bald; and *Action* is a variable which takes the value 1 if *Name* is stayed away from and 0 otherwise.

The steps in the test procedure are the following:

1. For each $Name_i$ in *RECORD*, $i = 1, \ldots, n$, find (a) the degree to which $Name_i$ is bald, and (b) the action taken:

   (a) $\qquad \mu Bald(Name_i) = {}_\mu RECORD$
   $$[Name = Name_i]$$

   (b) $\qquad Action(Name_i) = {}_{Action} RECORD$
   $$[Name = Name_i]. \qquad (6.41)$$

2. Using the relative sigma-count, find the proportion of cases in which the command is complied with:

$$\rho = \frac{\Sigma_i \, Action(Name_i)\mu Bald(Name_i)}{\Sigma_i \, \mu Bald(Name_i)}. \qquad (6.42)$$

3. Test the constraint induced by the fuzzy quantifier *most*:

$$\tau = {}_\mu MOST[Proportion = \rho]. \qquad (6.43)$$

The computed value of τ represents the degree of compliance over the period under consideration; and the test procedure which yields τ represents the meaning of the dispositional command (6.38).

The Concept of a Canonical Form and its Application to the Representation of Meaning

When the meaning of a proposition, p, is represented as a test procedure, it may be hard to discern in the description of the procedure the underlying structure of the process through which the meaning of p is constructed from the meanings of the composants of p.

A concept which makes it easier to perceive the logical structure of p and thus to develop a better understanding of the meaning representation process, is that of a canonical form of p, abbreviated as $cf(p)$ (Zadeh, 1981).

The concept of a canonical form relates to the basic idea which underlies test-score semantics, namely, that any semantic entity—and, in particular, a proposition—may be viewed as a system of elastic constraints whose domain is a collection of relations in the explanatory database. Equivalently, let $X_1, ..., X_n$ be a collection of variables which are constrained by p. Then, the canonical form of p may be expressed as

$$cf(p) \triangleq X \text{ is } F, \tag{7.1}$$

where $X = (X_1, ..., X_n)$ is the constrained variable which is usually implicit in p, and F is a fuzzy relation, likewise implicit in p, which plays the role of an elastic (or fuzzy) constraint on X. The relation between p and its canonical form will be expressed as

$$p \rightarrow X \text{ is } F, \tag{7.2}$$

signifying that the canonical form may be viewed as a representation of the meaning of p.

In general, the constrained variable X in $cf(p)$ is not uniquely determined by p, and is dependent on the focus of attention in the meaning-representation process. To place this in evidence, we shall refer to X as the *focal variable*.

As a simple illustration, consider the proposition

$$p \triangleq Janet \ has \ blue \ eyes. \tag{7.3}$$

In this case, the focal variable may be expressed as

$$X \triangleq Color \ (Eyes(Janet)),$$

and the elastic constraint is represented by the fuzzy relation *BLUE*. Thus, we can write

$$p \rightarrow Color \ (Eyes(Janet)) \ is \ BLUE. \qquad (7.4)$$

As an additional illustration, consider the proposition

$$p \triangleq Dick \ is \ much \ taller \ than \ Nina. \qquad (7.5)$$

Here, the focal variable has two components, $X = (X_1, X_2)$, where

$$X_1 = Height \ (Dick)$$

$$X_2 = Height \ (Nina);$$

and the elastic constraint is characterized by the fuzzy relation *MUCH.TALLER* [*Height*1; *Height*2; μ], in which μ is the degree to which *Height*1 is *much taller* than *Height*2. In this case, we have

$$p \rightarrow (Height \ (Dick), Height \ (Nina)) \ is \ MUCH.TALLER.$$
$$(7.6)$$

To make the meaning of a canonical form more precise, it is necessary to introduce the concept of a possibility distribution (Zadeh, 1978a). Specifically, let X be an n-ary variable $(X_1, ..., X_n)$ in which $X_i, i = 1, ..., n$, takes values in a universe of discourse U_i, implying that X takes values in $U = U_1 \times U_2 \times \cdots \times U_n$. Informally, the *possibility distribution* of X, expressed as Π_X, is the fuzzy set of possible values of X, with the understanding that the possibility that X may take a value $u \in U$, written as $Poss\{X = u\}$, is a number in the interval [0, 1].

In terms of the possibility distribution of X, the canonical form of p may be interpreted as the assignment of F to Π_X. Thus, we may write

$$p \rightarrow X \ is \ F \rightarrow \Pi_X = F, \qquad (7.7)$$

in which the equation

$$\Pi_X = F \qquad (7.8)$$

is termed the *possibility assignment equation* (Zadeh, 1978b). In effect, this equation signifies that the canonical form $cf(p) \triangleq X \ is \ F$ implies that

$$Poss\{X = u\} = \mu_F(u), u \in U, \qquad (7.9)$$

where μ_F is the membership function of F. It is in this sense that F, acting as an elastic constraint on X, restricts the possible values which X can take in U. An important implication of this observation is that a proposition, p, may be interpreted as an implicit assignment statement which characterizes the possibility distribution of the focal variable in p.

As an illustration, consider the disposition

$$d \triangleq Overeating \ causes \ obesity, \qquad (7.10)$$

which upon restoration becomes

$$p \triangleq Most \ of \ those \ who \ overeat \ are \ obese. \quad (7.11)$$

If the focal variable in this case is chosen to be the relative sigma-count of those who are obese among those who overeat, the canonical form of p (7.11) becomes

$$\Sigma Count(OBESE/OVEREAT) \ is \ MOST, \quad (7.12)$$

which in virtue of (7.9) implies that

$$Poss\{\Sigma Count \ (OBESE/OVEREAT) = u\} = \mu_{MOST}(u),$$
$$(7.13)$$

where μ_{MOST} is the membership function of $MOST$. What is important to note is that (7.13) is equivalent to the assertion that the overall test score for p is expressed by

$$\tau = \mu_{MOST}(\Sigma Count(OBESE/OVEREAT)), \quad (7.14)$$

in which $OBESE$, $OVEREAT$ and $MOST$ play the roles of composants of p.

It is of interest to observe that the notion of a semantic network may be viewed as a special case of the concept of a canonical form. As a simple illustration, consider the proposition

$$p \triangleq Ron \ gave \ Shelley \ a \ red \ pin. \qquad (7.15)$$

As a semantic network, this proposition may be represented in the standard form:

$$Agent \ (GIVE) = Ron$$
$$Recipient \ (GIVE) = Shelley$$
$$Time \ (GIVE) = Past$$
$$Object \ (GIVE) = Pin$$
$$Color \ (Pin) = Red \qquad (7.16)$$

Now, if we identify X_1 with $Agent \ (GIVE)$, X_2 with $Recipient \ (GIVE)$, etc., the semantic network representation (7.16) may be regarded as a canonical form in which $X = (X_1, ..., X_5)$, and

$$X_1 = Ron$$
$$X_2 = Shelley$$
$$X_3 = Past$$
$$X_4 = Pin$$
$$X_5 = Red \qquad (7.17)$$

More generally, since any semantic network may be expressed as a collection of triples of the form (Object, Attribute, Attribute Value), it can be transformed at once into a canonical form. However, since a canonical form has a much greater expressive power than a semantic network, it may be difficult to transform a canonical form into a semantic network. A simple example of a proposition for which this is true is

$p \triangleq$ *Over the past several years the combined income of Patricia's close friends was about half a million dollars.*

(7.18)

Concluding Remark

We have described some of the basic ideas underlying test-score semantics and illustrated its application to the representation of the meaning of propositions, predicates, dispositions and commands. These ideas are simple in nature and, with a little practice, it is easy to learn how to use test-score semantics to represent the meaning of almost any semantic entity through the construction of an explanatory database and a test procedure. What is much more difficult, however, is to write a program which could construct an explanatory database and a test procedure without human assistance. This is a longer range problem whose complete solution must await the development of a substantially better understanding of natural languages and knowledge representation than we have at this juncture.

Notes

1. It should be noted that a quantifier may be viewed as a second order predicate.
2. As will be seen presently, this mode of counting yields the so-called *sigma-count* of large balls. An alternative way of counting the number of large balls is described in Zadeh (1981).

3. Standardized rules for aggregation and combination may be likened to ready-made clothing.
4. A more detailed discussion of various types of counts may be found in Zadeh (1981, 1983a) and Dubois (1982). Note that the power or the sigma-count of a fuzzy set may be viewed as a special case of its *measure* (Zadeh, 1968).
5. This proposition is a fuzzy version of the familiar example '*The King of France is bald*' which is associated with the presupposition '*There exists the King of France*'.

References and Relevant Publications

Atlas, J. D., How linguistics matters to philosophy, presupposition, truth and meaning, in: *Syntax and Semantics 11* Oh, C. K. and Dinneen, D. (eds.). New York: Academic Press, 265–281, 1979.

Ballmer, T. T. and Pinkal, M. (eds.), *Approaching Vagueness.* Amsterdam: North Holland, 1983.

Bartsch, R. and Vennemann, T., *Semantic Structures.* Frankfurt: Attenaum Verlag, 1972.

Barwise, J. and Cooper, R., Generalized quantifiers and natural language, *Linguistics and Philosophy 4* (1981) 159–219.

Barwise, J. and Perry, J., *Situations and Attitudes.* Cambridge: MIT Press, 1983.

Bergmann, M., Presupposition and two-dimensional logic, *J. of Philos. Logic 10* (1981) 27–53.

Black, M., Reasoning with loose concepts, *Dialogue 2* (1963) 1–12.

Bosch, P., Vagueness, ambiguity and all the rest, in: *Sprachstruktur, Individuum und Gesselschaft*, Van de Velde, M. and Vandeweghe, W. (eds.). Tubingen: Niemeyer, 1978.

Bosch, P., The role of propositions in natural language semantics, in: *Proceedings of the 6th Intern. Wiggenstein Symposium on Ontology and Language*, Leinfellner, W. *et al* (eds.), Kirchberg, 1982.

Carnap, R., *Meaning and Necessity.* Chicago: University of Chicago Press, 1952.

Chomsky, N., *Rules and Representations.* New York: Columbia University Press, 1980.

Cooper, W. S., *Logical Linguistics.* Dordrecht: Reidel, 1978.

Cresswell, M. J., *Logic and Languages.* London: Methuen, 1973.

Cushing, S., *Quantifier Meanings—A Study in the Dimensions of Semantic Competence.* Amsterdam: North Holland, 1982.

Davidson, D. and Gilbert, N., *Semantics of Natural Language.* Dordrecht: Reidel, 1972.

DeLuca, A. and Termini, S., A definition of non-probabilistic entropy in the setting of fuzzy sets theory, *Information and Control 20* (1972) 301–312.

Dubois, D. and Prade, H., *Fuzzy Sets and Systems: Theory and Applications.* New York: Academic Press, 1980.

Dubois, D., Properties de la cardinalite floue d'un ensemble flou fini, *BUSEFAL 8* (1981) 11–12.

Etchemendy, J., *Tarski, Model Theory and Logical Truth.* Ph.D. Dissertation, Stanford University, Stanford, CA, 1982.

Fine, K., Vagueness, truth and logic. *Synthese 30* (1975) 265–300.

Gallin, D., *Intensional and Higher-Order Modal Logic.* Amsterdam: North Holland, 1975.

Gazdar, G., *Pragmatics: Implicature, Presupposition, and Logical Form.* New York: Academic Press, 1979.

Goguen, J. A., The logic of inexact concepts, *Synthese 19* (1969) 325–373.

Groenendijk, J., Janssen, T. and Stokhoff, M. (eds.), *Formal Methods in the Study of Language.* Mathematical Centre: Amsterdam, 1981.

Grosz, B. J., Focusing and description in natural language dialogues. Tech. Note, AI Center, SRI International, Menlo Park, CA, 1979.

Haack, S., *Philosophy of Logics.* Cambridge: Cambridge University Press, 1978.

Halvorson, P. K., An interpretation procedure for functional structures. Cognitive Science Center, MIT, Cambridge, MA, 1981.

Hausser, R. R., A constructive approach to intensional contexts, *Language Research 18* (1982) 337–372.

Herzberger, H. G., Dimensions of truth, *J. of Philos. Logic 2* (1973) 535–556.

Hintikka, J. K., *Logic, Language-Games, and Information: Kantian Themes in the Philosophy of Logic.* Oxford: Oxford University Press, 1973.

Hintikka, J., Game-theoretical semantics: insights and prospects, *Notre Dame Journal of Formal Logic 23* (1982) 219–241.

Hobbs, J. R., Making computational sense of Montague's intensional logic, *Artificial Intelligence 9* (1978) 287–306.

Janssen, T., Compositionality and the form of rules in Montague grammar, in: *Proceedings of the Second Amsterdam Symposium on Montague Grammar and Related Topics,* Groenendijk, J. and Stokhoff, M. (eds.), 1978.

Janssen, T., Relative Clause Constructions in Montague Grammar and Compositional Semantics, in: *Formal Methods in the Study of Language,* Groenendijk, J., Janssen, T. and Stokhoff, M. (eds.), Mathematical Centre, Amsterdam, 1981.

Johnson-Laird, P. N., Procedural Semantics, *Cognition 5* (1977) 189–214.

Kamp, H., A theory of truth and semantic representation, in: *Formal Methods in the Study of Language,* Groenendijk, J. A. et al, (eds.), Mathematical Centre, Amsterdam, Tract 135, 1981.

Keenan, E., (ed.), *Formal Semantics of Natural Language.* Cambridge: Cambridge University Press, 1975.

Keenan, E. and Fultz, L., Logical types for natural languages, *UCLA Occasional Papers in Linguistics 3*, 1978.

Khatchadourian, H., Vagueness, meaning and absurdity, *Amer. Phil. Quarterly 2* (1965) 119–129.

Kindt, W., Vagueness as a topological problem, Ballmer T. and Pinkal, M., (eds.), *Approaching Vagueness*. Amsterdam: North Holland, 1982.

Klement, E. P., An axiomatic theory of operations on fuzzy sets, Institut fur Mathematik, Johannes Kepler Universitat Linz, Institutsbericht 159, 1981.

Lakoff, G., Hedges: A study in meaning criteria and the logic of fuzzy concepts, *J. Phil. Logic 2* (1973) 458–508. Also in: *Contemporary Research in Philosophical Logic and Linguistic Semantics*, Jockney, D., Harper, W. and Freed, B. (eds.). Dordrecht: Reidel, 221–271, 1973.

Lambert, K. and Van Fraassen, B. C., Meaning relations, possible objects and possible worlds, *Philosophical Problems in Logic* (1970) 1–19.

Landman, F. and Meerdijk, I., Compositionality and the analysis of anaphora, *Linguistics and Philosophy 6* (1983) 89–114.

Lewis, D., General semantics, *Synthese 22* (1970) 18–67.

Main, M. G. and Benson, D. B., Denotational semantics for 'natural' language question-answering programs, *Amer. Journ. of Computational Linguistics 9* (1983) 11–21.

Mamdani, E. H. and Gaines, B. R., (eds.), *Fuzzy Reasoning and its Applications*. London: Academic Press, 1981.

McCarthy, J., Circumscription: A non-monotonic inference rule, *Artificial Intelligence 13* (1980) 27–40.

McCawley, J. D., *Everything that Linguists have Always Wanted to Know about Logic*. Chicago: University of Chicago Press, 1981.

Miller, G. A. and Johnson-Laird, P. N., *Language and Perception*. Cambridge: Harvard University Press, 1976.

Montague, R., *Formal Philosophy*, in: *Selected Papers* Thomason, R., (ed.). New Haven: Yale University Press, 1974.

Moore, R. C., The role of logic in knowledge representation and commonsense reasoning, *Proceedings of the National Conference on Artificial Intelligence*, 428–433, 1982.

Morgenstern, C. F., The measure quantifier, *Journal of Symbolic Logic 44* (1979) 103–108.

Osherson, D. N. and Smith, E. E., Gradedness and conceptual combination, *Cognition 12* (1982) 299–318.

Partee, B., *Montague Grammar*. New York: Academic Press, 1976.

Peterson, P., On the logic of *few, many* and *most, Notre Dame J. of Formal Logic 20* (1979) 155–179.

Peterson, P. L., Philosophy of Language, *Social Research 47* (1980) 749–774.

Rescher, N., *Plausible Reasoning*. Amsterdam: Van Gorcum, 1976.

Sag, I., *Deletion and Logical Form*, Ph.D. Dissertation, MIT, Cambridge, 1976.

Sager, N., *Natural Language Information Processing*. Reading: Addison-Wesley, 1981.

Sandewall, E. J., Representing natural language information in predicate calculus, in: *Machine Intelligence 6*, Meltzer, B. and Michie, D., (eds.). New York: American Elsevier, 255–277, 1971.

Schank, R. C., *Conceptual Information Processing*. Amsterdam: North-Holland, 1975.

Scheffler, I., *A Philosophical Inquiry into Ambiguity, Vagueness and Metaphor in Language*. London: Routledge & Kegan Paul, 1981.

Schubert, L. K., Goebel, R. G. and Cercone, N., The structure and organization of a semantic net for comprehension and inference, in: *Associative Networks*, Findler, N. V., (ed.). New York: Academic Press, 122–178, 1979.

Schubert, L. K. and Pelletier, F. J., From English to logic: context-free computation of 'conventional' logical translations, *Amer. Jour. of Computational Linguistics 8* (1982) 26–44.

Searle, J. (ed.). *The Philosophy of Language*. Oxford: Oxford University Press, 1971.

Simmons, R. F., *Computations from the English*. Englewood Cliffs: Prentice-Hall, 1983.

Staal, J. F., Formal logic and natural languages, *Foundations of Language 5* (1969) 256–284.

Stitch, S. P., Logical form and natural language, *Phil. Studies 28* (1975) 397–418.

Suppes, P., *A Probabilistic Theory of Causality*. Amsterdam: North-Holland, 1970.

Tarski, A., *Logic, Semantics, Metamathematics*. Oxford: Clarendon Press, 1956.

Van Fraassen, B. C., *Formal Semantics and Logic*. New York: Macmillan, 1971.

Wahlster, W., Hahn, W. V., Hoeppner, W. and Jameson, A., The anatomy of the natural language dialog system HAM-RPM, in: *Natural Language Computer Systems*, Bolc, L., (ed.). Munchen: Hanser/Macmillan, 1980.

Wilks, Y., Philosophy of language, in: *Computational Linguistics*, Charniak, E. and Wilks, Y. (eds.). Amsterdam: North Holland, 205–233, 1976.

Winograd, T., Towards a procedural understanding of semantics, *SAIL Memo AIM-292*. Stanford University, Stanford, CA, 1976.

Woods, W. A., Procedural Semantics for a Question-Answering Machine, *Proceedings of AFIP*, Washington, 457–471, 1968.

Yager, R. R., Quantified propositions in a linguistic logic, in: *Proceedings of the 2nd International Seminar on Fuzzy Set Theory*, Klement, E. P., (ed.). Johannes Kepler University, Linz, Austria, 1980.

Yager, R. R., Some procedures for selecting fuzzy-set-theoretic operators, *Inter. Jour. of General Systems 8* (1982) 115–124.

Zadeh, L. A., Probability measures of fuzzy events, *Jour. Math. Anal. and Applications 23* (1968) 421–427.

Zadeh, L. A., Quantitative fuzzy semantics, *Inf. Sci. 3* (1971) 159–176.

Zadeh, L. A., A fuzzy-set-theoretic interpretation of linguistic hedges, *Jour. of Cybernetics 2* (1972) 4–34.

Zadeh, L. A., Fuzzy languages and their relation to human and machine intelligence, *Proc. of Intl. Conf. on Man and Computer*, Bordeaux, France, S. Karger, Basel, 130–165, 1972.

Zadeh, L. A., The concept of a linguistic variable and its application to approximate reasoning, *Information Sciences 8 and 9* (1975), 199–249, 301–357, 43–80.

Zadeh, L. A., Fuzzy sets as a basis for a theory of possibility, *Fuzzy Sets and Systems 1* (1978a), 3–28.

Zadeh, L. A., PRUF—a meaning representation language for natural languages, *Int. J. Man-Machine Studies 10* (1978b) 395–460.

Zadeh, L. A., Test-score semantics for natural languages and meaning-representation via PRUF, *Tech. Note 247, AI Center, SRI International*, Menlo Park, CA, 1981. Also in: *Empirical Semantics*, Rieger, B. B., (ed.). Bochum: Brockmeyer, 281–349, 1981.

Zadeh, L. A., Precisiation of meaning via translation into PRUF, in: *Cognitive Constraints on Communication, Representations and Processes*, Vaina, L. and Hinikka, J., (eds.). Dordrecht: Reidel, 1984.

Zadeh, L. A., Possibility theory as a basis for meaning representation, in: *Proc. of the Sixth Wittgenstein Symposium*, Leinfellner, W., *et al*, (eds.), Kirchberg, 253–261, 1982.

Zadeh, L. A., A Computational Approach to Fuzzy Quantifiers in Natural Languages, *Computers and Mathematics 9* (1983a) 149–184.

Zadeh, L. A., Linguistic variables, approximate reasoning and dispositions, *Medical Informatics 8* (1983b) 173–186.

Zadeh, L. A., A fuzzy-set-theoretic approach to the compositionality of meaning: propositions, dispositions and canonical forms, ERL Memorandum No. M83/24, 1983c. Also in the *Journal of Semantics 3* (1983), 253–272.

Zadeh, L. A., A theory of commonsense knowledge, ERL Memorandum No. M83/26, 1983d. Also in *Aspects of Vagueness*, Skala, H. J., Termini, S. and Trillas, E., (eds.). Dordrecht: Reidel, 257–296, 1984.

Zimmer, A., Some experiments concerning the fuzzy meaning of logical quantifiers, in: *General Surveys of Systems Methodology*, Troncoli, L., (ed.). Louisville: Society for General Systems Research, 435–441, 1982.

Zimmermann, H.-J. and Zysno, P., Latent connectives in human decision making, *Fuzzy Sets and Systems 4* (1980) 37–52.